Virtual Product Creation in Industry

Rainer Stark

Virtual Product Creation in Industry

The Difficult Transformation from IT Enabler Technology to Core Engineering Competence

Springer

Rainer Stark
Industrial Information Technology
Berlin Institute of Technology
Berlin, Germany

ISBN 978-3-662-64299-3 ISBN 978-3-662-64301-3 (eBook)
https://doi.org/10.1007/978-3-662-64301-3

This Springer imprint is published by the registered company Springer-Verlag GmbH, DE part of Springer Nature.
The registered company address is: Heidelberger Platz 3, 14197 Berlin, Germany

Contents

Abbreviations

2D	Two-dimensional
3D	Three-dimensional
A&U	Analyze and Understand
ABS	Anti-lock braking system
ADAMS	Automatic Dynamic Analysis of Mechanical Systems
AGV	Automated Guided Vehicle
AI	Artificial Intelligence
ALM	Application Lifecycle Management
AMQP	Advanced Message Queuing Protocol
AOIS	Automated Optical Inspection System
API	Application Programming Interface
AR	Augmented Reality
ARCNET	Attached Resource Computer NETwork
ASE	Advanced Systems Engineering
Automation ML	Automation Markup Language
AutoML	Automated Machine Learning
B2B	Business-to-Business
B2C	Business-to-Costumer/Client
BCE	Before Common Era
BEM	Boundary Element Method
BoL	Begin of Life
BOM	Bill of Materials
BPI	Business Practice Investment
BRep	Boundary Representation
BSD	Berkeley Software Distribution
C3P	CAD, CAM, CAE and Product Information Management
CAD	Computer-Aided Design
CAE	Computer-Aided Engineering
CAM	Computer-Aided Manufacturing
CAPP	Computer-Aided Process Planning
CASE	Computer-Aided Software Engineering

CAx	Computer-Aided x
CentOS	Community Enterprise Operating System
CERN	European Council for Nuclear Research
CfB	Call for Bids
CFD	Computational Fluid Dynamics
CMM	Capability and Maturation Matrix
CNC	Computerized Numerical Control
CoAP	Constrained Application Protocol
CP/M	Control Program for Microcomputers
CPPS	Cyber Physical Production System
CPS	Chyber Physical System
CPU	Central Processing Unit
CRISP DM	Cross Industry Standard Process for Data Mining
CRM	Customer Relationship Management
CSG	Constructive Solid Geometry
CSL	Control and Simulation Language
D&C	Drive and Control
DA	Data Analytics
DAc	Data Acquisition
DC	Data Collection
DCx	Data Contextualization
DEEP	Digital Engineering Excellence Provider
DII	Data Identification and Interpretation
DM	Digital Master
DMDA	Data Modeling and Data Analytics
DMM	Data Modeling and Management
DMS	Document Management Systems
DMU	Digital Mock-Up
DNC	Distributed Numerical Control
DND	Data Need Definition
DOS	Disk Operating Systems
DP	Digital Prototype
DPE	Digital Platform Engineering
DRC	Design Rule Checking
DS	Digital Shadow
DSE	Data Science and Engineering
DT	Digital Twin
DTE	Digital Twin Engineering
DTV	Digital Technology Vendor
DV	Data Visualization
DVU	Data Value Understanding
EAI	Enterprise Architecture Integration
E-CAD	Electrical Computer-Aided Design
ECM	Engineering Change Management
ECU	Electronic Control Unit

EDM	Electronic Document Management
ENIAC	Electronic Numerical Integrator and Calculator
EoL	End of Life
EOS	Engineering Operating System
ERP	Enterprise Resource Planning
ESC	Electronic Stability Control
ETL	Extract Translate Load
FAT	File Allocation Table
FEA	Finite-Element Analysis
FMI	Functional Mock-Up Interface
FMS	Fortran Monitoring System
FMU	Functional Mock-Up
GD&T	Geometric Dimensioning and Tolerancing
GDPR	General Data Protection Regulation
GNU	GNU's Not Unix
GPGPU	General Purpose Computing on Graphics Processing Units
GPL	General Public License
GPSS	General Purpose Systems Simulator
GPU	Graphic Processing Unit
GSP	General Simulation Program
GUI	Graphical User Interface
HIL	Hardware in the Loop
HMD	Head-Mounted Display
HMI	Human-Machine Interface
HTTP	Hyper-Text Transfer Protocol
IA	Information Activity
IaaS	Infrastructure as a Service
IBM	International Business Machines Corporation
ICT	Information and Communication Technologies
IDE	Integrated Development Environment
IfB	Invitation for Bid
IGES	Initial Graphics Exchange Specification
IIO	Intelligent Information Object
IIoT	Industrial Internet of Things
IL	Information Logistics
IoT	Internet of Things
IP	Intellectual Property
IP	Internet Protocol
ISP	Internet Service Providers
IT	Information Technology
ITIL	Information Technology Infrastructure Library
JT	Jupiter Tesselation
KBE	Knowledge-Based Engineering
LSI	Large-Scale Integrated Circuit
Mac OS X	Macintosh Operating System (today: operating system of Apple)

MBD	Multibody Dynamics
MBE	Model-Based Engineering
MBS	Multi-Body Simulation
MBSE	Model-Based Systems Engineering
MES	Manufacturing Execution System
MoC	Management of Change
MOC	Manufacturing Operations Center
MoL	Mid of Life
MQTT	Message Queuing Telemetry Protocol
MRO	Maintenance Repair and Overhaul
MRP	Material Requirements Planning
MS-DOS	Microsoft Disk Operating System
MULTICS	Multiplexed Information and Computing Service
NC	Numerical Control
NURBS	Non-Uniform Rational B-Splines
OCM	Organizational Change Management
OEM	Original Equipment Manufacturer
OLP	Offline Programming
OSLC	Open Services for Lifecycle Collaboration
OTA	Over the Air
PaaS	Platform as a Service
PADL	Part and Assembly Description Language
PC	Personal Computer
PCB	Printed Circuit Boards
PD	Product Development
PDAs	Personal Digital Assistances
PDGS	Product Design Graphic System
PDM	Product Data Management
PDP	Product Development Process
PLC	Product Lifecycle
PLC	Programmable Logic Control
PLM	Product Lifecycle Management
PMTI	Process, Method, Tools and Information standard
POC	Prove of Concept
PPSE	Product and Production Systems Engineering
PSS	Product-Service Systems
PVM	Parallel Virtual Machine
QPL	Quick Part Locator
R&D	Research and Development
RAM	Read Access Memory
RE	Requirements Engineering
REST	Representational State Transfer
RfQ	Request for Quotation
RISC	Reduced Instruction Set Computer
RM	Requirements Management

ROI	Return of Investment
ROM	Read Only Memory
RTLS	Real-Time Location System
RUP	Rational Unified Process
SaaS	Software as a Service
SAM	Software Asset Management
SDD	Systems Definition and Derivation
SDK	Software Development Kit
SEA	Systems Environment Analytics
SIL	Software in the Loop
SIM	Systems Interaction Modeling
SLE	Systems Lifecycle Engineering
SME	Small and Medium Enterprises
SOA	Service-Oriented Architecture
SoS	Systems of Systems
SpecIF	Specification Integration Facility
SQL	Standard Query Language
STEP	Standard for the Exchange of Product model data
SW	Software
SysML	Systems Modeling Language
TCO	Total Cost Ownership
TCP	Transmission Control Protocol
TDM	Team Data Management
TET	Twinning Engine Type
UDP	User Datagram Protocol
UML	Unified Modeling Language
UNIVAC	Universal Automatic Calculator/Computer
USB	Universal Serial Bus
V&V	Validation and Verification
VFS	Virtual File System
ViC	Virtual Commissioning
VLSI	Very Large-Scale Integration (circuits)
VPC	Virtual Product Creation
VR	Virtual Reality
XaaS	Everything as a Service

List of Figures

List of Tables

Chapter 1
Motivation and Approach

This book was published because the power of *Virtual Product Creation (VPC)* and its solution elements still are significantly underestimated and hence underutilized within industry. In many companies, significant investments have been made into *PLM (Product Lifecycle Management) technologies*—a term that has been introduced in the second half of the 90s of the last century to designate information applications, which should help to manage the product throughout its entire lifecycle. Nevertheless, no robust working environment has yet been achieved to support digital engineering activities consistently. Today's industrial situation is characterized by a number of factors:

- a clear separation between physical (prototyping and rig) testing and IT-based digital (prototype) simulation,
- difficulties to adopt new digital working practices across all disciplines; enterprise-wide and cross-enterprise,
- pending decisions on core competence in computer-aided design, analysis, validation and verification versus outsourcing digital work packages with the idea of "plug & play" engineering,
- unclear ownership for new digital engineering tasks and opportunities and
- obsolete business models which exclude positive digital value creation (in terms of models, investigations, functional validations and process verifications etc.) and only charge for IT license and server cost as well as for human resource cost.

The role of *Virtual Product Creation* as a fully recognized engineering discipline is to serve as key business enabler for digital value creation synchronized with strategic and operational business plans. In today's best-case scenario, however, *Virtual Product Creation* is still only understood as a joint work stream or project between IT and Engineering departments to launch new digital IT-tools with direct applications and to develop and deploy related IT tool oriented methods for a certain group of digital design engineers and analysts. The dilemma resulting from this limited *Virtual Product Creation* approach is that core-engineering processes remain

R. Stark, *Virtual Product Creation in Industry*,
https://doi.org/10.1007/978-3-662-64301-3_1

mostly untouched. This book aims to address the need, as well as practical and scientific ways to seriously extend today's limited Virtual Product Creation footprint to a full engineering discipline in order to achieve business goals from today and of tomorrow. The future extension of Virtual Product Creation has to take seriously into consideration the interplay between digital product function activities of customers and users in the market as well as maintenance/functional service during the product use lifecycle back with digital activities in product and production engineering.

In silent conviction, even ordinary business managers admit that today's product creation can no longer exist without the engineering discipline *Virtual Product Creation*. However, there still are major misunderstandings in how to use those technology elements consistently, effectively and most efficiently. The biggest dilemma is the question of how to integrate digital activities within the company's business value creation network. Senior business management and operational middle management still regard digital engineering rather as a type of IT technology and a wider spectrum of Virtual Product Creation and PLM processes, methods and solution elements as cost contributors rather than as company competence assets.

The author of this book has spent fourteen years in automotive industry as part of a major global OEM from the first half of the nineties up to the end of the first decade of this century. After having done research in the field of CAD/CAM and Engineering Design in academia, he started in industry to gain substantial experiences as system engineer in body engineering within several global vehicle development projects. During this time, he was responsible for design, test and release of automotive body systems manufactured around the globe. With this product and production development experience he started to build up a new department in the area of C3P (CAD, CAM, CAE and Product Information Management) and PLM. During the last eleven years of his industrial career, he was responsible in various technical expert and manager positions to develop, implement and operationally ensure digital engineering technologies as growing part of the automotive development activities, both in product development and for manufacturing engineering.

Today he holds responsibilities as full-time University professor of the chair *Industrial Information Technology*—belonging to the *Institute of Machine Tools and Factory Management (IWF)* in the faculty *Mechanical Engineering and Transport Systems* at the *Technische Universität Berlin (TU Berlin)*. In this position, he is able to enhance digital solutions of the current generation and has excellent opportunities to define and influence next generation Virtual Product Creation solutions. Both R&D activities are accomplished together with his research teams in academia and in industry, with national and international students and in close cooperation with other research institutes, software vendors and industrial companies.

What exactly is this technical book about? Overall, it provides a unique view into the realities of Virtual Product Creation activities in todays' industrial engineering execution. The book's content structure is divided into three major parts. Part one starts with a prologue to lay out the differences in approach in the young technical field of IT supported Engineering and Virtual Product Creation throughout the last 50 years. Part 1 also explains the transition of information technology and its organizational set-up in enterprises as a baseline. The second part of the book (Chaps. 7

through 16) provides refined technical understanding on Virtual Product Creation technologies and methods as well as their practical business use and today's limitations. The third section of the book (Chaps. 17 Through 21) deals predominantly with business integration challenges of IT based digital engineering working practices as part of Virtual Product Creation and with the outlook on future digital engineering approaches, business models and solutions.

At the present time, digital technologies represent an absolute necessity for the creation of new products; however, the day-to-day product development operations have not yet embraced more than forty to fifty percent of the "digital potential". The obvious question is why? This book delivers honest and detailed answers to this question. In addition, it aims at delivering good understanding to industrial managers and technical experts as well as to academia, researchers and students on the weaknesses and strength of digital engineering solutions and the corrective actions which have to be taken in order to make virtual product creation set-ups eligible to solve striking engineering challenges of the future.

Each chapter provides a quick essential guide in the beginning followed by a detailed review with corresponding analysis material. The overall target for the author is:

- to explain the difference between IT/PLM technologies and true digital engineering working solutions,
- to detect and describe best practices and fundamental flaws of today's Virtual Product Creation solutions in industrial practice,
- to motivate the young generation of students and engineers to learn and engage with digital engineering technology appropriately and
- to stimulate industry business leaders to more pro-actively push the full potential of the Virtual Product Creation and to establish new organizational set-ups for the next generation of Virtual Product Creation engineers.

Despite all challenges in making the most effective use of Virtual Product Creation solutions today, as well as likely in the future, there is no other option for modern and future engineering than to rely on appropriate Information and Communication Technologies (ICT). However, it is indispensable to enhance the way these fast evolving technologies are actually integrated into engineering execution processes.

Chapter 2
Prologue—Understanding the Difference in Approach

Lisa Weber, Li Wang and Jonathan Lee,[1] three Master Students in Product and Production Systems Engineering (PPSE) within the integrated Global Engineering School that is jointly offered by three leading technical Universities in Europe, USA and Asia are in their final year of graduation. It is the year of 2030 and they have to finish their project assignment within ten working days. This project is part of their interactive exploration curriculum to learn essentials of Virtual Product Creation working patterns and the associated digital transformations. The new way of self-learning comprises the Knowledge and Information Subject Survey (the new KISS[2] approach) by studying, in this particular case, the history of IT (Information Technology) supported digital engineering approaches. Those digital engineering principles should be assessed starting from the 60s and 70s of the twentieth century and the emerging approaches in the first and second decade of the twenty-first century up to the presence in 2030 and beyond.

Lisa Weber, Li Wang and *Jonathan Lee* are all happy that they belong to a privileged generation of students who can use a special configured virtual research assistant. This is a kind of science avatar who actively uses a premium "research search service" for documents, knowledge, best practices and information retrieval and interpretation across all available online databases within the global internet networks (and it could be used within enterprise to interrogate intranet sources and data lakes, if permissions are given!). All three students won the right to use this avatar for one month, as part of an award package by the University consortium due to their excellent research project results from the previous semester. They can use this science avatar exclusively in

[1] All person names formatted in italic format are used to illustrate fundamental scenarios to explain principles and styles of Virtual Product Creation solutions within this book and are entirely fictional with no relation to individuals in present or past reality. Real individuals in presence and past are designated by normal formatting.

[2] In today's industrial application scenarios, complex IT-technologies are confronted with the request to follow the KISS principle, which in the western world is referred to as: *keep it simple and short/stupid*. This requirement expresses the social-organizational desire to keep digital solutions manageable for the majority of users by avoiding difficult and complex case differentiation.

R. Stark, *Virtual Product Creation in Industry*,
https://doi.org/10.1007/978-3-662-64301-3_2

order to calibrate the community based relevant knowledge it can gain via state-of-the art excessive social engineering wiki networks which were formerly known as web 2.0 technologies.

After two days of refined configured research by their research avatar *Li Wang* and *Jonathan Lee* both feel amazed about the significant difference in executing Virtual Product Creation within the last 50 years. It is similar to what their grandfathers used to tell them about the industrial application scenarios around the significantly evolving manufacturing technology throughout the middle part of the twentieth century. *Jonathan* analyses the first three typical scenarios of Virtual Product Creation analyzed sorted and interpreted by his research avatar, whereas *Li* concentrates at the same time on the use of deep learning methods to predict future perspectives of technology and their applications. *Lisa Weber* holds special knowledge on sustainable engineering approaches: her father, a Professor in Europe, had written several papers on human technical intelligence recognition to ensure true sustainability of products and technical systems. This new engineering intelligence research direction has been started after the machine learning (Artificial Intelligence) hype had left too many disappointments in industry in the first part of the 2020th. The results of their analysis look as follows.

2.1 Pioneering in Self-made Mode by Technical Experts

In the very first beginning of the availability of computing machines in the middle of the twentieth century a relatively small group of Mathematicians and Scientific Engineers had recognized that digitization of calculation did offer new opportunities to assist the execution of tedious scientific and engineering reckoning activities. Due to the fact that the interaction with such computing machines was tedious and error prone—punching cards were to be used as programming interface originally—only the scientific oriented experts were attracted to develop “engineering application programs” such as calculation routines to solve differential equations in the case of dynamic structures.

Based on this mathematical scientific oriented passion the first generation of “0-level digital users” (in the sense of digital calculation and drawing methods) developed their own applications. They used algorithmic methods followed by subsequent coding activities with the help of technical programming languages such as FORTRAN, PASCAL or C. This gene-ration of “Digital Engineers” integrated the role of software developers and application users in one person. It was more or less a club of its own and mainstream Engineers did not get involved at all. They continued to refine traditional physical test and development methods without paying real attention to the new emerging engineering discipline.

2.2 Scaling-Up by Growing Digital Design and Analysis Groups with Customized Solutions

In the end of the 70's large industrial companies and innovative companies with close relationships to universities and research institutes attracted the new generation of pioneering digital experts. Subsequently, those teams began to grow steadily. With the help of the first commercially available CAD/CAM systems[3] growing out of the lab environments of universities and aerospace, aviation and automotive companies a new era started to evolve in the early 80's. Later on, specialized CAD/CAM companies started to offer the first generation of digital engineering tool sets, which were initially tightly linked to the special type of hardware. In the second half of the 80's these new digital capabilities of modeling, designing, drafting, calculating and analyzing mechanical structures hit the arena of mainstream engineering: predominantly young generation designers started to use 2D and first 3D CAD/CAE[4] and CAM-systems for their ordinary engineering work. It was the time of split teams, still significant numbers of drawing board-oriented draftsman and the new generation of "CAD jockeys", young design professionals with profound modeling but limited engineering skills. Due to the rare numbers of specialized computer hardware and vector screens, engineering managers were forced to set up shift systems for the growing number of the CAD design teams. The introduction of UNIX based workstations in the second phase helped to significantly lower cost and to provide workstation usage for every digital engineer.

At the end of this phase (up to the early 90s) the clear set-up of separation between software developers and "ordinary technology applying" CAD designers was finally established in industry. In the CAE and CAM area, however, application experts and analysts still developed major parts of their own tool sets and they continued software maintenance and improvement. The new era of specialized CAD service providers started to evolve.

2.3 Desire to Restrict—The Dilemma of Limited Understanding of the Role of Virtual Product Creation

From the early 90s onwards, organizational changes and the introduction of project management approaches were necessary to put in place simultaneous and concurrent working principles in industrial companies. As consequence of *time to market* reduction needs, the digital workforce was asked to deliver more design and analysis models in shorter times. The number of digital models was growing fast enough to convince companies and solution providers to put more efforts into data management

[3] CAD: Computer Aided Design (initially Computer Aided Drafting); CAM: Computer Aided Manufacturing (initially dedicated to computer assisted generation of numeric controls for tooling machines).

[4] CAE: Computer Aided Engineering, special tool sets to calculate and analyze product properties.

solutions such as EDM and PDM.[5] Within the companies themselves, it was no longer sufficient just to use CAD, CAE and CAM as individual tool sets. Data exchange solutions between the individual CA-applications as well as more powerful collaboration and information management environments became the bottleneck. Significant investments were made to introduce the next level of virtual product creation tools sets.

It took until the millennium before leading companies in virtual product creation operation recognized that digital processes and methods represent the real competence that needs to be established in engineering instead of digital tool functionality alone. In this phase of entering the "adult status of virtual product creation" most of the development projects were still led through ordinary physical artifacts driven business managers. This generation of business managers still could afford—according to the company's internal reward and recognition values—to keep their digital knowledge limited. However, they were unable to control and correct fundamental flaws in digital processes and methods and, therefore, had to rely on a new generation of digital experts.

Those groups of digital process and method experts were partly organized in backbone engineering organizations or still in "process IT" organizations. Those technical competent experts and managers usually were limited in political power unless they were made responsible also for operational engineering working teams (which did not happen often at that time).

The gap between ordinary engineers and managers who constantly asked for digital solution simplifications and evidence of verification prove-outs, and the new group of "full digitally convinced" engineers became bigger from year to year. A crisis in form of missing operational robustness in digital engineering did occur in many organizations, especially in those where research and development expenses for virtual product creation was cut significantly, although new challenges in mechatronic system development, e.g., did ask for significant new investments towards new PLM-based (Model-based) Systems Engineering. Prove out of improved solutions no longer could take place and necessary research actions to integrate digital solutions (process, method, tools and information standards) were scaled down as soon as economic downturns occurred.

The era of traditional business case driven IT enabled engineering started to decline. Consequently, limiting and restricting virtual product creation to enabling technologies only had failed. The PLM vendor landscape to the same time was reduced to an oligopoly scenario: 2–3 major PLM vendors controlled technological innovation and all associated digital business matters.

[5] EDM: Engineering Data Management; PDM: Product Data Management.

2.4 The New Digital Presence—Living a New Understanding of Information Based Value Creation

During the financial crisis and more than 10 years later during the pandemic crisis driven business recession again in the beginning of the third decade of the twenty-first century a group of experienced Virtual Product Creation researches and PLM and industry experts started to define staggered plans how new alliances between research, industry and PLM vendors could be shaped up. New business models, engineering principals and engineering execution models (such as model-based systems engineering) put *Information Activity* (IA) and the related neutral process and method solutions into the focus of digital value creation. Gradually, however, it became also evident that new harmonization efforts were necessary in industry to meaningfully steer innovations in

- *Information Technology* (IT) leveraging smart infrastructures as well as information backbone technologies and networks
- *Data platforms and data values* to offer data contextualization and training pools for machine learning support and new data driven smart services
- *Information Logistics* (IL), the new discipline to exchange and transfer information objects to the entry points of engineers' needs, engineering process activities and model based engineering execution points.

These efforts aim at reducing the drastically increased risks of "information outages" and "information tsunamis" in global engineering operations and business collaborations.

From 2015 onwards, it became possible to introduce the capabilities of the matured WEB 2.0 technologies into the new day-to-day digital engineering task execution within a systematic approach—IT departments reverted their development approach to integrated "agile" development and operations (DevOps) progression. Social network capabilities could be made available within the new digital engineering work pattern, mash-up technologies started to help providing new intuitive interfaces in order to improve comprehending complex product and lifecycle views as well as delivering intelligence of visual analytics to decision makers. The dream of involving all partners of product creation within the virtual process became true and the dominance of a PLM vendor oligopoly had vanished naturally. New start-ups, inspired from new the WEB 2.0 and Apple/Android apps technology opportunities (leveraging the new hyped approach of "microservices), as well as business professionals from the Business Intelligence and Big Data expert groups started to create new paradigms of digital engineering. Within such new innovative solutions it became possible to dynamically offer task oriented engineering apps based on online search engines, not only screening classically structured data but also the vast majority of unstructured data within and across companies. Google, Baidu and other IT and Web intelligence companies all over the world started to inspire the new generation of start-up companies, digital lab providers, Digital Life Cycle Engineers and Cross Innovation Managers. A new idea about the "future PLM" was born, the real starting point of true Virtual Product Creation for all engineers, planers, managers, users and interested customers.

A new generation of virtual service providers in close cooperation with a wide group of digital service developers had grown up until 2030 in addition to the "classical" PLM vendors. From then onwards a new virtual eco system was formed which consists of this new group of digital economy business partners, neutral modern virtual product creation research institutions, the "transformed" PLM vendors, meanwhile nameplated as Digital Technology Vendors (DTV), and the industrial partners themselves. From then onwards, all four partners were equal in importance and well connected in order to design and set up virtual product creation solutions in the industry. Closed shop virtual product creation was successfully transformed into an open, trust and competence oriented digital business. This was possible due to the introduced value and price system for data, information sets and digital models which was invented in Europe based on the desire to extend the classical production factors *land, work and capital funds*. The thinking of data souverenity had started around 2020.

Almost in parallel a growing number of researchers, engineers and computational experts did develop new ways of engineering by introducing the 2nd generation of digital models for the description of technical system core modules (beyond the traditional first generation geometric based CAD, CAE and DMU models). Those 2nd generation models did evolve from an initial set of models, which were only used for control problems. The new generation of model-based (systems') engineering tools and methods allowed to co-simulate and to integrate various perspectives of the technical system model: mechanical behavior, control behavior, functional and logic behavior as well as time- and state based dynamic model transitions and interaction up to human machine interactions (including cognitive elements). Towards the end of the third decade of the twenty-first century, it became possible to use real-time full digital model-in-the-loop validation of technical systems. Engineers needed to get a skill set-up update every 2 years in order to keep in synchronization with the exploding world of system modeling opportunities! Organizations, however, were not in the position to integrate such new working methods quickly enough, which resulted in a growing number of technical system modeling experts working for projects rather than companies on a pure contractual base.

It had therefore taken approx. 70 years to close the circle back to where the digital innovation had started: engineering competence uses technological commodity in a robust and innovative style. In 2030, however, it is the *IX-technology* (*Information Authoring, Information Logistics, Information Learning, Information Technology, Information Retrieval, Information Mash-Up, Information Models* etc.) which is the base for virtual product creation and PLM based systems engineering instead of the physical tool shop that used to be the base for component and prototype engineering in the middle of the twentieth century.

Jonathan Lee is intrigued by the results of the research avatar. For him it is amazing to understand the difficult way of virtual product creation from initial invention of basic IT technology up to the deep immersion of digital engineering solutions into almost every angle of (virtual) product value creation. He halts for a couple of seconds of wondering and then switches back on to his real assignment: picking a 2–3 year period between 1960 and 2030 and explaining virtual product creation in industry from the understanding out of this period. He remembers the time frame when he left kinder garden and all of a sudden became aware of new digital computing possibilities in school: it was the time frame around 2011/2012, the time right after the major finance turbulence in the world. This would be a good period, he thinks. His research avatar—unfortunately the older software load still limits the search retrieval and interpretation capability to information and knowledge in English language only—suggests an interesting book for further study on this subject: *Virtual Product Creation in Industry—the difficult transformation from IT enabler technology to core engineering competence.* The book you are currently reading …

Li Wang meanwhile got many attractive job offers out of his project assignment since he was able to demonstrate and showcase in a set of global digital pitches the application of new sets of artificial intelligence methods to assemble a perfect mix of digital models and methods for any engineering task. Li Wang will be on world tour for the next two weeks to decide whether he wants to work in Asia, Australia, Europe, Africa or America. Due to new super-fast networks and emerging digital technologies it is no longer necessary to be physically present for personal interaction with teams and technology. *Li Wang* will be part of the first generation digital entrepreneur, Chief Digital Engineer and University Professor in one person – a total new career opportunity of the digital future!

Lisa Weber has started first steps of her new career to become the Deputy Governor of the EU (European Union) *digital sustainability program* that is central part of the *Green Deal Act* in Europe. The avatar analysis results were not yet sufficient for her to directly derive the new sustainable business and technology value elements of the future. However, it helped her significantly to recognize shortfalls and to deduct clear conclusions how to accelerate open digital data business to offer circular economy solutions as basis for any 2050 business model. Digitalization to drive sustainability started to become the new critical political and technological power in the world, perfect for Lisa in her future role as Deputy Gonvernor.

Lisa, Li and Jonathan continue to stay connected with each other through their personal digital avatars. Their personal avatars remind all three of them not only to stay interested in the individual achievements (as it was popular in social networks already before…) but also to learn from their individual shortfalls. All three agreed to use the *sustainability mode* of their avatars alerting them mutually if they do not leverage the new digital opportunities of Virtual Product Creation hard enough to drive for the new sustainable balance of industry/technology/business, society/humans and ecology.

Chapter 3
The Big Picture—Information Technology in Enterprises

Executive Summary

This chapter deals with the following topics:

- information technology and its technological history,
- important evolutions and major milestones of modern information technology (incl. desktop/workstation operating systems and mobile operating systems) and
- the IT setup within enterprise organizations.

Quick Reader Orientation and Motivation

The intention of this chapter is:

- to provide a solid understanding of the roots of modern information technology.
- to describe the technological hardware development from calculation machines up to modern super computer farms.
- to explain the importance of desktop/workstation and mobile operating systems to provide the basis for application-oriented software.
- to gain insight into how the technology and the organization of IT has been developed from middle of the twentieth century until today.
- to comprehend that the execution of information technology has reached meanwhile an (IT) factory level with its own rules and services.
- to gain insight into Industry Senior Manager expert assessments with respect to changes of the future IT landscape.

3.1 Introduction and Basics

The term "information technology" comes with two meanings today: firstly, the original meaning in terms of tools and applications which create, transform or convert sets of electronically storable data and information. The basis for such technology comprises basic computer architecture, binary coding techniques, data models as

R. Stark, *Virtual Product Creation in Industry*,
https://doi.org/10.1007/978-3-662-64301-3_3

well as related software, hardware solutions. Nowadays, the term *ICT (Information and Communication Technology)* is often used for this first meaning and also includes message and network technologies. Secondly, the term *IT (Information Technology)* designates a department or an organization within an enterprise which is held responsible for the operation of computers, servers and related "IT services and solutions". IT as a department or enterprise organization obviously is in charge to coordinate all general aspects of information technology in its original sense. In contrast to the original information technology development role, IT departments nowadays are oftentimes exhausted to put significant focus on policy aspects, user administration, data storage tasks, operational readiness of hard- and software equipment and general help desk services.

The obvious question is why such a difference in using the term IT really exists or what might have changed in the last twenty years of IT perception within the enterprise. The following sections will provide a deeper understanding, starting from the history of information technology up to the current set-ups of IT in modern enterprises and will also come to conclusive statements with respect to Virtual Product Creation.

3.2 History of Information Technology (IT)

Information Technology (IT) is one of the crucial enablers for modern digital engineering which is designated in this book consistently as Virtual Product Creation (VPC). A detail explanation of VPC is given in Chap. 4. To understand IT better, it is necessary to reflect the origin and the different stages of computation and related tools.

3.2.1 Hardware: From Numbering and Mechanics Towards Electronics

The development of the first calculators started with the invention of numbers and numbering systems. Most numbering systems are based on the counting of the fingers of the hands. Therefore, it is not astonishing that the Sumerian, Egyptian and Babylonian numbering systems are based on the number ten. Our decimal system was invented in India and arrived in Europe via the Middle East. But the first known calculation tool is the abacus, dated around 2700–2300 BCE [1]. The abacus was invented in Babylon a city-state of ancient Mesopotamia [2]. The earliest archaeological evidence for the use of an abacus was in Greece and dates to the fifth century BC. Primarily the cultures of the near and far east (Mesopotamia, Egypt, Persia, Greece, Rome, China, India and Russia) made effective use of this first calculation machine, interestingly enough, based on different number systems with base numbers 60, 16, 5 and 10 [3].

From the point of view of motivation, the "mathematical number and calculation machines" started with the need of mankind to use arithmetic calculations for day-to-day activities such as for selling goods on the market or for building complex stone or wood constructions which extended the easy enumeration capabilities of the fingers or the mental memory of human beings.

The next step forward concerning calculation tools was the invention of wooden sticks to enable easy multiplication and division of numbers by the Scottish Lord J. Napier. He developed the sticks around 1600, the user had to cut out paper slips and glue them on wooden sticks. Astonishingly, the paper slips were available until 1920 [1].

In the year 1623 W. Schickard designed a calculating machine (calculating clock) for addition, subtraction, multiplication and division based on sophisticated mechanics (see Fig. 3.1): it is the first mentioned machine with spur gearing. Its special feature was a gear-driven carry mechanism, which aided in multiplication of multi-digit numbers. However, multiplication and division was only possible with the help of the user. During multiplication the user had to determine sub-products with the earlier mentioned sticks from Napier. Afterwards the sub-products are manually added to the six-digit calculation unit to be summarized. The only implemented model got lost during the confusion of the Thirty Years' War. A second model which Schickard asked his friend J. Kepler for construction was destroyed by fire. It took a few centuries until B. v. Freytag-Löringhoff could prove the operational reliability by reconstructing the machine in the years 1957 to 1960 [1].

In 1642 the French Mathematician B. Pascal designed a gear-driven semi-automatic adding machine the "Pascaline", which is the first known full functioning

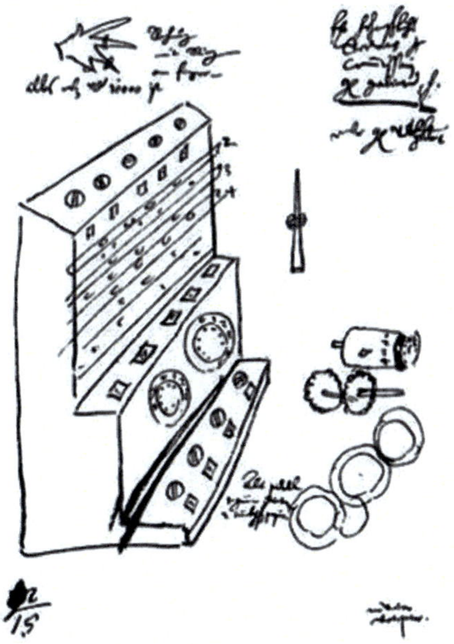

Fig. 3.1 Original drawing by W. Schickard (*Source* Wikipedia)

mechanical adding machine. The goal of Pascal was to relieve the user from the burden of trivial but labor-intensive calculations. Pascal built up to 50 prototypes, but sold just a few due to the cost and complexity of the machine. Furthermore, the Pascaline was difficult to use and could only add and subtract, so that it did not find widespread use [4].

In 1674 Gottfried Leibniz built the "Stepped Reckoner," a calculator using a stepped cylindrical gear (see Fig. 3.2). The Stepped Reckoner was the first machine which could perform all four basic arithmetic operations, addition, subtraction, multiplication and division; it was more complex than the Pascaline. One of his greatest developments was the invention of a binary notation. He even developed a plan for constructing a machine which uses binary arithmetic, but he could never finish it [4].

A further contribution with an enormous impact on the development of computers was the invention of a punch card driven loom around 1801 by J.-M. Jacquard. The loom as illustrated in Fig. 3.3 had to fulfill the task of weaving of patterns and was controlled by a sequence of punch cards.

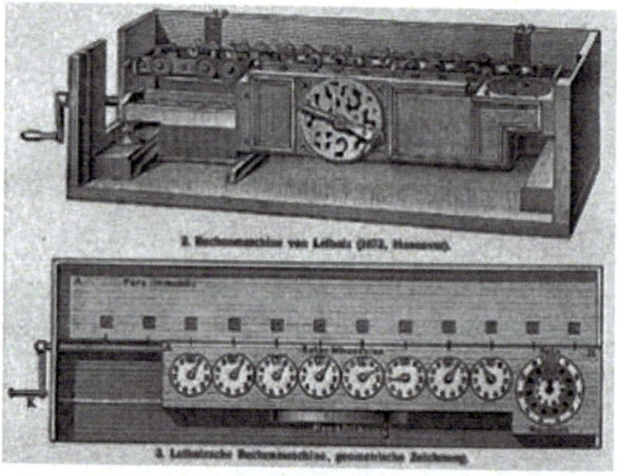

Fig. 3.2 Stepped reckoner from Leibniz (*Source* Meyers Konversations-lexikon, Ed. 6)

Fig. 3.3 Jacquard's loom at Manchester Museum of Science and Industry (*Source* photograph by G. H. Williams)

Fig. 3.4 Babbage's analytic engine (The National Museum of Science, London)

The idea of using punch cards to control the pattern and most notably the ability to change the pattern by using different punch cards can be seen as a conceptual pioneer for later computer programs.

The first concept of a modern computer was drafted by the British mathematician C. Babbage in the year 1822 (see Fig. 3.4). He was influenced by Jacuard's loom and planned to use punch cards as storage for his machine. The outline of his Difference Engine already included all important parts of a modern computer, as input devices, memory unit, arithmetic and logic controller unit and output devices.

In the 1880s H. Hollerith developed a calculation machine which could count, compare and sort information on punch cards. The punch card was mechanically scanned, whereas by discovering a hole a circuit was closed. The first usage of the machine was during the census of population in the United States of America. With the help of the machine the time for evaluation could be reduced from 7½ years to just 6 weeks. In 1896 Hollerith founded the *Tabulating Machine Company*», for the construction of similar machines. Later on the name of the company changed to «International Business Machines Corporation (IBM)» in the year 1924. Since then until the late 60s IBM produced punch cards for machines in offices, which found widespread use.

In the year 1919 Eccles and Jordan, both US physicists, invented the flip-flop electronic switching circuit [5]. The invention was critical to high-speed electronic counting systems.

During World War II the building of modern computer was intensified on all sides, since the military was in need of fast ballistic calculation. The British cryptographs needed machines to break the secret code of the Germans.

Fig. 3.5 Reconstruction of the Z3 computer of K. Zuse in the Deutsche Museum in Munich

In the year 1941 the German engineer K. Zuse developed an operating computer, which was used in the aircraft industry. The first three models named Z1, Z2 and Z3 were all destroyed during the war. The Z3 was the first fully functional general-purpose computer which was controlled by a program (see Fig. 3.5).

Beside the basic arithmetic operations Z3 could also find square roots. However, the computer never reached its full potential, because of disagreements with the German regime at that time. Z3 counted to the 0 generation of computers [6].

Another development which became operational during World War II (1943) was the Colossus, a British vacuum tube computer. Colossus was designed as part of the British crypto analysis program at Blechly Park for the purpose of deciphering messages from the German Army. It was the first electronic calculation device which was programmable; but it was still not possible to store a program. The first version had 1500 vacuum tubes and was a room filling machine. The second version had 2500 tubes and could process five characters at the same time and can be seen in Fig. 3.6. Astonishingly, the existence of the machine was kept under concealment until 1976 [7].

In 1943 construction on the ENIAC (*Electronic Numerical Integrator and Calculator*) has started, it was the first modern computer for solving general problems (see Fig. 3.7). It filled a 140 m^2 room and was developed by J. W. Mauchly and J. P. Eckert of the University of Pennsylvania, where it was used in 1946. The construction was financed by the US Army during World War II. ENIAC had a weight of 27 t, was

Fig. 3.6 Improved Colossus Mark II (1944)

Fig. 3.7 Programmers operate at ENIAC's main control panel (US Army photo from the archives of the ARL Technical Library)

5.5 m in height and 24 m long and contained 17,468 vacuum tubes. ENIAC was able to calculate 100,000 operations/s, the first ones contained calculations to prove the concept of building a hydrogen bomb.

The theory behind general-purpose computer was first drafted by A. Turing with his 1937 published paper "On computable Numbers" which presented the concept of Turing machines. Later on, the theory was extended by J. v. Neumann, he introduced the concept of a storable program in the year 1945. This was one of the major developments towards nowadays computers, which became known as the von Neumann architecture. Data and instructions to manipulate the data could be stored in the same place. This design was inspired by the work of the computer pioneers J. P. Eckert and J. Mauchly and is still the basis for computers today [8].

H. Aiken developed in the year 1944 at Harvard University a calculating machine called "Automatic Sequence Controlled Computer" also known as Mark I, which was a lot bigger than the one from Zuse. It was an electro-mechanical construction based on the ideas of Charles Babbage. The proportion of the machine was astonishing, she was 15 m long, 2.5 m high and was composed of 700,000 individual parts, with 3000 ball bearings and 80 km of line wire. Mark I had the capability to add in 0.3 s, multiply in 6 s and divide in 11 s, wherefore 72 addition counter with 23 decimal places had been used [1].

The first computer with real-time capabilities was developed in 1949 by J. Forrester at MIT and became known as Whirlwind. At that time, it was the largest computer project with an annual budget of $1 million and a team of 175 people working at it. Whirlwind could multiply in twenty seconds and was therefore the fastest computer of the early 1950s. But it was not always reliable as Whirlwind was out of order for a few hours each day. Furthermore, Whirlwind had storage problems, as the storage tubes lasted only one month the costs summarized to an enormous monthly amount [9].

The invention of the first transistor in the year 1947 at the Bell Laboratories resulted in a revolutionary development of computers. The invention was published in a scientific paper that appeared in the Physical Review of 1948 and was announced in a press conference at Bell Laboratories. At that time in history the excitement of the inventors about the transistor was not shared by the public. Not until the end of 1951, the manufacturing of the transistors started. And from there on it still took a long time until hot and unreliable vacuum tubes could be replaced by small transistors. At the end in the year 1972, the three inventors J. Bardeen, W. Brattain and W. Shockley received the Nobel Prize in physics for the transistor [10].

All the new developed machines have to be operated and this opened a new field for inventions. Therefore, on the software side new developments took place. G. Hopper, a professor in mathematics and later on an army admiral, developed the first compiler, named A-0 in the year 1952 for the programming language FLOW-MATIC. The term "debugging" for searching errors in computer programs could also be traced back to her. During her work on Mark I she detected the failure of a relay caused by a moth. Whereupon she stuck the moth in her logbook, see Fig. 3.8, with the accompanying sentence: "First actual case of bug being found."

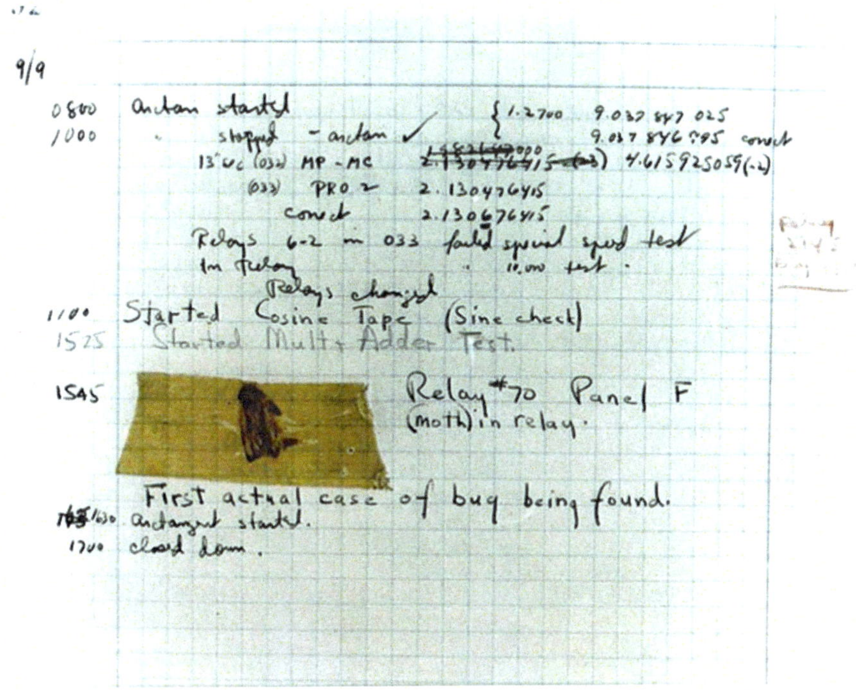

Fig. 3.8 The first "computer bug"; *Source* U.S. Naval Historical Center Online Library Photograph

From there on the term bug was used for software errors and not only by engineers for malfunctions in machines. She developed a lot of software which was still in use in 1999 and a lot of people feared that this software will cause a big disaster, because Hopper reserved only two digits for the date in her Cobol-Libraries [11].

The greatest breakthrough in terms of miniaturization could be reached in 1958 by the production of the first microchip. Whereas the introduction of the first commercial computer with a monitor and keyboard input by DEC is an important milestone in the direction of nowadays computer.

In January 1963 I. Sutherland at MIT introduced Sketchpad, the first commercial computer-aided design (CAD) software, developed as part of his Ph.D. thesis. Sketchpad allowed the user to draw directly on the computer screen with the help of a light pen. During this period CAD research started in many European countries. In France, for example, the research in 3-D surface geometry computation was funded by the car companies Citroen and Renault. Before the 60s ended a lot of companies like GM, Ford, Lockheed began to adopt and adapt the new technology. Nowadays it is difficult to imagine a manufacturing firm without CAx-systems and the capability to transfer digital data to CNC machine tools. However, at the beginning it was not easy to accomplish the use of CAx in industry. However, until today there still exist a great gap between the state of the art and the real usage of CAx-systems.

Another important invention was the mouse by D. Engelbart in 1964. Although his invention was not really appreciated at that time, but later on this device has served all kinds of users till today.

The next era (1974–today) is called the microcomputer era and was divided into five periods by Remi et al. [1]. The era is characterized by the application of large-scale integrated circuit (LSI) and very large-scale integration circuits (VLSI). The fundamental technique is used till today, what has changed is the packing density of the circuits. This development from 1971 until 2010 can be seen in Table 3.1.

Period 1 (1974–1982)

In this period hard- and software merged into an entity as well as commercial data processing with manufacturing systems. First local area networks (LAN) were created and first CAD/CAM installations were established in many companies.

In the year 1977 the most important computer companies for end user computers were funded. Steve Jobs and Steve Wozniak corporate Apple Computer, and Bill Gates and Paul Allen found Microsoft. The influence of Microsoft was increased by the decision of IBM to select PC-DOS as operating systems for their new PC. Furthermore, the first Apple was designed, which consisted mostly of a circuit board. In addition, the open-architecture IBM PC was launched in 1981 and started the area for home computers.

Table 3.1 Miniaturization of microchips

Year	Name of the chip	Number of transistors on chip	Size of chip
1971	Intel 8080	2300	10 μm
1974	Intel 8080	4500	6 μm
1978	Intel 8086	29,000	3 μm
1985	Intel 386	275,000	1.5 μm
1995	Intel Pentium Pro	5,500,000	0.6 μm
2002	Intel Itanium	220,000,000	0.13 μm
2006	Quad-Core Intel Xeon	291,000,000	65 nm
2007	Quad-Core Intel extreme (Penryn)	820,000,000	45 nm
2010	Six-core Core i7 (Gulftown)	1,170,000,000	32 nm
2012	Quad-core + GPU Core i7 Ivy Bridge	1,400,000,000	22 nm
2015	Quad-core + GPU GT2 Core i7 Skylake K	1,750,000,000	14 nm
2017	28-core Xeon Platinum 8180	8,000,000,000	14 nm
2018	Apple A12X Bionic (octa-core ARM64 "mobile SoC")	10,000,000,000	7 nm
2019	HiSilicon Kirin 990 5G	10,300,000,000	7 nm
2020	Nvidia's GA100 Ampere	54,000,000,000	7 nm

Source https://en.wikipedia.org/wiki/Transistor_count

Period 2 VLSI, sequential architecture (1982–1990)

In this period the usage of computers for automatization in industry was pronounced. On the software side standard solution for specific classes of problems came into existence. Therewith a new programming paradigm, the modular programming paradigm, came into being. Modules, which could be integrated in existing software solutions, were developed. Later on these developments led to the object oriented paradigm. Using such paradigm, the source code was structured in communicating objects rather than modules. This new paradigm enabled an abstract view on the software development process which facilitated a discussion over software design for a broader group of experts besides software developers. Another invention was the provision of graphical user interfaces for interactive systems, which enormously increased the user friendliness of applications. At the end of this period in the year 1989 Tim Berners-Lee proposed the World Wide Web project during his stay as researcher at CERN (European Council for Nuclear Research).

Period 3 VLSI, parallel architecture (1990–2010)

From 1990 on the PC started to be a mass product in the private and business sector, due to the ease of learning and ease of handling of the new generation of computers. Moreover, the internet started to be a medium for the masses and had more and more impact on the life of everyone. On the hardware and software side the era of parallelization started, Dual und Quad-Core CPU 's were developed at the same time as multithreaded software which could take advantage of the multi cores. Linux-clusters with PCs and Open Source were introduced, which could achieve billions of operations per second. Furthermore, 64-bit architectures were also available on PCs, before this architecture was only used for super computers. The advantage of this architecture is a simpler calculation of big integer values, whereof algorithms like encryption algorithms could benefit from. Moreover, with the new architecture the usage of more than 4 GB main memory was possible. Further milestones in this area were the development of distributed embedded software engineering and the invention of portable computers like laptops and handheld devices like smartphones. Furthermore, a new generation of programmable graphic cards for general purpose programming (GPGPU) on the GPU of the card has been developed. Therewith, for time consuming simulations like physics-based simulations, the calculation time could be massively accelerated.

Future hardware developments

In the future, asynchronous chips could be used to boost operations per second even further. In asynchronous systems the data flow could be controlled by switching networks for local coordination instead of tact cycles. With this concept not only the computation could be accelerated, but also the power consumption could be reduced.

Another new development could be the use of quantum computers. With this new type of computer the dual system will get obsolete, because more than two states can be represented. Once it will be possible to get a quantum computer to work robustly, a range of new applications will emerge. As of today, the quantum computer operation is still not stable. However, hardware lab research in this direction is booming and in computer science the theoretical concept has already reached sophisticated levels.

3.2.2 *Software: The Key Role of Operating Systems of Modern Computers*

In order to make functional use of computers, it is essential to provide an *operating system* that controls the link between the base resources of modern computer architecture. The operating system serves as control linkage between the various sets of application software (office software, communication software, engineering applications etc.) and the key hardware components of a computer such as

- central processing unit (CPU),
- different types of memory such as cache memory (very fast memory to reduce loading times of operations into the CPU processing routines), RAM (read access memory), ROM (read only memory), disk memory and external drives (incl. USB sticks) etc.
- internal communication bus system and
- external devices such as monitor, mouse, key pad, touch screen, printer, plotter, USB driven devices, network node components such as switches, bridges, hubs, repeater, proxies, router etc.

The evolution of operating systems meanwhile has a history of more than 60 years. The following sub-section gives an exclusive summary of both the "traditional" operating systems for computers (desktop and laptop) and the "new" operating systems for mobile devices.

The timeline of computer operating systems

The era of operating systems started in the 50s, before that time operating systems were unknown, even programming languages did not exist. In Fig. 3.9 the timeline of the main operating systems from the 50s up to the year 2020/1 is shown.

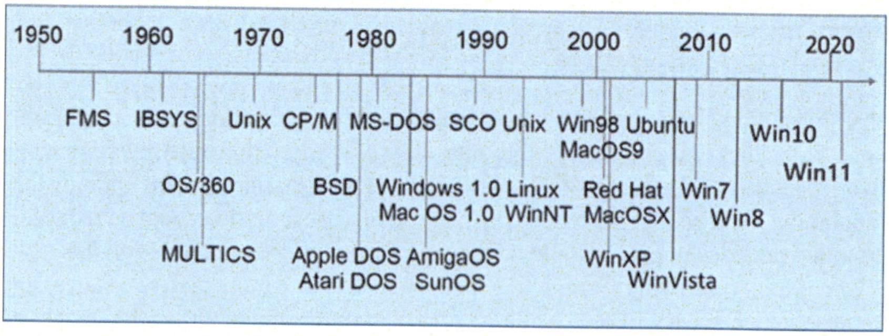

Fig. 3.9 Timeline of operating systems

Operating systems in the 1950s

In the 50s simple batch operating systems-were developed. These operating systems could only run one job at a time. The first operating systems were characterized by their diversity. They were strongly hardware dependent. Even in one firm, there existed different operating systems for different computer architectures. The user known in today's IT operation—mainly a practitioner of applications—was not known at that time: a user was the operator of the system, which would be called today a system admin or super user. Such an operator had to load cards to start a job and unload the card when the job was done. A computer program consisted of a lot of cards. The main one in this era was the Fortran Monitoring System (FMS) for the IBM 709 by North American Aviation [12].

Operating systems in the 1960s

In the 60s many companies started to provide their computers with operating systems. The main inventions in this era were the introduction of multiprogramming and time sharing systems. Multiprogramming stands for the capability to load different programs in the main memory of the computer. This enables the execution of more than one program "simultaneously". The simplest multiprogramming systems were so called spooling batch systems. Spooling (Simultaneous Peripheral Operation On-Line) was a technique that enabled the operating system to start a new job if the current running job had been finished.

An operating system developed by IBM for its 7090 and 7094 computer called IBSYS had multiprogramming capabilities. IBSYS was based on the FMS and was used with control cards. The IBM OS/360 enhanced multiprogramming capabilities further by [13].

An operating system with time sharing capabilities allows different users to share time on the same machine during their programs execution. The development of this technique was inspired by the desire to achieve shorter response times.

One of the first timesharing operating systems was the Multiplexed Information and Computing Service (MULTICS). The development of MULTICS was a combined research project by MIT, Bell Labs and General Electric. Apparently, MULTICS was a kind of failure, because of the high expenses during development, which drove Bell Labs to finally withdraw from the project. Nevertheless, it had a high impact on the development of subsequent operating systems, because all important concepts of algorithms had already been tested on the system. MULTICS could support hundreds of simultaneous time-sharing computers. Therefore, the MULTICS developer at Bell Labs started to rewrite MULTICS and called it UNICS and later on UNIX [14].

In early 1969 UNIX was developed by computer scientists at Bell Labs and AT&T. With UNIX a new conceptual view was developed on how operating systems should work. With the appearance of UNIX the era to design a unique operating system for each computer architecture was over. As a consequence, for many firms it was easier and cheaper to adopt UNIX as their "standard" computer architecture rather than continuing to develop their own. UNIX became the dominant time-sharing operating system used on all kinds of computers. Bell Labs licensed the source code of UNIX to Universities almost freely, which is the reason for the quick improvement of UNIX and its' widespread usage. Over the years UNIX developed further and is still used up to now, for example in its Berkeley version BSD (Berkeley Software Distribution) also now known as FreeBSD. UNIX most impact on mainstream computing lies in the 80s. Afterwards UNIX builds the basis for many modern operating systems for PCs like MacOSX, and the Linux-family and for many embedded operating systems like Android and others.

Operating systems in the 1970s

In this era the disk operating systems were developed. The first one was CP/M (Control Program for Microcomputers), developed by Intel and Digital Research as an operating system for microcomputer. In the beginning it was kept very simple. Later on other utilities as editors and debuggers were added. It was a single user system and can be seen as a milestone for the later on coming availability of personal computers. With this invention the era of personal computers has started.

(Other disk operating systems in this era Apple DOS, Atari DOS, PC-DOS).

Operating systems in the 1980s

One operating system which quickly came to dominate the IBM PC market was MS-DOS (Microsoft Disk Operating System). When Bill Gates bought PC-DOS, he asked the developer of the operating system to join his new founded firm Microsoft. Afterwards, he renamed the operating system to MS-DOS and licensed it to IBM. From this point in time all IBM PCs were shipped out with MS-DOS preinstalled.

An important invention in this era was the development of a GUI (Graphical User Interface) by Douglas C. Engelbart at Stanford University. Engelbart also invented the mouse in the year 1963, but at that time without GUIs there was no real usage for it. These ideas were adopted by researchers at Xerox PARC (an innovative "office of the future" laboratory). Steve Jobs, student and hardware developer in the garage of his parents, saw the GUI, while visiting PARC and realized its potential value. Jobs then had the plan to build an Apple with a GUI, which then led to Apple Macintosh. The operating system on the Apple Macintosh was known as the most user-friendly operating system at that time. From then onwards computers came along with an operating system which had not only a GUI but also a new interaction device: the mouse. This was the first computer which could be used by people with no computing background [14].

Microsoft followed shortly after Apple Macintosh by implementing a GUI in their new operating system which was called Windows.

From the mid 80s onwards computer network and distributed operating systems were developed. With a network operating system the computer of one user could be connected to computers of other users. Resources like files and printers could be shared amongst those users. Furthermore, the user could log into remote machines. One of the first network operating systems was Netware, developed by Novell. A distributed operating systems goes one step further, it adds an abstraction layer to the system. Herewith, the operating system and all resources appear to be local for the user. Consequently, the user no longer has to bother with which operating system he/she is connected in a network since he/she has the availability to access all resources in this system [13].

Other popular operating systems in this era were: Commodore DOS, SunOS, and Windows 2.0.

Operating systems in the 1990s

In this era also, Microsoft operating system developed networking capabilities. Windows NT (NT stands for New Technology) describes a new family of operating systems. The first one was released in the year 1993. Newer operating systems from Microsoft such as Windows XP, Windows 7 or Windows Server 2008 also belong to this family of operating systems. They meanwhile have the same capabilities as the UNIX family: multiprogramming, multiuser and processor independencies.

The main invention in this era was the development of PC cluster computing. A computer cluster is a compound of computers linked to each other for example by local area networks.

The first Beowulf-class PC cluster was developed at NASA's Goddard Space Flight Center in 1994 using early releases of the Linux operating system and PVM (Parallel Virtual Machines) running on 16 Intel 100-MHz 80486-based PCs connected by dual 10-Mbps Ethernet LANs. The Beowulf project developed the necessary Ethernet driver software for Linux and additional low-level cluster management tools and demonstrated the performance and cost effectiveness of Beowulf systems for real-world scientific applications. All Linux systems are developed under the GNU General public license (GPL). The acronym GNU is self-referring and stands for "GNU's Not UNIX", implying that GNU software was not developed from UNIX code. The first kernel for a linux operating system was developed by Linus Torvalds.

(Other operating systems in this era: Windows 95, Palm OS, Windows 98).

Operating systems in the 2000s

In the year 2002 Apple started with Mac OS X a new generation of operating systems switching from Motorola CPUs to Intel CPUs. From then onwards Apples operating systems also run on PCs and not only on PowerPC. Apple committed itself to this switch, because of the stagnation in the development of Motorola CPUs. In the beginning, Mac OS X was built with cross-platform capabilities, but from 2009 onwards Motorola CPUs were no longer supported. The transfer to Intel CPUs helped to spread the operating system to new user communities. Meanwhile Mac OS X is the second most used operating system after the Windows family.

In this era one of the major developments were new human computer interaction possibilities without the usage of a mouse. Almost all operating systems meanwhile include support for multi-touch screens. One example is Windows 7, which was released in 2009, resp. Windows 10, which was released in 2015. Other popular operating systems in this era wre: Windows 2000, Windows ME, Red Hat Linux, Solaris, FreeBSD, Suse Linux, Debian, Novell Netware, Ubuntu, Windows Vista.

In Table 3.2 the usage of operating systems for PCs in January 2020 is shown.

Table 3.2 Usage of desktop operating systems according to statista https://www.statista.com/statistics/218089/global-market-share-of-windows-7/

Windows	MacOS X	Linux	Chrome OS	Others/unknown
77.7%	17.04%	1.9%	0.5%	1.83%

Table 3.3 Distribution of the 500 most powerful supercomputers worldwide from 2017 to 2020, by operating system (June 2020) https://www.statista.com/statistics/565080/distribution-of-leading-supercomputers-worldwide-by-operating-system-family/

Linux	CentOS	Cray Linux Environment	bullx SCS	Redhat Enterprise Linux	Others
54.2%	23.6%	6.8%	3.4%	1.8%	10.2%

In the area of supercomputer, however, it is Linux the commonly used operating system (see Table 3.3). Linux holds a market share comprising the top 500 supercomputers worldwide of 54.2% in June 2020, whereas Windows is not relevant anymore in this sector. Even Microsoft started to use Linux on its own Azure Cloud Services (https://www.cybersecurity-insiders.com/microsoft-uses-linux-instead-of-windows-for-its-azure-sphere/).

After the introduction of operating systems on PCs, it has to be mentioned that approximately 90% of CPUs are meanwhile used within embedded systems. Embedded systems represent algorithmic processing units and are built in mobile phones, digital cameras, DVD recorders, cars, washing machines, exercise machines, television sets etc. With the help of embedded systems such devices, products and machines are enabled to follow intelligent functional or control behaviours. Most of these modern devices, products and machines use a 32-Bit- or 64-Bit-operating system [14]. One of the most widespread operating system used on these devices is Symbian OS, which runs also on a lot of smartphones.

Hereinafter, this chapter of the book concentrates on operating systems for mobile communication inasmuch as those are one of the inventions with enormous impact on mankind with relation to future business operation and communication within society networks.

Mobile operating systems

A mobile operating system is an operating system that controls a mobile device. Typical examples are smartphones, personal digital assistance (PDAs) and tablet computers. Nowadays the operating systems running on smartphones are the same as on PDAs and tablet computers. Today, smartphones include the capabilities of PDAs and tablet computers.

First generation mobile operating systems (1992–2006)

This first generation of mobile operating systems was dominated by Symbian OS, Blackberrry RIM and Windows Mobile.

The first actual smartphone was developed by IBM in the year 1992. It was called Simon, had a touchscreen, mail, calendar, address book and a sketch pad. The operating system running on Simon was Zaurus OS. Since then operating systems for smartphones evolved rapidly.

In the year 2000 Microsoft released its Pocket PC 2000 with Windows CE 3.0 running on it. It offered applications like Pocket Office, which included trimmed-down versions of Word, Excel and Outlook. The successor of Windows CE was Windows Mobile 2003 with also supported add-on keyboards and Bluetooth connections to other devices.

Already two years before, in 1998, the three-telecommunication companies Ericsson, Motorola and Nokia founded the new company called Symbian LTD. Under this new umbrella, the formerly known mobile operating system prototype EPOC32 was relabelled to Symbian. Between 1998 and 2006 the use of the Symbian operating system was growing up to 67% market share amongst the first generation smart phones (Palm OS and Windows CE were trailering significantly behind). However, it was the reputation of Symbian that was difficult to be software coded. In June 2008, Nokia announced the take-over of the Symbian LTD and at the same time to establish a Symbian Foundation in order to provide an open platform for all companies with respect to develop their mobile operating system derivatives. Finally, the Symbian Foundation was established in April 2009 and the availability of the Symbian platform could be realized in February 2010 (with a remaining part of close sourced components under the ownership of Symbinan LTD).

During the peak usage periods of Symbian in Q1 of 2007 almost 16 million smart phones were equipped with this operating system. Overall, more than 126 million smart phones were sold with Symbian OS. As lessons learned it became clear that the number of Symbian OS developer was by far exceeding the opportunity to recruit them from the job market. Consequently, the community thinking was the right one, but came 3 years too late in order to catch up with the growing role of the second generation mobile operating systems (see next section).

Finally, Nokia closed all activities in Symbian development in January 2014 after several years of constant decline in overall mobile phone business. Meanwhile Nokia has started to concentrate on Google Android operating system as part of their re-entering to the market after some experimentation with the latest Windows Phone mobile operating system.

Similar to the technology history of Symbian, the operating system of RIM (originally owned by Research in Motion Limited and later known as BlackBerry Operating System (OS) had its major starting point in the late 90s. RIM itself existed already since 1985 as part of a technological cooperation with Ericsson in order to develop a two-way paging service and new wireless networking capabilities. BlackBerry OS was specializing in secure communications and mobile productivity. Nevertheless, the peak of BlackBerry OS was in 2008 and declined with the Lehmann Brother crisis and the emergence of Android. Similar to Nokia, also BlackBerry used for several years (2010–2015) another operating system kernel—QNX a commercial Unix-like real-time operating system, originally aimed at the embedded systems market and originally developed in the early 1980s by Canadian company Quantum Software Systems—before moving to the Android platform (see next section).

Second generation mobile operating systems (2007–2018)

The second generation became highly influenced by Google Android and Apples iOS. The starting point of the second generation was the release of the first iPhone by Apple. It offered as main innovation multi-touch capabilities. But the main invention was the speed of its internet browser, Safari provided by the operating system. Currently Safari still is the fastest browsers on smartphones.

One year after Android another operating system was released. It has been developed by the Open Handset Alliance under the supervision of Google. The Open Handset Alliance is a consortium of different firms like Intel, HTC, Samsung, ARM and so on. Dissimilar to the trend in operating systems for computers the number of operating systems for mobile communication devices still grows and many smartphone and tablet vendors even create their own add-ons to the base operating system architectures.

A new paradigm has been created by Apple and Google by providing standards for the global developer community to develop applications based on their mobile communication operating systems. Those applications are simply called "Apps". The apps technology has the high potential significantly penetrate into the future business and engineering world and to revolutionize the interaction with information. Enterprises will face the challenge with the young generation of employees to satisfy their expectations to handle official company information as easy as it is enabled by the *"apps-based"* smart phone generation.

3.3 The Set-Up of IT in Industrial Companies

In order to understand today's set-up of IT in industrial companies a dedicated section will provide a quick historic analysis of the main drivers for the different technical and organizational foundations. Subsequently the state of the art of today's IT factory set-up will be briefly discussed as well as the resulting problems and limitations which are subject for major changes in the future.

3.3.1 History of IT Technical and Organizational Drivers in the Twentieth and Twenty-First Century

Information technology started out as a research activity at several universities and research institutes in the mid of the twentieth century, predominantly in the US, in England, in France and in Germany. It took until the 60s in the twentieth century before the new information technology could be used within enterprises in a wider sense: it was necessary that generalized hardware became available to enable the execution of "custom" oriented calculation applications.

The first official set-ups of Information Technology departments in industrial enterprises occurred more consistently after the Second World War, however, not before the second part of the 50s. Around that time the rather small, central IT groups were located high up in the organization, close to the company owner or president. As seen earlier in IT history (compare the section before) it was necessary that business drivers were found in industry to promote the use of "calculation machines": in the early days of IT in enterprises in the mid of the twentieth century it was the need for better organizational, sales, marketing and accounting work execution which served as catalyst for calculator machines and computers. Especially sales revenue and related tax calculations in the financial sector of big enterprises were traditionally established close to Senior Management offices in order to guarantee close loop and secret interaction between the small dedicated groups of analysis experts and the cadence of leadership meetings. Quick turnover of calculation tasks no longer could be ensured efficiently and precisely enough by manual reckoning methods and activities. Consequently, the business and accounting tasks did drive the innovation, set-up and usage of calculation and computer machines within first "unofficial central" IT departments as illustrated in Fig. 3.10.

Technical computation activities initially started in the 50s of the twentieth century due to the first digital control driven milling as well as other manufacturing, drafting and calculation machines [15]. From then onwards, more and more engineering functions in product development and in manufacturing were built up on calculation machines (compare more details in the Chap. 5 "Technology History"). New critical skill sets were provided by engineers and mathematicians who tried to help themselves in resolving tedious calculation problems to analyze test results, determining

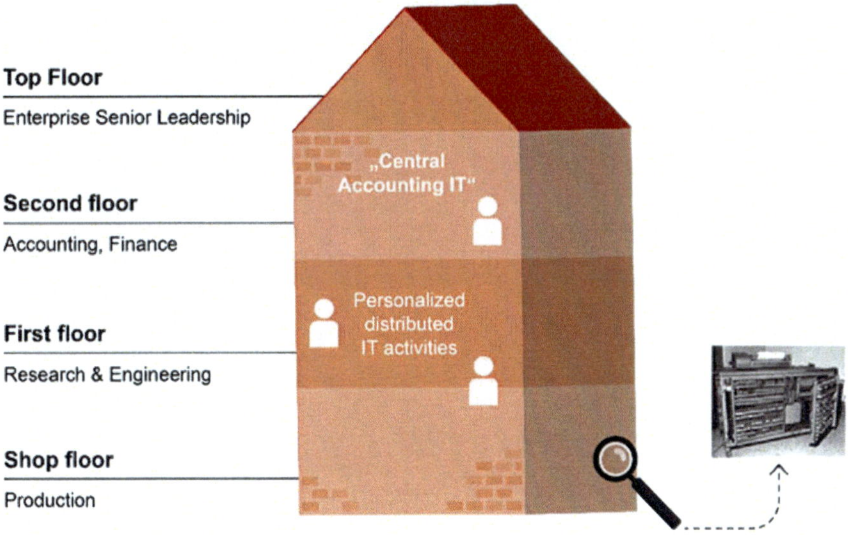

Fig. 3.10 Early organizational/local IT set-up within industrial enterprises (end of 1960s to the beginning of the 1970s)

mechanical structures or dynamic problems as well as time and quality measurements in manufacturing.

During the initial times of setting up IT departments there was no special architectural preparation of building space and offices to accommodate for these new calculation machines. Initially no special IT infrastructure drivers existed. Nevertheless, over the first ten years of industrial use it was also recognized that the enormous amount of space to install bigger computation machines such as already known from military and research center installation would need special space and utility preparations (electric power, cooling) within the old, traditional office and factory buildings.

Traditionally, the ground floor areas and later on also locations near elevators on higher floors of newer buildings were chosen to provide space for mainframe computational hardware. Special purpose and centrally set-up main computer facilities were operated by special technical staff with special skill sets to maintain and repair the still rather fragile hardware (relays, transistors etc.).

As shown in Fig. 3.11 new layouts of office and workshop/manufacturing buildings in the second phase of IT organizations in industrial companies did lend themselves to a natural separation of different IT sections each of which belonged an individual finance, engineering and manufacturing department. In addition, a first kind of "central", general IT laboratory was established to host general purpose compute power in form of mainframe and workstation clusters.

Between the early 70s and the mid 90s of the last century step by step a growing need to connect terminals to mainframe and workstations to compute clusters was

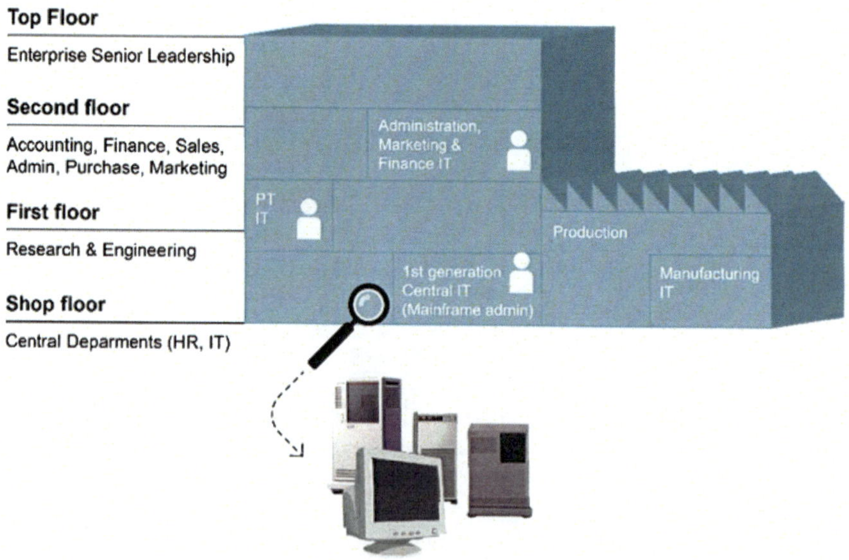

Fig. 3.11 The 2nd phase of IT organization in enterprises—the majority as local set-up belonging to functional organizations (mid-1970s to mid-1990s)

recognized. At that time, however, this trend was mainly driven to better utilize hardware for individual computation tasks.

Starting in the 70s and then sharply increasing during the 80s of the last century professional IT experts were established in big enterprises with the sole responsibility to set-up, install, maintain and improve computer installations. Small and Medium Enterprises (SME) could not yet afford those experts and waited in many cases until the mid-nineties to allow a full set-up of those IT departments with the right skill mix of hardware set-up and maintenance, software installation, database technology, network and related infrastructure, data center and store operation and customer support resp. help desk execution. In line with the sharply increasing number of users, IT applications and required skill sets to run an IT operation it became necessary for enterprise IT organizations to grow, both in size and individual array of skills.

During the 20 years between 1980 and the millennium, industrial companies could benefit from the first wave of young professionals with university degrees: those IT literate experts brought fresh information technology knowledge from university majors such as computer science, informatics and computational engineering into the so far "IT on the job self-learning" community of engineers, physicians and mathematicians.

With the evolving network technology and cross enterprise engineering and production evolution big sized enterprises and subsequently also SME had to establish in addition to their local (building-to-building) IT set-up also regional (city-to-city) IT set-ups and cross-regional and global IT set-up (see Fig. 3.12).

Information technology became a critical enabler for global economy operation. It evolved from a supporting equipment industry to an own mainstream business sector by providing bundled services out of hard- and software, telecommunication, mobile and cloud computing, IT operations and helpdesk service as well as associated policies and compliances. Despite the .com hype and crash right after the millennium all enterprises did accelerate their infrastructural IT set-up and prepared themselves for a location independent IT service provision from anywhere in the world. Therefore, the effective control over shared internet networks, related bandwidth and latency characteristics, secure virtual private network connections and most efficient software license utilization over the enterprise network has become a major business operation in industry. Those operations are executed more and more by contractually time-bonded IT-departments inside and outside the actual enterprise. Low cost country sourcing in India, China and other countries in Eastern Europe, South East Asia and Africa is still one of the major answers to a year-over-year cost saving task which many of the IT executives have to deliver as part of the yearly commitments to company's profits. From this standpoint Information Technology has been established as commodity.

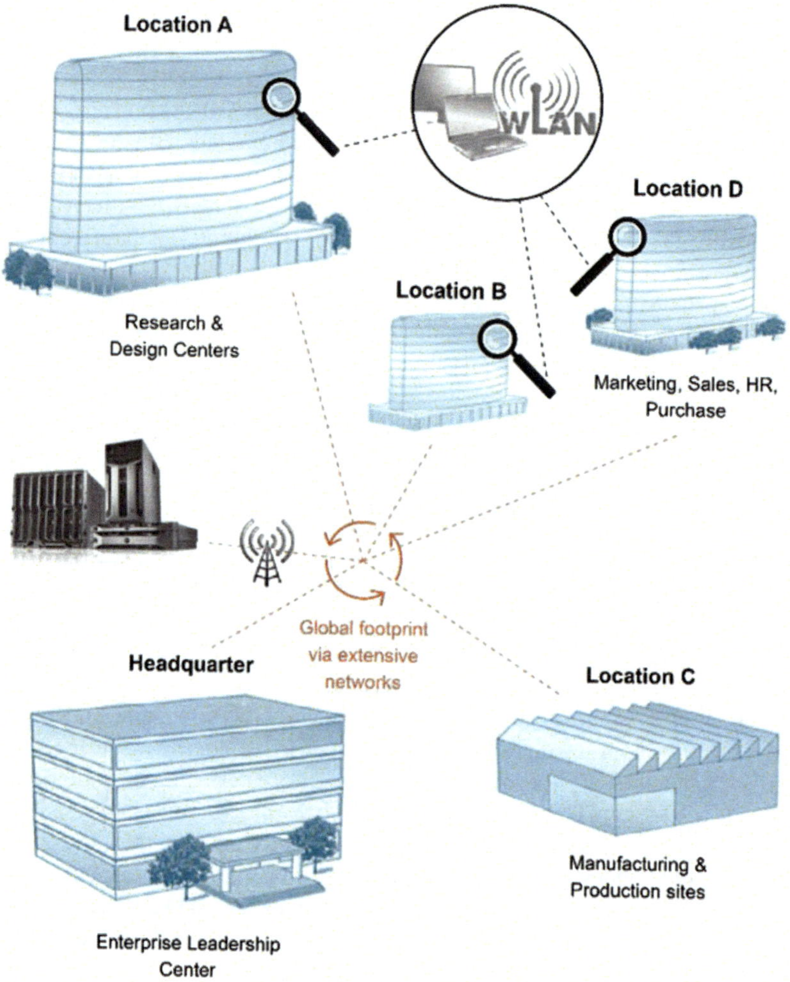

Fig. 3.12 The 3rd phase of IT organization in enterprises—global services and policies + outsourced operation (mid of 1990s to 2010+)

3.3.2 *Today's IT Factory Set-Up and Future Business Concepts*

Meanwhile each single work place in administration, accounting, product planning, project management, marketing and sales, research and development, manufacturing engineering and production control is dependent on the network connected computational intelligence with the help of hardware such as desktop computers, workstations as well as mobile connecting and computing devices and with the help of task dependent software applications. For each of these "digital value creation activities", however, it is necessary to make investments upfront to be able to make effective use of the IT tools and applications. The investments need to be divided into two parts, the obvious part, i.e. the direct IT related investment, and the non-obvious part, i.e. the "hidden" business and work competence related investment:

Information Technology Investment (ITI)

Obvious direct IT related investment to provide the base working system:

- IT architecture planning, application architecture integration
- Data model development and data base configuration and customization
- Computer Hardware and software purchase
- Network infrastructure and data center equipment (e.g. back-ups and storage)
- System administrator/user help desk training.

Business practice investment (BPI)

Non-obvious, "hidden" business and work competence related investment to establish the working capability for IT enabled value creation:

- Active support in the process of determining *enterprise key business processes* driven and actively steered by Business Senior Management in order to drive commitment and alignment of the type and degree of information technologies; such key processes are e.g.:
  - Product Process (lifecycle such as planning, engineering, manufacturing, logistic, production, use, recycle, reuse etc., compare [16])
  - Costumer Order Process (offer, order, build, delivery etc.)
  - Sales, Service and Maintenance Process
  - Key Support Processes (finance, procurement, marketing etc.)
- Research and analysis to determine the *core IT enabled and supported value creation work packages* within those *enterprise key business processes* (see above under a), such as information and digital model authoring, storage, retrieval, distribution, routing and delivery, visualization, analysis, synthesis build etc.
- Development of effective and efficient *key information process logic* to ensure robust execution of the above determined *key IT enabled and supported value creation packages*. Such logic determines the process for individuals and clarifies how often these individuals should be delivered with up-to-date information and related data or whether they should be automatically triggered by specific digital model update status. This clarification becomes increasingly important within business operations since the timely delivery of information and data sets come with effort, cost and infrastructure impact.
- Modification of traditional working practices and behaviors in order to match them with the business affordable *key information process logic*. It might well be that turn-around cycles of 24 h which might be desirable for quick engineering progress are not affordable and engineers need to find different ways of interaction and reasoning.

- Engineering of suitable conceptual data models, business/IT application methods, use cases and templates in order to realize *efficient information logistics* (data transport, distribution and information flow) and *information intelligence* (analysis of existing and synthesis of new information objects)
- Piloting of and educating the new business (process/method) approach in a phased manner—this investment needs to include testing of the corresponding IT application developments in order to finally justify and confirm the direct IT investment.

Whereas the first type of investment (ITI) is an accepted—but very controversially discussed—investment schema in enterprises the second one (BPI) is merely established as a fundamental, regular investment schema. As Information Technology can no longer provide business value without this second type of investment (BPI) most companies are in a crisis towards business operation innovation.

From this standpoint, Information Technology is only in the beginning to get fully established as a key competence and indispensable skill set in professional life, especially for engineers in Virtual Product Creation. Industrial enterprises in today's world have not yet sufficiently understood how to establish an investment schema to guarantee the Business Practice Investment (BPI), which constitutes the sufficient condition to guarantee robust digital/IT related value creation. The reason for this "currently increasing lack" is twofold:

- The non-existence of a digital value creation model and consequently the limitation of today's business controller models to only accept ITI funding.
- Limited sensitivity and knowledge of enterprise leadership management on/about the critical dependency between business and collaboration competencies and information technology.

A written survey/telephone interviews based on 16 key questions/assessments conducted in the first half of 2011 amongst more than 10 senior IT business managers in automotive (BMW, Daimler, Ford, Mazda, Continental, both headquarter and within the regions) and technology (energy) industry (Siemens) as well as additional single expert interviews within the mechanical engineering, railway and aviation/aerospace industry (experts from Airbus, Bombardier, MAN, Rolls-Royce) have revealed the following interesting set of characteristics:

The top three IT budget spending line items relate to:

- Operational business support such as executing data centers, help desks, data base administration etc. together with dedicated projects:
- to reduce the number of IT applications (*IT consolidation*) and hence to realize a scalable reduction of license cost;
- to use similar or even the same IT applications and server installations where possible (*IT harmonization*);
- to restructure IT "back-office" functions by building up scalable "enterprise architecture integration" (EAI) solutions by introducing SOA (service-oriented

architecture) interfaces between individual business process critical business IT applications and the enterprise digital backbone (web service platform);

- to reduce license cost, improve compliance and policy control and streamline internal operational footprint by outsourcing ITIL based services to special IT companies.

The top three business tasks of the internal IT organization in terms of effective use of internal IT headcount relate to:

- Acting as "IT front office" in terms of:
- Following strategic decision by Senior Management on business processes and organization as well as consulting Senior Business Management on IT solutions as part of future business strategies and
- on specific IT enabled business functions with regard to upcoming business process needs and opportunities.
- Project Management execution with respect to new IT solution requirements management, development and coding coordination as well as test and deployment tasks.
- Governance, compliance control and capability tracking coordination work.

Obviously, there exist a tendency in global acting enterprises to balance out differences between regional business needs, availability and cost of human expert competences and expenditures for recruitment and knowledge ramp-up. Based on the individual IT business set-ups the organization IT decides on in- and outsourcing strategies. The level of outsourcing has common aspects across the interviewed enterprises and branches, but also significant differences. As a common business approach in information technology divisions a range of tasks is outsourced.

The first group of "highly outsourced IT business tasks" comprise the following activities:

Data center operation and data base administration: on average the outsourcing rate is higher than 80%, the distribution, however varies between 50 and 100%. Two interesting observations are visible:

For companies with a significant global footprint the degree of outsourcing for this task seems to be higher in the European and US locations.

The type of outsourcing differs from *full operation outsourcing* (i.e. outsourcing incl. management of the operation to an outside company) to *mainly labor outsourcing*, i.e. keeping the management control within the enterprise but outsource the actual "doing work" to agency headcount. The later one is more popular in Anglo-American and some far east companies since legislative work regulations in those countries provide the opportunities for mid to long term arrangements of such outsourcing set-ups in contrast of different working laws in many European countries and Japan.

User help desk support, incl. first line support (dispatcher of the central user help desk), second line support (knowledgeable contact persons or "on the job" support personnel to provide the direct user support at the working desk) and third line support (full technical experts and method consultants who back-up the second line support. Common budget practice today is central funding for those operations—service oriented budget contributions from the various business and engineering disciplines have not yet been implemented!

Technical execution of IT application development projects: surprisingly enough, the interview/questionnaire activity revealed that all Automotive OEMs heavily rely on outside companies to operationally execute and technically lead the customization and configuration work of existing market IT solutions and to develop new/additional IT application solutions. In the meanwhile, the majority of automotive, railway and aviation/aerospace OEMs no longer keep the core expertise for such IT development tasks and therefore, in many cases the OEM personnel concentrates on operating on the project-steering level. There seems to exist a difference to other industries (such as Siemens in the technology/energy sector) and even to some automotive suppliers where the level of outsourcing of those tasks is limited to 30–50%.

A second group of outsourcing levels is formed by IT business tasks where the outsourcing rate is usually smaller than 20% at all companies.

This second group of "less intensive" IT business outsourcing consists of the following two activities:

Consulting support of Senior Management for IT enablers in enterprise business strategies. Outsourcing for such as task is limited to special assessments and expert back-ups but not for the entire range of consulting activities.

Project lead of IT application development tasks—please also compare the statements under the third task of highly outsourced task above!

There exist a third group of miscellaneous outsourcing levels with no clear trend across the companies and industry branches.

This third group comprises the following IT business tasks:

IT infrastructure and IT enterprise (architecture) integration: here the level of outsourcing differs between 30 and 100%. It seems that US based companies still keep a higher internal rate within such tasks.

Research studies, new innovations and related pilot projects: there is a significant difference between German based OEMs which favors a significant (at least 30–50% rate!) of outsourced work, based on their excellent and fully trusted network with German universities and research institutes versus Anglo-American and Japanese companies with less than 15% and not necessarily using public/independent but private expertise from outside. Due to a high appreciation of application-oriented research in major parts of Europe (such as Austria, Germany, Italy, Sweden, Switzerland) it is possible that universities and research institutes (e.g. such as Fraunhofer) have established firm and clear intellectual property agreements with industry and still remain well established within the science community. In the Anglo-American world science and research has been treated by industry and universities more as an academic exercise with a clear separation line as compared to industrial praxis. As an additional observation by the author in having exchange with Chinese authorities, China seems to follow the central European way of tight interaction between company and university/research institute innovation work in accelerating innovation schema.

Traditionally, budget spent on Information Technology within an enterprise is subject for discussion, classification and ranking. The expert interviews revealed interesting trends on IT budget spending which can be summarized as follows (in comparison of the period between 1990 and 2010 with reference points in 1990, 2000 and 2010):

The overall IT budget in percentage to the overall enterprise budget has decreased from 2.5 to 4% in 1990 down to 1 to 2% in 2010. Some companies had significant higher IT investments around 2000 and did reduce them again towards 2010. No generic rule exist how to measure IT budget percentage on overall budget (due to the individual definition how to count e.g. BPI, i.e. business practice investment: as IT budget or business budget).

There exists a general trend that operational IT expenditures such as for data center and help desk operation are cut by using outsourcing opportunities and by establishing commodity solution instead of special enterprise customized solutions. IT project budget with a close relation to business initiatives usually remain on a similar level.

Globally acting enterprises have invested intensively within the Asia–pacific region. Hence, those enterprise locations have sharply increased their IT budget by more than 40 to 70% over the period between 1990 and 2010.

There exists a significant difference between the companies on how they use their IT budget with respect to new IT solution development and to the related Business Practice Investment (BPI): companies such as BMW, Daimler and Siemens use double digit percentages of their overall IT budget for such activities, whereas Japanese companies such as Mazda and automotive suppliers such as Continental keep it limited on a low single digit percentage.

The percentages of budget spent on Virtual Product Creation (VPC) respectively Product Lifecycle Management (PLM) solutions versus entire IT budget vary across the enterprises and highly depend on the replacement and innovation cycle of the individual company solution. In all enterprises, however, license cost is the dominant factor and accounts for 70 to 90% of the operational IT budget for this field. Second largest element of the running cost is represented by purchased service for experts, IT consultants and application specialists.

The next focus of the interviews was centered on the question which internal model is used within the enterprise to determine, define, drive, develop and deploy new Virtual Product Creation (VPC) and/or Product Lifecycle Management (PLM) solutions. Interestingly enough, there is the following difference: today's traditional model within matured European and American enterprises favors a lineup of two separated groups, one central expert and consulting team in IT (sometimes also named Process IT) and another central or networked digital process and method department within Engineering. The Asian model still sees a concentration of such responsibility within the IT organization. Due to the rather fragmented business setup the Energy, Machine and Technology sector (e.g. Siemens, MAN Diesel and Turbo) as well as the railway sector (e.g. Bombardier) run all of their major Virtual Product Creation and PLM projects out of the IT organizations and Senior Management encourages/enforces a link to subject matter experts and/or key users from key engineering departments on a project level only. Today's Senior Engineering Management oftentimes still believe that leadership in driving new Virtual Product Solution might distract Engineering Middle Management from their ordinary operational task and that therefore IT Managers "own this task" and should get supported by Engineering Project Leaders only.

In summary, it is noticeable that the majority of *today's* enterprises still struggle with the fact that Virtual Product Creation and PLM innovation is mainly driven out of IT departments rather than by engineering itself. This explains the dilemma of many companies: new ways of engineering oftentimes are missing the right business motivation and targeted execution path. IT experts and process/method project leaders and consultants are no longer in the position to drive such business conversion if engineering management and experts are not taking up the leadership.

As a forecast into the future, the IT Managers were asked the question how they see the changing landscape in the future, i.e. within next five years. They were given the following three Virtual Product Creation competency set-ups to choose from for a best future set-up:

- Central organization as a combined PMTI (Process, Method, Tools and Information standard) competence team;
- Two central organizations, one information technology competence team within the IT-department, one as best practice Business PMTI team in Engineering;
- Several qualified decentralized information technology and PMTI expert teams under central coordination.

The majority indicated that their set-up of today follows set-up #1 but that they probably will transition to a set-up #2. All managers, however, have shown significant interest in a set-up #3, especially since a set-up mix might be necessary for the future in order to drive the differences of IT enabled business according to the following suggested way:

- Set-up #1 for IT commodities and IT maintenance;
- Set-up #2 to allow best user segmentation whilst keeping a high degree of communization for the backbone solutions;
- Set-up #3 to become capable of following a true business value driven approach to stay competitive in a fast-changing business environment.

From 2015 onwards, the term "bi-modal" IT became popular amongst IT Management. On the one hand companies are handicapped by legacy IT technologies which force them to run on old data base platforms and mainframe computing technology especially in their back-end solutions. On the other hand, new IT development approaches (agile software development, open community design) and new technology elements (e.g. converged databases, html5 visualization etc.) make it possible to quickly establish customer front-end solutions. It is, however, controversially discussed what the best approach will be to merge them or to guarantee smooth co-existence to each other.

The following chapters will provide dedicated insights to the Virtual Product Creation history as well as to the Virtual Product Creation technology and solution landscape.

References

1. Levi P, Rembold U (2002) Einführung in die Informatik für Naturwissenschaftler und Ingenieure. Hanser Verlag
2. Carruccio E (2006) Mathematics and logic in history and in contemporary thought. Aldine Transaction
3. Ihrah G (2001) The universal history of computing: from the Abacus to the quantum computer. Wiley, New York
4. Lister M, Dovey J, Giddings S, Grant I, Kelly K (2003) New media: a critical introduction second edition. First published 2003, second edition published 2009 by Routledge, Abingdon. ISBN13: 978-0-203-88482
5. Eccles WH, Jordan bWF (1919) A trigger relay utilising three-electrode thermionic vacuum tubes. The Radio Review, p 298
6. Zuse K (1970) Der Computer. Mein Lebenswerk. Verlag Moderne Industrie, München
7. Randell B (1976) The colossus, technical report series No. 90. Computing Laboratory, University of Newcastle upon Tyme, UK
8. Dale N, Lewis J (2009) Computer science illuminated, 4th edn. Jones and Bartlett Publishers, Inc.
9. Slater R (1987) Portraits in silicon. MIT Press, Massachusetts Institute of Technology
10. Bernstein J (1984) Three degrees above zero. 1. Bell Telephone Laboratories - History. Press Syndicate of the University of Cambridge, Great Britain
11. Beyer KW (2009) Grace hopper and the invention of the information age. Massachusetts Institute of Technology, USA
12. Godbole AS (2005) Operating systems, 2nd edn. Tata McGraw-Hill Publishing Company Limited, New Delhi, India. ISBN 0-07-059113-X
13. Ceruzzi PE (2003) A history of modern computing, 2nd edn. MIT, USA. ISBN 0-262-53203-4
14. Tannenbaum A (2009) Moderne Betriebssysteme, 3rd edn. Imprint der Person Education Deutschland GmbH, Germany. ISBN 978-3-8273-7342-7
15. Spur G, Krause FL (1997) Das virtuelle Produkt: management der CAD-Technik (The virtual product: management of the CAD Technology). Fachbuchverlag Leipzig. ISBN-13 978-3446191761
16. Carlson B, Burgess A, Miller CH (1996) Timeline of computing history. The IEEE Computer Society. Accessed 17 June 2010. http://www.computer.org/cms/Computer.org/Publications/

Chapter 4
Virtual Product Creation (VPC) Explained

Executive Summary

This chapter deals with the following topics:

- Understanding of the term and development element *virtual product*,
- Explanation the activities within and the capabilities of *Virtual Product Creation* and
- The difference and relationship between *Virtual Product Creation* and the overall concept *PLM (Product Lifecycle Management).*

Quick Reader Orientation and Motivation

The intention of this chapter is:

- to provide an overall understanding of the discipline *Virtual Product Creation (VPC)* as part of the product lifecycle concept and to map the VPC activities throughout the lifecycle engineering spectrum;
- to explain the principle elements of a virtual product;
- to introduce the variety of digital models as part of the virtual product elements;
- to provide an understanding how *Virtual Product Creation* differs from other digital enterprise capabilities;
- to introduce the major capabilities of *Virtual Product Creation* in the context of technical system development, digital manufacturing and service development.

4.1 The New Engineering Discipline *Virtual Product Creation*

Virtual Product Creation is a discipline for digital engineering in Product Development (PD) as well as in Manufacturing/Production System Development and Process

R. Stark, *Virtual Product Creation in Industry*,
https://doi.org/10.1007/978-3-662-64301-3_4

Planning. During the last ten years, it became evident that the ongoing digital activities of Maintenance Repair and Overhaul (MRO) and ongoing product updates have also become part of Virtual Product Creation activities. Especially functional product updates via the delivery of software updates as well as algorithmic self-learning aspects as part of autonomous operations will drive regular and intense Virtual Product Creation processes. Virtual Product Creation, therefore, represents a major part of the overall digital enterprise capabilities.

Virtual Product Creation encompasses many engineering activities, which in contrast to the traditional engineering approach no longer uses physical but virtual elements as a "model for progression".

The term *virtual* implies that those elements do not (yet) exist in the traditional physical world and therefore only exist in a specific transient or seeming format, i.e. they are not yet real. In order to avoid that such virtual models only exist in the minds of individual human beings in their specific roles such as engineers, planers, managers, professors, students, workers etc., the power of digitization and standardization helps to make them "existent, visible and executable" on computers. Hence, the Webster dictionary explains the term virtual as, amongst others, "being on or simulated on a computer or computer network".

The term *product* means traditionally "a thing that is produced or created by labor". Nowadays, however, a product can also be a service or a software. A product is offered on a market and has a lifecycle. A product in general can be used in many different real-world environments but typically, the field of use must already be anticipated early on during its ideation. Typical product examples are:

- a specific cutting machine for usage in a factory,
- a sedan automobile for passenger transportation vehicle as part of traffic system in certain global regions,
- a coffee machine for private use in a household or as a vendor machine in a cafeteria, etc.

Since the beginning of the millennium, the concept of a product-service system evolves as a further evolution of traditional products: product service systems treat products and their offered service as an integrated offering to the market.

The term *virtual product* was formally introduced in a comprehensive manner by the researchers Krause and Spur in 1997 (Das virtuelle Produkt, the virtual product, compare [1], p. 307):

- The *virtual product* is the central information carrier for a complete computer-aided product development. The interaction with the *virtual product* is only possible via the help of modeling and verification functionalities [of a (PLM-type) software].
- The *virtual product* can be defined as a realistic computer-aided representation of a product, with all functions of the product for all life cycle phases of the product. Specifically, the activities of product planning, design/styling, design engineering, manufacturing engineering, production operation, recycling and service are addressed.

- The *virtual product*, therefore, is a computing model, which describes the product with the help of information sets and product functions as realistic as possible.

Figure 4.1 shows the definition and illustrates the basic understanding of a *virtual product*: the virtual product today is represented by a series of computer-aided models and information sets which requires the existence of corresponding modeling and simulation algorithms usually represented by different sets of IT-applications (such as CAD, CAE, CAM etc.).

The cascade of different elements associated with a virtual product is shown in Fig. 4.2: first, it is necessary to provide a *designation* or sometimes even an explanation of the virtual product in mind. If a new virtual product—or a sub-system or component of a virtual product—is "born" in the mind of a developer, oftentimes just a "nickname" or a general term is being used before assigning an official part or product designation to it. Strict naming and numbering data base regulations as part of the overall PDM or PLM environment also kick-in and require such specific designations in order to guarantee precise and reliable collaboration with other engineers and departments.

It is important that all members of the virtual product creation community associate the proper *imagination* with the virtual representation of the product. For such goal, it is essential to use symbols for recognition (e.g. icons) and simple illustrations together with short explanation of the main function, the architecture and the context of a virtual product.

In addition, it is necessary that the virtual product is represented in a certain way with the help of product model representations. As outlined in Fig. 4.2 the most common product model representations are the following ones:

- Design models (e.g. CAD, i.e. Computer Aided Drafting/Design): partial models, which describe the geometric shape and topology of a product, meanwhile often enriched by additional semantic data such as technological parameters (tolerances, reference system), functional information and manufacturing and quality inspection data.
- Behavioral models (3D CAE models) of the virtual product such as FEA (Finite Element Analysis), MBS (Multi Body Simulation) or other 3D CAE model types.
- Physics models, also called 1D CAE models: they represent the functional and/or behavioral models in a specific mathematic way within out explicit 3D geometry.
- Full selection of a high number of different virtual product models in specific light weight visualization representation, also known as DMU (Digital Mock-Up) model or even in 3D stereoscopic format known as Virtual Reality scene model.
- Various specific models to represent GD&T (Geometric Dimension and Tolerancing), or CAM (Computer Aided Manufacturing) or CAPP (Computer Aided Process Planning) for manufacturing process and tool path information capturing.

The term *product creation* describes all business, planning and engineering activities of product planning, product development or engineering and manufacturing/production system development in order to guarantee form fit and function of

What is a Virtual Product?

Definition:

A virtual product represents all - or at least major - capabilities of a real functional product by integrating various sets of digital product models with corresponding simulation algorithms. Although it exists only virtually it can be configured, used and tested as part of different physical, functional and interactive conditions and environments.

Understanding:

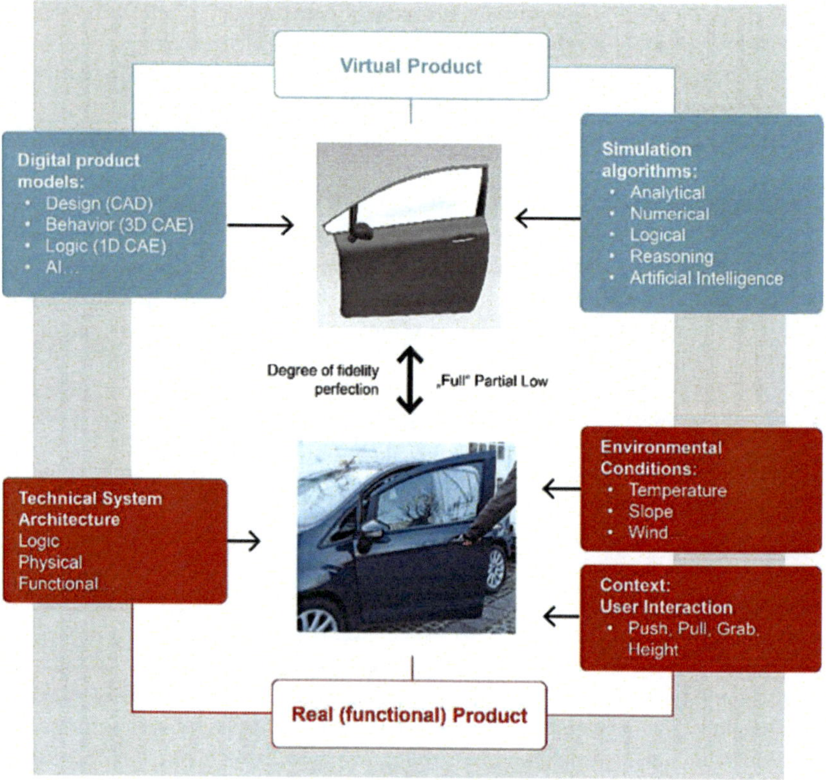

→ **Consequences:**

The degree of fidelity perfection match between the real product and the virtual product depends on:

A) avilable digital modeling capabilities
B) acceptable digital modeling efforts
C) needs and priorities to virtually validate, test and use real (functional) product behaviours

Fig. 4.1 What is a virtual product? Definition, understanding and consequences

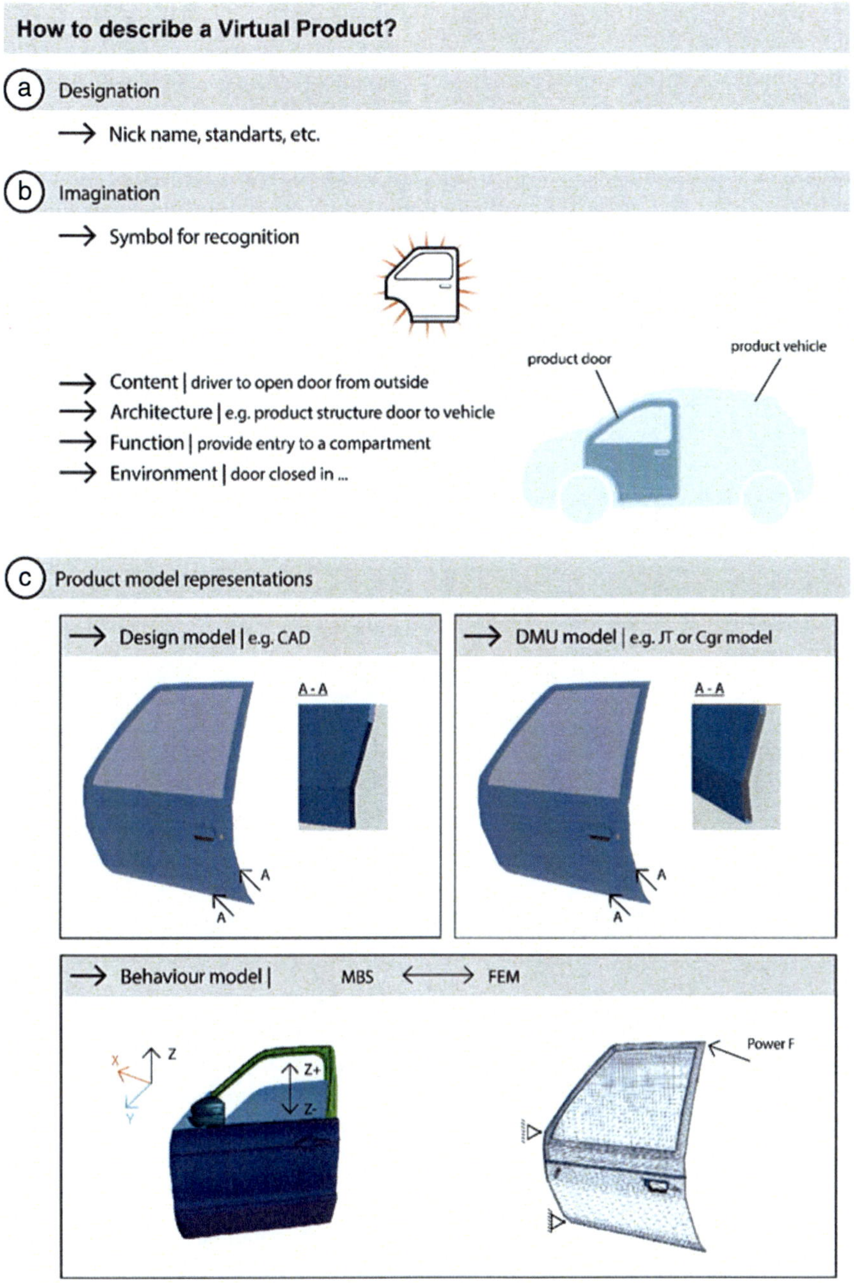

Fig. 4.2 How to describe a Virtual Product? Designation, imagination, product model representation and meta-information

the product itself and its production and operational readiness (compare [2]). In the English-speaking world there exists a synonym to the term *product creation* that is called *product realization*. However, the term product realization implies oftentimes also the notion of "putting something into reality" and, therefore, is oftentimes understood as the tail end *of product creation*. Therefore, the author recommends using the term product creation, which also nicely fits into the German term *"Entstehung"*: here, it carries a notion of "following an implicit process of growing something to its final existence".

Definition of *Virtual Product Creation*:
Virtual Product Creation constitutes of all process steps and engineering activities (and their iterations) that use digital applications, IT tool functions, software algorithms, working methods and assessment and decision capabilities to create, modify, simulate, analyze, test, validate, verify, sign-off, release and exchange *virtual products* and their derivations.

Therefore, *Virtual Product Creation* is determined by the explicit tasks and activities of engineers and the specific capabilities of computer algorithms and software, in order to define, change and use CAx models and other digital models (such as algebraic or function models) in the context of other digital data and information.

Virtual Product Creation therefore is pivotal to determine shape, function and characteristics as well as logical and physical behavior of real products under *"adjustable virtual conditions"*.

4.2 Virtual Product Creation Capabilities and Activities

The activities of *Virtual Product Creation* must be understood in the context of the overall lifecycle of products and related services. Figure 4.3 shows typical *Virtual Product Creation activities* along the product life cycle. In the beginning, when a new product idea and plan is born, the full set of system engineering and/or product development activities is started and followed up very stringently, including all associated project management and gateway execution schemas. In general, there exist different styles of development processes:

- from traditional concept design, embodiment design, detail design and physical prototype engineering and testing;
- via V-shape illustrated and phased mechatronic and systems engineering approaches including formal development phases for requirements engineering, functional modeling and networks, system architecture and behavior simulation,

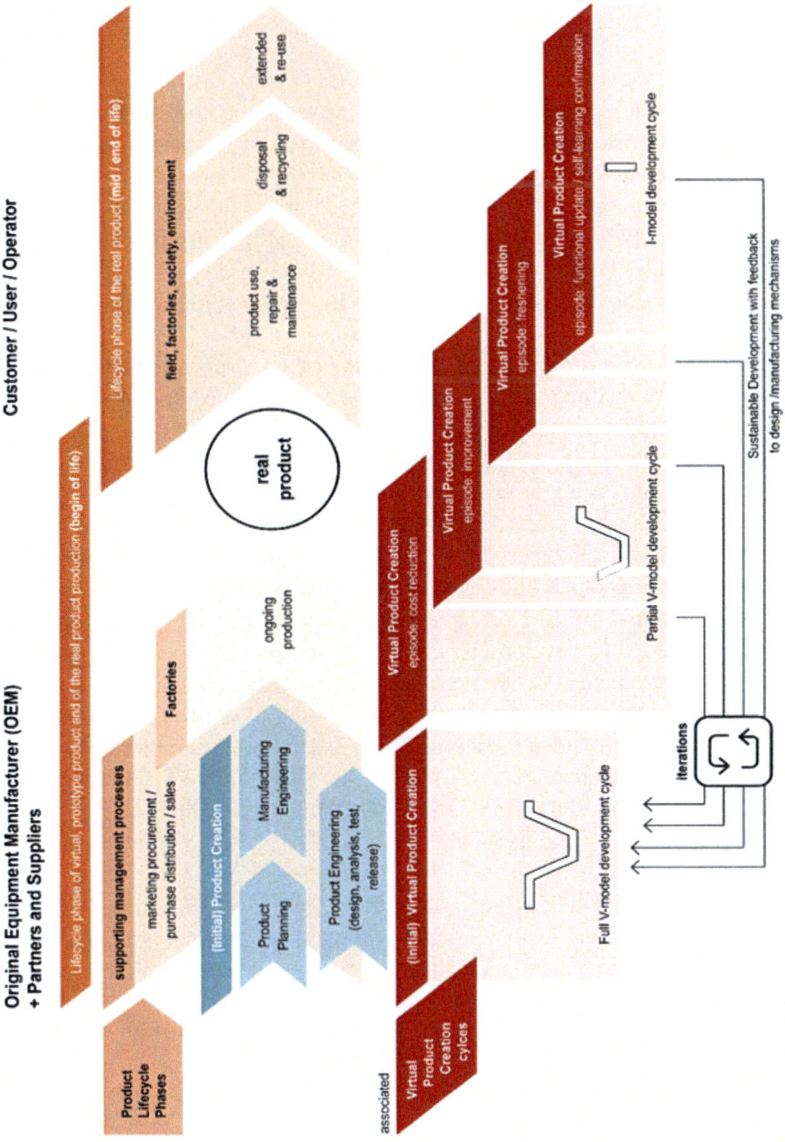

Fig. 4.3 Initial Virtual Product Creation (VPC) and VPC episodes later in the life cycle

disciplinary detail design (E/E, mechanics and software) up to different kind of prototyping, sign-off and release management;
- up to agile development activities based on incidents and/or small development task arrangements with stakeholders and costumers.

Meanwhile, all of those development and engineering styles are deeply relying on *Virtual Product Creation activities* and related capabilities such as *methods and tools*. The degree of *Virtual Product Creation execution* is a matter of *affordability* (cost and resource wise), *necessity* (need for upfront and ongoing simulation and confirmation to avoid risk and gain of clarity during the development process and its stage gates) and *complexity* (part, assembly, sub-system, full product or system and interface to other systems). Unfortunately, standards for the degrees of *Virtual Product Creation capabilities* in industrial companies and industry branches don´t yet exist. Consequently, there is a high degree of uncertainty regarding the questions of which technologies and methodologies need to be deployed in order to become and stay competitive.

Hence each company has the burden to define their own virtual product creation destiny and capability roadmap that is difficult due to the limited knowledge ordinary management holds in the subject of *Virtual Product Creation*. By the end of the 90s software vendors in Europe and in the US started to (mis-) use the term product lifecycle: originally, the new engineering discipline *Virtual Product Creation* encouraged designers and analysts to substantially create models for design and behavior simulations. Over the years, however, those activities started to create challenges and problems due to missing capabilities to store and manage computer-aided models and files consistently and safely. Consequently, throughout the 90s of last century, software vendors started to develop IT-systems for engineering data management (EDM), team data management (TDM) and finally product data management (PDM, the next level up to EDM and TDM), which provided all tool sets to name, number, store and manage computer-aided models and their associated files in the context of projects and product structures.

Around the millennium, the term PLM (Product Lifecycle Management[1]) was originally introduced by IT vendors in order to gain more traction in the mindset of business processes and the strong push of re-engineering efforts strongly supported by Senior Management and the new business sector of Business Consulting. PLM in its neutral meaning carries the idea of describing the full life cycle of a product and its surroundings such as the factory, its production systems and cells, other resources and the usage in the field. The lifecycle of a product, therefore, is divided into three phases Begin of Life (BoL), i.e. from the first idea of a product until its readiness to be manufactured, Mid of Life (MoL), i.e. the production of the product as well as its usage and maintenance in the field, and End of Life (EoL). Nevertheless, the core driver for all of the PLM embedded digital engineering and simulation activities and associated business and IT solutions are represented by the appropriate *Virtual Product Creation tasks*.

---

[1] Please compare detail descriptions and explanations of the term and the disciplines of PLM in [3–5] and follow the major technologies of PLM in Chap. 11.

In addition, the new influence of IoT (Internet of Things) driven data streams from already existing live products will need to get connected to the upfront models of the BoL (Begin of Life) phase of the entire product lifecycle. Such connections are established with the help of the new generation of virtual product models, called *Digital Twins.*

As depicted in Fig. 4.3 the style and intensity of virtual product creation activities might differ along the needs in the engineering and development cycles. During the last years, it became clear that more and more development episodes and specific aims of development need to be supported by virtual product creation activities. If indeed, the degree of self-learning and autonomous products and technical systems will substantially and steadily grow—as the strong efforts in the mobility and industry sector indicate—then the virtual product creation skill set needs to be transformed in order to "on the fly" simulate the consequences of AI (Artificial Intelligence) driven software control on technical products and systems in the context of the environment.

The core competence *Virtual Product Creation* provides a rich and complex mix of capabilities that consists of digital processes, methods, tools and model/information/data objects. Figure 4.4 depicts the situation for technical system development, which includes product development. *Virtual Product Creation* is executed within a *domain layer*, the *system development layer* and the *application layer* with specific digital modes that correspond to specific development tools. In

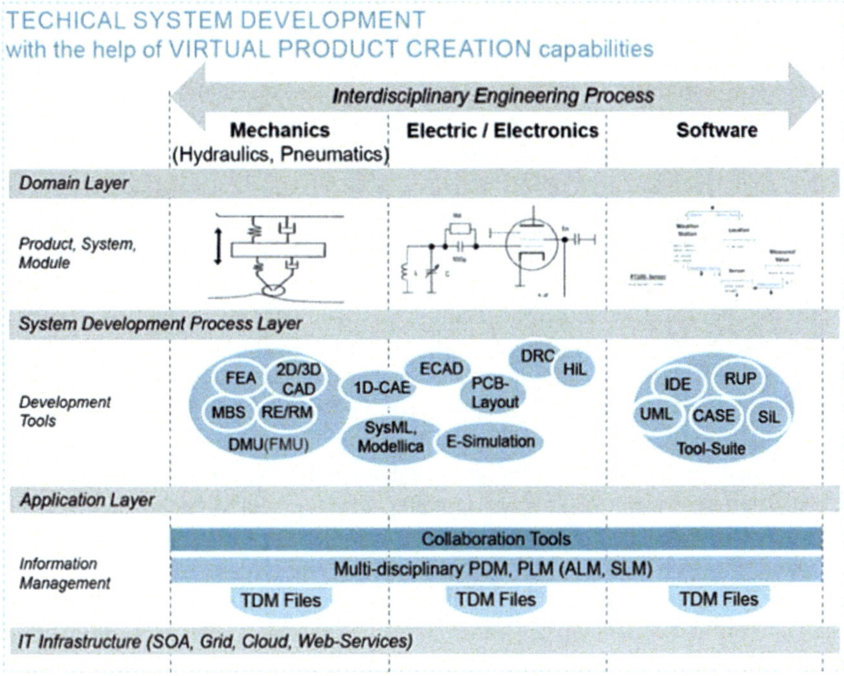

Fig. 4.4 Layer view of Virtual Product Creation capabilities for technical systems development

order to maintain, manage and share all of those models and information it is essential to maintain and organize a substantial information management environment based on a robust *information technology (IT) infrastructure layer*. Information and data management environments do offer a wide range of handling of documents, models meta data and administrative data within the context of user-oriented functions and underlying data bases: starting from *Team Data Management (TDM)*, mainly architected for file-based data repositories of teams of 20–30 individuals. Those environments have been extended to *PDM (Product Data Management)* and *PLM (Product Lifecycle Management)* environments, which offer globally available full configuration and structure rich repositories and exchange platforms. Nevertheless, still today, there are different environments for the "physical" model world of hardware and for line of code environments for software.

As the *domain layer* at the top of Fig. 4.4 shows, it is essential to differentiate between the three major technical domain areas mechanics (hydraulics, pneumatics), electric/electronics and (embedded) software. For each one of those major domains different sets of modeling and development environments and tools have been developed.

The mechanics VPC portfolio offers development tools such as *Computer Aided Design* (CAD; design of components and assemblies), *(3D) Computer Aided Engineering* (CAE; algorithms to analyze shapes and behaviors with the help of discrete volumetric or surface elements and components) tools such as *FEA* (Finite Element Analysis) and *MBS* (Multi Body Simulation) but also simplified mathematical equation based *1D CAE* approaches. From 2010 onwards, official tools for requirements management (RM) and requirements engineering (RE) have been also added to the mechanics-oriented development world (before it was mainly used in software development). *DMU (Digital Mock-Up)* provides a digital environment which can visualize and analyze dynamically complex products in real time manner.

The next level up from DMU is called FMU (Functional Mock-up). There currently exist two basic understandings of the FMU: the first one concentrates on the upfront representation of all pure (neutral) functions and their interplay, whereas the second one does encompass all logical and behavioral characteristics, i.e. not just the mechanical (hydraulic/pneumatic) ones. The later one obviously requires the descriptions and behavior models of the electrical/electronic domain and of the software control domain in addition to the mechanical (hydraulic, pneumatic) one.

The entire mechanic VPC portfolio has its foundation on the mechanical physics models based on masses, forces, inertia, momentums and torques, displacements, velocities and accelerations, both in static and dynamic circumstances.

The electric/electronics VPC portfolio offers development tools and environment such as E-CAD and PCB (Printed Circuit Boards) layouts and simulations (e.g. schematics of electric circuits and networks). In addition, there exist Design Rule Checking (DRC) tools for determining the best parameter set-up for semiconductor manufacturing and complex HIL (Hardware in the Loop) test environments in order to provide test beds for embedded software testing against a full set of the functional behavior of the overall technical system under development. The physical parameters and models of current, voltage, magnetic flux and associated electronic functions and networks represent the basis of this VPC portfolio.

The relatively young discipline of Software Engineering as part of Computer Science made its way into the Virtual Product Creation portfolio from the 2000s onwards as part of the sharply increased embedded software control of modern mechatronic products. Here the tools suite embraces a range of different capabilities in order to develop robustly and test software code and to verify its function within the bigger context of the technical product/system. *Integrated Development Environments* (IDE) as well as *CASE* (Computer Aided Software Engineering) provide connected engineering tools for code editing, code testing, software compilation and linkage and build automation as well as error reporting and explanation. CASE also provides orientation for software development approaches such as waterfall or scrum development. It also provides options for requirements management for information system and software development as well use case specifications, e.g. in the context of *UML* (Unified Modeling Language). Some software companies, such as IBM, have established specific environments such as *RUP* (Rational Unified Process) in order to provide adaptable process frameworks of the entire software development innovation and development operation and testing. Finally, there exists a range of *SIL* (Software in the Loop) engineering and test environments, describing a test methodology where executable code such as algorithms (or even an entire controller strategy), usually written for a particular mechatronic system, can be tested within a modeling environment that can help prove or test the software.

Figure 4.5 illustrates the Virtual Product Creation capabilities within the context of digital manufacturing. Principally, there exists a similar technology world as in

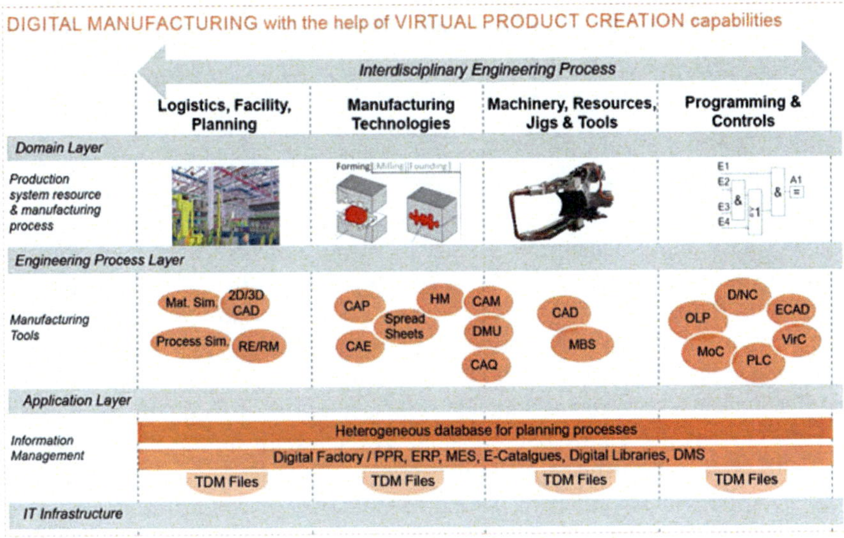

Fig. 4.5 Layer view of Virtual Product Creation capabilities for digital manufacturing

the technical system development arena; however, the viewpoint in manufacturing is different. Typically, the following four areas of expertise prevail with respect to Virtual Product Creation for digital manufacturing:

- logistics, facilities, overall "digital factory and process" planning;
- simulation of specific manufacturing technologies and physics;
- modeling, simulation and management of machinery, resources, jigs and fixtures;
- programming of robots and machines, control engineering.

In all these VPC disciplines, a robust data management regime needs to be orchestrated through the application of specific data management solutions like TDM (Team Data Manager) as well as within overall PDM (Product Data Management) environments. Different kinds of underlying data bases (files based, client server based) exist in order to safely store and provide data and information of the final product itself, the manufacturing process steps and related technical details as well as the wide variety of resources (tools, fixtures, jigs, machines, workers etc.).

As depicted in Fig. 4.5, the *Virtual Product Creation* capabilities *for logistics, facilities and overall production systems and lines* provide a range of mathematical and digital process planning and simulation techniques. They are necessary in order to describe, simulate and analyze the interaction of the manufacturing objects within the *Digital Manufacturing (Factory)* set-up, i.e. the digital representation of the real production environment in a factory.

The physical behavior prediction of material cutting, flow and forming as part of specific manufacturing technologies such as milling, late work, casting, forming, deep drawing, drilling, welding, brazing, coating etc. does require computer aided engineering tools (FEA, MBS etc.) and methods. Interactions with the machines themselves do incorporate more and more *HMI (Human Machine Interface)* simulations. Overall configurations are oftentimes stored in type of spreadsheets linking all critical machine process parameters.

Automation of discrete manufacturing cells with robots and jigs require a solid simulation of dynamic forces, momentum and torques as well as efficient motion control, which is achieved by traditional modeling and simulation packages such as CAD (Computer Aided Design) and MBS (Multi Body Simulation). Automation within an overall manufacturing cell or production line, however, needs software and hardware control. Hence, a wide range of *Virtual Product Creation solutions* is offered for such tasks: *DNC (Distributed Numerical Control)*, *PLC (Programmable Logic Control)* and *OLP (Offline Programming)* make it possible to create, simulate and analyze/test upfront—as part of ViC (Virtual Commissioning) within the *Digital Factory* environment—and to finally provide control programs for all numerically controlled machines and robots within the real factory.

As part of manufacturing engineering, the VPC capability *MoC (Management of Change)* is key in order to manage all engineering changes during the development of the product, the manufacturing process and all related manufacturing resources. Later on during real production, the *MOC (Manufacturing Operations Center)* is critical in order to control ongoing job batches as part of *MES (Manufacturing Execution Systems)*.

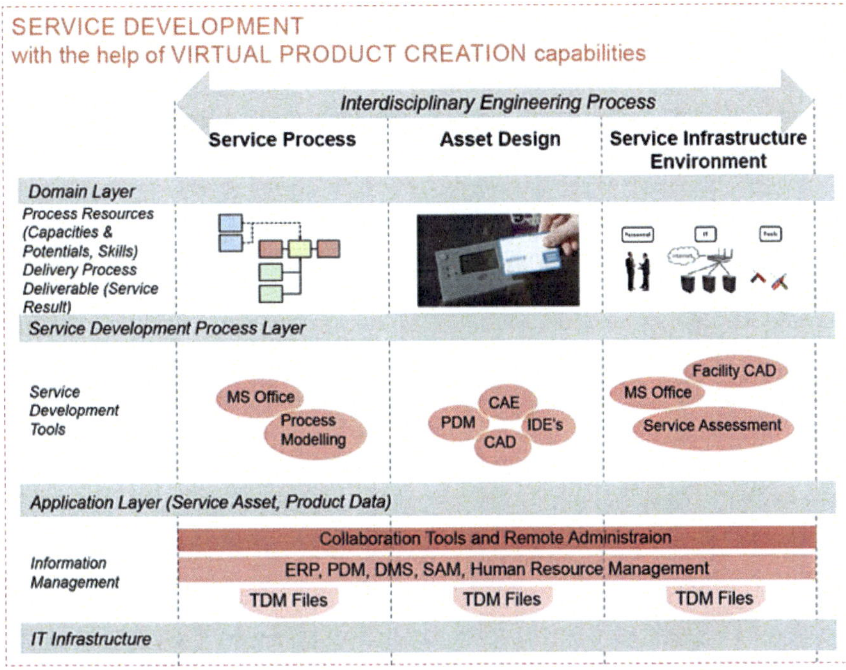

Fig. 4.6 Layer view of Virtual Product Creation capabilities for service development

Finally, Virtual Product Creation capabilities should be explained in the context of *service development* (see Fig. 4.6). From 2010 onwards as part of the increased importance of maintenance and customer service as well as of product-service systems (PSS) *Virtual Product Creation capabilities* are increasingly requested within the context of the *service process design* itself and in close conjunction with the related *asset design* (i.e. products, machines, gadgets etc.) and the service infrastructure environment (compare Fig. 4.6).

One recognizes the use of similar core VPC capabilities also here in the layer of service development tools but within a different context. The usage of modeling paradigms and tools such as CAD (and in some advanced research approaches even already VR) is still in the beginning since service process and its interaction with products, infrastructure elements and processes still miss a substantial library of relevant objects, their notations and the common engineering semantic.

However, it is worth noting that service development is highly dependent on the availability of information from both, PDM and ERP systems, as well as from specialized environments such as *SAM (Software Asset Management)* and Human Resource Management.

All of the above introduced Virtual Product Creation activities and capabilities will be explained in detail in Chaps. 7 through 16 (all major technologies).

Virtual Product Creation of today and its linkage to the different phases of Product Lifecycle Management keep a clear distinction, however, to other digital enterprise capabilities. Examples of other digital enterprise capabilities are represented by the event-based order to delivery management world in the factory, where the following IT solutions are widely used:

- ERP (Enterprise Resource Management), with its integrated Product Planning System (PPS) capabilities and
- MES (Manufacturing Execution Systems) together with other shop floor IT systems.

These technologies are not subject of this book but should be mentioned as an important boundary condition for *Virtual Product Creation* concentrating on the operational side of the factory business.

The concept and the solution sets of the Digital Twin, however, might close the separation of lifecycle and event based digital business in the future! (See more details in Chaps. 20 and 21).

References

1. Spur G, Krause FL (1997) Das virtuelle Produkt: management der CAD-Technik (The virtual product: management of the CAD Technology). Fachbuchverlag Leipzig. ISBN-13: 978-3446191761
2. Anderl R, Eigner R, Sendler U, Stark R, Smart Engineering (2012) Interdisziplinäre Produktentstehung (Interdisciplinary product creation). Acatech, Springer Vieweg. ISBN 978-3-642-20371-9
3. Saaksvuori A, Immonen A (2008) Product lifecycle management. Springer, Berlin. ISBN 978-3-540-78173-8
4. Stark J (2005) Product lifecycle management—21st century paradigm for product realization. Springer, Berlin. ISBN 1-85233-810-5
5. Eigner M, Stelzer R (2008) Product lifecycle management—Ein Leitfaden für product development und life cycle management (A guideline for product development and lifecycle management). Springer, Berlin. ISBN 978-3-540-44373-5

Chapter 5
The Technology History of Virtual Product Creation

Executive Summary

This chapter deals with the following topics:

- description of the evolution from shop floor to modern Virtual Product Creation and beyond, focusing on three different application fields: geometric modeling, verification and validation and product data management
- understanding of the interrelation between working technologies and traditions, knowledge about products and processes and collaboration aspects (local, regional, global) on the one hand, and the fast IT evolution on the other hand.

Quick Reader Orientation and Motivation

The intention of this chapter is:

- To gain a first insight into how technologies have emerged over time and in which sequence.
- To provide the necessary background to assess the origin and the maturity of virtual product creation tools and methods.
- To introduce the fundamental concepts of virtual geometric modeling.
- To give an overview over different computer simulation technologies for verification and validation activities.
- To present the core concepts and functionalities of Product Data Management (PDM) systems.

5.1 The History of Computer Aided Design (CAD) Systems and Geometric Modeling

Before geometries were modeled in Computer-Aided Design (CAD) software tools, geometries were drawn manually on paper. Since the tradition of manual drawings

R. Stark, *Virtual Product Creation in Industry*,
https://doi.org/10.1007/978-3-662-64301-3_5

dates back several hundreds of years, it has evolved and improved over time. In order to understand why such optimized practices have been replaced by computer-aided tools, one has to understand the advantages of using CAD over the traditional methods.

The first wave of CAD systems improved what had already been done before: the 2-dimensional drawing of geometries. But products have 3-dimensional shapes and thus multiple drawings of one product have to be done, that show the product from different perspectives and in different cross-sections. The high amount of drawings that have to be done results in a strong need for drawing creation productivity. If only a few minutes can be saved in creating one 2-dimensional drawing, then many hours of work could be saved.

Drawing geometries in a software provided several advantages over manual drawings. Digital drawings allow for more efficient ways for editing, storing, copying and distributing. The possibility to edit a digital drawing allows for removing mistakes or adjusting details without having to redo the entire drawing. Since design models need to be provided to multiple stakeholders (i.e. other engineers who need to align their design, and manufacturing experts) they must be copied and distributed which both is much easier with data rather than with paper. Furthermore, the storage of design models can be realized in many different ways. For example, they could be organized by their affiliation to different components, different engineering teams or different areas of manufacturing. While one single digital model may be referenced from different 'views' (i.e. data models that represent different structures for organizing engineering items), a drawing would have to be copied several times, if it was to be stored in different structures. Hence, the management of design models is more efficient for CAD models. All these advantages increase overall productivity and were the main reasons for the first development of CAD systems.

The first (2D) CAD system, named Sketchpad, was introduced in 1962 by Sutherland, a researcher of the MIT. All following systems were developed by significantly sized manufacturers (>10.000 employees) for use within their own company. Together the aerospace and the automotive industry were pioneering the field with notable systems such as DAC at General Motors in 1964 [1], CADD at McDonnel-Douglas in 1966 [2], PDGS at Ford in 1967 [3] or CADAM at Lockheed in 1967 [4].

After this first wave of CAD systems a new feature was introduced into CAD systems: the possibility to generate and modify 3-dimensional geometries. The advantage over traditional drawing approaches was evident: if 3-dimensional shapes are directly modeled in 3-dimensional space, they only need to be modeled once. This is also why design models were often modeled using clay, wood or other materials. Unfortunately, such approaches cannot provide the exactness that a drawing can provide, and drawings were still required. By modeling 3-dimensional shapes in a CAD tool, they need to be modeled only once and the resulting model is precise enough. Thus, the time for creating multiple drawings could be saved. Furthermore, new technologies allowed for precisely defining complex surfaces. With 2-dimensional drawings such precision could theoretically only be realized by creating an infinite number of fine-granular cross-sections of a 3-dimensional shape or—depending on the topology of the geometry shape—by a few representative cross-sections in case of simple prismatic or rotational parts.

The first generation of 3D-CAD systems were mostly developed at research institutes, like in Europe most notably BUILD (University of Cambridge, 1978), PROREN (Ruhr-Universität Bochum, 1978) and Compac (Technische Universität Berlin) [5]. These systems built on research results that laid the foundations for modern 3-dimensional modeling: the mathematical concepts of non-uniform rational B-splines (NURBS), boundary representations (B-Rep), constructive solid geometry (CSG) models and wire frame models and specific 3D modelling languages such as the Part and Assembly Description Language (PADL) by Voelcker. It is important to understand that the first 3-dimensional modeling system did not support graphical modeling but instead modeling had to be done by writing mathematical formulas and code scripts.

Shortly after the first 3D-CAD systems were presented by academia, OEM would introduce these new technologies into their existing CAD solutions, subsequently replacing 2D modeling by 3D modeling. Examples comprise GEOMAP at Toyota [6, 7], PDGS at Ford or CADD at McDonnel-Douglas. While some of these solutions where custom-developed by external software companies, several OEM-internal developers would start their own businesses, leading to a wave of CAD vendors introducing ready-to-use CAD offerings for small and medium sized companies, too. During that period many of the CAD systems that still exist nowadays were born: PE CAD from HP (in 1980), UniSolids from Unigraphics (1981), CATIA from Dassault (in 1982), SDRC from I-DEAS (in 1982), InterAct (1983) and IGDS (1984) from Intergraph and Euclid from Matra (1985).

It was also during that time, the early 1980s, that the computer hardware market was shaken by the introduction of RISC processors and the first workstation computers, most notably UNIX workstations. While CAD systems usually ran on computer hardware that was built for the single purpose of running CAD systems, the new workstation concept allowed for different usage scenarios. In the mid-1980s, graphical processing power also allowed for more advanced graphical editing for the first time. The company PTC profited from that development first by introducing their CAD system Pro/Engineer in 1987 that revolutionized the way 3D modeling was done in a graphical user interface.

This second wave laid the fundamentals for today's CAD-systems: the approach of solid modeling by which shape generation is done by sequentially adding basic shapes such as cuboids, spheres or cylinders (often represented as B-Reps) to a 3D model and combining them (using CSG models) in order to build more complex shapes. It was only in 1996 when a new approach to modeling was presented by Lüddemann that suggested to virtually imitate the process of clay modeling [8]. Nevertheless, solids modeling established itself as the most widely adopted approach and it can be found in every current CAD system.

While the basic modeling kernel remained stable for many years, CAD systems provided new functionality in other ways. In 1976 Grayer introduced an approach that allows for automatically generating machine control code for a milling machine directly from a CAD model. The concept of programming the machine's routines, using so-called numerical control (NC) code, had already been introduced in 1952 by Parson [9]. This new approach allowed for automating the task of manually writing such NC code. Other approaches followed thereafter and nowadays the generation of NC code is supported by almost all CAD systems for many types of machines (plate work, grinders, etc.).

In 1987 Pratt and Wilson presented the concept of features [10] and parametric models that drastically increased engineering productivity. The basic idea of features and parametric models is to allow engineers to model design intent like 'hole' or 'thickness' explicitly with semantics instead of doing so indirectly via geometry only. The concept of a 'hole' can be selected from a set of reusable 'features' and the engineer only needs to place the 'hole' in the given coordinate system and specify its main parameters (such as diameter and depth). This is especially useful for standardized shapes such as screw threads and it saves time and ensures correctness by automating modeling tasks. In 1992 Schulte and Stark suggested using features as "higher level primitives" in order to transfer manufacturing relevant information about geometries from a CAD system to a CAPP System [11]. In 1994 Rieger developed one of the first feature modeling editors [12]. In 1998 Dassault Systems introduced its new CAD System Catia v5 that implemented the idea of features allowing the user to specify parameterized templates for parts [13, 14].

While CAD systems were thus providing ever more useful functionality many product models were still available as drawings and a very pragmatic question arose: how to convert these drawings into 3D models? In 1981 Jansen developed an approach for automatically converting technical drawings into 3D models [15–17]. Later, in 1995 and in 1997 Liu and Luth improved this approach by also generating more complex splines and semantic information in the resulting 3D models [18].

Until today, the CAD system market has been heavily consolidated and only few of the former system vendors have survived. The main competitors in the 2010s were Dassault's CATIA and Solid Works, Siemens NX (former Unigraphics) and Solid Edge, PTC's Creo (former Pro/Engineer) and Autodesk's AutoCAD (2D) and Inventor (3D).

While early CAD system offerings subsequently introduced substantially new modeling approaches, the focus of CAD vendors today is on iteratively improving productivity. Approaches like "shape morphing" in CATIA v5 [19] allow users to easily reshape freeform surfaces based on fixed feature points. Approaches like "direct modeling" from Spaceclaim (2007) or "synchronous technology" from Siemens NX (2007) aim at allowing the user to resize and reshape geometries more easily by intuitively pulling or pushing them with simple mouse-movements instead of typing parameter values into the forms, and by partially recognizing dependencies between parameters automatically. In the 2010th first digital platform (*Software as a Service, SaaS*) based CAD modeling environments were founded, such as on shape in 2012, which was acquired by PTC Inc. in 2019.

5.2 Digital Product Validation and Verification

While the previous section described the historical development of technologies for modeling geometries of a product, this section presents the development of validation and verification technologies.

5.2.1 *Introduction into Validation and Verification (V&V)*

Verification is the process of confirming that a technical system or a digital model (of a technical system) complies with all its specifications ("Did we build it right?"). Validation, on the other hand, is the process of confirming that a technical system complies with the customers' and all relevant stakeholders' expectations ("Did we built the right thing?") [20–25].

It is important to understand the difference between both. If a specification fully reflected all customers' and all stakeholders' expectations, then the process of verification would, at the same time, validate a technical system. However, this is usually never the case. Therefore, both processes must be performed along the product development process.

Verification happens at several stages in the product development process. It usually beginsg when the first digital models (e.g. geometries, simulation models, etc.) have been created. While the digital models only constitute parts of the whole technical system and while they are not physically built yet, they can be compared with specifications. When all partial digital models have been created, they should ideally be integrated (e.g. as virtual assemblies or co-simulation models) and then again be compared to the specification. Finally, when the real physical system has actually been built, it should again be compared to its specification. Verification can be performed by engineers completely and does not require the involvement of the customer or other stakeholders.

Validation can only be truly performed when a prototype of the technical system exists or when it has been finally physically built. Before this exists, validation can only be performed against a set of assumed performances of a product without sufficient confidence that this can actually be achieved (as it is, for example, the case in the quality function deployment approach). Virtual prototypes allow for partially validating a product before it is physically built. Nevertheless, the final physically built system that exists must be validated again. Validation must always involve customers and/or other stakeholders.

The specifications relevant for verification and building prototypes usually consist of requirements and digital models that describe a technical system's behavior. Requirements are first specified at product/system level and are subsequently broken down into detailed specifications for its subcomponents and parts. The discipline of requirements management provides methodologies for collecting and detailing requirements but is not focused upon in this section. It is thus assumed that detailed requirements and related digital models already exist and need to be checked against each other. Requirements that are verified mainly comprise the following aspects:

- spatial constraints (boundaries) for geometries,
- kinematic behavior of parts,
- physical behavior of parts and forces,
- behavior of interrelated processes, and
- user experience.

Boundaries for geometries can refer to a static geometry or to the space that a geometry occupies considering its possible movement or positioning. The latter is tightly related to the kinematic behavior of parts that analyses how a set of parts that share geometric interfaces or simply a common space can be moved or positioned. A kinematic analysis thus provides the foundation for the verification of spatial constraints that consider kinematics.

The physical behavior of a part focuses on the interrelation between geometries (with specific material characteristics) and physical forces and movements that are applied to it. Typical examples comprise material deformation under different pressures applied, vibration of bodies, or movement of air or water on surfaces.

The behavior of processes is relevant when many different physical or digital processes are dependent on each other. For example, the technical execution of the physical function "braking a vehicle" involves many different system interactions such as the physical behavior between the ground and the tire and between the brake disc and the brake pad, the behavior of sensors that measure the forces that work on the brake and the behavior of the software that reacts to the sensors signals and that may control the brake pad in return. A separate analysis of all these system interactions without taking into considering the cross-effects may result in unforeseen behavior of the complete technical system. Therefore, this aspect is very important with regard to validation.

Finally, user experience is an important factor in order to focus on how a human user (the customer) perceives a product when interacting with it. It mainly focuses on the effects of product characteristics that may directly affect human sensory perception. These may comprise noise, haptics like textures of surfaces, odor, visible shapes, colors and different aspects of dynamic interactions.

In order to verify the different aspects mentioned before, different computer-based simulation technologies have been developed and evolved over the last decades. They are presented in the following sections.

5.2.2 *Evolution of V&V Technologies and Computer Aided Engineering (CAE)*

The development of the first algorithms, languages and theoretical approaches to simulate physical aspects of a system or process flows dates back to the 1930s. During that time Enrico Fermi used Monte Carlo algorithms to calculate the properties of neutrons and presented according to numerical methods for investigating statistical problems. In the 1940s Jon Von Neumann and Stanislaw Ulam presented the roulette wheel technique that was applied to the same problem. The simulation of physical phenomena was thus one of the first fields for applied simulation approaches. At this time though computer technology was not available and hence the presented approaches were not yet implemented as software.

In the 1950s discrete event computer simulation was introduced. The IBM 650 computer was used and the algorithms were implemented in assembler language (i.e. not a high-level programming language as commonly used nowadays). In this case, no physical phenomena were investigated but abstract process flows. At this point they were not yet applied to engineering use cases.

In the 1960s computer simulation gained strong momentum and many different formal simulation languages (for describing simulation models and setups) were presented. Carl Adam Petri presented petri-nets as an approach to model process flows. Geoffrey Gordon presented the General Purpose Systems Simulator (GPSS), also an approach for simulating process flows, and applied it to the problem of weather prediction. Harry Markowitz, Bernard Hausner, and Herbert Karr presented SIMSCRIPT, a language for modeling and simulating events and schedules, and used it to simulate inventory problems. Ole-Johan Dahl and Kristen Nygaard presented the programming language SIMULA that was also used for modeling object flows through processes [26]. SIMULA build the foundation for later programming languages such as Smalltalk and thus introduced basic concepts for the object-oriented programming paradigm that is one of the most commonly applied approaches in software development nowadays. Further simulation languages comprised SOL (A symbolic Language for General Purpose System Simulation) from Don Knuth and J. McNeley, the General Simulation Program (GSP) by Keith Douglas Tocker and CSL (control and simulation language) from John Buxton and John Laski.

With all these new simulation approaches and technologies openly available manufacturer's interest in simulation increased. Companies like Boeing, Martin Marietta, General Dynamics, Raytheon, or Southern Railway built simulation groups that investigated the applicability of these approaches to their engineering-specific problems. At the same time, computer manufacturers like IBM, Control Data, and UNIVAC focused on providing suitable hardware solutions allowing the industrial application of simulation languages. Computer performance was limited at this point of time though, thus limiting the complexity of simulation models that could be simulated.

The 1970s continued where the 1960s ended, and further event- and process-centered simulation approaches and languages were presented at scientific conferences. Alan Pritsker presented multiple event simulation languages such as GASP IV, SLAM or SAINT [27]. Parkin and Coats presented a new algorithm for event-based discrete simulation [28].

But the 1970s also marked the advent of the first computer aided engineering (CAE) systems. This term summarizes software systems for finite-element analysis

(FEA), for computational fluid dynamics (CFD), and multibody dynamics (MBD). In the early 1970s the first three dimensional models for calculating fluid flows were introduced at Boeing [29]. In the late 1970s the FEA systems ANSYS and Abaqus were developed, providing means to companies to model, simulate and analyze material deformation or heat transfer problems. In 1977, Orlandea et al. [30] introduced the MBD system ADAMS (automatic dynamic analysis of mechanical systems) that allowed for calculating the kinematics of three-dimensional objects.

The introduction of CAE plays a major role from a product verification perspective. Process- and event-centered simulation approaches can be used to analyze abstract system behavior models, while CAE can be used to analyze geometry models. Hence, both approaches support different engineering activities at different stages of the product development process. Furthermore, the analysis of geometry models always matters when developing a (physical) technical system, while the analysis of abstract system behavior is only relevant for rather complex systems such as airplanes. Hence, CAE put simulation technologies on the map of many more companies, from tool machining companies to car manufacturers.

The 1980s mark an important change in the history of simulation technology. Computer hardware became significantly cheaper, thus also allowing smaller companies to profit from simulation software without having to commit to massive financial investments. With cheaper hardware, more powerful computers could be afforded and more complex simulations became possible. Furthermore, an increasing number of off-the-shelf software solutions was offered on the market, on one hand in the area of material requirements planning (MRP) for manufacturing and Computer Aided Process Planning (CAPP), and on the other hand for solving complex mathematical equations. While MRP and CAPP represent solutions focused on specialized engineering tasks, toolboxes such as Matlab, which was introduced in 1984, were generic solutions that could be applied for solving simulation tasks for different purposes. In addition to advanced math functionality, Matlab also provided a graphical user interface for modeling data flow and visualizing simulation results. Thanks to its large acceptance and deep market penetration it still is an important offering on today's market (marketed as SIMULINK since 1992).

MATLAB marks a cut in the way simulation software was used. While earlier, simulation models were programmed in a specific language, MATLAB allowed users to create simulation models graphically. Computer simulation thus became more accessible to a wider range of non-expert users. This trend continued in the 1990s and nowadays all important simulation software systems provide such graphical modeling interfaces.

Another noteworthy innovation that happened in the 1980s was the first introduction of a virtual reality setup with a head-mounted display (HMD) that included a motion tracking system (at the University of North Carolina). In 1989 VPL Research spawned the first commercial offer of such an HMD, called the "EyePhone".

In the 1990s simulation systems and computer hardware became increasingly powerful, yet no substantial theoretical innovations were introduced. Manufacturing planning was the most common application scenario for process-centered simulation approaches. The market of off-the-shelf software solutions for computer simulation expanded and consolidated. Systems such as GPSS, EXTEND, MAST, Micro Saint were developed, replacing former solutions that required programming. It is important to understand though that even such graphical simulation modeling systems still require the programming of scripts to some extent. Since different systems employed different proprietary scripting languages, Hilding Elmqvist introduced Modelica in 1997. That is an object-oriented language for the modeling of technical systems providing a standardized format for reusing and exchanging dynamic system models. Modelica is still used today in many simulation software systems such as SimulationX or Dymola.

The 2000s marked the advent of hardware-in-the-loop (HIL) simulation where real, physical electronical/or mechanical components are connected to a simulation software (through sensors and actuators that are connected to computer interfaces). This allows for verifying the interaction of multiple electronical, mechanical and software components where a part of the components already exists physically and other parts are still under development. Such functionality is often provided by development tools for modeling and programming the data flow between electronical components, such as LabView (first introduced in 1983) or dSPACE (first introduced in 1988). HIL is widely used in the development of cars and trucks but also in the development of all other kinds of mechatronic systems.

Basic Explanation of simulation approaches and technologies

Simulation Technologies have evolved with one main goal in mind: minimizing the efforts of testing physical prototypes. Instead of building a costly physical prototype, simulation software allows for testing a virtual prototype instead. Since the second half of the 90s an overall Digital Mock-Up (DMU) can be created if all geometries are well structured in a product information database, and a broad range of different digital models exist to allow for specific virtual prototype simulations.

This approach also allows for testing a product (or one of its components) early in the design process, i.e. even before aspects such as manufacturing need to be considered. Hence, problems can be discovered earlier and the duration of development iterations can be shortened. Finally, manual testing tasks can be automated, further lowering testing costs.

As emphasized in the previous section there exist different simulation approaches, each one suited for different verification purposes.

Spatial constraints (boundaries) for geometries are usually verified directly in a CAD environment and do not require additional simulation software. Modern CAD environments meanwhile provide easy-to-use clash analysis functionality, which for a long time was a privilege of specialized DMU tools only. When an engineer places multiple CAD parts in one shared space, the CAD environment is able to analyze the resulting assembly and identify all spots where parts 'collide'. If parts are moveable then their kinematics can be modeled in the CAD environment, too, and the clash analysis functionality will consider the whole space that each moveable object may occupy in any of its possible positions.

Usually though, CAD environments do not provide means for modeling physical behavior. While they can detect clashes of parts they cannot compute what exactly happens if these parts interact with each other with specific forces applied to them. Analyzing the physical behavior of a product or its parts thus requires specialized CAE simulation software.

In order to analyze physical product behavior, continuous dynamic simulation approaches are applied. This is what FEA models are used for, which have been mentioned in the previous section. Examples for physical behavior comprise:

- the way a car body deforms when it crashes into another object,
- the turbulences resulting from a current of wind meeting an airplane's wings or
- the vibration required to make a building structure collapse.

In the continuous dynamic simulation, a geometry model is translated into a set of differential–algebraic equations modeling continuum mechanics. Since solids and fluids behave differently, different models are used for describing solid mechanics and fluid mechanics. The car body deformation and the collapsing building structure are both examples for solid mechanics. The air turbulences are an example for fluid mechanics.

The algebraic equations are then solved by mathematical algorithms and the results are reflected back into the geometry model. This allows for visualizing them in a geometrical representation. Often, physical behavior (such as the degree of deformation measured in millimeters or the range of movement during vibration) is also visualized in charts and diagrams.

While the continuous dynamic simulation focuses on geometry and physics it is often also desirable to analyze the behavior of disembodied things such as signals or data flow. This is especially interesting in electronical and mechatronics engineering where components usually do not interact through the application of physical forces but the sending and receiving of electronic signals. This is what the process-centered simulation approaches are used for, that have been presented in the previous section.

In process-centric simulation models, functional components of a system are modeled as a graph of nodes that are interconnected through edges that transfer quantifiable signals (e.g. in software such as Dymola, LabView or SimulationX). Each node may have multiple input and output edges and it processes inputs into outputs. For each edge, a direction and a signal type (e.g. a visual signal such as light at a specific luminosity or a data input stream of digits) is specified. Each node can be modeled as a mathematical function with the signals from the incoming edges signals as its parameters.

Therefore, each functional component's behavior can be modeled separately and finally all functional components can be simulated in their aggregated behavior. Usually, such simulations reveal where components may receive input signals that are not out of their accepted range of values (e.g. a light signal that is too dark or too bright and that can thus not be measured properly by a photometer) or where functional components fail (either because they generate wrong outputs or because they do not generate outputs at all, e.g. in case of unsolvable mathematical equations). Such simulations are also used to optimize the behavior of functional components (by fine-tuning the mathematical function that represents their behavior).

As mentioned in the previous section, process-centric simulation models can be developed before the functional components themselves are developed and they can provide useful insights on the constraints for interfaces between functional components. At a later stage of the product development process, when functional components have been designed, the virtual models of the functional components can be used as inputs for continuous dynamic simulation software. That other simulation software (with one specific virtual model of a functional component) can then be linked to the process-centric simulation model, replacing the mathematical function of one node (that was only based on an assumption earlier in the product development process). Hence, the initially assumed behavior of one functional component can be replaced by its actual behavior (assuming that the continuous dynamic simulation model is valid). Finally, all developed functional components can be "co-simulated" and their real interplay can be analyzed and validated.

While process-centric simulation models are used for modeling the behavior of a system with respect to the input and output signals that its different components receive and generate, such models often do not provide any insight on the temporal aspects of a system's behavior. Functional components of a system send signals from one to another but sometimes it is essential to know at which point in time these signals are sent and how long one component needs to wait for another to send a specific signal. This is a very similar problem to that in business process or project planning where one wants to minimize idle times in the process/project but also wants to ensure that single activities have enough buffer time in case of unforeseen events.

In such cases, state machines (or process models with underlying state machines) are used for modeling the system behavior. Similar to process-centric simulation models, functional components are modeled as a graph of nodes that are interconnected through edges. Each node is modeled as a set of attributes, such as duration or likelihood of failure. Often, minimum, maximum and average values can be specified for such attributes. In addition to nodes that represent functional components, there also exist nodes that guide the process flow (decision, parallelization or synchronization points). This allows for modeling parallelization and iteration.

5.3 Product Data Management (PDM)

When the first 2D CAD systems were introduced in the late 60s no file systems existed yet and data could not be transferred through a computer network. That means the created drawings could not be saved as a local file in a folder on a computer and they could not be sent to a server that provided storage functionality. Instead, they could either be plotted/printed or saved on a magnetic tape. The management of the created models thus involved manual tasks dealing with physical objects (i.e. plots or magnetic tapes) that had to be stored in some physical storage place. Since drawings were made manually for decades before the introduction of the first CAD systems, approaches for the storage, indexing and access existed already. Nevertheless, these approaches relied on human users for indexing, searching and securing engineering results, they required physical storage space, and the distribution of engineering results required physical copying and distribution through classical postal or delivery services.

When the first wave of 3D CAD systems was introduced in the late 70s, the first file systems (e.g. System VFS, FAT) and the first data exchange protocols for computer networks (e.g. Ethernet, ARCNET) had already been presented by researchers, but were not yet widely adopted in business practice. Drawings were still managed physically.

It was only in the 1980s that the physical management of drawings (and all other kinds of documents) started being partially replaced by the digital management of files. More powerful file systems like the Berkeley Fast File System were introduced that allowed for storing drawings and documents directly on a computer. Also, the first relational database systems (e.g. POSTGRES, INGRES) were introduced. While they did not allow for managing complex data like documents or drawings, they could manage huge amounts of small data, and could thus be used to store and manage information about suppliers, orders, customers, etc. Data bases were used in software systems (introduced in the beginning of the 1980s) that allowed for managing metadata about paper-based documents digitally. That means that metadata (i.e. information about creation data, document type, author, version, etc.) was managed digitally while the corresponding documents were still stored physically. Later on these systems evolved into so-called Electronic Document Management (EDM) [31] systems and could then also manage documents digitally (in a file system), hence rendering physical storage obsolete. Examples for early EDM systems comprise SoftSolutions (1979), Saros Mezzanine (1986) and PC Docs (1989) [32–35]. Today such systems are called Document Management Systems (DMS).

In order to allow companies to "migrate" older documents, that only existed physically, into digital documents these systems also provided document imaging (i.e. scanning) functionality. Furthermore, text-analysis algorithms would allow for indexing text-based documents semi-automatically thus saving indexing efforts. And finally, documents could be searched using full-text search.

While the advances in file systems, network protocols and the introduction of EDM systems provided significant advantages for managing documents, they did only address "generic" data management challenges (e.g. indexing, searching, storing, etc.). Product development faced specific challenges though, that these systems did not address, mainly revision and configuration control and the management of the lifecycle of product data.

Version, revision and configuration control is an important field of activities in product development because one component can be used in multiple different product versions or configurations. Hence, one single version of a CAD model could be a part of different assemblies or a part in different bills of materials (BOM). While simple version control usually only allows for saving consecutive versions of one document (i.e. 1, 2, 3, etc.), in product development one document may exist in different versions in different "contexts" (e.g. assemblies, BOM, etc.). This complexity could not be handled with early EDM systems.

The other challenge with respect to product data is its life cycle. Different engineering artifacts (i.e. requirements, system models, CAD models, etc.) often go through different development and release stages (e.g. idea, concept, design, released design, etc.). Each of these stages may affect the access rights for the corresponding product data and the way it is stored and versioned. Early EDM systems did not allow for specifying such characteristics and all documents were simply treated the same way.

While the amount of product data increased steadily in the 1980s, the challenges mentioned before showed that there was a need for specific IT support for the management of product data. As a logical consequence, especially large manufacturers like Boeing or Ford with strong in-house research and development departments would develop their own company-specific PDM systems. For example, Ford's PDGS system would feature a component called Data Collector that provided PDM functionality connecting globally distributed development centers [3]. Smaller companies, on the other hand, were less affected by the challenge of overwhelming amounts of product data, but first and foremost they simply could not afford to develop their individual solutions.

The first PDM software that was sold on the market was SherpaWorks from Sherpa, that was released in 1984 [36]. In 1989 IBM introduced a PDM software called ProductManager [4, 37, 38]. But it was only in the 1990s that the market for PDM systems grew significantly. In the early 1990s Unigraphics and SDRC, two companies that already offered CAD solutions at that time, released respective PDM offerings (Unigraphics iMan in 1991 and SDRC Metaphase in 1992). In the late 1990s other CAD vendors followed their example, and in 1998 PTC released Windchill and Dassault Systemes released Enovia. BAAN introduced BAAN PDM in 1996 [39] and Eigner + Partner introduced CADIM/EDM in [40]. Hence, the "new" PDM market was (mainly) shared among CAD vendors, thus explaining the initial focus of most PDM systems on the management of CAD models. Many other types of product data, such as requirements, simulation models and results or factory layouts would still be managed outside of PDM systems. The CAD vendors, realizing this maladjustment, would thus redefine their image from CAD vendors to "Product Life-cycle management (PLM) solution providers" in the late 1990s, and enhance their products with corresponding, additional functionality.

It should be noted that the term PLM is not only limited to the management of product lifecycle information or data within a specialized IT system. PLM also comprises the management of information and information flow between processes at a more general level. Eigner and Stelzer even refer to PLM as a solution strategy [41]. A PLM system alone can thus not cover all aspects of PLM.

Today, PLM systems support the management of almost any kind of product data. Their typical components are:

- a central data vault where all product data is securely stored,
- a workflow engine for controlling product data centric processes such as release processes,
- user interface components for handling,

- bills of materials/product structures,
- product configurations,
- version management,
- (standard) parts management, and
- project management.

Additional functionalities typically provided by PLM systems comprise [42] advanced search, file conversion, secure file transfer, task management and/or change notifications.

Together with a suitable PLM strategy, these functionalities aim to provide the following business benefits [43]:

- to save development time and cost through the reuse of parts, modules, platforms, etc.;
- to reduce the amount of engineering changes after the start of production through better support of V&V activities in the early PDP stages;
- to improve collaboration through well-defined processes and responsibilities;
- to confidently ensure the availability of relevant data;
- to increase the amount of time engineers spend on innovative and value-creating activities through reducing the efforts for laborious data management activities, and
- to provide continuous support of business processes through the reduction of information gaps between heterogeneous IT systems.

While initially PLM systems provided mostly data management functionality, current PDM systems support all kinds of processes, either through workflow functionality or through specialized, task-oriented plugins (e.g. requirements management views) for the graphical user interface. Nevertheless, they usually do not cover data and processes management from the entire product lifecycle, but only from the beginning of a products' life, the product development phase.

There exist multiple reasons for the PLM systems' focus on product development, the most important one being that later phases of the product lifecycle are often managed not by the same company that develops the product, but by external partners. Reaching an agreement on a common PLM approach in such an Extended Enterprise setting can be time-consuming and challenging [44]. Hence, traditionally, IT systems are used by one company only and companies do not interlink their IT systems or use shared IT systems. Instead, the different companies that are involved in the product lifecycle manage their own data and processes separately. As a result, there exists no single IT system that supports data management and process support for all phases of the product lifecycle, but a variety of specialized IT systems in each different phase of the life cycle.

Another practical reason stated by Grieves [45] is the fact the whole lifecycle of a product may last up to 100 years which is much more than the typical lifetime of an IT system. At the start of production of a product (often after multiple years of development) the initially introduced PLM system may already be out of date. If for that reason another IT system is introduced for managing information from the later phases of the product lifecycle, then it makes no sense for PLM vendors to cover these phases in the first place.

Nevertheless, PLM system vendors are still aiming to provide solutions for later phases of the product life cycle, too. Currently, data from the use and the end-of-life phase of a product is often managed only in ERP (Enterprise Resource Planning) systems, if at all. Since ERP systems are focused on business-centric topics such as sales numbers, parts supply, logistics, etc. important information relevant for engineering is often not collected in them. Hence, there exists a demand for managing engineering-relevant information in these lifecycle phases that is likely to be addressed by PLM vendors in the near future. While there already exist partial solutions for supply-chain management and factory data management, the management of information about a product's usage, wearing, maintenance and disposal is still poorly covered.

Existing IT solution offers from competitors (for later product lifecycle phases), such as ERP, pose a practical challenge for this extension of PLM system's functionality though. PLM system vendors must penetrate new market areas facing stiff competition. It thus remains to be seen whether or not this will hold PLM systems back from actually covering the entire product lifecycle somewhere in the future. Please refer to Chap. 11 to gain more insight to PDM/BOM and PLM.

References

1. Krull F (1994) The origin of computer graphics within General Motors. IEEE Annals Hist Comput 16(3):40. https://doi.org/10.1109/MAHC.1994.298419
2. Rake W (1973) Computer-Aided Design and Drafting (CADD): an advanced designer's tool. SAE Technical Paper 730934
3. Whitney DE (1993) History of CAD/CAM at Ford
4. Weisberg DE (2008) The engineering design revolution. the people, companies and computer systems that changed forever the practice of engineering
5. Spur G, Krause FL, Harder JJ (1982) The Compaq solid modeler. In: Computers in mechanical engineering, New York, pp 44–53
6. Whitney DE (1991) Visits to Prof Kimura's CAD research Lab July 5 and July 19 for discussions about product realization, IMS, and Product Development Cycles
7. Kimura F (2009) Geomap-III: designing solids with free-form surfaces. Comput Graph 4(6):58–72
8. Krause JL (1997) Virtual clay modeling. In: Pratt MJ, Sriram RD, Wozny MJ (eds) Product modeling for computer integrated design and manufacture. Chapman & Hall, London, pp 162–175
9. Reintjes JF (1991) Numerical control. Making a new technology. Oxford series on advanced manufacturing, vol 9. Oxford University Press, New York
10. Shah JJ, Mäntylä M (1995) Parametric and feature-based CAD CAM. Concepts, techniques, and applications. a Wiley-Interscience publication. Wiley, New York
11. Bernardi A, Klauck C, Legleitner R, Schulte M, Stark R (1992) Feature based integration of CAD and CAPP. In: Krause F, Ruland D, Jansen H (eds) Informatik Aktuell. CAD'92 Neue Konzepte zur Realisierung anwendungsorientierter CAD-Systeme, Berlin, pp 295–311
12. Rieger E (1994) Semantikorientierte Features zur kontinuierlichen Unterstützung der Produktgestaltung. Produktionstechnik - Berlin, vol 158. Hanser, München, Berlin

13. Kent JR, Carlson WE, Parent RE (1992) Shape transformation for polyhedral objects. In: Proceedings of the 19th annual conference on computer graphics and interactive techniques. ACM, New York, NY
14. Brill M (2006) Parametrische Konstruktion mit CATIA V5. Methoden und Strategien für den Fahrzeugbau. Hanser, München [u.a.]
15. Krause F, Jansen H (1993) Luth N (1993) Neue Methoden der automatischen Zeichnungsinterpretation. Zeitschrift für wirtschaftliche Fertigung und Automatisierung 88(12):589–592
16. Spur G, Jansen H, Krause F (1986) Automatische Digitalisierung und Interpretation technischer Zeichnungen für CAD-Prozesse. ZWF CIM 81 (1986) 5:235–241
17. Spur G, Jansen H, Krause F (1986) Verarbeitungstechniken zur automatischen Zeichnungserfassung für CAD-Prozesse. ZWF CIM 81(5):235–241, 9:460–466
18. Luth N (1997) Eine Methode zur automatischen Strukturinterpretation in digitalisierten technischen Zeichnungen. Berichte aus dem Produktionstechnischen Zentrum Berlin. IPK, Berlin, Berlin
19. Ali A, Brebbia CA (2006) Digital architecture and construction. WIT transactions on the built environment, vol 90. WIT, Southampton
20. Aiaa guide for the verification and validation of computational fluid dynamics simulations. Aiaa, [S.l.]
21. Author Unknown (1979) Schlesinger terminology for model credibility. In: Simulation, vol 32, No 3, pp 103–104
22. American Society for Quality (ASQC) (1978) Quality systems terminology (A3-1978)
23. ASME (2006) Guide for verification and validation in computational solid mechanics
24. IEEE (1990) Standard glossary of software engineering terminology (Std 610.12-1990)
25. Kapurch SJ (2007) NASA systems engineering handbook. Retrieved from https://www.nasa.gov/sites/default/files/atoms/files/nasa_systems_engineering_handbook.pdf. Accessed on 07 Nov 2017
26. Nance RE (2002) RGS Perspectives on the evolution of simulation. Oper Res 50(1):161–172
27. Rossetti MD, Hill RR, Johansson B, Dunkin A, Ingalls RG, Goldsman D, Nance RE, Wilson JR (2009) A brief history of simulation. In: Dunkin A (ed) Winter simulation conference. Winter simulation conference, s.l, pp 310–313
28. Nance RE (1993) A history of discrete event simulation programming languages. The second ACM SIGPLAN conference on History of programming languages. ACM, New York, NY, pp 149–175
29. Robert PE, Saaris GR (1972) Review and evaluation of a three-dimensional lifting potential flow computational method for arbitrary configurations. Boeing Co.
30. Orlandea N, Chace MA, Calahan DA (1977) A Sparsity-oriented approach to the dynamic analysis and design of mechanical systems. J Eng Industry
31. ISO Electronic document management -- Vocabulary -- Part 1: Electronic document imaging (12651-1:2012)
32. Author Unknown (1990) UTAH software firm forms new marketing company. Deseret News
33. Klaproth F (1998) Lossau N. The document management system Saros Mezzanine and the new product AGORA as key component in a digital library architecture at Göttingen University Library 1513:685–687
34. Delphi Consulting Group (1995) In its newest market research service, document management: the paperless revolution. Delphi Finds Three Vendors Dominate Market
35. Chen SCM (1994) The document masters: PC DOCS, Saros, and SoftSolutions. PC Magazine, pp 316–317
36. CADAZZ (2004) CAD software - history of CAD CAM. http://www.cadazz.com/cad-software-history.htm
37. IBM (1997) IBM ProductManager Version 3 Release 2
38. Sendler U (2009) Das PLM-Kompendium. Referenzbuch des Produkt-Lebenszyklus-Managements. Xpert.press, Springer, Berlin, Heidelberg
39. Baan Development B.V. (1998) BAAN IVc4. BaanPDM 5.1.3 DBA Guide. Retrieved online from: http://baansupport.com/docs/baan/U7175AUS.pdf

40. Schmidt S, Wierschin H (1998) Global Engineering auf der Basis weltweit verteilter Daten und Dokumente. In: Industrie-management, 10/1998
41. Eigner M, Stelzer R (2013) Product lifecycle management. Ein Leitfaden für Product Development und Life Cycle Management, 2nd edn. VDI. Springer, Dordrecht
42. Saaksvuori A, Immonen A (2008) Product lifecycle management, 3rd edn. Springer-Verlag, s.l
43. Feldhusen J, Gebhardt B (2008) Product Lifecycle Management für die Praxis. Springer, Berlin, Ein Leitfaden zur modularen Einführung, Umsetzung und Anwendung
44. Stark J (2009) Product lifecycle management. 21st Century paradigm for product realisation, 1st edn. Springer, London Limited, s.l
45. Grieves M (2006) Product lifecycle management. Driving the next generation of lean thinking; [how GE, P&G, Ford, Toyota, and other leading companies achieved dramatic increases in productivity and profit]. McGraw-Hill, New York

Chapter 6
The Set-Up of Virtual Product Creation in Industry—Best Practices, Error Modes and Innovation Speed

Executive Summary

This chapter deals with the following topics:

- History and understanding how the *typical set-up* of *Virtual Product Creation* capabilities looks like in industry,
- Explanation of best practices in *Virtual Product Creation* capabilities and set-ups and
- Reasoning of inherit error modes and the challenge of keeping abreast with innovation speed

Quick Reader Orientation and Motivation

The intention of this chapter is:

- to give an overview and orientation of how the collective suite of *Virtual Product Creation capabilities* should be implemented within industrial operation and organizations
- to explain how best practices in Virtual Product Creation set-up needs are dependent from company culture, management style and collaboration needs
- to understand the new role of Virtual Product Creation beyond the traditional mindset of digital technology set-up
- to offer approaches to allow for high speed innovation without losing control on consistent deployment and scalability of Virtual Product Creation operations.

6.1 Basics Awareness and Sense for Change

In order to understand the role and the set-up of what may summarily called the *Virtual Product Creation in Industry*, one has to look at how engineering traditionally was set-up, how it evolved and how it works today. Roughly speaking, until the nineteenth century is was common for engineers to work on product prototypes in the workshop

R. Stark, *Virtual Product Creation in Industry*,
https://doi.org/10.1007/978-3-662-64301-3_6

directly. Over time they realized that is was much more beneficial to think ahead and to develop sketches, drawings, or blue prints or even a master plan firstly every time an alteration to the design of the product was introduced and then to actually work possibly on a specific prototype to conduct prove-outs and tests.

With developing and producing an ever-bigger variety of products, it became clear that engineering needed to develop its own methodology regarding how to become effective, efficient and overall successful, not only technically but also business wise. Technical drawings were an excellent conduit to prove a certain concept of maturity and completeness. Over time, they became an important "tool" and indeed the method of choice how to describe a product in both design and in "functional carrier" terms by adding appropriate notes as text annotations and descriptions onto a drawing.

However, it rapidly became obvious that there was a need to standardize such an approach. At the very least, the technical drawings should be interchangeable within one company. Better still, if they were interchangeable and in fact standardized across an entire industry or market.

Therefore, standards for technical drawings were defined: first among certain companies, then on a national and finally on an international level. However, technical drawings had (and still have) their limitations. As a consequence, product development still needed a physical prototype or at least a mock-up to understand and comprehend whether the idea that an engineer initially had in mind really came into realization and finally into operational mode. The technical drawing could not—and was never intended to—represent things such as strength, fatigue or all functional aspects of a product. At this point, the notion of the process arose as something that needed to be represented in abstract terms. Processes were needed in order to make sure that the different engineering experts could work closely together. Processes were understood as "a description or prescription to bring various working steps into a specific sequence" so that highly skilled workers were able to work hand in hand with lesser skilled and even with unskilled workers on developing a product from the very first idea into a rough plan and finally into a functional prototype. From a manufacturing point of view, it was also necessary to include requirements, feasibility aspects, standards and tolerances onto the technical drawings.

In the 1960s and 70s of last century researchers especially in the German-speaking countries envisaged a methodology of how to design engineering products starting from functional and technical physical perspectives. There were scientists and professors such as Pahl and Beitz [1], Hubka [2], Hansen [3], Rodenacker [4] and many more who developed theoretical models of what design is all about, how synthesis and analysis engineering activities should evolve and how maturity and/or compatibility driven progression of engineering development should work. Similar trends arose in the Anglo-American world with different foci: design research was much stronger driven by innovation aspects and stage-gate oriented design progression with the help of project management.

Only selected industrial companies, however, did pick up the *functional and technical-physical perspectives* directly into their development process practices whereas the stage-gate and project management influences into development process guidance were implemented intensively and consistently by the majority of the industrial companies. Consequently, still today many companies are suffering from the fact that they still have to convert their pragmatic activity driven design progression into more consistently function and system driven design, development, validation and verification.

In addition, due to quickly spreading product variations in the nineties of last century, the *functional and technical-physical* driven design science theories were rather ousted by project management models and practices from the Anglo-American business and management schools in order to find practical ways of steering standardized stage gate development approaches (incl. the introduction of firm gateway and milestone events).

In parallel, virtual product creation ideas, concepts and solutions based on digital artifacts such as CAD models were introduced allowing extending a 2D drawing to a 3D model of not only distinct mechanical parts but of the entire (virtual) product. However, in order to describe deliverables technical drawings with their limitation to geometrical shapes had to be enhanced by adding internal data models and data structures. In addition, the introduction of finite elements methods and other kinds of CAE models made it possible to describe the intermediate steps of the virtual design progression rather than being limited to a representation of just the physical constituents and elements. The view on deliverables rapidly moved from a static one, i.e. from progression in terms of concatenated steps to a truly dynamic understanding of the entire engineering process. It became possible to determine which elements should be at which maturity level and at which point in time along the product and manufacturing engineering processes. The entire product, including all simulation aspects such as NVH, crash, fatigue, durability, thermal and acoustic behavior etc. should ideally be represented in a standardized model. Industry and Digital Technology Providers (also known as PLM vendors) did miss out chances of converting to similar digital model structures, identical theoretical foundations and consistent and mutually acceptable data exchange agreements. Driven through pressing IT innovations and associated rollout of virtual product creation technologies new fields such as digital manufacturing and technical systems integration did cause a lack of consistent deployment of the originally developed theoretical models of design and systems engineering (functional and technical-physical perspectives) as described above.

In the last 30 years, the industrial praxis had to cope with a sharp acceleration of introducing new digital tools sets and associated digital models as shown in Fig. 6.1. Therefore, many engineering practices had to primarily introduce appropriate data and project related management solutions in order to keep control on the various digital model types. The content and theory integration of such distinct digital technologies and approaches, however, did not take place yet!

As a logical consequence, the modern research teams *Industrial Information Technology* of TU Berlin and *Virtual Product Creation* of Fraunhofer IPK took on such challenge by extending this evolutionary progression a step ahead with the description of an overall *engineering operating system.* The focus, therefore, is on expanding

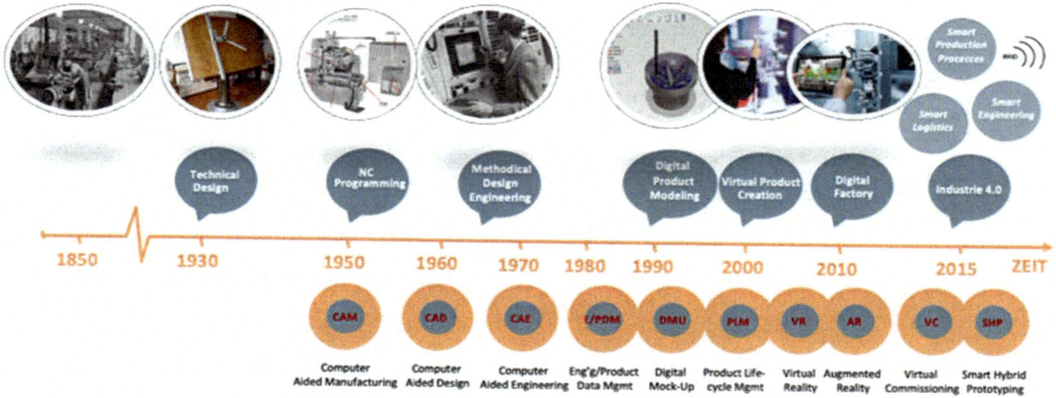

Fig. 6.1 Evolution of digital technologies in engineering

the view from a limited engineering process viewpoint to one overall systemic guidance that puts at the center the «*human engineering activities*» and views these as interacting with three dimensions, namely «*processes and organization*», «*virtual and physical artefacts*», and «*tools and IT systems*». This system is designated *as Engineering Operating System* (EOS), compare details in [5, 6]. The core principle of the EOS places human engineering activities in the center of the model in interaction with processes and organization, different types of artefacts (physical and digital), as well as with tools and IT systems (compare Fig. 6.2). The resulting full EOS model is illustrated in Fig. 6.3. The three sets of "processes and organization", "tools and IT systems" and "virtual and physical artifacts" form the basis of such interactive model, which should help to reshape the rather limited digital engineering approaches in industry.

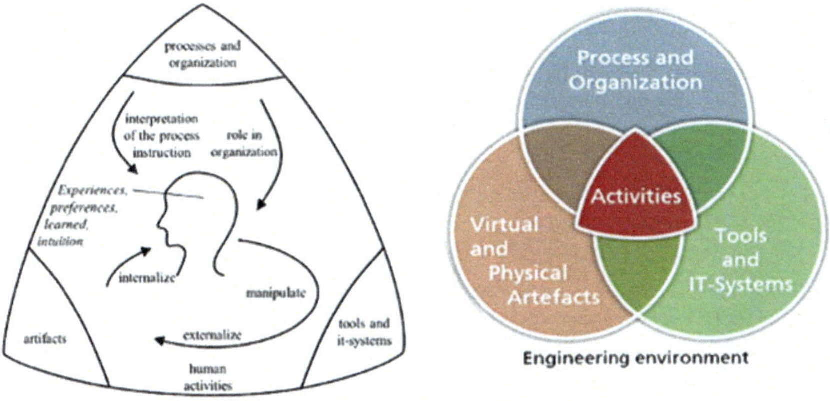

Fig. 6.2 Core principles of the EOS (compare [5, 6])

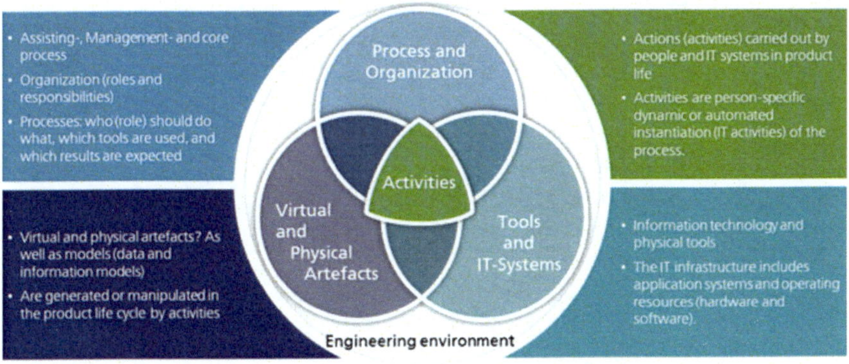

Fig. 6.3 Detailed explanation of the Engineering Operation System (EOS)

Based on these three sets and the interplay between them, the actual value-creating activities are carried out by humans within their roles (e.g. design engineer, manufacturing engineer, CAE analyst, data engineering etc.). A deeper insight to the EOS principles and working mechanisms is provided in [5, 6].

6.2 Understanding Ownership of Skillset—The Difference Between Traditional Engineering Skillset and New Digital Skillset

It is important to understand the difference between the traditional engineer skillset and the new digital skillset. Both are needed, but in a steadily evolving priority mix! Traditional engineer skillsets were centered mostly on knowing the technical parameters of how to describe a certain mechanism, a physical layout, and then bringing it into physical realization. Engineers always had to do the testing themselves in terms of proving out whether their ideas were feasible. Therefore, traditionally engineers were in ownership of their own verification and validation procedures, methods and tool sets. It was the nature of an engineer to have an idea, to build it with certain skills of how to build prototype and to test it accordingly with an idea of how the test can be successful.

In a next phase, testing became a separate discipline by itself because it is not simply just testing in order to prove out whether the object can withstand the stress level induced by an outer load of external forces or moments. Modern testing transformed itself into answering the following question: How long can it last intact without failure and how much reserve does it still carry?

Consequently, the engineering discipline was further progressed and developed in trials of having refined engineering testing procedures and methodologies. Up to the point where quality management methodologies like Taguchi and others added additional skillsets in terms of doing clever testing, reducing the number of tests and finding out about the nature of the design and its potential to be optimized (e.g. in terms of light weight). It rather became important to analyze why something was failing and not just the fact that it failed or not. Traditional bogey testing evolved towards reliability and key live testing procedures. Latest research foundations to combine the traditional test based engineering verification approaches with new virtual product creation based digital engineering modeling approaches have been published by Gerhorst in his dissertation work in 2017 based on his rich quality, robustness and engineering experiences at Ford Motor Company (compare [7]).

Digital skillset in product engineering became necessary in the design phase to produce a drawing on a computer; that is why CAD originally was called *Computer Aided Drafting*. Then, gradually, this skill set in product engineering evolved in using richer sets of commands within a modeling environment called CAD (Computer Aided Design). Consequently, designers and CAD engineers were in the position to describe the shape and the geometry of a component with respect to later on producing it in the production hall on a machine. Therefore, also the work planning was implicitly supported in order to ensure the manufacturing of the part. This digital skill set stage was followed by a more frequent interaction between digital modeling in design resp. product engineering and making sure that the risks to produce the designed parts in the factory were understood early enough. Consequently, the digital skill set was expanded into manufacturing engineering by starting digital checks to prove and ensure manufacturing feasibility. This then was the starting point to implement digital process planning activities and digital CAM (Computer Aided Manufacturing) capabilities.

Similarly, the evolution on the product verification and early validations side took place. Inspired through specialists in larger companies who developed mathematical procedures to automate stress and fatigue problem calculations and driven by younger generation engineers who transported new type of calculation toolsets into mainstream engineering industrial companies systematically implemented CAE (Computer Aided Engineering) digital skill sets within Product Engineering first. However, the speed and intensity differed substantially. Those industries that saw opportunities to save physical prove-out prototypes were amongst the most ambitious ones; others tailored a long and missed the chance to establish those digital skill sets early enough.

In addition, CAD engineers and especially component and system engineers needed to become clear on the digital skillsets of naming and numbering to make products unique in terms of their description and how they could be referred to in something called a Bill of Materials (BOM). All these formerly treated analog working elements on paper—traditionally executed by supporting workforces rather than by engineers—were transferred into digital toolsets. First, this was achieved by simply introducing digital spreadsheets, before more sophisticated workflow and database approaches were introduced which required a deeper understanding of object uniqueness in a larger data model and data base schema.

Over time, it became necessary to teach engineers in understanding basic digital concepts but primarily in using specific digital toolsets and applications. It was still quite similar kind of sequence of steps of how those elements were done until the point came no longer part by part and digital description by digital description were sufficient. In the nineties of last century and in the beginning of the two thousand, the overall product and mock-ups became a "demanding deliverable" and the skill set of how to organize and deliver in a day-by-day activity became a competitive edge. The digital models themselves have become a new additional digital skillset compared to the IT application skill set, and those companies who are able introduce those new digital skill sets quickly and robustly are having significant advantage in terms of fast, robust and system compliant engineering turn around also across locations! This important challenge and opportunity, however, is not yet understood by many companies and their leaders, which create permanent stress and malfunction in delivering engineering and production of products and technical systems.

There still exist the major misunderstanding that digital skill sets are good enough to be represented within the *Information Technology* departments rather than in Engineering and Manufacturing. This is still today the big challenge of successfully

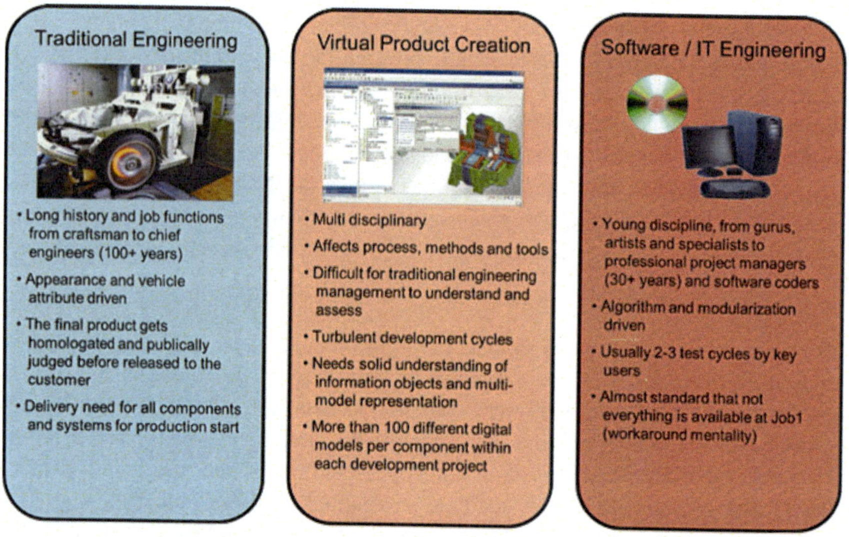

Fig. 6.4 "New" Virtual Product Creation skill set in between the typical company competence traditional engineering and Software/IT

running Virtual Product Creation in Industry! Naively enough, today's discussion is centered around providing as much as possible Software Engineering digital skill set due to the high significance of software intelligence of modern products, also the overall technical system knowledge, functional layout and above all, system interaction remains the most challenging part of future Virtual Product Creation!

Summarizing the evolution and today's overall situation of establishing digital skill sets as part of a new core *virtual product creation* discipline within industrial product and manufacturing engineering Fig. 6.4 shows the biggest challenge in striding forward towards a fully recognized and appreciated core *virtual product creation* competence.

6.3 Understanding the Nature of *Virtual Product Creation Collaboration* in Development Project Execution

The set-up of Virtual Product Creation activities in industry is characterized by an overall ideas and project progression, which uses as core model an implicit control behavior of the involved members and resources as outlined in Fig. 6.5.

As starting point for an engineering project or undertaking, the explicit reflection of a significant set of requirements (oftentimes dozens, hundreds or even thousands) lays the foundation as input to the virtual product creation activities (input W in Fig. 6.5). Before the engineering activities can start according to the overall EOS (Engineering Operating System, compare Figs. 6.2 and 6.3) all internal procedures

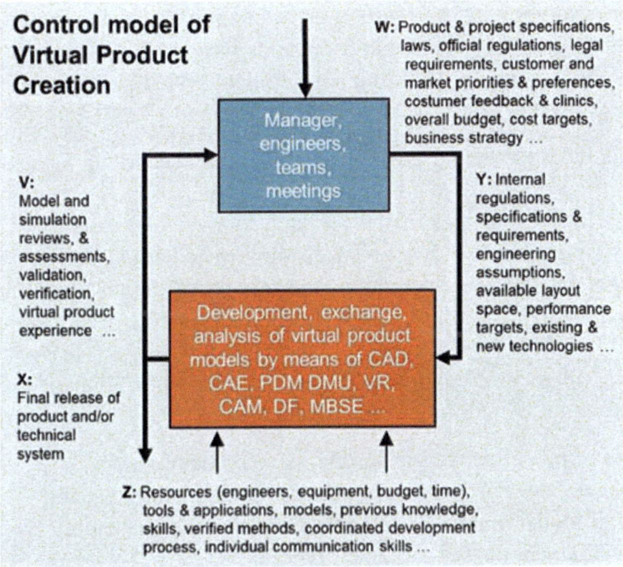

Fig. 6.5 Control model of virtual product creation

and targets as well as any outer constraints to the product or technical system should be known, understood and analyzed (see reflection step Y in Fig. 6.5).

Such reflection step might already require substantial digital capabilities as part of requirements management or early conceptual solution reasoning.

Another key ingredient of the Virtual Product Creation control loop constitutes the set-up of engineers, designers, digital tools and digital models (see input Z in Fig. 6.5). This is the expertise pool and knowledge foundation of a company in order to start the virtual product creation activities such as modeling, assessing, analyzing and modifying. If this set-up is not appropriately developed and provided beforehand, the control loop is severely handicapped right from the beginning, i.e. many unnecessary iterations might be the consequence! The progressing in iterations might depend on how a company plans to divide and to structure the design problem into individual solution components and how certain solution assemblies are treated from functional, overall layout and supply chain point of view. The conceptual design loop describes and tries to determine architectures of products and their components. A digital mock-up based on the first CAD model layouts might support such elements of describing and modeling entities and bringing them together for discussion and review. Using them for CAE, proving them out analytically, studying results of static and dynamic behaviors, clarifying questions on component manufacturing and assembly capability for manufacturing feasibility are other key activities of design reviews (see the validation and verification activity V in Fig. 6.5).

In order to become effective and efficient in applying such a *control model for Virtual Product Creation* industrial companies have to establish key capabilities within the product and manufacturing development system:

1. *Virtual Engineering delivery phases* within the overall development process which includes the definition of clear deliverables according to the overall engineering progression (like e.g. maturity and compatibility targets for the individual delivery packages of the product components, assemblies and partial systems). Figure 6.6 shows as an example of a virtual delivery process for an electric bicycle, Fig. 6.7 shows the respective digital model network and progression plan, whereas Fig. 6.8 shows the principle of an automotive body shell CAD deliverables plan.

   These three different levels of descriptions show the different kind of clarification needs within virtual product creation business in industry:

   a. type of model delivery regarding development phases,
   b. linkage between model types to ensure compatibilities and
   c. detail model maturity according to components/assembly type as well as development phases and gateways.

2. A common understanding amongst all involved development partners how long iteration cycles should last according to the VPC control model should last is key. The term "digital engineering turn-around cycle" might be the best explanation for this core competence: how quickly can a new design proposal or the analysis of a recognized problem be executed within the digital engineering collaboration across all sites and partners? This subject usually causes a delicate and oftentimes implicit discussion since most of the companies have not yet established responsibilities for such digital progression/collaboration arbitration and determination amongst all partners. Traditionally, the process re-engineering internal experts or outside consultants do have neither the digital knowledge nor the engineering expertise in order to lead such a critical undertaking. IT departments might to assess and determine the IT service and operation in terms of IT processing time, e.g. for data base replication and data service delivery (e.g. conversion), but have no insight to the engineering delivery constraints. Traditional engineering departments usually have limited digital process knowledge and not enough IT technology know how to act in leading function. This is the reason why special *digital innovation and operations* departments in product development and manufacturing engineering are getting established to hold such knowledge mix and to play this critical leadership role. Such a strategic decision, however, needs full senior management understanding, endorsement and active support!
3. A solid set-up of appropriate digital design reviews that are best suited to engage the right level of experts, partners, suppliers, management and ultimate decision makers in line with different development situations and types of decisions.

From the 90s onwards, companies started to set up digital design reviews as a new class of engineering excellence. There exist different types of classical design

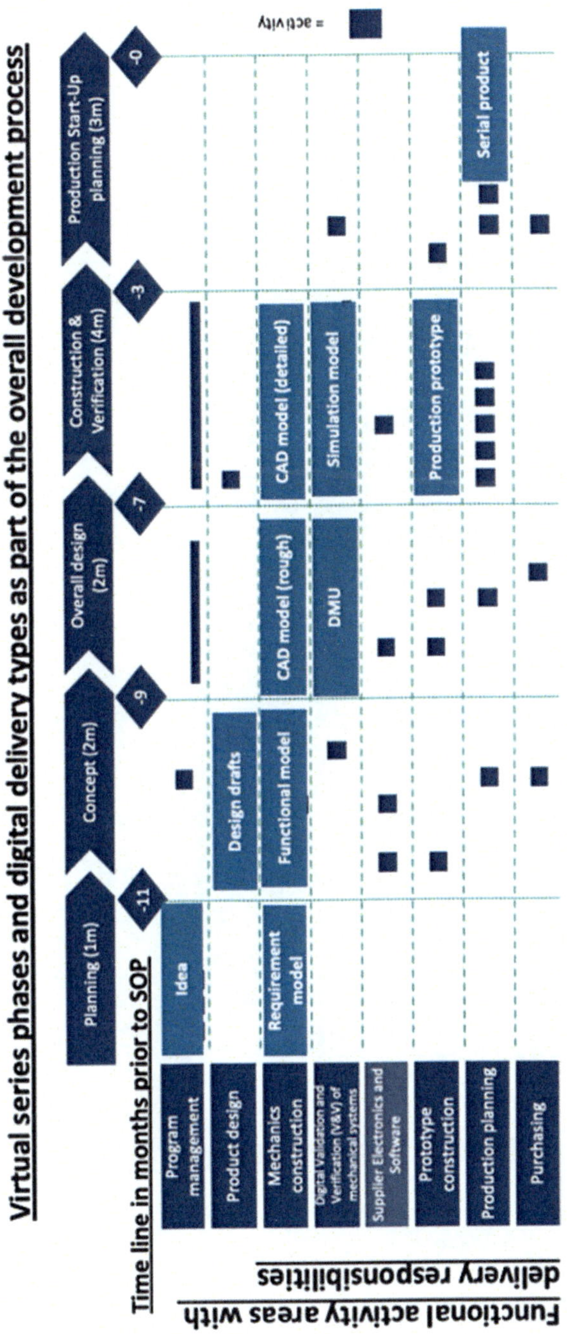

Fig. 6.6 Virtual delivery process for an electric bicycle

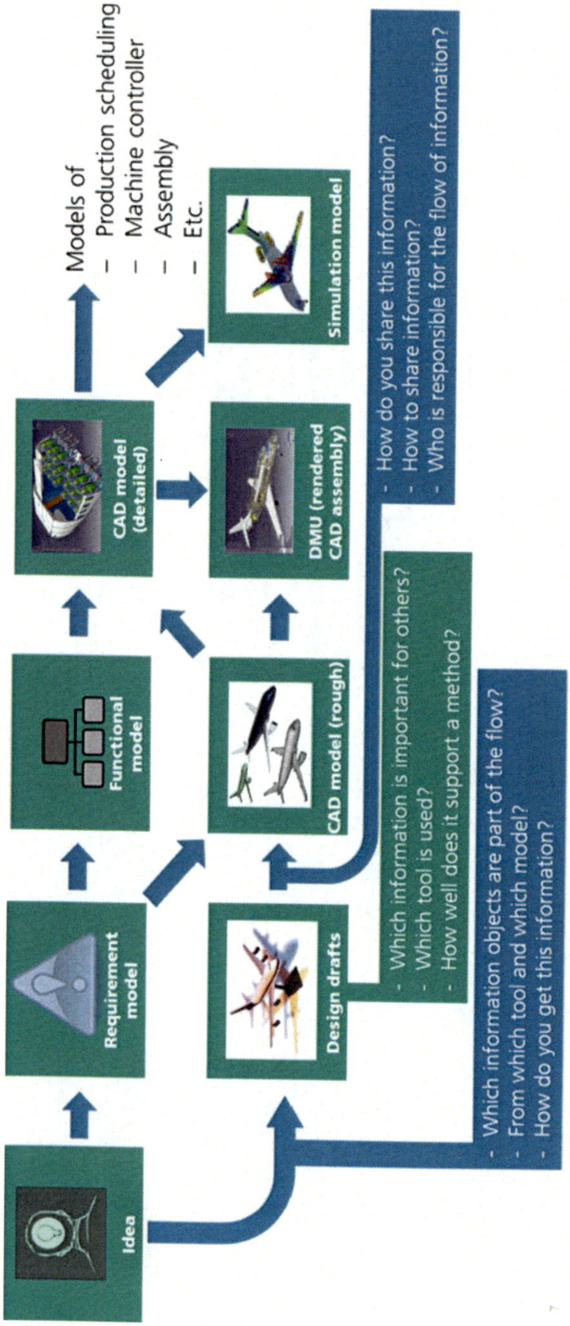

Fig. 6.7 Digital model network and progression plan

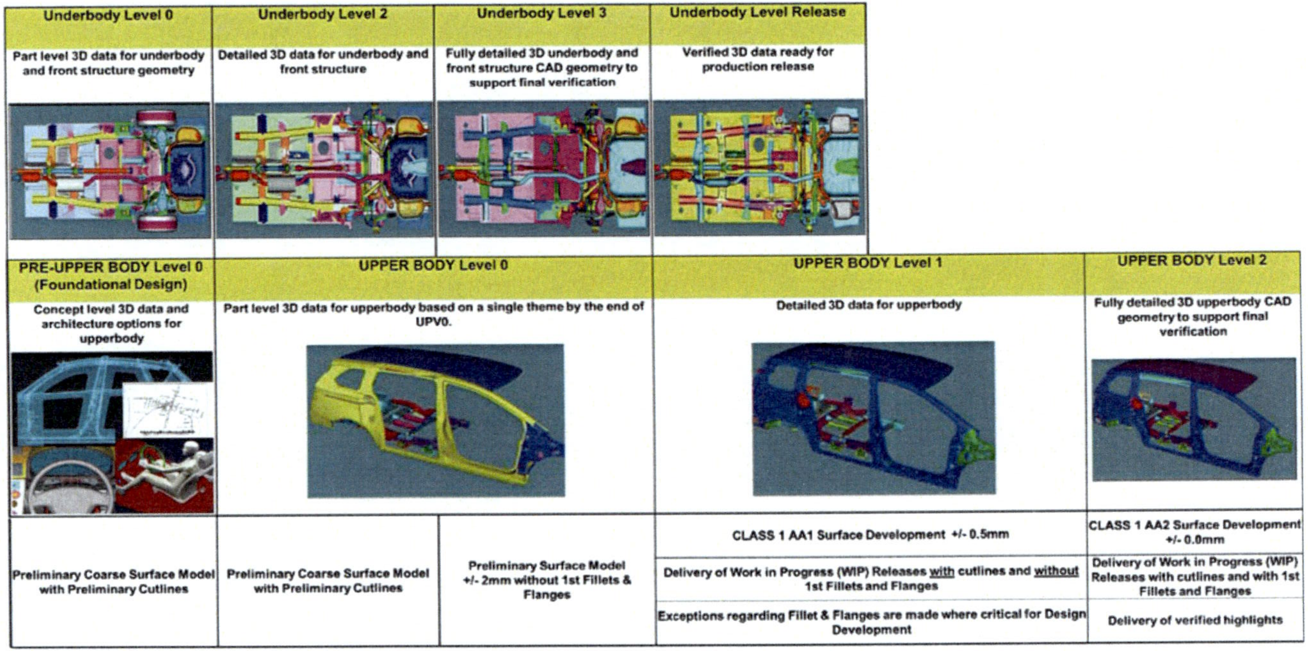

Fig. 6.8 Principle of CAD deliverables plan (example body shell)

reviews in industry, partial with overlapping content, which had to get modified in order to make them digitally effective and efficient:

- **System Compatibility/Integration Reviews (SCR/SIR)**: The purpose of a CR is to review the readiness of the interfaces of the design components and partial systems to each to ensure system integration readiness.
- **Preliminary Design Review (PDR)**: The purpose of PDR is to review the conceptual design to ensure that the planned technical approach will meet the requirements.
- **Critical Design Review (CDR)**: The purpose of CDR is to review the detailed design to ensure that the design implementation has met the requirements.
- **Production Readiness Review (PRR)**: The purpose of PRR is to ensure that the design is completely and accurately documented and ready for manufacturing release.
- **System Acceptance Review (SAR)**: SAR verifies the completeness of the specific end products in relation to their expected maturity level and assesses compliance to stakeholder expectations.
- Other special types of design reviews: Operational Readiness Review (ORR), Periodic Technical Review (PTR), Technical Peer Reviews (TPR)...

Generically, a design review is a documented, comprehensive and systematic examination of a design. The aim is to assess its ability to meet quality and attribute requirements, to identify potential problems and to lay down the development of problem resolution. A design review may be carried out at any stage of the design process, can be applied on physical artefacts, but, nowadays, is mainly applied on digital models that represent form, fit, function, behaviors and other control intelligence of products, devices, machines, technical systems and product-service-systems (PSS).

According to the official academic design methodology, design reviews do not play any important role! Why? Group discussion, critique, and problem solving in teams are not at all an appreciated part of the traditional synthesis and analytical method steps; design progression is mainly build-up on personal and intrinsic analytical thinking, robust methodology and determined process execution. However, the reality in industrial practice is different:

- Development progression is highly driven by experience and compromise! → This triggers reviews with other experts.
- The timely compression of development execution from the beginning of the 90s in the twentieth century (≈30 to 50%) has driven milestone-oriented program and project management. This naturally triggers reviews amongst all development partners to meet milestone timing with best possible solutions.
- Technically more complex and diverse solution elements (mechanical, hydraulic, pneumatic, electronic, software control) make product synthesis and integration more dependent on holistic systems engineering thinking, which calls for multi-domain product and system compatibility and validation reviews.

- Different development partner interests and boundary conditions (OEM, supplier, engineering service provider, final customer service provider etc.).

The execution goal of digital design reviews is usually oriented towards:

- ***Status*** (Where are we compared to the initial plan and to the models/data we reviewed last time?)
- ***Assessment of goal achievement*** (Are we still well on track to achieve the overall goal) and
- ***Resolution and next steps*** (What do we have to do next to deliver?)

In order to ensure a sound definition of the *digital* review scope and engagement of participants the following elements are key:

- ***Determination of the review type and review leader*** (preparation, conduction and result preparation)
- ***Establishing a way of review execution and operational behavior*** (on which technical depth level detail expert discussions are allowed, which solution ideas and issues resolution steps are possible, which external partners need to get involved etc.?)
- ***Determining all necessary digital preparation steps***, e.g. in terms of delivery and storages dates of digital models in databases, type of configurations to be populated in order to allow for sound review execution during the reviews, preparation of predefined digital camera views on pre-analyzed issues in the right sequence etc.
- ***Selecting the most appropriate digital review tools and IT-technologies*** to support the live execution of the review (incl. multi-site collaboration, broadcasting and streaming technologies).

Similarly, to the traditional paper-based design reviews, significant preparation time is necessary to prepare and conduct such digital reviews in order to ensure successful outcome and result delivery. Obviously, it depends on how many technical design objects are subject for which type of review and how many different experts, partners and management personnel are involved. As outlined in Fig. 6.9 a small digital design review might engage 5–20 persons and is executed in a local centric mode with some extend remote experts. Such set-up eases the preparation time and the digital execution effort. Usually such team reviews are half-formal, coordinator initiated and follow an atmosphere and spirit to deliver official outcome rather than informal outcome in engineer-to-engineer reviews. Usually, such a digital design review follows an official invitation one week in advance and is executed on a regular base (every second week or monthly). Efforts are taken to pre-prepare analysis investigations based on working and analysis model layouts prior to the review in order to enable a smooth drive through the review content. Figure 6.10 shows the setup of a full program or project (compatibility) digital design review. Such a review is rather formal and puts focus on a correct behavior by all attending job roles and authority levels. Between 20 and 50 concurrent participants and delegates as well as

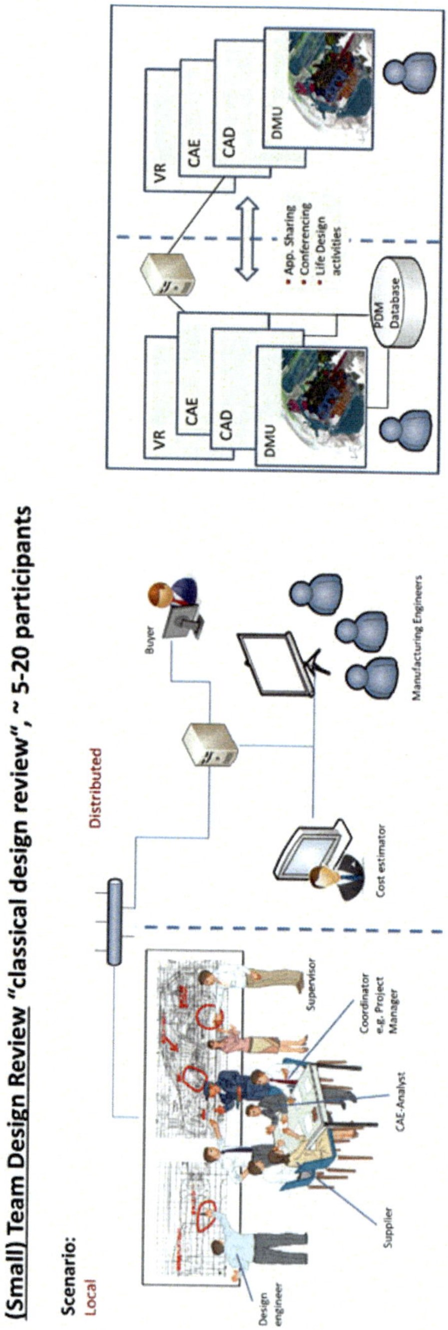

Fig. 6.9 Set-up of (small) digital team design reviews

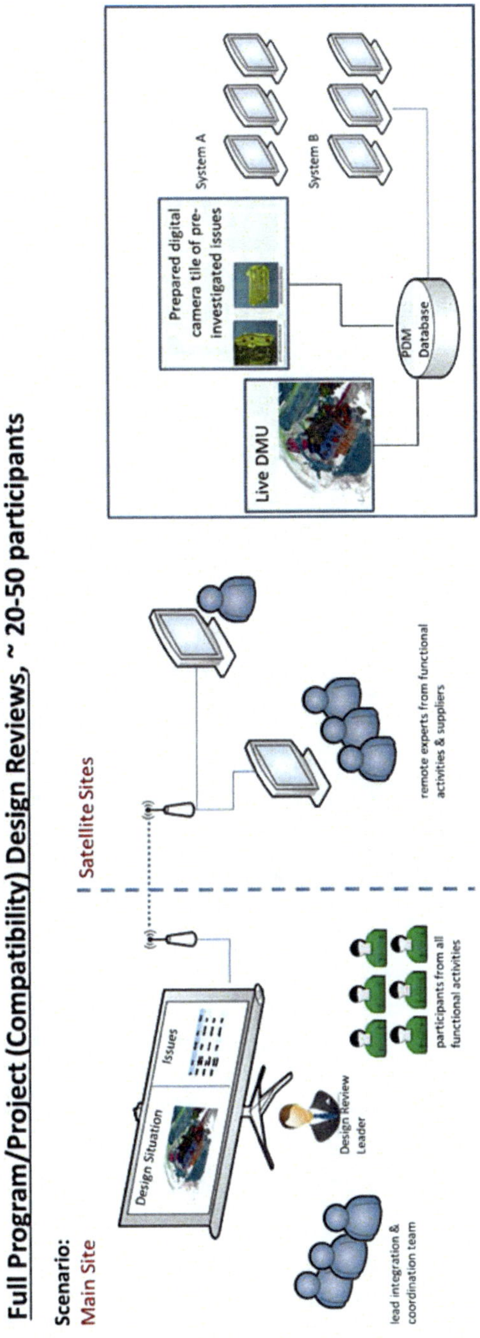

Fig. 6.10 Set-up of full program/project (compatibility) digital design reviews

experts from each functional activity are attending the review following a technical and development policy relevant atmosphere. The review time lasts 1½–4 h.

Under the clear leadership of review leader ("commander in charge"), all participants follow a rather formal review agenda according to chosen lead review criteria. The review cadence is officially linked to the operational development plan as well as to respective milestones/gateway deliveries. The review coordination and integration team needs 3–5 days active preparation after an official and thoroughly followed design freeze! The typical review duration is between 3 days and a full working week.

Virtual Product Creation Experience in Industry (*Best Practice*): Proactively Utilizing the Power of Digital Design Reviews

The development situation

A typical day in product development in automotive industry: According to the overall milestone process, the vehicle program team is just ten days away from the critical gateway called *"design theme approval and full investment commitment"*. The vehicle program director Paul N. is in trouble and has too many open issues and no consolidated understanding yet of the *real* development status of his 250 person team to be sure that he can defend solid program delivery plan for 3 derivatives and 5 powertrain combinations within 1.5 billion $ within the next 36 months. Therefore, he has determined in a crisis meeting 3 weeks ago to conduct a full vehicle design and system compatibility review today, lasting 3 working days. All relevant engineering, manufacturing, purchase, and program controlling activities need to be involved; the lead coordination is with the overall Vehicle Engineering Package and DMU (VEP-DMU) organization, which has formed a Vehicle Engineering Review Core team.

The Vehicle Engineering Review Core team is ready to go!

Michael B., the Package Leader in the vehicle program team, and *Dr. Ryan*, the Digital Engineering and Innovation Leader in the Research and Development Center, have geared up their teams during the last 5 working days. Both gave "green light" last night at 9 pm to the Vehicle Engineering Manager Joe T. who is determined to deliver a successful digital design and compatibility review to his direct boss Paul N. Now it is 10 a.m. in the next morning, all key stakeholders also from the other development site, 90 min flight time away, are onsite to be most responsive in review practice and quick sub-team get together. The digital review begins.

Digital Technology, problem transparency and determined actions

The core team had instructed all functional activities 2 weeks prior to the digital review to publish the current design intent status of their digital models according to the precise rules of product structure, variants and maturity. Overall, more than 2000 updates have been recorded in the PDM data base until 3 working days prior to the review, a tracking of recognized issues have been done in parallel with links to the updated product models. For the last 3 working days the core team has found approximately additional 400 issues, piling up to approx. 1100 recognized issues (compatibility problems, technical questions, unclarities in design, clashes, missing parts etc.) before starting the review.

After 2 h in the review, already having addressed 30 issues with 6 immediate solutions, 15 potential solutions options and 9 issues where the severity of the issues are confirmed to be of highest priority with probably 2 critical ones, the following situation started to escalate (immediately prior to lunch):

- One of the design team proposals requires a low and smooth glass shut line incl. a low cowl area with a relative "fast" slanted windscreen.
- According to Program Management the functional activity Body Engineering as attribute leader of the wiper system has to deliver a system solution with 0$ on cost to an existing wiper architecture solution of a SUV type vehicle as part of the modular commodity plan.
- The Design Engineering department of the Design Studio has created a digital issue prior to the review with the following text: *wiper system concept is "inappropriate" and cannot be assessed.*

Michael B. calls up this issue and asks one of his DMU engineers to show the current situation in the DMU. Zoran Z. quickly invokes one of the prepared camera tiles, which shows the 3D design dilemma: the wiper system with motor, gear linkages and wiper blades flies approx. 500 mm above the cowl area in the air without any meaningful attachment schema …. Zoran Z. can express this issue with just one sentence! Vehicle Manager Joe T. confirms that this solution is absolutely inappropriate and ask the responsibly Body System Engineer to come forward to the major screen to explain the resolution of this issues which already exists for more than 4 months! Unfortunately, nobody stands up.

Now, the Chief Program Engineer Paul N. who used to be Body Engineering Chief prior to his new management position asks specifically for the report out of Reiner U. …, a voice from behind just utters: Reiner U. is not in today; he is most likely in the assembly plant today, 1 ½ hours car drive away, unreachable. Paul N. reacts quickly and asks specifically for Andrew F., the responsible Body Supervisor to come immediately to the review to explain the situation. The team decides to break for lunch and to reconvene after lunch break by starting directly with this very issue. Before leaving for lunch break, the Program Chief Paul N. and the Vehicle Manager Joe T. directly ask Dr. Ryan, the Digital Engineering and Innovation Leader, if he can check the history of the PDM database for additional intelligence of the wiper system history. Dr. Ryan confirms that his team can do that quickly within the break and will be ready to report out at the restart after lunch.

Restart of the digital design review at 1 p.m.: the focus right away on the "flying" wiper system still 500 mm in the air above the cowl (hood area close to the windscreen). 2 min after 1 p.m., Andrew F., Body Supervisor, appears in the meeting together with his direct boss, Steve D., the Body Engineering Program Manager. Both take a seat in the first row of the review room, on the opposite side of Paul. N. and Joe T. All 58 participants are eagerly interested in the new round of finding clarity on the "flying" wiper system; Michael B. starts the session by quickly recalling the issues content: *wiper system concept is "inappropriate" and cannot be assessed.*

Paul N. directly addresses the following question to Andreas F.: "Andreas, we are wasting time in this review due to the non proactiveness of your team, I expect immediate clarification from you personally!" Andrew F. stays cool and determined and responds in the following way: "Paul, I am sorry that you think that we waste time here, but my team is not guilty for it. I assume that this new digital engineering solution is not capable of showing the latest design, you should make sure that you have the right IT and method support in place for the entire team, rather than potentially wasting time of us in Management and of my engineering team …" Paul N. looks to Joe T. and both are directing the attention to Dr. Ryan. He took on the assignment to clarify this point proactively with his team during the lunch break. Therefore, Paul M. simply responds to Andy F.: "No problem, Andrew, we do know that the digital solution works, if it is used correctly in terms of working methods and the right responsibility is applied! Hence, Dr. Ryan will provide some insight of this issue to all of us, Dr Ryan, please do so!"

Dr. Ryan stands up and reports out on the point: "There only exist two versions of the wiper system assembly in the PDM data base, the first one from 5 months ago, already in the wrong vehicle position, 500 mm above the cowl, loaded up by an agency Body CAD Designer called Ray O. The second version from 3 months ago, just with a name change to this vehicle program without any further modification. That is all activity by Body Engineering!"

Andrew F. frowned on this quick and clear report out by Dr. Ryan and looks with a slide hope in his eyes to his direct boss Steve D. Steve mumbles unknown words to himself, stands-up and states the following striking words to the Program Chief Paul M. in front of all 58 participants:

> It is time to call crisis on this non-technical but management issue…, and not from us, you Paul should rather do it as Program Chief! According to our new development policy of making suppliers fully responsible for the development of purchased systems, it is no longer our OEM responsibility to take care of functional and package issues and development tasks; it is the task of the full service supplier. We have no

longer budgeted for those engineering activities! And the purchase department did not yet chose a supplier due to competitive edge, still waiting for a best offer until the next gateway…; you better put pressure on the Purchasing Manger Brigit E.!" He looks around to all participants in the room, points to the screen with the "flying" wiper system and says in an elegant British manner; "These new digital review capabilities based on the new digital engineering toolset are superb! They guide us, what really needs attention. We may all appreciate it!

Paul N., the Chief Program Engineer, quickly understood the dimension behind it. He states to Michael B., the core Vehicle Package Leader running the entire design review: "Michael, please note down that the "flying" wiper system has provided enough evidence to a root cause issue of our current development process and system: the non-alignment of development responsibilities and budget provision in the absence of a sourced supplier. I will personally set-up a meeting tomorrow with Brigit E., the Purchase Manager, to get this dilemma addressed!"

Lessons learned, best practice and follow-up

- If professional transparency is applied to well-prepared digital design reviews, even hidden problems will get detected.
- The right virtual product creation competence in digital design reviews help to avoid potential story telling around the issues.
- The right culture of digital design reviews is key to deliver the right results and follow-up actions.

The incident of the "flying" wiper system as reviewed in the design review helped the vehicle manufacturer to quickly resolve the process issue of "not yet sourced full service suppliers". In short term, extra budget was given to Body Engineering to cover the wiper system task in a responsible way for the period until the supplier was sourced. Generically, the determination of gateway deliverables was changed and the subsequent vehicle programs no longer had to experiences this trouble.

Chief Program Engineer Paul N. and his entire engineering team did successfully defend the entire vehicle program in the following senior management review and did deliver the vehicle program within the next 3 years with more than 15 digital design and compatibility reviews in time and budget. The vehicle could eventually successfully launched to the market and did perform well in terms of sales figures, business revenues and customer satisfaction.

6.4 The Traditional Set-Up of Virtual Product Creation and Its Flaws

Industrial companies have traditionally established functional and organizational divisions such as marketing, product planning, product development, purchase, controlling, manufacturing planning, production, logistics or service operations and have added technology and service-oriented departments to it such as information technology, infrastructure maintenance, human resource and finance. Start-ups and small enterprises run their business much leaner by combined responsibilities. Medium sized companies (more than 250 employees) and big companies incl. major OEMs have added project organizations to such a set-up in order to be more nimble in project execution. The skills needed in such projects are provided by the classical functional organizations within such companies by setting up a matrix reporting line system or temporarily by partners and (system) suppliers from outside the companies.

Due to the above-described situations, three principle approaches are common to establish new, additional and/or experienced and expert's skill set for virtual product creations in industry:

- **Option 1**: Implementing digital engineering skill set for virtual product creation primarily within the existing functional divisions and activities *without* any fundamental changes in the underlying roles and responsibilities set-up.
- **Option 2**: Innovating digital engineering and overall digital process knowledge primarily via cross-functional organizations as *first mover and driver* for subsequent mainstream boost up of VPC skills in the existing functional divisions.
- **Option 3**: Pushing new and additional virtual product creation technologies through (process) IT or dedicated temporary digital technology project organizations in order to overhaul and (re-) establish deeper digital skillsets in the existing functional divisions.

In all of the above three options the generic challenge exist that a difficult separation exists for a longer time period (at least 3 years) in terms of joint or competitive ownership between:

- IT infrastructure, hardware/software competence and IT operations service for VPC, usually owned by IT departments
- Digital process and application competence in engineering, analysis/verification and manufacturing within the functional activities in product development or manufacturing.

Germany, middle and northern Europe traditionally, i.e. in most of the cases, prefer option 1 as a starting point. If bigger digital transformations are needed, they rather fail to modify and innovate the traditional ways of digital working. Oftentimes, several years are wasted or only little progress is made to come to new and more proactive virtual product creation operations in business. This, however, is not a competence issue—skill set and knowledge usually improves significantly over time—but rather a new mindset and risk dilemma. Department leaders rather prefer to optimize their existing way of working and want to see *proven evidence* that the new virtual product creation solutions deliver better results. In addition, due to high outsourcing, many mainstream functional departments no longer have the leadership and key user knowledge to pro-actively change the as-usual situation towards speedy digital innovations without help from outside. This is the reason why companies in that region rely heavily on *well-trusted* consulting competence from the outside. If option 1 has finally flawed or continued to be non-effective most of the middle and northern European industrial companies favor or change to option 3 rather than considering and selecting option 2. French and southern European companies can act more flexible but usually have to resolve a complicated set-up and do need a strong and long-standing network approach as part of a hierarchical elite approach in order to drive innovation forward, also in case of virtual product creation skillset and innovations. Small and medium companies usually experience much higher challenges due to their limits in terms of number of experts available within the organizations.

The Anglo-American way in industry appreciates and favors speed, innovation and change much more, and looks for rapid and more visible transformational change. Back to the nineties of last century, the American way of deploying and implementing virtual product creation innovations started from option 3 and move gradually over to option two from the early 2000th onwards as long as sufficient management leadership support and top down momentum were available. Promoting and pushing is a strong element for such transformations but long-term skill set development oftentimes is compromised or even not possible due to high turnover in employees and high-outsourced work force. The British way orients itself—also in case of virtual product creation—very strongly towards a well-refined project management approach, which usually earns higher value and appreciations than technical skills in virtual product creation technologies.

The approach of Japan still today is highly characterized by option 3. However, long-standing relations within the companies themselves and very tight relations to their dedicated suppliers' network make it possible to keep an evolutionary digital skill set progression via a parallel working set-up according to option 1. Only for very dedicated goals or in crisis modes option 2 is used deliberately. Therefore, digital transformations with new VPC skills take substantial time. South Korea usually has a similar set-up to North America, whereas China is more similar to the European approach and follow a very stringent *Made in China 2025*× plan, which enables quick and consistent steps to close the gap to world-class digital leaders.

All three options, however, can only mitigate but cannot avoid or even eliminate today's systemic flaws in the usual set-up of virtual product creation technologies, skills, competences, operational activities, implementation actions and associated leadership profiles in industry:

- Insufficient and non-stringent overall development, deployment and implementation management approach with respect to:
  - only coarsely developed virtual product creation architecture with respect to business and engineering target plan

  - quickly and not thoroughly derived use case portfolios for today's and future development and application scenarios
  - unclear deployment metrics, risky deployment plan and non or limitedly evaluated migration and implementation efforts within engineering and manufacturing departments
  - unclear leadership and responsibility split between project organization, IT departments and functional activities.

- Limited know-how and skillset within the industrial enterprises, both from digital technology expertise and from digital engineering resp. manufacturing operations know-how perspective; as shown in Fig. 6.4 such new VPC skillset is necessary and key to bridge the traditional engineering discipline and the genuine discipline of software and information technology.
- Non-appropriate leadership control, foreign domain language and obsolete review formats to cope with the high number of technical challenges in the virtual product creation solution portfolio, especially with respect to:

  - IT application bug fixing and IT infrastructure robustness,
  - developing and implementing new digital engineering responsibilities and working methods,
  - difficulties to determine appropriate data models, explanation of IT tool error states as well as creation and delivery of the right type and enough matured digital models of technical systems and components,
  - no experience in solving "hidden" personnel and organizational problems within the digital transformation,
  - irritations about "bug fix prioritization" listings, low reliability on impact analysis and resolution plans from digital experts resulting in limited buy-in from middle and senior management
  - confusion and fears in middle/senior management about rather "techy language blended" report outs of virtual product creation architects and experts in formal reviews.

- oftentimes high-level agreements only on the best compromise between effectiveness, efficiency, robustness and simplicity of digital process and solution execution resulting in many *on-the-fly* process modifications during roll-out and implementation causing churn, frustration, delivery risks and unstable project situations.
- wrong hopes and expectation in big bang implementations with rapid one-off know how transfer rather than ramp-up deployments with learning and skillset development over a substantial time line.
- unclear leadership calls to establish new digital task ownership within organizations, especially if their nature is cross- or multi-disciplinary.

In order to realize the full potential of virtual product creation and its digital solutions sets it becomes essential to provide the right understanding of new and future jobs in Virtual Product Creation beyond the traditional CAD/CAE enabled designers

Fig. 6.11 New and evolving job roles and functions

and analysts. A number of new job functions such as DMU (Digital Mock-up) engineer, PDM expert, data engineers, data analysts etc. have already been established in industry, at least partly. Moving into the future, however, there will exist more stringent needs to overhaul and change existing job roles and functions or to create even new ones more rapidly (see Fig. 6.11):

- **Sustainability mentor**: use of data analytics, digital scenario and value creation composition for connected sustainability networks
- **Product-Service-System architect**: integrated and balanced development of products and services as a joint offering
- **Systems architect**: owner to oversee, synthesize, control and protect the cross-domain system architecture of technical systems
- **Systems engineer**: design, validation, testing and verification of technical (sub) systems and components with models and data
- **Sustainability Factory Lifecycle Engineer**: design and analysis of factory system design within sustainable and circular economy constraints
- **Global Production Network Designer**: modeling and optimizing global production networks to achieve best operation resilience
- **Digital Factory Model Integrator**: integration engineer to digitally build, super compose, align and test all digital manufacturing models for digital factory simulation
- **Professional in Product Lifecycle Management** (see Fig. 6.12).

Unlike the long lasting cycles of waiting many weeks or even several months before a next substantial progress iteration was finalized within the overall development process the introduction of substantial virtual product creation solutions made it possible to significantly reduce such "engineering turn-around cycles" (see VPC control model in Fig. 6.5) down to one week or even less. This potential is seducing with respect to the overall reduction of development time and time to market and

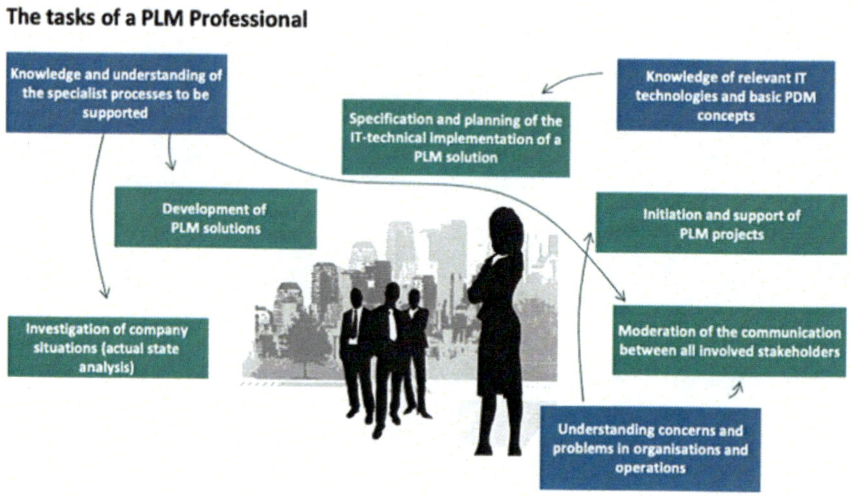

Fig. 6.12 The tasks of a PLM Professional according to Fraunhofer

obviously creates new demanding requests to the engineering community in the mindset of management.

It, however, uncovers also a wide range of more fundamental issues and challenges of virtual product creation operations. Once central question need to be answered in any case: who (person, team, empowered role) in the oftentimes widely dispersed engineering project team holds the responsibility, the accountability and the willing power for the daily collaboration problems by executing virtual product creation?

The following issues are somewhat related to this question, are widely known and create massive churn in the digital engineering operations but oftentimes are not solved or not even understood due to missing leadership and/or profound knowledge in virtual product creation deployment:

- Ownership of and execution duties for digital engineering processes and related VPC development tasks (what does it mean in detail, what is agreed, appreciated, or even valued?)
- The traditional conflict between traditional physical oriented engineering approaches and new or emerging digital engineering approaches are not clarified strategically and generically (i.e. independent of development projects) and/or not tactically within the development project. The believe, the trust and the personal commitment to use virtual product creation solutions and skill wherever possible remain critical elements which in many companies are underdeveloped and not yet part of career plans and execution power.
- The digital capability to support the "expected" digital engineering turn-around target by management does not exist. This limited digital capability can have many reasons such as insufficient IT architecture of the underlying PLM solution set, underperforming PLM architecture due to limited data base operations and replications, error prone conversion and translation services, wrongly used digital

working methods, high number of different IT applications and associated data model formats etc.
- High degree of technological interfaces and organizational interfaces (confusion how to resolve these different types of interface issues)
- Politics around information ownership, access and deliveries
- Missing business models for digital value creation and non-appreciation of data and model values.

Beyond those day-to-day related technical, process, ownership and execution challenges and limitations, there exist a phenomenon in most of the modern industrial companies which can be called *"unknowingness"* or even *"innocence"* of leadership levels how to sustainable set-up virtual product creation intelligence in functional engineering activities. The following syndrome types do characterize such dangerous attitudes from different angles:

- The *"is not my problem"* syndrome; nobody told me that this is my responsibility, why should I care? Even if it affects me, I am innocent, somebody has to fix it and help me!
- The *"selling syndicate"* syndrome *("vanilla pie the sky")*—those teams who do feel ownership for virtual product creation are often politically forced by their senior initiative or project chiefs to always sketch an absolute positive picture of the virtual product creation architecture, solution set and associated capabilities. If those expert teams nevertheless treat negative arguments or complaints from the digital user side seriously and openly, they might be treated as betrayers!
- The *"overcharged victim"* syndrome; many user areas in engineering and manufacturing as well as in further downstream areas refuse using new virtual product creation solutions and the associated tasks and activities due to the excuse that this cannot be handled any longer due to many other tasks already. In some case this might be justified but in many cased this argument is used a scapegoat excuse not get measured against this new digital task and therefore to simply refuse a professional acceptance.
- The *"job stopper and SWAT team"* syndrome If the pressure gets too high in using new digital solutions due to inherit software bugs and digital process and activity flaws, group leaders, key users and direct supervisory leaders call crisis declaring a "job stopper" situation. This is combined with the demanding request of establishing a SWAT (Special Weapon and Tactics) team approach to help, mitigate and resolve those problems.
- The *"not recognizing and innocent dig deeper"* syndrome; due to fears that a political crisis within the company or project need to be avoided in any case, teams and leaders try to resolve partial, smaller and more detailed problems in order to get the positive feeling to help and improve the situation. This behavior first need to be appreciated and honored, however, can lead to many inefficiencies and problems for the overall situation since the assessment of the bigger problem is avoided, ignored or simply pushed out. Technical leadership is needed and critical in such situations since typically the "complaint storm" of many other leaders might be in the way (do we really have a bigger problem?). In any case, this

technical leadership behavior requires the right personal character and typically a well-earned trust and authority within the organization to act in such a way.

Virtual Product Creation Experience in Industry (How to Deal Meaningfully with *Bad Practice*):

Blame IT in order to disguise error modes in using digital modeling methods in engineering

The product development situation

Another quite stressful day in automotive engineering: According to the overall milestone process, one vehicle program team is under major pressure to deliver all digital model deliveries for the important gateway "Model delivery for final design theme decision" in a few weeks. This gateway is to confirm all major functional attributes, i.e. the developed system design of the body structure need to be virtually verified and signed-off. To be able to conduct all major CAE attribute simulations like crash, NVH and durability the Body Engineering design teams have to deliver all 420 sheet metal components with their final weld flanges and all 4450 spot-welds. The other vehicle program is already in pre-launch mode and has to make fast changes in order to solve urgent issues of manufacturing feasibility in pre-launch mode approx. 15 months prior to Job 1.

The situation of the overall Virtual Product Creation situation

Dr. Ryan and his global digital method and deployment team have major challenges to deliver new digital modeling, data management and CAE analysis solutions to the dispersed engineering team across 3 development sites in Germany, England and Japan as part of a major new global platform development. This global team consists of ~350 engineers and started to use the new Virtual Product Creation architecture (new CAD system, new PDM system, new product structures, several new CAE and Digital Factory solutions and a ran principles) 18 months ago.

To the same time, the new solution has to support another fast running SUV vehicle program that is the first runner of the new gateway development plan reducing time to market delivery by 9 months! This team just started to use the new virtual product creation solution 6 months ago.

The job 1 of the new digital solution set followed a development period of 2 ½ years as part of a global major initiative called *Digital Development System Next Generation (DDS Next Gen).*

The day of calling "digital crisis"

It is 10:30 am in the morning of a sunny Wednesday. Dr. Ryan receives a demanding phone call directly out of the major design review room in the Product Development center.

The Engineering Director, in person, calls him from his mobile phone. Five minutes ago, the Body Engineering team as part of the Senior Management preparation event prior to the critical gateway review of the new development plan called out a major crisis. The accusation is simple but tough: Dr. Ryan and his team rolled out a digital solution that does not work and will kill vehicle program timing with major consequences for the company and individuals!

Dr. Ryan immediately knows that this is prime time now, again, and that he has to hurry over to the conference room of the vehicle program team: "You have to go into the arena of *critical energy* to understand the real dilemma". He had experienced those situations numerous times, yet another major crisis, again a long day until 10 pm at work, probably …

Five minutes later, he enters the room, sticky air and cold faces, ~40 person in the room, fully crowed. It seems that everybody is waiting for him…, one person stands up and offers a chair for him, and no other seat is available anyhow. Dr. Ryan finds himself positioned directly opposite to the senior management team, consisting of the Engineering director Frank L., the Body Chief Steven V. and the Body Manager Andrew F. and a number of other Vehicle and Body Management members. Amongst them, the former colleague of Dr. Ryan, Jeff R. from the Body Engineering department. The Engineering Director Frank T. immediately nails the problem for Dr. Ryan:

> Dr. Ryan, today it became clear in the review that the ***IT solution of the new virtual product creation solution is not working*** and that the ***new CAD modeling methods are absolutely unacceptable***; hence, your ***digital solution does cause the delay of the entire vehicle program. The damages could be immense, in the range of 20 million € lost revenues!*** Management expects from you immediate action, a resolution plan until tonight and your team members might have to become responsible for the work themselves!

Dr. Ryan stays calm and remains concentrated while he is asking the following loaded question back to Senior Management: "May somebody explain this conclusion in more detail to me, please! I am also highly interested in the root cause of this severe accusation to my team, especially since the tests of the new digital modeling; approach was successfully signed-off by the Body Engineering key user team 4 weeks ago after a 3 weeks stress test across the UK and Germany! This dilemma might be rather an operational problem of the agency designers not understanding the method or executing them wrongly! We have heard from the office floor chat that designers are assigned on short notice, do not take appropriate training classes and do not even attend the offered awareness sessions!"

After 5 to 6 s of interested mumbling in the room the Body Manager Andrew F. finally speaks up: "This does not matter any longer Dr. Ryan, we are in deep sh… now, and we need immediate action from you! It cannot be that we need 20 min to model just one f… spot- weld! Back in the old days, this was done in 20 s on a drawing board! This is all digital crap". Dr. Ryan asks immediately back to him: "Let us stay professional Andy; which person told you that modeling of one spot weld takes 20 min?" Dr. Ryan S. knows that this can only come from the immediate supervision of the CAD Designers, hence he looks over to Jeff R., his former colleague and spotted that Jeff feels really uncomfortable and unhappy.

Dr. Ryan recognized that he needs to talk straight to him, but after the meeting. Setting him up in front of the meeting does not help at all! Alliance in crisis is key within digital business. Dr. Ryan S. added the following commitment to the Senior Management team: "I fully understand the severity of the problem, we will work immediately on it and you will get a plan by tonight! Is this o.k. with you?"

The Engineering Director Frank T. and the Body Chief Steven V. become relieved and thank Dr. Ryan for his personal commitment. They expect readiness of a personal report out in a smaller Senior Management team meeting after 8 pm; a meeting invite will come within the next 30 min.

The meeting closes immediately. On the way out, Dr. Ryan asks his former colleague Jeff R. for a coffee together to discuss the situation very informally… On the way to the onsite coffee shop Jeff opens up to Dr. Ryan: "Sorry Ryan, but I did not have any other choice, the pressure was too high! Senior Management did not provide budget for all needed CAD Designers. I know that my team has difficulties to cope with the new modeling approach. I wanted to discuss this with you anyhow today, but during the meeting, I was urged to report out on the problem…. Therefore, I just restated what one of my CAD designers told me last night: they need 20 min on average including all working aspects to get one spot weld modeled and properly documented and registered in all necessary data repositories! I did not expect this management explosion at all; it seems that they need an excuse for other problems!"

The immediate SWAT Team action

Dr. Ryan calls a crisis meeting with his own international team right after the meeting with the vehicle program team to get prepared for the night meeting with Senior Management. During the next 3–4 h, the dilemma starts to become more transparent after interviewing two of the CAD Designers of the Body Engineering team together with Jeff:

- time consuming solution to create/update spot welds in the new digital solutions (complex, difficult to learn, limited assistance)
- missing spot-welds occur regularly
- mispositioned spot welds (not on the flange, somewhere in space)
- wrong spot-welds spread across the flange (too dense, too coarse, major differences in pitch distance without engineering reason).

Dr. Ryan starts to develop a plan for the late night meeting to calm down all parties but to get to the truth of the matter.

Taking on the tough assignment in the late night Management review

In the meeting Dr. Ryan presents a two-fold plan which gets full support:

- Immediate full method hands-on on for all CAD Designers in charge
- Six Sigma root cause analysis with support by Senior Management.

Action in crisis mode, staying professional with rationale (4 weeks)

- Best methods engineers support the CAD Designers in identifying errors
- IT departments get involved to investigate CAD workstation and long waiting times to connect to the PDM data base
- Root cause analysis of the weld flange delivered by the part designers
- Two extra CAD Designers to protect gateway deliveries short term.

Further course of actions (3 months)

- Six Sigma project by Dr. Ryan S. with Body Chief as project champion
- Stringent follow up of the DMAIC process for the modeling methods (Define, Measure, Analyze, Improve, Control)
- Deeper investigation of the CAD tool application with the PLM vendor
- Investigations of IT fundamentals on the workstation staging area.

The results and lesson learned

- The description in the crisis meeting was irritating and exaggerating since the statement that modeling takes 20 min was wrongly referred to just *one spot weld* rather than to a *full weld joint*, consisting of approx. 5–7 spot welds, as shown below!

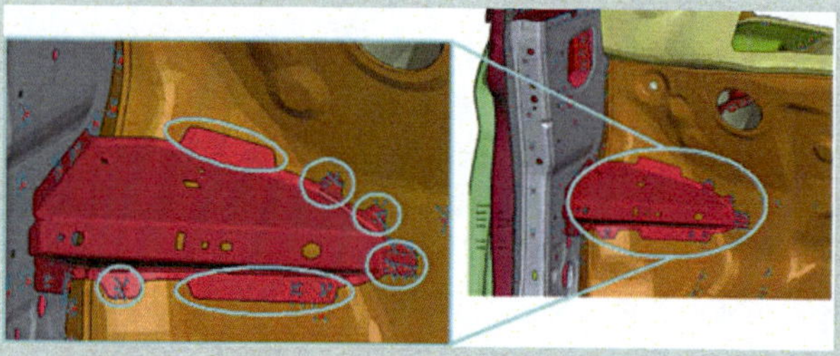

- The major problem was *not the IT solution* but the problem was indeed caused to 80% by bad digital modeling not following the methods during the spot weld creation and by already badly designed weld flanges (also not following stringently the methods) as part of the body assembly modeling *within Engineering*!
- The neutral Six Sigma effort did help tremendously to identify a range of different and additional weaknesses in the overall digital solution set incl. the CAD application itself, the IT operations set-up for workstation to PDM database communication and the various options to ease, streamline and improve the modeling methods themselves. The project did deliver timing improvements by 50% down from 9.3 min to 4.7 min for a weld joint consisting of 5–7 welds quality improvements (reduced rework) by a factor of 20! A yearly benefit from more > € 100,000- could be achieved.
- No program timing delays, no disaster situations due to professional and transparent problem solving
- Virtual Product Creation needs deep expert level leaders like Dr. Ryan who are able to communicate and interact with Senior Management as well as working closely with deep digital technologies and various expert levels.
- Staying rational and socially clever to mitigate critical energy in meetings and staying open for criticism, such attributes help to gain and keep trust during difficult times in virtual product creation deployment and operations.

6.5 The New Role of Virtual Product Creation—Evolving from IT Technology Towards Engineering and Lifecycle Competence

During the last 10 to 15 years, it became apparent that Virtual Product Creation needs a robust IT architecture but that IT by itself cannot guarantee any business success of using the digital solutions in the right way as part of engineering and manufacturing of technical systems.

Each industrial company has to be able to determine the complete IT solution architecture for Virtual Product Creation (oftentimes also referred to as PLM solution architecture by PLM vendors) by itself (incl. neutral consulting partners) before getting into pre-sales contact to PLM or Digital Solutions providers. This is a critical capability since this forms the basis for the foundation to evolve towards business critical digital engineering, business and lifecycle activities within the enterprise and as part of the associated wider partner and supplier networks. The IT solution architecture, however, is highly dependent on the intended digital working solutions across the company. It is, therefore, not to be mixed up with a sub-task such as data model configuration or customization. In addition, it is not just a simple derivation from a couple of use cases which oftentimes are subject of an early POC (prove of concept) activity with PLM/Digital Solution providers!

Please note in the following **Part 1** of the **Virtual Product Creation architecture framework**. This part 1 consists of the major building blocks of the *IT oriented Virtual Product Creation architecture* and comprises the following 10 elements:

1. The core principles and extent of centralized versus decentralized digital authoring environment(s)
2. The single versus multiple logical data models; this is not be confused with the specific data models as part of individual applications and data micro services
3. Data, information and model management and interaction principles; this distinction become critical since the increase of different enabling services on top of them will be decisive for the use pattern in business and engineering processes and activities (modeling, analyzing, simulating, reviewing, signing-off etc.)
4. Data, information and model distribution; which transactional services are necessary within the architecture and to the outside resp. at input
5. Supplier and partner connectivities
6. Business system integrations (linkage to BOM, ERP, MES, SCM, CRM etc.) incl. all backend data base services
7. Legacy design compatibilities, interoperability capabilities and migration services
8. Linkages and services to data analytic, semantic data linking, data lake repositories and AI architectures (such as lambda)
9. IoT (Internet of thinks) and Industry 4.0 WEB services
10. Front end visualization, cockpits and interaction intelligences.

Experiences from previous years show that it becomes critical for business success to put much higher attention to the digital working patterns of the target behavior of the *Virtual Product Creation architecture* beyond the classical use case determinations. Agile software development demands a concentration on a number of prioritized use cases in terms of guaranteeing robust error free software coding, testing, packaging and delivery for execution (bundling etc.). This is the right view from software engineering perspective only, but does not satisfy the business needs in engineering, manufacturing, service and field operations. This has become a major dilemma in many companies during the last ten years that created decreasing trust in IT solutions. The *error free rate* of software might increase but the *effectiveness to use* the software application for engineering and manufacturing of technical systems might not increase at all or might even be diminished!

Hence, the **Part 2** of the **Virtual Product Creation architecture framework**, which drives aspects of the *business orientation,* becomes essential if not more important than the ones from IT perspective (compare part 1):

1. Determination of the overall digital transformation goals from business perspective, levels and major target areas incl. the identification of the "best and most appropriate" data and information flows.
2. Identifying of critical connections between the EOS (Engineering Operating System) dimensions and influences onto the intended activity types (see Figs. 6.2 and 6.3) to avoid late findings in collaboration execution.
3. Product data structuring and partitioning with respect to industry branch and company specific product types and configurations.
4. Data and model types to describe and represent technical systems, product-service-systems and for smart products (connected by internet services) incl. all related model parameters.
5. Data and model delivery and maturity set-up plans, compare Figs. 6.6, 6.7 and 6.8.
6. Agreements on Virtual Product Creation control model principles and determination of supported engineering turn-around cycles (Fig. 6.5).
7. Determination of *model-based (systems) engineering degree* for the various *Virtual Product Creation architecture levels.*
8. Digital Design review and collaboration patterns (e.g. as shown in Figs. 6.9 and 6.10).
9. Extending and modifying existing skills and job functions as well as the creation of new job functions (see Figs. 6.4, 6.11 and 6.12).
10. Introduction of new data and model values as incentive to motivate and control Virtual Product Creation value creation.

From business value, proposition point of view it meanwhile became clear that companies have to start moving away from *Information Technology (IT)* focus towards *Information Logistics (IL)* and *Information Activity (IA)*:

- *Information Technology (IT)* for platform infrastructure, hardware commodities (server, network, nodes, transport protocols, infrastructure recovery) and basic data base operations,
- *Information Logistics (IL)* as set-up for IT application-oriented services such as data distribution and replication, exchange and consistency/completeness checks, back up and archiving, broadcasting, streaming etc.
- *Information Activity (IA)* as process/data/information/model based digital value creation, i.e. determination of content and solutions to enable and resolve engineering, manufacturing and in-use or in-field services as well as digital tasks and activities incl. engineering modeling, collaboration, review, assessment and analysis etc., i.e. the new digital working competence.

It, therefore, becomes critical that *Virtual Product Creation* needs engineering experiment environments and test beds for all aspects of IT, IL and IA rather than "just" a combination of IT applications and administrative databases.

Furthermore, the following principal shortcomings and related improvement needs of today's VPC and PLM solutions have been recognized but not resolved. They need to be addressed more stringently in research and in each digital transformation project in industry:

- better support of human expert task and role interaction, reasoning and knowledge processing,
- adjustable degree of simplicity and complexity task execution,
- less software function, more task assistance etc.
- PLM and VPC architecture maturation rather than customizing
- early prototyping of new digital work patterns
- alignment of product and factory lifecycle needs with VPC/PLM architecture lifecycles via identifying core elements of data/model types and potential micro service IT elements.

6.6 Best Practices of Integrating Virtual Product Creation into Mainstreaming Engineering

Driving forward with increased levels of consistent virtual product creation activities it essential to use the power of the next levels of the digital transformations. Besides technology opportunities, the real focus must lie on the proactive penetration and usage of virtual product creation along the engineering processes and associated activities. The question may have to be raised how this can be done and who (which teams) feel responsible for it?

It all starts from the *business needs* versus *individual wants* dilemma. Traditionally, high-level business process logics and processes are changed or rebuilt and then a special team has to find out how this can be "best integrated" in the day-to-day digital engineering work practice. Such team usually is under high pressure to deliver a "new" or "just adapted" digital solution and then to explain it to the workforce how to use it. This causes much churn, frustration, objection and fears which delays the transformation significantly. Therefore, this approach is no longer appropriate or even "best practice" moving forward!

The new, modern and more participative approach builds on *incentive and active participative models* rather than the classical *persuading, pressure, directing and pushing* models to set-up an ownership readiness and pro-activity approach.

This new *incentive and active participation approach* to drive the future of Virtual Product Creation uses the following different phases of preparation and engagement:

- Understanding and appreciating the roadmap of change, not just for the anonymous company but also for the individuals or teams; this has to include an active involvement with respect to the reflection of the good and bad existing practices and the ideas and desire for change and improvement
- Constant education and on-the-job training to better support the recognition why, what and how to change and in which time blocks
- Explaining the new intended way of working by using key users from functional engineering activities in close cooperation with VPC experts and IT personnel to showcase and then to prototype future engineering and manufacturing work scenarios
- Allowing all interested individuals to engage into such an approach, obviously within defined time allocations and with the duty to deliver detailed ideas, proposals and prototyping results
- Offering incentives for engineers and designers to engage within new Virtual Product Creation projects. Incentives can be granted e.g. as part of technical responsibility opportunities, special training activities, free time allocations or with respect to future career opportunities.

Based on such a proactive engagement model it is key that the central VPC development team and the associated trusted key user advisors keep closely involved in these activities in order to conclude the best future working and digital technology mix as well as any impact on the VPC architecture.

It is extremely important to engage different management and engineering workforce levels proactively, i.e. not just to report out to them but to use their experience and "design capability" to reflect, change and modify the intended digital working modes. The following best practices exist to get management and engineers to live ownership of day-to-day virtual product creation:

- Officially down cascade data and model ownership with full accountabilities (i.e. all related budget duties for IT services and for supporting data/model workforce!) and make management and lead engineers responsible to protect the team's digital work deliveries in terms of designing, modeling, analyzing, storing, structuring etc. (like it is usually done for testing of physical objects).
- Justify and approve access rights for any kind of collaboration partner regarding data and models.

- Show ownership in daily digital engineering activities and act as leaders to enable easy design review set-up and progression (with intuitive but safe access and preview capabilities via modern visual analytics such as clever color-coding).

The above best practices show that major changes are necessary in the behavior how Virtual Product Creation as new discipline in industry. This requires, however, the right mindset in the companies and a much higher understanding of Virtual Product Creation. In Chap. 17, the hidden demands of the engineering community and in Chap. 18 the challenge of modifying Management leadership behavior is investigated in detail. The next ten chapters (7 through 16) will describe all major technologies of Virtual Product Creation in detail.

References

1. Pahl G, Beitz W (2007) Engineering design. Springer Print ISBN 978-1-84628-318-5; Online ISBN 978-1-84628-319-2
2. Hubka V (1994) Theorie Technischer Systeme - Grundlagen einer wissenschaftlichen Konstruktionslehre (Theory of technical systems—basics of a scientific design school). ISBN 978-3-662-10446-0
3. Hansen F (1974) Konstruktionswissenschaft: Grundlagen und Methoden (Design science: basics and methods) http://www.dmg-lib.org/dmglib/handler?docum=4885009
4. Rodenacker WG (1991) Methodisches Konstruieren (Methodical design); Springer ISBN-13: 9783540539773; ISBN-10: 3540539778
5. Lünnemann P, Wang WM, Ibánez Manteca P, Stark R (2017) Engineering activities—considering value creation from a holistic perspective. In: Proceedings of the 2017 International conference on engineering, technology and innovation (ICE/ITMC), Funchal, Madeira Island, Portugal. 27–29 June 2017. ISBN 978-1-5386-0774-9/17/$31.00
6. Stark R, Lindow K, Reich D, Vorsatz Th (2018) Smart engineering capabilities for new generation product offerings. In: 23rd International seminar on high technology, Piracicaba, Sao Paulo, Brazil; pp 29–60; ISSN: 2175-9960
7. Gerhorst F (2017) Resource-optimized verification planning for mechatronic systems in the virtual stage of product creation. In: Stuttgart: Fraunhofer Verlag, 2017, XV, 165 pp. Berlin, TU, Diss., 2017 Berichte aus dem Produktionstechnischen Zentrum Berlin; ISBN: 978-3-8396-1255-2

Chapter 7
Major Technology 1: Computer Aided Design—CAD

Executive Summary

This chapter deals with the following topics:

- Basics and advanced techniques of Computer Aided Design
- Providing insight into how engineers benefit from using CAD technologies
- Describing functioning, benefits, and limitations of CAD technologies in practice.

Quick Reader Orientation and Motivation

The intention of this chapter is:

- to give an overview of CAD technology in Virtual Product Creation as driver and enablers for Digital Transformation in engineering
- to present CAD technology as part of Virtual Product Creation from a practitioner's point of view to analyze the need and usefulness for day-to-day industrial work practice
- to give instructions on how to use CAD technology
- to explain models, frameworks, and mathematical representations that help to grasp the internal working modes of CAD technology.

The purpose of computer-aided design (CAD) is to determine and modify two and three-dimensional objects and to specify part manufacturing information. In the beginning of CAD, it started with 2D drawings that were similar to conventional technical hand drawings. Subsequently, CAD technology started to offer 3D-modeling where 3D shapes could be created directly. Nowadays it is even possible to derive a technical drawing from a 3D model and to convert 3D models into virtual products with functional behavior.

R. Stark, *Virtual Product Creation in Industry*,
https://doi.org/10.1007/978-3-662-64301-3_7

7.1 Engineering Understanding of CAD

The engineer uses CAD systems to conduct the following design and engineering tasks:

- sketching of principal sections of components and deriving first product concepts out of it,
- modeling all details of components and assemblies in the phase of embodiment design,
- validating product basic functions, behaviors and producibility, with the help of calculations,
- detailing and documenting specifications for manufacturing and testing as part of virtual prototyping and prior to physical prototyping.

7.1.1 *Why Does an Engineer Use CAD?*

As shown in Fig. 7.1a, 3D geometry master models created by CAD are used as the basis for styling, design, simulations and verification. Authoring of geometric models constitutes a major engineering activity within the product development process. CAD models get converted into special models and file formats in order to be used as a source for Digital Mock-Ups (DMU), Rapid Prototyping and Numerical Control (NC) programming tasks in digital manufacturing.

7.1.2 *What Does CAD Do for an Engineer?*

The main task of CAD is to enable and support activities an engineer conducts to develop the shape and function of a product and/or a technical (sub-) system. CAD provides the user with tangible model entities, for instance, solid primitives or features which accomplish an efficient modeling process. The CAD system itself needs to be involved by the design engineer via end-user functions and executes automatically substantial mathematical operations, model checks and other background tasks, as depicted in Fig. 7.1b. By using inherent knowledge about certain types of model intelligence, the CAD system can even propose new design options, as well as design adaptions or alternatives.

Figure 7.1b indicates another task of CAD: supporting engineers. For instance, engineers can be supported in designing sketches or drafting preliminary design concepts. It takes part in different design stages such as new design, design adaption or diversifying and it even helps developing entire design models with all the details. On the way to the final model, it also ensures design activities like informing, modeling, calculating, drawing, interpreting, changing and documenting. In addition, CAD

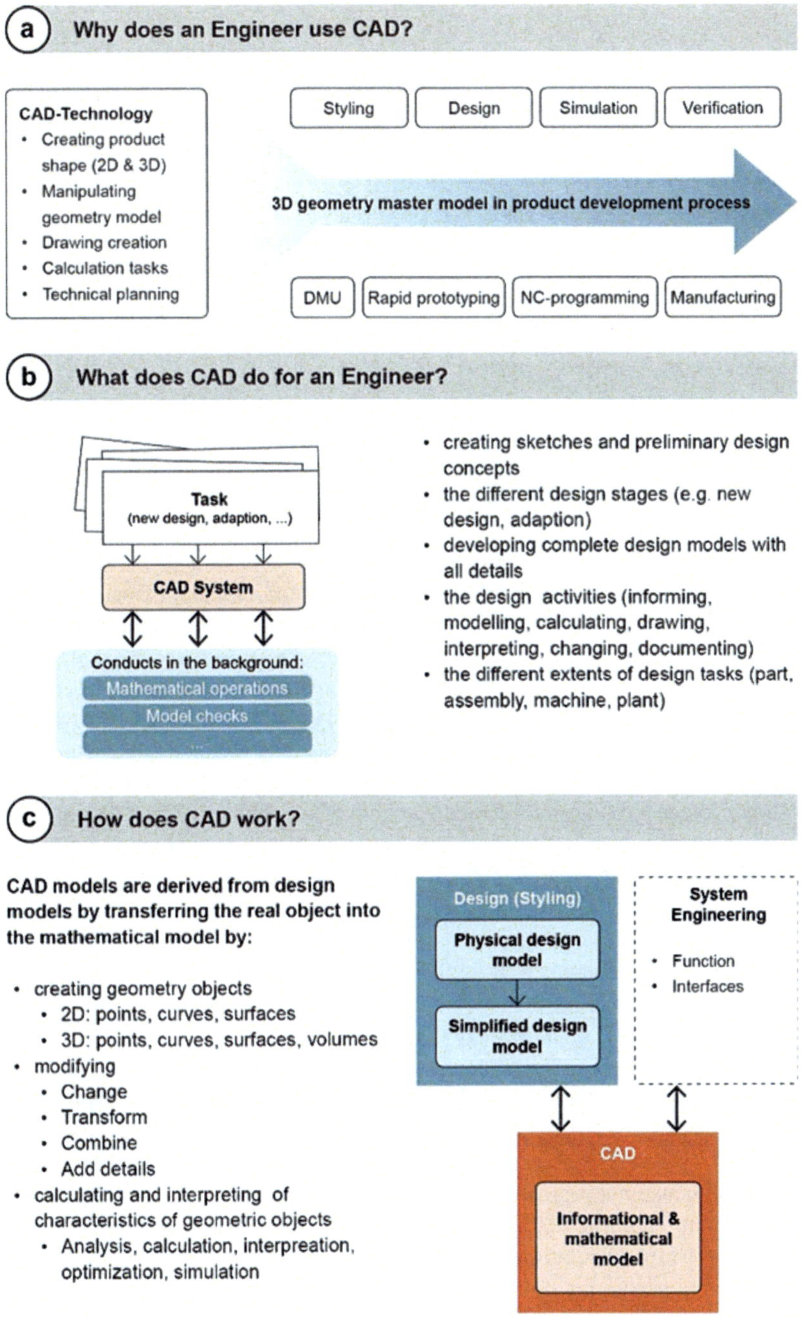

Fig. 7.1 Engineering application of CAD

assists the engineer in conducting additional design tasks, namely, arranging parts, assemblies, machines or even entire plants.

As a consequence, CAD models contain a range of explicit and implicit information sets. Due to this fact, it is possible to save information and knowledge an engineer is integrating into a product. Such information can be product structures or partitions (what is the product or technical system composed of) and calculation results such as volumes and surfaces. Hence, such information can be reused and variants or changes can be easily realized without rebuilding the entire design. In conclusion, CAD supports the engineer in every step of design: by describing a shape, defining details, developing variants, testing, analyzing and validating a model. Nevertheless, CAD requires substantial interaction with the engineer in order to make accurate decisions.

7.2 How Does CAD Work?

CAD technology enables the creation of a mathematical model of the design of a part or an assembly. Accordingly, there are different manners of using CAD. It is possible to create new objects by transferring ideas and considerations into a virtual model. However, it also helps a lot to transfer real objects into virtual models, either by modeling manual or, much easier, by using 3D scanning and subsequently reverse engineering technologies. The other way around, there is a way of directly producing real models based on CAD data. This procedure is called rapid prototyping and transforms the CAD model into a defined set of design layers for rapid machine operations. The previously mentioned mathematical model represents the design model throughout the entire development process. Within this process, interaction between design, styling, system engineering and CAD can be supported. Figure 7.1c shows the interplay of information exchange about physical and simplified design models, functions, interfaces as well as informational and mathematical models. Nowadays, engineers increasingly rely on the robust representation of the proper information and mathematical models when they drive characteristics of products through the higher-level systems engineering approach.

Further, as shown in Fig. 7.1c, there exist three types of engineering applications of CAD: creating, modifying and calculating as well as consequential interpreting of characteristics of geometry objects. Creating geometry objects comprises certain elements; in a 2-dimensional area, those elements include points, curves and surfaces, whereas, 3-dimensional representations are additionally based on volumes. Modifications contain changes, transformations, combinations and added details. Analysis, calculations, interpretations, optimizations and simulations, in turn, belong to calculating and interpreting of characteristics of geometry objects.

7.2.1 System Architecture of a CAD System

Figure 7.2 illustrates, that a CAD system consists of different system modules including user interface, graphical system, application interface, geometric modeling kernel, database management system and database. In addition, there are several extensions of CAD systems such as structural analysis, NC-programming, application modules, feature modeler and data exchange.

The *graphical user interface* belongs to the *user interface.* It creates a connection between a computer and a person through graphic symbols. More precisely, a software module realizes computer controlling by a computer mouse or, in case of a touch screen, by a movement of fingers. The *graphical system* offers interactive functions such as measurements, control modes, adjustment-settings and geometric modeling modes. Furthermore, the *application interface* allows integrating new features into present applications or providing data for other applications.

With the help of a *geometric modeling kernel,* a model can be created, modified and analyzed as well as it allows access to the internal data model. It forms the algorithmic part of a CAD system. Hence, its tasks are a registration of new elements into the database, manipulating, linking, and deleting existing elements and conducting diverse global functions. Registration implies, for instance, adding a full circle in a sketch. Manipulating can be extending a distance, linking implies determining intersection points and deleting can be erasing a circular arc. Examples for global functions are calculating the extremal values of geometry or evaluating surface and volume areas (see [1], pages 75 ff.).

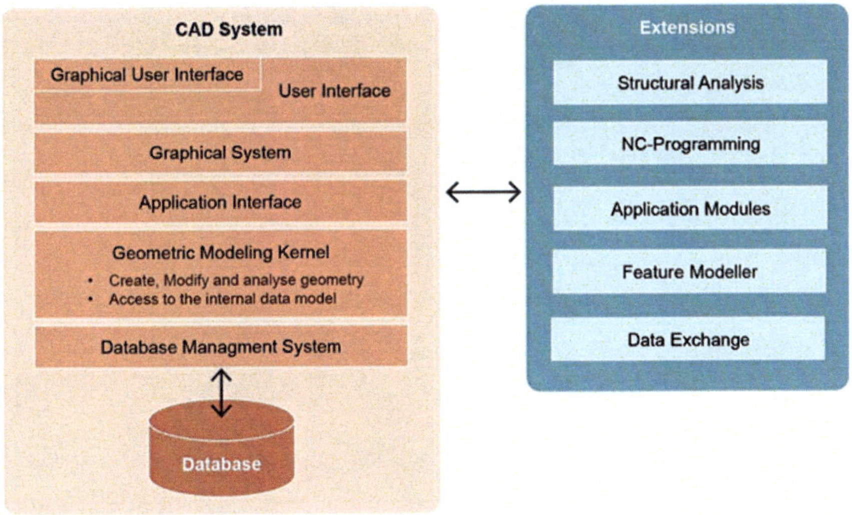

Fig. 7.2 System architecture of a CAD system

In most of the modern CAD systems there exist additional capabilities concerning storing engineering application knowledge such as design feature, templates and best practice processes. Those capabilities are often encapsulated in separate modules (feature modeler, knowledge rule, etc.). Those capabilities enable real engineering rather than just design modeling task execution.

Finally, the *database management system* facilitates administrating the database. The *database* is the core of every CAD system because it organizes and collects geometric elements and their topological relation. The content and structure of the database influences the performance of a CAD system. Typical storing of data includes geometrical data, drawing data, attributes of those two data types and organizational data. The stringent database integration of all design entities makes the difference between the traditional drawing board and Computer Aided Design. Manual design activities are focused on representations of design in documents whereas CAD creates digital models (compare [1], pages 77 ff.)

7.2.2 CAD Modeling Technologies

Figure 7.3 defines different kinds of modeling in CAD. As shown, the two widely used types of geometry processing are Constructive Solid Geometry (CSG) and

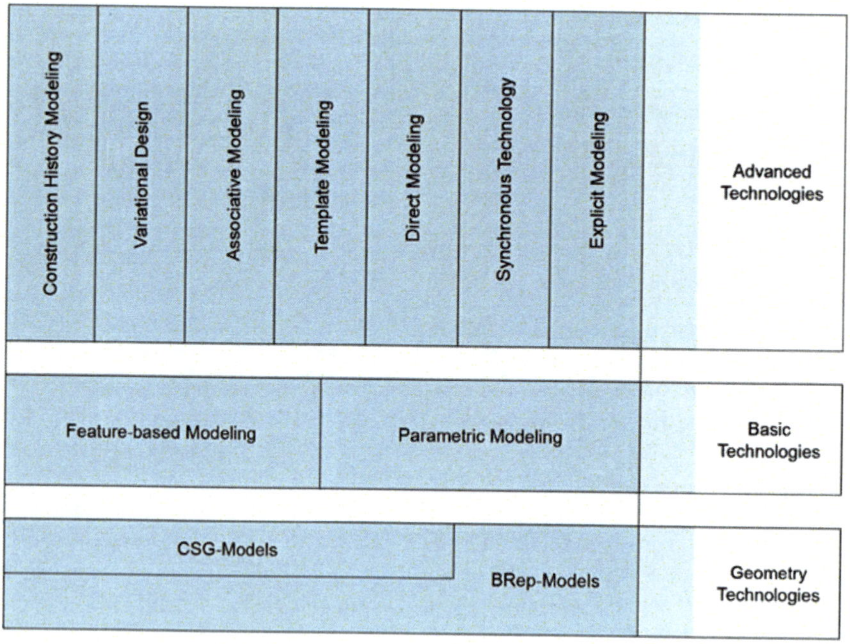

Fig. 7.3 Basic IT technology of CAD

Boundary Representation (BRep). Their explanations follow in the sub-chapter *Geometry Processing and Topology*. Firstly, basic and advanced technologies get exemplified.

CAD modeling, in general, builds geometric shapes that describe a geometric object. CAD modeling determines location, orientation and shape of solids, surfaces and other geometric elements in 2- or 3-dimensional space. Such space is related to a coordinate system. In addition, a Euclidean metric is defined. This metric enables measurability of distances that are implicated in a model. The shape comprises information about the scale and reflection of a geometric shape. These shapes can be either two-dimensional or three-dimensional.

7.2.3 Geometry Processing and Topology

The main types of 3D-models are wireframe, surface and volume models (Fig. 7.4a). The *wireframe model* only describes edges and points of geometric shapes. It was the first and simplest representation of geometric shapes. Due to the fact that computer resources were limited long time, it was still used up to the end of the 80´s of the last century to describe geometric shapes of products. It is the easiest representation of a 3D-object that carries the least information and hence is the least realistic type of three models (refer to [2], pages 21 ff.)

After having realized the limitations of wireframe modeling (no exact representation of surface information, difficulties to view those models due to no hidden line removals of edges, limitations toward part manufacturing etc.) engineers and mathematicians started to develop *surface model representations*. It was firstly used for the description of analytical difficult describable surfaces, which are common in body shells of vehicles, ships and airplanes or in fluid dynamic machine parts. Such surfaces have different bends in several directions and mostly an important aesthetic aspect (according to [2], pages 33 ff.)

The most ambitious of those three main types is the *volume model*, also referred to as a solid model. It shows real physical objects and allows users to create, safe, calculate and modify solids. Its main sub-models embrace BRep, CSG, hybrid, cell and other models. CSG and BRep will be explained more detailed in the following sub-chapter. A *Hybrid model* is a model where different model types are combined. For instance, surface models get connected to wireframe models; hence, they are non-homogeneous models with different forms of representation in one system. In relation to Fig. 7.4b, hybrid model implies the combination of CSG and BRep (see [2], page 122) In addition, *cell models* exist. They are used to compose an entire volume of a 3D object. i.e., cell model is an object represented by an arrangement of neighboring cells in a three-dimensional room (compare [2], page 128).

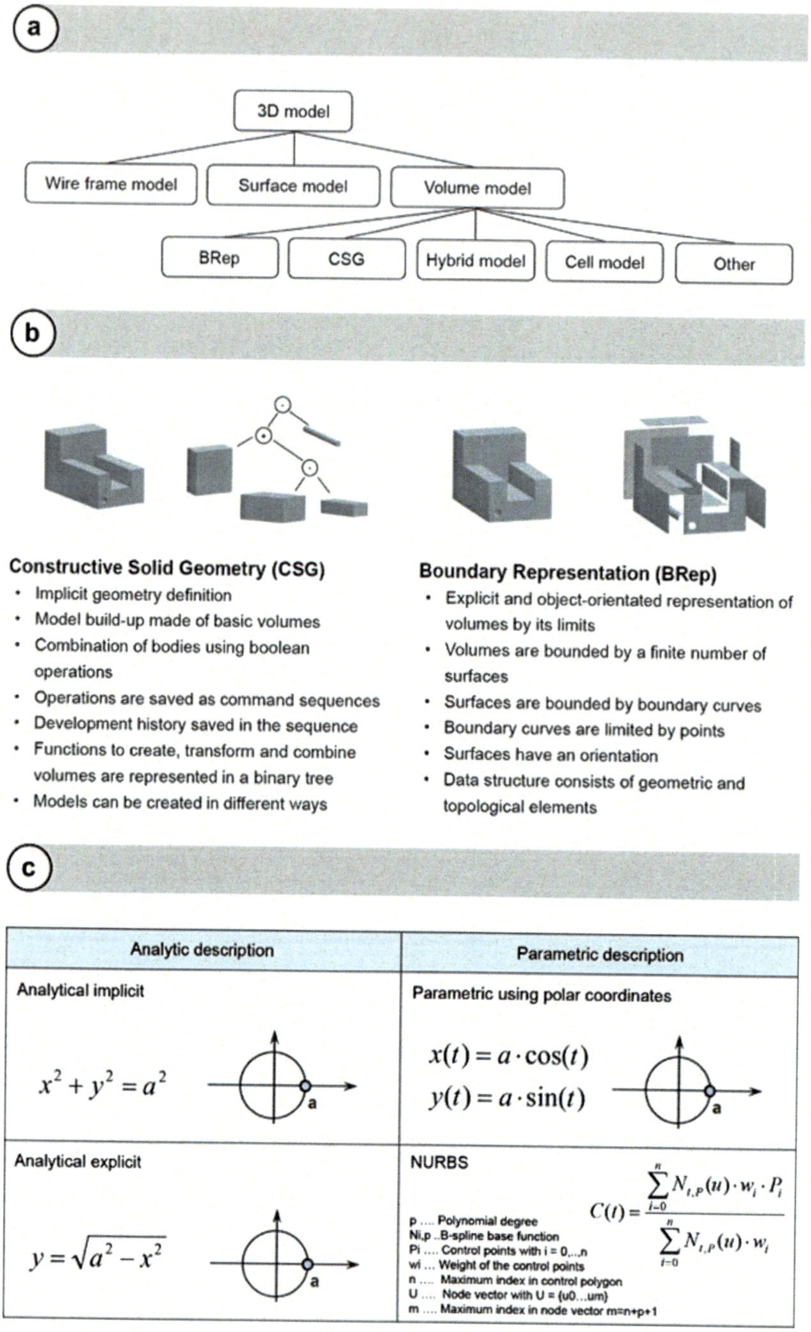

Fig. 7.4 Classification and geometric- and topological-processing of 3D modeling

7.2.4 Volume Model Types

Figure 7.4b defines that *Constructive Solid Geometry (CSG)* represents a process of building solids. Within this process, there are different steps in which predefined bodies get connected through Boolean operations. Those operations include *unite, subtract* and *intersect*. The numbers and types of predefined bodies vary depending on the certain CAD-system. In every CAD system, there exist simple shapes. Such shapes can be blocks, bowls, cylinders, cones, prisms and pyramids.

The connection between single basic volumes and their definition is established by the notation of command sequences. The development history of solids is saved consistently and it is represented in a so-called binary CSG-tree. Since there exist many possibilities how to create, transform and combine volumes, a high variety of model creation cannot be avoided. Consequently, CAD systems have to deal with non-unique computer internal representations in the case of CSG models. Hence, *CSG* is an implicit geometry definition that implies no information of real surfaces and edges of an object (refer to [3], pages 4 ff. and [4], pages 69). In turn, as shown in Fig. 7.4b, *Boundary Representation (BRep)* is an explicit geometry definition that owns an object-orientated representation of solids by its limits. Single volumes are defined by a limited amount of orientated surfaces and the surfaces themselves are bordered by boundary curves.

In the same way, certain points limit boundary curves. This representation hierarchy is shown in detail in Fig. 7.5. This figure also explains the relationship between geometric and topological elements as well as it explains the elements more precise. Such elements build the data structure of BRep-models. Geometry defines the shape of a logical element in a particular mathematical way. Topology comprises the hierarchical belonging and the finiteness of single elements as well as the neighborhood relations between them. Consequential, such a BRep network structure is complex. Possibilities exist to add technological information to every single element of an object. In contrast, no information about the construction history is given. Therefore, a BRep-model does not contain any information on what kind of volume primitives it is based on (according to [4], page 69).

7.2.5 Mathematical Representation

The analytic and parametric description of geometric elements is shown in Fig. 7.4c. Analytic elements can be either implicit or explicit. An advantage of explicit description is, that it is easier to control whether a Euclidian point is part of a 3D element or not. By comparison, the parametric description needs iteration to identify a point. If implicit equations are used, it is not possible to calculate the point of an element directly. In this case, the complexity of solving depends on the type of implicit description (compare [5], pages 134 f.). The mathematical description of a circle is

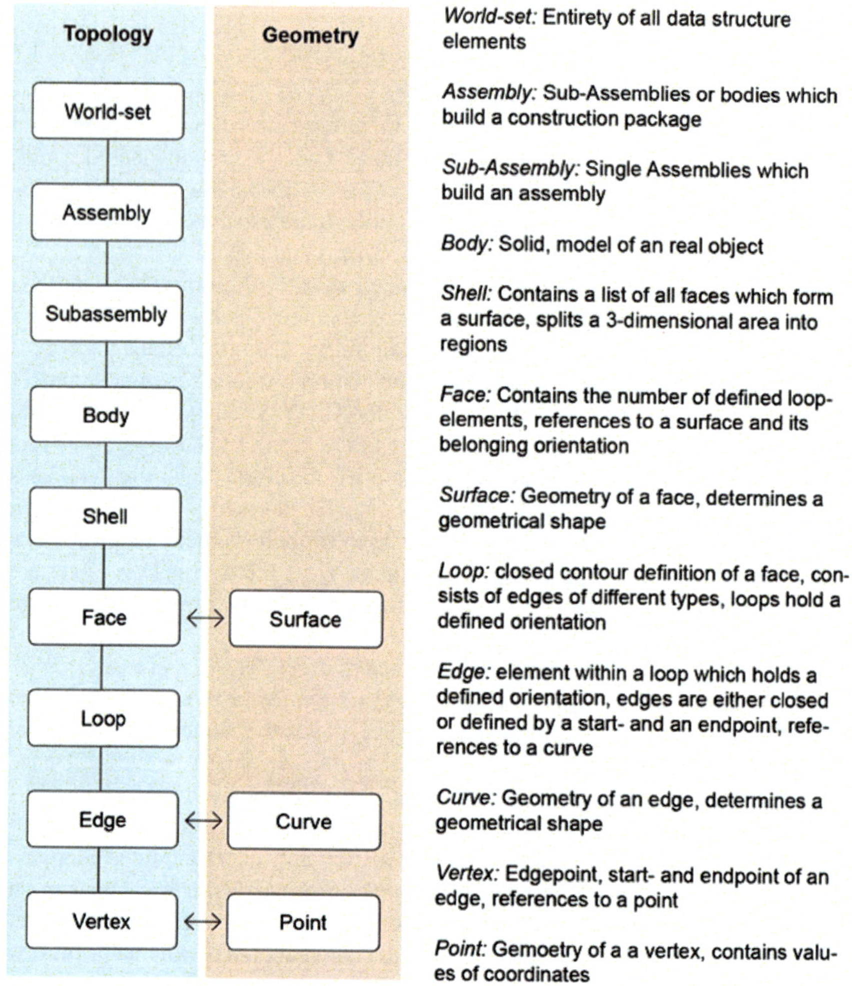

Fig. 7.5 Description of topology and geometry (see [2])

shown in Fig. 7.4c. Besides analytic and parametric equations there is also a description that is implemented by NURBS. Those are non-uniform rational B-Splines, they can describe freeform shapes in an analytic way, but at the same time, they need a very high computer capacity (refer to [5], page 151).

7.3 Basic Technologies

The most common basic modeling technologies of CAD include feature-based modeling and parametric modeling, as shown in Fig. 7.3. The following two sub-sections will explain these modeling technologies and give an example of how they can be used by design engineers. As an example-model for the explanation, a shaft with two sections, a chamfer and a drilled hole will be used (see Fig. 7.6).

7.3.1 Feature-Based Modeling

Feature-based modeling is a CAD modeling methodology that uses *features*—design entities with specific engineering meaning and relevance- to allow integrating design information into product development processes. For that matter, it is important to distinguish the meanings of the expression feature.

A feature is represented by the following two elements:

- Syntax comprises all notations to explicitly or implicitly describe the geometric shape of a feature.
- Semantic is associated with the feature syntax and represents the engineering meaning of a feature.

Therefore, differentiation between manufacturing, design and assembly features is most common (see Fig. 7.7). Design features provide inherit knowledge about functional and property intentions and targets. Manufacturing features represent linkages to manufacturing operation plans and details. Assembly features build constraints

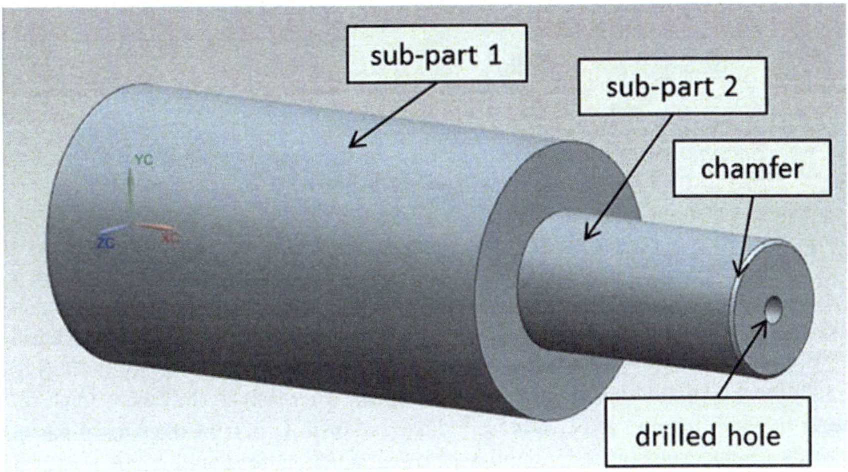

Fig. 7.6 Shaft as example model in Siemens NX

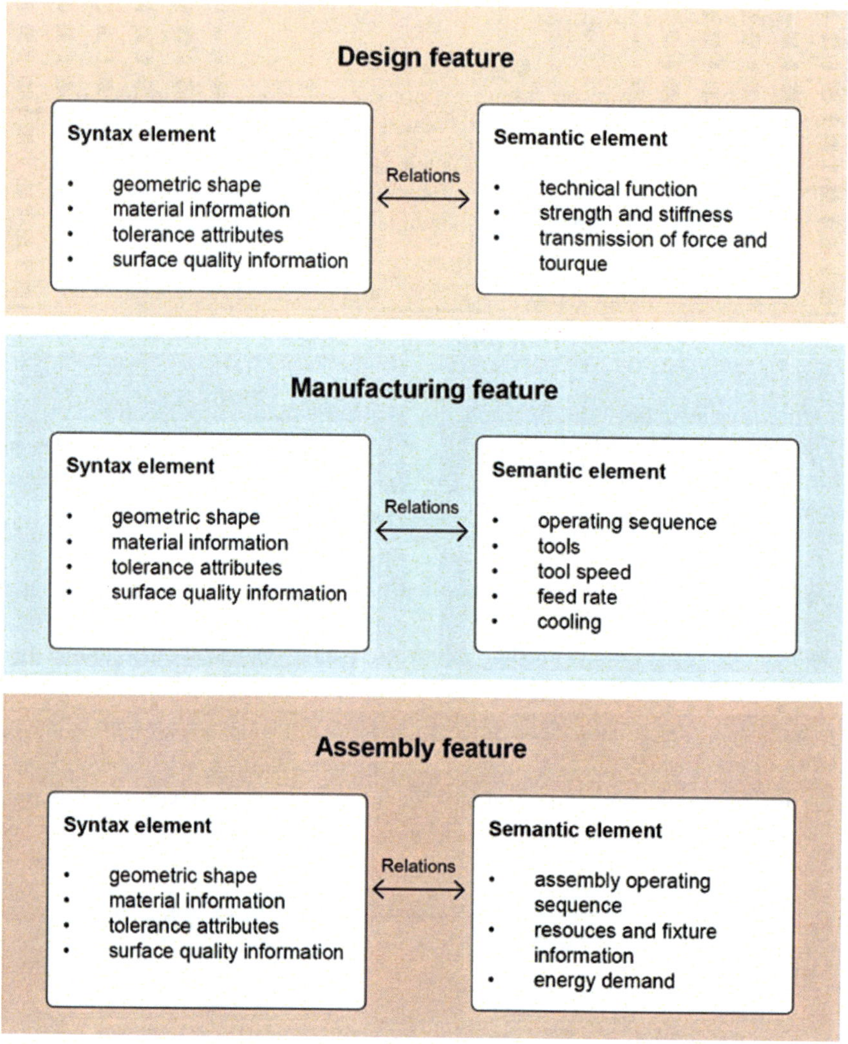

Fig. 7.7 Examples for syntax- and semantic elements of different types of features

between design features of different components and therefore should help to address assembly sequence planning. With the help of features, there are possibilities to create bosses, pockets, chamfers, holes and many more design primitives. Still today, the use of design features is mostly limited to quickly create geometric shapes, the authoring of additional semantic information has not yet been intensively offered. Industrial companies will have to make more effective use of it as part of the Model-based Systems Engineering (MBSE) approach, by which functions will get mapped to features in the CAD world.

Features need to be organized in databases which enable data reuse for designing. An important point of using feature-based modeling is to reference features correctly. In case original relationships are not kept, regenerating a model can cause a movement of features and the complete design model will appear in an incorrect state. In addition, the possibility of feature recognition exists.

Subsequent digital planning environments either use direct design features from CAD or, as in most cases, only receive non-intelligent geometric models, and subsequently use a parser tool to recognize relevant manufacturing and assembly features. In the following section, an example is shown to illustrate the approach of feature-based modeling. The same model will be used later on to declare the procedure of parametric modeling and to create a direct comparison between both modeling technologies.

Using Feature-Based Modeling

For the shaft that is going to be built, the first step is to create a cylinder by a design feature called *cylinder*. Figure 7.8 illustrates this approach and it also shows how to define or change the dimensions of the *cylinder*. In the CAD-software *NX* features enable creating simple solids as well as creating solids based on sketches. Since the model was created with *NX*, the feature *cylinder* belongs to the category design feature. In comparison, the CAD-software *CATIA V6* does not offer such an equivalent feature. However, there exists another method of creating a shape like that based on a sketch. In *NX* this feature is named *extrude*, as it is shown in Fig. 7.9. The representative feature in *CATIA V6* is named *pad*.

First, a sketch of the shape, which builds the base for the solid, needs to be created. In the example case, this sketch contains a circle which has the same diameter as

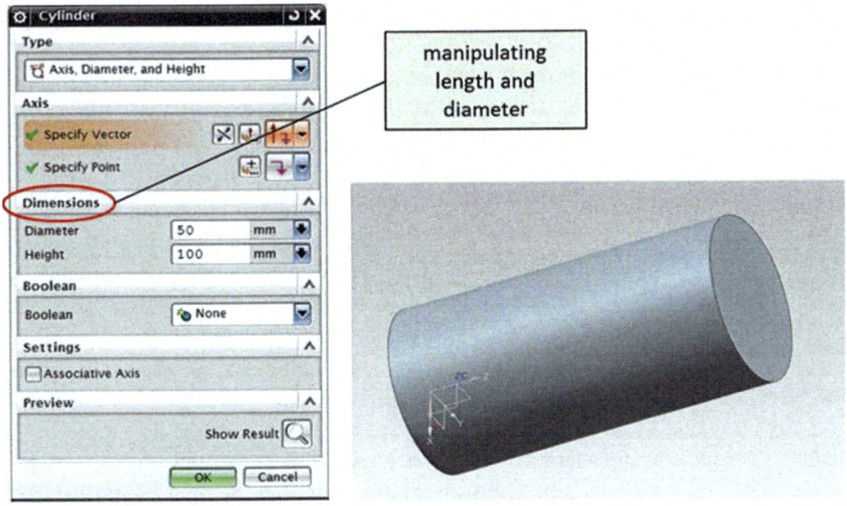

Fig. 7.8 Design feature cylinder in NX

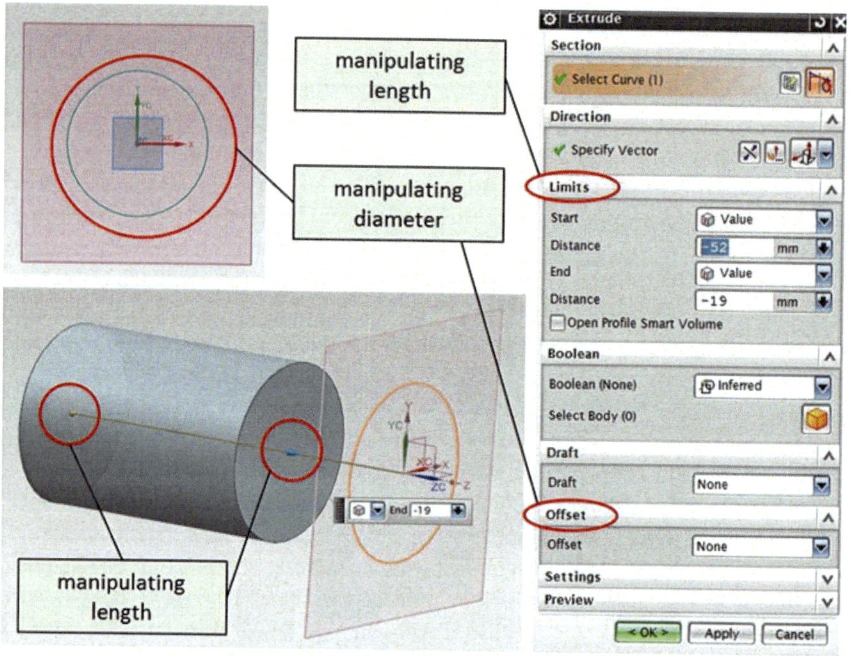

Fig. 7.9 Design feature extrude in NX

the cylinder. While using the feature *extrude*, the sketch needs to be selected and its length has to be determined, as it is shown in Fig. 7.9. Manipulating the cylinder's measurements can be realized by changing the sketch, i.e. by altering *limits* or *offset* in the feature window or by pulling the knots of the solid. Directly, the second cylinder of the shaft design can be created accordingly.

Since there exist two solids the subsequent operation has to connect both solids by using the feature combine function *unite*. Figure 7.10 illustrates this methodology. In *CATIA V6* there is no need of combining feature because solids unite automatically while creating new sub-solids within the context of a part.

Figure 7.11 shows the application of *chamfer* and *hole,* which belong to category detail features in *NX*. In *CATIA V6* names of these features and the category, which they belong to, are called the same. Manipulation of the size and other properties of those features can be realized through different options. The size can be changed by pulling the arrow of the chamfer and by changing the value of distance in the belonging feature window. The position of the feature *hole* can be varied directly in the relevant sketch. Size as well as the shape can be defined in feature window *hole.* In *CATIA V6* it is even feasible to pull the end of the hole to change its depth.

There exist features for many construction steps. As a result, features can be used for basic geometries, the combination of solids as well as for final model details. Additional feature types exist for modeling in specific modeling environments. Such

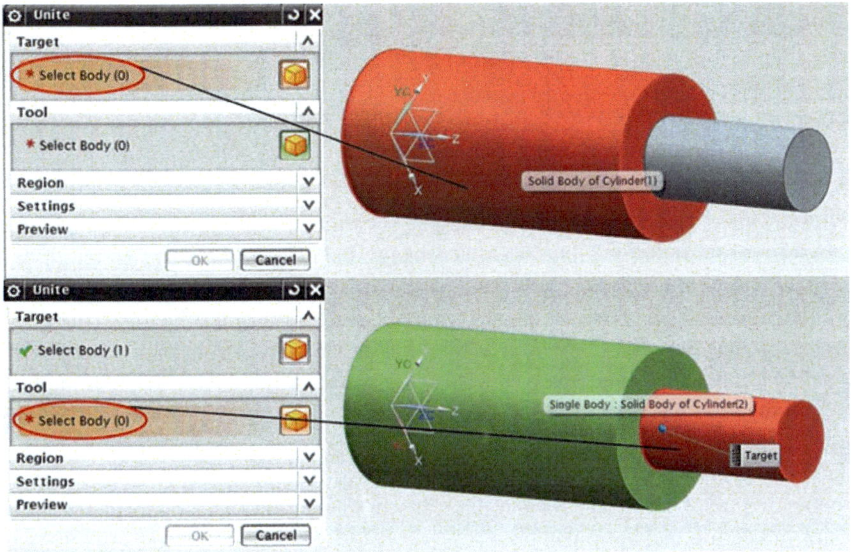

Fig. 7.10 Combined feature unite in NX

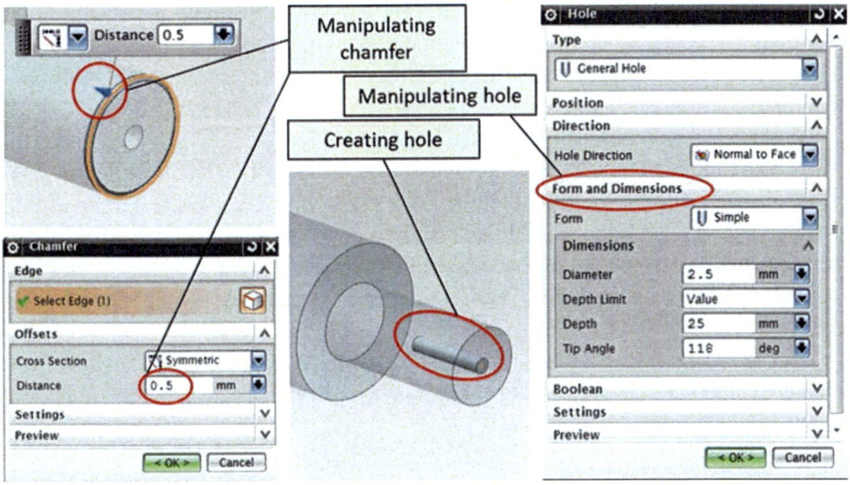

Fig. 7.11 Detail feature chamfer and hole in NX

features comprise, for example, modeling entities in welding, sheet metal or surface applications.

7.3.2 *Parametric Modeling*

The idea of *parametric modeling* is to create models which are mathematically stable and can be altered easily. It is based on a mathematical description of a computer internal model. The realization happens through mathematical relationships between single measurements like lengths, angles or other geometry determining factors. In addition, those constraints and relationships get connected through systems of equations. From this it follows that a single change of measurement causes an entire modification of a model in a stable and precise manner. Another important point of parametric design is the way of equation solving. Equations are solved in sequence; this means that they are solved one after another. Resulting from that, *parametric modeling* entails a lot of advantages such as flexibility which is of particular importance in the product development process because a high number of alterations are usually made in this engineering activity. Moreover, it ensures variational design, it keeps information from the design engineer, so it becomes comprehensible and documented for other designers and, last but not least, it provides opportunities for model simulation and optimization.

Using Parametric Modeling

The section illustrates an example of how to use parametric modeling. The illustration is based on the CAD-software *CATIA V6R*. Using the shaft example as in the section "*Using feature-based modeling*" this section concentrates on creating a parametric model step by step (Similarities with and differences to the CAD system *NX* are mentioned).

First of all, a sketch needs to be created (see Fig. 7.12). The sketch does not need to be fully constrained in this first step, because the measurements will be replaced with parameters. In *NX* the course of action is the same as in *CATIA V6*, but some of the functions have different names. After designing this sketch and before referencing measurements, a list of parameters needs to be generated. In *NX* parameters are called expressions. This list, as shown in Fig. 7.13, comprises parameters that belong to the particular parametric design.

First, it is necessary to choose *user parameters* in *CATIA V6*, in *NX* this function is called *user-defined*. After that a new parameter can be created, in *CATIA V6* by

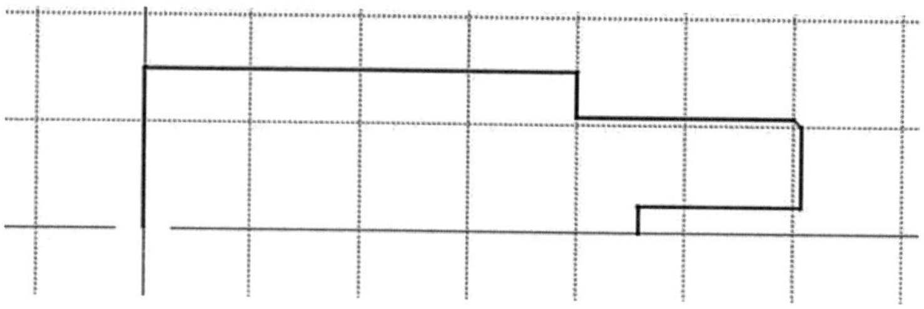

Fig. 7.12 Sketch of the model in CATIA V6

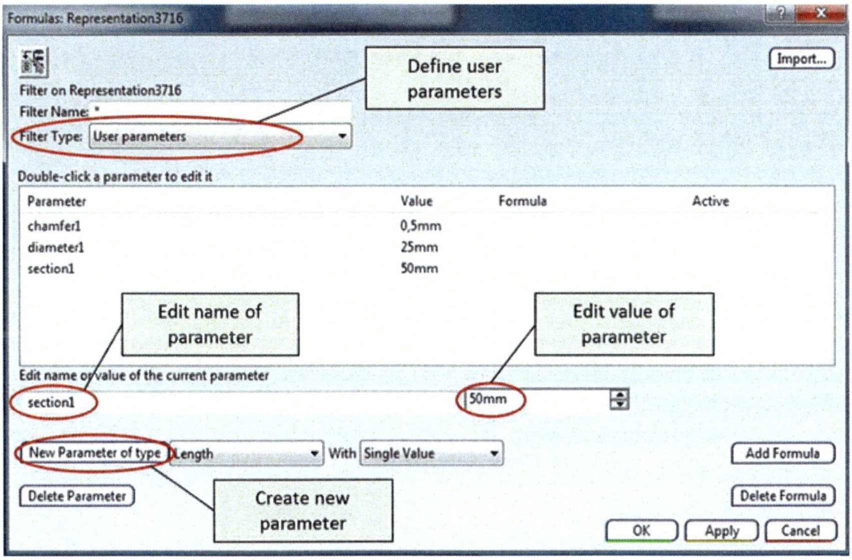

Fig. 7.13 List of parameters in CATIA V6

selecting a *new parameter of type*, accordingly selecting an empty column in *NX*. Then it is possible to edit name, type and value of a parameter. Besides length, types can be also angle, volume, mass and many more. Figure 7.13 illustrates this application. Hence, there exists a high range of opportunities to alter a model by manipulating the belonging list of parameters.

Subsequently, the activity of referencing needs to get started by using parameters which were automatically created before, without direct measurement values. In that way, it is relatively easy to change parameters and alter the model through relations which were created before. How this could look like can be seen in Fig. 7.14.

Once parameters are set in a sketch or in the part design environment, parameters of the model can be varied through the list of parameters. Last of all, as illustrated in Fig. 7.15, this sketch gets revolved by design feature shaft and the whole solid construction is completed. The equivalent feature in NX is revolved. In contrast to feature-based modeling, every shape is saved in the sketch and alterations can be easily made by changing the parameters or changing details in the appropriate sketch.

7.4 Advanced Technologies

Due to the fact that CAD is meanwhile an industry standard tool used by many engineers nowadays, and that design tasks just as products become more complex, the need and use of new CAD-technologies steadily increases. Therefore, certain

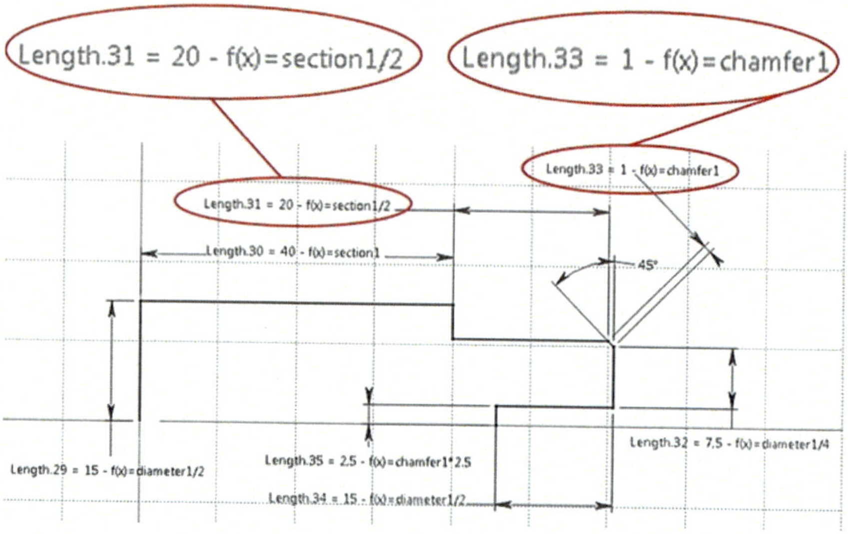

Fig. 7.14 Referencing a sketch using parameters in CATIA V6

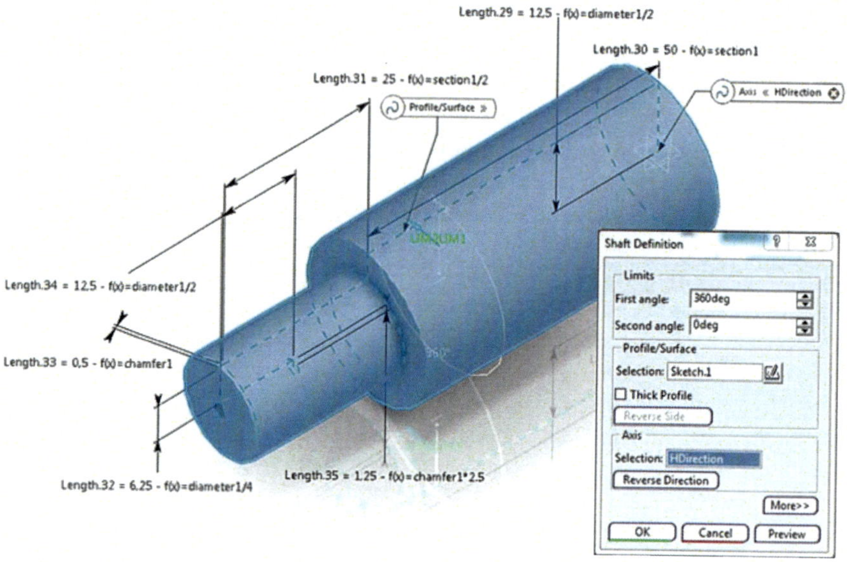

Fig. 7.15 Design feature shaft in CATIA V6

advanced technologies are offered. Those technologies, as shown in Fig. 7.16, are construction history modeling, variational design, associative modeling, template modeling, synchronous technology, direct modeling and explicit modeling.

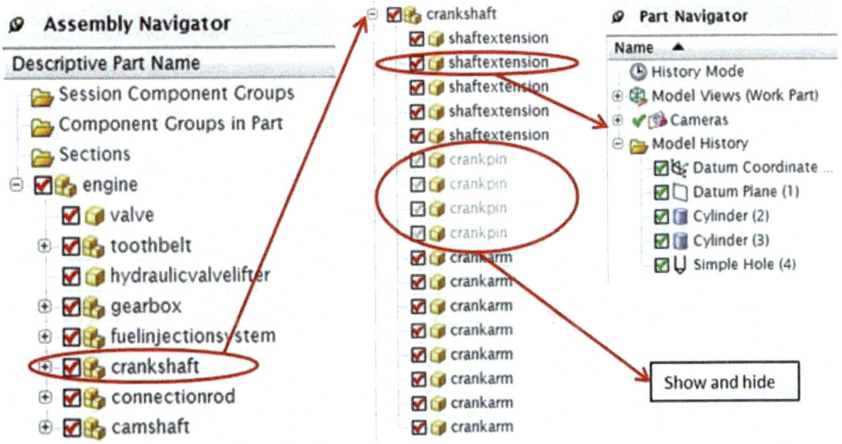

Fig. 7.16 Construction history tree in NX

Construction history modeling has been introduced in the first half of the 90´s of the last century and comprises the documentation of a sequence of model creation steps. The visualization is realized through a history tree structure, as it is shown in Fig. 7.16. Each assembly consists of different parts or subassemblies. This representation is based on *NX*. In contrast, *CATIA V6* does not split the tree into different single trees. It integrates the features of a part and the belonging subassemblies in one construction history tree. In *CATIA V6*, an assembly is named *product* and a component is called *part*. Figure 7.16 illustrates an NX example of the engine that is composed of parts and subassemblies, which consist of numerous parts. Those are shown in the assembly navigator on the left side of the figure. These parts have their own construction history as well. In the right area of the figure, features which build a certain part are visualized. Features, Boolean operations and other steps of construction are listed in the part navigator in the order as they were used. Consequently, it is possible to change this order by manipulating the tree. Another possibility is to change the properties of a feature or other step by double clicking the certain icon in the tree. Through the construction history tree, parts and assemblies can be hidden and shown and other characteristics of parts, like weight or count, can be displayed.

As industrial practice has shown, it still remains critical to control changes at the origin of the tree, right through the entire tree without problems. As a rule of thumb, such history should not exceed 150–200 steps in order to stay robust and flexible enough to introduce such a history model to another engineer. This documentation enables traceability of how a model was built. If the history steps are not too complex, it even alleviates correctly changing the model.

Variational design is a different type of parametric modeling. In contrast to parametric modeling, the equations get solved simultaneously. This means that there is no order in presetting the constraint evaluation and, accordingly, more than one solution is formed. From this it follows that parameterization in variational design does not have to be complete. This can result in nonlinear equation systems which require a high computational effort to be solved, and even impossible solutions can be the output. Additionally, it might be difficult to explain a failing equation solving process.

The main idea of associative modeling is creating dependencies among different geometric objects of a single model or even among different models. This enables opportunities to drive design sections and surfaces from one design component to another one and creates a boost in update effectively in case of change propagation. Managing relationships between design models is also known as *multi-model-links* in the CAD-software *CATIA*. Such dependencies or relations between two geometric objects in associative design are directional relations. An example of associative modeling helps understanding this definition more precisely. The shape of a sealing flange of a vehicle door is depending on the shape of the related windowpane curve and the related body shell. If there are any changes made on the windowpane curve and/or the body shell, the shape of the sealing flange will be altered as well.

Furthermore, parametric-associative modeling represents the connection of parametric and associative modeling. This enables simple and fast adjustments and variants of models by varying parameters and replacing reference geometries. Alterations of the model size imply updates of the model. In this case, the CAD system keeps the idea of the construction of the engineer because it saves the design intent and the geometric shape (according to [6], pages 19 ff.)

Template modeling has been introduced with the existence of *CATIA V5* in the beginning of the 2000 years. In order to improve productivity in 3D associative and parametric modeling structures for robustly managing parameters and reference links were offered. The industry has heavily introduced design templates between 2000 and 2010, mainly in the vehicle body, chassis and power train design. Based on the author´s expert insight into the industry, an investment of car industry in the order of 2000 person-years à 150,000 USD (i.e. $\approx$ 300 million USD) has been made in the first 15 years of the 2000 years for intelligent geometric templates only.

Template modeling enables productivity increase and shortening of product development time. A standardization of iterative activities and modeling processes is possible through templates; this allows the re-use of those shapes just as well as it reduces and economizes design-engineering work. Modeling time savings up to 50–60% are realistic for specific design tasks. Consequently, template modeling implicates a basic model which can be modified in numerous ways and it delivers a substantial set of construction modeling information.

Furthermore, templates create stable creation processes and storage of construction information. Templates support the integration of design guidelines, concepts and methods. Nevertheless, they can also entail some disadvantages, for example, high complexity in design, complication of expansive alterations, further education of engineers and just as well as the necessary continuous maintenance and adjustment. But most of those problems should only appear during the implementation of process within a company.

In detail, templates are used in different ways. They appear in technical drawings, *PowerCopies*, *UserFeatures*, part and assembly reuse as well as in *Knowledge-Based Engineering* applications like *Business Process Knowledge Templates* (see [6], page 393). The terms and description of functionality which are used hereafter originate from *CATIA V5*. The equivalent terms in *NX 10.0* (and later versions) are *WaveLinks* for *PowerCopies*, *User Defined Features* for *UserFeatures* and *Template Based Design* instead of *Business Process Knowledge Templates*.

Templates in technical drawings consist of the layout in general and the title block, which adopts modeling information like tolerances and weight into its output. *PowerCopies* are suitable for geometry elements which are either used more than once in the exact same way or which build the basis for related shapes. The user is free in modeling alterations in terms of linked copies. If the original model gets modified, it depends on the chosen settings, which are usually done before working with linked copies, whether they change as well or stay as they were copied (compare [6], pages 396f.) *UserFeatures* allow operators to create their own features as well as to save and share them with other users (refer to [6], pages 414f.). Part and assembly templates can be defined as a consistent extension of *PowerCopies* onto a level of assembly design, which means it is possible to link whole parts and even assemblies into the new assemblies (according to [6], pages 419f.).

Due to the fact that *Business Process Templates* possess the highest level of complexity and that they are the advancement of the template applications mentioned before, they will be explained as an example for template structures in general. Hereunto Fig. 7.17 shows the single elements and structure of *Business Process Templates*. Subsequently, those devices will be explained elaborately: as shown in Fig. 7.17, parameters, constraints and formula belong to the CAD modeling process. Thereby parameters represent certain characteristics within CAD files. A formula, also known as relation, defines those parameters and has the ability to use them as mathematical arguments. Constraints describe geometric boundary conditions within

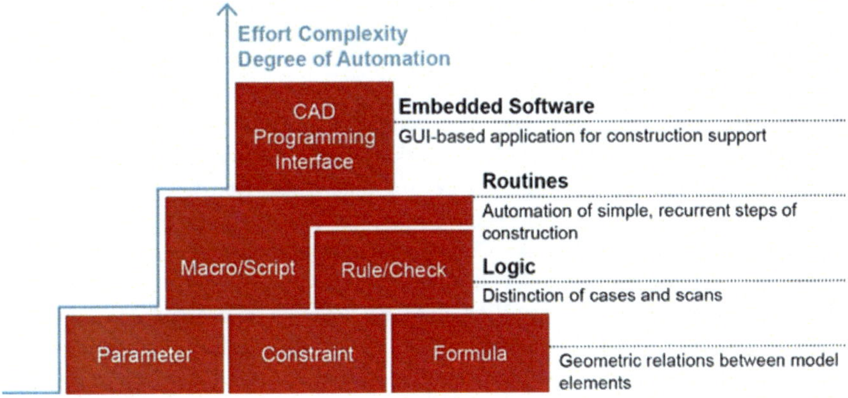

Fig. 7.17 Structure of business process knowledge templates

sketches or assemblies. Macros or scripts belong to the level of information technology and are connected within the CAD system. Either the input of a script occurs through automatic translation during the modeling process or it can be done by direct programming with the aid of CAD system integrated editors. Furthermore, macros or scripts are controlled by rules and checks. An example for a determined rule can be the maximum weight of a component which gets controlled by the check system. It does only allow following the defined rules and gives commands to change if there is a need for it. Through their CAD Programming Interface they even enable users without modeling and construction knowledge to make changes in ways of simply typing information e.g. mass, measurements, number of drill holes or size of bolts into it. The Interface can be served in external ways from the particular CAD-system but it still needs a license for the connection and interaction between the actual model, the script and itself.

Figure 7.18 shows an example of the structure of parametric associative design in template modeling. This method ensures stability of updates. Stability is important because there are high requirements of update processes within the use of templates. For this purpose, geometrical information needs to be exchanged between different parts in multiple ways. As shown in Fig. 7.18 there are three levels which comprise a base model, an adapter and the connected components. The base model builds a strict specification for the structure of the single elements. It contains concepts, edge geometry like planes, axes, vertexes, surfaces and freeform shapes. This geometry gets transferred into the adapter by different types of links.

Those links serve the task of depositing geometrical information and parameters of origin elements into other CAD components, also referred to as target parts. The issued information can be handled like a "normal" geometry and the size of the target part gets reduced as well. Additional, publications are mostly necessary for an explicit release of geometry elements. Publications support the feature of giving names to either parts or sketches. This property provides repositioning as well as creating new parts without direct access to the original geometry and independent from relevant product structure at once. In practical use, it is often prescribed to work with publications because they help to gain a clearly arranged product structure and simplification of geometry exchange.

The adapter (see Fig. 7.18) prepares the input geometry of the base model and remits it to separate component templates. Within the processors of the adapters and parts formulas and rules are used to assimilate the given information and geometry. Furthermore, the individual parts get referenced and transferred. In the case of structures with high complexity the update processes are partly operated by particular, embedded software like *Component Application Archticture* (CAA) or *Visual-Basic-for-Application* (VBA) in *CATIA V5* and *CATIA V6*.

Through the use of **direct modeling**, engineers can manipulate a CAD model in a direct way such as twisting, pulling or pushing it. Direct modeling is an intuitive approach and can be easily operated. Accordingly, it is often used for composing freeform surfaces or other desirable product shapes. Significant time (more than 20%) can be saved through direct modeling and engineers need less time to learn how to handle this way of modeling compared to other major technologies in CAD.

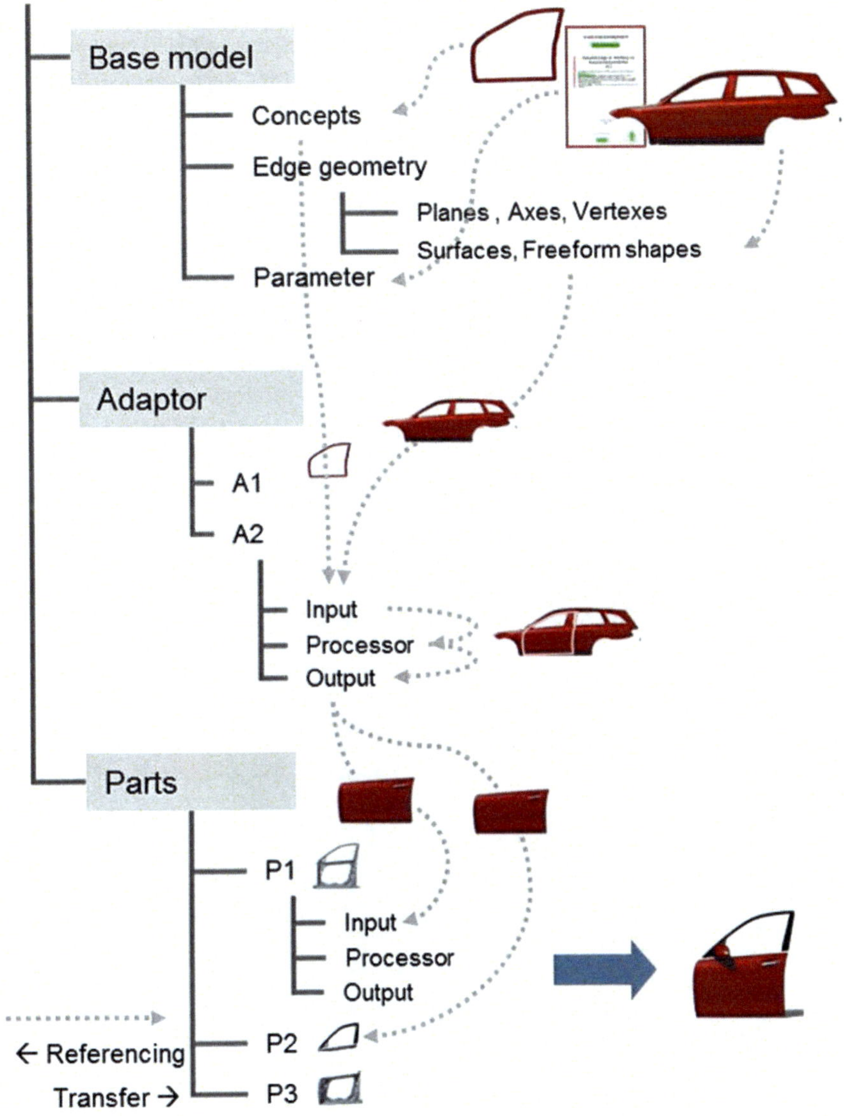

Fig. 7.18 Exemplification of parametric associative template modeling

Moreover, it is possible to adjust parts which are already in existence without any complications (compare [7]).

The latest innovation in direct modeling is called **Synchronous Technology** and finds its use in different application fields. E.g. it allows the engineer to alter a model or assembly completely independent from its construction history but still keeps it. This offers a range of advantages. On the one hand, there is the possibility of working with models from external CAD software without losing important information. In the same way, it permits easy-to-use workability for engineers who do not know how a certain model was built originally. Furthermore, it represents a new type of a point of intersection between CAD and Computer-aided manufacturing (CAM). Production engineers or Numerical Control (NC) programmers do often work separate from design engineers and consequently, they are often not informed well enough about the construction structure which causes problems in preparing for manufacturing. With Synchronous Technology it is much easier to make changes absent from doing them through the history tree which often causes failures (refer to [8], pages 3f., 7).

Figure 7.19 shows one way of altering a model under use of Synchronous Technology in *NX 10.0 (and later versions)*. Hereto the relevant faces of a model get selected and moved by either pulling the geometry directly or input the value of distance. In certain CAD-software this technology has different names, so it is called *Live Shape* in *CATIA V6* (according to [9]) and *synchronous technology* in *NX* (see [10]). *Creo Elements* is a fusion of *CoCreate2007* and *Pro/ENGINEER* which got developed especially for the purpose of direct modeling (compare [11]) and with

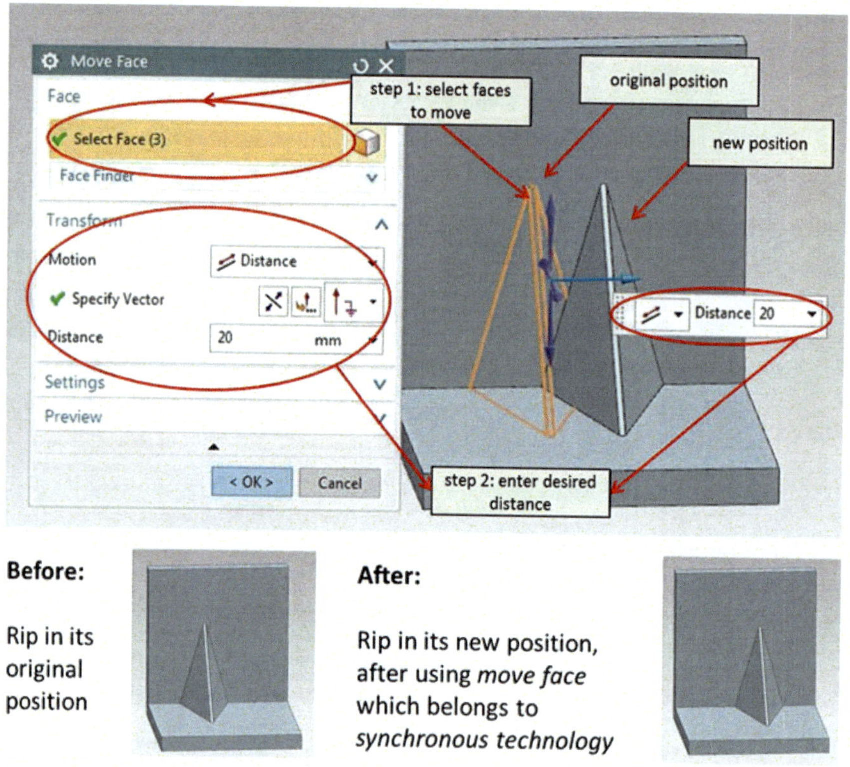

Fig. 7.19 Synchronous Technology in NX

ANSYS it is possible to interact with *SpaceClaim* as a direct modeling surface (refer to [12]).

Another technology that offers easy to use interaction and intuitive application is called **explicit modeling**. Especially for repurposing and varying a model it constitutes a useful tool. Via explicit modeling, it is possible to change a single body like a block into a complex geometry by adding, copying, cutting and pasting single shapes. Just as well as direct modeling it ensures to save a lot of time and increases flexibility of adaption and editing (according to [13]).

A completely new type of CAD systems are **cloud CAD systems**. Hereafter, such systems will be explained through the example of *Onshape* which got developed by key members of *SolidWorks* and other experts. *Onshape* is the first full-cloud CAD system and its beta version was released in March 2015 by the *Onshape* Company. The term full-cloud means that the system is running in a web browser, uses cloud-native documents instead of files for its saving processes and there is no need for installation. At the same time, this means that internet access is obligatory. It offers part- and assembly design as well as 2D-drawings, direct modeling applications, data management and it even allows working with imported data. Such data can consist of common exchange formats like *Parasolid*, *IGES* and *STEP* or even native files from other CAD systems like *SolidWorks*, *CATIA*, *CREO*, *Autodesk Inventor*, *JT* and *AutoCAD* [14].

Furthermore, the new system enables the user to expand working on mobile devices like smartphones or tablets through a mobile app. *Onshape* offers a free version and a professional version which is much cheaper than common CAD-licenses. This fact facilitates access to CAD for private users and small companies. Another achievement consists of easy sharing and simultaneous working on the same document by different users which even provides merging various solutions of a part into one integrated document. The CAD system or modeling kernel gets updated every few weeks without any effect on earlier created data which leads to high update stability. Users and software developers are connected through an online forum with direct contact. So it is possible to improve *Onshape* by the user in the short term by reporting problems or giving advices for new feature types [14].

Other cloud CAD systems which are not fully-cloud systems include *CATIA 3DEXPERIENCE on the Cloud* from *Dassault Systemes* [15], *Fusion 360* from *Autodesk* [16] and a private cloud from *Siemens PLM* which is meanwhile branded under the name of Xcelerator Cloud [17].

References

1. Vanja S, Weber C, Schlingensiepen J et al (1994) CAD/CAM für Ingenieure: hardware, software, strategien. Vieweg, Braunschweig, Wiesbaden
2. Grätz JF (1989) Handbuch der 3D-CAD-Technik: Modellierung mit 3D-Volumensystemen. Siemens-Aktiengesellschaft, Berlin, München
3. Gyurecz G, Renner G (2011) Correcting fine structure of surfaces by genetic algorithm. Acta Polytechnica Hungarica

4. Abulawi J (2012) Ansatz zur Beherrschung der Komplexität von vernetzten 3D-CAD-Modellen. Dissertation, Helmut Schmidt Universität
5. Spur G, Krause FL (1997) Das virtuelle Produkt: Management der CAD-Technik. Carl Hanser Verlag, München, Wien
6. Braß E (2005) Konstruieren mit CATIA V5: Methodik der parametrisch-assoziativen Flächenmodellierung. Carl Hanser Verlag, München, Wien
7. Rudeck E (2009–2015) Concurrent engineering blog: what CAD users need to know about direct modeling. http://www.concurrent-engineering.co.uk/Blog/bid/93709/What-CAD-users-need-to-know-about-direct-modeling. Accessed on 3 Aug 2021
8. Siemens PLM Software (2011) NC programming with synchronous technology: white paper. Part model editing made easy for NC programmers; NC Programming with Synchronous Technology White Paper (geoplm.com)
9. Ault HK, Phillips A (2021) Direct modeling: easy changes in CAD? Direct modeling: easy changes in CAD? (erau.edu). Accessed on 3 Aug 2021
10. NXTutorials Plus (2020) NX synchronous modeling demonstration. NX synchronous modeling demonstration—YouTube. Accessed on 3 Aug 2021
11. Parametric Technology Corporation (2018) Introducing PTC Creo direct. https://www.youtube.com/watch?v=B7FN6XePsZI. Accessed on 3 Aug 2021
12. ANSYS (2021) ANSYS spaceclaim direct modeler. Ansys spaceclaim|3D CAD modeling software. Accessed on 3 Aug 2021
13. PTC CoCreate Explicit Modeling. PTC CoCreate explicit modeling—YouTube. Accessed on 3 Aug 2021
14. Onshape Inc. (2017) Onshape´s 3D CAD system. Beginner tutorial 1/5—onshape 3D CAD—creating sketches and objects—YouTube. Accessed on 3 Aug 2021
15. Dassault Systèmes (2021) Catia 3D experience on the cloud. The 3DEXPERIENCE platform, a game changer for business and innovation|Dassault Systèmes (3ds.com). Accessed on 3 Aug 2021
16. Autodesk Inc. (2021) Fusion 360 is the next-generation tool for the next generation. http://www.autodesk.com/products/fusion-360/students-teachers-educators. Accessed on 3 Aug 2021
17. Siemens Product Lifecycle Management Software Inc. (2016) Siemens' NX software now available in cost-effective private cloud environment: optimizes IT resources, enhances deployment flexibility, further expands platform-of-choice options. https://www.sw.siemens.com/en-US/?Component=222055&amp%3BComponentTemplate=822. Accessed on 3 Aug 2021

Chapter 8
Major Technology 2: Computer-Aided Industrial Design—CAID

Executive Summary

This chapter deals with the following topics:

- Basics and advanced techniques of Computer Aided Industrial Design (CAID)
- Providing insight into how engineers benefit from using CAID technologies
- Describing functioning, benefits, and limitations of CAID technologies in practice.

Quick Reader Orientation and Motivation

The intention of this chapter is:

- to give an overview of CAID technology in Virtual Product Creation as driver and enablers for Digital Transformation in engineering
- to present CAID technology as part of Virtual Product Creation from a practitioner's point of view to analyze the need and usefulness for day-to-day industrial work practice
- to give instructions on how to use CAID technology
- to explain models, frameworks, and mathematical representations that help to grasp the internal working modes of CAID technology.

Designers in the automotive industry—in the sense of designing the shape/style—most intensively use Computer-Aided Industrial Design (CAID), which is also known as Computer-Aided Styling (CAS). It supports the design/styling process much better than CAD systems because it is more intuitive to work with and allows inaccuracies in sketching as well as in modeling. Techniques, which belong to CAID, are e.g. creating high quality freeform surfaces, 3D sketching, virtual clay modeling and high-end rendering. In the automotive industry, CAID is used intensively for car body and interior design. In general, however, it finds its use in every industrial design sector from consumer goods, via design shapes of other transport systems such as trains and planes up to machine box design.

R. Stark, *Virtual Product Creation in Industry*,
https://doi.org/10.1007/978-3-662-64301-3_8

8.1 Engineering Understanding of CAID

This section explains why and in which phases designers (styling experts, shape modelers) use CAID instead of CAD. Furthermore, it will be shown how CAID works in reference to the context of digital and physical product development as part of the overall product creation process.

8.1.1 Why Does an Engineer Use CAID Instead of CAD?

As shown in Fig. 8.1a CAID allows a much higher level of detail in sketches than CAD and by doing so it even allows for time savings in digital modeling. CAID finds its use primarily in the design sector because there it is important to create different eligible solutions in a short time and thereby support the creative process. If CAD was used for this phase significant more time would be needed: the reason is that CAD systems offer less stringent surface modeling capabilities compared to CAID capabilities. Amongst several modeling laminations, CAD does not allow easily for any modeling inaccuracies and it is deemed to be difficult to work intuitively with it within the highly creative phase of modeling. Another advantage of CAID consists of the possibility of making quick alterations in a sketch by simply deleting curves without causing inadvertent effects as it would happen in a CAD system. At the same time, it is even not necessary to change an entire sketch.

In conclusion, CAID enables considerable time savings and it reduces the efforts during modifications, which comes into interplay a lot during a design & development process.

8.1.2 Where is CAID Being Used?

CAID is mainly used in the first stages of product development. As shown in Fig. 8.1b these stages are the concept, design and verification phase. Within this context, designers use CAID partly in the concept and in the verification process (during the rework of the shape after design modifications). Its main usage, however, is part of the design phase. CAID therefore builds an intersection between different product development stages and between engineers and designers. The two main fields of application for CAID consist of the design study as a starting point for the technical development and of the design process of creative and aesthetic product modules. The design phase includes different engineering tasks beginning with the search for solution principles followed by the separation into feasible solutions or modules, the design of the chosen module itself and finally the integration of the module design into the overall design.

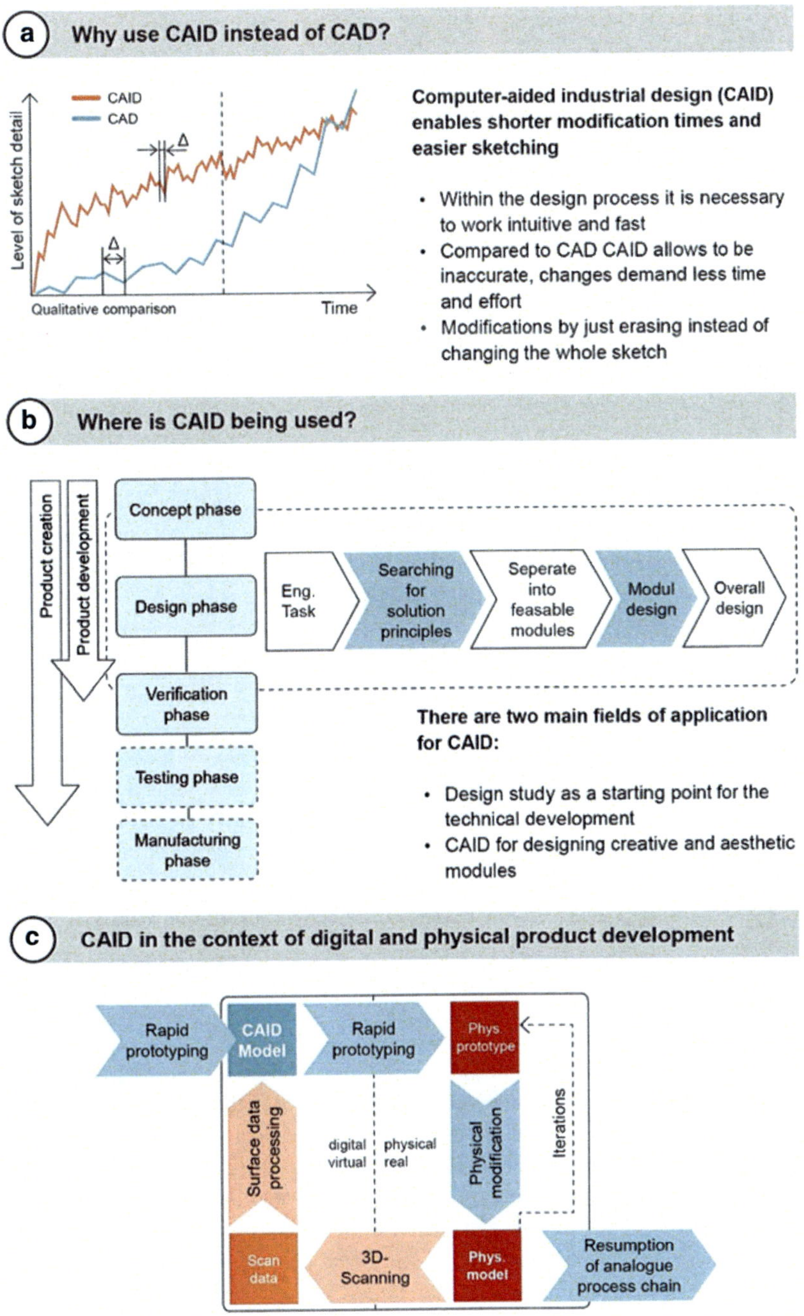

Fig. 8.1 Engineering application of CAID

8.1.2.1 CAID in the Context of Digital and Physical Product Development

The modeling process of CAID consists of two areas, the digital or virtual environment and the physical or real environment. As shown in Fig. 8.1c, the modeling process usually starts with the creation of a virtual model with the help of CAID software. As part of rapid prototyping, the digital model becomes a physical model. Rapid Prototyping is also known as "3D Printing" or more accurately stated as *Additive Manufacturing* and it uses techniques such as *Stereolithography* (STL), *Selective Laser Sintering (SLS), Selective Laser Melting (SLM)* or *Fused Deposition Modeling (FDM)* (see more details in [1]). Afterwards the manufactured prototype might get modified until the desired shape is achieved. There is no limit to the number of alterations during the physical modification and it is also possible to resume this analogue modification process chain after having passed through the following "digital" steps. Those "digital" steps comprise a 3D Scan that converts the physical into a virtual model again and generates scan data within the scanning process. The gained data can be imported into CAID software and there the surface data gets processed until the desired shape is reached—or this process chain can be continued until this aim is finally achieved.

8.2 How Does CAID Work?

This section defines the basic IT technology of CAID and explains how a classical design process works with CAID. Furthermore, it will clarify two input devices of CAID as well as the technique of three-dimensional immersive modeling.

8.2.1 How Does a Classical Design Process Use CAID?

As shown in Fig. 8.2a, there are three stages of the engineering process which comprise the concept, design and detail modeling phase. Within this context, the detail modeling phase is to be understood as a sublevel of the design phase.

CAID is used in all of these phases but in the detail modeling phase, it is used only in a specific way. First, designers (styling experts, shape modelers) need to work out a first concept based on the given appearance assignment and the principal idea of style. After this initial design phase in most of the cases, still a physical mock-up gets build up by means of rapid prototyping. The shape data of the physical part can be added to CAD via the technology process called *reverse engineering*. Through the fact that CAID and CAD can be both managed in product data management systems (PDM) it is possible to work in both areas and exchange data from one to the other software. As shown in Fig. 8.2, CAID gives information of shaped surfaces to the CAD system and CAD transfers information in the form of engineering constraints

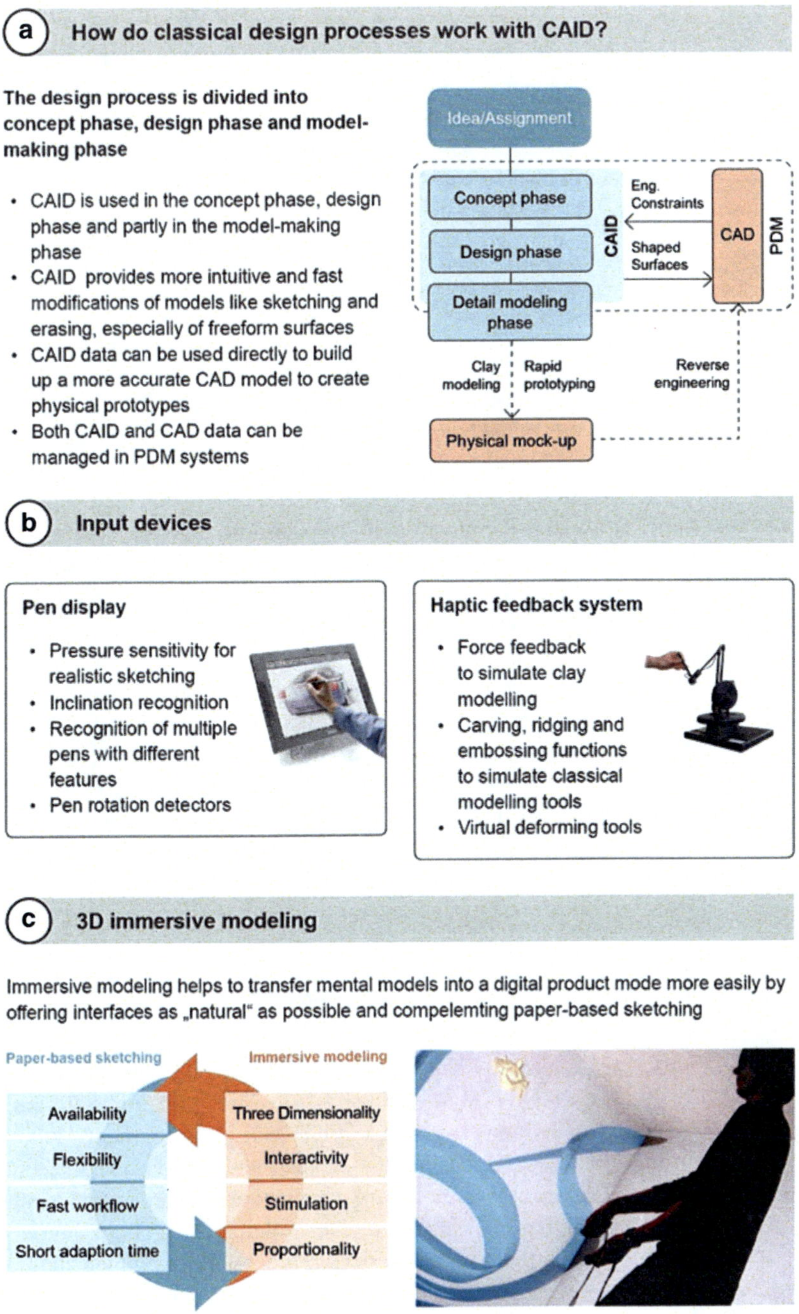

Fig. 8.2 Basic IT technology of CAID

to the CAID system. The reason for extending work outside CAD is the specific set of extra surface related modeling capabilities of CAID to better support the styling design process of product shapes. The main advantage of CAID is that it provides more intuitive and faster modifications of models than CAD, especially in the field of freeform surfaces. Specialized CAID modeling capabilities are also key during the final class. A (visible styling surface of the outer shapes of products) high precision generation.

8.2.2 *Input Devices*

Figure 8.2b shows two typical input devices of CAID, a pen display and a haptic feedback system. The pen display on the left side in Fig. 8.2b provides 3D sketching. Principally 3D sketching works like 2D sketching but the designer needs to employ the method of sketch rotation in order to work in different views to create the three-dimensional shape. To support the function of virtual rotation the pen comes with a rotation detector. Furthermore, the pen has a pressure sensitivity and an inclination recognition for realistic sketching and it can recognize multiple pens with different features. The display itself can be connected to a computer like any other device (e.g. keyboard or mouse).

The second technology shown in Fig. 8.2b on the right side is the haptic feedback system, which helps to imitate the classical process of clay modeling in a virtual environment. This technology is mainly used in design studios in the automotive industry so far. With the help of such a device, the designer receives a volume dependent force feedback during the modeling operation. Through its various functions in the virtual environment, it is possible to simulate classical modeling tools like carving, ridging, embossing or other deforming tools.

8.2.3 *Three-Dimensional Immersive Modeling*

During the last 10 years, intensive research has been conducted in order to achieve the next level of 3D immersive design (see [2], 3). As shown in Fig. 8.2c three-dimensional immersive modeling is part of a Virtual Reality environment. With the help of different input devices, the designer is enabled to create virtual shapes by moving the device in 3D space—all active movements are recorded and mathematically translated into digital line, surface or even volume data. Those interfaces support the efficiency and naturalness of the design process through their intuitive handling and interaction. Moreover, three-dimensional immersive modeling can be understood as a complementation of the traditional two-dimensional paper-based sketching. As the schema in Fig. 8.2c shows, it increases quality characteristics of 2D manual sketching and it offers additional properties of 3D modeling. Consequently, 3D immersive modeling supports traditional paper-based sketching qualities like availability, flexibility, faster workflows and shorter adaption times. In addition, it enables new elements such as three-dimensionality, interactivity, stimulation and proportionality.

The user in Fig. 8.2c works with special input devices like brushes which can be provided in different sizes with various functionalities and grippers. In general, there are four types of input devices: discrete, continuous and hybrid devices as well as miscellaneous input like speech. The operator in the figure uses continuous input devices.

Artists, designers and animators mainly use the technology but the application also starts to be popular for engineers. By offering new ways of modeling, immersive technologies help to merge the traditional disciplines of art, design and engineering to a new comprehensive skill set. This circumstance paves the way for distinctly easy communication between designers and engineers in the product development process. Within this context, it provides 3D sketching, 3D modeling as well as it helps to present and to alter virtual models in a spatial environment.

8.3 Advanced Technology of CAID

In order to provide intuitive human interfaces for CAID Designer, it is necessary to implement refined mathematical algorithms, which enable traditional sketching and drawing techniques of the analog designer work practice into digital CAID environments.

Figure 8.3a explains the principle of virtual tape drawing, one specific way of realizing a successful analog/physical way of using tapes to express feature lines of outer shape design into the digital/mathematical interaction world of CAID. This interaction method provides the notion of a fixed and loose tape that can be activated and fixed to each other by using two mouse buttons. Usually, there exists a degree of freedom for the CAID designer in terms of which configured input device should be used: besides the traditional mouse device (and its different buttons) other 3D interaction devices might be used.

Figure 8.3b presents another rather intuitive CAID functionality: the spline curve creation (a free from 3D line in space or on the surface) based on multi-stroke sketching with a time-dependent drying ink metaphor. The interaction cycle allows using the successful interaction technique "multi-stroke" of a traditional painter and sketcher from the analog world to describe lines of proportions. This rather innovative digital technique enables the use of an intelligent "latency mechanism metaphor" of drying ink to give the opportunity to modify and change the line sketch before making a final commitment to it in the digital CAID modelling environment.

Figure 8.3c describes the principle of how simplified light models are used to create highlight lines (left-hand side, independent from point-of-view) and reflection lines (right-hand side, dependent on point-of-view). Since CAID tools are used to create aesthetic free form surfaces it is necessary to provide special tools to CAID designers. One key function of such toolset enables the control of the quality of the

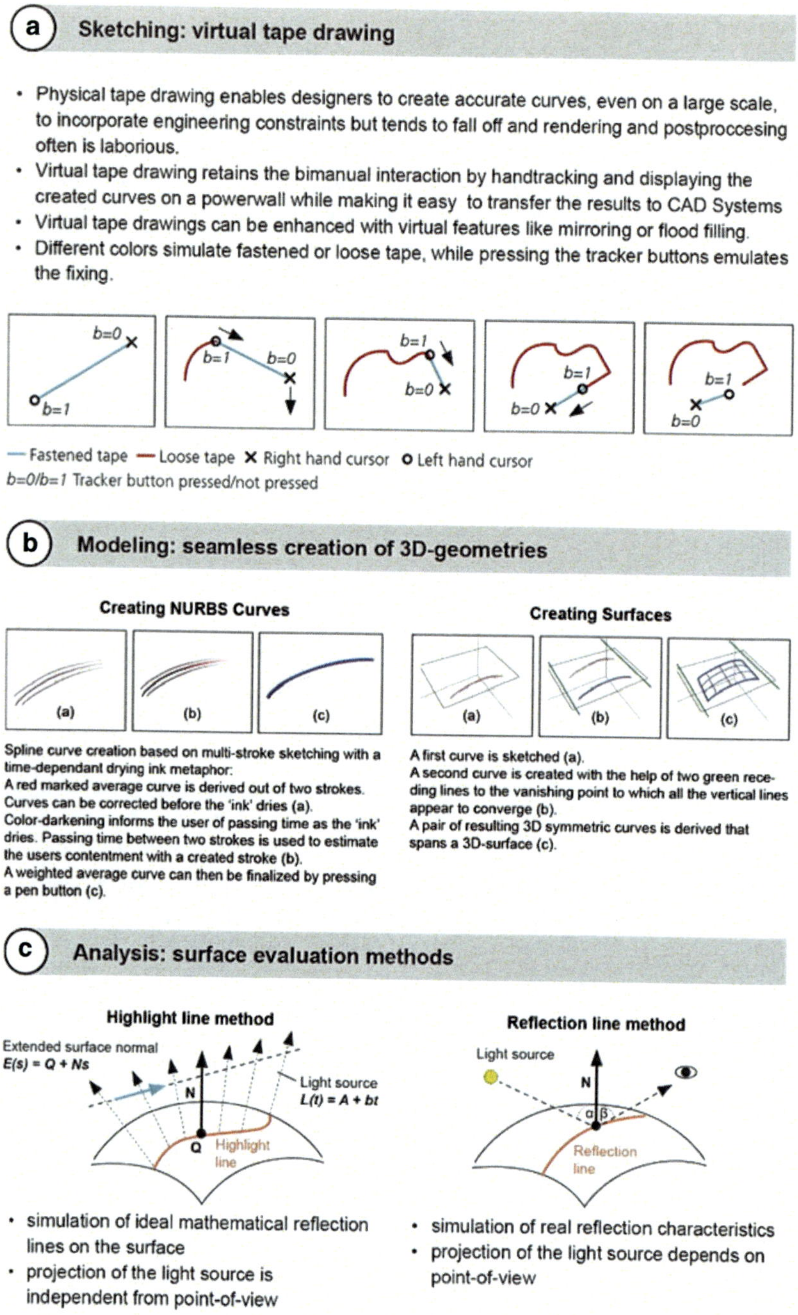

Fig. 8.3 Advanced technology of CAID

surface flow: in order to avoid unwanted light reflections at those aesthetic surfaces the continuity of the surface curvature control (2nd derivative of the mathematical notation of the free form surface) needs to be ensured. CAID designers use such "highlight and reflection analyzing" tools in order to test and modify the original free form surface definition, i.e. the NURBS (Nun Uniform Rational B-Splines) polygon itself and other NURBS geometrical parameter settings.

Highlight Lines

The highlight line is one of the surface evaluation methods. It is a simplified form of the reflection line by eliminating the point-of-view factor for the assessment. Independency from the point-of-view allows the surface to be rotated for further inspection, without changing the properties of the highlight line.

A highlight line is created by positioning a light source over the assessed surface in consideration of the highlight line's position/location and its orientation. The projected light source on the surface is a set of points, whose extended surface normal intersects with the light source. The equations for both, the light source and the extended surface normal, are as follows (compare [4]):

$$L(t) = A + Bt \tag{8.1}$$

L(t) the idealized linear light source, which is placed over the assessed surface (see Fig. 8.4).
A a point, which is located in the idealized linear light source/the point of origin of the light source.
B describes the direction vector of the linear light source *L(t).*
t a parameter, which determines/"limits" the range of the light source.

$$E(s) = Q + Ns \tag{8.2}$$

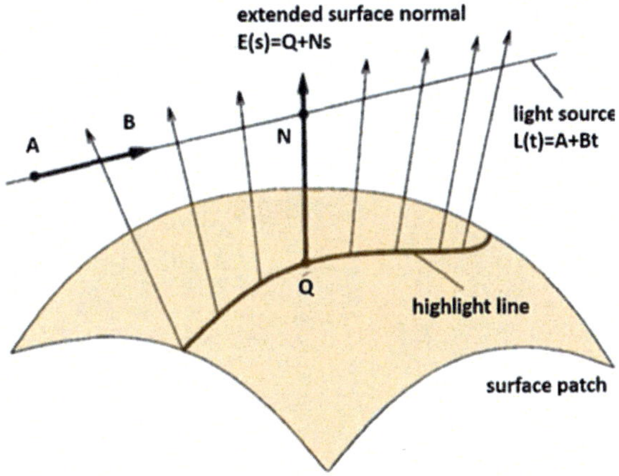

Fig. 8.4 Surface evaluation method: highlight line. Original source [4]

E describes the extended surface-normal of the observed point Q.
Q the observed point on the assessed surface.
N the normal of the observed point Q.
s the parameter which dictates the extension of the normal of the observed point Q.

If point Q belongs to the highlight line, the extended surface normal $E(s)$ would then intersect with the light source $L(t)$, or in other words, the signed distance between the extended surface normal and the light source would be zero. As the name suggests, the light source is in the form of a line or does not possess any radius. The equation for the signed distance between extended surface-normal and the light source and the conditions that are to be met for the highlight line are explained in Eqs. 8.3 and 8.4 and in Fig. 8.5 (compare [4]).

$$d = \frac{(B \times N) \cdot (A - Q)}{\|(B \times N)\|} \tag{8.3}$$

A light source with a radius r = 0 results in a highlight line with

$$d = 0 \tag{8.4}$$

A light source with a radius r > 0 results in a highlight band with

$$d \leq r \tag{8.5}$$

As shown in Fig. 8.5, d$(u, v) \neq 0$ since point S(u, v) does not belong to the highlight line.

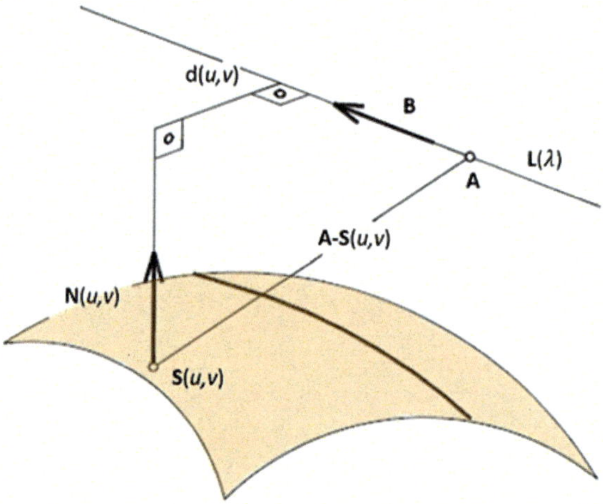

Fig. 8.5 Illustration of the perpendicular distance of the observed point and the light source. *Note* notations are different from the equation written above. *Source* [5]

Even though the highlight line method is independent from the observer's point of view, the placement/location of the light source, the orientation and the shape of the surface influence the resulting highlight line.

Placing a highlight line over a planar face would result in a single and uninterrupted projected curve on the face. However, placing/ using highlight lines on a non-planar face may produce various results. Placing a highlight line over a convex side of a non-planar surface still produces a single and uninterrupted curve on the face. This is shown in Fig. 8.6.

Placing a highlight line over a concave side of a non-planar surface would however result in a loop, interrupted or intersecting projection on the face. These occurrences are shown in Fig. 8.7.

This happens because the surface-normal directions on the planar and convex face are scattered out. The surface normal directions on the concave surface, however, are

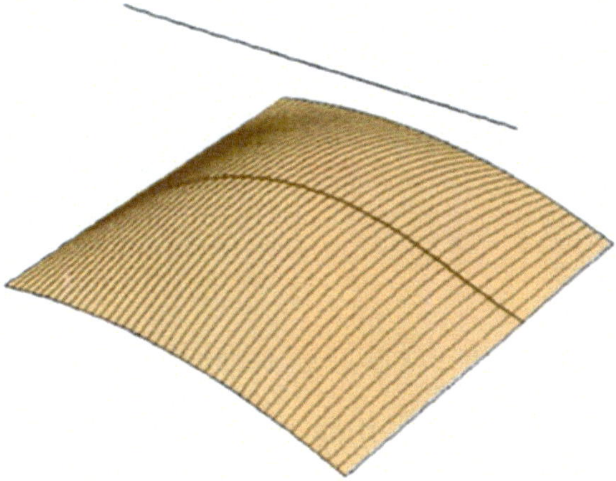

Fig. 8.6 Application of a highlight line placed over the convex side of a surface (single uninterrupted curve). Original source [4]

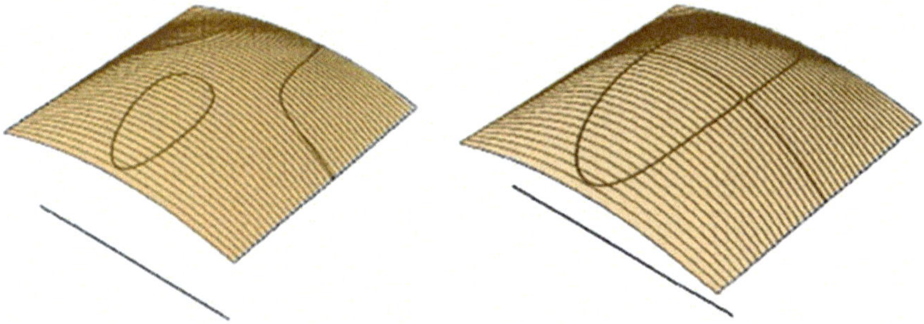

Fig. 8.7 Application of highlight line place over the concave side of a surface. Loops, discontinuity and intersections occur. Original source [4]

more "focused" in a certain direction and intersecting with each other, thus creating the projected light would result in a loop, intersections and/or interrupted curve.

Figure 8.8 shows two examples for interactive surface analysis functions provided by the CAID module of the CAD system CATIA from Dassault Systèmes for Designers.

Figure 8.9 shows the direction of the normal vector of a concave surface. Multiple parallel highlight lines placed over a surface can also be used for a more thorough and complete assessment of the surface quality.

The light source is generally described with a single line from a given point A with the direction B. However, this light source model can be expanded into a boundary

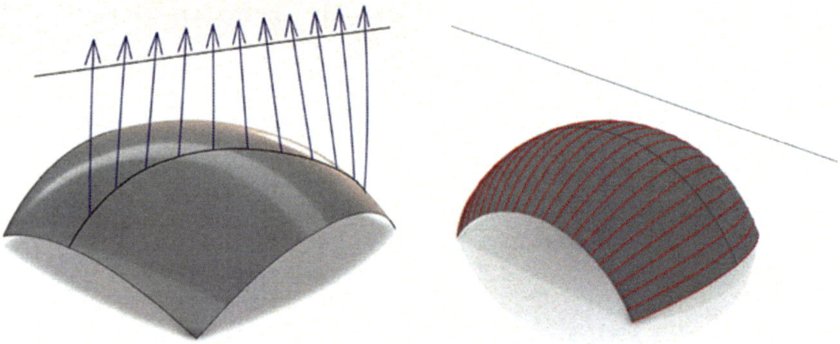

Fig. 8.8 Implemented highlight analysis functions in the CAID module of the CAD system CATIA; left side: vectors to highlight curvature change, right side: highlight line created point projection onto the surface to establish redlined curves through individual cutting planes. *Source* Dassault Systèmes

Fig. 8.9 Surface-normal vectors of the concave surface. Original source [6]

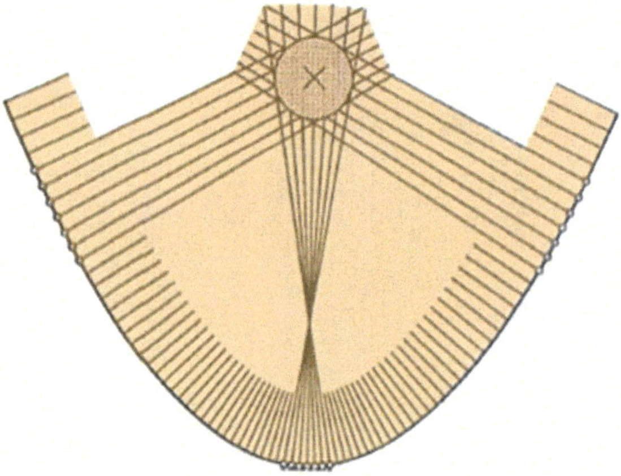

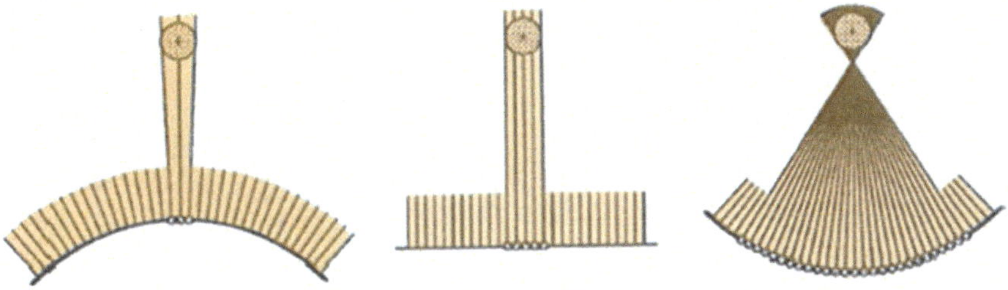

Fig. 8.10 Application of highlight band on planar and non-planar surfaces and illustration of normals on different surfaces. *Source* [6]

band model, when an idealized light source is given a certain radius. This model would better simulate a real light source (compare cases in Fig. 8.10).

The light source model can be expanded by giving the light source a certain radius r. The condition, which has to be met for a point on a surface to belong in the highlight band, also changes. With the former light source model (highlight line) the distance between the extended surface-normal and the light source has to be zero, in order for the surface point to be included in the highlight line.

For the highlight band, the distance between the extended surface-normal of a point on the surface Q and the centre axis of the light source has to be smaller than the radius of the light source. The equation and condition which are needed to be met for the highlight band are shown in Eqs. 8.3 and 8.5. A highlight band provides additional information to the surface quality assessment aside from the detection of irregularities. As seen in Fig. 8.10, light source with the same radius projects highlight bands with different widths on different types of surfaces. Therefore, the observation of changes in the width of the highlight band enables the detection of changes in contours.

In practical CAID modeling work it is important to be competent in specific CAID system functionalities to ensure proper surface generation and testing. Based on the principals explained earlier in this sub-chapter the following figures show various applications of surface quality analysis functionalities.

Figure 8.11 shows two methods based on the *angle pitch value* in the CAID module of the CAD System CATIA of Dassault Systèmes.

Working mode of the *normal to the surface approach* (left side of Fig. 8.11): the system draws a series of curves on the surface. At all points along a given curve, the angle between the *local normal to the surface* and the coordinate axis Z is constant. The spacing between each curve represents a change in the angle of the local normal given by the *angle pitch value.* The origins of the curves are the points on the surface at which the coordinate axis Z is normal.

Working mode of the *tangent to the surface approach* (right side of Fig. 8.11): the system draws a series of curves on the surface. At all points along a given curve, the angle between the *local principal tangent to the surface* and the coordinate axis Z is constant. The angular spacing between each curve is given by the *angle pitch*

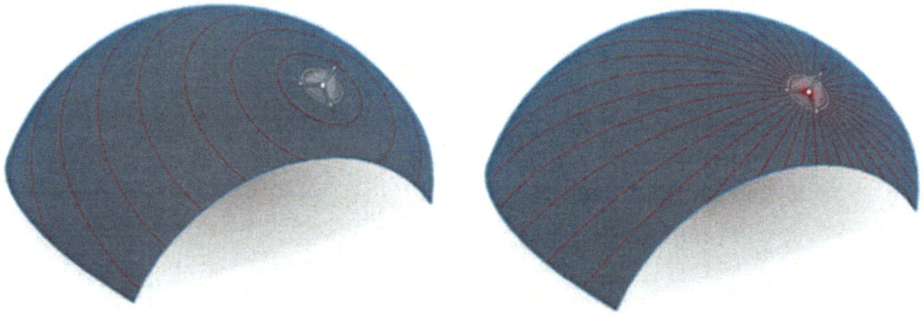

Fig. 8.11 *Angle pitch value* between the local normal to the surface (left side) and the local principal tangent to the surface with respect to the determined (in this case outgoing) axis of the coordinate system at the top. *Source* Dassault Systèmes

value. The origins of the curves are the points on the surface at which the coordinate axis Z is normal.

Figure 8.12 shows the *explicitely controllable intersection method*. Within such *surface quality analysis method* intersections (red lines) are ceated onto the target surface, always referred to defined cutting planes. The cutting planes can be explicitely controlled in the following ways:

1. Parallel based on the orientation of the black plane as shown in Fig. 8.12
2. Normal in relation to any curve
3. Based on an already existing plane in the CAID model.

The amount and the stepwidth of the cutting planes can be flexibely controlled and determined, as well as the start and end points (boundaries) relative to the reference plane.

Fig. 8.12 Intersection method. *Source* Dassault Systèmes

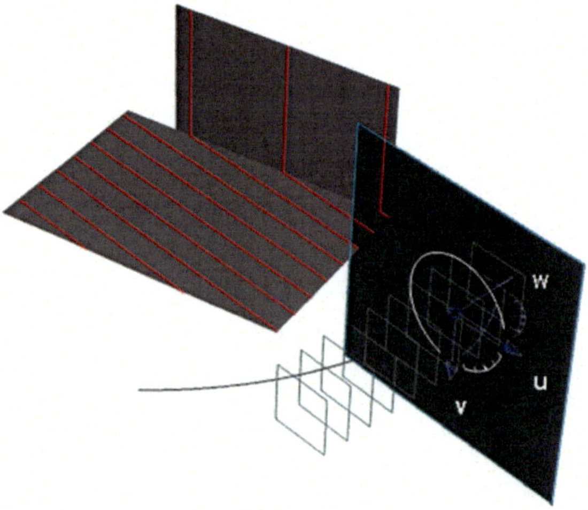

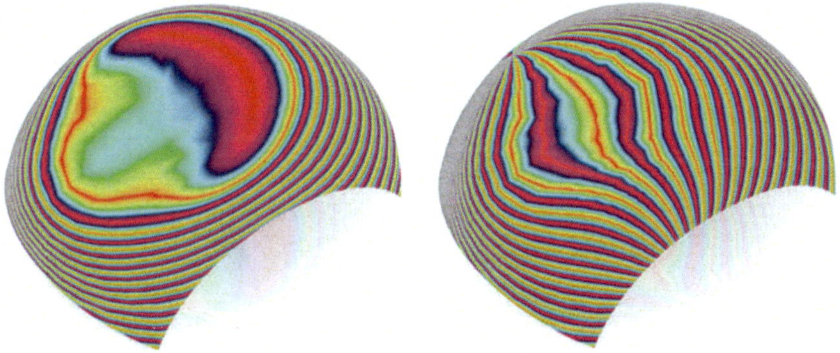

Fig. 8.13 Color coded high light lines; left side: lines directly from the light beam of the parallel light source, right side: by reflection of stripes on the interior side of an imaginary cylinder around the geometry. *Source* Dassault Systèmes

Finally, Fig. 8.13 illustrates the method application of highlight lines based on the principle of *color coded "zebra" line projections* onto the target surface.

Overall, there exist a wide range of additional CAID surface modeling functionalities with integrated testing procedures, both similar across different CAID/CAD systems but also with specific custom oriented features and unique solution implementations in different CAID/CAD systems.

References

1. Thompson MK et al (2016) Design for additive manufacturing. Trends, opportunities, considerations, and constraints. In: CIRP annals—manufacturing technology. https://doi.org/10.1016/j.cirp.2016.05.004
2. Israel JH, Wiese E, Mateescu M, Zöllner C, Stark R (2009) Investigating three-dimensional sketching for early conceptual design—results from expert discussions and user studies. In: Computers and graphics.https://doi.org/10.1016/j.cag.2009.05.005
3. Stark R, Israel JH, Wöhler T (2010) Towards hybrid modelling environments-Merging desktop-CAD and virtual reality-technologies. In: CIRP general assembly, Pisa, Italien, 22.-28.08.2010. CIRP annals—manufacturing technology 1/59, pp 179–182
4. Beier KP, Chen Y (1994). Highlight-line algorithm for realtime surface quality assessment. Michigan: department of naval architecture and marine engineering. Comput Aided Des 26. Online available under: PII: 0010-4485(94)90073-6 (psu.edu)
5. Balster C (2014) Übertragbarkeit von Modellierungs- und Parametrikkonzepten in der Flächenkonstruktion zwischen verschiedenen CAD-Systemen. Diplomarbeit (Transfer capabilities of modeling and parametric concepts in surface design between different CAD systems; Master Thesis), TU Berlin, Germany
6. Beier KP, Chen Y (1994) The highlight band, a simplified reflection model for interactive smoothness evaluation. In: Sapidis NS (ed) Designing fair curves and surfaces: shape quality in geometric modeling and computer-aided design. SIAM, Athens, pp 213–230

Chapter 9
Major Technology 3: CAPP, CAM and NC Technology

Executive Summary

This chapter deals with the following topics:

- Basics and advanced techniques of Computer Aided Process Planing (CAPP), Computer Aided Manufacturing (CAM) and Numerical Control (NC)
- Providing insight into how engineers benefit from using CAPP, CAM and NC technologies
- Describing functioning, benefits, and limitations of CAPP, CAM and NC technologies in practice.

Quick Reader Orientation and Motivation

The intention of this chapter is:

- to give an overview of CAPP, CAM and NC technology in Virtual Product Creation as driver and enablers for Digital Transformation in engineering
- to present CAPP, CAM and NC technology as part of Virtual Product Creation from a practitioner's point of view to analyze the need and usefulness for day-to-day industrial work practice
- to give instructions on how to use CAPP, CAM and NC technology
- to explain models, frameworks, and representations that help to grasp the internal working modes of CAPP, CAM and NC technology.

This sub-chapter explains all *Virtual Product Creation* technologies which are essential to transform the digital delivery of product development (represented by CAD models) into digital artefacts which can formally and completely describe the working elements to eventually realize the physical shape of the product as part of the digital manufacturing process. The explanation starts with Computer-Aided Process Planning (CAPP) and is followed by Computer-Aided Manufacturing (CAM) and Numerical Control (NC).

R. Stark, *Virtual Product Creation in Industry*,
https://doi.org/10.1007/978-3-662-64301-3_9

9.1 Computer-Aided Process Planning—CAPP

Computer-aided process planning (CAPP) is the generic term for software tools that assist in the planning of manufacturing processes. CAPP serves as a bridge between CAD and CAM. CAPP is used to determine how a design will be manufactured in a production system via digital planning methods. Without a successful CAPP, it is impossible to transform complex design information into manufacturing.

9.1.1 Engineering Understanding of CAPP

The following subsection will explain the basics of CAPP. For this purpose, the functionality and benefits of CAPP in Virtual Product Creation are demonstrated.

9.1.1.1 Why Does an Engineer Use CAPP?

Process planning of manufacturing operations takes place organizationally between product design, manufacturing engineering and operational production. In process planning, the manufacturing processes, their sequences and the manufacturing conditions are determined. Process planning for production is a complex task with many variables from different areas and departments such as component design, manufacturing engineering, task sequence determination, ergonomics, material supply and logistics. All of those aspects need to be considered as part of process planning. The goal is to convert the virtually designed product (assemblies with individual components), in economic and competitive terms, into a physical component with minimum resource investments and high delivery robustness and quality of the product itself.

Process planning is traditionally associated with high manual effort and requires substantial heuristic knowledge. CAPP is used to assist manufacturing engineers in process planning and to make decisions as objective as possible instead of just relying on the knowledge and experience of individual experts. CAPP, therefore, allows engineers to systematically develop appropriate methods for the manufacturing of single components and/or full products with a reduced manual effort and to fulfill efficiency targets of the manufacturing process, the production system or even the entire factory.

9.1.1.2 What is CAPP Doing for an Engineer?

Engineers, in general, and Manufacturing planners, in specific, use CAPP for tasks like:

- Process selection
- Operation sequencing

- Machine and tool selection
- Process scheduling
- Manufacturing condition determination
- Manufacturing time and cost estimation.

Two of the most complex challenges in manufacturing planning are *process selection* and *operational sequencing*. CAPP, therefore, engages algorithms to optimise process sequencing via mathematical methods [1].

By using CAPP, constraints such as the technical priority graph and capabilities of resources can be automatically checked and problems and restrictions in the production system can be displayed. Through simulations in the virtual production system, engineers can secure and optimize the manufacturing of new products before the production system actually exists.

In summary, CAPP is used to match, secure and optimize the connection between manufacturing processes and resources, demonstrated in Fig. 9.1.

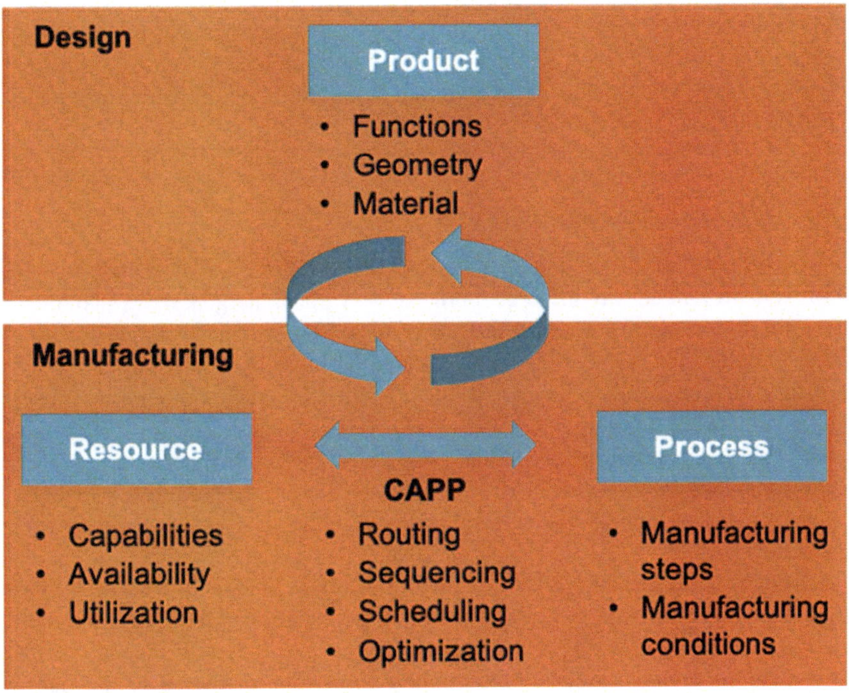

Fig. 9.1 Connection between product, process, resource and CAPP

9.1.1.3 What Are the Benefits of CAPP?

CAPP is used to reduce the time and effort required to create consistent process plans. Furthermore, CAPP is used as an automated interface between CAD and CAM to achieve full integration of structural and process data into the manufacturing system design.

The use of CAPP in virtual product creation leads to the following advantages:

- The process plans can be generated automatically from the product data by using feature recognition technology.
- The manual effort of scheduling is reduced because algorithms can be used to automatically create and compare different planning variants.
- The time required for process planning can be reduced because steps as the generation of process plans and scheduling can be automated.
- The planning quality can be increased, since the processes in production can be simulated virtually to identify and correct problems (such as material flow) before they are implemented.
- The utilization of manufacturing workplaces and machines can be increased, because in planning many different variants can be created and easily compared with acceptable manual effort. In addition, target fulfilment of planning variants can be checked simultaneously.

9.1.2 How Does CAPP Work?

CAPP systems pursue two different approaches—variant and generative—to translate design information into manufacturing steps. The variant approach uses the similarity of parts to segment them into different groups. There are master process plans for each part family which are used and edited to match the requirements of a certain component. Group technology (GT) code is widely used for the classification of parts into families of similar ones. In contrast, the generative approach creates a process plan for each part from scratch without manual effort. Manufacturing databases and appropriate part descriptions are used to generate a process plan for a certain part [2].

A lot of research in the area of CAPP took place in the last decades. The current research on CAPP systems focuses on the generic approach. The feature-based technology—originally a major research topic in the early nineties of the last century—is used to translate the implicit knowledge of the developer into a computer-interpretable way to automate and optimize the planning of manufacturing processes. A feature is described by a compilation of characteristics and/or properties of a product. The description of a feature consists of the relevant property itself (for example geometric shape and topology to identify machining features, compare Fig. 9.2), its value and its relation and constraints. Features can be used as integration elements over the entire life cycle of a product.

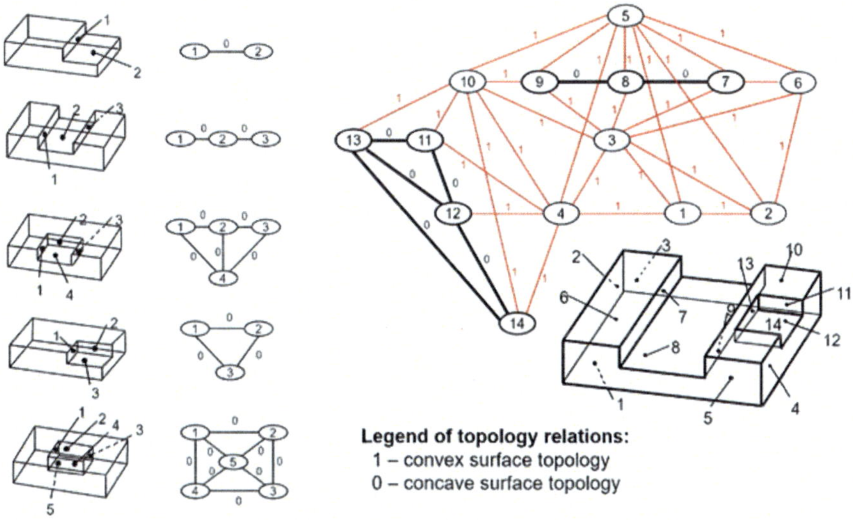

Fig. 9.2 Identification of milling machining features for process planning by recognizing topological regions of a digital design component

In addition to the feature-based technology, AI technologies such as knowledge-based systems, genetic algorithms and also artificial neural networks are used in CAPP systems to create and optimize manufacturing processes. The process plans for new products can be generated based on the geometry description, the material and other variables that influence the manufacturing decision [3].

However, in some areas like assembly planning of complex products such as vehicles, process planning still requires a lot of manual effort. Complex products and complex production systems lead to an enormous complexity in process planning that cannot be handled by a fully automated solution these days. The progress in CAPP is significantly slower compared to CAD and CAM, although much research has been done in recent years. The multidisciplinary nature of process planning makes it difficult to automate CAPP in the industrial praxis [4].

9.1.3 CAPP Methodology and Technology

In the following section, the steps by which a methodological and technological description of the process of manufacturing a part/product is built will be described. In addition, the main problems when using CAPP are to be considered.

As stated above, CAPP is the bridge between CAD and CAM. In this regard, CAPP has a large number of methods and steps in the planning of the production process.

The main approach of CAPP works as follows: according to a given model of a digital product (designed with CAD), a work plan is created for its production or assembly sequenced process steps in a way that it can be altered in terms of sequence of the process steps, type of working steps and duration of the working steps. In order to virtually simulate, test and modify an entire production process or, like in a specific case here, of an assembly sequence it is necessary to leverage a full digital model that consists of the following elements:

- the digital component which needs to be assembled to another part or sub-assembly (both to be represented with a 3D CAD or visualization model),
- the relevant digital representations of assembly resources (tools, fixtures, transport systems, safety devices etc.) and
- the digital representation and interaction model of a human worker (in case of a manual assembly task).

The description of one specific digital assembly sequence step, for example, includes information about the following core elements:

1. The 3D representation, the sequence and the trajectory of the components during the assembly operations.
2. The equipment which are used to perform the assembly tasks and the worker actions used in each operation.
3. The digital simulation tool which mathematically controls the processing and timing of all operations within the 3D environment.

Figure 9.3 displays the preparations for the production process using the Technomatix software (Siemens).

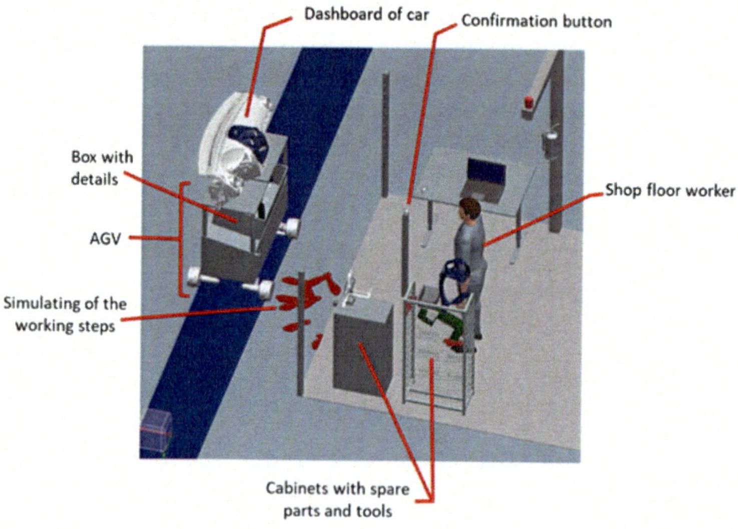

Fig. 9.3 CAPP: sequence of operations and operation elements in 3D

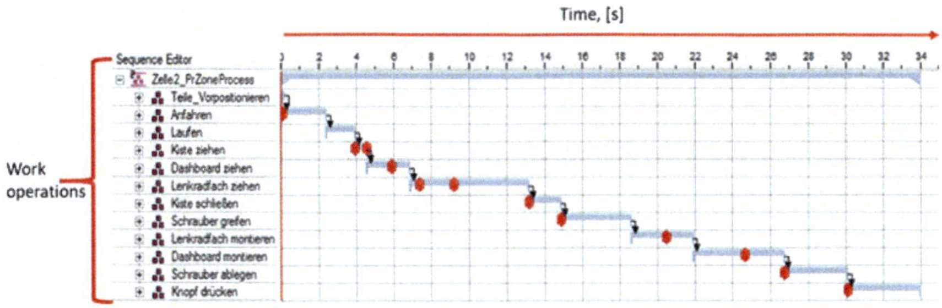

Fig. 9.4 Time sequence simulation of worker actions at the station

This example shows how the production of the car dashboard is planned. Each part of the assembly process is programmed with a specific action of the virtual model of the worker with the necessary cycle time.

In addition, a wide range of advanced programming functions for production planning, such as the visual components, typical interactions and assembly steps, is provided. A process engineer, therefore, can evaluate the movement of a manufacturing worker on the shop floor and his speed of work (compare the time chart of the assembly sequenced steps see Fig. 9.4).

In the automotive assembly example shown in Figs. 9.3 and 9.4, the simulation begins with a preset of parts. This operation ensures the preservation of the initial positions of all parts. An automated guided vehicle (AGV) is then launched. The AGV travels through the AGV line to the middle of the assembly station. The worker then walks to the AGV, opens the box under the AGV and takes the parts (a trim cover in this case) that need installation into the dashboard. Then the worker transfers the box to its original position and walks to the cabinets to get the relevant assembly tools (screwdriver, bolts etc.). The worker takes the tool and screws the part (cover) to the dashboard. Afterwards, the worker puts the tools back to the cabinet, presses the confirmation button, and allows the AGV to go to the next workstation. Accordingly, at this station, it is necessary to perform 12 different operations (see Fig. 9.4) in order to fasten the part (cover) with all necessary details. Such an assembly process sequence allows for optimization, both on behalf of the worker activities and movements as well as on behalf of the part installation sequence itself.

9.1.4 *Requirements for CAPP*

Production planning in general is a trustful, responsible, sometimes even complicated and time-consuming process, since reliable information connections to different stages of production, technology and software have to be implicitly and explicitly established. As a consequence, detailed requirements are imposed on CAPP to ensure the quality and integrity of the planning process in order to achieve high robustness for the subsequent production system operation:

- A process plan must ensure all quality requirements of the part defined in the part drawing or with the help of digital annotations at the 3D CAD model itself.
- A process plan should deliver on overall goals such as reasonably high production efficiency and high yield in order to fulfill production quota according to due dates.
- A process plan should ensure low production costs (both piece price and investments).
- A process plan should help to improve the working conditions and promote uninterrupted development of manufacturing technology [5].

In Fig. 9.5, a traditional paper-based process plan is shown as it has been used for more than 50 years. It includes many elements, such as: materials of parts, work equipment, number of operations, etc. Such documents are usually prepared for each operation of the production system. With the help of computer aided support

NJA COMPANY	ROUTING SHEET		Product No.			Page __ of __	
			Part No.				
			Part Name				
material		Blank		Number of part per unit		Quantity	
OP No.	Name of operation	Equipment	Jig or fixture	Tool	Gage	Standard Hours	Remarks
			Planner		Approved		
			Date		Date		
Alteration No.	Approved	Date	Issue No.		Issue Date		

Fig. 9.5 Example of routing sheet [5]

for production work planning those final documentations are just an output of the internal digital process data model which is generated, altered, simulated and tested within the CAPP environment as already shown in Figs. 9.3 and 9.4.

9.1.5 CAPP Challenges and Problems

Production planning and digital planning activities as part of CAPP pose mainly the following challenges:

- to ensure the principal technological, the business and the necessary geometrical accuracy to manufacture and assemble single part, (sub-) assemblies and/or full product in the given working conditions of the production cell, the entire production system or even within the complete factory. These challenges need constant attention and they are usually addressed within industry via different stages of feasibility studies and heuristic determinations which do include all manufacturing partners and suppliers.

  Unfortunately, up-to-date there still exist many interface challenges for the exchange of digital models of digital product and process models: besides general data exchange formats like STEP and IGES more specific data exchange models for manufacturing resources such as AutomationML have been developed and introduced over the last couple of years.
- to establish frameworks and platforms for flexible cost planning and competitive production price estimations across companies.
- to develop simulation methods for automatic efficiency analysis of resources and equipment in production and for adapting flexible production concepts for existing production line and machine regimes in factories.
- to develop production plans for low cost production based on frugal manufacturing principles.
- to maintain and adapt process plans to organize uninterrupted production of products with the possibility of further improvement under given conditions.
- using new *Industrie 4.0* design and planning systems for describing networked *Cyber Physical Production Systems* (CPPS) to enable future aspects such as (see more details in Chap. 20):

  - Modeling of digital twins and the associated digital analysis streams
  - Design and implementation of tools which enable automatic digital twin creation and process integration
  - Development and establishment of approaches for validating digital twins as part of digital production planning
  - Development of methods for validating real-world pro-ducts, plants etc. by using the digital twin at an early stage
  - Reverse design of the data analysis required for determining minimum sensor population and managed basic AI modules.

All these challenges and problems are typical and exist (or will exist in the future) at every enterprise. Achieving a balance among them will allow process and manufacturing engineers to adapt production for each specific purpose. Some problems are solved by purchasing new equipment, which oftentimes drive major adjustments of the manufacturing process themselves. However, one should always keep in mind that the main task of industrial CAPP is to develop such a manufacturing process for products, which will take the lowest effort or price in resources, working hours and equipment of the enterprise.

9.2 Computer-Aided Manufacturing—CAM

Modern digital manufacturing engineering is developing at a rapid pace in the direction of production automation with the widespread use of the latest robots, CNC (Computerized Numerical Control) technology and, of course, additional control and analysis software like PLC (Programming Logical Controller) or Operational Data Acquisition as part of Manufacturing Execution Systems (MES). Flexible technologies provided by software allow engineers to quickly and efficiently adjust the production process of parts for each individual request from the client. Automation of the product design process using CAD/CAPP/CAM technologies is a way to increase production efficiency and product quality. In this regard, this sub-chapter will consider CAM technology as well as the advantages it provides for engineers and its capabilities. A typical process of an engineer working with a CAM system will also be presented.

First of all, let us get acquainted with the definition of CAM.

According to Nageswara Rao [6], *CAM generally refers to the computer software used to develop the computer numerical control part programs for machining and other processing applications.*

Another definition, according to Alavala [7] states: *CAM is a computer system that helps to manage, plan, and design production operations in terms of resources and time.*

In summary, the author recommends the following definition of CAM: *CAM (Computer-Aided Manufacturing) provides methods, tools and information standards to help engineers carry out automated calculations of tool paths for processing on CNC (Computerized Numerical Control) and DNC (Distributed Computerized Control) machines digitally and provide the distribution and load of such digital control programs to production machines using computer and digital networks.*

9.2.1 CAD/CAM Integration

It is important to understand the relationship between CAM and CAD representations. CAD-systems are developed to create geometric and topologic models (2D and 3D) of parts, components and assemblies with the help of internal digital (and mathematical) representations and to create design documentation such as drawings (compare Chap. 7 CAD Modeling Techniques). Basically, modern CAD systems include all the necessary modules for modeling a three-dimensional part and the design of all the necessary supporting documents (specifications, sketches, etc.).

In turn, CAM-systems are designed to devise the processing of products on machines with computerized numerical control (CNC machines) and to transfer and load operating programs for these machines (milling, drilling, erosion, punching, turning, grinding, etc.). CAM systems are also called pre-production systems. Currently, they are almost the only way to manufacture complex parts and shorten their production cycle since the old days of manual settings with a tool machine are no longer state-of-the-art in industrial practice. CAM systems use a three-dimensional part model created in a CAD system.

How does the relationship between these two software solutions look like? The manufacturing engineer needs to take a series of actions in order to check the CAD design and to prepare the resulting control sequences for the execution of the individual machine operations at a given CNC machine before starting the operational production. For this, the engineer uses CAD-CAM systems in the following way (see Fig. 9.6):

- By engaging upfront in the development, design and analysis of the product with the help of a CAD system that provides active manufacturing feasibility support (e.g. applying rules of manufacturability for specific geometric shapes under given tool characteristics) to the CAD designer.
- By verifying the topology, shape and dimension of the CAD model and by simulating its manufacturing processing and its material behavior with the help of CAM knowledge and simulation according to the following aspects at a given CNC machine:
  - Recognizing specific machining features (see Fig. 9.2)
  - Minimizing tool clamping
  - Automating tool holdings
  - Simultaneous multi-axis kinematic operation
- Under appropriate physical conditions (i.e. by using verified CAD/CAM parameter settings of used materials, tool wear, cutting speed, tool forward speed, lubricant cooling, surface conditions, tool-workpiece angle etc.).
- By creating a final optimized control program for CNC machines via indicating the sequence of product processing processes for each surface area, such as, for instance: lathing, milling, grinding, slicing, drilling, etc., or, in other words, by using the CAM system.
- By issuing control commands to each individual CNC machine (using a CAM system) as part of the Distributed Numerical Control (DNC) network.
- If necessary, by carrying out verification at each stage of production of the part and adjusting the program for each individual CNC machine.

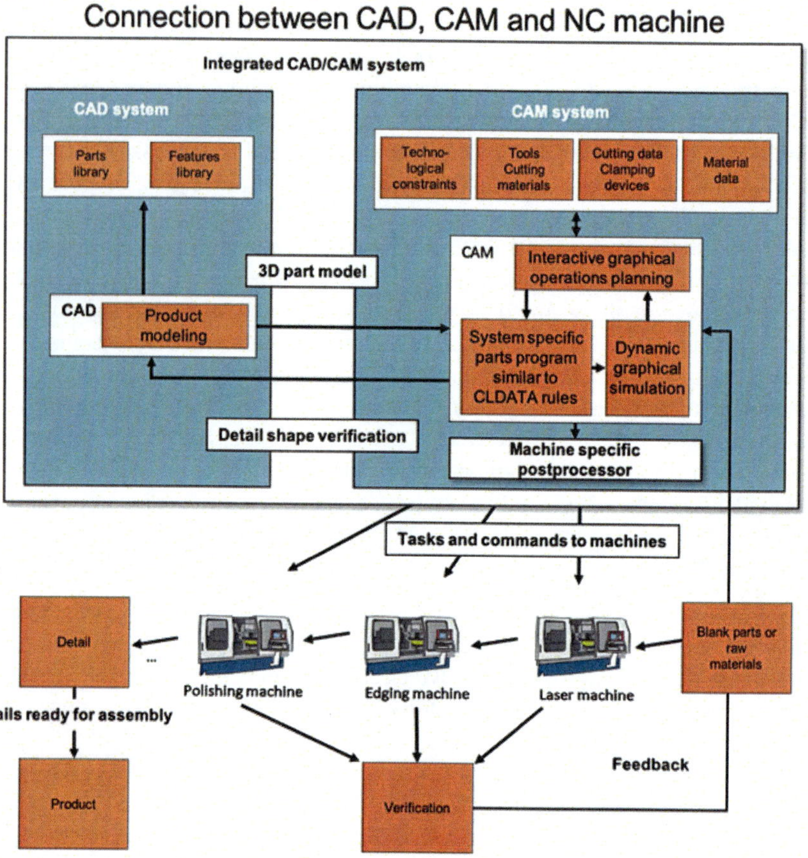

Fig. 9.6 Connection between CAD, CAM and NC machine

With the CAM graphical system, the engineer receives visual feedback for each step of the work piece processing if the appropriate settings of the individual process steps have been accomplished with the necessary manufacturing knowledge beforehand.

Another useful aspect of the CAM graphics system is its ability to simulate toolpaths. This is a computer animation that shows exactly how the program will work on a specific CNC machine with the given verified parameter settings of the physical behavior as described above. If the analysis shows that the machine operation does not work properly or according to anticipated work plan assumptions, the settings before the actual processing of the physical machine on the shop floor.

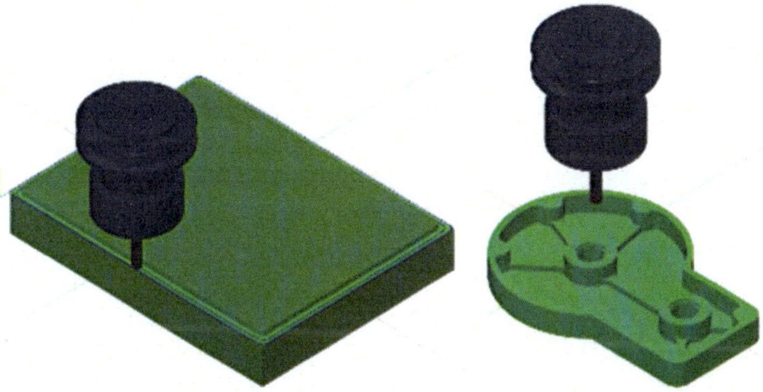

Fig. 9.7 Part manufacturing using CAM tools—milling operation with tool and tool holder (in black color), raw part on the left side and finished part on the right side, both in green color

9.2.2 Engineering Understanding of CAM

This section explains how and why engineers use CAM systems. By using CAM software, an engineer is able to develop and analyze the process of creating a part at each stage of the CNC operation. Each company produces products of varying complexity, so CAM tools are an indispensable technology to ensure proper and automatic production compliance.

Figure 9.7 shows a typical example of the result of an engineer working with CAM software. After receiving the 3D model, the CAM engineer must create the correct sequence of commands for its manufacture. To accomplish this, the engineer determines the necessary form of the raw work piece, selects the material and the appropriate tools for the job (Fig. 9.7, left side). Figure 9.7 on the right side, shows the final stage of the milling workpiece after the full machine operation simulation in the CAM software. Visual simulation during the processing shows how the manufacturing process evolves and which problem might occur (e.g. clash with tool clamping and fixtures which are not shown in Fig. 9.7).

9.2.3 Why Does an Engineer Use CAM?

Due to the general increase in production and the speed of delivery of the finished product to the buyer, it is necessary to respond more quickly to changes in production and produce the same products in a shorter time.

Therefore, best practices give better results. In order to be one of the best companies, process and manufacturing engineers must adhere to planned targets. Similarly, in order to increase the productivity of production processes, manufacturing and process engineers should use CAM software to simulate the manufacturing of a product on a tool machine and to finally transfer commands to the CNC machine (Fig. 9.8a).

CAM software allows process engineers to analyse and test the performance of 3D models even before the appearance of a physical prototype. By using these programs, they test the CAD model, speed and production capabilities on specific machines (Fig. 9.8b).

Designers will be able to perform simulations at any time during product design and development. However, engineers often model details at the concept stage and continue to refine them throughout the development cycle (Fig. 9.8c). This allows to evaluate the manufacturing characteristics of the product precisely and to optimize it in terms of cost and quality, leaving time for innovation and error correction.

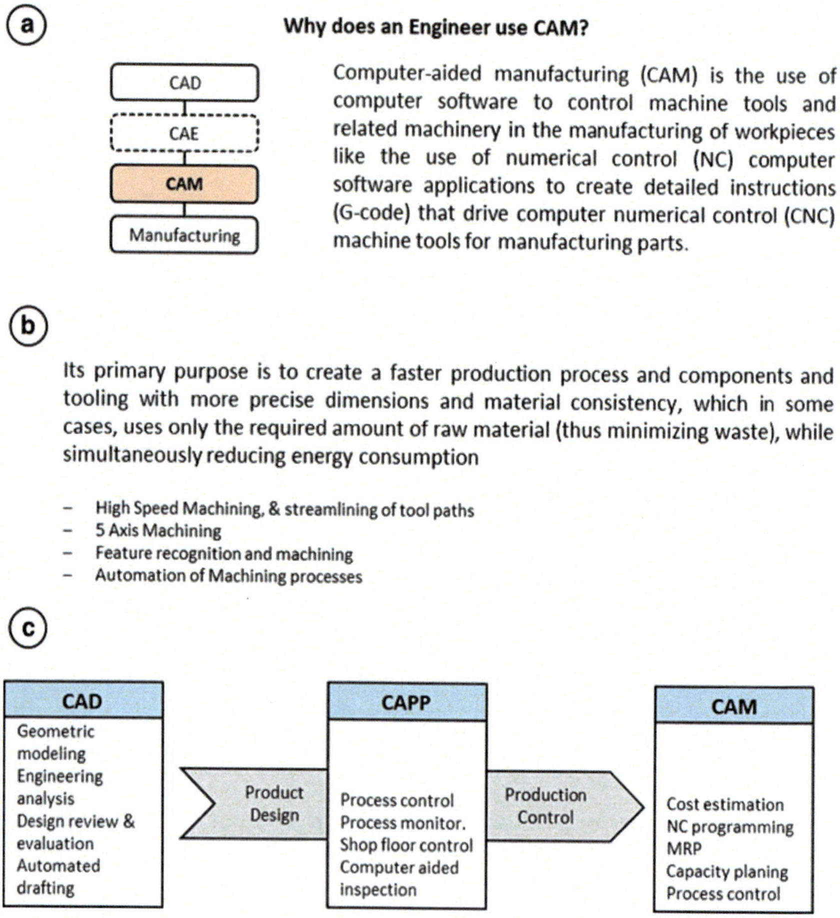

Fig. 9.8 CAM as final step in the overall process flow to accomplish a CAD design model within manufacturing (picture was provided)

Overall, it is worth noting that the full benefits of computer aided tools can only be applied if the interplay of leveraging the 3D CAD model description in CAD in the context of appropriate CAPP activities will lead into the final processing of the specific CAM working steps. Hence CAM can also be seen as the final process step to convert a CAD design model proposal into a manufactural work piece on a specific tool machine as part of a general manufacturing process.

9.2.4 *What Are the Benefits of CAM?*

CAM systems for planning, preparing and creating the process for manufacturing CNC parts are much faster than when doing this work in the traditional way.

The processes of preparing the control program by means of a computer application and by manufacturing a desired part on a CNC machine represents the first advantage of CAM. The second advantage of utilizing a CAM system and CNC machines is that a higher precision can be achieved in part manufacturing. Without such an approach, it would be impossible in today's competitive business environment to accomplish manufacturing operations of such a high amount of new and modified products in a fast, accurate and cost-efficient quality-controlled manner.

In addition, the ability to create and analyse a virtual three-dimensional model of a complex part before production starts allows, in many cases, avoiding design and technological errors even at the stage of production preparation. In this regard, it can be concluded that a modern engineering company can competitively and successfully occupy its niche in the market if it meets the following three conditions:

- Reducing the period of preparation for production and launch of products on the market to a minimum.
- Achieving lower cost of production compared to main competitors in due course of production optimization.
- Ensuring "best competitive" quality products.

Further to this, engineers have a number of significant advantages by using CAM systems, compared with machines under manual control, when it comes to creating high-precision and complex parts:

- High speed in component production.
- Lowering GD&T (Geometric Dimensioning and Tolerancing) deviations and higher consistency with each component or finished product in a predictable deviation range.
- Reaching greater operational efficiency due to computer-controlled machine operation, which do not need to take breaks as physical machine fitters.
- Achieving high complexity machining operations and high operating times of machines.

There exist some limitations, however. CAM-enabled machines are usually designed for a specific task and are not incredibly versatile in adapting to product design modifications as part of the on-going product refreshing cycles. Hence it is necessary to deploy solid and extended library concepts for already used, modified, derived and totally new CAM control programs in the context of associated CAPP working plans.

9.2.5 CAM Technology and Process

This section explains how a typical design process works by using CAM systems and it shows the associated steps and features of simulating the part (workpiece) processing process. The main features of CAM systems in engineering are also considered.

9.2.5.1 CAM Technology Features

CAM systems are designed to automatically create control programs based on geometric information prepared within the CAD system, CAM systems offer the capability and choice of working tools loadable at a specific tooling machine, a range of physical and kinematic parameters (which could be loaded from pre-prepared libraries as part of the tool and tool machine management database) and the tool working paths as created in the CAM system itself. The main advantages for an engineer when interacting with CAM are the visibility of the work, the convenience of choosing a geometry, the high speed of calculations and the ability to check and edit the created tool paths.

Different CAM systems may differ from each other in scope and capabilities. For example, there are systems for turning, milling, woodworking and engraving. Despite the fact that most modern CAM systems are able to create control programs for any type of production, such a separation by field of application remains relevant.

If an engineer needs to use milling, then they need to purchase a milling module for the corresponding CAM system. If only turning is needed, then it is enough to purchase a turning module of the same system. The modularity of building CAM systems is part of the marketing policy of software vendors of CAM systems and allows manufacturing enterprises to save significant funds by acquiring only the necessary design and technological capabilities.

9.2.5.2 Typical Design Process with CAM

Now we shall take a look at a simple example of how an engineer works in a CAM system. Typically, such a process involves four main steps:

Step 1: The CAD engineer develops a three-dimensional CAD-model of the detail with certain parameters, material and other features which should all have already

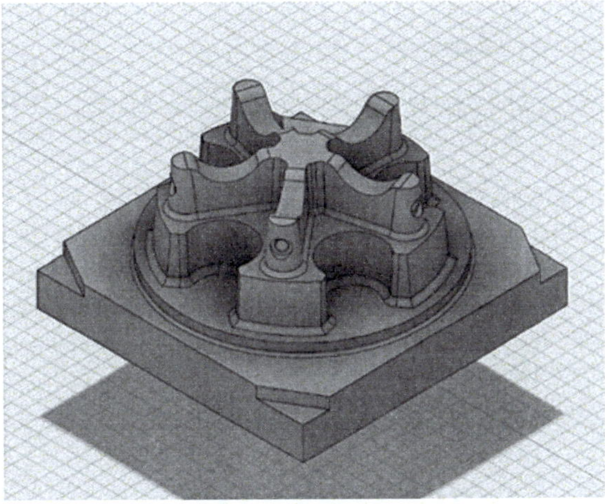

Fig. 9.9 CAD model of a component

considered Design for Manufacturing (DfM) rules. For instance, Fig. 9.9 shows a typical part considered to be manufactured on a CNC mill.

Step 2: The 3D model of the part is imported into the CAM system. The manufacturing engineer, who should have knowledge as production technologist and programmer, determines the surfaces and geometrical elements necessary for processing, makes the choice of the processing strategy as well as the cutting tool and finally sets the cutting mode. The system is then able to calculate the tool paths (Fig. 9.10, on the left side).

Step 3: In the CAM-system, a visual check of the arising trajectories is performed, the CAM programmer has the ability to quite easily correct errors that may appear, for example tool path correction or cutter change. In the Fig. 9.10 (on the right side), a detail within the production simulation is shown (potential clash). According to the

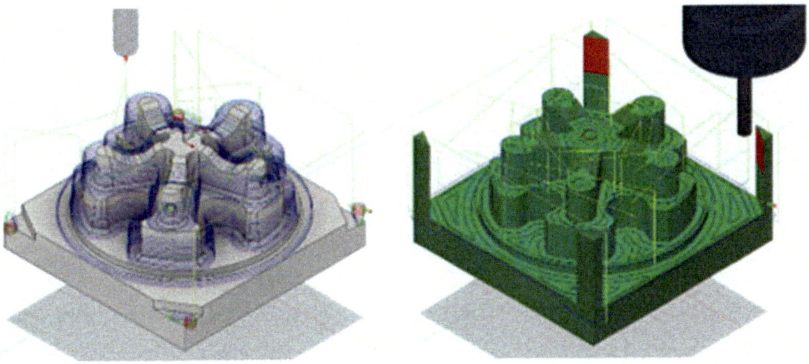

Fig. 9.10 CAM process features of a component

results of the simulation, the engineer can estimate the material removal and further optimize the process.

Step 4: The final delivery of the CAM system is the control program code. This code is created using the postprocessor, which in turn customizes the control program to the characteristics of a specific machine and the CNC system. Examples of such code are shown in Fig. 9.20. According to this code, the cutting head will pass from point to point throughout the entire workpiece, giving it the desired shape.

The post processor is a unique driver that converts the developed plan for the movement of the cutter (in the CAM program) and the technological commands into machine code. Such a code is developed in strict accordance with the capabilities of a particular CNC machine [8].

9.3 Numerical Control—NC

Numerical Control (NC) is an electronic method of controlling machine tools (CNC machines). CNC machines execute the individual processing steps for manufacturing automatically. The machining steps are defined in the NC program which is read by the CNC machine from a data carrier or data storage device (data base, server as part of a DNC network). Afterwards, the controller of the CNC machine evaluates and executes the work instructions.

9.3.1 Engineering Understanding of NC

This section explains why NC is used in production and what NC is doing for engineers. Moreover, it will be demonstrated how NC **programs** are used to manufacture individual workpieces and components.

9.3.1.1 Why Does an Engineer Use NC?

NC is a common control method in production because it leads to the following advantages:

- Automated, fast and high precise manufacturing
- Capability to automated recurring tasks
- Automatic generation of NC on the basis of CAD data
- Reduction of errors in the translation of CAD data into NC code due to automation
- Offering deeper understanding of the work process by graphical representation support
- Optimization of tool paths through upfront simulation
- Collision detecting using the NC program

- Conducting of NC programming comfortably in the office rather than at the tool machine on the shop floor
- Providing higher uptimes of CNC machines and higher overall productivity due to working on the NC program off-floor
- Offering free form surfaces manufactured through multi-axis control capabilities through NC.

Therefore, engineers can rely on NC technology for the physical production of a CAD model if all digital transformation steps and the right parameter settings in the NC code of a specific tool machine for a specific material can be ensured.

9.3.1.2 What is NC Doing for an Engineer?

Engineers use NC for automated, fast and high precise manufacturing of individual designed work pieces. Engineers do not have to control CNC machines during manufacturing. Numerical controlled machines execute machining steps automatically and can be adapted very quickly to another product by replacing the data carrier. The execution of NC programs can be repeated to manufacture additional work pieces without any further effort. The possibility of off-floor programming ensures that the machine tool is not blocked during the creation of NC programs, which enables a high utilization of machine tools. Moreover, engineers can create the NC programs automatically on the basis of CAD data. As a consequence, NC is nowadays used on almost all machine tools. Manual adjustment control of the machine itself is only used in machine experimentation and ad-hoc fabrication cases, complete shop-floor oriented tool machine programming is reduced more and more even in small enterprises.

9.3.2 How Does NC Work?

All work piece specific control information like path, feed and speed are written in the NC program. The NC program contains all necessary working steps block by block in the right order. There are different kinds of CNC machines and it is necessary to convert neutral code which is generated by CAM software to the exact code dialect used by a specific CNC machine. The NC postprocessor is computer software which converts general code of the geometry (CLDATA) to the code dialect for a specific CNC machine.

Cutter Location Data (CLDATA) is a programming language for NC processor output data which is standardized in DIN 66215 [9]. CLDATA describes the manufacturing operation in absolute terms. Every NC postprocessor can convert the standardized CLDATA files to CNC machine specific code. The structure of CLDATA is demonstrated in Fig. 9.11.

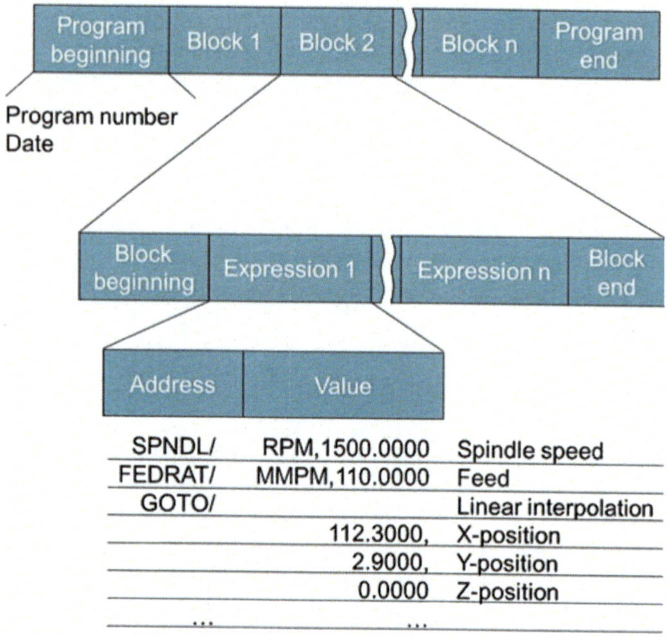

Fig. 9.11 Structure of CLDATA standardized in DIN 66215

The structure of machine specific NC programs is standardized in DIN 66025 sheets 1–3. Each block includes a number of expressions to parametrize different functions. The position information is stored as a numeric value for each axis. It is important to consider that it is the contour of the workpiece and not the tool path the one that is programmed. The tool path is calculated automatically at the machine. Figure 9.12 illustrates the structure of NC programs.

The NC program is stored on electronic data carriers or transmitted directly from the computer to the CNC machine. An industrial computer at the machine reads the NC program and executes the machining steps chronologically. An interpolation program in the CNC calculates intermediate positions because the path does not have to be parallel to the axes. The interpolation points set the position values for a simultaneous movement of all needed axes.

Closed position control loops are used to control the positions of the axes precisely. The actual positions of the axes are measured continuously. The differences between the position set values and the actual positions are amplified by the position controller and used to regulate the motors. Figure 9.13 demonstrates the functionality of the closed position control loop of a translational axis.

The relative movement between the work piece and the tool is used for automated machining of individual designed work pieces.

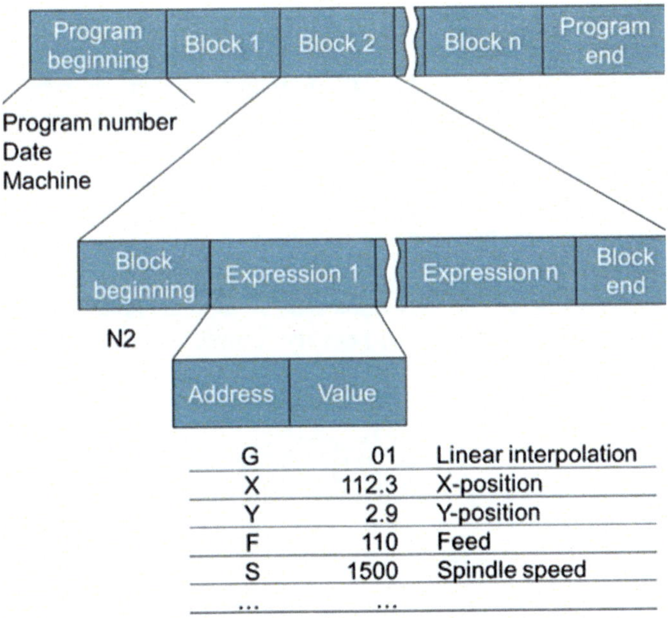

Fig. 9.12 Structure of machine specific NC programs standardized in DIN 66025 (representation based on [10], p. 525)

Fig. 9.13 Functionality of the closed position control loop (representation based on [10])

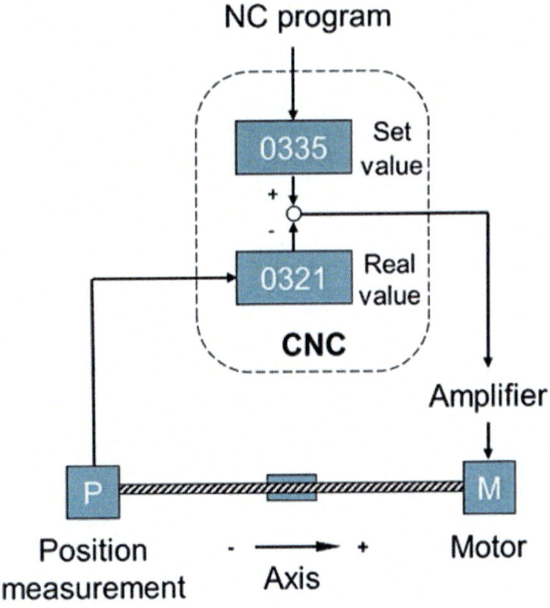

9.3.2.1 Where in the Product Development Process is NC Used?

CAx software is an integral part of today's product development process. Every product starts with an idea. CAD, CAE, CAPP, CAM and NC are used to turn the idea into a real product.

The first step in the process chain is design and development. In design and development CAD and CAE are used for design, digital mockup, simulations and optimizations. Afterwards, CAPP and CAM are utilized in production planning (Manufacturing Engineering) to assign resources, schedule operations, select tools and create NC programs. The N programs are transferred to the production and NC is used to execute the manufacturing steps automatically. The linkage and application areas of CAx systems in product development are shown in Fig. 9.14.

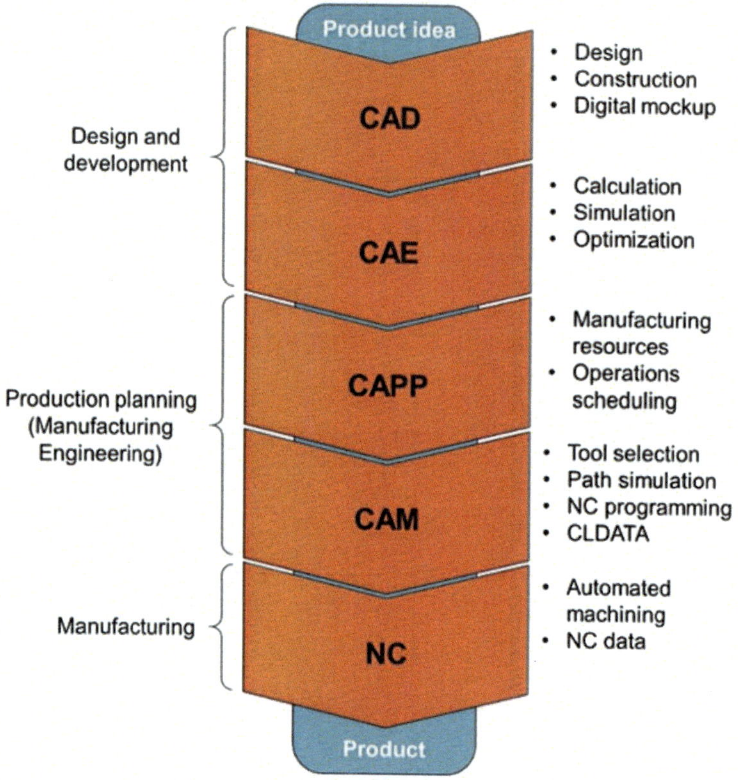

Fig. 9.14 CAx linkage in the product development process (representation based on [3])

9.3.2.2 What Is the Application Area of NC?

There is a wide area of application for NC. NC programs can be used to control and automate processing steps for manufacturing, assembly, equipping and inspection. Machine tools that use NC are for instance milling machines, lathes, drills, sawing machines, grinders, laser machines, 3D printer as well as combinations of these machine types.

9.3.2.3 NC Programming Technology

The following subsection will explain the features of NC and the NC programming process. NC program examples for a lathe process (see Fig. 9.19) and a milling process (see Fig. 9.20) are shown to demonstrate the standardized data input format of NC programs.

CNC machines are freely programmable machines, which usually consist of a combination of translational and rotational axes. Each axis is equipped with an electronic position measuring system and has a controllable drive. The measurements of the axes take place with a resolution of 0.0001 mm or 0.00001° and even finer to achieve precise work piece surfaces. The motion sequences and the technological information like feed (F), speed (S), tool (T) and miscellaneous functions (M) such as tool changes are specified in the exchangeable NC programs. Complex manufacturing steps such as high-speed milling would not be possible without NC [10].

There are three different control modes for NC as shown in Fig. 9.15. Point-to-Point control is only used for positioning when no tools are utilized. All programmed axes start simultaneously at rapid traverse until each axis has reached its target position. It is the fastest way possible to reach a certain position. With the line control, one individual axis can be traversed at a defined feed. So, the path is always parallel to the axes. The third and most relevant control mode is the continuous path control. Any two-dimensional and also three-dimensional path can be realized with the continuous path control. The movements of two or more axes are synchronized by the use of interpolation points in order to achieve the smallest possible deviation from the programmed path.

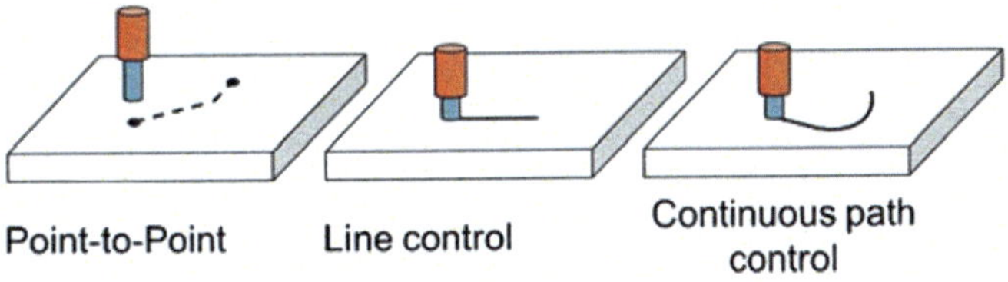

Fig. 9.15 Three different path control modes for NC

The NC technology enables:

- Easy program shifts
- Easy program modifications
- Short set-up times
- High flexibility
- High manufacturing accuracy
- Direct use of CAD data for programming [10].

9.3.2.4 Process Chain for a Milling Process

The geometries of the designed workpiece and the blank are created in a CAD system. CAD-NC modules or independent NC systems are used to generate the process steps by taking into account the tool geometry and technological information as cutting speed. Subsequently, the tool path is created and checked if necessary. A NC postprocessor is used to convert the machine-neutral tool path (CLDATA) into a machine-specific NC program to take into account the specific geometry, kinematic and switching functions of the CNC machine. In the last step of the process chain, the NC program is transferred to the milling machine to manufacture the workpiece. Figure 9.16 shows the NC process chain for a CNC milling process.

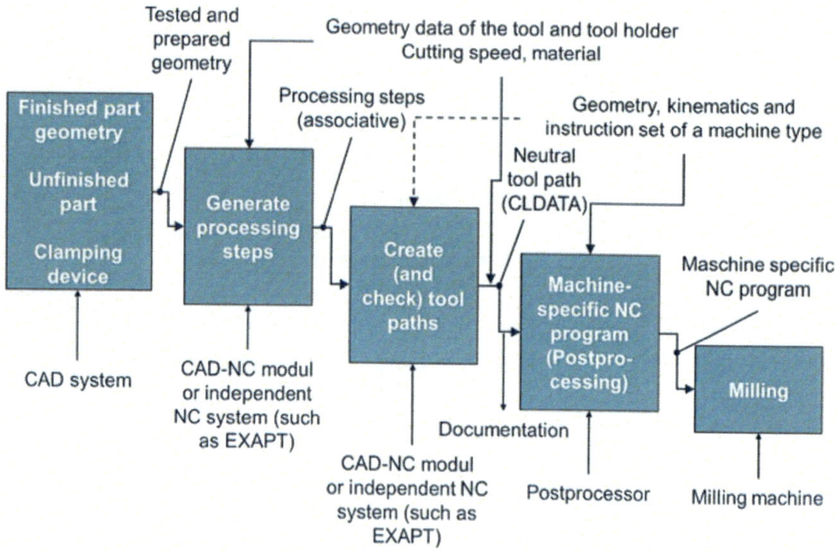

Fig. 9.16 Process chain for a CNC milling process

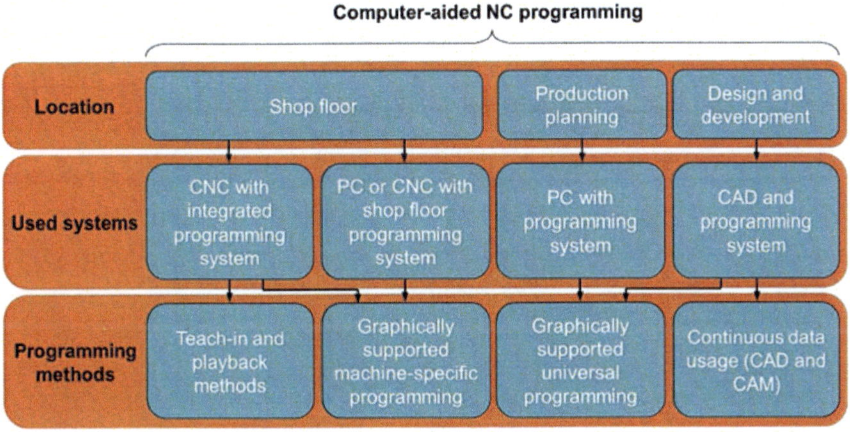

Fig. 9.17 Types of computer-aided NC programming (representation based on [10])

9.3.2.5 Typical Programming Process with NC

It is important to know that all common ways to create NC programs are computer-aided. There are different ways for computer-aided NC programming which can be distinguished by location, used systems and programming methods as shown in Fig. 9.17.

One option is to create the NC program directly on the shop floor. CNC machines often include an integrated programming system which can be used to create and edit NC programs. The advantage of programming on the shop floor is that the workers constantly monitor the progress in production, so, they can use their expertise to eliminate mistakes and optimize the process themselves. In order not to block the CNC machines during programming, in most cases, the NC programs are created externally and transferred to the CNC machine via a storage medium or a network connection.

In addition to shop floor programming, the NC programs can already be created in design and development or production planning. In these cases, the NC programs can be generated directly on the basis of CAD data. So, there is no need to create technical drawings and instructions for the operators as the NC program contains all the necessary geometric and technological information. It should be noted that these generated programs use a universal code that needs to be translated to the machine specific code dialect by means of a postprocessor.

9.3.2.6 Linear and Circular Interpolation

NC programs are not created manually and even changes are not inserted manually in the machine code these days. However, it may be advantageous to know and

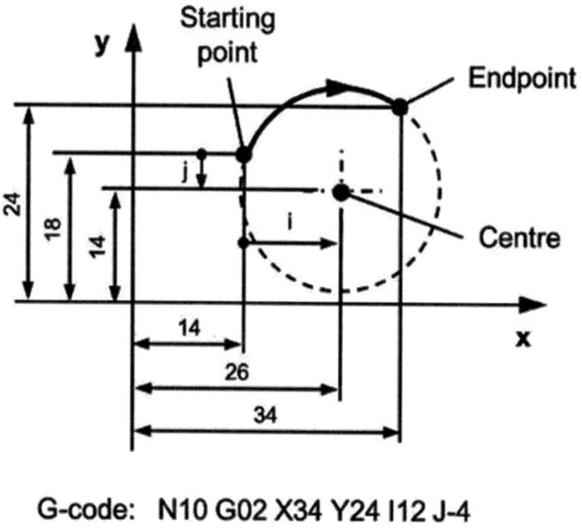

Fig. 9.18 Programming circular interpolation

understand the standardized data input format of NC programs because it is used by almost all modern CNC machines.

Important tasks are linear and circular interpolations. The command G01 is used for a linear interpolation between a start and end point. A high number of intermediate points are calculated automatically in order to get an accurately linear path. Circular interpolations are programmed with the commands G02 (clockwise) and G03 (counter clockwise). A programing example for a clockwise circular interpolation is illustrated in Fig. 9.18.

9.3.2.7 NC Code Examples

In the following two NC program examples are shown to demonstrate the standardized data input format G-code (standardized in ISO 6893) of NC programs. Figure 9.19 shows the G-code for a lathe process. When creating NC programs for lathes, it is important to consider that the X-values are always programmed using diameters. Figure 9.20 shows the G-code for a milling process.

9.3.2.8 Step-NC

The most common programming language for CNC machines (G-code) has remained essentially unchanged since the early 1950s. Back then, paper tape was the most popular medium for data transfer between computers. Although the capabilities of computers and CNC machines have improved considerably since then, the same programming language is still used to control CNC machines [11].

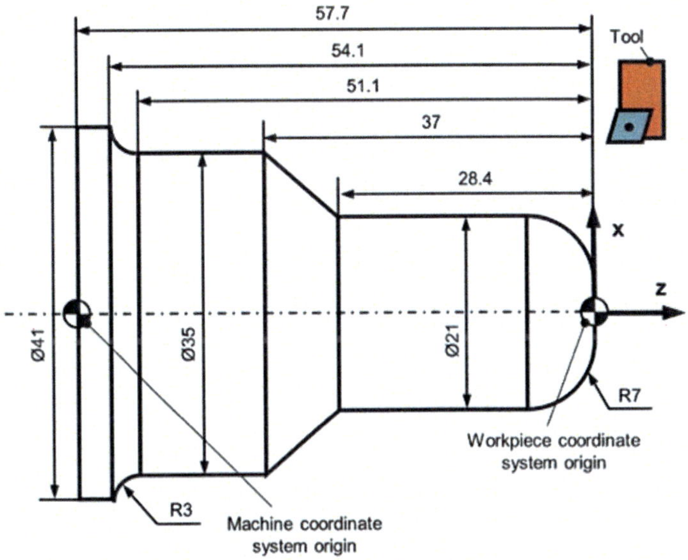

Block						Description
N1	G54		Z57.7			Offset
N2	G90					Absolute positioning mode
N3	M04				S600	Spindle start counterclockwise
N4	G00	X-4	Z4			Rapid to starting point
N5	G01		Z1		F500	Prepositioning
N6	G02	X-2	Z0	I1	K0	Tangential approach
N7	G01	X7				Move along contour
N8	G03	X21	Z-7	I0	K-7	
N9	G01		Z-28.4			
N10		X35	Z-37			
N11			Z-51.1			
N12	G02	X41	Z-54.1	I3	K0	
N13	G01	X60				Tangential move off
N14	M05					Spindle stop
N15	M30					End of program

Fig. 9.19 G-code example for a lathe process

The control language STEP-NC (standardized in ISO 14649) was developed to replace G-code with a modern, associative communication protocol that connects the process data with the product description of the component. The control language STEP-NC is therefore not limited to axis movement commands of the machine tool. The machine tool can be provided with information about the desired result of the machining. The use of STEP-NC is intended to enable faster, more accurate and more autonomous machine tools that can access product and process models [11].

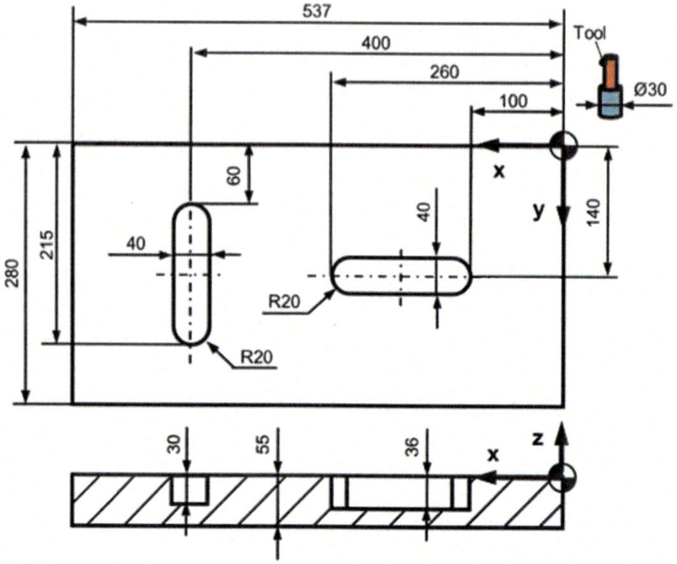

Block							Description
N1	M03					S1000	Spindle start clockwise
N2	G41						Start cutter radius compensation left
N3	G00	X181	Y159	Z2			Rapid above groove
N4	G01			Z-36		F1500	Move to cutting depth
N5	G03	X180	Y160		I1	J0	Tangential approach
N6	G01	X120					Move along contour
N7	G03		Y120		I0	J-20	
N8	G01	X240					
N9	G03		Y160		I0	J20	
N10	G01	X179					
N11				Z2			Move off
N12	G00	X419	Y139				Rapid above groove
N13	G01			Z-30			Move to cutting depth
N14	G03	X420	Y140		I0	J1	Tangential approach
N15	G01		Y195				Move along contour
N16	G03	X380			I-20	J0	
N17	G01		Y80				
N18	G03	X420			I20		
N19	G01		Y141				
N20				Z20			Move off
N21	M05						Spindle stop
N22	M30						End of program

Fig. 9.20 G-code example for a milling process

References

1. Deja M, Siemiatkowski MS (2018) Machining process sequencing and machine assignment in generative feature-based CAPP for mill-turn parts. J Manuf Syst 48:49–62. https://doi.org/10.1016/j.jmsy.2018.06.001

2. Yusof Y, Latif K (2014) Survey on computer-aided process planning. Int J Adv Manuf Technol 75:77–89. https://doi.org/10.1007/s00170-014-6073-3
3. Kumar SPL (2017) State of the art-intense review on artificial intelligence systems application in process planning and manufacturing. Eng Appl Artif Intel 65:294–329. https://doi.org/10.1016/j.engappai.2017.08.005
4. Al-wswasi M, Ivanov A, Makatsoris H (2018) A survey on smart automated computer-aided process planning (ACAPP) techniques. Int J Adv Manuf Technol 97:809–832. https://doi.org/10.1007/s00170-018-1966-1
5. Wang HP, Li JK (1991) Computer-aided process planning. Elsevier Science, Oxford
6. Rao PN (2004) CAD/CAM: principles and applications. Tata McGraw-Hill, New Delhi
7. Alavala CR (2015) CAD/CAM: concepts and applications. PHI Learning Private Limited, Delhi
8. Kief HB, Roschiwal HA (2007/2008) NC/CNC Handbuch 2007/2008: CNC, DNC, CAD, CAM, CIM, FFS, SPS, RPD, LAN, NC-Maschinen, NC-Roboter, Antriebe, Simulation, Fach- und Stichwortverzeichnis. Hanser, München
9. Deutsches Institut für Normung (1974) DIN 66215-1: programming of numerically controlled machines; CLDATA, general structure and record types. https://doi.org/10.31030/1301864
10. Kief HB, Roschiwal HA, Schwarz K (2025/2016) CNC-Handbuch 2015/2016: CNC, DNC, CAD, CAM, FFS, SPS, RPD, LAN, CNC-Maschinen, CNC-Roboter, Antriebe, Energieeffizienz, Werkzeuge, Industrie 4.0, Fertigungstechnik, Richtlinien, Normen, Simulation, Fachwortverzeichnis. Hanser, München
11. Hardwick M, Zhao YF, Proctor FM et al (2013) A roadmap for STEP-NC-enabled interoperable manufacturing. Int J Adv Manuf Technol 68:1023–1037. https://doi.org/10.1007/s00170-013-48940

Chapter 10
Major Technology 4: Computer Aided Engineering—CAE

Executive Summary

This chapter deals with the following topics:

- Basics and advanced techniques of Computer Aided Engineering (CAE)
- Providing insight into how engineers benefit from using CAE technologies
- Describing functioning, benefits, and limitations of CAE technologies in practice.

Quick Reader Orientation and Motivation

The intention of this chapter is:

- to give an overview of CAE technology in Virtual Product Creation as driver and enablers for Digital Transformation in engineering
- to present CAE technology as part of Virtual Product Creation from a practitioner's point of view to analyze the need and usefulness for day-to-day industrial work practice
- to give instructions on how to use CAE technology
- to explain models, frameworks, and
- representations that help to grasp the internal working modes of CAE technology.

In modern virtual product creation, functional and behavior simulations of products (and associated services) play a foundational and constantly increasing role. The need for not only creating a virtual model, but also "studying its behavior in reality" as well as "improving and streamlining its structure" [1] motivated scientists and engineers to develop advanced methods and software tools. Whenever a product has to be designed, there exist a long range of different kind of functional and behavioral requirements this product needs to fulfil. Furthermore, the entire manufacturing process is highly influenced by the shape and assembly structure of the product. Therefore, it is vital to predict if the product behaves as functionally intended, but also how it can be manufactured and assembled. Computer Aided Engineering (CAE) designates the state-of-the-art tool and methods which are available to conduct such analysis and prediction tasks.

R. Stark, *Virtual Product Creation in Industry*,
https://doi.org/10.1007/978-3-662-64301-3_10

10.1 Background and Evolution of CAE

Managing and controlling execution efficiency and availability of resources as part of virtual product creation is a major task and delivery of PLM (Product Lifecycle Management) IT solutions. Hence, it was introduced broadly in industry and became indispensable for virtual product creation in industry during the last two decades. CAE can be considered as an integral part of PLM and Virtual Product Creation (compare Chap. 4 *"Virtual Product Creation—what is it"*) that provides methods for engineers to simulate a product's behavior under real conditions. Originally, *CAE was a term used to describe the procedure of the entire product engineering process, from design and virtual testing with sophisticated analytical algorithms to the planning of manufacturing*. However, it became clear after some years, that the nature and diversity of both, model and file management as well as analysis and simulation tasks, do differ significantly from those of the design and product structure, BOM (Bill of Material) and the manufacturing process. Consequently, the terms *PLM* and *Virtual Product Creation* were created to extend the original CAE procedure idea to the full landscape of the entire product lifecycle (compare Chap. 4 *"Virtual Product Creation—what is it"*).

With the rise of powerful computers in the late 70s, it became increasingly possible to calculate large numerical problems. The development of the Boeing 777 in the early 1990s can be seen as one of such corner milestones. This was the first extremely complex product entirely virtually designed, where also a digital mock-up (DMU) was developed (compare Chap. 12 "Digital Mock-Up"). In automotive industry, e.g., extensive CAE simulation was driven by highly increasing occupant safety requirements, which led to early technology, compute centers in the 70ties and 80ties and then to a major thrust in the 90ties by intruding powerful UNIX workstations and Cray supercomputing.

Today's CAE software landscape is partially traceable to the development of application-specific tools by specific corporations within the IT and PLM vendor market place. Unlike in the past, even large OEM corporations no longer develop their own CAE codes but rely on the tool competence of IT and PLM vendors. Interestingly enough, there exist still today many hidden CAE kernel applications which were programmed initially by universities institutes or by CAE expert teams of companies in the 70ties, 80ties and 90ties of the twentieth century: some of them remain within their originally code (e.g. Fortran) and would need major refactoring in order to be transferred to modern software code architectures.

Today there also exist highly specialized service providers, who concentrate on the development of CAE software for different technology branches, without actually being active in product development themselves.

10.2 Engineering Understanding of CAE

The use of computer-assisted simulation reduces financial risks, which are generally connected to product development. According to [2], product development from a corporate perspective is mainly centered on the action of making an investment, and the corresponding expectation of future profit. CAE promises an efficient and goal-oriented mode of working, as well as monitoring and increased control over the development process.

10.2.1 Why Does an Engineer Use CAE?

Unlike in the past, when engineers were dependent on building physical prototypes in order to test the future final product properties and behaviors, engineers nowadays can rely on the capabilities of CAE solutions to simulate upfront the relevant product functional performance capabilities. Such front loading of computational engineering capabilities with the help of CAE solutions helps engineers to avoid unnecessary and costly engineering iteration cycles and costly modifications to physical prototypes (see Figs. 10.1 and 10.4).

The process of product development begins with the idea and its drafting in CAD as depicted in Fig. 10.1. To reach the final goal of physical manufacturing, the product needs to be designed and evaluated in terms of all critical requirements such as static loads, durability, heat resistance etc.). Without the use of CAE, calculations were traditionally done manually (e.g. based on algebraic mathematics) incl. heuristics and experiences from respective literature or company knowledge. Once all preparation methods have been completed, a prototype is made, which is then tested under realistic conditions. The insights gained here then possibly result in the prototype being redesigned, in which case the iterative design process would start from the beginning until all conditions are satisfied before then serial production can begin. In such traditional process, all tests had to be done with the physical prototype, which has the following disadvantages:

- Significant time delay due to the lead-time to manufacture and assembly physical prototypes.
- Costly operations to procure and product all physical components.
- Significant efforts and time delays to loop back findings of the test results at the physical test stand into the digital master file (CAD).

The early use of CAE (see Fig. 10.1 on the right side) significantly improves the conception and embodiment phase prior to prototyping, as well as speeding up the entire process. The calculation and behavioral prediction in realistic conditions allows for optimization before the prototype is even created. The numerical methods used by CAE can make assertions that are more precise and that help to check and verify individual scenarios and circumstances early on. It is, however, necessary to be

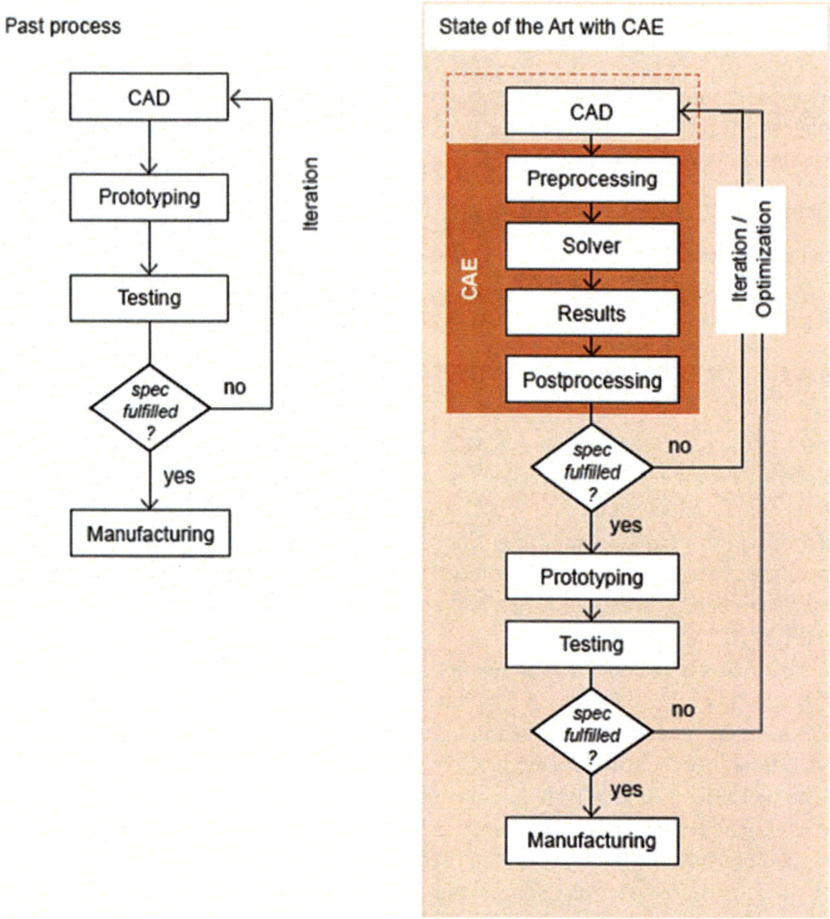

Fig. 10.1 Front-loading with upfront CAE to reduce iterative engineering cycles

proficient in all CAE methodological steps and procedures (such as all four steps in the orange box: pre-processing, solver and simulations control, results management and post-processing). The earlier insights can be generated how a component or specific feature will function within a construction unit, the more effective the development process can be (see the principle in picture in Fig. 10.2).

Timely knowledge prevents making last-minute stressful and costly changes in the end, which could even impact on other areas of the project, too. An analysis of the dependencies and interconnectedness of components as part of the CAE model is important to be determined early on.

Early decisions regarding the general direction or the type of problem-solving approach (compare [3]) affect more than 80% of the product life cycle costs. This is why it is important to be able to predict costs and technical feasibility (function,

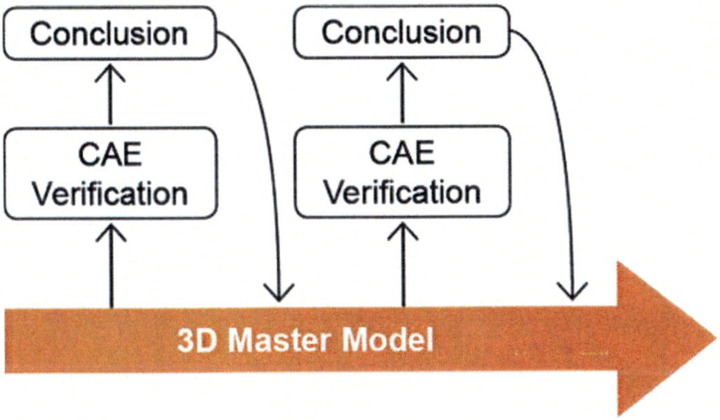

Fig. 10.2 Effect of CAE conclusions on the product development process

behavior, performance, manufacturability etc.) of a product concept early on. The growing complexity of technical systems, as well as the desire for higher levels of employee productivity, require increasingly more powerful digital and virtual solutions based on IT resources (compare Chap. 5 *"The technology history of Virtual Product Creation"*). The continuous advancement of information technology with faster and more robust digital processes, as well as higher degrees of virtual model details and network data rates, provide businesses with innovative solutions to build and use simulation models. The potential for numerical calculations of scientific or technical problems is increasingly being discovered and will continue to evolve. Simulations follow the idea that *a complex problem reduces to a series of greatly simplified problems* [4]. For this reason, the computer simulation of such processes and engineering problems is particularly well-suited and has meanwhile reached a full mature level. Therefore, CAE has been evolving from an exclusive expert skill set to a widely used engineering capability across different degrees of engineers, designers and analysts.

10.2.2 What is CAE Doing for an Engineer?

CAE is meanwhile used as a standard engineering validation and verification solution in different industries and technical applications: in classic machine-construction industries, in automotive, aviation, aerospace and maritime industries, a diverse set of products such as vehicles, aircrafts, ships, machine tools, pumps etc. a high variety of CAE analysis templates exist. CAE is used both, for the product development itself, as well as for production planning of the products. Production planning stretches to industrial areas such as material sourcing, storage, logistics and disposal. This is why product data management (PDM) and product lifecycle management (PLM) are becoming increasingly important for manufacturing enterprises (compare the

Sect. 5.3 *"Product Data Management"* and Chap. 11 *"Product Data Management and Bill of Materials"*). The result is a desire for more integrated software solutions that can deal with the wide spectrum of tasks [5]. A market overview for CAE software is for instance provided by [6].

The CAE software application landscape and its associated model fundamentals are diverse and offer different technology foundations. Due to the fact that several scientific disciplines and technical branches have developed their own solutions sets based on individual demands for numerical calculation methods, different CAE product and research prototypes have been developed and are still under new development. Figure 10.3 illustrates the different CAE disciplines and simulation types:

As shown in Fig. 10.3, structural analysis is one of the major CAE disciplines. This field of expertise is focused on the simulation of components or structure regarding specific physical phenomena. This can include analysis of components under static load, acoustic analyses, and questions regarding thermodynamics, fluid mechanics or electromagnetism. Here it is important to document the behavior of a material or a continuum in a particular state of aggregation. The structures are generally set in a three dimensional space, meaning that the calculations and models are based on a 3-D case.

In solid mechanics, materials are generally analyzed that correspond to the Hook's law. Distortions in the purely elastic area can be solved with linear numerical methods. Plastic distortions can be calculated within limits in a reasonable way. However, outside those limits non-linearity of material laws lead to a significant increase of the associated calculation efforts. In fracture mechanics other laws apply, which require their own specific methods. Variables to be calculated include tension, elongation and displacement. For solidity analysis, usually the *Finite Element Analysis* (FEA)—here the mathematical principles of virtual displacements of small finite elements are applied—is used, but the *Boundary Element Method* (BEM)—using the mathematical equations of integrals—can also be leveraged if the volume is rather thin.

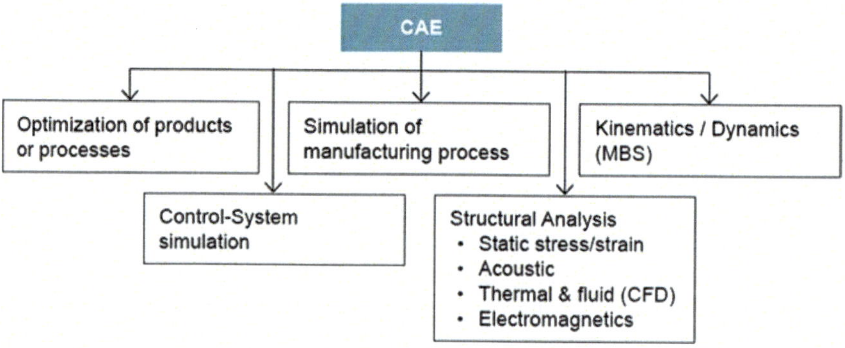

Fig. 10.3 The main disciplines of CAE

Closely combined with the solidity analysis, is the *eigenmode* analysis. The goal is to determine the *eigenfrequency* and the *eigenmode* of a structure. Resonance vibrations present a common problem in machine construction, as many systems are subject to the vibration generated by internal and external forces. This is where FEA or BEM are then employed.

In the numerical *Computational Fluid Dynamics* (CFD) fluids and gases are analyzed. State variables include pressure, flow velocity and density, among other things. The base equations (typically modelled after Navier–Stokes) are calculated or approximated using the *Finite Difference Method* (FDM), the *Finite Volume Method* (FVM)—both are based on the mathematical principle of differential equations—or the FEA approach.

In the area of thermodynamics, processes are examined where energy, in the form of heat, flows through a medium and is radiated off or transferred into a different form of energy. There are close correlations to computational fluid dynamics, as warming processes are often viewed in terms of fluid dynamics. One application is the combustion process in a gas motor, for example.

With numerical methods, electromagnetic effects are also studied. Calculated state variables may be electric field strength, eddy currents, etc.

In the subdomain of kinetics and dynamics the movement of components or assemblies and the dynamic forces and momentum being created are analyzed. The technical term for this area is called *Multi-Body Dynamics* or *Multi-Body Systems*. With the control systems simulation, technical systems are viewed on a global scale, and the flow of energy and materials is analyzed. With help of the physical and control technological dependencies between components of a system, a kind of circuit diagram is established (not a spatial diagram), which is why the field is also called a 1D simulation. The simulation of manufacturing processes and the optimization of products or processes helps itself to methods and technologies from the CAE subdomains, but can be viewed as separate branches due to their content and praxis (compare Chap. 9 "*CAPP, CAM and NC Technology*"). For example, the simulation of a welding robot may include the structural analysis (in the form of thermodynamic analyses) and kinetics (for the movement of the robot). Optimization processes can be applied to structural analyses as well as control systems.

In numerous technical applications the movement of a body in space, and/or the relative movement of components in relation to one-another one are subject of analysis. Such components can be connected via joints, springs or dampers. Force, momentum or acceleration occur which are dependent on the movement, or affect the multi-body system. Sketching out behavior of such kinetic or dynamic systems is the goal of the *Multi-body Simulation* (MBS). The objects can ideally be shown as rigid or flexible. In the kinetic simulation, there are globally no open degrees of freedom, but with a dynamic one, there are. The latter is numerically more difficult to solve.

By following the three-dimensional modeling of the system in the CAE software, the physical attributes and dependencies can be defined. The corresponding material attributes, connections, starting and border conditions (forces, momentum, bearings/support) are defined. As such, a system of equations can be established and calculated.

The *multi-domain and control systems simulation* is used when energy or material sizes need to be viewed on a global scale. In technical systems, the connections between individual components are shown as a circuit diagram. Cause and effect are clearly defined [7]. As the structure of a three-dimensional component is not directly considered unlike to the previous simulation varieties, only the physical relationships between the actors are considered, the method is also known as a 1D CAE simulation. When dependencies between different physical quantities can be generated, the simulation is called a multi-domain simulation.

With the help of the above described various CAE methods Engineers can create digital prototypes in order to reduce or totally replace physical prototypes. In the world of technical system development, physical prototypes serve traditionally as:

1. Selected models or artefacts to prove out (*validate*) constructional, functional and behavioral ideas and intentions with the help of their special physical realization (in non-production mode)—in automotive industry, e.g., the terms *working horse* or *mule* exist for early full vehicle prototypes.
2. A rather complete set of physical systems and components at a certain development gateway stage (mostly prior to and shortly after the production release gateway) under almost production ready circumstances in order to *verify* a determined set of constructional, functional and behavioral attributes of the product according to a sign-off list of requirements and engineering/ performance targets.

A physical prototype, therefore, might still be needed despite of powerful CAE simulations for a number of reasons: especially in early CAE capability ramp-up phases in enterprises "invisible" human errors in CAE model build and simulation set-up and execution might occur during engineering development. This can lead to late surprises such as component collisions, fatigue problems, thermal failure etc. Due to late notification until physical prototype testing cost intensive rectifications is then unavoidable and causes "unnecessary" churn, readiness delays and on-cost. Another source of error could be the simulation itself: if a mistake is made in the model assumption or the planned calculation procedures of the component or product environments, this fault can distort the entire findings of the CAE project. This is more often the case than expected, since realistic conditions can be best anticipated as outer boundary conditions for simulation, but they cannot be perfectly imitated due to limited knowledge about complex physical interrelations. Therefore, CAE needs careful verification of the used model types. Another common reason for still constructing a prototype is the desire to see and feel the actual product with all human senses. This can affect the tactility of the product, or the comfort of a product (like a seat). Current VR technologies do attempt to bridge this gap in a virtual space, but options are still limited regarding other human sensibilities such as force feedback and tactile experiences.

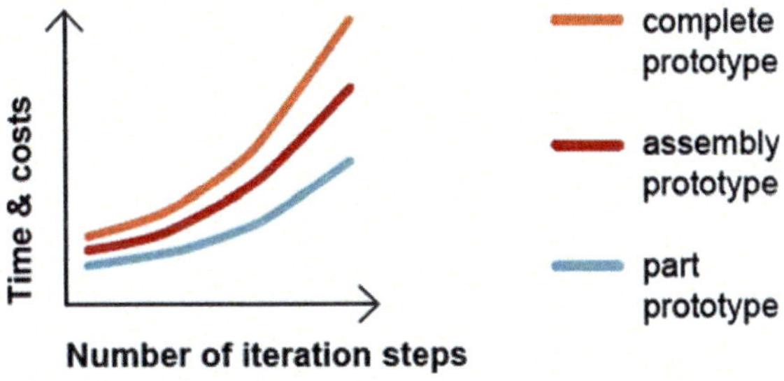

Fig. 10.4 CAE target to reduce development costs and expensive physical prototypes

Nevertheless, the better a CAE simulation of a component or product is, the less reliant engineers have to be on physical prototypes. This is desirable as a physical prototype can be expensive and time-consuming, as qualitatively shown in Fig. 10.4.

Each iteration of the prototype costs additional time and budget. For example, if a company considers the development of a new train generation, it is hardly financially possible to build more than one prototype of such train, before arriving at the final product for the customer. In aviation industry even the first full flight prototype is also finally sold to a specific customer after integrating all updated packages as part of the development completion. For substructures, such as seats or armrests, a higher number of physical prototypes might be possible. Thus, the CAE is in a position to replace physical prototypes by digital ones and consequently helps engineers to improve product development on multiple levels by realizing a systematic validation and verification of components, sub-systems products and complete technical systems consisting of multiple, interacting products.

10.3 How Does CAE Work?

With the help of CAE and its simulation techniques, the behavior of a (technical) system that either already exists or is under development can be analyzed with respect to certain system or product attributes/properties. The simulation can be understood as an experiment on a digital model: the results of such "simulation experiment" can be used to drive conclusions regarding the behavior of a real technical system (product, machine, production system etc.). In order to trust the outcome of a specific CAE simulation, it is necessary that the underlying digital model type and the simulation algorithms have been generically validated against tested behavior of an equivalent physical realized technical system or product.

Please note below the major ten steps of a successful CAE analysis project:

1. Detail scope analysis of the (real/physical) technical system set-up and clarification of the technical system performance targets.
2. Determination of the purpose of the CAE analysis (and simulation types) and clarification of the intended simulation type requirements.
3. Selection of the appropriate CAE discipline(s) (compare Fig. 10.3).

4. Digital model formulation (incl. usage of the correct model elements).
5. Digital model generation and implementation (model build).
6. Verification of the digital model according to certain model criteria.
7. Simulation run, i.e. conduction of the simulation on an appropriate computer or computer cluster.
8. Validation of the simulation results according to engineering and model theory knowledge.
9. Evaluation of the simulation results towards the technical system performance targets (with potential modification proposals to improve the original digital model, e.g. CAD model).
10. Closure and book shelfing: report out to and discussion with engineering partners and stakeholders, documentation of the results (incl. design modification proposals) and lessons learned conclusions with potential improvements of CAE analysis procedures.

Within the first step, the scope analysis of the technical system, the subdivision of the system into its sub-systems and components takes place. This step is often directly combined with or preceded by a formulation of the requirements and should determine what will be depicted for which purpose. Following the determination of the CAE analysis type and its underlying core method/procedure in steps 2 and 3 it will be decided which results are expected in which type of format. As a result, these decisions largely determine which objects receive black-box characters, as they may require special control algorithms, which the analyst and costumer of the CAE analysis project may not need or want to view.

In steps 4 and 5 the active choice needs to be made which model elements should be used within the model formulation and how exactly those model elements need to be linked to each other in order to achieve most realistic simulation behaviors of the digital model. In addition, the corresponding assembly model is determined as part of all connected component models. At times, there is academic differentiation between theoretical and experimental model building. The former describes characteristics of the system to be displayed by utilizing (physical) laws and hypotheses and corresponding mathematical formulation and equations. The experimental model development, however, concentrate on taking observations from physical test stand experiments into account, generally by using and incorporating measured test data into physical model assumptions. The test measurements are, however, generally approximated or fitted on a curve for simplified usage in order to enable the linking to the solving process of the theoretical models.

The implementation part of step 5 explains the conversion of a digital preprocessor model into a format that is machine-readable, i.e. this is the transition from a model preprocessor to the simulation model in a specific solver environment. Generally, specific for representational languages, procedures or even programs are used for it. This work step can efficiently decide which models are effective and feasible, using effective representation or programming methods. If models are not documented at all or are only poorly documented, the verification of the digital model in step 6 becomes tedious and time consuming.

Step 6, the verification of the digital model, is in fact the last stage where domain knowledge can and should be integrated into the digital model, these steps should, therefore, provide ease-of-use for the user and can help to avoid modeling errors. The more complicated and layered the data entry options are for the user, the more likely it is that errors can occur, which can lead to results that do not accurately reflect the behavior of the technical system to be represented. The split between usability and accessibility for less simulation-competent users, the desire for efficient programming, as well as shorter development times will always stay in competition to each other and should be clarified upfront during the requirements phase of the simulation project.

Step 7, the execution of the simulation, finally describes the experiment based on the implemented model. Generally, this involves the distribution of parameter values, which are subject to change in repeated instances. The solver environment steers the execution of solving all inherit mathematical equation tables in a timely fashion. Depending on the type of computer or computer clusters in usage it might take between seconds, minutes or hours (for complex problems with several hundreds of thousands or even millions of mathematical equations it might even take weeks). The results of a simulation are affected by the choice of the mathematical solution algorithm as well as by its parameters. These approaches should be documented in order to make the experiment reproducible. There exist multiple approaches to achieve simulation results. The simplest is to change the accuracy and the time step distance until no significant variations appear in the results. It is advisable to compare the results of different solution algorithms.

Step 8, the validation of the simulation results, represents one of the biggest challenges. It does offer, however, significant potential to improve the results of a CAE analysis project by validating the soundness and the correctness of the results. In the validation process, it is determined whether the results really accurately resemble the original system. The question is often difficult to answer with a simple yes or no. The simulation can never perfectly represent the behavior of the original systems, which is why deviations are often found. To assign limitations on deviations is usually difficult to prove. Often, a qualitative process for specific values over time with determined deviation limits can provide a good indicator for the validity of a model regarding specific characteristics even without performing a quantitative proof.

The evaluation of the simulation results (step 9) constitutes an extensive and oftentimes difficult task, since it is necessary to apply domain knowledge and CAE model build/simulation knowledge at the same time. Therefore, the CAE analyst needs to engage closely with the System, Component and Design Engineers in order to conclude meaningful results. Oftentimes it also remains invisible whether numerical particularities of the solving process itself (e.g. rounding occasions) might also have an influence on specific final results of structural analysis (such as stress and displacements), thermal analysis (such as temperatures or thermal flux) or acoustic and vibration analysis (*eigenvalues, eigenfrequency* etc.). In order to close the loop at the end of the CAE analysis (step 10) it is essential to include the final results of the CAE simulation run but also details from the preparation steps beforehand. The following result types are of high interest at the end:

- the official report and the associated oral explanation as part of the virtual product creation collaboration,
- the discussion with other experts and stakeholders,
- the creation of lessons learned on modeling practices and
- the final documentation of the entire simulation project.

This final part of the process is oftentimes neglected by CAE analysts in industrial practices as well as by researchers in science: leaving a gap-riddled documentation of the models and of the verification methods often leads to issues and impacts, when the models are to be reused or adapted for a different project.

The steps 3–6 are iterative in their execution (please also compare Fig. 10.5). The goal is to have the model describe the desired characteristics as accurately as possible, after all. As a result, the process should begin with simple models, which can be refined by comparing measured values, for example. This commonly requires detailed mapping of sub-components, but does not preclude the need for more specific parameter values.

10.4 CAE in Product Development

The calculation/simulation of technical systems takes place in one of the following three phases of product development, fulfilling a specific function within each one of those phases:

1. In the first phase, at the concept design phase, i.e. during the preliminary calculations to drive major physics of the design it is important to establish major requirements of a product (e.g. approximate number of components, required operating power, material alternatives etc.). As the design is generally not finalized during this stage, the calculation is often performed analytically along experiences or guidelines or with the help of concept CAE models. Figure 10.6 shows an example of a car body shell CAE model during the concept phase of the vehicle development. In order to support the target setting of the overall vehicle performance targets with respect to vehicle package, NVH behavior (noise, vibration, harshness) and crash, it is decisive to use *CAE concept models* in various degrees to determine major dimensions and topological design principles of the body structure. The upper part of Fig. 10.6 depicts a simple concept model whereas the lower part shows a refined one based on input from other different digital model sources.

2. When the product is (partially) constructed and it is necessary to make decisions about component separation as part of the embodiment development phase, the concept design is transformed into a proper system, product and component design. This second phase needs the help of specific CAE verification calculations. Within this context, the use of CAE software tools becomes mandatory,

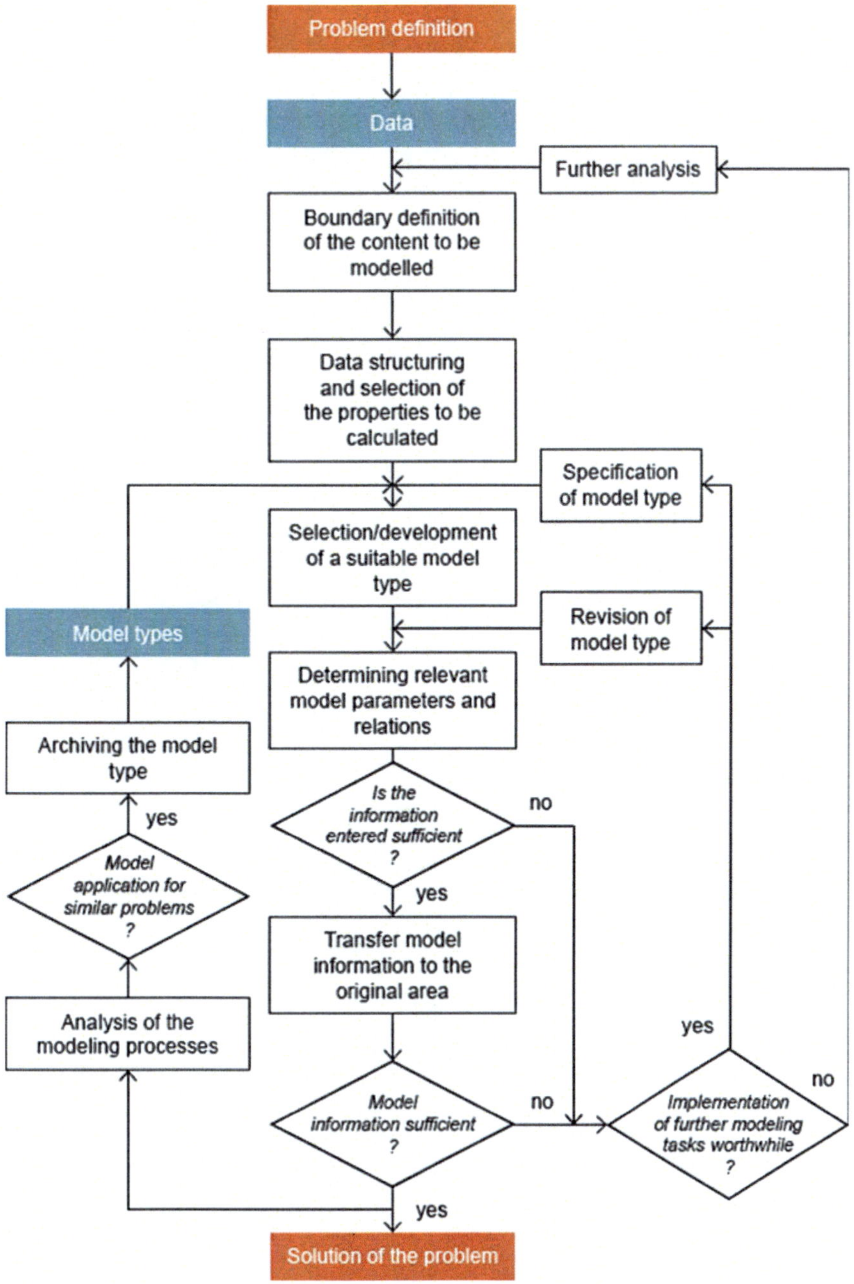

Fig. 10.5 Iterative flow of activities in order to achieve a robust CAE model build (formulation, generation and implementation)

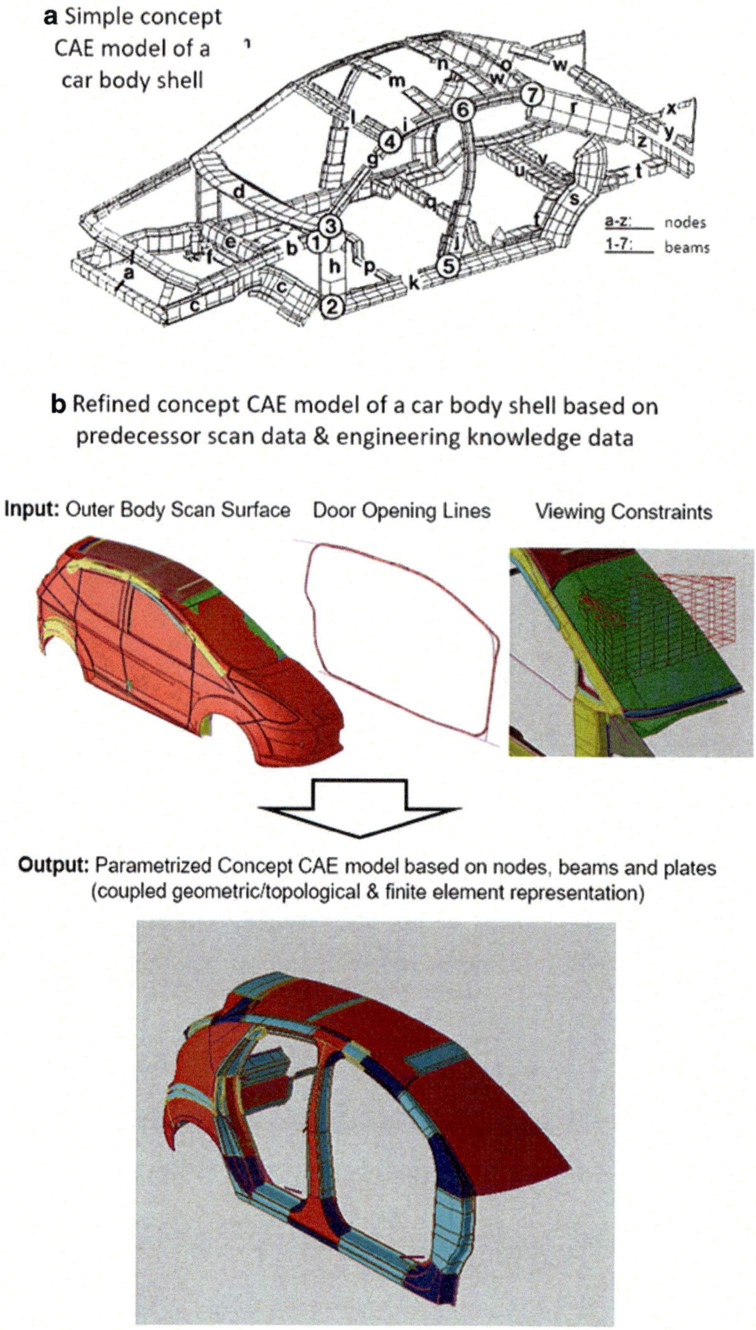

Fig. 10.6 Different types of concept models as part of car body engineering

as the overall technical and/or product system is already too complex for traditional analytical calculations. Oftentimes design details and crucial development decision scenarios need the extensive support of CAE analysis. Figure 10.7 shows examples of a detailed body shell FE (finite element) mesh model (a), the connecting FE assembly of an engine mount mesh model to a rail assembly mesh model (b)—both are used for vehicle crash simulation purposes)—and detailed FE mesh models for the durability analysis of a wheel hub (c).

The third development phase of CAE calculation is driven by optimization goals. If a product does not yet fulfill certain product performance targets, or if there exist ample potential for design characteristics improvement, specific CAE analysis support is desired (please compare example in Fig. 10.8).

10.4.1 From CAD to CAE—CAE Model Build

In classical virtual product creation process, the 3D design is developed with the help of CAD software. All relevant product, design, function and manufacturing information is stored in the CAD model or as associated meta data (as part of the data storage environment such as PDM). That includes the design (geometric information), normal or supply components (screws, glue etc.), manufacturing information as well as a material list. If CAE is integrated in the CAD application, the model generally can be immediately transferred to the simulation (see Fig. 10.9).

If the CAE software is in a separate application, the information needs to be transferred. If the CAE software cannot process the native CAD file format, instead only working with exchange formats, the 3D data has to be converted [5]. If appropriate interfaces are present between the CAE, CAD and database software, the files can be accessed simultaneously. The CAD model is then exported into a neutral format (like STEP or IGES) and subsequently imported into the CAE software.

Before components are transferred to the simulation software, first it is evaluated whether the construction details are relevant for the simulation. Often construction details (for example chamfers, drillings) are not important to a simulation, and can unnecessarily increase the complexity of the calculation grid, and the resulting calculations [8]. As safety factors are generally used for permissible values (e.g. max. stress or deformation), higher real stresses are covered by omitting features.

One possibility to evaluate the influence of suppressed features and to include them in the results of a simulation is the use of benchmarking. In this case, general cases are simulated with and without a feature, and results are compared afterwards.

To take drilling as an example, if it is known that the maximum stress on a bore increases by a factor of 2 compared to the same design without a bore at a certain type of load, this factor can be applied in similar applications. A possible discrepancy is covered by safety factors. It is to be ensured that the benchmark can be transferred to a specific case.

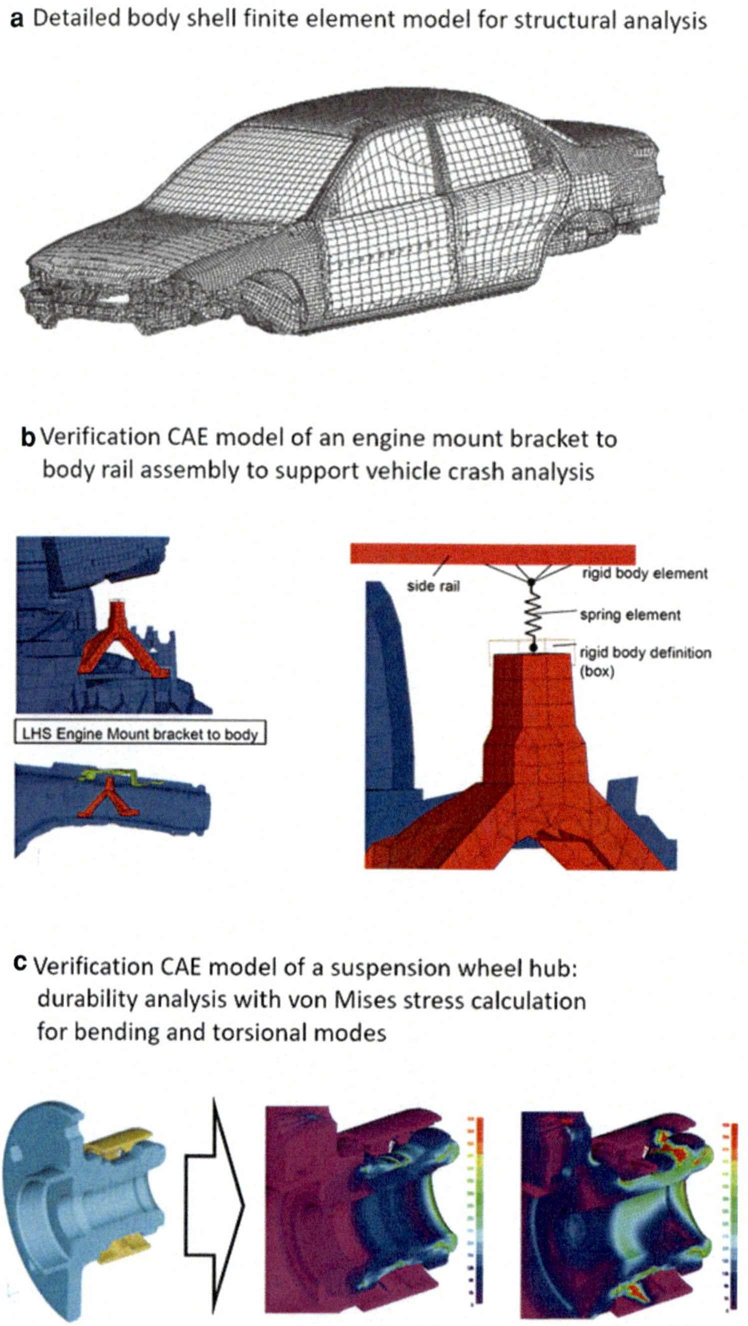

Fig. 10.7 Different types of verification CAE analysis in vehicle development

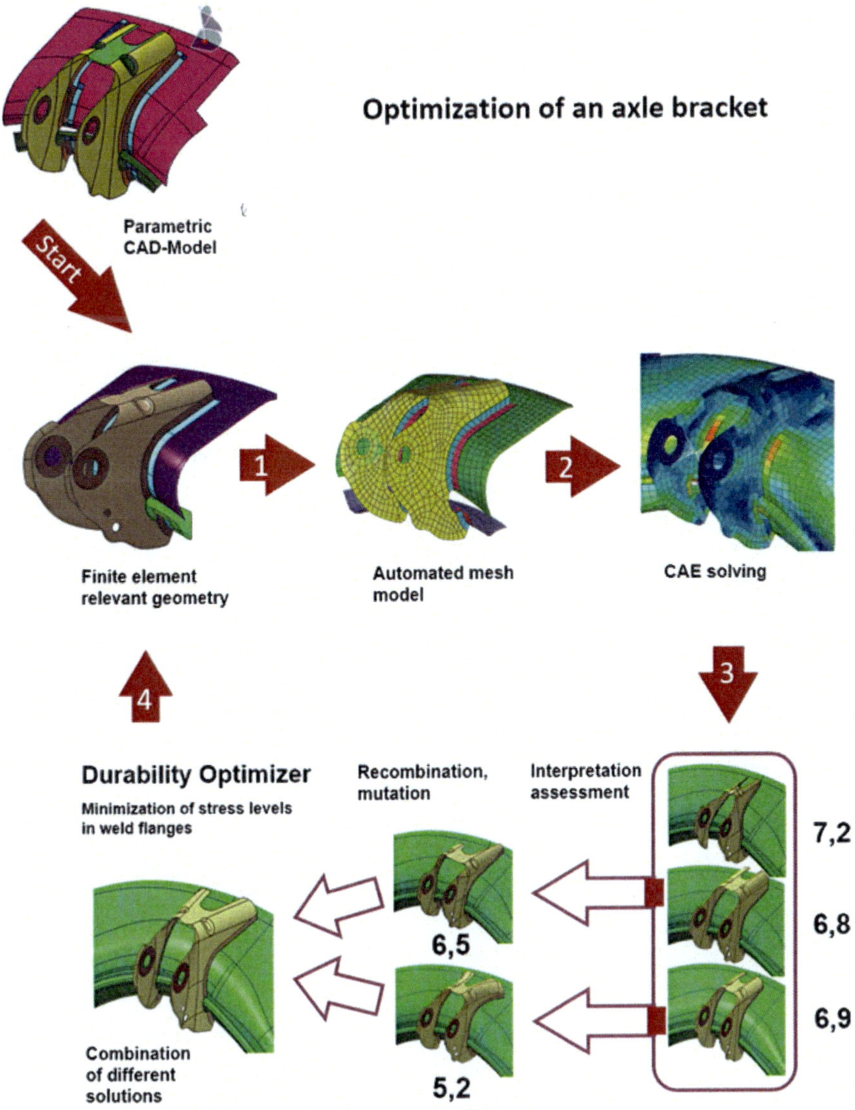

Fig. 10.8 Design topology optimization by using CAE analysis

This is where further CAE model build preprocessing takes place to set up the mathematical model (structure, boundary conditions, loads etc.) and resulting numerical equation system. After having performed the simulation run the post-processing to analyze the component or assembly also takes place within the CAE system environment. Analysis result files from the computing process are typically displayed in the same CAE software. The insights gained then flow back to the construction,

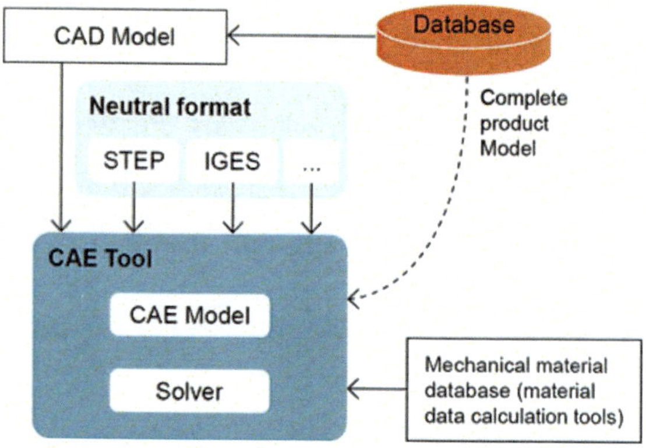

Fig. 10.9 Model transformation from CAD to CAE

which is how the component can be improved. This overall model build process for the Finite Element (FE) analysis is illustrated in Fig. 10.10.

Figure 10.11 shows the core relations between CAE software modules, simulation phases and result types. CAE software packages offer specific functions for specific tasks such as creating a model environment to pre-process the geometric model with all necessary engineering boundary conditions such as loads, forces, inertia, moments etc. and specific finite element connectivity conditions (compare e.g. illustration B in Fig. 10.7 with respect to the engine mount bracket integration into the rail structure).

As a result of the pre-processing stage, the simulation model is created. The solver then solves the mathematical model numerically and generates a result file. The post-processor then looks at the results, and creates a representation for analysis. After critical areas and zones at the design component or assembly are found, the

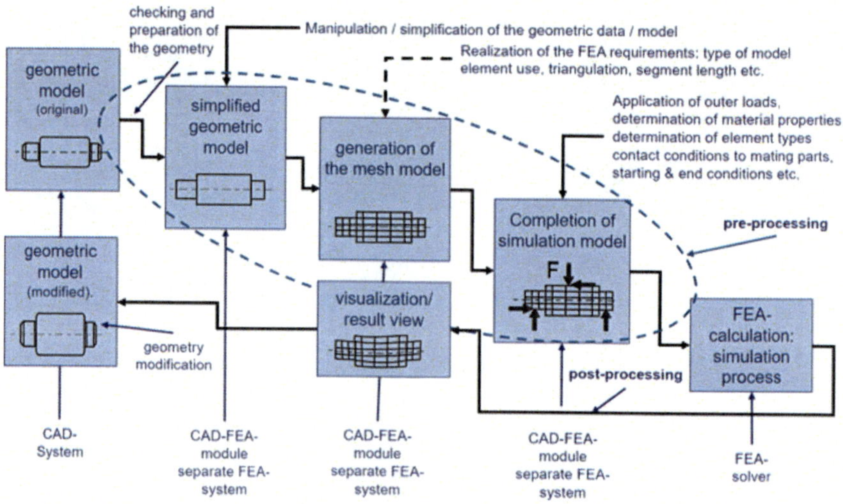

Fig. 10.10 General model build process for a finite element (FE) analysis

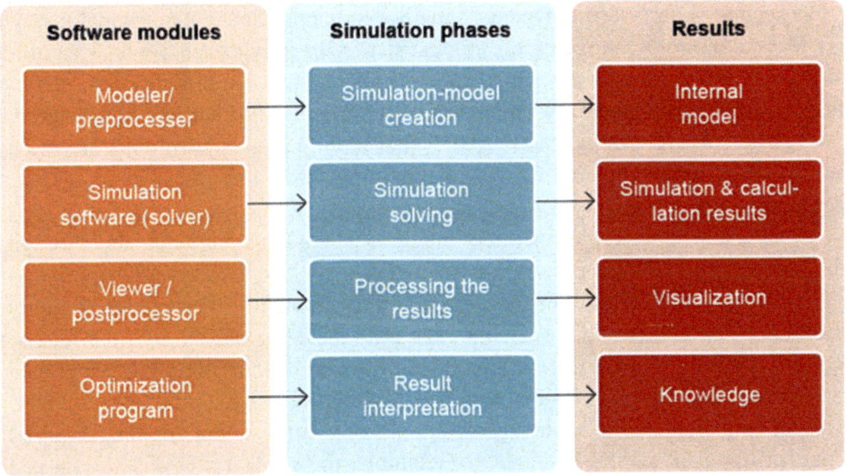

Fig. 10.11 Core relations between CAE software modules, simulation phases and result types

model undergoes an optimization process. For this, the results need to be correctly interpreted. Root causes for critical areas are also studied in order to improve them. Finally, the new insights and knowledge can be (re-) used in future projects as well.

The following explanations focus on the CAE Build process with respect to CAD-CAE transfer and necessary neutral exchange file formats, where a structural or kinematic simulation of 3D data is the aim—not a 1D simulation. This is why the use of neutral formats and its capabilities becomes important. In practical application, the creation of a CAD model is generally done by the engineering design department, while the preparation of the simulation model belongs to the calculation/CAE departments. This clear separation of responsibilities within a company is mainly due to competencies. The following steps need to be followed during the transition phase.

10.4.2 Interfaces/Formats to Transfer CAD Models to CAE

When converting CAD files into a neutral file format, two types are differentiated [5]: the first one represents geometrically exact systems, which can replicate data without loss in geometrical precision with the help of mathematical methods. Examples are STEP, IGES or JT. Currently, some formats also allow the storage or product information, which goes beyond pure geometry. That includes assembly structures, material data or different configurations.

The second neutral format type uses methods to approximate the original geometry. Most commonly, surfaces are represented with polyhedrons, which significantly reduces the amount of data. This does allow for some inaccuracies, particularly on freeform surfaces. Normally, only geometric information will be stored, no assembly structure or other metadata. That is why these file formats are often used for DMU-applications or for visualizing large assemblies rather than for CAE calculations.

CAD models, which represent three-dimensional volumes, can be created through two different modeling methods (compare Chap. 7 "*Computer-Aided Design—CAD*"): a surface model defines the volume only implicitly through its (volume) surrounding topological connectivity of the poly surfaces and through the mathematics of the individual surfaces. This type, however, cannot directly be leveraged to check whether a point is inside or outside of the implicitly defined volume body.

Depending on the CAE application, a surface or volume model might be required as part of the CAE model build. If, for example, the goal of a CAE analysis is a strength calculation for a tin construction, the surface model suffices, as the surface mesh is only set up with 2D shell elements anyways. If a complex casting is needed or the liquid flow in a pipe is to be simulated, the entire volume body needs to be meshed. Here it is very helpful if the information about the interior and surface of the object is known beforehand.

The neutral CAD format IGES (Initial Graphics Exchange Specification) enables to exchange geometric information as well as metadata like assembly structure or material. It was published by the U.S. National Bureau of Standards in 1980 (nowadays National Institute for Standards and Technology, NIST) and, therefore, the code is standardized. The rules how to convert the original geometry are not clearly defined, this is why the representation differ from software to software. With the conversion, the original shape does not lose its geometric accuracy [9]. IGES saves data as surface model without the information about volumes.

Similarly, to the IGES format, STEP (STandard for the Exchange of Product data) belongs to the geometrically exact exchange formats. It is standardized according to ISO 10303 and specialized with different application protocols for specific technology branches (for example AP203 for general mechanical engineering, AP224 for manufacturing purposes. The construction history as well as features and geometric constraints are lost. For this reason, it is difficult to edit the STEP model at later stages. Due to its versatility and performance capacity STEP is widely recognized. For visualization applications, the format is not considered the first choice due to its complexity [9].

The exchange format STL (Surface Tessellation Language) exists since 1988 meanwhile is a wide-spread geometric interface option. The original geometry is approximated with triangles, and the degree of accuracy can be influenced or set by the user individually. Information regarding geometric features, component assembly structure or construction history are also not saved in the STL format. The format can compress data sets efficiently and its application spans from simple geometries without free-form surfaces up to complex models or models with a high accuracy requirement producing large amounts of data. A clear disadvantage of the STL format represents the fact that the STL format lacks standardized data representation modes (numerical representation, header etc.). The facetted representation, however, is well suited for visualization applications (compare [10]).

With the JT format (Jupiter Tessellation), geometry is either approximated with triangles (like STL) or displayed geometrically exact (like STEP or IGES). Advantages of JT are the standardization according to ISO 14306 and the continuous functional updates to satisfy the requirements of different technological branches. Object and meta data as well as assembly structure and geometry features can be stored and safely exchanged. These characteristics make JT one of the versatile and sustainable exchange format.

10.4.3 Pre-processing of a FEA Model

The finite element analysis (FEA) is a standard method for the calculation of continuums, as already introduced in sub-chapter 10.2.2. As shown in Fig. 10.12 the structure of the CAE/FEA model is separated into spatial elements, in order to reduce a global problem to its basic physical calculations in each element. On the elementary level with defined properties, the state variables and the local behavior can be evaluated with approximating functions [4]. The process in an FEA is generally a good example also for the workflows of other CAE simulations. Similarly to the general CAE process, the FE analysis begins with the import a CAD model, some CAE applications, however, also offer a modeling environment of geometry within the CAE application itself. The structure is idealized, meaning components irrelevant to the simulation are eliminated from the calculation model and features that are less important are suppressed. The further preprocessing then contains the following steps:

- The material values are defined according to the construction and the features of the design and are then assigned to individual components.
- Afterwards, the element type is selected: 2D shell elements, volume elements or their subtypes and the corresponding mesh based on those elements gets generated. CAE engineers have to use substantial heuristics to refine the mesh according to geometric shape characteristics and/or areas of physical load applications (auto meshing algorithms might be limited in assuming the right engineering knowledge).
- The completed mesh is then assigned with boundary conditions in the form of load cases, which contain clamping and fixtures, the forces and momentum affecting the component and other conditions such as symmetry, contacts etc.
- The solver then generates the individual elements according to the discrete structure transfer functions, assigning them to the total stiffness matrix.
- The results, meaning the component stiffness and resistance to bending or yield, is then calculated in each element and displayed in color coding as part of the post-processing. The component behavior can then be predicted using the plots,

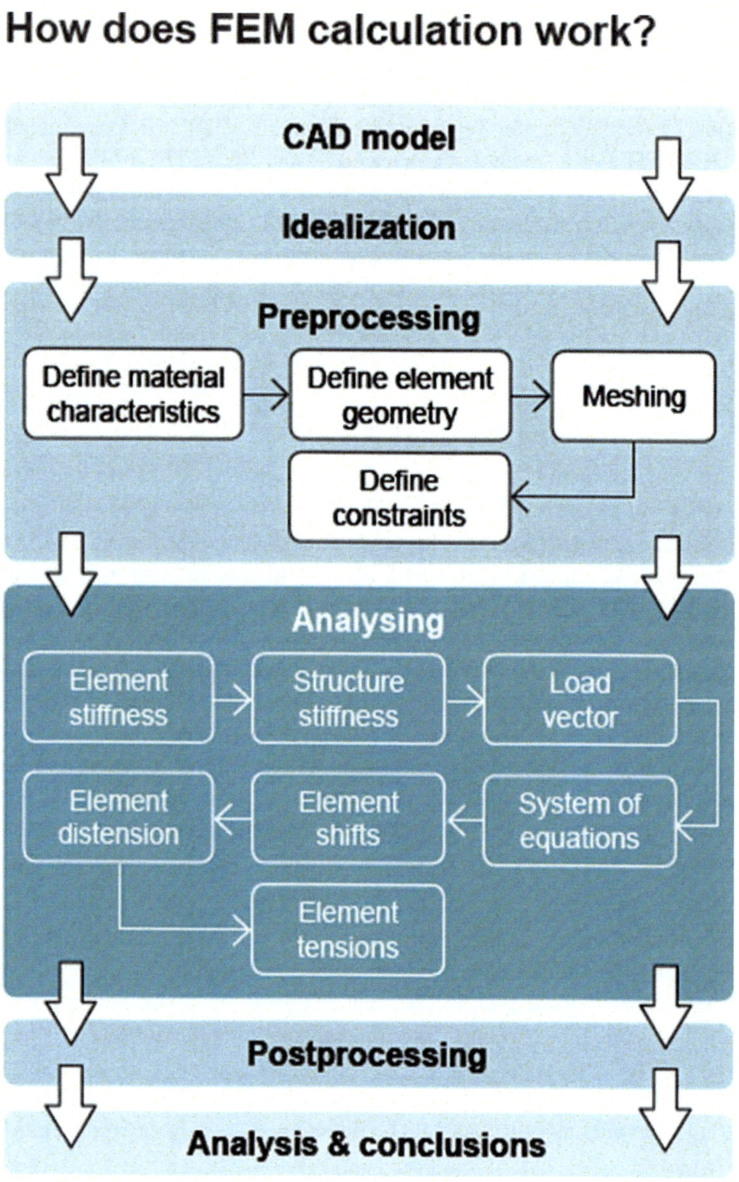

Fig. 10.12 How does FEA calculation work?

tables or graphs for analysis and evaluation (compare [10] and [4]) as well as Fig. 10.12).

The generation of the FE mesh, itself, has a strong influence on the quality and accuracy of the simulation. Depending on different quality markers (for example skewness, aspect ratio, Jacobian matrix and determinant structure, warpage etc., compare also [11]) it can be analyzed whether an element can be calculated numerically in a reproducible way, or if distortions, inaccuracies or singularities are to be expected. The mesh is generated manually with the help of the meshing tool of the preprocessor and the knowledge of the CAE analyst. Methods such as the "feature approach" allow for a clean mesh establishment and allow for optimized meshes under consideration of geometric and functional elements (features). The condition is that the basic CAD data set contains these geometric elements.

The general rule is that the more refined the mesh, the more precise the solution approach for differential equations can be expected. Areas in which critical component strain, high gradients or rigidity deflection is expected, need to be meshed more finely [10]. Since the manual adaption of the mesh takes time and is expensive, it is important to test how much detail needs to be evaluated beforehand, in order to achieve a sufficiently accurate calculation result [12].

A technique called *adaptive meshing* exists to automate CAE model build: where higher stresses are located the mesh is refined automatically by the solver to get more accurate results. Figure 10.13 explains this principle in more detail: in the example of a tool holder first a fixed or automated mesh is being applied by using standard meshing algorithms which take in to account the curvature related mesh element size rules. After having applied all outer constraints and loads the numerical solve of the finite element problem is executed and the high stress areas are detected. Based on such results the refinement of the mesh in the areas of high stress is applied in order to receive refined results which can differ from the initial results by 10–25%. The upper example a of Fig. 10.13 shows the example of the original tool holder design whereas the example b (lower areas) shows the same principle at the already beforehand *topology optimized* or *generatively designed* tool holder. The approaches of *topology optimization* and *generative design* are explained later in this section and represents a specific approach (compare also Figs. 10.14 and 10.15).

There is a wide variety of element types to display the complex structure of a technical application. In the following, the main types, their features and their uses are elaborated [12]. In principle, they are categorized into 1D, 2D and 3D elements, depending on the number of dimensions in which the principle force transfer takes place. Elements can have different amounts of integration points, depending on the degree of approximated transfer functions (linear, quadratic etc.).

The rod element belongs to the 1D-Elements. The main distension takes place via the longitudinal axis and is much larger than the measurements in perpendicular directions. It can only handle forces or line loads in direction of its longitudinal axis (tension and compression). A typical use would be the representation of steering linkage for wheel suspension.

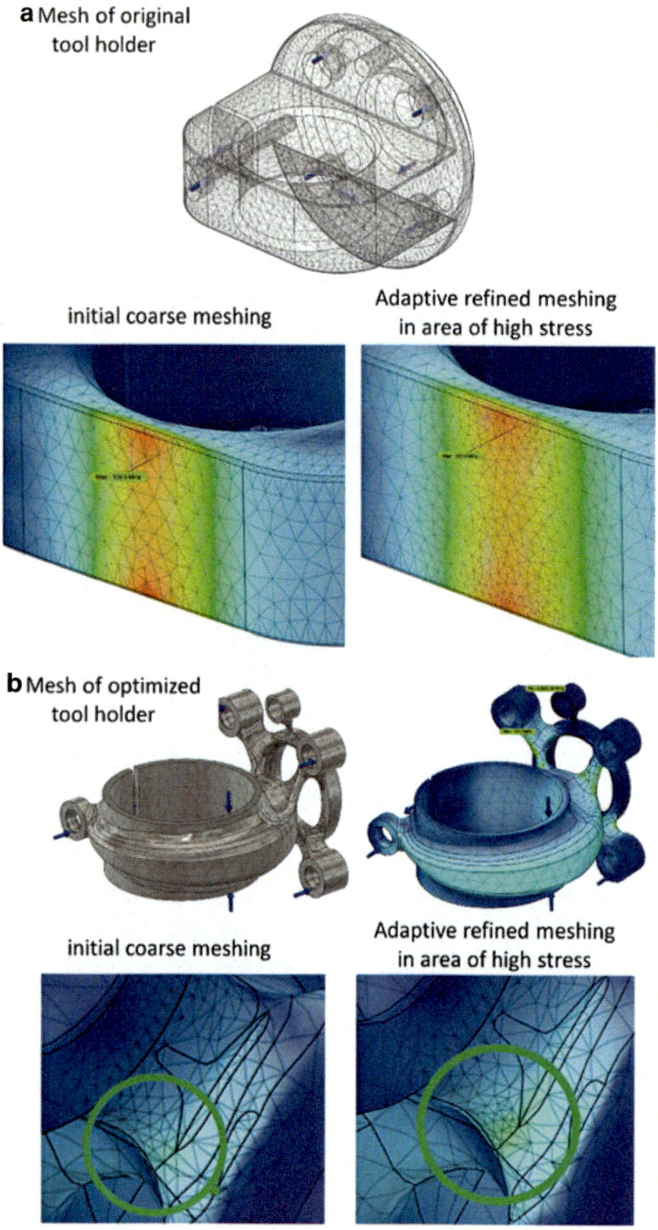

Fig. 10.13 Principle of adaptive meshing in areas of high stress (courtesy support by Autodesk Inc.)

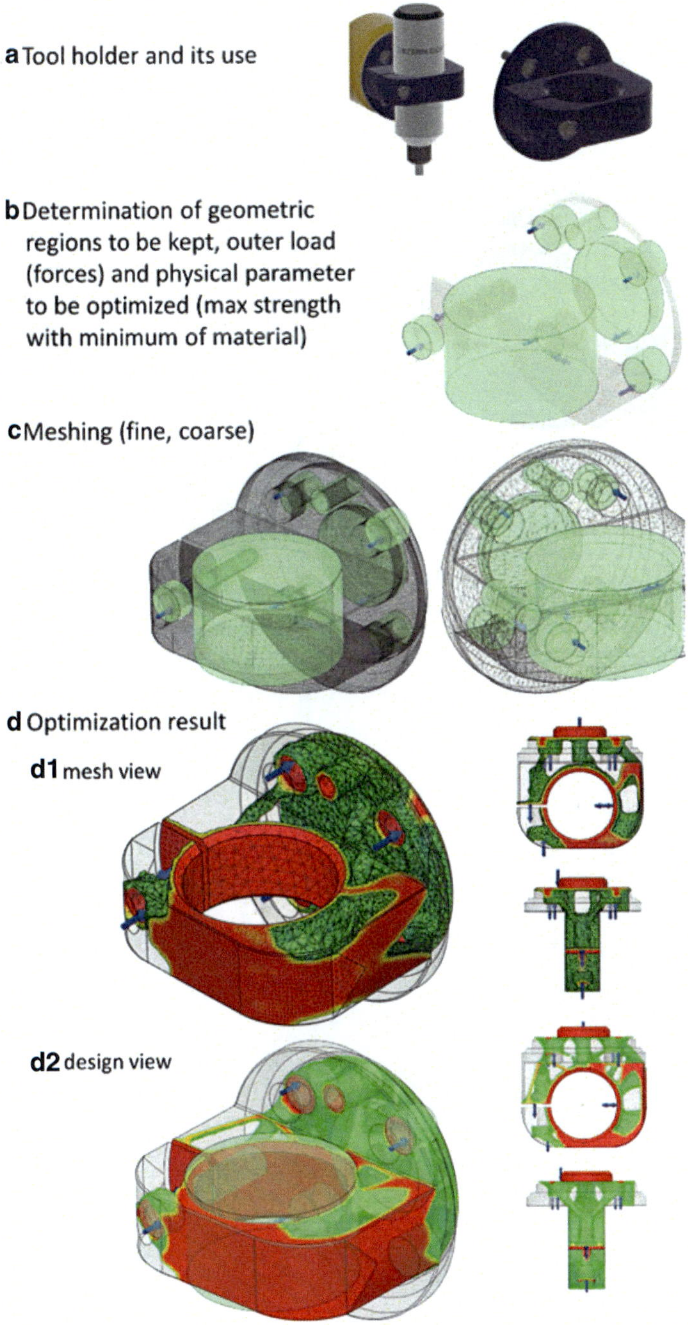

Fig. 10.14 Topology optimization (courtesy support by Autodesk Inc.)

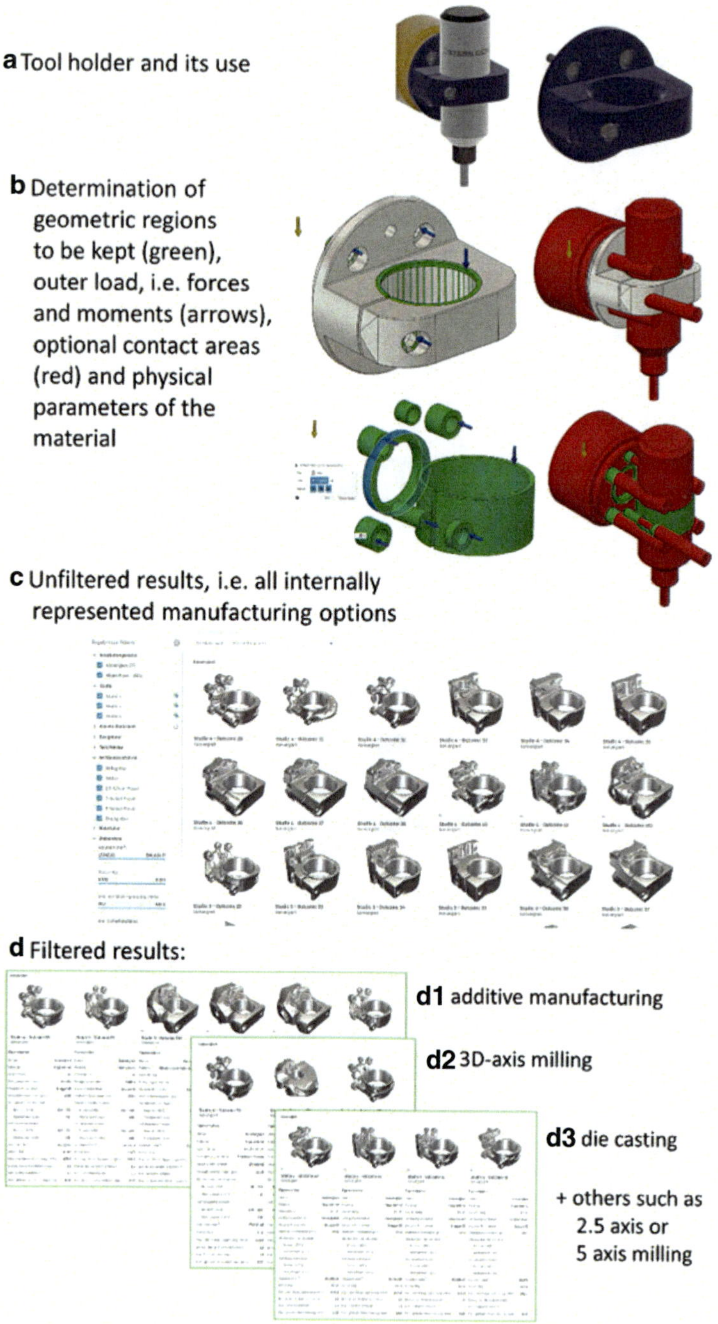

Fig. 10.15 Principle of generative design (courtesy support by Autodesk Inc.)

The beam element is a 3D-body. It has the same properties like the rod but it is also capable of transferring twisting and bending torque. The Bernoulli Hypothesis applies, that is why no shear deformation are taken into account. The bends of the beam are smaller than the height of the beam. Different cross-sectional areas can be defined. Because of its characteristics the beam element may be used for screws or even tubes.

Shell elements are created by layering disk and plate elements. This is why forces and moments in the plane of the element as well as perpendicular to the plane can be transferred. The combination of planar, shear and bending loads result in principal stress, which is to be considered for analysis. The thickness of the shell element is virtually added, for meshing only a middle surface is needed. It is used for sheet metal, in general thin structures with a defined thickness. In addition, composite or anisotropic materials may be represented. Shell elements can be triangles or squares. However, the theory of shell elements are rather complex and cannot be understood by engineers easily (recommended references are: [13, 14]).

With volume elements, 3D-structures are meshed. They are able to represent and reflect the behavior of continua, no matter which geometry. The shapes of the volume elements are hexahedrons, pentahedrons or tetrahedrons.

There exist several coupling elements, like distribution or rigid couplings, that are used to connect elements, for example to a single fixed point. Another application is the connection of a point mass to the structure. The operating degrees of freedom can be adjusted. Furthermore, spring or damper elements or contact definitions may be set up.

10.4.4 Utilizing FEA Models Within Optimization Problems

With the help of CAE tools, products and processes can be optimized for specific features and towards specific behaviors. Each type of optimization calculation is based on the principle that certain necessary constraints need to be kept and criteria for the optimal desired state are defined (via the *objective* or *target function*). In order to reach this optimal state, design variables need to be flexible (see [15]). The goal of the objective/target function is dependent on the design, converging on a local or global extremum, the desired state has been reached. Whether a global optimum can be reached is oftentimes also dependent on the starting conditions. Gradient based methods often lead to local extrema, which is why the influence of starting conditions on the optimization calculation should always be checked thoroughly (see [16]).

There is a basic distinction between the optimization of a control system and a 3D structural optimization. In the first case, a system with mathematical/physical dependencies is present, for example a combustion motor with a transmission. The goal of a system optimization can be the optimal working point, meaning the least consumptive state of the system. In the structural optimization, one optimizes a three-dimensional structure for specific features or towards specific behaviors. These would include factors such as minimal weight, reduction of component tension or minimal material use for the same constructive functionality. The three main techniques of structure optimization are topology optimization, parameter optimization and form optimization.

The outcome of the topology optimization (compare Figs. 10.8 and 10.14) is simply based on an available design space, including load introductions (for example, storage, forces, static analysis, thermic analysis etc.). Boundary conditions are defined, such as maximal tension or material usage, as well as symmetry conditions. The solver then calculates at which points of the design space material is required and where it is not necessarily needed. The result is an analysis of the ideal usage of the design space. The phenomena is also visible in nature: for example, tree branches or bone marrow only grow in places where material strength is needed. The result is a more or less defined design recommendation, which is then converted into a constructible format.

Complex structures with irregular geometries are particularly suited for additive-generative or casting procedures. Figure 10.14 shows an example of topology optimization of a tool holder: a given tool holder CAD design (see step a) is constraint in step b by determining design topology areas to be kept (due to the assembly design constraints), by applying loads and by setting the optimization criterion. After applying the mesh (step c), the algorithms can solve the constraint optimization problem, which result in the optimization outcomes (see step d).

In contrast to topology optimization, parameter optimization is based on an already existing design. Design parameters (e.g. wall thickness, length of a girder, thickness of an axle) are defined as alterable in the objective/target function. This often has the advantage that (in the case of an FEA) the FE mesh is retained and can be adapted in the frame of smaller geometric changes, without requiring the generation of a new mesh.

Shape optimization improves the local geometry of a component. A common use of FEA is to minimize tension on transfer points (radials, rigidity deflections on cross section modifications etc.). The requirement is—similar to the parameter optimization—that a certain design has already been established, and only details need to be adapted. The variable can be the position of nodes on the surface of the component, for example. If the mesh is not deformed too drastically during the shape optimization, it is sufficient to perform an automatic mesh smoothing after each step without re-meshing.

10.5 Advanced CAE Technologies

The simulation types named in this sub-chapter are meanwhile highly developed and matured. Therefore, applied with the right engineering and CAE method knowledge, they can achieve precise results in their specialized fields. A current challenge is the so-called “flexible body dynamics”, the combination of multi body dynamics and finite element analysis, oftentimes using two types of CAE simulation based on a co-simulation framework. This means that the bodies of a MBD model are no longer treated as rigid, but can be replaced by moldable FE models of the actual component. The technique of integrating an FE model is called floating frame of reference (FFR), representing one method for the co-simulation approach. For example, the movement and expansion of the piston rod in a combustion engine can be simulated. Due to the reciprocal influence of a MBS and a FEA model the problem is non-linear and analytically as well as numerically difficult to solve [17].

A further focus of study is the combination of multi domain simulation (MDS) and MDB/FEA. A possible application is the simulation of an electric window motor in a car: The control of motors or sensors is set in the field of mechatronics. The entire system can be displayed in a multi domain system. At the same time, the dynamic strain on the windowpane or the fixture for it on the body can be simulated via MDS. Furthermore, it is of interest to implement such a system as Hardware-in-the-Loop (HiL). This involves combining several real components with the virtual system. Another application is the digital twin: here a real system is digitally mapped and fed with measured data (such as movement, duration etc.). This can then be used to draw conclusions regarding the current or future state of the system.

The real-time simulation of a flexible thin-walled component and cables or hoses also presents a unique challenge. The difficulty lies in the numerous conditions: cables and hoses are sometimes comprised of multiple layers of different materials. In addition, a cross-section with twisted wire strands does not clearly show deformations, further complicating behavioral predictions. The assembly of thin-walled objects, for example covers in the automobile industry, often takes advantage of their flexibility, which is why it is manually warped to achieve its correct installation position. This deformation is difficult to simulate due to the human component and is generally analyzed in practical testing.

The field of research topology optimization is not fully developed and has much potential. For problems such as effectively used assembly space, highest degree of rigidity etc. there is generally not only one solution, but a variety of local optima with different specifications. As the algorithms work with evolutionary and partially heuristic principles, the question whether a problem was optimally solved is often not easy to answer.

Acoustic phenomena in systems are examined using NVH analyses. They are needed in the automobile industry, for example, where attention must be paid to motoring experience and comfort. The complex automobile system contains a number of oscillation sources (motor, transmission, chassis etc.) and resonance bodies (covers, body etc.). The system as a whole is impossible to simulate in a structural analysis due to its complexity. As a result, simulation procedures such as FEA, BEM and MBS are combined. With the transfer path analysis (TPA) the paths on which sound and vibrations are transferred from their originator to the recipient (generally the human) are analyzed [18].

Due to new compute power with the help of grid and cloud computing it is nowadays possible to combined FEA analysis with design synthesis. This new type of hybrid approach is called *Generative Design*. There does not yet exist an absolute clear scientific or normative industry wide standard definition for the term and field of *Generative Design*, however, the common understanding can be expressed by the current explanation by AUTODESK Inc. (https://www.autodesk.com/solutions/generative-design# visited in September 2020):

Generative design is a design exploration process. Designers or engineers input design goals into the generative design software, along with parameters such as performance or spatial requirements, materials, manufacturing methods, and cost constraints. The software explores all the possible permutations of a solution, quickly generating design alternatives. It tests and learns from each iteration what works and what doesn't.

In actual facts *Generative Design* combines engineering design and analysis tasks with cost estimation and manufacturing feasibility work which usually are handled separately from each other:

- 3D Design of individual components
- 3D Assembly and constraint product modeling
- CAE analysis and shape/topology optimization
- Design for Manufacturing (DFM) and manufacturing feasibility analysis as well as
- Cost engineering.

In order to achieve meaningful design exploration and associated design proposal offering with additional information sets the user is guided through a systematic approach. Figure 10.15 gives an insight into the generative design principle: starting from the overall assembly situation (see part a) and the interface design knowledge to the adjacent part (see part b) the design and analysis engineer receives a set of design alternatives (see part c).

In the following, the design and analysis engineer has to provide additional information sets for the algorithms to resolve the underconstraint mathematical problem according to the options at the pareto front (i.e. to achieve the improvement of one design factor without deteriorating another design factor). Hence, in order to reduce the high number of solution proposals offered by the compute algorithms it is necessary to further filter or select the manufacturing technology options (see part d) by the design and analysis engineer. Based on AI (artificial intelligence) heuristics cost estimations are provided to the use of such Generative Design environment in order to discuss and or decide on the option to take. This design proposal then needs to be further refined and executed by classical design (CAD) and analysis (CAE) methods and tools.

10.6 Exemplary Automotive FEA Project Cases

This section illustrates a typical FEA analysis as part of an automotive body shell use case. The example deals with the digital (CAE based) design verification of a body shell rear end in combination with a towing hook design. It shows the assumptions and simplification which are made based on engineering knowledge and heuristics in the automotive body shell development. The purpose of this example is to show the tight interaction between product/technical system know-how and specific CAE method and tool knowledge. Only if both viewpoints are aligned and supported by prior system validation test (using hardware) such CAE (FEA) based design verification can be applied successfully in industrial practice.

This industrial project case is divided into four parts: Fig. 10.16 illustrates the overall situation of the FEA analysis case by describing the load case (towing of

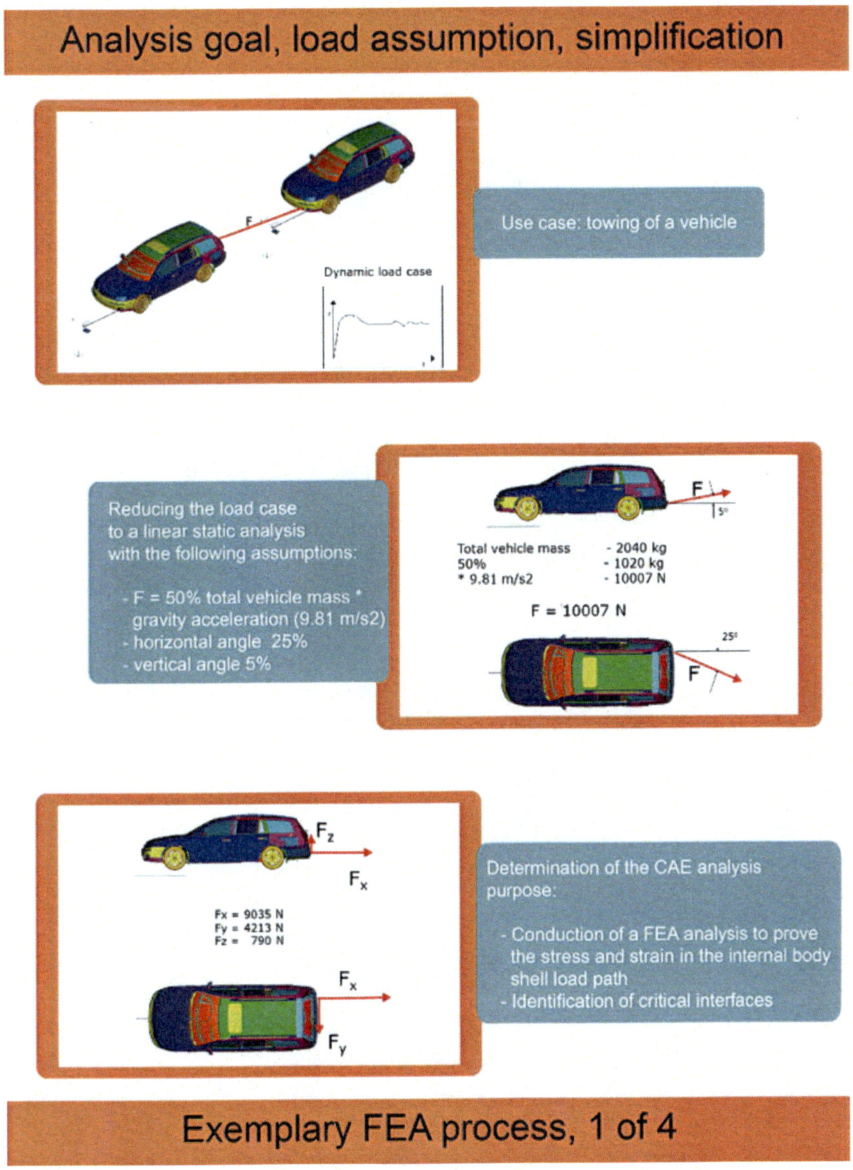

Fig. 10.16 Exemplary automotive FEA project case, part one

another car), the first simplification (reducing the dynamic load case to an equivalent static one with the help of higher forces) and the clarification of the angle of direction of the force elements). Figure 10.17 shows how a CAE Engineer has to make the right decision to cut the overall body CAE mesh model and to reduce it to a (relevant) one for the vehicle rear end. The aim is to calculate the load path and stress analysis

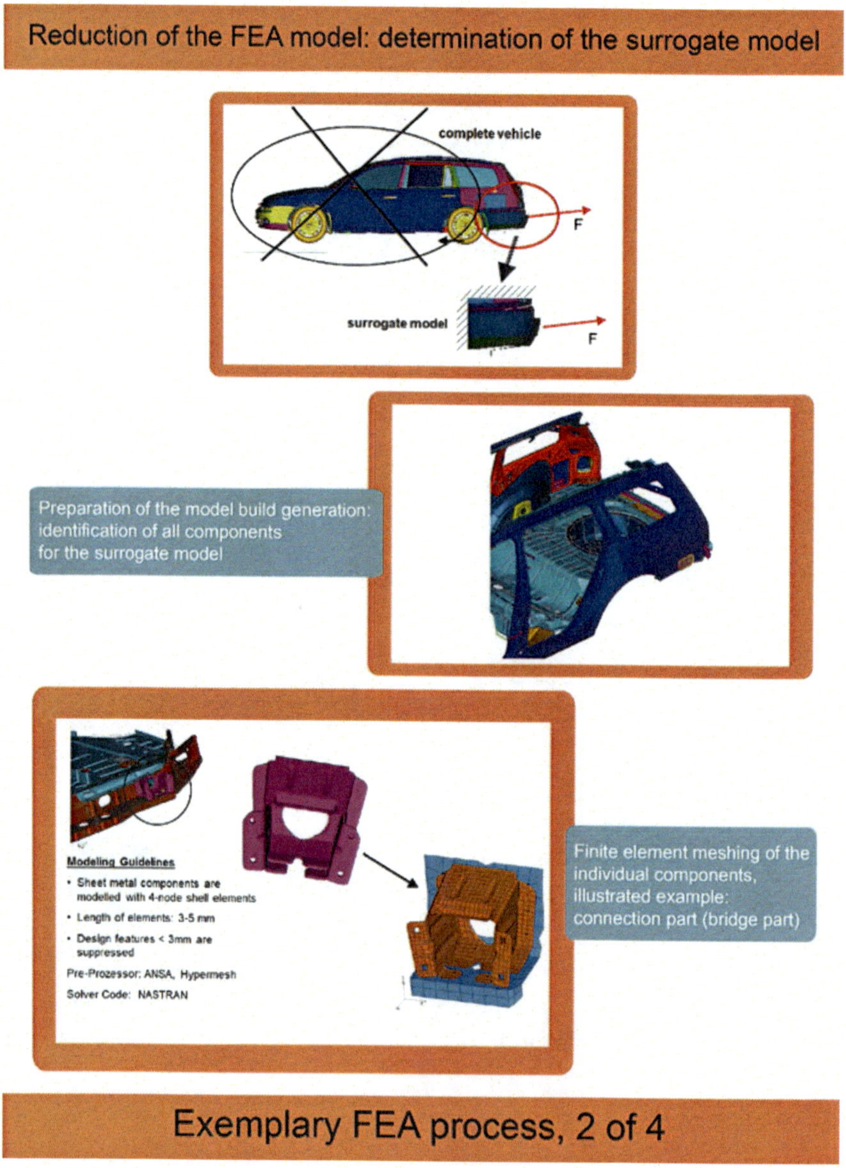

Fig. 10.17 Exemplary automotive FEA project case, part two

based on the outer load assumptions (as already clarified in Fig. 10.16) and the correct boundary conditions. Figure 10.18 shows the library of mesh model types which are proven and recommended within the company's best CAE practice guidelines ("CAE cookbook") and how the element connections need to be modeled. Especially the interface between rigid machinery parts (like the screw in towing hook assembly

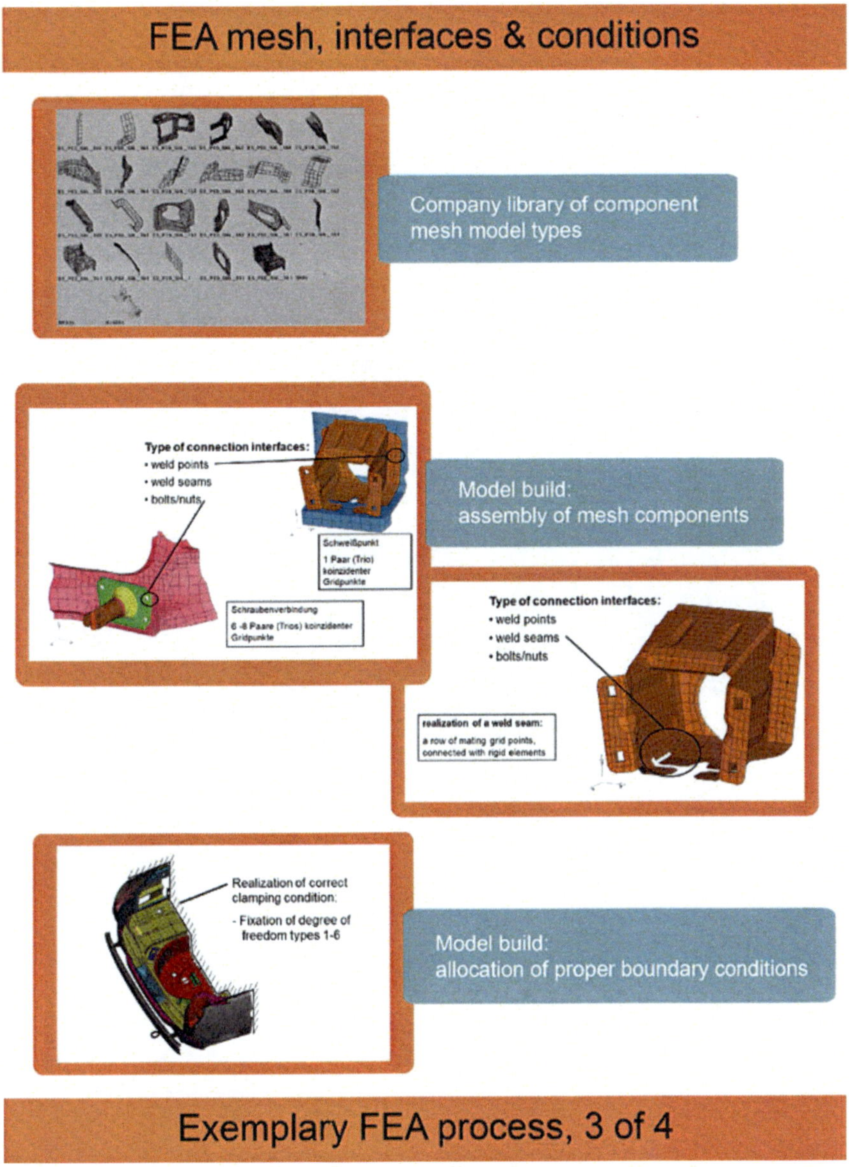

Fig. 10.18 Exemplary automotive FEA project case, part three

plate) and sheet metal components (of the body shell itself) need to be modeled on purpose with the help of specific grid points. The bottom picture illustrates the need to position and fix the body shell rear end assembly in the right way. Figure 10.19 shows the necessary final digital model verification before the simulation run can be started.

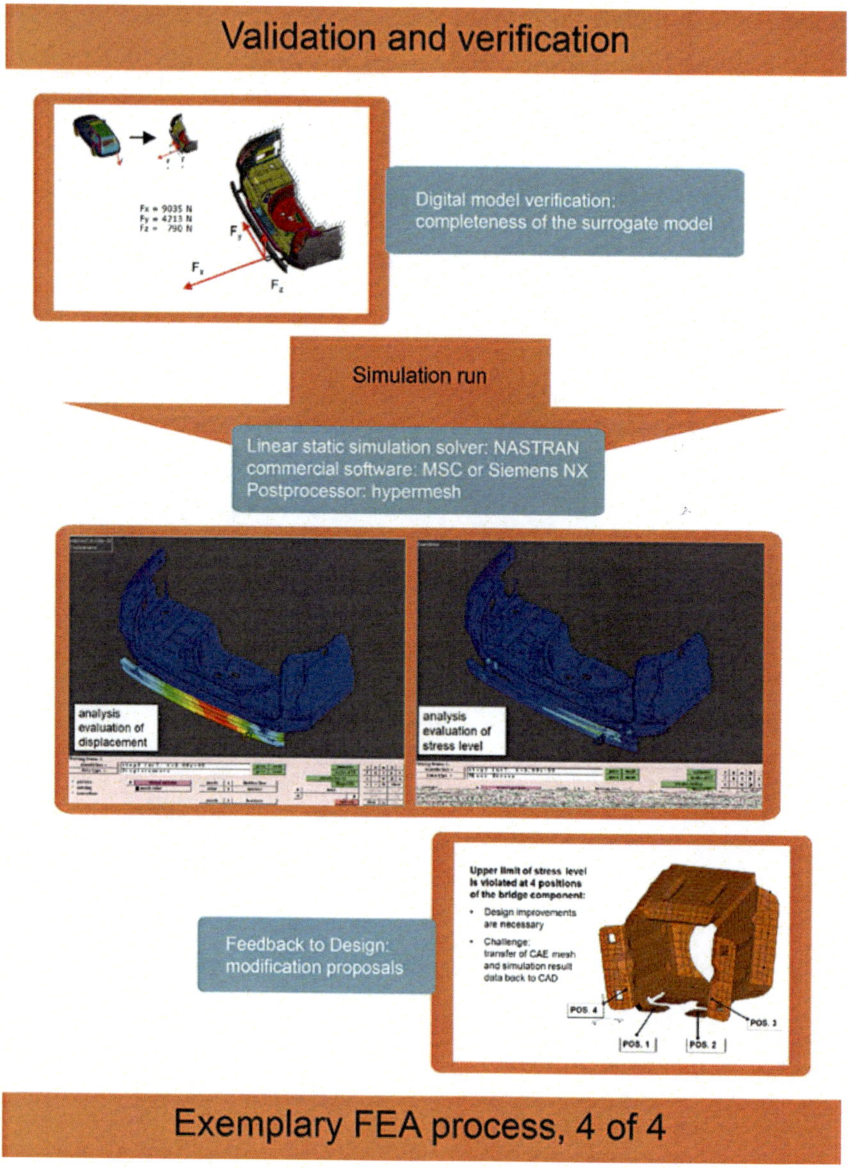

Fig. 10.19 Exemplary automotive FEA project case, part four

It also illustrates the post-processing results of the FEA analysis. The graphical visualization of the displacement (left side) and of the stress distribution within the sheet metal assembly (based on von Mises stress type) are indicators for the CAE analysts to finally judge whether the results are acceptable or do cause problems over time. CAE analysts then prepare an analysis report based on specific components and their embedment within the overall technical system or assembly.

Finally, improvement proposals are made and an overall assessment is discussed with the Design Engineers with respect to possible design modifications. As a consequence within this project case, the bridge component (bracket), shown in the bottom of Fig. 10.19 had to be redesigned since the stress levels at the folded flanges were too high. A deep drawn bracket would usually provide more stiffness but it does it would not usually require more efforts to provide an extra tool set for its manufacturing compared to a sheet metal folding part).

A second example deals with a modular CAE model build to support efficient CAE analysis for front crash investigations as shown in Fig. 10.20. There exist two different CAD models as design alternatives for the front bumper (part a). The CAE mesh assemblies (part b) are divided into different domains, which are individually meshed depending on specific strategies and accuracy requirements. The interface conditions for the individual connection types of the mesh elements can also be maintained if associate mesh model build is deployed: this included the connection type, the intelligence of determining the number of connection types (based on rule) and the time step settings for the simulation itself. The only difference between the CAE mesh assemblies is the mesh of the bumper itself, the other mesh parts of the assembly stay exactly the same. The lower part of Fig. 10.20 shows the simulation results for the full-frontal crash (c1) versus the one of the partial frontal crash (c2).

10.7 Final Remarks

Computer Aided Engineering (CAE) meanwhile constitutes a major digital engineering discipline which is indispensable to validate and verify early design concepts, functional layouts and final release ready design proposals before any physical prototype evaluations take place. Due to the fact that all associated skills in tools and technology, product and CAE models, modeling methods as well as in the underlying mechanical laws and mathematical formulations require substantial knowledge of engineers and analysts make it extremely challenging to distribute those skills widespread in the organizations. Therefore, especially Small and Medium Size Enterprises (SME) are still dependent on massive support by specialized engineering service providers instead of building up internal CAE skilled engineers.

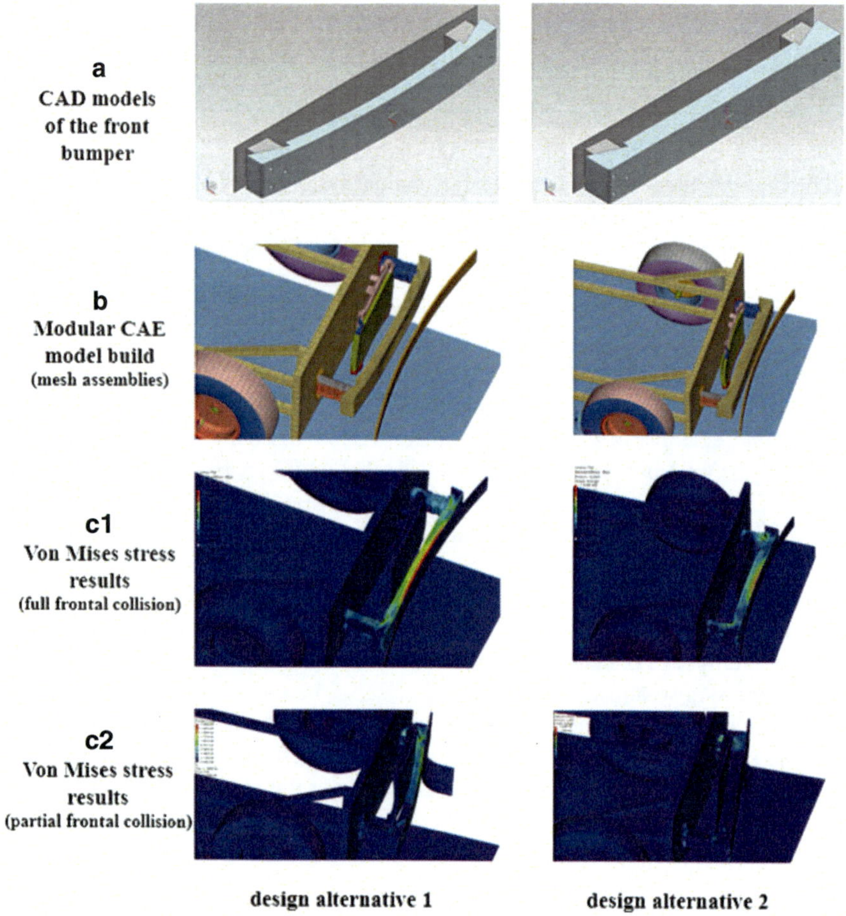

Fig. 10.20 Modular CAE model build as part of a conceptual front-end crash evaluation

Two trends are currently visible and evident:

- Bigger companies try to mix the skill set between design engineers and CAE analysts in a new way in order to start with CAE rather than using it as after-the-fact verification solution only
- The rising complexity and connectivity of technical systems under development do require new advanced CAE modeling and analytical skill sets of engineers and analysts as part of the comprehensive Advanced Systems Engineering (ASE) development capability. This does require, however, new assistance for engineers in system modelling, systems integration and system validation/verification.

As of today, CAE is still seen as an expert group development operation and not yet as an engineering skill set which needs to be support by every engineer. Therefore, middle and upper management need to be re-educated in virtual product creation capabilities in order to drive the critical new functional and behavior analytic skill sets within major product development processes and company organizational set-ups.

References

1. Kolbasin A, Husu O (2018) Computer-aided design and computer-aided engineering. MATEC web of conferences, vol 170. https://doi.org/10.1051/matecconf/201817001115
2. McHaney R (2009) Understanding computer simulation. Roger McHaney & Ventus Publishing ApS; ISBN 978-87-7681-505-9
3. Matsuura Y, Kaminishi K, Matsuura Y (n.d.) (2018) The impact of CAE advancement on product development process. In: International conference on innovation and management. http://www4.pucsp.br/icim/ingles/downloads/papers/TL_060.pdf. Accessed 27 Aug 2018
4. Jagota V, Sethi AP, Kumar K (2013) Finite element method: an overview. Walailak J Sci Tech 2013:1–8
5. Hirz M, Dietrich W, Gferrer, Lang J (2013) Overview of virtual product development. In: Hirz M, Dietrich W, Gferrer A, Lang J (eds.) Integrated computer-aided design in automotive development (p. Chapter 2). ISBN 978-3-642-11940-8
6. Orbis Research (2018) Global (North America, Europe and Asia-Pacific, South America, Middle East and Africa) CAE software market 2017 forecast to 2022. http://www.orbisresearch.com/reports/index/global-north-america-europe-and-asia-pacific-south-america-middle-east-and-africa-cae-software-market-2017-forecast-to-2022. Accessed on 27 Aug 2018
7. Brunberg A, Abel D, Autschbach R (2009) An object-oriented model of the cardiovascular system with a focus on physiological control loops. In; IFMBE proceedings, vol 22, pp 2600–2603. Springer, Berlin, Heidelberg
8. Suresh G (2007) A formal theory for estimating defeaturing-induced engineering analysis errors. Comput Aided Des 39:60–68. https://doi.org/10.1016/j.cad.2006.09.006
9. Pfouga A, Stjepandic J (2018) Leveraging 3D geometric knowledge in the product lifecycle based on industrial standards. J Comput Design Eng 5:54–67. https://doi.org/10.1016/j.jcde.2017.11.002
10. Erhunmwun I, Ikponmwosa U (2017) Review on finite element method. J Appl Sci Environ Manage 21:999–1002. eISSN: 2659-1502; print ISSN: 1119-8362
11. Altair Engineering Inc. (2012) Element quality and checks. http://altairuniversity.com/wp-content/uploads/2012/04/Student_Guide_211-233.pdf. Accessed on 23 Aug 2018
12. Steinke P (2007) Finite-Elemente-Methode - Rechnergestützte Einführung. Springer. ISBN 978-3-642-53937-4
13. Timoshenko S, Woinowsky-Krieger S (1959) Theory of plates and shells. Mc Graw-Hill Book Company, Singapore
14. Rieg F, Hackenschmidt R, Alber-Laukant B Finite Elemente Analysis für Ingenieure (Finite Element Analysis for Engineering). Hanser. ISBN 978-3-446-43469-1
15. Larsson R (2016) Methodology for topology and shape optimization: application to a rear lower control arm. Chalmers University of Technology, Göteborg, Department of Applied Mechanics, Sweden
16. de Weck O (2005) Design optimization—structural design optimization. Massachusetts Institute of Technology (MIT). http://citeseerx.ist.psu.edu/viewdoc/download?doi=10.1.1.207.9728&rep=rep1&type=pdf. Accessed on 17 Nov 2019
17. Shabana AA, Wang G, Kulkarni S (2018) Further investigation on the coupling between the reference and elastic displacements in flexible body dynamics. J Sound Vib 427:159–177
18. Siemens AG (2020) Transfer path analysis—a guideline to qualifying and quantifying vibro-acoustic transfer paths. https://www.plm.automation.siemens.com/en_us/Images/What_is_transfer_path_analysis_tcm1023-220904.pdf. Accessed on 27 Sept 2020

Chapter 11
Major Technology 5: Product Data Management and Bill of Materials—PDM/BOM

Executive Summary

This chapter deals with the following topics:

- Basics and advanced techniques of Product Data Management (PDM) and Bill of Materials (BOM)
- Providing insight into how engineers benefit from using PDM and BOM technologies in the context of PLM (Product Lifecycle Management)
- Describing functioning, benefits, and limitations of PDM and BOM technologies in practice.

Quick Reader Orientation and Motivation

The intention of this chapter is:

- to give an overview of PDM/BOM technology in Virtual Product Creation as driver and enablers for Digital Transformation in engineering
- to present PDM/BOM technology as part of Virtual Product Creation from a practitioner's point of view to analyze the need and usefulness for day-to-day industrial work practice
- to give instructions on how to use PDM/BOM technology
- to explain models, frameworks, and digital representations that help to grasp the internal working modes of PDM/BOM technology.

The basic need for and the role of data management in Virtual Product Creation, both have already been introduced in Chap. 4 (*Virtual Product Creation (VPC) Explained*). The history of Product Data Management (PDM) has been explained in Sect. 5.3 as part of the overall Chap. 5 (*The Technology History of Virtual Product Creation*).

This chapter defines and explains the use of Product Data Management (PDM) and Bill of Materials (BOM), based on given standards and the technological state of the art. Engineering and administrative tasks, carried out in specific functions of Product Data and BOM Management, as well as the related architectures are both

R. Stark, *Virtual Product Creation in Industry*,
https://doi.org/10.1007/978-3-662-64301-3_11

described in detail. Furthermore, this chapter provides an insight and a critical analysis on how engineers benefit from using PDM and BOM and how the IT technology behind those solutions in principle works. This chapter also covers crucial integration aspects of PDM/BOM solutions within the bigger scope of PLM (Product Lifecycle Management) and ends with an outlook on expected further development in the area.

11.1 Introduction of PDM and BOM

As Virtual Product Creation has evolved, the number of data and information sets which were created has been growing exponentially and engineers, designers and analysts started to spend many hours, days and sometimes weeks to search for the individual data models, documents and meta data in digital repositories such as shared drives, local repositories and specific IT application data collectors. Ford Motor Company had recognized this dilemma in the mid 90ties of last century already early on in the first wave of digitalization and started to address this "search syndrome" and called it "Sam Schwartz's Search for Data" (see Fig. 11.1).

To address the "Sam Schwartz's Search for Data" issue Ford Motor Company created an overall solution approach which was called *Product Information Management (PIM)* in 1996, which put already the information set, i.e. the semantic meaning of data within a context, into the core of an appropriate management environment. The PIM solution provided a range of solution sets to find, organize and safely store

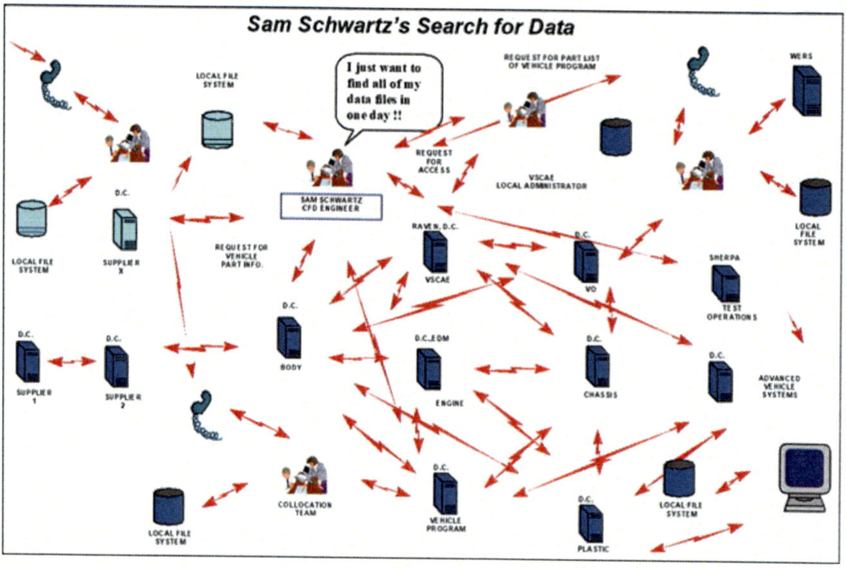

Fig. 11.1 The "Sam Schwartz syndrome" of wasting time in searching for digital data sets in a company IT data storage landscape (see [1])

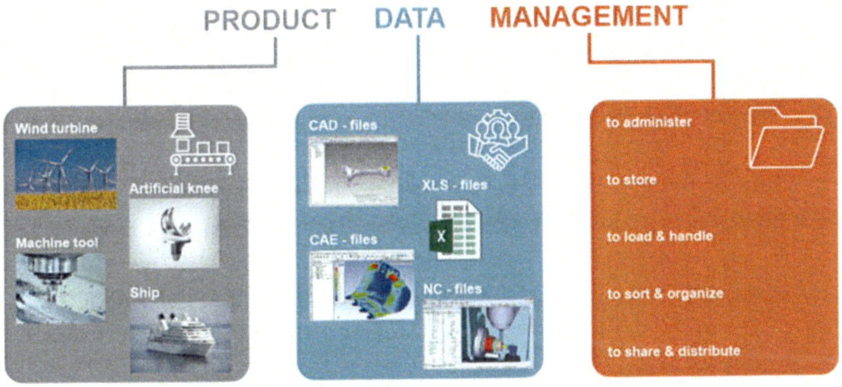

Fig. 11.2 The three elements of PDM (product data management)

different kinds of information sets and the associated files. The term PIM, however, did not gain the buy-in from the global community and towards the end of the 90ties the European research community and some technology vendors formed the term *Product Data Management (PDM)* as evolution from the limited and oftentimes localized approaches *EDM (Engineering Data Management)* or *TDM (Team Data Management)*. It is important for engineers to understand what actually is meant by Product Data Management. Figure 11.2 illustrates the three PDM core elements *Product*, *Data* and *Management* with the help of examples. PDM concentrated first on structured data sets and file types.

Product Data Management (PDM) is the digital solution environment to store and manage product defining and representing data. PDM manages product data and process-related information in a central software along the product creation process. This information includes authoring data and models such as CAD models, CAE models, parts information, manu-facturing instructions, requirements, manufacturing notes, meta data of model creation, change or storage and related documents such as norms and standards. Within PDM, the focus is on managing and tracking creation, change and archiving of all information related to products in their widest sense. This notion is extended to production environments, i.e. how to manufacture and assemble products, and to services in the field associated to the product use (please compare the information management tasks and characteristics in Figs. 4.4, , 4.5 and 4.6). PDM, therefore, makes this data available in downstream phases of the product lifecycle [1].

The VDI guideline 2219 [4] defines PDM as the consistent storage, provision and management of information about products and the associated development processes [3] and designates its system integration into the wider landscape of virtual product creation (compare to Fig. 11.3).

Essentially, a PDM system provides solutions for secure data management, engineering workflow execution and configuration management. The ideal PDM system

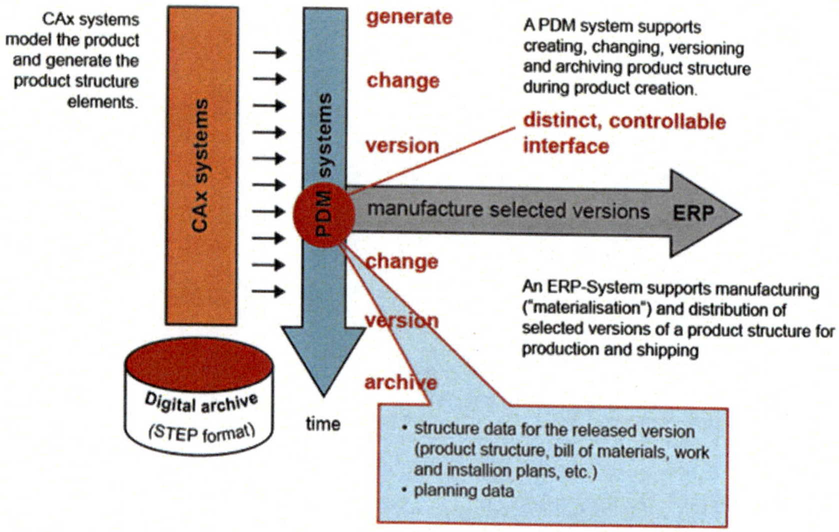

Fig. 11.3 PDM interaction with systems in virtual product creation [4] and [6]

is accessible to multiple applications and teams [1]. The term PIM (Product Information Management) as precursor for PDM has evolved from the end of the last century: today, it designates rather enterprise outbound oriented product information system for e-commerce. PIM systems visualize, offer, complete and enhance product data for costumers.

In addition to PDM, *Product Lifecycle Management (PLM)* is a systematic approach to managing the various transitions that a product undergoes throughout its lifecycle. That way, the affected systems also include production planning (plant and technology) and PPS, as well as sales planning, sales, distribution logistics, end-of-life management, including service and, in some cases, even recycling issues. In general, the PLM management system covers the following phases (see Fig. 11.4):

- **Begin of Life (BOL)**: involves the development of products and design processes
- **Mid of Life (MOL)**: includes e.g. supplier collaboration, product information management and service management
- **End of Life (EOL)**: includes strategies for the disposal or recycling of products

The classic product life cycle

- goes through several phases (BoL, MoL, EoL),
- which are partly overlapping and
- which are coordinated with each other.

In each phase (BoL, MoL, EoL) …

- different specialist processes take place,
- different experts are involved and

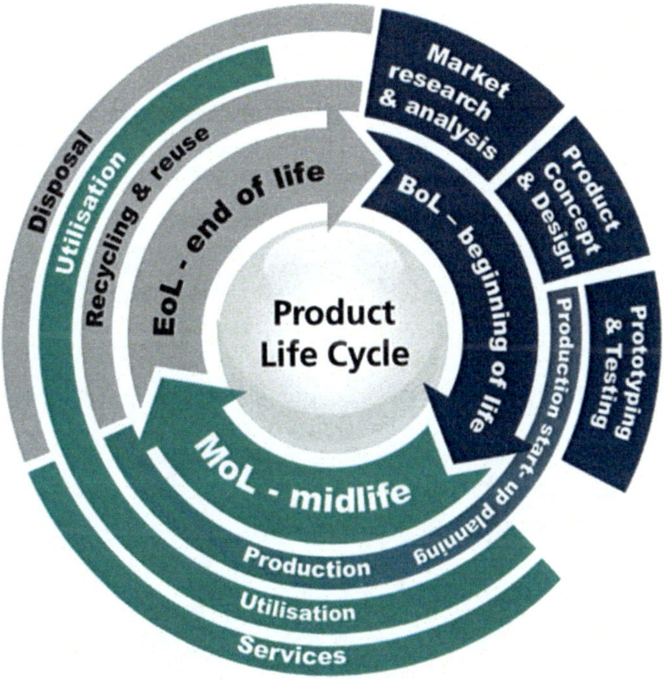

Fig. 11.4 Product life cycle model of Fraunhofer IPK[1]

- different models are created and different tools are used.

The explicit representation of product life cycles as part of the Virtual Product Creation and PLM solution portfolio comprises the following facts:

- Products are usually not completely new, but are further developments (new versions, facelifts) of already existing products
- The innovation cycles of the used technologies behave differently
- Software updates, for example, can often be carried out faster than the replacement cycles of mechanical and electronic components
- Information, documents and models (e.g. CAD models, structures, parts lists etc.) are (ideally) reused in the following products.

With such a life cycle argumentation in mind, PLM and Digital Technology vendors started to create and offer their solution portfolio around the corresponding holistic data and model lifecycle progression. Amongst them, in the early 2000th, EDS was one of the first vendors which started to offer such a solution portfolio based on the core PDM application suite *Teamcenter* which was built up on the

[1] The Fraunhofer Institute for Production Systems and Design Technology (IPK) in Berlin, Germany, offers various learning modules for *Professional in Product Lifecycle Management* under: https://www.ipk.fraunhofer.de/en/events/mehr-koennen/seminars/plm-professional.html.

PDM tool box Metaphase (originally jointly developed by SDRC and Control Data) as *Teamcenter Enterprise* and on the CAD Manager system IMAN originally from Unigraphics, also known as *Teamcenter Engineering* (compare to Fig. 11.5).

The entire Teamcenter PDM application suite was further sold to UGS Corporation and finally merged into the Teamcenter UA (Unified Architecture) by the new owner Siemens PLM, known today as Siemens Digital Industry Software, in the

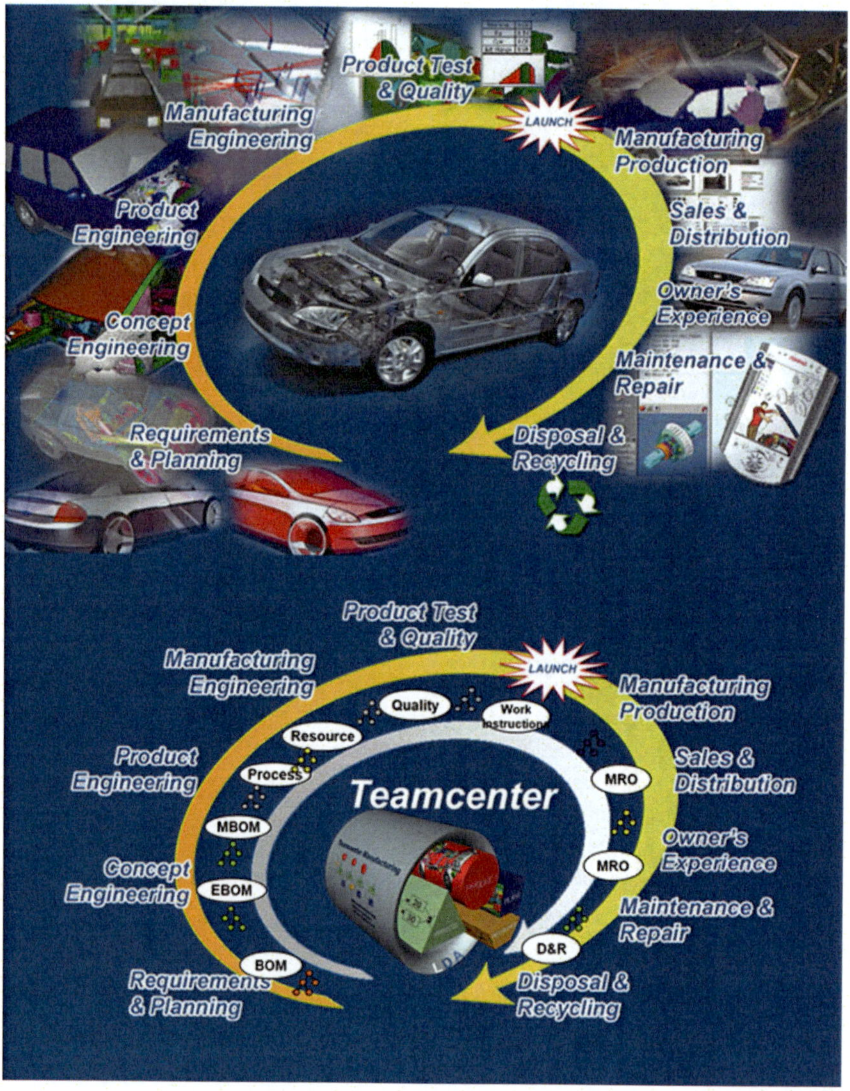

Fig. 11.5 Holistic PLM mindset and corresponding PLM backbone concept (outlined by [8] as part of their paper presentation in Stuttgart)

end of the 2000th. At the same time in the DACH (Germany, Austria, Switzerland) and European region, similar important PDM evolutions took place: in Germany the EDM/PDM extension of Contact Software (founded in 1990) towards the CIM Database architecture was driven organically inside the company rather than venture capital funded. The Germany PDM solution from Eigner + Partner (founded in 1985) took a merger and acquisition route, originally called *CADIM/EDB (CAD Informations Management/Engineering Data Base)*, later known as *axalant* and *e6* (after merged with Agile 2001) and finally taken over by ORACLE in 2007 (compare to [7]. A third solution is offered by the company PROCAD under the brand name PRO.FILE, which is positioned as a meanwhile extended product data backbone. Finally, it was interesting that SAP as an ERP (Enterprise Resource Planning) and MRP (Manufacturing Requirements Planning) solution provider also entered the PDM market as a fourth German company with its own solution SAP PLM [7]. In France the data management solutions of Dassault Système has evolved a long way from the original partnerships with IBM (USA) data management solutions, acquiring the IBM Product Data Management solution—Product Manager—to create the new company ENOVIA Corp. to develop ENOVIA LCA, the neutral SmarTeam PDM application up to the MatrixOne based new architecture of the new ENOVIA solution which is now and integral part of 3DEXPERIENCE). Please also compare with other earlier historic PDM evolutions in Chap. 5.3.

Product data are basically divided into *use-data* and *meta-data* over the entire life cycle. From a historical perspective, *use-data* includes all data that contains application executable models, calculations, simulations, work plans, etc. These are data sets that are created and used in the context of digital value creation—always with the aim of enabling the function, behavior and/or production of a product or the provision of a service. *Metadata* describe the use-data sets and provide additional information about their creation, usage and logistics. This includes, for example, date information about the creation, storage, modification and provision to others, versioning, persons and ownerships, status information, structural information, time effectivity, etc. In addition, they are used to describe the "state" of the product as well as the history of its digital development.

Product Data Management includes the collection and provision of product-related data. This includes both, the use- and meta-data as described above. Overall, the management of product data covers a wide range of tasks [1, 2, 3 and 8] such as:

- Store and order product data.
- Prepare data for search and use.
- Product Structure Management: composition of the product from assemblies and individual parts, BOMs (such as quantity, structure, development, or production BOMs). Traceability.
- Variant and Configuration Management: Management of different product characteristics using validity checks.
- Characteristics Lists/Classification: definition of characterizing and classifying product attributes, e.g. for reuse.

- Materials Management: use of defined materials or exclusion of unwanted materials in production.
- Change Management: traceability of technical changes over the entire product life cycle.
- Status Management and Release Systems: traceability of the validity of individual objects, so that only valid documents or released components are used in the next development step.
- Manage part versions.
- Deliver views on the product.
- Distribute and exchange product data to users, stake holders and external partners, suppliers and authorities.

CAx-systems are available for generating *use-data* in the product creation phase. The CAx-systems and the resulting CAx-models (*use-data*) can be subdivided into the following sub-classes (also compare to Chap. 7):

- Computer-aided design of mechanical products (M-CAD)
- Computer-aided design of electrical products (E-CAD)
- Computer-aided design of electronic products (EDA)
- Computer-aided engineering (CAE)
- Computer-aided NC programming (CAM).

EDA systems refers to systems that support the design of printed circuit boards (PCB). The software is created in special programming systems, which are outside CAx, summarized as IDEs (Integrated Development Environments) and ALM (Application Lifecycle Management) systems.

These systems have in common that they create different types of digital mock-up of the product and that they derive the documentation for manufacturing of products and their components. In other words, these systems generate data for the downstream process steps of the product life cycle. Furthermore, they also provide the basis for digital twins through the CAx models (compare to Chaps. 20 and 21). The intense use of CAx in product development addresses an increased variability, broader functionality and rising complexity of industrial products and an associated increased model complexity and data volume.

In the traditional paper-based world, each technical drawing was supported and accompanied by a parts list that defines parts that need to be ordered and assembled for a dedicated purpose. The DIN199 and ISO7573 define parts lists as specifications of all constituents of an assembled part by part reference number, quantity, part number, technical data, etc. [4]. Different types of parts lists are defined according to the corresponding purposes and use such as the distinction of variants, the description of building blocks or recording the amount of parts across variants. The wider, yet firm term *Bill of Materials (BOM)* comprises in its definition any kind of material according to a dedicated purpose. One example for a BOM content is additional material such as oil or glue, required during the assembly, but not mentioned in the parts list.

The *Bill of Materials (BOM)* is something that often considered a simple thing. After all, what can be complex in a list of parts? It used to be a simple list of parts on a drawing or later on in an excel spreadsheet. However, as we can see in practice, manufacturing companies are usually underestimating the complexity of making the list and fail to manage it properly. The results are usually painful and severe problems might occur quicker than you can think—missed components, wrong cost estimation, delayed product deliveries, frustrated engineers and customers who run around trying to figure out where the last and most updated bill of materials is. In addition, many OEMs in industry branches such as aerospace, automotive and rail have already invested heavily in the range from $50 to $100 million per enterprise into the development and deployment of new, upgraded or refurbished BOM technologies and applications in the first twenty years of this millennium! This can range from a replacement of traditional BOM systems by modern PDM system based BOM applications up to new WEB front ends to traditional mainframe BOM backbones. The discipline of defining specific "BOM methodologies and repositories" for software applications and to the sharply rising embedded software content in physical products has emerged and is also know under the technology terms ALM (Application Lifecycle Management) or Software BOM.

As a consequence, in modern product development, the *Bill of Materials (BOMs)* solution set represents a complex multi-disciplinary information structure combining mechanical, electronic, software, and system information. Openness and simplicity are the main drivers to help companies to get up to speed with data management science behind the Bill of Materials. It should be easy to create a Bill of Materials and manage all aspects of it. However, it oftentimes becomes complex to robustly manage BOMs due to many aspects, attributes, variants, configurations, parts & assemblies, component types and stakeholders involved.

PDM Systems that hold the product structure, offer the possibility of dispensing BOM. By following certain criteria, the product structure is top-down traversed and the resulting filtered BOM is displayed. The BOM may be exported for further use. Figure 11.6 shows different types of structurally different BOMs for development phases from early planning until the support and sales.

BOMs have different composition in the respective IT system at different points in time. At present, most industrial companies use different IT systems in the various phases to meet the specific requirements. This is usually efficient in the individual process steps, but brings with it considerable disadvantages in the overall product data management. Therefore, consistency in many cases is not given on both, the process and data level.

11.2 Engineering Understanding of PDM and BOM

It can be stated, that the concept of product data management is always subject to a company-specific, inhomogeneous and often traditional understanding. The range extends from a synonymous use with the term drawing management (IT system) to

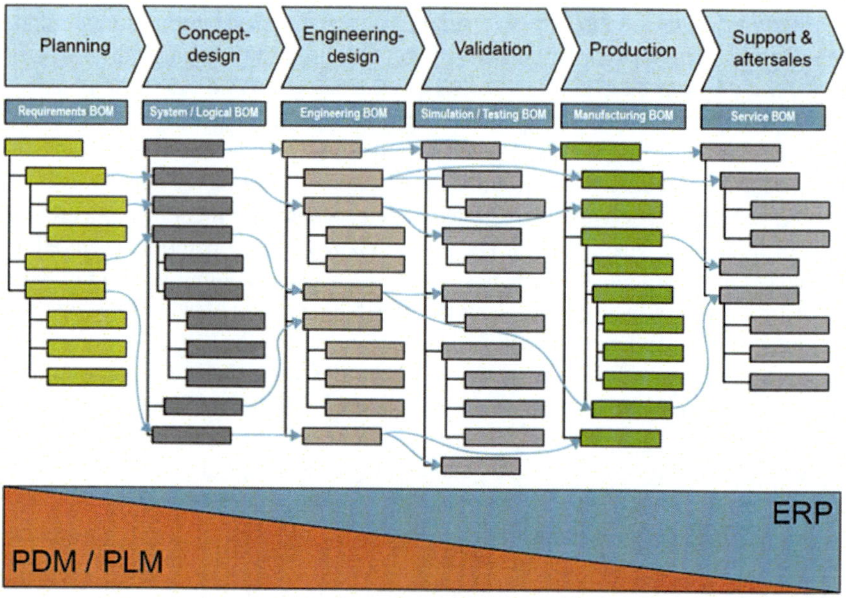

Fig. 11.6 BOM-management through different phases of the lifecycle [10]

interdisciplinary product life cycle management (process perspective). In science, the most diverse definitions and understandings are found, which are in constant change. However, the software by itself is not sufficient to serve an engineer need.

For a more comprehensive view of product data management, the levels of processes, methods, IT systems and information standards need to be considered in an integrated manner. Figure 11.7 depicts an extended view on VDI 4499 PMT Model [7], which are required for the management of an engineering application of PDM.

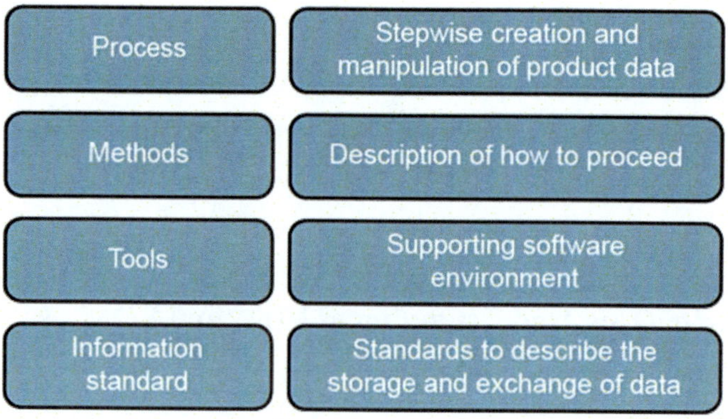

Fig. 11.7 PMTI (process, methods, tools, information standard) model

The added information layer permits the secure data storage and standards for data exchange. Furthermore, the data models of how to store data in a certain composition builds the information basis for engineering work. The tool layer provides the engineer with a data manipulation environment, one of them is the PDM tool. A further layer is the methods understanding of how to manipulate data. This is required in order to work in accordance with the tool functionality, process framework and information standards. The business process layer provides guidelines on how to order activities.

Engineering work of product data management comprises five practical dimensions. These dimensions are:

- Object Versions: Object versions describe the maturity level of product descriptive data generated during the development process.
- Domain Views: The domain views describe the different models of an object that are created in the development process (e.g. mechanics, electronics, software).
- Product structure: The product structure describes the structural relationships and interdependencies of objects that arise in the development process.
- Release status: The release status describes the status of an object in the development process.
- Product variants: Product variants describe alternative objects for an object.

The Bill of Materials is a structure and/or listing representing all items needed to make a product. Depending on the product and business it includes a different level of details about how the product is engineered, manufactured, assembled, sold and maintained. Every engineering and manufacturing business should be concerned if parts and assemblies, as well as any other related data, are not managed properly. PDM technologies and solutions can be leveraged to describe Bill of Materials.

11.2.1 What is PDM Doing for an Engineer?

The benefits for an engineer who applies PDM result from the supported collaboration within a company or even across companies. Three main aspects are provided by the PDM system: data storage in a centralized repository, metadata management and process management.

The CAx systems mentioned above generate a large amount of data/information in the individual phases of the product life cycle. Today, PDM systems are predominantly used for their management. The *use-data* (CAx-model) storage in the vault provides a centralized secure and save repository. Access right is permitted according to company settings, so that co-workers may further use the data. An engineer may use the repository as source of information. It might be relevant for engineers to find out which *use-data* (CAx-models) like CAD design models, CAE model builds and simulation runs or Digital Factory models [see 11] and control programs might exist already and which other design artefacts are available for the technical system and product context. Extended search functionalities that include design shape similarity search (via vector tuples), cartesian space and proximity search via bounding boxes as well as text and index search support engineers in allocating, accessing and using data. In addition, it is possible to learn about design history if engineers load sequences of versioned *use data* models and compare them in respective authoring applications. This opportunity, however, is not well supported today and will require much better functionality in the future.

The increased product complexity is caused by the fact that, in addition to the classic mechanical components (hardware), components from electrical engineering or electronics and software are increasingly being used to perform product functions. These types of products are called mechatronic products and need to get supported by an integrated product structure (=structured parts list). Due to the mutual influence of individual components throughout the change process, the effective management of this product structure is only possible via the functional capabilities of PDM systems.

The *metadata management* includes variant management, revising and classifying use data, setting a status as well as relating individual use-data to other use-data and to system/product context. These metadata thus indicate and control the further application of the use-data, e.g. if the part and its design data are ready and/or mature enough to

- be published to others (e.g. to enable engineering collaboration),
- be released for manufacturing,
- be purchased,
- be used in a dedicated product variant, etc.

For each of the above-mentioned metadata dimensions dedicated functionalities are included into the PDM tool. Once an object is generated in the PDM database all metadata indications of further use may be added and modified. Therefore, an engineer benefits from a professional and supported database managing metadata for design artefacts. Metadata are stored and managed centrally and are available for further use within the PDM system. Not only the engineer who authors and registers the data, but also the co-workers in Engineering and other down-stream users and costumers benefit from such data repository. One main benefit for an engineer on the data delivering and consuming side is the possibility to visualize and browse the product structure and other linked data sets. By doing so, engineers and down-stream users such as buyers, cost estimates and service personnel will be able to gain an overview and understanding of the product decomposition and characteristics.

The *engineering and management collaboration process* is supported by an integrated PDM workflow-management system. Figure 11.8 shows the example of an engineering change workflow, that involves different stakeholders such as fitters, engineers, designers, managers, production planners, suppliers etc.. The PDM system provides different digital change document with links to the relevant use-data, metadata and workflows. The stakeholders can get involved and connected via the change mechanisms and workflows of the PDM tool according to the company specific change process. The stakeholders complete and modify the content of the change

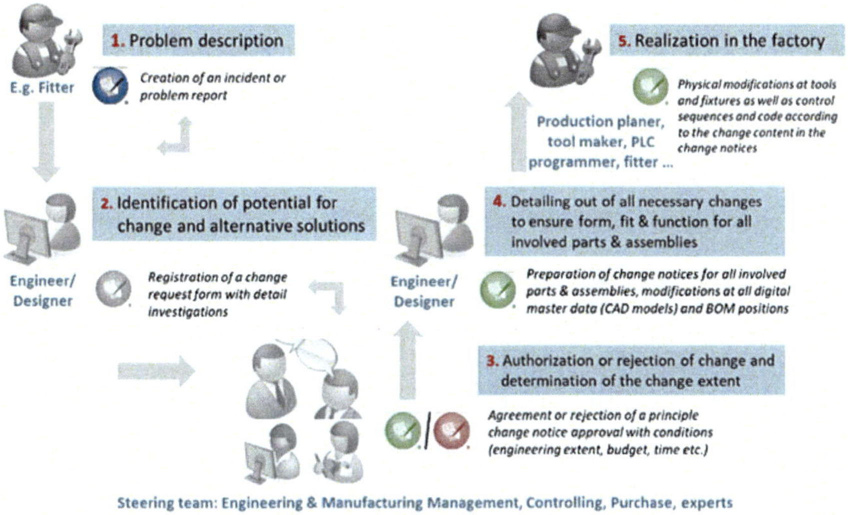

Fig. 11.8 Example of a workflow supported change process

documents and link use data to it. They authorize the generation of new or the modification of existing digital models of the parts in the change content. The status of the change documents, the transfer to the next stakeholder in the workflow as well as the status information is executed by the PDM system. An overview, statistics and transparency are provided to all change workflow users.

Overall, PDM systems support engineers in data management and process management in the overall product life cycle (compare to Fig. 11.4) in particular. Data management is divided into the following modules:

- Master data management (incl. vaulting)
- Product structure management
- Document management
- Classification of products, components and articles
- Integration of CAx authoring tools
- Collaboration of companies, partners, departments and engineers.

Release and change management are counted as part of the PDM process management and linked to methods of versioning and configuration management. Another aspect for the use of PDM in the company is the integration with the help of the integrated product structure across the disciplinary boundaries, which helps reducing or even eliminating separated data silos that arose in the past. PDM organizationally merges the different domains or disciplines (mechanics; electronics including electrical engineering and software) into one common data management approach. This is done by integrating the respective authoring tools into PDM. The result is a common product structure. PDM maintains relationships and structures of the individual business objects. However, this pre-supposes that business objects/items are maintained in a timely manner (incl. version control).

In the following some core PDM functions are described. They are offered to help engineers in order to coordinate, modify and interpret dynamic and causal modifications at *use- and meta-data.*

Version Management of Information Objects

All information objects (items) change through time. In order to manage and coordinate such information objects (items), this continuum of change must be broken into discrete units that can be identified, tracked, configured, retained, shared and distributed accordingly. Therefore, the *version* is a discrete unit of change. All types of information objects that need to be tracked are versioned. This includes part objects, design representations, placeholders and placeholder solutions. A *version number* identifies each *version.* The version number usually does not have any "intelligence" implied in it. It is systematically assigned and simply identifies the *version.* The *version number* does usually not have any significance and, therefore, does not get influenced by *part numbering.*

An example of a version sequence could be:

Part AA version 001

Part AA version 002

Part AA version 003 ...part AA version n.

State Management of Information Objects

State is the attribute that identifies whether the version is frozen or not. In an ideal digital working world, "*Work In Progress*" and "*Frozen*" would be the states required. When the work on a version is complete, it is frozen and distributed to all locations, all co-workers and partners in order to be used as the new "default" version. However, in the real world, things are not quite that simple. Sometimes a version must be exchanged or distributed even though it is not yet the new "default" version. Such situations include:

- An exchange of versions that are not yet complete for collaboration
- A data exchange to an engineering expert team which pre-checks the versions according to certain packaging or functional attributes
- A distribution of versions that are complete for review and approval.

To cover these cases, the *state attribute* oftentimes carries three values. Figure 11.9 shows the state values and explains their possible meanings in the context of product data management rules and methods within a company and their extended enterprise collaboration system.

It is important to note that there usually exists a connection between *versioning practice* and *application of state attributes* in Virtual Product Creation. So, please pay attention to some possible or typical design working rules related to *versioning*

State	Description	Frozen	Distributed	Default
Work in Progress	The version is still being worked on. The information can change at any time. It is not distributed and is not the default version.	N	N	N
Review	No more work will be done on this version. The information will not change again. It is distributed, but is not the default version.	Y	Y	N
Published	No more work will be done on this version. The information will not change again. It is distributed and is th new default version.	Y	Y	Y

The table below illustrates the use of stat with versions:

State	Version						
	1	2	3	4	5	6	7
Work in progress	✓	✓	✓	✓	✓	✓	✓
Review	✓	✓	✓	✓	✓	✓	
Published	✓		✓			✓	

Fig. 11.9 Example of the meaning/application of the *state attribute* in PDM

and *state* that need to be agreed amongst the engineering teams (usually lead by method experts):

- A new version can only be created if the previous version is *Review* or *Published.*
- In the example in Fig. 11.9 (lower part), Version 8 cannot be created until version 7 is moved to *Review* or *Published*.
- Once a version is *Published*, any previous versions that are still in *Review* state cannot be *Published.*
- After version 6 was *published*, neither version 4 nor version 5 could be *published* (compare to Fig. 11.9).
- Once a version is promoted to *Review* or *Published*, it cannot be demoted back to *Work in Progress* or *Review*.
- After version 4 was *rejected*, version 5 was *created to progress the design.*

Interestingly enough, the majority of industrial companies do not necessarily follow such stringent rules (yet). The traditional loose practice of CAD designers in the early creative part of engineering development never made it obligatory to work stringently to such collaboration methods—if it comes to the "ordinary change managed" design progression towards releasing such stringent rules, however, becomes mandatory!

#	Maturity	Description
1	Development	This version is sufficient only for development work.
2	Sourcing	This version can be shared with the supplier for bidding.
3	Prototype	This version can be used to create prototype tooling or parts.
4	Tooling	This version can be used to order production tooling.
5	Production	This version can be used to produce production parts.

Fig. 11.10 Using *maturity levels* for lifecycle situations of information objects

Maturity or Status Management of Information Objects

During the development and manufacturing of products and their sub-assemblies and parts/components, their digital models and associated meta-data representations pass through various *statuses* or *maturity levels*, which are documented in the master data by letters, numbers or a combination of both. The *statuses* or *maturity levels* thus reflect their current status and can even form a status network among themselves. Typical *statuses* for lifecycle situations are:

- *W(ork),* ***W***: currently still worked on by an individual or a team,
- *P(re) R(elease),* ***PR***: officially suggested for release
- *A(proved),* ***A***: approved for release but not yet sent to downstream
- *R(eleased)* ***R***: authorized for release incl. downstream usage (ERP)
- *B(locked),* ***B***: blocked for any modification and usage
- *C(hange),* ***C***: in change, i.e. modifications are under way
- *E(ol),* ***E***: end of life, i.e. no further modifications are possible.

As an alternative, other groups of industrial companies prefer to orient themselves towards *maturity levels* of lifecycle situations as shown in Fig. 11.10.

In comparison to the *state attribute*, the *statuses* and *maturity levels* do not concentrate on the *collaborative aspects* of the information object but rather on the level of the *life-cycle* of the information object and the associated *authorization or modification situations*. As an information object goes through its lifecycle it achieves different levels of *statuses* and *maturity*. *Maturity* is a measure of what the object is "good enough for". *Maturity* is an attribute of the object itself, not the use of it. *Statuses*, however, orient themselves rather to the allowance of how the information object can or should be used in the lifecycle (as opposed to how it is used in the engineering collaboration which is expressed by the *state attribute*).

The list of maturity values is an ordered list of increasing maturity. Using the example maturities in Fig. 11.10, a version with a *maturity value* of "Prototype", is also suitable for activities that require "*Development*" or "*Sourcing*" maturity. Like other attributes of an information object, *maturity* is modified over the lifecycle of an information object. Figure 11.11 shows a time sequence example and the way how

#	Date	Time	Event	Version	State	Maturity
1	4 / 24	09:10	The user creates the new object.	1	Work in progess	Development
2		17:15	The user freezes the object to share an intermediate review with his colleagues.	1	Frozen	Development
3	...	...	...	...	...	...
28	5 / 07	10:02	The user versions the object to make final change before sending it out for bids.	14	Work in progress	Development
29		11:50	The user completes the changes and sets the maturity value to Sourcing.	14	Work in progress	Sourcing
30		11:53	The user freezes the versions for sharing with the suppliers.	14	Work in progress	Sourcing

Fig. 11.11 Time sequence example how to apply *maturity levels* in combination with *versions* and *state attributes*

maturity levels are assigned.

Maturity as well as *statuses* are declared values (as opposed to derived or calculated values) that may span multiple versions of the object. When a new object is created, it defaults to the first maturity level value (row #1 in Fig. 11.11). When a new version of an object is created, the new version is assigned the same *maturity* as the previous version (row #28 in Fig. 11.11). The user can change the maturity value for a version while it is in WIP state (row #29 in Fig. 11.11). Once a version is frozen (either *Review* state or *Published* state), the *maturity* value cannot be changed. The list of maturity values may vary by object type, however. The list of maturity values will not be configurable. A visual cue on maturity could be implemented as an aid to design and visualization. An example of this might be the application of decreasing degrees of transparency to geometric representations of parts based on increasing levels of maturity. Within the CAD or visualization tool, mature parts would appear fully opaque and those in earlier stages of development would be more transparent by varying degrees. Such implementations, however, are not (yet) common in today's VPC industry practice. In theory, *statuses* and *maturity levels* can apply to all objects used in the PDM, such as documents, models, parts, articles, components product structures, design representations, etc. In addition, they must also be coordinated across IT-systems, for example with ERP (Enterprise Resource Planning) systems. However, it is up to the rules, working frameworks and methods within the individual company and their extended enterprise system to decide how wide and extensively they are used in daily practice.

Master Data Management (incl. vaulting)

One of the most important steps in populating PDM environments with digital data is the *Check-in* process of CAD models (as the prime example for any other authoring application). CAD models represent a specific status of the design process with

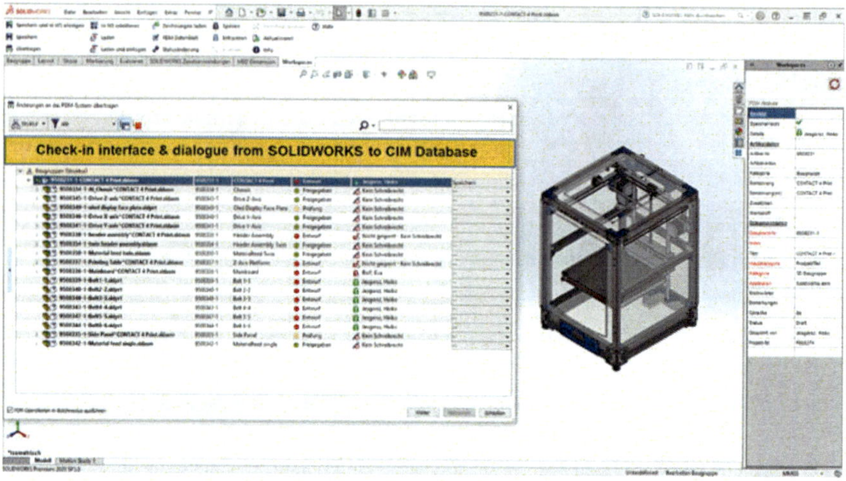

Fig. 11.12 Check-in of CAD models (SOLIDWORKS) to PDM (CIM database)

respect to a design concept or intent, to design details and to mature interface solutions in between parts and components as part of an entire CAD assembly. In order to "hand-over" the CAD model working state to the structured environment of *Product Data Management (PDM)* tools, a range of important "Check-in" concepts need to be understood by the user (e.g. CAD Designer) and need to be robustly managed by the interface between CAD and PDM system. As shown in Fig. 11.12 the check-in interface carries a number of critical information sets which need to be aligned between both digital model types.

First of all, it is highly desired to establish and to continuously defend the *single master principle* during virtual product creation and for the entire digital lifecycle: this principle guarantees that *meta data* information objects in PDM (also known as items in some PDM environments), which are created upon the check-in of CAD models in such a CAD-to-PDM bottom up "hand-over and save" operation, follow a unique identification schema to avoid any duplication of objects within the entire PDM system environment (i.e. even across site implementations and across company servers as part of digital collaborations between project partners). Consequently, each meta object or item newly created in PDM during the check-in of CAD models will generate a unique identifier (UID), which in turn is checked and maintained during any update of such object (e.g. during the next CAD check-in process).

The same, however, should also be applied to the CAD models themselves, which are handed over to the control of the PDM system. The CAD models are treated as *use data* in the PDM environment saved in a secured location on the PDM controlled disk space (which is often referred to as "vault"). Similar to the UID of metal objects, many PDM architectures use unique identifiers of CAD models below the UID of the meta objects. This however, needs to be ensured in precise correspondence with the identification schema of the individual CAD systems that handle such concepts within their internal regime of CAD assemblies as well. However, still many CAD system installations only use file names as identifiers. The attempts to embed global identifiers have unfortunately failed in many companies, but did succeed in single vendor CAD/PDM implementations. Modern open identification schemes, therefore, are based on URIs (Uniform Resource Identifiers) as known from WEB technologies.

The check-in of 18 CAD assembly objects (items) in Fig. 11.12, therefore, might cause hundreds of handling calls between the CAD client and the PDM staging area that ensures the correctness of all check-in items in the PDM environment. This is the reason why such check-in processes are handled by specific CAD-PDM interface broker applications. They are managed at the PDM server side in order to keep sensitive control of all handshake and process services sequences and potential error log handling. However, in case of modern workspaces it becomes reasonable to exchange complete change sets between PLM and the CAD environment, reducing and simplifying such interactions.

Secondly, the CAD to PDM check-in process might transfer a rich mix of additional meta data of the CAD assembly and component models to the respective PDM meta objects (items). As shown in Fig. 11.12, in the main picture (on the right), information sets such as numbering, naming, status of design work and responsible object owner are part of this transfer. As shown in the side panel on the right side, there are additional important data sets transferred upon check-in such as categories of classification, document #, assembly #, project name, etc. Additionally, model data has to be extracted: positioning of parts in an assembly, material used, the geometry itself etc. Such extraction can happen via the interface, or later in the backend system.

It should be point out, however, that there exists an alternative to the population of meta objects in PDM through the CAD check-in process: the *top down product structure and object creation* in PDM. In such a situation, the product data objects already created upfront in PDM (aligned to and usually maintained in product structure) do create respective objects within the CAD assembly file, which can be then populated with CAD models in the CAD system. Such top-down working style is often used in the early days of a development project by setting up meta data populated product structure. Later on, in the development project, mostly the CAD to PDM bottom up approach is used.

11.2.2 What is BOM Doing for an Engineer?

Everyone needs at least one *Bill of Material (BOM)*—if not several—in the engineering team and to be able to produce the product in the manufacturing company. In the following, please reflect a simple breakdown of why everyone needs a *BOM* and what specific needs are demanded by different people and teams:

- Product Development needs to track the information about the product to manage what is needed and what is a possible impact from changes. Product Development finally determines the content of a product via the *(Design or Engineering) BOM*.

- Manufacturing planners need to have a *(Manufacturing) BOM* to organize production planning, production and assembly processes.
- Procurement and Purchase usually need a BOM to make cost estimation and find the best suppliers and contractors.
- Sales departments usually need a Bill of Materials for sales configurations and offering features in the products.
- Depending on the type of manufacturing—*Engineering to Order*, *Configure to Order* or *Build to Stock sales* the companies need BOM for different reasons including order details, forecasts or project management.
- Support and Maintenance need a BOM to know exactly what was sold to the customer, how to maintain it and how to manage the ordering of spare parts.

In order to provide a deeper understanding of BOM solutions and their core conceptual elements, it is first of all necessary to give an insight into the different views engineers have with respect to the development and engineering of components, parts, assemblies, modules and their digital models. As shown in Fig. 11.13, the conceptual viewpoints of what to describe and how to describe the virtual product differ significantly between the *Bill of Materials* authoring and the *CAD modelling activities*: the BOM expert and Release Engineers mainly deal with part numbers, part types, part names, quantities, feature conditions and rules as well as with time effectivities for production start and end of production. This is absolutely key to determine the correctness and completeness of the parts lists for all product variants. Digital models (geometric shape, functional and behavioral simulation etc.) are expected to be generated and completed by others.

The CAD Designers as well as the Component and System Engineers primarily concentrate on the technical solution development, validation and verification side of the business. Therefore, they have to deal with modeling concepts, e.g. in CAD design, like solids, surfaces, attributes, technical parameters and linking mechanisms

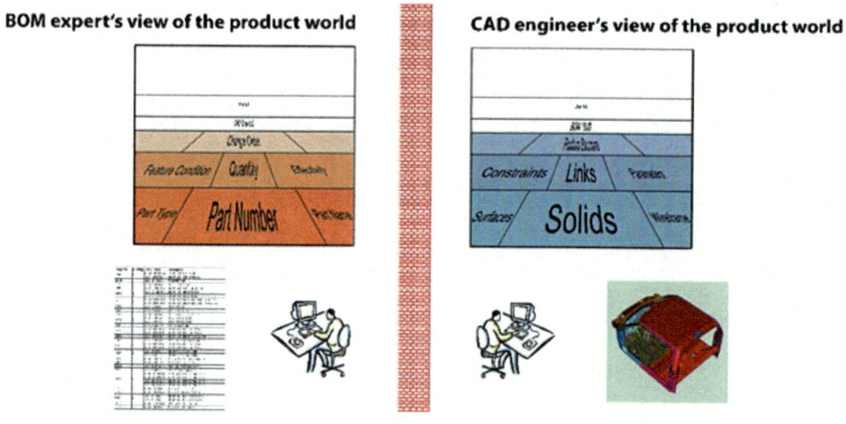

Fig. 11.13 Differences in understanding and expectation between a BOM viewpoint and a CAD modelling view point

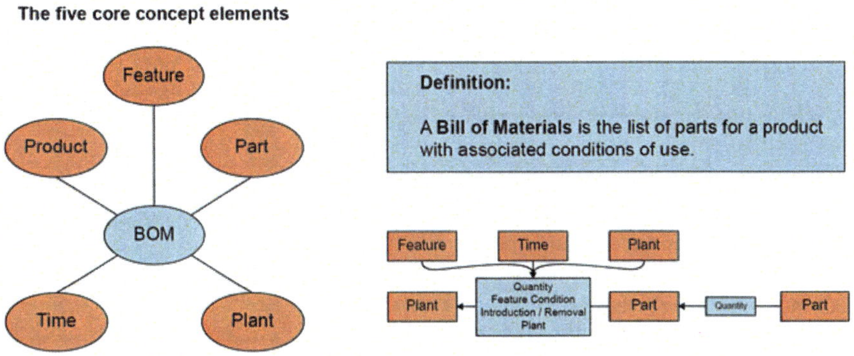

Fig. 11.14 The five core concept elements of BOM solutions

between them. Neither the correct naming nor numbering standards nor the necessary coding language for product families and variants are the main focus of their professional work. However, the correctness of those designations and the completeness of the product and technical system context also constitutes an absolute must for those engineers to become successful in finalizing their own deliveries.

As a consequence, *Bill of Materials* might be a significant help or even represent a challenging duty for a different kind of engineers. In order to provide better understanding, it is necessary to define and explain the main core concepts of *Bill of Materials*. As outlined in Fig. 11.14, the *Bill of Materials* (BOM) is defined *as a list of parts for a product with associated conditions of use* by leveraging the following 5 core elements:

1. The target entity of the BOM description, the *product*.
2. The elements which are called out by a BOM description, the *parts*.
3. The environment in which the BOM is used for the build, the *plant*.
4. The expressions to describe the part usage in the BOM, the *features*.
5. The relevance of the BOM with respect to *time*.

This understanding raises the question of which information sets should at all be captured within a BOM based on or related to those five core elements. The answer is not simple and it differs between industrial companies and industry branches significantly. However, the following information sets are usually amongst the most common ones:

- **Part Number**: This is a unique string that identifies every single component in the product. Not all part numbers are the same. There exists very little standardization across BOM management in companies. Therefore, *OEM*, *Manufacturing*, *Suppliers* and many other specific part numbers exist. Those represent additional pieces of information. Engineers might not use all of them. But, in order to have a unique *Part Number* for every single item it is a must to keep things organized.
- **Description (of the part)**: It usually gives the organization a simple way to look at the details about parts and components. Used for the search, simple identification

Description is not unique and must not be used to reference parts for any purposes except human interaction. *Part Number* is used for everything else.
- **Part Name**: *Names* make it easier for organizations to track components. Many systems have internal names which are specifically useful to *Operations*, *Test*, and *Engineering* teams. In many cases, companies have introduced official names as part of their internal information standards, but not all of them do it this way. A separate nomenclature might be used with customers in *Sales* organizations. Unlike part numbers, part names are not necessarily unique. Consistent use of part names like connector, adapter, and cable, can help keep everyone on the same page of (semantic) understanding.
- **Unit of Measure**: A very important piece of information that helps to identify how to measure the quantity of the part (which is essential to determine the number of all parts within the product BOM).
- **Type (s)**: Usually the *type attribute* absorbs different types of information to classify a part to help identifying how to manage the part. Organizations can apply multiple *types* to classify parts and how data about each part can be processed.
- **Revision, Version and Phase**: This is used to identify the level of maturity and specific changes. Companies are using various practices during the creation of new part numbers of revised already existing parts.

In addition to these parameters, organizations are usually including information about plants (factories) manufacturer, suppliers, cost and other pieces of data in BOMs. The information sets described above are usually the same about each item or object in the BOM, i.e. each part or purchased assembly. However, another group of information sets is typically included in BOMs and it represents a specific *part usage*:

- **Quantity**: It is the ultimate way to describe how many pieces of the part you need to have in a given product. Note, in some cases (like the "instance BOM"), quantity can only be calculated!
- **Reference Designator**: It is typically used in the electronic Bill of Materials. It usually represents a uniquely identified instance of the part (assembly).
- **Location**: It is typically used in mechanical BOMs to differentiate between instances of the part/assembly.
- **Effectivity**: It is an important element of BOMs. *Effectivity* usually represents specific characteristics by showing if a part (or assembly) can be used and under which conditions. There exist different types of effectivity—date, configuration and serial number.

In addition, there exist different types of BOMs principles for Engineers in order to enable the best use of BOMs within business processes and decision-making meetings:

- **Configurations, configured vs resolved BOM**: Last, but not least a specific piece of information is related to complex products that have multiple configurations. A BOM that contains information about multiple configurations usually is

called "*configured BOM*" and the BOM that represents information about a single configuration is called "*resolved BOM*".
- **Single-level, multi-level, flattened BOM**: A list of components represented as one level is usually the basic way to plan the BOM. However, components are connected into assemblies and those usually have sub-assemblies. This is a simple *single-* or a *multi-level structure.* A "*flattened BOM*" is a report that summarizes the usage across multiple levels.

It is essential for Engineers to be able to create and modify BOMs since the capabilities to conduct *BOM Data Management* become essential for product completeness and correctness. Flexibility and ability to configure application are two main things for Engineers and manufacturing to work with *Bills of Materials*, *Catalogs*, *Vendors* and *Order* planning. Therefore, to have a robust data model to support the *Bill of Materials Data Management* becomes essential. This capability is called the *reference-instance model.* The idea of such a model is to create a granularity around the elements *Item Reference* and *Item Instance* as part of the *Bill of Materials* data management principles. Let's use the example of a simple skateboard as shown in Fig. 11.15. You can think about a collection of items (components) that are available to build a product—wheels, boards, axles, screws, etc. You can build multiple products from the set of components.

Think that a skateboard is one of them. A *reference* is an abstract object that can be modeled, e.g. as shown in *OpenBOM* solutions, which has been developed and it is offered by the team around Oleg Shilovitsky. This object has a set of properties and represents any part, assembly, or component. In some companies, such objects are also called generic parts since they exist in many or even in all products. So, a *reference* can be a nut, a bolt, a screw, an electric motor, or any piece of equipment one can buy, manufacture, or outsource. The BOM solution *OpenBOM* uses, for instance, catalogs to define references and configure data properties. An instance is an actual part used in a specific product (engineering product or built product) or an entire product that should be built. Thus, if you create a skateboard, which has 4 wheels, then you have 1 reference of a wheel and 4 instances of a wheel. As shown in Fig. 11.15, references are represented as nodes, while instances are represented as links. *OpenBOM* allows to create and customize properties on both reference and instance models. As an example, you can add a property to a catalog (e.g. cost) and all Items of this catalog will get this property immediately regardless of which BOM this item is used for. At the same time, it is possible to add instance property (e.g. *Reference Designator*) to a BOM, which will be only used for electronic BOMs and will not be available and used for mechanical parts. Properties are building blocks of information. You describe any characteristics of the product by using properties of different types—text, numbers, references, cost, lists, date, etc. Engineers use BOMs in order to display and monitor product development progress, to provide an overview of certain product properties such as cost or weight, to deliver a domain specific view of the product composition e.g. in modules or to serve as a baseline, for example, to finalize the production extend or the hand-over of a product to a customer. As elucidated already above, the PDM product structure is traversed, according to

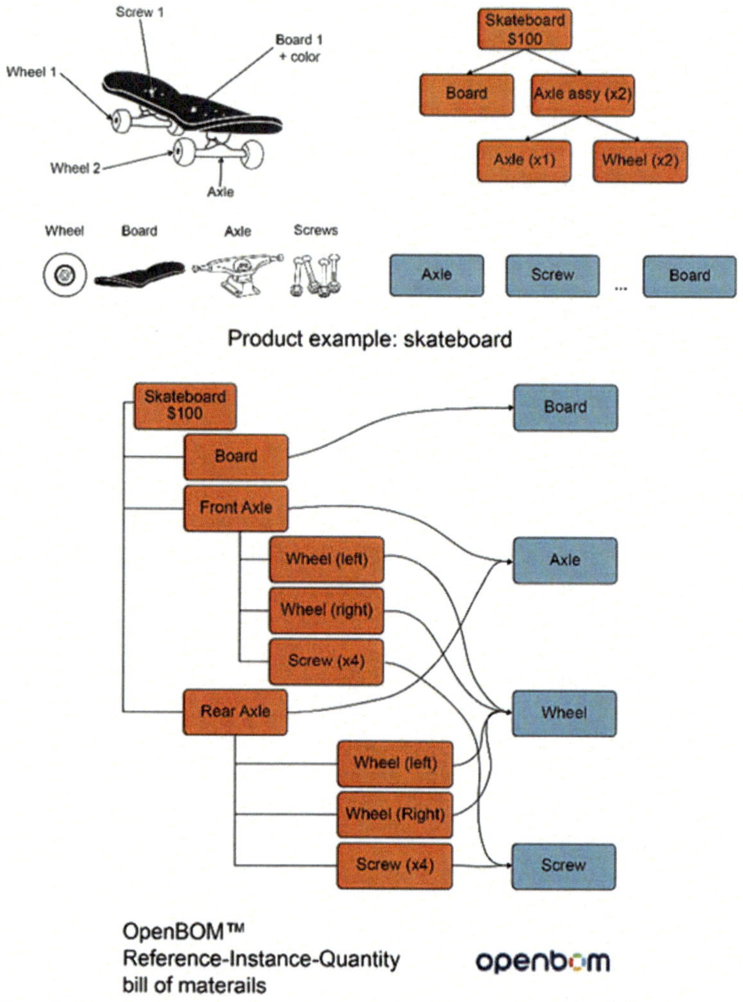

Fig. 11.15 *Reference-instance model* in BOM management

relevant filters for status, variants, etc. and the resulting BOM is visualized in a table-like view.

In accordance to the product development process, certain milestones or quality gates are defined, where BOMs are derived or diverted from PDM. The subsequent management of the engineering progression and the related decision processes heavily rely on the BOM existence and on the status of the BOM with respect to completeness and content. These management and decision processes such as “Does the product meet the costing boundaries?” exhibit the exceptional practical significance of BOM and, therefore, underline the importance for both, Engineering and Planners as well as for Managers. According to the different materials that compose a product or a subset of a product, and the order in which the material is displayed, BOMs show different views on the product. When using the metadata in combination with the material, BOMs support the cost- and weight management of products. When exporting and saving the BOM, it serves as a frozen baseline and supports the baselining process. The range of BOMs might differ according to the product's complexity and size. In the aerospace industry entire products consist of ~1billion objects, cars are built within a range of up to 100.000 and cell phones some 1.000 assembled objects. Figure 11.16 shows an example of a PDM E-BOM structure of a machining center consisting of several thousands of individual BOM lines in the PDM system CIM Database from Contact Software.

It is challenging for engineering teams to ensure and declare BOM completion throughout the engineering iterations and at specific gateways of the development process. The reasons for such challenge are manifold: changing assumptions about development project content due to product planning adjustments, unclear ownership due to increased level of partner and supplier involvement, new team members with limited experiences, new technology solutions with different number of sub-system components and assemblies, changing manufacturing assumptions, pending decisions on make or buy determinations with direct influence on purchased assembly situations etc. Hence, any help to easily recognize possible missing parts are welcome. As shown in Fig. 11.17, it is advantageous if Engineers can use the linkage to a PDM system that contains a generic product structure and represents the source for the Digital Mock-up (DMU).

Within both views (PDM product structure, see the midsection of Fig. 11.17, and the graphical view of the DMU, see the right section of Fig. 11.17), it is easily visible that a *structural reinforcement member* is missing on the left side of the underbody.

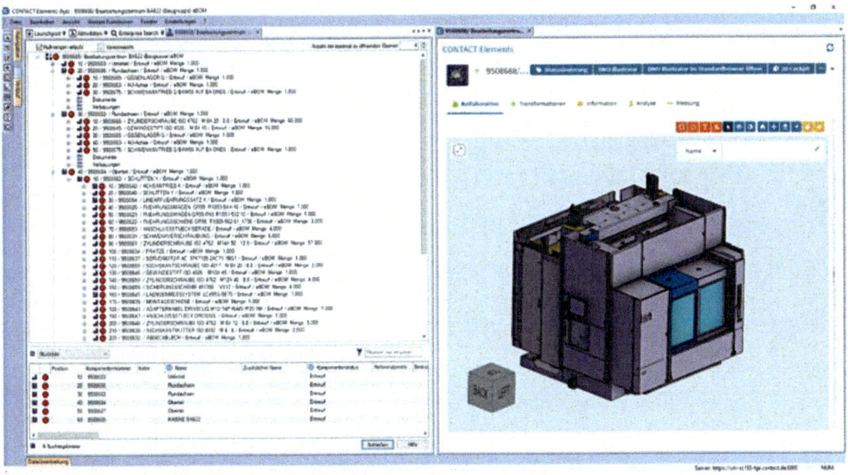

Fig. 11.16 PDM product structure based BOM of a SW machining center (*Source* CONTACT Software and Schwäbische Werkzeugmaschinen)

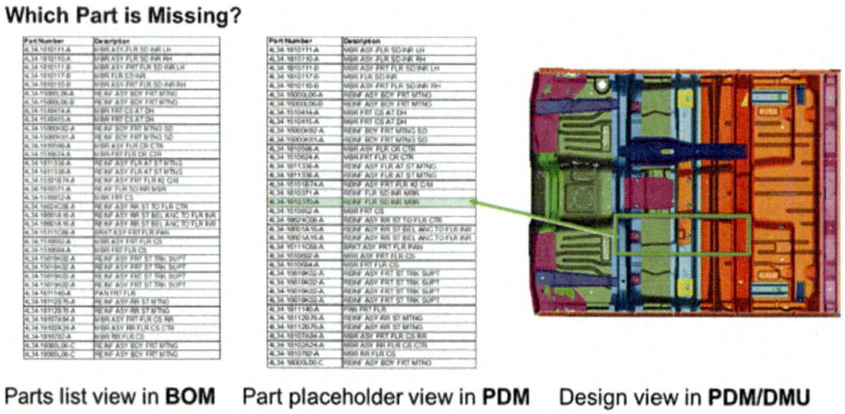

Fig. 11.17 Support by PDM structure and DMU to achieve BOM completion

The BOM Engineer can leverage the advantage of PDM product structure since it uses generic placeholders for all generic parts of a product and therefore can easily flag out a missing part instance. Figure 11.18 shows the other way around: if the parts list of the BOM is in correct shape it is possible to identify missing digital models (such as CAD) if a stringent linking exists between the BOM to the PDM product structure and the DMU.

How to Derive BOMs from PDM?

From the beginning of this millennium onwards an increasing number of manufacturing companies have understood how essential the introduction and use of a PDM system is in order to create, share, distribute and authorize product data in the context of the entire product, technical system and/or factory environment. As a consequence, PDM representations and functionalities have become decisive in order to directly

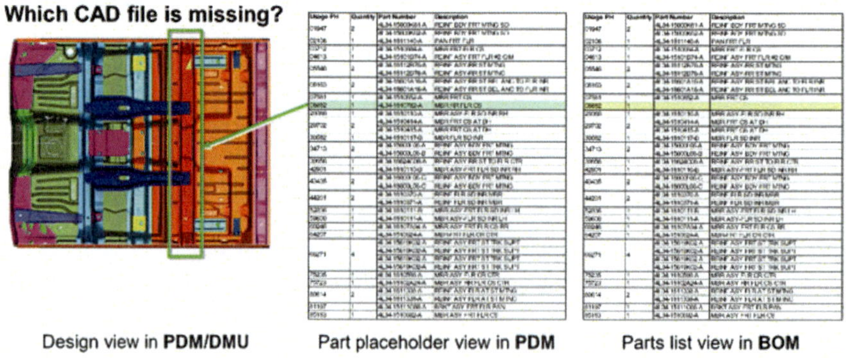

Fig. 11.18 Usage of BOM parts list to find out about missing CAD files

derive BOM from the product data records and structures in PDM or to be in tight interaction with the management of BOM (structures) in separate IT-systems.

As it was already indicated by Fig. 11.15 through Fig. 11.18 it becomes, therefore, essential to know and understand the principle ways of how to derive or divert a BOM from PDM environments. One of the core "digital context orientation" mechanisms in product development is the product structure, which ideally reflects the functional view of the product. Although there exist different views of what the product structure should represent *(functional view* vs. *organizational view* vs. *physical architecture view* vs. *manufacturing sequence view*, etc.), the product structure is in any way documented in form of the parts list in a certain order and with equivalent sub-structuring.

The *product structure*, therefore, represents a list for a product that contains all components, elements and/or groups belonging to the product. A *functional product structure* is usually a set up in Virtual Product Creation which is created for the first time in the CAD system (mechanical, electrical, electronic) as a structured assembly file. Such product structures are also called *Design BOMs*. Such an implicitly "bottom-up created" structured CAD assembly file-based product structure is transferred to the PDM system in order to create the PDM based product structure objects for further administration and supplementation.

This is done by means of the PDM functionality *Integration* and employs, e.g. the *CAD check-in process* (compare to section *Master Data Management* and Fig. 11.12). Such CAD driven *Design BOM* forms the basis for building the overall product structure, into which additional *Design or Analysis BOMs* from the other CAx systems can be collected and merged. It is then finalized to the *Engineering BOM* (E-BOM). If software is also understood as a new type machine element similar to the classic one, such as a screw, nut, etc., then the software representations (e.g. source code, or run-time software application) must also be included in the overall product structure and managed in PDM. The software itself, is generated outside of PDM in the corresponding CASE (Computer Aided Software Engineering) tools like the "traditional" components of the hardware or the components of the electrical engineering/electronics. Some companies also support a "top-down" based product structure object creation directly within PDM which can be downloaded to CAx-systems for further population.

The materials and/or assemblies required to manufacture the product are defined as part of the engineering progression in product development and are used to typically create a *function-oriented product structure*, also known as a *design bill of materials* or *Design BOM*. The *Engineering BOM* is built up in the PDM system on the basis of this product structure. In this *Engineering BOM*, non-geometric elements (such as oil) are also included as independent component objects and are subject for associated release and change processes in PDM. The *Manufacturing BOM* is based on the *Engineering BOM* and is created via manufacturing sequence and assembly stations oriented structuring principles. The Manufacturing BOM gets further expanded to include plant-specific and/or supplier-related information. Today, this is still done in most of the cases in the ERP system, where manufacturing routing and sequencing is also created.

The increased complexity of products, which—as mentioned—is reflected by a common product structure consisting of hardware, electronics and software, has also an impact on the functional capabilities of BOM management in PDM. In the past, *PDM based BOM management* focused—more or less—on the management of mechanical components in the form of individual parts or assemblies. Today, the aim is to build up and manage a common structure of the product in the PDM by integrating multiple sets of authoring tools. This aspect is particularly important with regard to change management processes and activities that need tight IT-architecture integration.

Managing Different Types of BOMs Across the Life Cycle

As part of the two main phases of product creation—*product development* and *manufacturing engineering*—different types of BOMs are created and used for business. In product development, the focus is on the *Engineering BOM*, which carries mainly the functional view of the product in a structured form to realize major product functions and building blocks. In addition to in-house production parts, the *Engineering BOM* also contains *purchased parts and assemblies* or parts that were developed internally and are manufactured externally. The *Manufacturing BOM* as ordered set of the manufacturing sequences of all *Engineering BOM* objects (parts, assemblies etc.) is not used in all companies and is meanwhile often directly replaced by the *"Bill of Process"*. The *"Bill of Process"* reflects work and build plans incl. the individual manufacturing and assembly process steps as well as linkages to manufacturing resources and tools in the (generic or specific) plant environment. *Quantity BOMs* are regularly used in product development and manufacturing engineering as specific type of listing. They reflect how often each component is contained in the entire product or in specific (sub) assemblies.

Figure 11.19 depicts a PDM tool supported comparison of two different table-like structures, the engineering BOM (E-BOM) and the manufacturing BOM (M-BOM) of a machining center (please also compare with the different lifecycle phases for the *Bill of Materials* in Fig. 11.6). This helps Product Engineers and Manufacturing Planners to distinguish between the functional and system viewpoint of product development and the manufacturing build sequence view in production planning. From the manufacturing perspective, it is important to understand exactly the correct assembly sequence order of any single item with respect to the already assembled unit as part of the manufacturing feasibility and build check-process. The PDM system provides E-BOM/M-BOM comparison capabilities with respect to BOM line order, exact quantity number and sub-structure identification and highlighting differences in 3D visualization.

Within the product life-cycle, additional types of BOMs are used. In *plant engineering*, for example, an "*as-built-BOM*" is encountered. In its first form, it reflects the construction status of the plant at the time of PAC (Provisional Acceptance Certificate). This acceptance of the plant takes place together with the customer and forms the starting point for the utilization phase of the plant at the customer's site.

At the same time, it is also the start of the warranty period. Within the scope of the warranty, a repair list of parts can be generated, for example, in which the

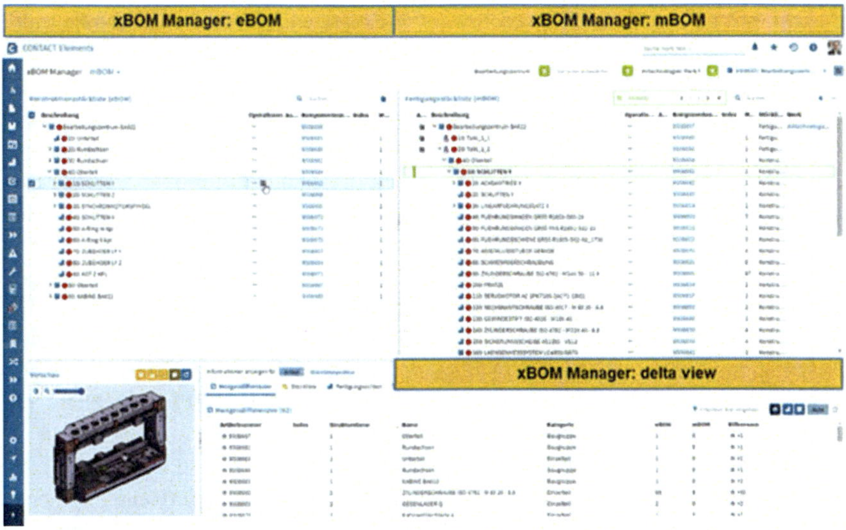

Fig. 11.19 E-BOM/M-BOM comparison capability shown for the example of a SW machining center in CIM database (*Source* Contact Software and Schwäbische Werkzeugmaschinen)

components to be manufactured for repair are listed. The final "*as-built-BOM*" is generated with the final acceptance of the system by the customer at the time of FAC (Final Acceptance Certificate). It represents the final construction status of the plant at that moment. The warranty period also ends and the plant enters the operating mode for which the customer is responsible. If work is to be carried out on the system during this phase, lists of parts must be created for it. Such listing, for example, may include the *"repair list of parts"* as already mentioned and/or the *"spare list of parts"*. The latter one collects all spare parts specified for the system. The identification of a component as to whether it is a *spare part* is done in the *Engineering BOM*. Unfortunately, it can happen, that a component in one product sub-system is declared as *spare part* and the same component as part of another product sub-system is not. For such purpose, an indicator is set in a corresponding column of the *Engineering BoM*. Since a component can be classified as a *spare part* for the one plant but not necessarily for another plant, the *spare part identification* does not take place directly in the master data of the component. After processing the replacement order or the repair order, the "as-built-BOM" must be updated. This is called the "*as maintained-BOM*" and documents the current construction status of the system.

The service list of parts is based on repairs and maintenance. It is set up during development and provides information on what can be removed, what needs to be repaired and maintained, and which components are specified as spare parts. In plant engineering, the *Engineering BOM* is mostly order-related, since orders are processed here according to the "engineering to order" (EtO) principle. For working with the "*as-built-BOM*" or the "*as maintained-BOM*", it is necessary that the BOMs always represent the current status of the product. PDM system capabilities meanwhile fully support their efficient management.

Domain views can be represented by classification of PDF information objects (e.g. part master) according to domain-specific interests. The objects' classification and master data then cover domain-specific information via attribute sets and values. Other views can also be derived from the product structure if appropriate indicators are set on their master data. For example, a listing can be created from the product structure that contains all purchased parts or all spare parts of the product.

In order to meet the increased customer demand for individual products, companies often use product configuration and variant management approaches based on a "*maximum bill of material (Max BOM)*". A "*maximum bill of materials*" is understood to be the one in which all possible versions of the product have been thought through, analyzed and defined in advance. Within a "*maximum bill of materials*", all available variants of the components/elements are simultaneously managed within a predefined hierarchy (using product structures) or listings (using flat partitions). In practice, however, it turns out that for products with high complexity and engineer-to-order components it is rather difficult to use the *"maximum BOM"*, since it is intellectually highly challenging to cover all product scenarios and variants in a single BOM. Therefore, this type of BOM only covers a certain size and a certain type of product. It follows the idea that several "*maximum BOMs*" are necessary for a wide range of products and that it is too risky to use a "*Super BOM*" for all together. This fact justifies an increased change and administration effort. The sales order then generates the concrete, order-related BOM from the maximum BOM.

Object dependencies for selecting the correct material are stored on the BOM item of the configurable material which can be represented in a configured product structure in PDM. The sales order deletes the parts that are not required from the overall configuration and this also creates the *"sales order-related BOM"*. In contrast to the *"maximum BOM"*, a *"modular BOM"* separates the product structure (=product hierarchy) from the component variants. Here, configurable components can be reused at module levels. A prerequisite for this is the development of configurable materials. From the previous considerations it becomes clear that the bill of materials is the original, central information carrier in the context of product creation and product manufacturing. It is supported by various documents (and in the future also by different kind of models), such as specifications, drawings, protocols, NC programs and calculation results. To ensure consistent management of all of them, PDM qualifies as a safe integration environment.

Companies are dependent on accurate management of overall costs and weight roll-ups, calculation and analysis as part of different product configurations, product architectures and overall product portfolios. This is critical not only during product planning and development but throughout the entire life cycle. If companies are successful in establishing robust and reliable roles and responsibilities within their workforce for such core information sets, PDM environments can be leveraged to provide a range of capabilities for authoring, tracking, dashboarding and analyzing cost and weight attributes in line with the product variant and design descriptions. Figure 11.20 shows an example for a cost roll-up and analysis of an automotive seat

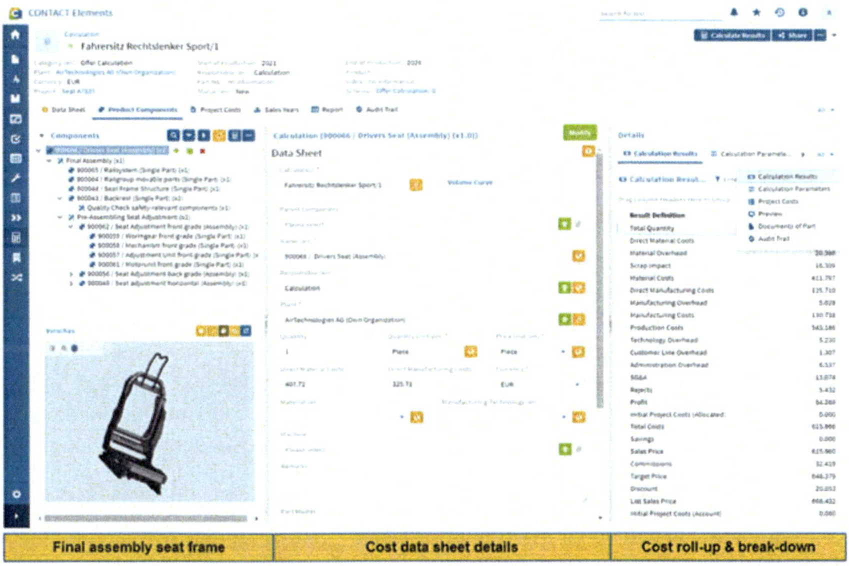

Fig. 11.20 Cost roll-up and analysis of an automotive seat frame in the PDM system CIM database (courtesy of Contact Software)

frame.

PDM provides to a large extent part-related data that enables advanced calculations and estimations aggregating such information on an upper-level product perspective. *Part master data* oftentimes includes weight information (calculated on density and volume in the PDM or taken over from CAD or typed in as estimated value by the engineer), the centre of gravity, translations (part locations in models), cost attributes, material information, supplier information or else. Some information is related to the part designs in the sense of its geometric specification, represented by CAD documents in PDM. Other information is related to parts in the sense of the producible unit, represented by the part master in PDM.

Manufacturing engineering data extends this part-wise available product data with aspects from e.g. work plans—information about the manufacturing process in terms of manufacturing sequence times per work place or operation. Information about parts, manufacturing processes and used resources allows PDM to calculate aggregated views for entire BOMs. Examples for such PDM based analytics are calculations of centres of gravity (e.g. for simulations), cost estimations for products, reports to fulfil requirements on material compliance, or time assumptions on production time and resources needed to manufacture a complex product.

The following example in Fig. 11.21 illustrates classification data of a battery component as used for EHS (environment, health, safety) compliance in clipping industry.

The example of PDM-based product costing applies, for instance, in RFQ (request for quotation) processes in automotive industries or machinery. A quick response to a

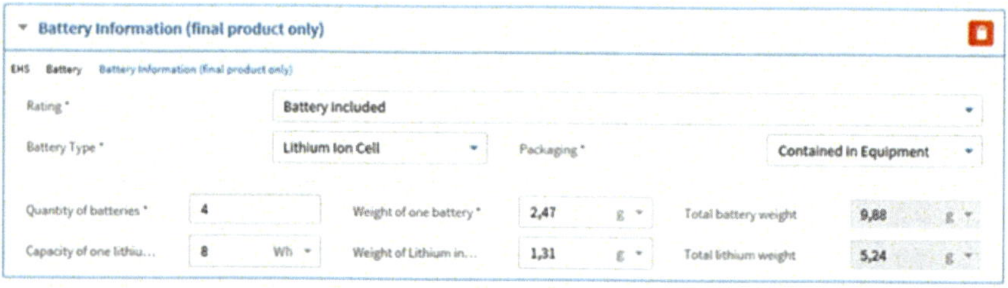

Fig. 11.21 Classification data for a battery component (example of a selected data subset, courtesy by Contact Software)

RFQ is particularly important in most ETO (engineering to order) and CTO (configure to order) scenarios. As for feasibility reasons, Engineers have to figure out and decide about changes and efforts for customization construction. A calculated product price will consist of recurring direct material and manufacturing costs, depending on the production volume, as well as from volume-independent project efforts. In most cases, a high level of reuse of existing components is desirable or the new product is an instantiation of an existing modular platform. In this phase, there might not be a complete BOM existing, as well as there is no certainty, if this product will ever be built. This means, that for such types of business processes (e.g. performing an offer calculation), it makes sense to manage even further material-centric information inside PLM. Figure 11.20 shows a product costing app in CIM Database PLM including a structure view (left), master data (middle) and cost calculations (right).

Although PDM/PLM offers this great capability, it is by far not leveraged to the level it could. Many companies suffer from badly maintained master data by historic reasons. Oftentimes master data management is perceived as non-liked or unfavourable additional burden by many engineers. Within many digitalization projects and PLM initiatives a professional approach to PDM-based MDM (Master Data Management) becomes inevitable to …

- … meet targets of "design-to-cost",
- … enable quick RFQ responses,
- … enhance planning robustness in manufacturing engineering (production process planning)
- … be prepared for material compliance requirements,
- … generate data for e-commerce platforms,
- … generate data for PIM (Product Information Management),
- … enable many more use cases in BOM management.

11.3 How Does PDM Work?

PDM Systems are complex software systems with a software architecture is based on a so called 4-tiers (layers) concept, as depicted in Fig. 11.22. This enables, first

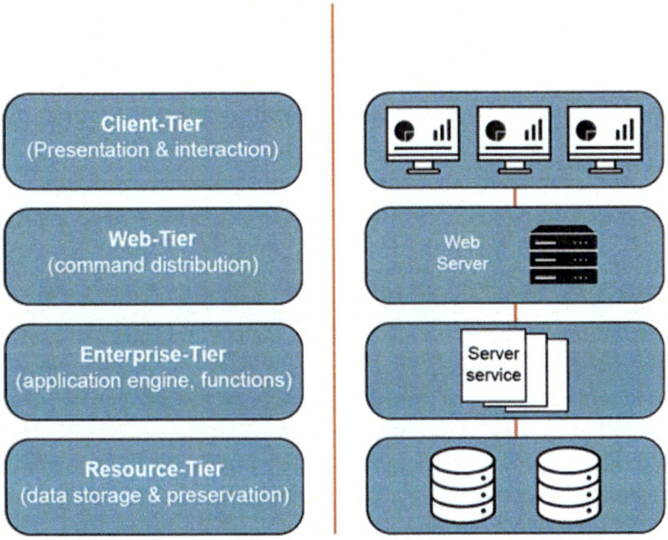

Fig. 11.22 4-tier layer of PDM systems

of all, a secure data storage of both use- and metadata in the base layer which is called *resource tier*. The separation between *use data* storage and *metadata* storage is included due to security aspects. The *use data* are saved as delivered from an authoring tool and it is not possible to modify them from the PDM application side. Due to the strict access rules this storage location is also called *vault*. The vault is a reserved and access-controlled area of the server storage disk. The *metadata* are created in the PDM environment and are usually stored in relational databases, which are defined through a vendor specific database schema. This schema defines how metadata are organized, in which way they might be combined or extended as well as conditions which need to be met so that they remain consistent.

Depending on the product, complex and comprehensive data sets need to be organized for a large number of different users. Therefore, a management of access and manipulation rules for both, use data and metadata, are required. The *Enterprise Tier* provides a large series of PDM applications as well as supporting roles and rights management as part of the PDM working policy. Furthermore, in this layer the business process integration is managed and executed, as well as all above mentioned metadata manipulation such as variants, versions, status, processes and workflows, data translations, data exchange etc. System administrators have to access this layer when adjusting system behavior of business rules. Usually, there exist a clear "separation of duties" regulation within industry between the *owning and viewing groups* of data in PDM (typically the users of the functional activities) which only receives access rights to certain projects or types of data and the *system administrator group* (part of the IT department) which does not own any data but are allowed to access all data areas (in case of problems) and to change access rules within the entire PDM environment.

The *Web Tier* addresses the connectivity of the PDM server to all computers, workstations and portable interfaces. This is required because of the distributed workshare within and across companies. All commands are distributed via WEB services and linkages across all network elements (nodes and connections) by using the different layers of IEEE protocols.

The top layer in the 4-Tier architecture, the *Client Tier*, represents the user interface. The user interface provides means for visualization of use and metadata and menu elements for data and command entries executed by the user. It acts as interaction between human beings and the PDM system intelligence. All user interface actions are funneled through the WEB tier down to the enterprise tier where the user induced PDM core functions are executed by the PDM server.

Other than today's standard 4-tiers architecture the original PDM/EDM systems started with a 1-tier architecture only: the expert user had to directly work in the data base command environment. As part of the expansion of PDM in enterprise to IT less-skilled ordinary users from functional activities the 2-tiers architecture in the second half of the 90ties and the 3-tier architecture with the beginning of the millennium were established: the 2-tiers architecture required a powerful client computer to control all data base operations, whereas the 3-tiers architecture already provided a dedicated PDM server as part of the *enterprise tier* to control and monitor all PDM function induced data base operations.

Figure 11.23 shows the extension of the 4-tier PDM architecture of the PDM vendor CONTACT Software in Germany to a *multi-tiers architecture* by adding a certain separation of layers within the client tier. In the lower area of Fig. 11.23 the different software language implementation technologies such as the Python or Java stacks of the *enterprise tier* are shown, as well as the data base query technologies (e.g. SQL, Standard Query Language) of the *resource tier* and the underlying operating system technologies (such as Windows, Linux etc.).

11.4 How to Integrate PDM in Large Scale PLM Environments?

The common PLM (Product Lifecycle Management) vision in companies focuses on establishing a company-wide source of data and information (compare to Chap. 4). The goal is to enable a continuous flow of data and information across all areas of the company. All data users should have access to the company's central data source, the company-wide "data backbone". All other existing IT systems should have the possibility to access the corresponding data backbone via PDM functionalities. With the help of a PDM system, all users will always have access to the latest data of the entire company. It is also important that the functionalities of PDM systems described in the previous sections of this chapter are retained.

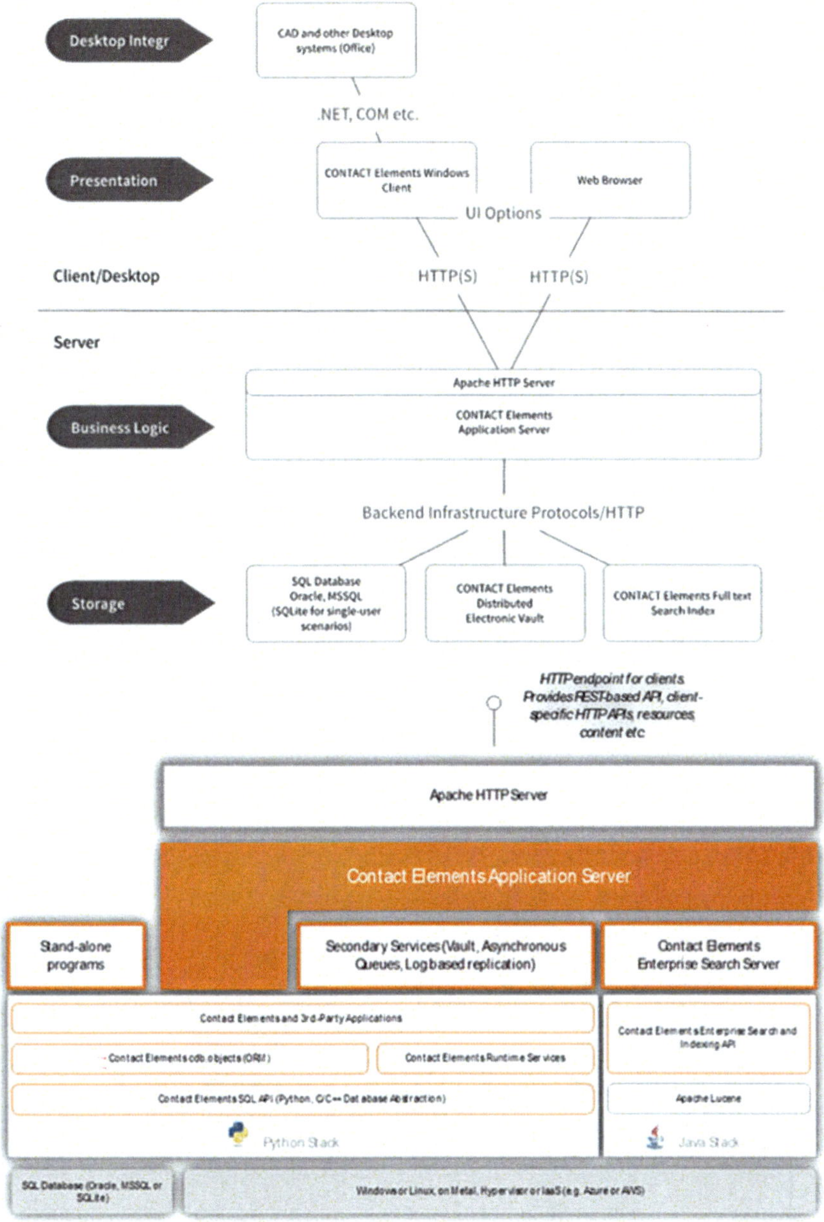

Fig. 11.23 The multi-tier architecture of Contact Software (top area: all tier elements; lower area: *enterprise tier* details)

To successfully implement PLM, it is necessary to establish the relationships between company-specific processes and product data in order to implement them within the IT landscape. Here, the product represents the central object for end-to-end data integration in the company. In this sense, it can also be mapped into a globally active value creation network (i.e. the cross-company integration of suppliers and customers). That way, PDM systems have to integrate certain company and cross-company specific processes and functionalities for supporting different active users and roles, e.g. engineers, designers, logistics, marketing, suppliers, economics.

Since a PDM server environment is often not based on a single IT server instance, a full PDM IT server architecture might have to be established within a company to ensure data availability across global regions. Two basic concepts are to be considered for such an architecture:

1. synchronicity and
2. heterogeneity.

Synchronicity ensures that data and information is kept synchronous between different sites of a company PDM server architecture or its relevant customers and suppliers. *Synchronicity* leverages PDM database queries and data exchange/transport technologies. *Synchronicity* serves the goal that data and information can be accessed at any time by authorized users at any site of the PDM server architecture, at the local instance (primary site), at the remote instance (remote site) or any other additional instance (additional site) as illustrated in Fig. 11.24.

Synchronicity requires a flexibility of execution options for the following core synchronization service scenarios:

- Regular data and information pushes from the primary site to the other (remote) sites in case of close inter-site collaboration needs,
- Synchronization of smaller updates through peer-2-peer interactions, usually triggered by end-users (such as engineers, designers, planers),

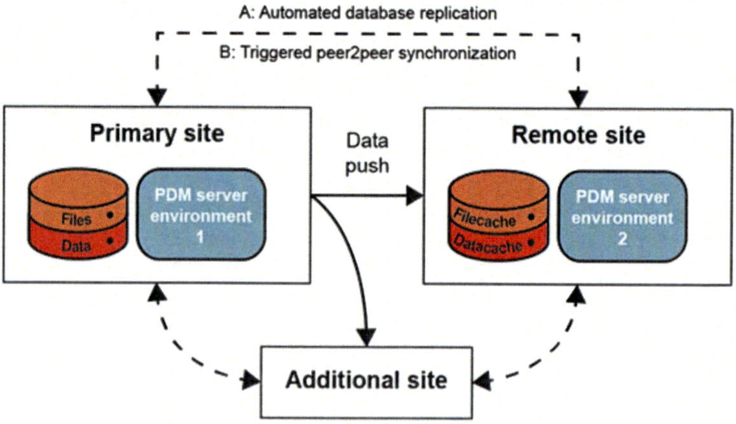

Fig. 11.24 PDM data base replication and synchronization mechanisms

- Pure peer-2-peer interactions in case of infrequent updates needs,
- On-demand pushes if different sites work on different data sets with non-regular exchange of common or standard information sets.

Heterogeneity, on the other hand, is a concept that allows the use of different PDM systems, services and other IT applications in a company or value network. In this case, the vision described above is served in such a way that PLM services ensure data and information exchange but need to be customized or interfaced to other major IT applications (see upper part of Fig. 11.25).

In addition, PDM and PLM solutions are getting integrated into the overall company IT architecture based on the overall EAI (Enterprise Architecture Integration) framework. Many business process driven IT services between major IT applications in modern industrial companies are meanwhile based on SOA (Service Oriented Architecture) principles (compare with the lower part of Fig. 11.25). Other IT services could be for example business data exchange to/from *ERP (Enterprise Resource Planning) system and BOM (Bill of Material) systems, CRM (Customer Relations Management) systems, Office and mailing systems, data lakes, Business Intelligence applications, CAx translation service environments* and so forth. SOA itself leverages ICT (Information and Communication) standards like RPC (Remote Procedure Call), CORBA (Common Object Request Broker Architecture), EIB (Enterprise Java Beans) and Web services based on SOAP (Simple Object Access Protocol) and REST (Representational State Transfer).

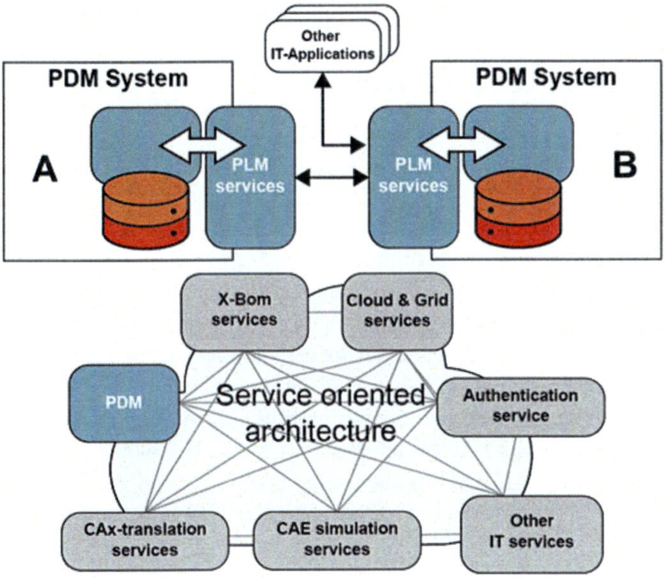

Fig. 11.25 SOA based IT integration principle of a company PDM system

Heterogeneity also requires open standards like PLM Services in order to give system providers independent solutions for PDM data exchange. In both cases, in synchronicity and in heterogeneity, a compromise has to be established between speed and cost of PDM database access (local or via long-distance network), degree of availability of data at each site and effort and quality for database synchronization.

The term *information logistics* has been created and suggested by the author of this book in order to describe the new discipline in information management that ensures that information needs of business and engineering partners are supported and realized by best possible selection, transport and delivery of available data sets within and between internal and external data base environments. Therefore, *information logistics* requires, as a fundamental base, a flexible technology stack

- between PDM and other major IT environments,
- across IT architectures,
- based on network availability and capabilities (such as sufficient bandwidth as well as low enough latency and round-trip time) and
- for an appropriate company or industry branch information model (which information sets are available in specific contexts and can be created based on targeted data analytics).

Within the context of data transport two major characteristics are noticeable:

- High network bandwidth is not sufficient to ensure high data transport in all cases; network latency affects transfer speed as well if many cross-chatting interactions between data services have to be executed.
- Unequal latency characteristics such as low latency at one site and high latency at the other site (e.g. at the other side of the world) cause reduced throughput independent from bandwidth availability.

Eigner and Stelzer [2, Chap. 8] describe in great detail the different concepts and alternatives in integrating authoring tools such as M-CAD, E-CAD, CASE and Office applications and explain what type of model exchanges need to be synchronized to which extent. In addition, they discuss a range of SOA based PDM/PLM IT integrations similar to the concepts introduced in Fig. 11.25.

In order to set up large scale environments, different roles have to work tightly together: IT solution architects, IT technicians, data architects, PDM method developers and PDM users, e.g. engineers, designers, suppliers as well as data engineers, data analysts and data scientists. What is oftentimes missing is a mid to long term plan how the individual information needs and the information standards match or need to be changed and aligned to each other. This situation will get exacerbated in the future due to sharply increasing data set availability form the operational field (compare with IoT and Industrie 4.0 solutions Chap. 20) and from the model rich and intense MBSE (Model based Systems Engineering) approach in product and manufacturing engineering (compare to Chap. 21).

11.5 How to Customize PDM/BOM to Company PLM and VPC Needs?

When implementing a PDM or BOM tool in a company environment it appears necessary to set the OOTB[2] or COTS[3] data model as well as function and behavior settings in accordance to company standards and processes. The adjustments might be driven e.g. by usability or process requirements or data security. Regularly, the implementation process is based on an investigation on the processes and methods of data management applied in the company. A fit-gap analysis identifies matches and lacks between processes and methods on the one hand and the intended tool behavior on the other hand. This process should be accompanied by company's domain experts, PDM vendor's technical sales or advisory staff as well as neutral consultants balancing the interest. Typically, a PDM implementation replaces an already running IT environment. Furthermore, the implementation process involves a large amount of the company's development staff. Therefore, it appears useful to guide the group by a rigid project management reporting on a C-level management circle or steering group.

Adaptions might be carried out both on the process and methods level as well as on the PDM application and IT services level. On the tool side several stages are applicable:

1. Since PDM tool capabilities have been growing over time, the digital technology vendors follow a modular approach, where customers choose the relevant modules (e.g. requirements management, change management, etc.) from the total given PDM tool portfolio. Thus, a first adoption to individual needs goes back to a module selection.
2. A change of settings or parametrizing is typically carried out in an administrative environment (compare to [8]) of the PDM tool in order to define administrative (e.g. roles and access rights) as well as logic behavior (e.g. workflows). The parametrization is accomplished without re-programming the PDM tool and thus supports regular software updates.
3. Companies even request adaptations to the tool behavior from software vendors including a re-programming of the functional behavior. These changes might be carried out deliberately, since these modifications might cause difficulties in software update, conformity to the OOTB resp. COTS PDM/BOM tool behavior etc.
4. In any case, PDM/BOM tools need their integration into the overall enterprise application environment. This causes extensive interface programming and related IT-service development and requests substantial PDM/PLM project attention between the host company and the related digital tool vendor. Substantial background info to those questions have been given in a range of publications in the first decade of the millennium:

[2] OOTB: Out of the box.

[3] COTS: Commercially over the shelf.

a. Eigner and Stelzer [2]: PDM integration as well as technical infrastructure and system functions in Chaps. 8 and 9,
b. John Stark [12]: addressing integration aspects as part of the PDM project in Chap. 30 as part of the overall PLM strategy,
c. Feldhusen and Gebhardt [13]: methodologies for PDM system integration (Chap. 6) which includes, e.g., aspects of analysis of the existing IT landscape and definition of the target environment incl. models and data types
d. Saaksvuori and Immonen [14]: integration of PLM systems with other applications (Chap. 5) and deployment of a PLM system (Chap. 6).

Parallel to IT tool modifications changes are necessary for process and method descriptions. Thorough tests of modifications need to be planned and accomplished until the intended business performance is reached. Moreover, training and people change management need to be prepared and carried out for successful roll-out (compare to Chap. 18).

One of the most challenging and controversial questions within a PDM/PLM project, however, remains the argumentation for, the justification as well as the final determination of the degree of customization of the OOTB and COTS PDM/PLM tool environments. Obviously, there exist different viewpoints between the Digital Tool Vendor (DTV), the IT department of the hosting company and the functional activities using PDM/PLM tool capabilities in actual business and engineering work processes. As shown in Fig. 11.26, the scope of such customization and the strategic focus need to be taken seriously into consideration, not only for the next 2–3 years of digital operation but for a longer time period, at least one decade ahead!

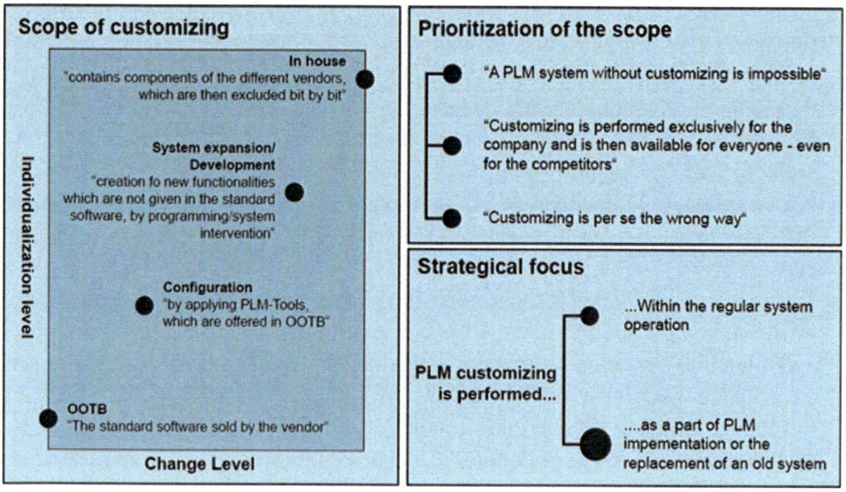

Fig. 11.26 Clarifying the strategic intent of PLM customizing

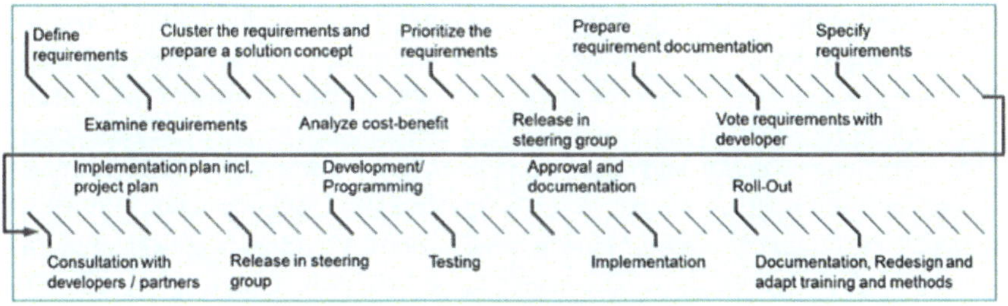

Fig. 11.27 PLM customizing leadership, development and implementation steps

Digital Tool Vendors have meanwhile painfully experienced that,

- their own configurations of OOTB and COTS PDM tool capabilities are difficult enough to be understood by business,
- the contracted customization of PDM tools and related integration projects usually harvest more stress and misunderstandings with their clients and might even hold up their clients from using new and advanced capabilities moving forward with new releases and
- it is difficult to ensure professional and target oriented project execution with their clients under immense time pressure.

Industrial companies who want to introduce, integrate and roll out PDM/PLM and BOM capabilities step-by-step within their organizations oftentimes miss to set-up the right PLM customizing leadership. In very few cases companies have an explicit PDM/PLM leadership element which stringently scrutinizes all aspects of PDM/PLM customization as shown in Fig. 11.27. The following weak elements in industry have been identified implicitly without being addressed explicitly yet in most of the cases:

- limited understanding of data flows and information needs within and across engineering and business processes and activities
- misunderstanding to which degrees PDM/PLM and BOM tool customizations will inhibit "easy integration" of future vendor tool functionalities and service capabilities
- nonexistence of a master plan for future IT application architectures and the related information model interface needs.

As a consequence, many companies suffer from quick PDM/PLM project decisions which in most of the cases are only valid in the context of a limited snapshot of a short-term tool integration rather than in the context of a full *information application* architecture (compare to Chap. 6). With respect to PDM/PLM customization needs assessment and PDM/PLM integration plans, the author of this book provides the following guideline to conduct appropriate preparation, analysis and decision steps which help to determine mid and long solution architectures (at least 5–10 years):

1. Establish and continuously maintain a comprehensive *information architecture* which describes information types, information needs for business and engineering process activities, information sub-sets in specific model types, mechanisms of information authoring, management and delivery. At the same time, *information ownership* and *information model ownership* need to be determined within the company governance structure. Conduct regular reviews in order to revisit the company *internal information standards* with respect to their readiness for *full digitalization.* This needs to be explicitly done in comparison to justifying stringent needs to continuing the support of long-lasting internal information standards which were originally created to support analog working styles based on printed lists!
2. For the areas of interest of (re-) establishing, refining or extending PDM/PLM environments, conduct a thorough *data flow analysis* for each of the major business or engineering process activities with respect to:
   a. Fulfillment of information needs with existing data flows and new/additional needs for the future
   b. Identifying shortcomings and limitations of existing data flows and corresponding IT interface characteristics (APIs, direct data import/export etc.)
   c. Derive a first set of an improved *information logistic* plan to ensure different degrees of digital care levels for the individual digital activities
3. Conduct a cross mapping of the *data flow and information logistic analysis* (step 2) within the EAI framework of all product master and structure leading IT applications (e.g. EDM, PDM, BOM, ERP etc.) in order to identify the intensity of current and future data repository characteristics with respect to:
   a. Today's and future *master repository* of information sets.
   b. Degree of *courtesy copy* needs of such master information sets within other repository environments due to critical business and engineering work collaborations.
   c. Degree of *information security levels* associated to the individual *information sets* to publish such information sets in cross repository data environments (e.g. data lakes, knowledge graphs, semantic networks etc.).
   d. Ease of *exchanging those information and data elements* across IT applications by using different types of *data exchange file formats and standards,* e.g. Standard for the Exchange of Product model data (STEP), Extensible Markup Language (XML) or Open Services for Lifecycle Collaboration (OSLC) (compare more details in [15].
4. Map out alternatives of *data model extension* to the *OOTB or COTS data models* from the Digital Vendor offerings to reach best possible information harmony across the EAI and the internal information architecture based on step 3. Put only those *information and data elements* onto PDM/PLM customization plans which have a robust realization chance based on data exchange standards and the certification to the *Code of PLM Openness* [16], compare Fig. 11.28.

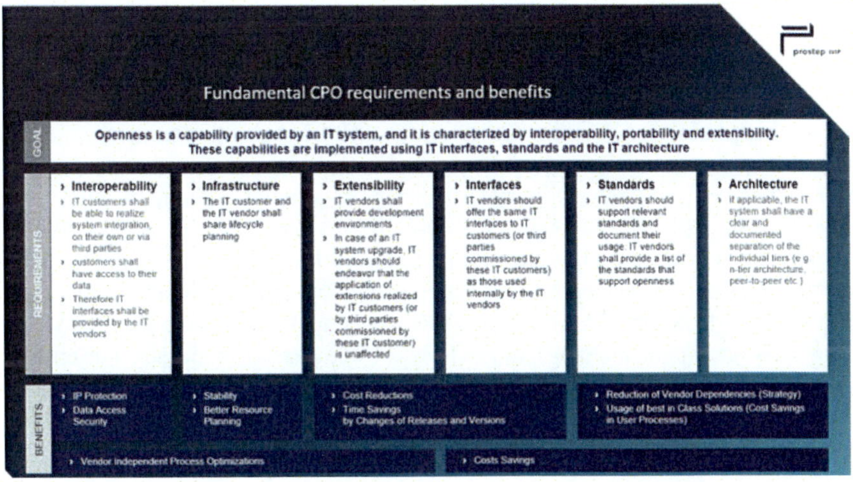

Fig. 11.28 Requirements and benefits of the code of PLM openness (CPO) according to [16]

5. Analyze specific tool functionalities that cannot and most likely will not be part of current and future OOTB and COTS software offerings of the Digital Vendors and add those to potential customization plans. However, apply stringent business justification rationale for those you want to move forward as *priority one* ("absolutely must have") customization elements!
6. Negotiate different models of customization with respect to:

   a. Stress the stringent support of the *Code of PLM Openness certification*[4] *(CPO)* [16] by Digital Vendors: as shown in Fig. 11.28, there exist six categories to be fully ranked as supportive in openness for PLM data and information. Digital Tool Vendors (DTV) acknowledging and following the CPO should be preferred since they can guarantee open standards based PDM/PLM solution integration to external IT architectures and applications.
   b. In case you have to rely on non-open data import/export solutions, make sure that you find agreement on how they might be best supported by Digital Tool Vendor (DTV) internal standards ensuring that future software releases provide a safe support of such customization content.
   c. For any tool specific functionality, try to arrange a specific disclosure period (usually 1–3 years) with the opportunity to open up this specific customization element to a fully supported OOTB or COTS capability of the software right after the defined disclosure time period (otherwise you run the risk

[4] The *Code of PLM Openness (CPO)* is an internationally unique initiative for open IT systems that was launched by leading German automotive manufacturers together with the prostep ivip association and which is sponsored by the German Federal Ministry for Economic Affairs and Energy (BMWi).

of paying ongoing maintenance fees for it, typically with a penalty of 3–6 months delay compared to the OOTB or COTS software release!).

Independent of the involvement of Digital Tool Vendors (DTV), the same approach needs to happen in case of internal company PDM/PLM self-development projects. Such scenarios are oftentimes even more complicated since significant content needs to be outsourced to IT development agencies or companies without specific code ownership for the lifecycle of the software! In any case, PDM/PLM and BOM customizations need to be considered and reflected not just for the start of a new PDM/PLM/BOM set-up, deployment and implementation but also for the long run. Most of PDM/PLM and BOM working environments last for more than 15 years before they get fundamentally adjusted again. The reason is simple: the efforts associated with such a change are immense and do require substantial investments, both in IT software customization and integration as well as in process & method development and human skills.

11.6 Expected Changes in Future Industrial PDM/PLM Operations

The "new value" of data and information across the lifecycle operation of technical systems and products will sharpen the "hidden value" of today's Product Data Management (PDM) and Bill of Material (BOM) solution environments. So far, it does not exist (yet) a value measurement schema for data and information sets mastered, managed and used in those repositories: it was expected to create those data sets in order to be able to proceed within various business and engineering process flows. The existence of such data, the linked semantic knowledge and the potential to learn from both was neither regarded as *"hidden knowledge about evidence of process execution"* nor treated as an *"asset to use it proactively for assistance in work and decision prediction"*. Unfortunately, it was so far simply treated as an "administrative thing" rather than a "true value".

The future, however, will make a substantial difference to this regard. Industrial companies will get under enormous pressure to drive digital transformation rebuilds of their *information architectures* and will get audited regularly on how they will conduct *"information quality and use assurance"* within their organizations and as part of their business and engineering process activities. At the same time, the pressure to link master and structure product data with "live data" from factory operations and from technical field use will grow substantially. PDM and BOM solutions need linkages to ETL (Extract Translate and Load) mechanisms of non-SQL (Structured Query Language) type of data bases hosting IoT (Internet of Things) raw and data analytics data. This, however, is a change for ordinary product data since it provides the base for Digital Twin applications (compare to Chaps. 20 and 21). Furthermore, future product data management environments will be integrated or at least linked to

Lambda architectures of AI (Artificial Intelligence) computing and their data analytic mechanisms and engines.

The author of this book has been conducting analysis work together with major OEM enterprises in the automotive sector in order to reveal the magnitude of order to product data sets generated and modified per calendar year and, therefore, to provide a base assumption for the hidden data assets within industry. It turned out that only three out of six companies were in the position to deliver meaningful data sets to this query. The other companies claimed to have currently no time, no resources and/or no method approach available to tackle such analysis task. The following motivation and query basics and principles have been given to the automotive OEMs:

The query request intends to find out to which levels data availabilities and dynamics within the well-established discipline of product data management have been developed in the 2nd decade of the twenty-first century. On a high level, this investigation of data extent and dynamics should help to understand and assess the potential of making effective use of AI (Artificial Intelligence) and new types of data analytics assistance within the fast-growing new discipline of Data Engineering and Analytics (DEA[5]) in virtual product creation. The anticipated advantages are manifold: improved and more robust interaction of Engineers with PDM information sets, automatic progress control for development control in virtual product creation, increased efficiencies in digital engineering processes and workflows, higher quality of products prior to product launches and market entries.

The Lead Principles for the Query:

- *Only what you can and will measure, can be improved!*
  (well established engineering principle that finally needs to be deployed for the digital data and model world in Engineering)
- *Data and models as well as their dynamic change behaviors do carry substantial implicit expertise, which, so far, was only measured, assessed, analyzed and used in very limited occasions for knowledge and intelligence creation!*
  (this is the core take-away from the rather small group of experienced experts in virtual product creation)
- *Let us start with such measurements—without compromising company knowledge. Encouragement will pay back!*
  (eventually pro-active drive will deliver new ideas and concepts for digital transformation, thread and traceability, DT3)
- *In case of interest and recognized needs let us build up bilateral or industry branch oriented deeper investigation projects!*
  (get into it, understand, recognize potentials, plan and act with further with dedicated activities)
- *Treat data governance in business seriously and demonstrate the right responsibility for it!*

[5] See explanations in Chap. 21.

The Target Metrics of Query:

1. *Number of established top nodes in PDM for the digital product or production line/station or plant (per vehicle project or factory environment)*
2. *Number of product structure of partition nodes below top node (all product structure nodes)*
3. *Total number of HW and SW product data items/models under the top node (to be separated between single parts and assemblies)*
4. *Number of stored iterations per year for each product data item on average (versioning, revisioning, baselining* etc.*).*

Three out of six automotive OEM (50%) recognized the importance of such an investigation for themselves and were glad to get into such engagement triggered by the outside. None of the OEMs (including the ones which did not participate) had done such analysis themselves and hence did not have any experienced approach to it. As a consequence, it was necessary to conduct intensive sessions with experts in IT and with data management experienced method engineers within the companies.

On average, it took 3 months to finally receive useful sets of numbers that were comparable across companies (it all happened in *isolated hub and spoke* approach through the author of this book; the companies involved do not know each other's engagement in the analysis, hence everything remains unanimous).

The query results (compare to Fig. 11.29):

First of all, none of the three OEMs (two from Europe, one from Asia) did use their PDM environments to master and store vehicle SW code. They use other repositories for vehicle SW firmware and controller code. The PDM usage timeframe for HW

Fig. 11.29 Results of product data authoring query in automotive industry (**a**–**c**: reference year 2016; **d**: average in time frame 2015–2020)

	OEM1 Europe[1]	OEM2 Europe[2]	OEM3 Asia[1]
a # of top nodes of the product	551	49	400
b # of product structure nodes below top node	1400	260	134
c # of product data objects in the PDM system (incl. multiple occurences)	2.042.007	60.164	1.786.400
d average # of product data iterations per year (change, addition, deletion)	2,1* per object => ~ 750.000	11,3** per object => ~ 1.360.000	2* per object => ~ 550.000
	Notes: 1 After 16 years of PDM usage 2 After 6 years of PDM usage * released product data only ** released + work in progress product data		

product data has been varied between 6 and 16 years. OEM 1 and 3 are comparable in sales and production figures, OEM 2 has higher numbers in sales and production compared to the other two.

One of the three OEMs does store "work-in progress" and "released" product data objects and use data (e.g. CAD) in the PDM repository, the other two OEMs concentrate only on "released" product data objects and use data (e.g. CAD) and keep entertaining a range of TDM (Team Data Manager) repositories for "work-in-progress" product data. Top nodes in PDM are added based on new vehicle projects or vehicle programs resp. architectures, factories and production lines are handled as separate top nodes. The growth of storing PDM data objects per years in each of the OEMs (compare line D in Fig. 11.29) differs between 550.00 and 1,360,000), depending on the number of work-in-progress iterations stored. All of such product data objects are generic, i.e. none of them belong to an individual vehicle instance.

In summary, this first query shows that there exist significant needs of further research to understand the opportunities of the existing data sets within the PDM repositories. At the same time, however, first research projects have already revealed that such PDM based data entries will become very useful in automatically predicting engineering progress (compare [17]).

The new generation of PDM systems will be able to manage and link a complex amount of information about the engineered product as well as from the live product in the factory or field/market. Digitalization thus opens up the possibility of using systems to simulate everything that could happen to the product in the real world. In other words, the product life cycle will be traceable in both the real and the digital world. The implementation of this solution approach will be time-consuming since the core foundations have to be built up within industry. Solutions such as *Digital Twins* and *Digitalization Platforms* for seamless data flows and analytic pipelines need to be established (compare Chaps. 20 and 21). In order to provide a solid base for *IoT* (Internet-of-things)-capable products, the backbone of structured and labeled data will also be essential for this next generation of digitalization. As a pre-requisite, robust data management is key and starts already with data generation of structured data at the beginning of the lifecycle. In addition, the question arises as to how legacy data of products can become part digital analytics. The maintenance of the "right data" will become the success factor of the future.

Another trend that has become increasingly apparent in the recent past is that production-related aspects, e.g. the creation of process sequences or assembly planning get shifted from the ERP (Enterprise Resource Planning) resp. PPS (Production Planning System) to the PDM system. The higher-level capacity planning will probably continue to take place in the classical ERP. The essential change in this paradigm shift, however, is that the *production BOM* derived from the *engineering BOM* is generated in the same system rather than in two different ones. In IT terms, engineering and work scheduling thus move closer together, as they work in the same system. It will be interesting to see how this ERP/PDM growing together will emerge over the next years.

The digital transformation allows for internal and external collaboration. Here, data can be retrieved directly by users in compliance with access rules without transfer processes specifically triggered by the data creator. However, this idea is counteracted by the current trend towards decentralization of IT applications, as this is accompanied by a certain degree of reduction of data traffic in the network. In contrast to change notifications, which are traditionally distributed manually and retrospectively, *block chain technology*, which can also be implemented in PDM technologies, may allow changes to be notified in real time to all affected parties, such as production, purchasing and suppliers.

At first glance, the fields of the "classical" application of PDM seem to be exhausted. But a deeper insight reveals that the expansion of PDM use seems to make sense, especially in the early phases of the product development process. The keyword here is "model-based system engineering" (compare to Chap. 21). To this end, company-specific concepts must first be developed with the help of external experts who guarantee a process-related view from the outside.

A general goal associated with PDM was the dissolution of the function-related and historically grown data silos in the companies. This was made possible by global data management. Within the field of *BIM (Building Information Modeling)*[6] this approach is referred to as *Common Data Environment (CDE)*. This CDE approach will further grow, especially towards the growing linkages of engineering data with operational data via Digital Twin technologies but also with respect to the aspired intelligences through data engineering, data analytics and data science (compare to Chaps. 20 and 21 for more details and insights).

The vision outlined by the author of this book and researches of Fraunhofer [18] predicts a convergence of the need towards *automation in engineering* and the ability to enhance next generation product data management with new assistance level and intelligence provision. As outlined in Fig. 11.30 it will be decisive to provide different degrees of PLM intelligences covering the following aspects:

- Configurable user centric interfaces to allow easy-to-use context information analysis and authoring for PDM users
- Information architecture concepts to open-up and support PLM wide data exchange and semantic data linkages based on open standards (e.g. through the CPO certification, details in [16])
- AI based supporting functions and assistance leveraging product data evolutions residing already within today's PDM environments—see PDM query results above
- Assistance and automation through semantic link technologies (WEB services such as REST), data contextualization, visual analytics and engineering intelligence formulations.

---

[6] The British Standards Institute defines BIM as follows: "the process of generating and managing information about a building during its entire life. BIM is a suite of technologies and processes that integrate to form the 'system' at the heart of which is a component-based 3D representation of each building element; this supersedes traditional design tools currently in use."

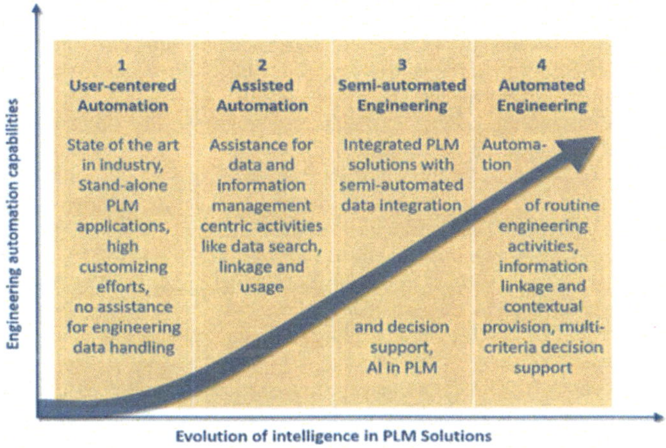

Fig. 11.30 The future "intelligence path" of PLM in Engineering [18]

In order to reach those new capabilities, industrial companies have to open up their implicit engineering heuristics and vast data repositories to researches from Universities and independent application-oriented research institutes. The Digital Tool Vendors need to start working more proactively with research on these new frontiers before building new applications (compare to Chap. 19 for more insight on Digital Tool Vendors). Overall, the new era of data engineering will grow substantially during the next years and will be explained in more detail in Chap. 21.

References

1. Stark R, Ammann D (1998) Integration von EDM und CAD als Voraussetzung für weltweite Fahrzeugentwicklung bei Ford (Integration of EDM and CAD as Requirement for Global Vehicle Development at Ford). In: 7th international PDM conference of CSC (Computer Science Corporation) Ploenzke: life cycle management—improving products and processes for tomorrow's markets; Mainz, Germany May 5–7, 1998
2. Eigner M, Stelzer R (2009) Product lifecycle management: Ein Leitfaden für product development und life cycle management. Springer, Berlin Heidelberg, A guideline for product development and life cycle management
3. Westkämper E (2013) Produktlebenszyklusmanagement. In: Westkämper E, Spath D, Constantinescu C, Lentes J (eds) Digitale Produktion. Springer, New York
4. Verein Deutscher Ingenieure (2016) VDI 2219 information technology in product development: introduction and usage of PDM systems. VDI-Gesellschaft Produkt- und Prozessgestaltung, Berlin
5. Deutsches Instituts für Normung e.V. (2002) DIN-Norm DIN 199. Ausgabe 3. DIN e.V., Berlin
6. Vajna S, Weber Chr, Bley H, Zeman K (2009) CAx für Ingenieure (CAx for Engineers), 2. Auflage. Springer, Berlin, Heidelberg
7. Eigner M (2021) System lifecycle management. Digitalisierung des Engineering, 1st edn. Springer Berlin; Springer Vieweg, Berlin. ISBN-13: 9783662621820

8. Stark R, Lietzau H-P, Klar A (2002) Management turbulenter Informationsflüsse in der virtuellen Produktentstehung (Management of turbulent information flows in Virtual Product Creation). In: VDI-Berichte Informationsverarbeitung in der Produktentwicklung: Von CAx zu PLM—Von der rechnerunterstützten Produktmodellierung zum integrierten Management des Produktlebenslaufs; Tagung zur CAT Engineering, Stuttgart, VDI-Gesellschaft EKV
9. International Organization for Standardization (2008) ISO 7573 technical product documentation—parts lists, 2nd edn. SO copyright office, Geneva
10. Abramovici M, Aidi Y (2013) Next generation product lifecycle management (PLM). In: Fathi M (eds) Integration of practice-oriented knowledge technology: trends and prospectives. Springer, Berlin, Heidelberg. https://doi.org/10.1007/978-3-642-34471-8_12
11. Verein Deutscher Ingenieure (2008) VDI 4499: 2008-2. In: Digitale Fabrik-Grundlagen. VDI-Gesellschaft Produkt- und Prozessgestaltung, Berlin
12. Stark J (2005) Product lifecycle management—21st century paradigm for product realization. Springer, London
13. Feldhusen J, Gebhardt B (2008) Product lifecycle management für die Praxis (product lifecycle management for practical use). Springer, Berlin Heidelberg
14. Saaksvuori A, Immonen A (2008) Product lifecycle management. Springer, Berlin Heidelberg
15. Prostep ivip association (2019) Collaborative systems engineering on the basis of engineering IT standards. ISBN 978-3-9817958-6-8. Available under: https://www.prostep.org/fileadmin/downloads/WhitePaper_SSB_final_EN.pdf
16. Prostep ivip association (2015) Code of PLM openness (CPO)—basic understanding of openness and associated requirements, version 1.4, 15.09.2015. ISBN 978-3-9817958-1-3. Available under https://www.prostep.org/fileadmin/downloads/PSI-16__CPO_V14_150915.pdf
17. Ebel H, Riedelsheimer T, Stark R (2021) Enabling automated engineering's project progress measurement by using data flow models and digital twins. Int J Eng Bus Manage
18. Damerau Th, Hayka H, Neumeyer S, Woll R, Stark R (2014) Intelligent information technologies to enable next generation PLM, product lifecycle management for a global market, Springer, Heidelberg, pp 485–495

Chapter 12
Major Technology 6: Digital Mock-Up—DMU

Executive Summary

This chapter deals with the following topics:

- Basics and advanced techniques of Digital Mock-Up
- Providing insight into how engineers benefit from using Digital Mock-Up (DMU) technologies
- Describing functioning, benefits, and limitations of DMU technologies in practice.

Quick Reader Orientation and Motivation

The intention of this chapter is:

- to give an overview of DMU technology in Virtual Product Creation as driver and enablers for Digital Transformation in engineering
- to present DMU technology as part of Virtual Product Creation from a practitioner's point of view to analyze the need and usefulness for day-to-day industrial work practice
- to give instructions on how to use DMU technology
- to explain models, frameworks, and mathematical representations that help to grasp the internal working modes of DMU technology.

A *Digital Mock-Up* (DMU) is a *digital collection of 3D models* that represents a comprehensive physical entity *with the help of a structured digital representation.* Usually, two different types of comprehensive physical entities are subject for such digital representation, *products* like e.g. machines, cars, aircrafts, trains or ships and *factory or production* line environments. In addition, a range of digital functionalities is offered by DMUs in order to investigate the digital models and their interplay. DMUs do also play another integration role: they are used where complex products or factories are developed and represented in heterogeneous 3D CAD environments like in automotive industry, aircraft design or plant manufacturing. Hence, DMUs represent (neutral) digital integration environments to avoid tedious CAD-to-CAD translation work for packaging and layout investigations, please compare Table 12.1.

R. Stark, *Virtual Product Creation in Industry*,
https://doi.org/10.1007/978-3-662-64301-3_12

Table 12.1 Application areas of hardware/physical and digital mock-ups [1]

	Engineering and manufacturing supportability	Training	Marketing
Primarily used	DMU for the product itself (airplane, ship, car…) but also for all its production means (factories, transportation equipment…) and verification of servicing procedures	HMU e.g. Space Shuttle Training Mock-up, International Space Station facilities, Fuselage/Cabin Mock-ups for workers being assigned to a new assembly line	HMU e.g. to provide customers a "touch and feel" impression, e.g. with fully functioning Cabin Interior—the "Sales Mock-up"; (scaled) Mock-ups for exhibitions
Secondarily used	HMU to validate particular risk areas, to cover certification relevant items, prove required functions (system tests) that are not yet reliably possible in a digital environment; examples: Design-, Production-, Engineering Mock-ups;	DMU supporting faster and better learning e.g. for Space Mission preparation; growing importance as computer performance and computer graphics advance (e.g. Virtual Reality)	DMU growing importance for external communication especially when coupled with virtual reality techniques, increased reactivity on customer need and requirements

However, even with modern computer power, traditional physical or hardware mockups (PMU or HMU)[1] still have their usage in product development [1].

Especially, when human product interaction comes into play, physical features like haptic and weight become important. This is the reason why Digital Mock-Ups are potentially not sufficient to represent a full engineering prove out environment for critical engineering investigations such as fitting clearance for manual assembly, service and dismantling tasks. Figure 12.1 shows the typical three types of representations (the final physical product and the physical mock-up resp. Digital Mock-Up during its development) in present engineering practice.

For distributed product development, a DMU can be used as a reference model, in which every new designed part or assembly can be directly implemented. This allows then for an easy check of the entire product assembly against other adjacent 3D models (e.g. space analysis) across multiple locations. This is not possible with physical mock-ups since they can only physically exist at one location and therefore cannot be virtually compiled to one representation over network and database connection as it is the case with Digital Mock-Ups.

[1] Hardware mock-ups (HMU) or physical mock-ups (PMU) are distinguished between the following types in aerospace and aviation industry [1]: Design Mock-ups, Sales Mock-ups, Production Mock-ups and Engineering Mock-ups (EMU). The latter represent the most "sophisticated ones" with functionalities usually evaluated using so-called "(System) Test Benches" or "Iron Birds".

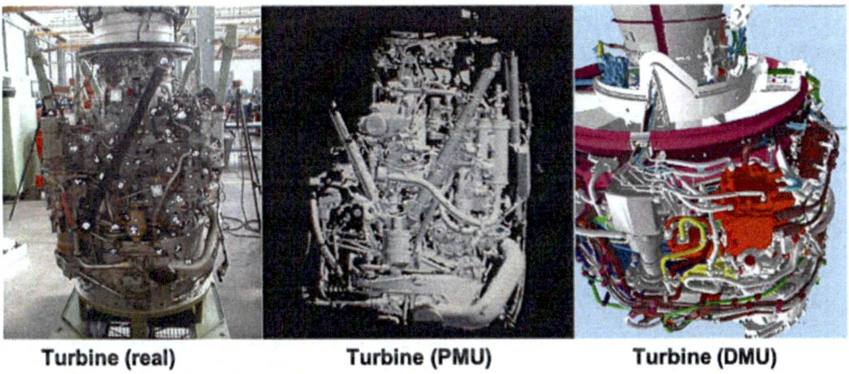

Fig. 12.1 Real aero engine, its physical and digital mock-up

12.1 Engineering Understanding of DMU

A Digital Mock-Up is a virtual representation of the entire product in all its variants, options and versions. It can be used throughout the product life cycle (if maintained consistently even after production starts!) and it supports validation, communication and decision-making processes. It is, therefore, a specific digital representation of a virtual prototype, which in turn is part of the virtual product.

Figure 12.2 illustrates different types of Digital Mock-Ups (DMU) of complex products and explains the core elements of a Digital Mock-Up. Hence, a Digital-Mock-Up (DMU) is defined and characterized as follows:

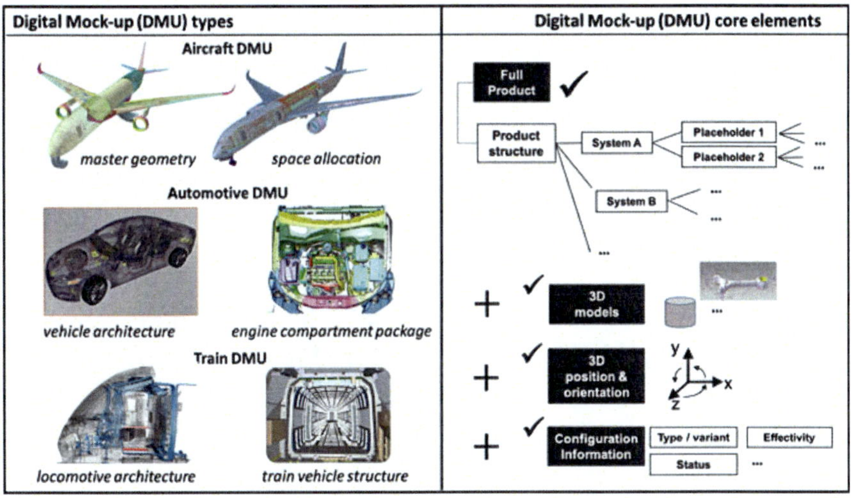

Fig. 12.2 Digital mock-up types and core elements

A Digital Mock-Up is a virtual representation of an entire product model of a (complex) product or other technical systems (e.g. production line, factory or service station). A Digital Mock-Up represents positioned and oriented 3D geometric models in a reference coordination system and offers configurations such as product type, variant, effective dates and/or other engineering/business attributes (e.g. weight class, temperature zone, cost level etc.*). A Digital Mock-Up serves as analysis, validation, communication and decision platform throughout the product/manufacturing engineering phase and—if possible—throughout the entire product lifecycle.*

Since the birth of the digital concept and environment DMU in the second half of the 90ties of last century substantial industrial experiences and further research developments have happened. Advanced engineering departments enrich the classical DMU by overlays of CAE analysis results (colored result file representations of FEA analysis, motion files of rigid motion simulations etc.) and transform the DMU into virtual prototypes in order to support the next level up to a fully virtual product. If elements such as logic functions, cross-domain interactions and behavioral modeling are getting added also the terms "Functional Mock-up (FMU)" or "Functional Digital Mock-Up" are increasingly used (compare [2] and [3]).

The DMU provides the geometric description, e.g., of an entire vehicle; the virtual prototype adds further data to the DMU so that functional and behavioral simulations get supported. The virtual product contains all information that is needed for the entire product lifecycle in different business sectors ([4], p. 9). A DMU consists of a (digital) product description and the collection of 3D models of all relevant parts in their correct spatial position. In particular, a DMU is used to detect collisions between parts and to simulate the assembly process [5], both in static and in dynamic situations.

For dynamic motion analysis, it is necessary to add motion behavior to the DMU by associating DMU components to motion files, which have been calculated before in Multi Body Simulation (MBS) software packages. The motion paths of DMU components are then driven by animated motion sequences in 3D space referenced to appropriate relative or absolute coordinate systems.

DMU structures and models are also leveraged in later phases of the product lifecycle than product development and manufacturing engineering: e.g. during review and analysis of problems and solution proposals in the mid of life activities such as ongoing production, product marketing and offering, during product use and as part of product enhancement activities.

12.1.1 Why Does an Engineer Use DMU Instead of CAD?

A DMU is used to perform investigations on assemblies of high complexity and with many components. Principally, it is possible to build these mock-ups from CAD models directly. However, CAD models usually contain data intensive, accurate and parametric geometrical information sets, which include a rich mix of meta data and supportive data structures. Oftentimes, such detailed information sets are not required for typical visual reviews, inspection work, clearance and collision analyses or kinematic simulations.

If approximated geometric data is used to an accuracy level of 0.5 mm with an overall product dimensional level of around 5 m, it is possible to reduce the size of the 3D geometric representation to only 30% of the original size. If precise data is then integrated for measuring purposes as part of the DMU representation the size is doubled again to an absolute level of around 60% of the original CAD file size. If then compression file technology is applied it can be reduced to the 30% of the original CAD file size again.

The reduction of 3D file size and clever applications of a range of algorithms for dynamic loading of visual data according to interactive viewing intentions supports swift and interactive working within DMU tool environments. CAD systems (either 3D or 2D) provide functionalities to construct and modify geometry of technical product models or drawings of mechanic and electronic parts. In cases where only a limited number of parts (max. up to 40 full 3D CAD models simultaneously) are in direct working interaction CAD models are o.k. to work with as long as the graphic and CPU (central processing unit) power of the computer hardware is good enough. For example, it is possible to support a scenario where a supplier who delivers bolts to an OEM, has to ensure the digital validation of the appropriate attachment situation. Here, a single CAD model for every bolt type carries all necessary information and the associated design/assembly situation for attach the right components via such bolts in appropriate design context can be supported well in a CAD environment.

In complex situations, however, where packaging, behavioral, assembly in context or other studies must be undertaken, a higher number of different 3D models need to be linked together in order to ensure speedy interaction without delays in human machine working modes. Here, the 3D CAD models are transformed into "lighter" DMU 3D representations (see explanation above concerning the geometric accuracy) to serve as basis for realistic visualization and simulation of the entire product or associated processes. These studies are, therefore, conducted with *Digital Mock-Up* (DMU) techniques using reduced data sets (i.e. lowered meta data richness and geometric accuracy) [6]. In order to handle these tasks, also non-relevant meta data information of the original CAD parts are deleted prior to the conversion into neutral 3D DMU file formats for the sake of speed.

A CAD design component may represent the left and the right tire of a vehicle, i.e. the same CAD file represents two instances of the same 3D CAD model. Digital Mock-Ups representations, however, create the visualization of the entire product in three dimensions and require, therefore, two separate entities of the instances in order to visualize the right and the left side position of the tire in vehicle position correctly.

Similarly, representations based on part records may use a single part number for an entire end-item assembly (like a full suspension subframe), that in fact consists of dozens of parts. In such cases, DMUs require unique identification of every entity. In addition, the whole configuration of the 3D DMU representations has to be able to resolve interferences, packaging and other design integration issues on an individual configuration level as well as on the combined cross configuration level.

12.1.2 What Does DMU Do for an Engineer?

Conceptual designers as well as System, Component and Manufacturing Engineers use Digital Mock-Ups to explore alternate product options, implications of various packaging scenarios and other information sets (like manufacturing process alternatives or ergonomic issues and consequences) across multiple configurations (see Fig. 12.3).

The Digital Mock-Up is used for packaging studies and engineering investigations throughout the product creation process. The DMU can also be leveraged for simulations such as package investigations, kinematics calculations or thermal simulations as well as simulations to plan the production process [9]. For example, modifying the air conditioning unit of a passenger car can affect the positioning and functioning of the steering devices. Similar system dependencies become apparent by studying the true 3D package positions in a front-wheel-drive engine compartment (compare Fig. 12.4):

- The transversal engine location drives the battery location
- The battery drives the brake booster location
- The brake booster drives the brake pedal location
- The steering column and the brake booster must not cross to allow column ride-down
- The steering column needs to be on the LHS (Left Hand Side) of the brake pedal
- The steering column cannot be further outboard due to the steering rack travel
- The brake pedal might end up too far inboard which causes trouble to be too near to the gas pedal and/or to limit the space for the inner compartment console.

Such a 3D based system integration and packaging analysis, however, does not work automatically and still needs substantial (automotive) system knowledge as well as the availability of a robust and reliable 3D representation of all technical systems in interplay of this vehicle zone in correct absolute and relative position. A specialist DMU or Packaging Engineer for engine compartment investigations leverages a full DMU with all core elements (product structure, 3D models, product position, product configuration), as outlined in Fig. 12.2. They have to conduct the

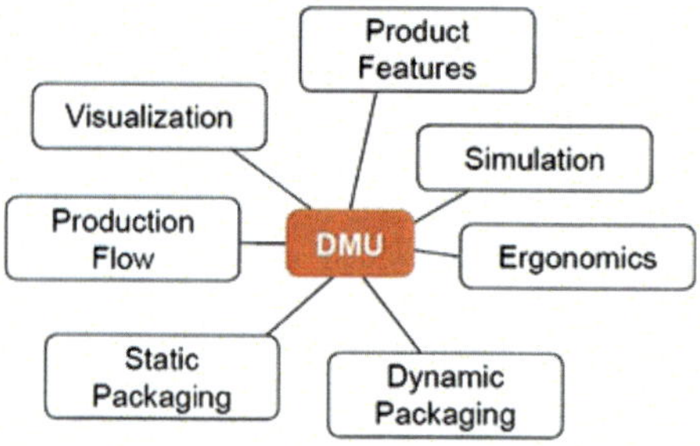

Fig. 12.3 DMU as central point for further applications

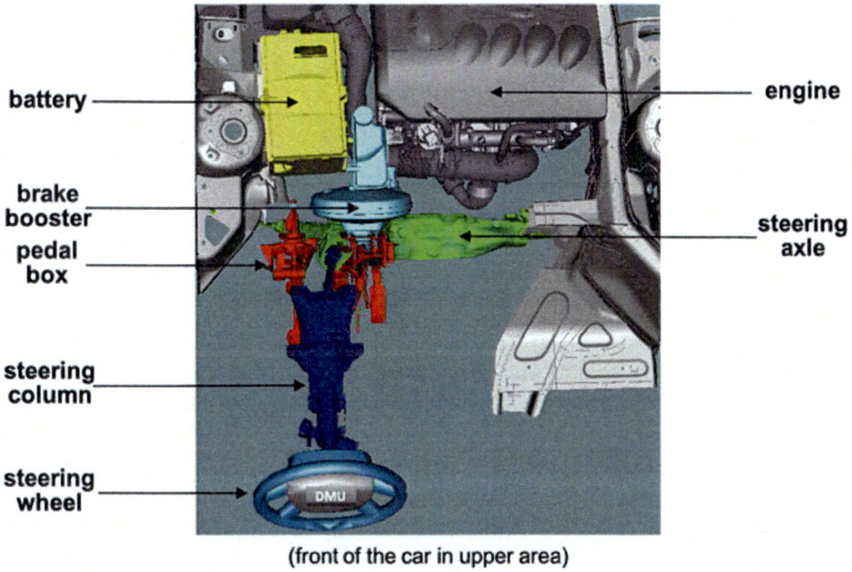

Fig. 12.4 DMU to investigate the mechanical package system dependencies between engine—battery—brake booster—steering column—pedal box positions

following methodological analysis steps in order to assess the situation as outlined above and shown in Fig. 12.4:

1. Determine the exact zone area of interest (usually done in xyz coordinate space or relative to a major sub-system). As it is shown in Fig. 12.5 (compare also the good match between the Digital and the Physical Mock-Up, which nowadays

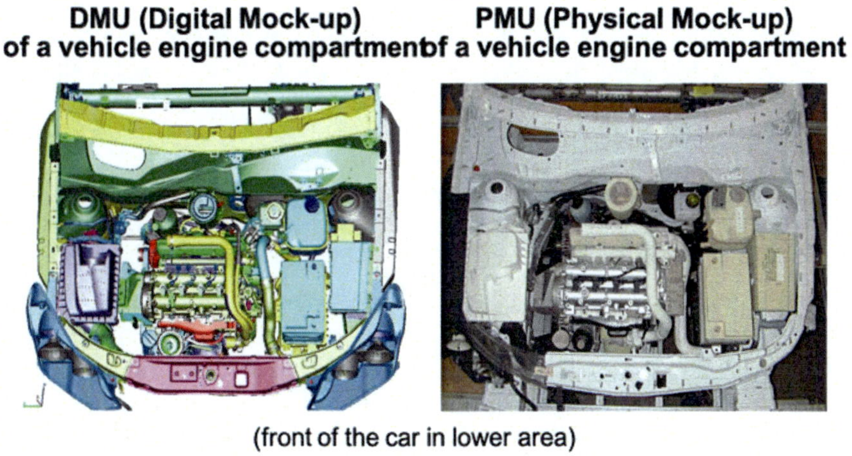

Fig. 12.5 DMU and PMU of an engine compartment of a passenger car

is no longer needed in most of the cases) the engine compartment zone can be limited by the car cross beam resp. dashboard panel (upper area) and the front grille opening (lower area). Such an engine compartment zone accounts for approximately 800–1000 individual parts, combined in different types of assemblies, incl. fasteners).
2. Perform an expert search for the systems of interest in the product structure, select a suitable component or assembly of such systems and conduct a proximity search in order to only show the possible sub-set of relevant components, assemblies and sub-systems for the specific system package analysis. Following this step, the number of different parts can usually be reduced by half or two thirds.
3. Down select those components and assemblies that follow the functional and package relations as shown in Fig. 12.4. For this step also significant technical knowledge is necessary by the DMU and Package Engineering since today's DMUs do not yet inherit functional and behavioral knowledge graphs. As result, the DMU or Package Engineer still has to work interactively with approx. 150 visual representations of different parts on screen, which explains why a lightweight representation of the geometry is necessary.

In any of such analysis cases, DMUs can help to assess effects of component movements and deformations or altered tolerances of components with respect to the entire product [7], e.g. with integrated static and dynamic analysis tools. However, any further detail analysis goal does trigger specific functional and behavioral models in outside simulations software packages (FEA, GD&T etc.) before those results are integrable into specific analytical type of DMUs.

12.2 The Role of a DMU in Product Development

DMUs start getting used for design and compatibility reviews at early phases of product development, long time before part records are released for production. In the old days of digital engineering development, up to the mid 90ties of last millennium, CAD models where composed in major CAD layouts every 4–8 weeks and afterwards, colored layout drawings where used to discuss the state of development progress. In contrast to these times, the appearance of Digital Mock-Ups drastically changed the advancements of digital engineering of major products such as aircrafts, cars, trains and big machineries. The constant building and delivery of DMUs every week or even every day became only possible due to the following three major advancements:

- Refinement of CAD assembly modeling in absolute product coordinate systems as part of the new digital modeling capability of Virtual Product Creation
- Set-up of solid publication processes from CAD to a companywide PDM environment as part of the new digital engineering process set-ups of Virtual Product Creation: unlike the traditional Team Data Management environments which only

provided data exchange insights for teams up to 20 or max 30 designers and engineers, the PDM environments provided data access and integration for thousands of designers and engineers

- Significant progress in IT technologies for Virtual Product Creation in order to:
- establish companywide intranet networks,
  - enable multi-lever IT server set-ups to constantly convert geometric and meta data formats into DMU type representations,
  - interrogate product and data structures via multi-branching structure traversal algorithms and
  - deliver rich and light weighed IT client interfaces to dynamically load and collaborate with data rich DMU representations of multiple Giga Bytes sizes.

With these new digital engineering capabilities of Virtual Product Creation DMU pioneering companies such as Boing, Ford, GM and Airbus started to set-up step-by-step a new regime of DMU based digital engineering methods to scale up Virtual Product Creation within their companies but also with outside partners and suppliers.

Figure 12.6 provides an overview on typical DMU usage pattern across the major phases of product development as they have been established since the end of last millennium. Quickly it became clear that the availability of DMUs are not only of advantage for CAD Designers and CAE Analysts. Manufacturing Process Planers, Engineering Managers, Buyers, Service Engineers and other product life cycle related personnel started to appreciate in the first decade of the 2000s the "easy to understand and consume" availability of DMU based design data nicely arranged in correct product position and step-by-step also in the most important product variant configurations. Obviously, it was difficult in every company to establish robust DMU creation

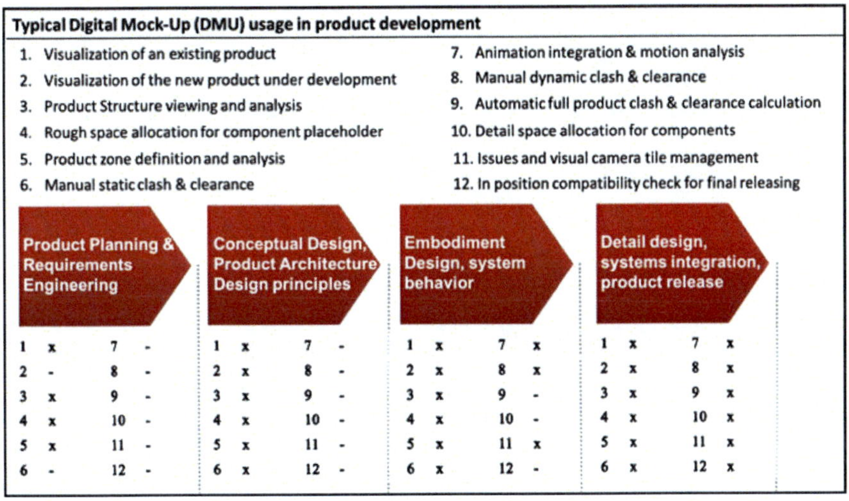

Fig. 12.6 Typical DMU usage types in different product development phases (each major usage type marked with an x at the usage instance)

processes as well as best fitting digital engineering process set-ups and associated digital skills. It was necessary to build up specific departments and business roles within the organizations to ensure scalable and active DMU based digital engineering processes. Consequently, DMUs may undergo a formal creation and release process as described in the next sub-chapter.

12.3 Usage of Different DMU Types

In order to understand the DMU creation process it is critical to determine the scope of usage as outlined in Fig. 12.6. E.g., static and dynamic DMUs may be distinguished and more advanced versions of dynamic DMUs called Functional DMUs are under development and in first deployments. All of them have their specific purposes and will be discussed in the following sub chapters. DMUs are created within various phases of the development process in alignment with the availability of CAD parts for mechanics or electronics, Functional DMUs are further enhanced by logic and software control features. It is essential that companies develop and release a CAD model progression plan with a schedule according to their virtual development process. Unfortunately, many companies are lacking such consistent model progression plan due to:

- missing ownership in individual departments or by classical engineering staff members
- missing understanding and awareness of urgency in Management
- unsorted digital policy rules within and across companies in order to steer digital engineering progression with suppliers and partners
- unavailability of robust process to track, review, reconcile & align matched delivery points in order to guarantee meaningful technical system development and compatibility
- unclear methods and easy-to-use IT tools for ongoing reporting and issues resolution.

12.3.1 Static Digital Mock-Up

In a static DMU, the digital representations of components and assemblies are non-movable and rigid. Static DMUs are used for collision detection, assembly and packaging studies as well as for complex design layout studies. These static DMUs are mainly related to mechanical design applications and often require mature state of the geometrical parts or at least representations that are easy to be interpreted by model consumers and reviewers. Therefore, the overall complexity of the represented 3D parts and the amount of data to handle are both very high. Large amounts of data can only be handled with the help of powerful computer technologies and dedicated data reduction software.

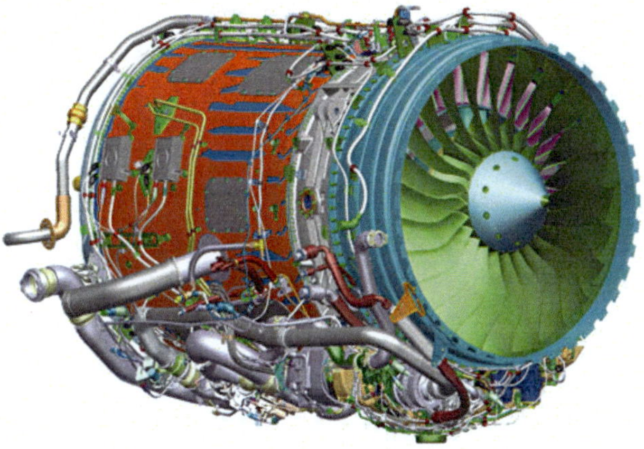

Fig. 12.7 DMU of an entire aircraft engine. Courtesy of Rolls-Royce Germany

The Digital Mock-Up of an aero engine (see Fig. 12.7) contains thousands of digital instances of visualization files, which are reduced in size from the original CAD master file with the help of data tessellation algorithms (i.e. data conversion operations which transform CAD models into triangulated approximations to reduce file size). Those data conversion operations are executed either at the CAD client side, that means at the CAD Designer's computer prior to the PDM upload or as part of the central data service of the overall PDM server regime (central server-side tessellation).

12.3.2 Dynamic Digital Mock-Up

The dynamic DMU is an enhanced version of a static DMU. It holds kinematic and dynamic elements and it is used for a variety of applications such as kinematic simulations, eigenmode analyses or for even more sophisticated simulations and investigations e.g. elastic or plastic deformations or behavior of hoses and cables. With a static DMU it is possible to ensure that parts and components of a product will not collide, meaning not occupying the same space while a dynamic DMU additionally allows for checking whether a specific part can really be assembled. For kinematic simulations one addresses information about degrees of freedom or motion capabilities to parts or components. This allows for virtual test of the functionality or for clearance and collision analyses.

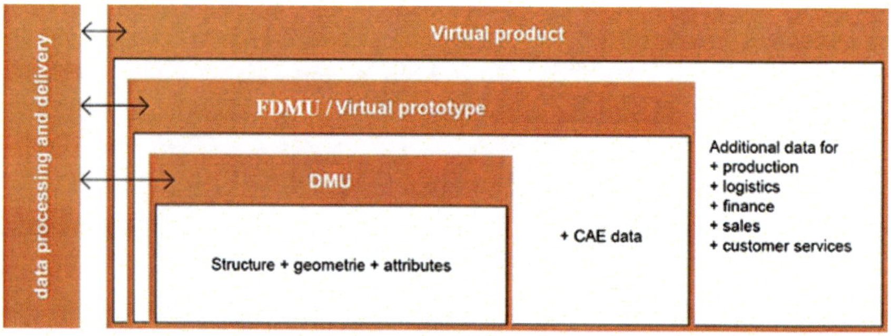

Fig. 12.8 Evolution from DMU to FDMU/virtual prototype and to (entire) virtual product

12.3.3 Functional Digital Mock-Up ("Functional Mock-Up")

Including more and more functions to products and considering larger and larger systems add also complexity to its digital representation. Functional Digital Mock-Up's (FDMU) driving the DMU concept further. In addition to the geometrical aspects of a DMU, a *FMDU* allows for simulation of the functional behaviors of such assemblies. *FDMUs* may couple control algorithms to CAD systems thereby allowing for a simultaneous functional testing of both together.

Let us understand the evolutionary positioning of the *FDMU* in the hierarchy of levels from DMU up the entire *Virtual Product* as outlined in Fig. 12.8: the Functional Digital Mock-Up (FDMU) belongs to the *Virtual Prototype level* and represents a DMU that is enriched with different types of CAE data (model characteristics, simulated behavior, etc.).

In order to achieve this evolution the concept of enrichment requires the incorporation of specific CAE models of mechanical components that are represented by visualization files of CAD models in current and correct position and configuration as part of the DMU, but also the integration of electric/electronic and software control model representations (compare Fig. 12.9).

A Functional Digital Mock-Up, therefore, adds the *functioning of the system under investigation* to the pure static geometrical aspects a conventional DMU. The Functional Digital Mock-Up (FDMU) is an extension of the well-established concept of a Digital Mock-Up. This approach constitutes a combination of traditional DMU with functional resp. behavioral simulation capabilities. Consequently, geometrical properties and functional aspects have to be considered simultaneously within a unique framework. This way, geometrical analysis/verification of a mechatronic system as well as functional/behavioral checkup of static, dynamic and logic functionality becomes possible. This combination constitutes an important step concerning the holistic validation and verification of mechatronic and cyber-physical systems (incl. the sharply growing cross product and technical system interactions).

FDMUs are intended to enable functional representations and analysis of a real technical system. As for all simulations, it is possible to use input data or boundary

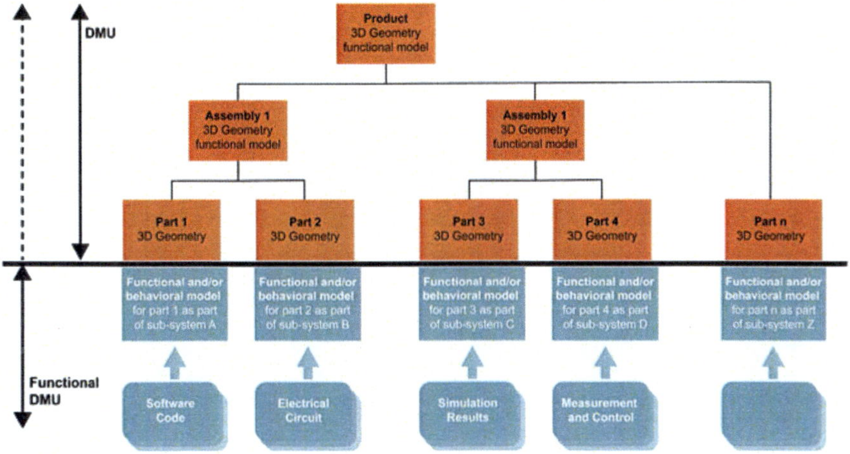

Fig. 12.9 FDMU and DMU

conditions obtained from real-life to conduct the simulations. In this sense, the FDMU takes over the role of the physical test object, as indicated in Fig. 12.10: although the full *Virtual Product* comprises additional information sets and characteristics (as shown already in Fig. 12.8).

The FMU/DMU has to align itself with the specific system analysis needs of the full *Virtual Product* in order to ensure that the three underlying FDMU elements model build-up, incorporated simulation models and virtual experiment assumptions are well aligned to each other.

An interesting property of the FDMU is the possible interaction between a visualization tool and numerical simulation in both directions. Hence, not only the simulation results can be presented e.g. within a VR-3D scene but the user can also actively influence a present simulation using the visualization environment.

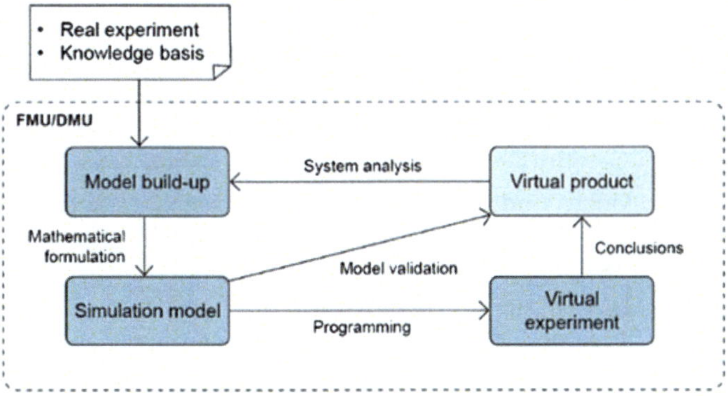

Fig. 12.10 Simulation set-up using FMU/DMU

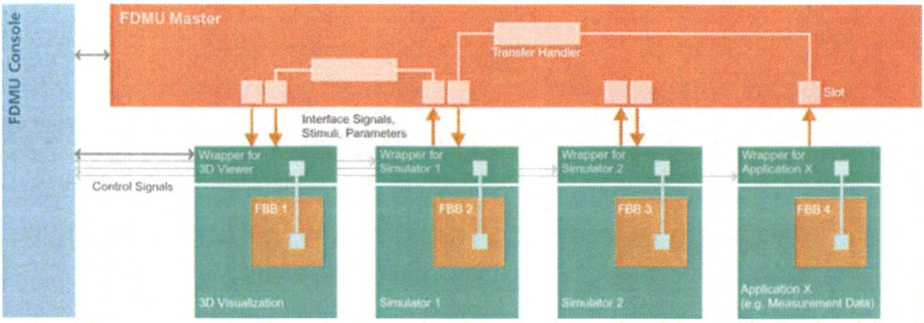

Fig. 12.11 FDMU principle based on a master simulator backbone according to [3]

Communication between visualization and simulation is carried out by a Master Simulator, as researched, developed and demonstrated within the Fraunhofer project "*functionalDMU*" between 2008 and 2011. This project has demonstrated that a FDMU allows for the functional interaction of a geometrical DMU with mechanics, electronics and soft- or firmware. Figure 12.11 shows the principle of organizing the individual functional simulation building blocks under the overall regime of the FDMU master (compare [2] and [3]).

Within the FDMU approach, the concept of a so-called Functional Building Block (FBB) is proposed by Enge-Rosenblatt et al. (compare [2]):

An FBB is an envelope summarizing geometric information (CAD models), behavioral models (e.g. described by differential–algebraic equations), and communication interfaces into one basic data module. Geometric information and behavioral information have to be created within their particular modeling tools. These models remain in their associated data files. Pointers to these files as well as all interface information and the mapping between geometrical data and the interface quantities of the behavioral model are collected within the FBB using the modeling language SysML for a unique description. Within every FBB, a simulator tool is also defined, which is capable to simulate the FBB's behavioral part.

A complete FDMU Simulation Model (FSM) consists of one or more FBB. Every input of an FBB must have an appropriate output belonging to another FBB. Furthermore, outputs can be propagated to the visualization to show simulation results using e.g. a geometric 3D model or some kind of plot versus time.

Obviously, there exist different kinds of technical realization options for Functional Building Blocks (FBB). Figure 12.12 shows a realization based on *Modelica*,[2] which is an acausal, object oriented technical notation language to describe physical models. *Modelica* uses mathematical algebraic and/or differential equations to describe the (inner) physical conditions and relations and offers interface connectors to couple the individual physical characteristics to other objects (or FBBs).

[2] Modelica is standardized, maintained and further developed by the Modelica Association. Please see details under: https://www.modelica.org/.

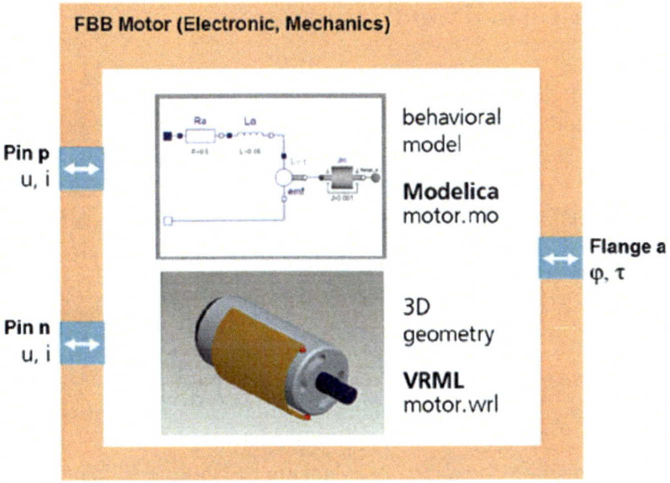

Fig. 12.12 Functional building block (FBB) example of an electric motor [2]

An alternative is a *causal* technical notation that directly uses so-called à-priori causalities of precedent objects with respect to the time-dependent influence on subsequent ones and that is mainly based on algebraic mathematical relations. The MATLAB[3] solution suite is, for instance, based on such a signal based causal notation with a wide range of other mathematical options to integrate technical-physical object behaviors.

A third alternative is represented by the system modeling language SysML which is maintained, explained and further developed by the *Object Management Group* (OMG)[4] and by *SysML.Org*.[5] The disadvantage of SysML, however, is that it cannot directly describe simulation models and, therefore, needs additional resources to realize the FDMU targets.

A fourth alternative is provided by direct couplings of encapsulated CAE simulation models by using the interface standard *Functional Mock-up Interface (fmi)*,[6] compare details in [7]: the Functional Mock-up Interface (FMI) is a free standard that defines a container and an interface to exchange dynamic models using a combination of XML files, binaries and C code zipped into a single file. It is supported by 100+ tools and maintained as a Modelica Association Project on GitHub.

[3] For more information please refer to: https://www.mathworks.com/products/matlab.html.

[4] For more information please refer to: https://www.omg.org/.

[5] For more information please refer to: https://sysml.org/.

[6] For more information please refer to: https://fmi-standard.org/.

12.4 DMU Set-Up and Model Building

Setting up a DMU starts with collecting geometrical, configurational, behavioral and other information and attributes from CAD, CAE and PLM systems, respectively. Using best practices and proven guidelines, the DMU is generated from those building blocks. Figure 12.13 shows the overall process of building up a DMU model and of providing it for DMU based analysis work. The "design" of the DMU, i.e. the DMU model build requires a solid information base of all core elements that are candidates to be incorporated into a DMU (compare Fig. 12.2). This can only be guaranteed consistently if the following rules and conditions can be met within the overall Virtual Product Creation environment of a company as already outlined in sub-chapter *The Role of a DMU in Product Development.*

To generate a DMU automatically, the designer has to assign the 3D CAD model to a node in the PDM/PLM product structure and has to position it in every relevant car configuration. Positioning can be done either in absolute coordinates or in relative coordinates describing the position with respect to another part.

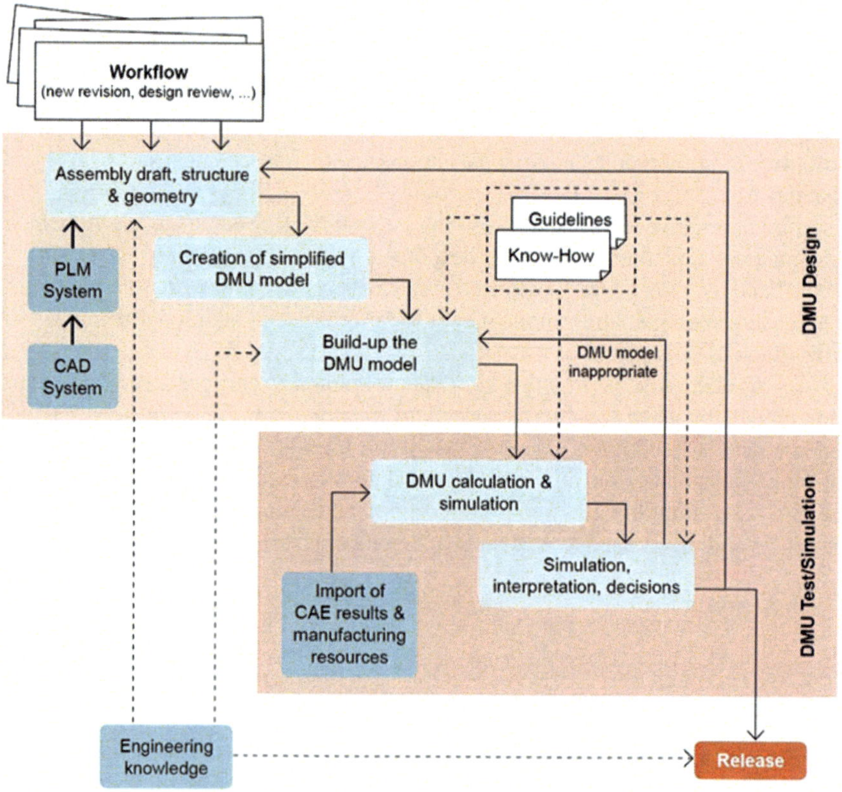

Fig. 12.13 DMU creation and use process

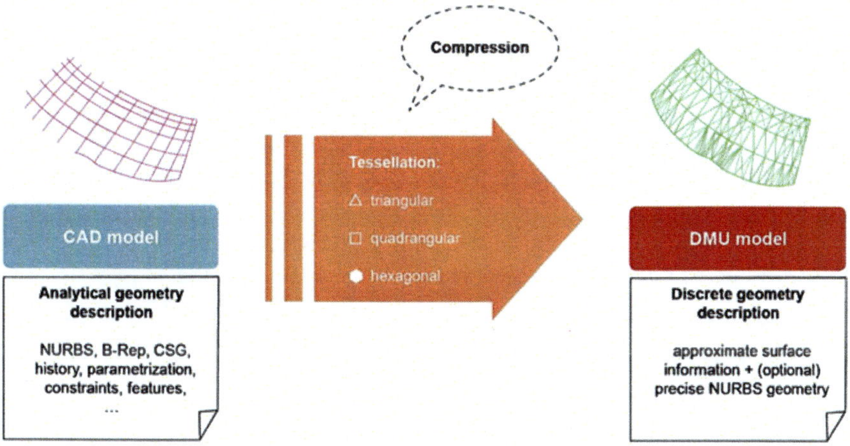

Fig. 12.14 Conversion of geometry from CAD model to DMU model

Based on a DMU model test and simulations such as geometrical layout or packaging clearance and clash analysis can be performed. For higher order simulations, the DMU model serves as basis. Depending on the aim of the DMU type (see sub-chapter before) the DMU model may be revised and enriched accordingly. Eventually, the DMU is released and can be used as single-source of truth for subsequent studies.

One of the core model build steps is the conversion of the exact (analytical and/or numeric) mathematical description of geometric entities such as surfaces and volumes into approximated ones. As shown in Fig. 12.14, the tessellation process converts the geometry descriptions of CAD models into approximated ones by using triangular (most common), quadrangular or hexagonal surface patches. In case of adding the original exact geometry representations from CAD to those approximated geometry representations in DMU compression technology is used to keep the file size small (compare explanations in sub-chapter "*Why does an engineer use DMU instead of CAD?*").

DMU models require only a reduced set of information compared to CAD part models. In principle, mainly the surface representation of the initial CAD parts is required to construct a sufficient geometry model in DMU. This information can be further reduced by simplification the surface modeling by a suited approximation, also called *facet representation*: it approximated the initial complex surface by more but simpler connected planes.

A process called tessellation, which means the approximation of any given surface by a set of geometrically simpler elements, realizes such approximation. These planes are polygon-shaped elements, where triangles are often used, as they are the simplest form of polygons. The process of approximating any surface by triangles is called *triangulation* (compare Fig. 12.14). For such purpose, various algorithms exist such as Delaunay algorithm or Watson algorithm [8].

Compared to the initial object, its tessellated surface model has significantly smaller data size. The DMU of the Smart Tripelec, shown in Fig. 12.15, has only 5.5% of the data size as the original CAD model. Another practical implementation offers different type of reduced and pre-configures partial DMU models which have been derived from a full product DMU. The lower illustrations of Fig. 12.15 show this effect. As starting point a full *multi variant vehicle DMU* based on ~9000–10.000 different visualization models in approx. 50.000–60.000 position instances would account for approx. 5 GByte of size (with an approximation accuracy of ±0.5 mm). Reduced vehicle DMUs (reduced by systems and variants) are offered like

- a *partial vehicle DMU* model with engine shrink wrap files only, without interior components, with limited underbody powertrain and electrical components

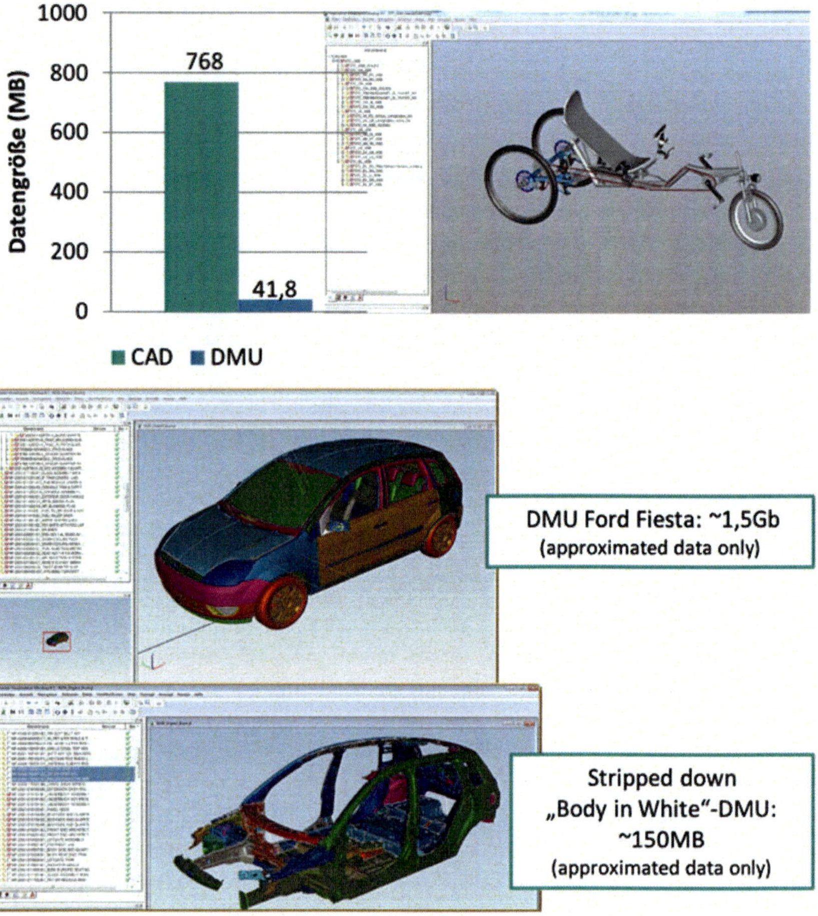

Fig. 12.15 File size of DMU models (CAD versus DMU) and partial vehicle DMUs, all based on JT

(approximated data only without underlying exact NURBS representations) with a file size of ~1.5 GByte and

- a reduced *Body-in-White* sub-*DMU model that* only accounts of a file size of 150 MByes (with ~500 part instances).

However, discretizing the initial object and replacing it by its tessellated derivative means that this deduced object has lost the information about history, relationships to its neighbors, parameterization, constraints and other features of the initial object. Accordingly, the positioning of the components is not included in the discretized geometry description and must be specified separately. The geometry models from different 3D CAD systems are transferred to the DMU. Thereby, the surfaces described according to exact mathematical rules are replaced by surface meshes from flat surfaces in order to accelerate the visualization of the models in the DMU system.

The transferred models from the various 3D CAD systems are integrated into the DMU models. Examples for often used DMU-data formats are

- *JT* (Jupiter Tessellation), originally a proprietary standard from Siemens, meanwhile ISO standardized, compare Fig. 12.16

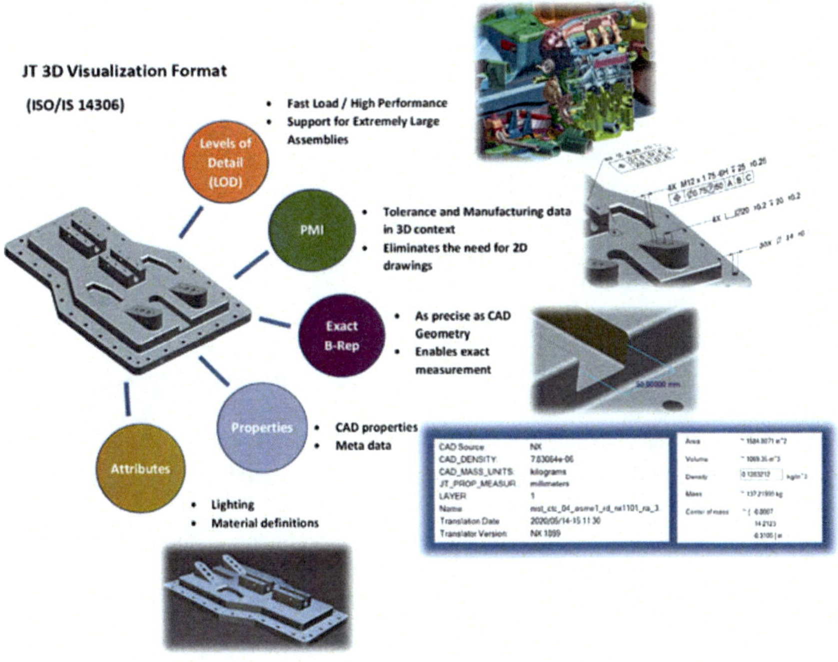

Fig. 12.16 Elements and capabilities of the 3D visualization format JT (courtesy support by Siemens Digital Industry Software)

- *3D XML*, proprietary format from Dassault Système, successor of the popular original Dassault Système DMU format *CGR* (Convergence Graphical Representation)
- *VRML* (Virtual Reality Modeling Language), ISO standard, meanwhile superseded by
- *X3D (Extensible 3D Graphics)*, ISO standard.

Originally the 3D visualization format JT (Jupiter Tessellation) was specified 1998 in a close collaboration between Hewlett-Packard and Engineering Animation, Inc. (EAI) and was developed based on the "Direct Model Toolkit". After having purchased EAI by UGS (1999) and after a later acquisition of UGS by Siemens AG (2007) JT nowadays belongs to Siemens Digital Industries Software.

JT is meanwhile an openly published data format as it has been accepted by ISO as an International Standard for 3D visualization. It is widely used for communication, visualization, digital mockup and a variety of other purposes at a majority of the world's leading manufacturing companies. In addition to visualization, many JT adopters use JT as a process format for workflows such as data exchange, supplier collaboration, and long-term data retention. As described in Fig. 12.16, JT supports characteristics such as approximated and exact mathematical (NURBS) description of geometric entities, level of detail (flexible mode of geometry visualization accuracy), PMI (Product and Manufacturing Information, i.e. support of engineering attribute data linked to geometrical entities) as well as other CAD properties, attributes and meta data.

For the positioning of parts and assemblies an absolute coordinate system is to be used, which all participants must comply with from the start. The product structure (optimally represented in the PDM/PLM systems), which must be defined and communicated at the beginning of the product development, is depicted in the reference model. The access of different areas to the DMU data can be regulated by the use of PDM systems [6].

However, the use of DMU is associated with a certain effort for the organization and usage. Furthermore, DMU geometries cannot be returned to the 3D CAD system because their structure must be changed when converting to the DMU system. The collision checks and measurements based on DMU analyzes also depend on the quality of the used geometry data. The overall creation of a DMU model, which is also called *DMU model build*, and its offering for DMU analysis work consists of six principal steps that are outlined in Fig. 12.17.

The overall DMU model build and the associated engineering support are owned by special DMU departments, which are meanwhile part of the engineering organizations. Such departments are not owner of the CAx data models and they are also not in charge of the underlying IT infrastructure (this is owned by the IT departments) which is needed to ensure robust DMU model build generation. The DMU departments, however, are the corresponding experts in order to build, verify, judge and analyze DMU models with respect to:

Process chain from CAD to DMU	Content	Possible methods and formats
1. Description of the process chain CAD → DMU	Development of a Digital Mock-Up, which serves as basis for different calculations and validations; can be run on screen or in background processes (e.g. for clearance & collision analysis)	Deduction of an assembly model with approximated geometry (triangulation of surface)
2. Starting data: CAD	Product structure and geometric model (solid or surface geometry)	CAD - native STEP, IGES, VDAFS, STL
3. Target model	Product structure and approximated geometry (polygon surface model)	Approximated geometry (polygon surface)
4. Data transformation	Deduction of an assembly model out of the product structure. Triangulation of the CAD geometry	STEP AP 214 JT
5. Additional information	Assembly paths, configurations	Interactive graphic simulation
6. Interpretation of results and feedback; process-chain DMU → CAD	Verification of the product configuration, checking according to collisions, optimization of assemblies	Analysis and Simulation, manual feedback of results into CAD

Fig. 12.17 Detailed steps of the overall process to create and use DMU models

- Setting scope and content of DMU configurations
- Orchestrating all IT services to perform the underlying data base interrogations, model conversion and model linking operations
- Enriching specific attributes and analysis results
- Keeping close contact to the CAD model authors in case of tessellation and or position/orientation errors.

- Providing hands-on navigation, analysis and documentation support in interactive design reviews incl. on-the-fly issues management
- Conducting and controlling off-line and automated clearance and collision analysis including the associated result management.

12.5 DMU Based Engineering Analysis Work

The DMU model build as described in the sub-chapter before must be aligned with the analysis and simulations tasks that belong to the identified DMU engineering investigation portfolio. Figure 12.6 introduced already an overview of typical DMU usage types in different product development phases. This sub-chapter will provide more insight into the immediate preparation of the appropriate DMU for such investigations and provides examples how the DMU based engineering analysis works.

To construct a DMU, it is necessary to have the *engineering bill of materials* and—which is even of more important during the development process—the *reliably configured product structure* of the product in question. Firstly, a DMU can be instrumental to analyze whether all required part geometries have been provided or which parts are missing just by visual representation of all linked visualization models (components, parts, assemblies) linked to the respective product structure nodes.

As in many other industries, the automobile industry, e.g., develops a number of different car variants of every model (left- and right-hand drive, automatic and manual transmission, different engines, and different customer-specific equipment) and so forth. Here, packaging is used to define the most critical configurations in terms of special constraints and requirements. These configurations then will be used to check the fitting of all parts with respect to each other. Clearly, a trade-off has to me made between the effort for these procedures and the number of variants to analyze. This trade-off is often made by experience and time constraints.

Figure 12.18 shows how the winglet sub-system of the overall aircraft wing system gets configured in the product structure which is available in a PDM System (in this case the PDM System Teamcenter Engineering of Siemens).

Usually, it is the task of a system and/or design engineer who owns this sub-system (winglet sub-system in this case), of a specific product modeler or of a BOM specialist to ensure that all components that are handled *as end-item assemblies* in the final assembly plant get variant coded with the correct company internal expression (feature code) language at the right level of the product structure. This task needs specific knowledge about:

- the generic product architecture of the aircraft OEM as shown in the upper part a (compare details in [1])
- the design models of the sub-system of interest (in this case the winglet sub-system) as shown in the moddle part b and
- the correct coding expressions as part of the company specific expression (feature code) language.

As of today, there do not exist any variant coding standards within industry sectors or cross-company consortiums, which makes it difficult to align the variant coding for the product configurations. Even the type of variant expressions (see Fig. 12.18 part c) do differ significantly: there might exist rules, family expressions or even numerical value assignments.

An aircraft DMU may contain 150,000 to 1,000,000 instances of visualization models depending on type (business jet, commercial aircraft or military jet or transporter), size (length, width, resp. single/double aisle), equipment levels and variant/configuration richness. Typically, such aircraft DMUs are used to represent different kind of development architectures through the virtual product creation process.

In the early development phases, the *Master Geometry DMU (MGD)* is created which contains all outer surface models of the fuselage, the wings, the empennage, power plant, and landing gear as well as the major aircraft axis. The MGD mainly serves as master to support basic airflow investigations with the help of Computational Fluid Dynamics (CFD) analysis, which constitutes a branch of fluid mechanics that uses numerical analysis and data structures to analyze and solve problems that

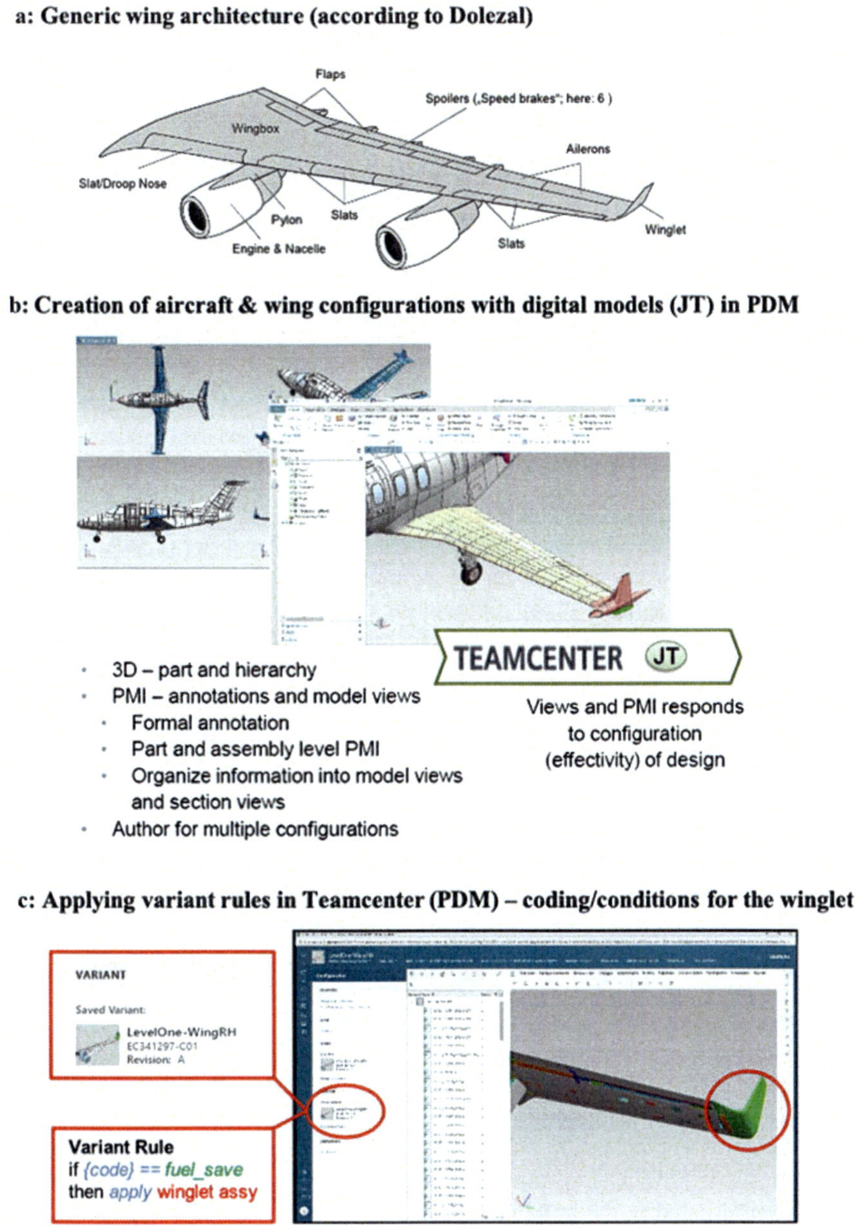

Fig. 12.18 Target oriented DMU model build preparation with the example of a configured wing system of an aircraft

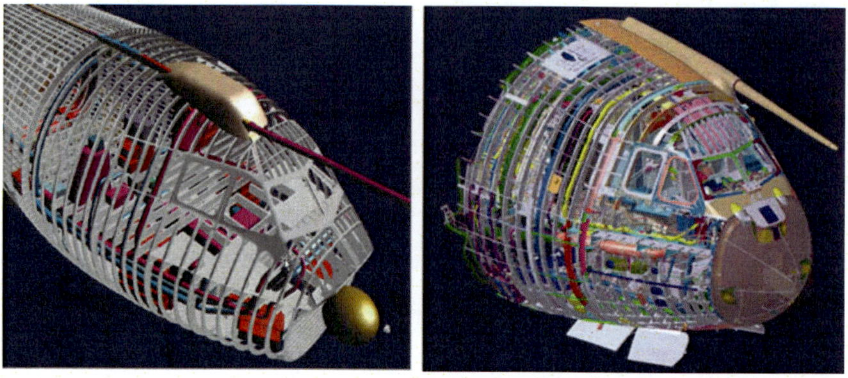

Fig. 12.19 Space allocation mock-up (left) and definition mock-up (right) of the nose fuselage section of the airbus A400M military transport aircraft based on Dassault Système CATIA CGR visualization files [1]

involve fluid flows. The second DMU often used in aircraft industry is called *Global Architecture Mock-up* (GAM): it is the master of the system architecture and support system installation engineering by representing package space and influence zones for fuel, electrics, hydraulics, air conditioning, flight controls etc. by simple cuboids, cylinders and flexibles. The third DMU type is called *Space Allocation Mock-up* (SAM) and represents major components in their allocated zones. The fourth type and most detailed DMU in aviation industry is called *Definition Mock-up (Definition DMU)* and contains all components throughout the final aircraft definition phase.

Figure 12.19 shows the difference between the *Space Allocation Mock-up (SAM)* and the *Definition Mock-up (Definition DMU)*. According to [1], those illustrations show two snapshots of aircraft development with a time span of about five years: while the *SAM* (left side) shows rough geometry of major structural elements and first space volume "claims" by systems and equipment the *Definition DMU* (right side) represent a densely packed nose section with all sorts of structures, systems and equipment fully detailed ready to be released for production.

If a consistent variant coded product exist within the PDM product structure it is then possible to traverse the product structure according to configuration sets that have been defined for the overall product. Figure 12.20 shows the situation of such a configured product—a full aircraft in this example—in the upper part a.

For such fully configures aircraft the aero engine systems usually are only represented in a „light" shrink wrap mode, i.e. only the outer shape and attachment points are fully represented in the propulsion DMU model. All internal combustion and other technical system representations might be suppressed before providing it as supplied system to the full aircraft DMU.

In order to support the engineering progression of a product development program the DMU analysis plays a vital role: althrouhg designers and engineers might have received an official *space allocation* (i.e. a firmly claimed and "contractually" assigned virtual 3D x/y/z coordinate system space volume) for their own technical

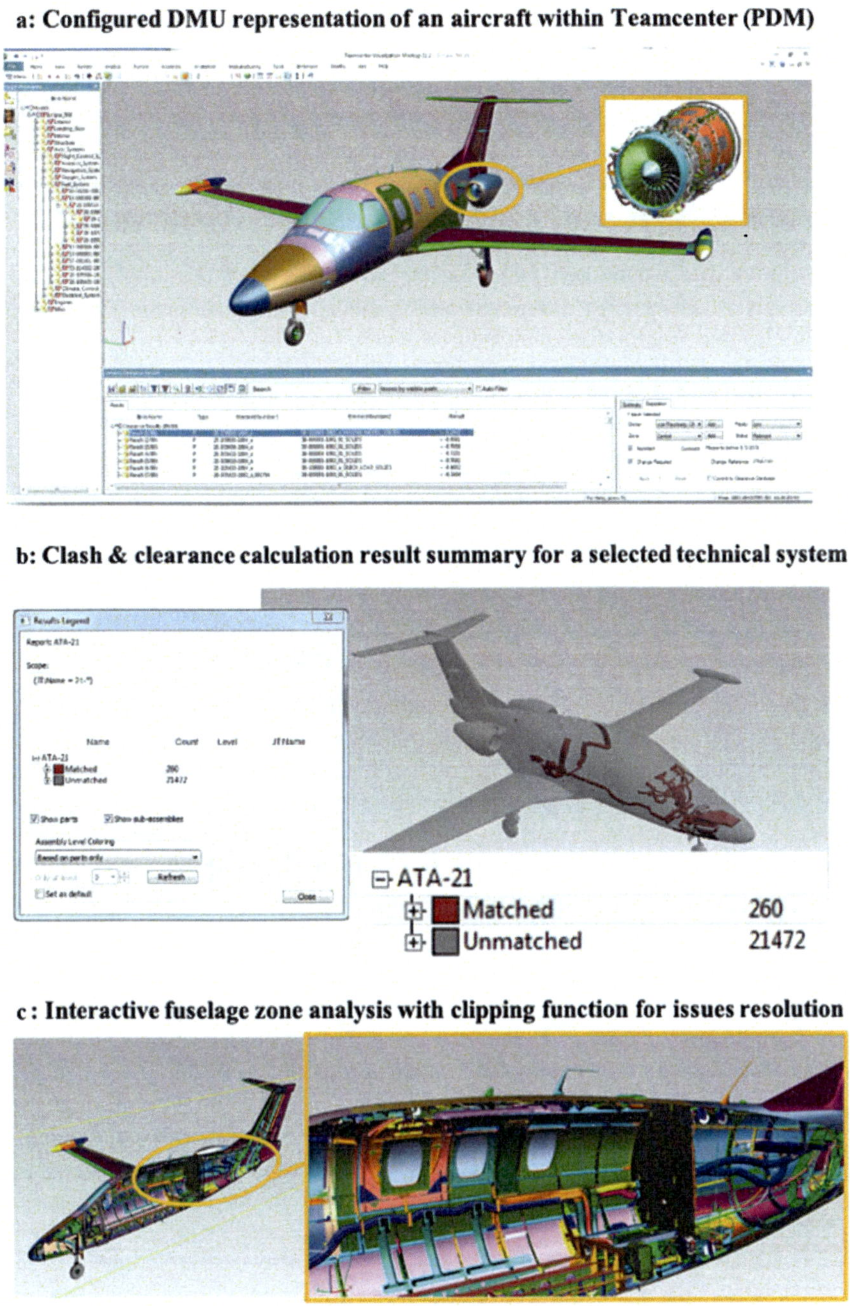

Fig. 12.20 Different types of engineering analysis based on a configured DMU

system components as part of the overall engineering steering process or mutually agreed one as part of the collaboration work amongst the neighbor system owners and the progam leadership team, there exist a high chance of mutually violating or distubring space violations.

In order to detect and manage those and to prevent system interacting noise distrubance the clash and clearance analysis process has been introduced as part of the overall DMU analysis activities. As starting point, it is executed based on static DMUs. As a second step, after having reached process maturity for the clash and clearance of the static DMU, also enriched dynamic DMU instances are added to the analysis. There exist different working modes within companies how such clash & clearance studies are conducted:

1. Automated calculation in a specific database (see results in Fig. 12.20, part b)
2. Interactive analysis of a DMU engineer or analyst (see Fig. 12.20, part c).

In order to be able to conduct automatic calculations within a DMU clash and clearance database, the following preparations must be done, both methodologically and technologically. Only very few companies in the world have meanwhile achieved this high maturity level of robustly executing such *automated* clash and clearance without disturbance and/or troublesome discussions around the daily, weekly etc. delivered results files with the respective system or design owners. Please note the pre-requisites for achieving such maturity level:

- Each configured DMU model will be additionally registered in an additional database that manages the used geometric space consumption of a component, assembly or system measured within a grid of volumetric elements called *voxels*. A voxel data base entry of a geometric entity, therefore, represents a value on a regular volumetric grid in three-dimensional space.
- Rules need to be defined together by the DMU analyst and the system/design engineers and owners with respect to:
  - Defining *clearance rules* with meaningful target values to guarantee sound technical system performance; e.g. within automotive package engineering a general clearance rule exists. This rule prescribes that a static clearance should not fall below a value of 15 mm between different components, which should not interact with each other. If such value is, lower a specific engineering analysis need to be conducted. For technical systems with specific dynamic interactions or possible disturbing error modes, the identifiable technical requirements need to be included into such specific clearance rules.
  - Defining *clash exclusion rules* in order to sort out all unintended clashes between components that cause clashes in their modelled shape due to standard CAD design modeling practices. This allows in many cases to reduce design modeling efforts of not showing the exact final shape of a components (e.g. in case of flexible or conformable material) or to only show the "as manufactured" but not the "as installed" design situation (this does include situations where rubber sealing and other interface parts are not shown in compressed form).

  - Allowing for specific *"non buildable worst case engineering configuration rules"* in order to accommodate for product variant spanning, invariable design situations or protection (often platform based engineering needs to protect for future, upcoming design situations which are not yet represented by appropriate 3D representations).

- Customized and tailored clash and clearance report generation for specific zones, product systems and system/design engineers and owners in order to meaningfully reduce the high number of first run reported clash and clearance issues from thousands (for a DMU consisting of 60.000–100.000 model instances) down to dozens or to a maximum of low hundreds. Since this requires substantial technical system development know-how, there exist a great potential of AI support for that task in the future!

It becomes obvious, that, still today, the majority of the clash and clearance analysis needs to be conducted interactively by DMU and system/design engineers by using 3D analysis methods in combination with inherit design and engineering knowledge of individuals or within an engineering team. For such interactive analysis, the following approach is being taken:

- The DMU resp. design/system engineer selects the relevant configuration from the list of offered DMU variants, manually selects the relevant DMU parts from the DMU product structure or uses different kind of search algorithms within the DMU software in order to finally determine the intended *collection of DMU parts for the analysis*, please note, e.g., the following three ones:
  - Neighbor search
  - Proximity search (all parts in the nearness of a bounding box with a certain size around the selected components)
  - Attribute search (selecting all components that carry a certain engineering or variant attribute such as released, in work, in configuration etc.)
- Having arranged the analysis collection of parts, now the interactive clash and clearance function of the DMU software can be activated; such algorithms works internally as follows:
  - For each of the DMU part the bounding box representation is used in order to detect which of the bounding boxed do interfere to each other (based on the target value of clash, i.e. less then ~0 mm distance, or the clearance value of interest). If a positive case is detected all facets of the approximated representation of the individual DMU parts are cross-analyzed to each other in order to determine where exactly between the parts the clash or clearance situation exist and the shortest distance is measured, represented with a line vector and stored.
  - After all analysis calculations are done within seconds or minutes (depending on the size of the collection of DMU parts for the analysis) a report is created which can be interactively reviewed and studied.

- Finally, it is decisive how the results are presented back to the DMU resp. design/system engineer in order to trigger a direct and smooth analysis interpretation and potential resolution steps. As shown in Fig. 12.21 there exist a couple of different visual aid opportunities to increase the immediate understanding of the engineering interpretation of the results:
  - Highlighted colored explanations of the clash volume, the penetration path and the overall intersection curve of the involved components and/or assemblies (see Fig. 12.21, part a)
  - Color-coding for the clearance zones between the involved components and/or assemblies (see Fig. 12.21, part b).

a: Presentation of a clash result within the DMU application 3DViewStation

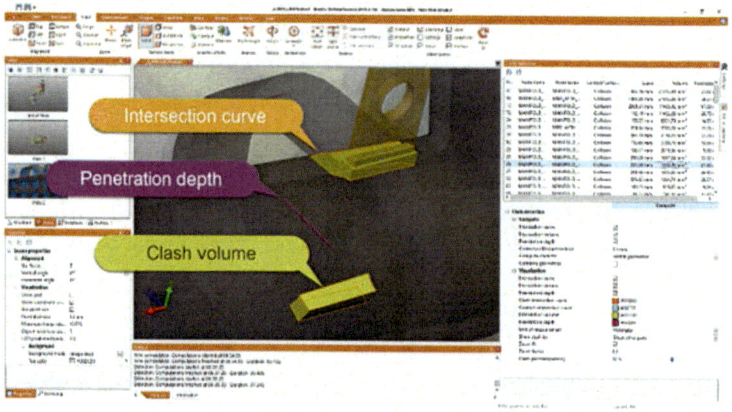

b: Presentation of a clearance investigation within the DMU application 3DViewStation

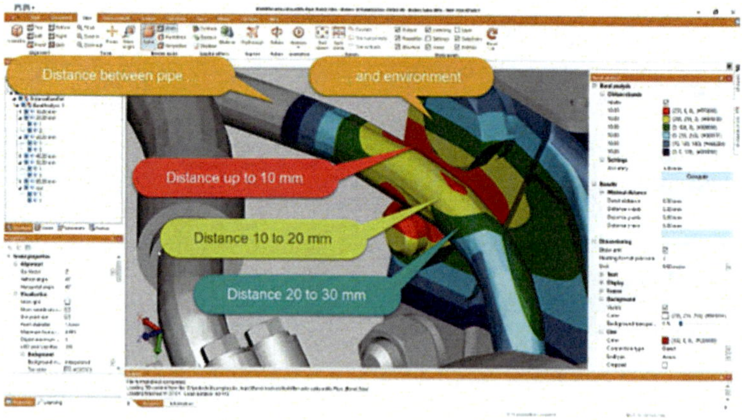

Fig. 12.21 Clash and clearance result presentation in the 3DViewStation of KISTERS

3D DMU models are also used for a range of other typical engineering investigations in order to judge the overall product system characteristics, behaviors and design opportunities. As shown in Fig. 12.22, the DMU engineers or users get trained to interactively use a set of sequenced DMU methods in order to drill down to the design and engineering point of interest.

Applying overall product system knowledge of the entire DMU model ((a) the DMU engineer or user selects the appropriate viewing angle and plane (b) in order to further applying down select methods such as part (de-) selection (c) and clipping plane application (d) in order to reach the free view on internal key characteristics of the product.

The power of Digital Mock-Up has meanwhile been transferred from aerospace/aviation and automotive industry also to other industries such as maritime and to new application fields such as city navigation and development.

As shown in Fig. 12.23, the number of elements and instances shown in such DMUs extends those of automotive DMUs by a factor of 10–20. In order to allow for smooth interaction within such environments Dassault Système has developed a service as part of the 3DEXPERINCE platform in order to dynamically configure, visualize and analyze large sets of DMU model instances. As part of the data representation schema all DMU objects within 3DEXPERIENCE are indexed (i.e. even the individual facets of the 3D models) in order to shown one or several facets at a time, given the user scenario (system engineer, layout engineer, manufacturing engineer, design engineer etc.). As an example, 3D faceted representation plus the semantic representation such as functional tolerances plus the links between objects can be displayed at once, for a manufacturing engineering role. As a designer role, the exact geometric representation can be dynamically added for such geometry to be modified. The semantic indexing technology deployed for such a scenario allows for extremely fast working sessions setup.

The DMUs in 3DEXPERIENCE support different kinds of views from sources such as 3D CAD, 3D laser scans, panoramic pictures, piping and instrumentation diagram (P&ID) and systems or functional diagrams, among others. The overarching data standard for such capabilities is called 3DXML, which exists in two types:

- 3DXML for authoring (design and modeling)
- 3DXML for experience (viewing, simulation results, animations…).

The '3D visualization' part of 3DXML is based on a derivative of the CGR (Convergence Graphical Representation) technology. In summary, Digital Mock-Ups (DMUs) are powerful environments for engineers, designers, planners and managers in order to provide consistent virtual prototyping capabilities to support the following types of investigations as part of the following three classes of DMUs:

1. Technical system, assembly and component interactions as part of a *Product DMU*
2. Machinery, tools and fixtures, handling and logistics as well as plant infrastructure interactions as part of a *Factory DMU* (compare Chap. 15)

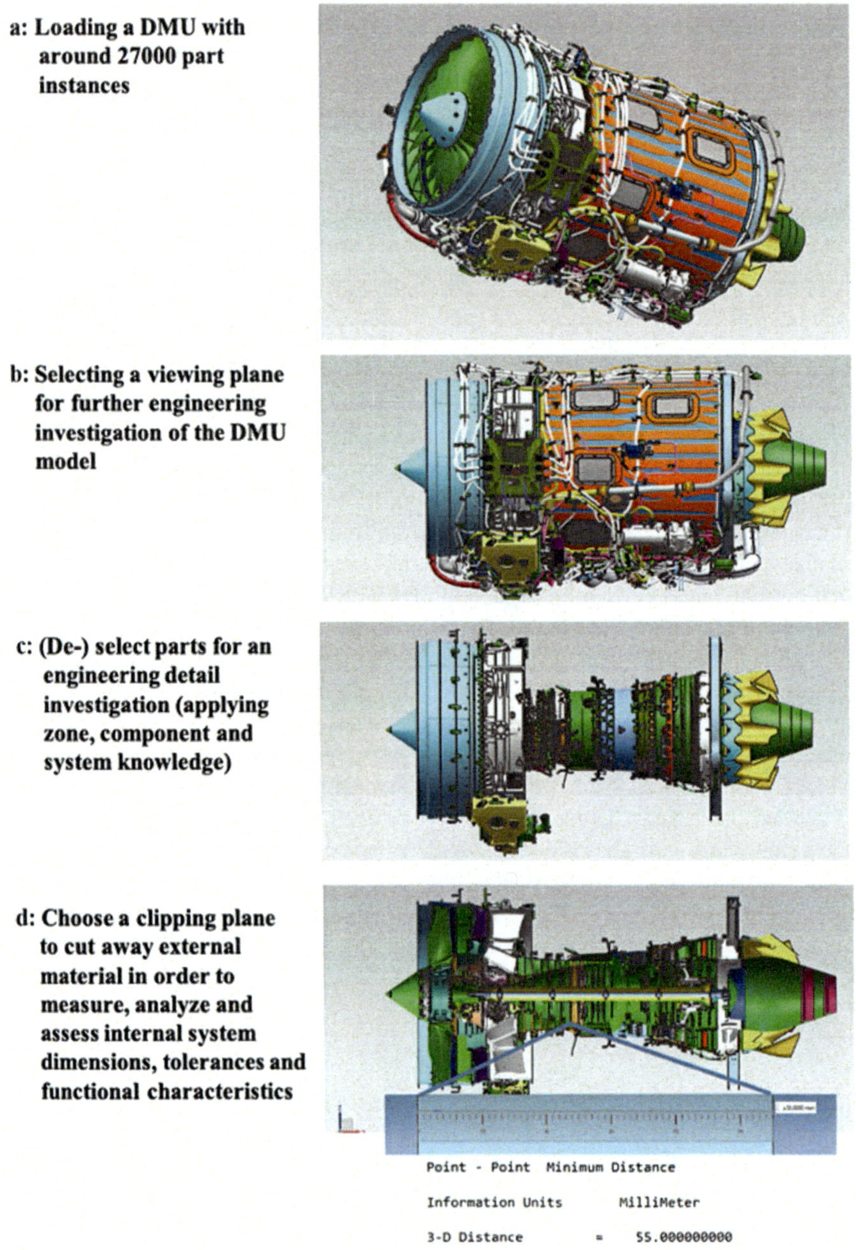

Fig. 12.22 Interactive DMU aero engine investigation using a series of 3D DMU methods

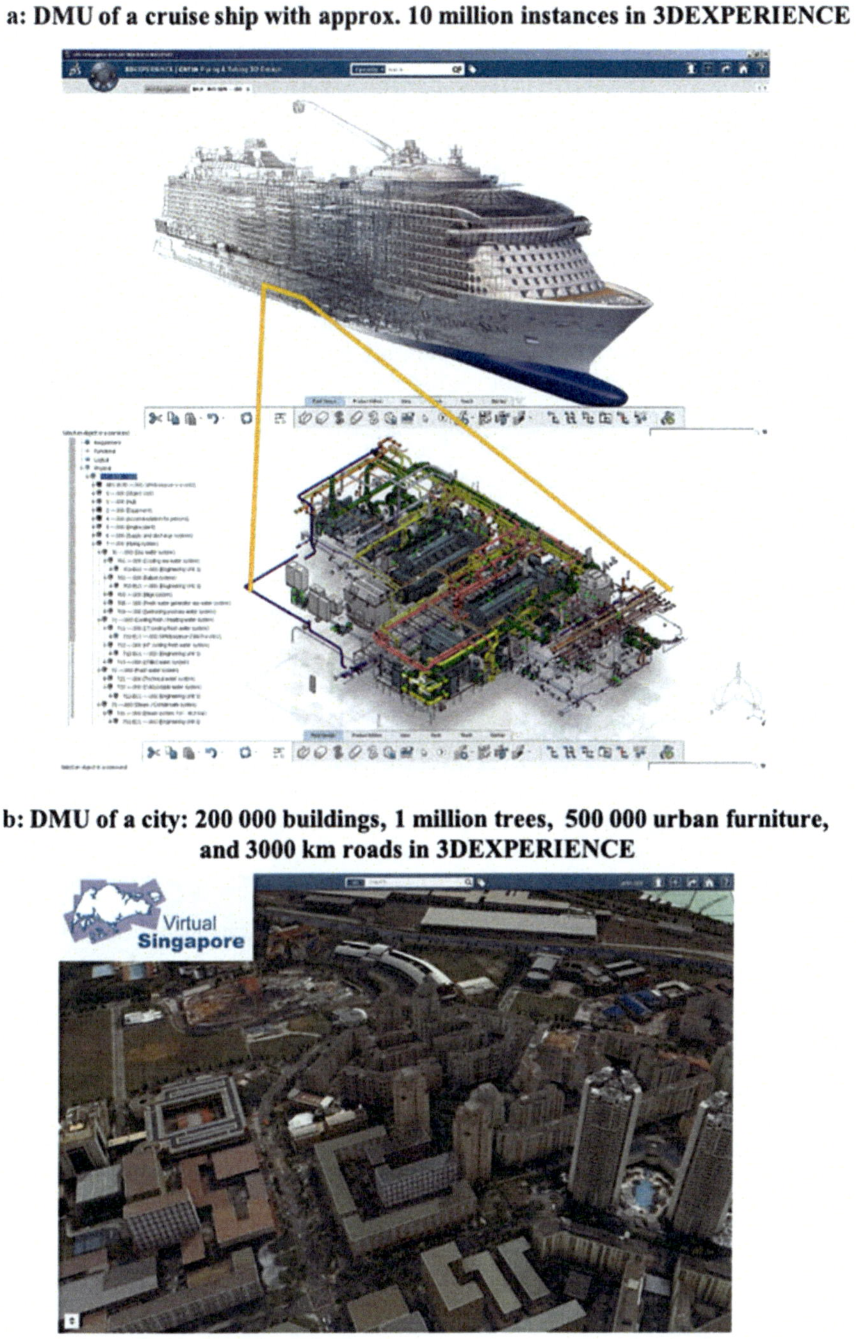

Fig. 12.23 Mega size DMUs in the 3DEXPERIENCE platform of Dassault Système

3. Building, technical building infrastructure (water, electricity, data networks, heating etc.), roads, squares, energy infrastructure, traffic systems (rail, road, aviation etc.) interactions as part of *Real World* (City, Agglomeration areas etc.) *DMUs.*

The International Organization for Standardization (ISO) has released in 2019 the world's first international standards for BIM (Building Information Modeling), the ISO 19650-1 and ISO 19650–2, which become pivotal for the set-up and exchange across companies and organisations of the DMUs of class 2 and 3 (see listing above). The full name for the standards is *"Organization and digitization of information about buildings and civil engineering works, including building information modelling (BIM)—Information management using building information modelling,"* and the organization has already released Part 1: Concepts and principles and Part 2: Delivery phase of the assets. The ISO plans to release a Part 3 (on the operational phase of assets) and a Part 5 (security-minded BIM, digital built environments, and smart asset management) within the following years (compare [10]).

References

1. Dolezal WR (2008) Success factors for digital mock-ups (DMU) in complex aerospace product development. PhD thesis; Technical University Munich
2. Enge-Rosenblatt O, Schneider P, Clauß C, Schneider A (2011) Functional DMU—a user-friendly interactive concept for the design of mechatronic systems. In: Bertram T, Dresden TU (eds) VDE/VDI-Gesellschaft Mikroelektronik -GME Fachtagung Mechatronik: 31.03.-01.04.2011, Dresden, pp 291–296. ISBN: 978-3-00-033892-2
3. Enge-Rosenblatt O, Clauß C, Schneider A, Schneider P (2011) Functional digital mock-up and the functional mock-up interface—two complementary approaches for a comprehensive investigation of heterogeneous systems. In: 8th International modelica conference—modelica'2011. Dresden, Germany
4. Göhlich D (2015) CAD in automotive industry and mechanical engineering; Lecture script at Technische Universität Berlin
5. Grote K-H, Antonsson EK (eds) (2009) Springer handbook of mechanical engineering. Springer. Available online at, Berlin. https://doi.org/10.1007/978-3-540-30738-9
6. Feldhusen J, Grote K-H (eds) (2013) Pahl/Beitz Konstruktionslehre. Methoden und Anwendung erfolgreicher Produktentwicklung *(Methods and Application of successful product development).* In: 8th new edition. Springer Vieweg, Berlin, Heidelberg. Available online at https://doi.org/10.1007/978-3-642-29569-0
7. Abel A, Blochwitz T, Eichberger A, Hamann P, Rein U (2019) Functional mock-up interface in mechatronic gearshift simulation for commercial vehicles. In: Proceedings of the 9th international modelica conference September 3–5, 2012. Munich, Germany
8. Anderl R, Grabowski H (2011) Virtuelle Produktentstehung. In: Grote K-H, Feldhusen J (eds) Dubbel: Taschenbuch für den Maschinenbau. Springer Berlin Heidelberg, pp Y15–Y29
9. Gausemeier J (2019) Rechnerunterstützte Entwicklung (Computer aided Development), in Enzyklopädie der Wirtschaftsinformatik. http://www.enzyklopaedie-der-wirtschaftsinformatik.de/
10. ISO Organization. Online information under: https://www.iso.org/obp/ui/#iso:std:iso:19650:-1:ed-1:v1:en. Accessed on 21st June 2020

Chapter 13
Major Technology 7: Virtual Reality—VR

Executive Summary

This chapter deals with the following topics:

- Basics and advanced techniques of Virtual Reality
- Providing insight into how engineers benefit from using Virtual Reality (VR) technologies
- Describing functioning, benefits, and limitations of VR technologies in practice.

Quick Reader Orientation and Motivation

The intention of this chapter is:

- to give an overview of VR technology in Virtual Product Creation as driver and enablers for Digital Transformation in engineering
- to present VR technology as part of Virtual Product Creation from a practitioner's point of view to analyze the need and usefulness for day-to-day industrial work practice
- to give instructions on how to use VR technology
- to explain models, frameworks, and mathematical representations that help to grasp the internal working modes of VR technology.

In 1995 Milgram et al. [1] asked the following question: What is the relationship between augmented reality and virtual reality? As an answer, they defined the *Reality-Virtuality* (RV) continuum as shown in Fig. 13.1. This represents a seamless continuum between a real environment and a virtual environment. The space between the two extreme environments is called *Mixed Reality*.

The *real environment*, also designated simply as reality, is the concept at the left end of the continuum. This includes any environment that consists of real objects, regardless of whether a person is present in this environment or if it is viewed from outside, i.e. on a screen. Such an environment is always bound to the rules of physics.

The opposite right end of the continuum is called *virtual environment* and can be represented by the concept of *Virtual Reality* (VR). The virtual environment

R. Stark, *Virtual Product Creation in Industry*,
https://doi.org/10.1007/978-3-662-64301-3_13

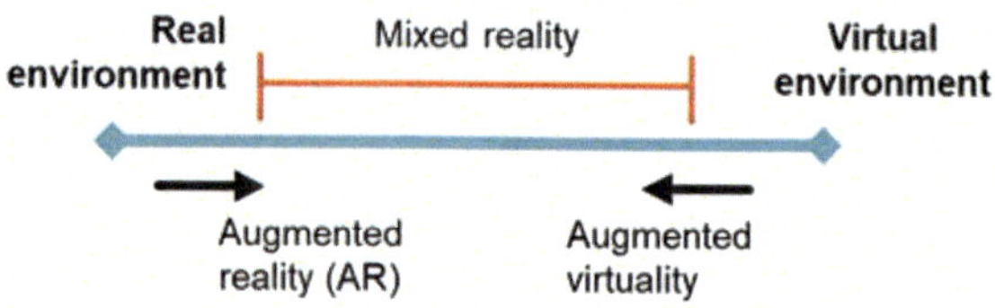

Fig. 13.1 Reality-virtuality continuum as defined by Milgram et al. [1]

provides an artificial "world" in which the user is completely submerged. A virtual environment is an environment that consists of virtual objects that are simulated by computer graphics or specific simulation engines, which might emulate some aspects of the real physical world by means of digital models and algorithms. Such an environment, however, is usually not bound to all stringent rules of physics.

Any environment between these two extremes is called a *Mixed Reality* (MR) environment in which real and virtual objects coexist. *Mixed realities* can be subdivided into *Augmented Reality* and *Augmented Virtuality. Augmented Reality* (AR) augments a real environment with virtual objects. *Augmented Virtuality* (AV) adds real components to a virtual environment [1].

According to LaValle, the term *Virtual Reality* (VR) refers to the induction of a targeted behavior in an organism by means of artificial-sensory stimulation, while the organism has little or no awareness of an illusion felt [2]. He also defined the following four key components:

- *Targeted/desired behavior*: the organism has an "experience" designed by the originator. Such an experience may be, for example, flying, walking or watching a movie.
- *Organism*: potentially any kind of living being that can be stimulated by artificial stimuli and thus put into a virtual reality, be it a human, fish, monkey or a fruit fly.
- *Artificial-sensory stimulation*: through the power of technology, one or more senses of the organism are hijacked and their normal input replaced by artificial stimulation.
- *Awareness*: during the experience, the organism is unaware of the stimulation and feels present in the virtual world. This is accepted as normal.

This definition includes any artificial stimulation. This takes cases such as watching a movie, listening to music or looking at a painted picture into account. Today's understanding of *virtual realities,* however, tends to exclude such cases and refers exclusively to those realities created by computer simulation. Sherman et al. [3] framed the following definition, which is here assumed to be the valid definition of *Virtual Reality*:

"*Virtual reality* is a medium composed of interactive computer simulations that sense the participant's position and actions and replace or augment the feedback to one or more senses, giving the feeling of being mentally immersed or present in the simulation (a virtual world)" [3].

They also defined four key elements of a *Virtual Reality* experience: a virtual world/environment, immersion, sensory feedback in response to user input, and interactivity [3].

Immersion refers to the (subjective) feeling of a person to be in a fictional environment. This can be an environment that produces reading a book or watching a movie in the minds of readers or viewers. However, these areas do not allow the person involved to interact with the environment, which is why one speaks here above all of a mental state.

In *Virtual Reality (VR)*, however, the user has possibilities for interaction, all-round vision and movement in the virtual environment. Therefore, immersion in this area is defined more precisely as a deceptive effect triggered by a technical, artificial stimulation of the sensors or sensory organs, which the user is not conscious of and which produces a real sense of the virtual environment. Immersion can be subdivided into mental and physical immersion in this area. The mental immersion describes the above-described state of strong involvement, the absence of doubt/disbelief, the perception of the environment as real. Physical immersion, on the other hand, describes the physical entry into a medium, the synthetic stimulation of the senses of the body through technology [3].

However, the terms differ in the bibliographical references found. For example, Slater describes immersion as only physical and he defines it as a measurable property of a system to the extent in which the sensory information corresponds to reality. On the other hand, he describes mental immersion as presence, a subjective feeling of the human that is triggered by their immersion in the virtual world. Thus, immersion or physical immersion enhances the mental immersion or sense of presence of a human in a virtual environment [4].

Another essential part of VR systems is the ability to interact with the virtual environment and its objects. Depending on whether the stimulation of the human senses can be influenced by the actions of the user, one distinguishes between interactive VR systems (closed-loop) and non-interactive VR systems (open-loop). In the case of a closed-loop system, the user has the ability to interact with the system, apart from using movements. For example, by voice commands, heart rate or body temperature [2].

Virtual Reality was invented by Ivan Sutherland who developed the first head-mounted display in 1968. 20 years later, virtual realities were first extensively introduced to research and film culture as an admired vision of the first digital revolution in the 90 s. With the use of a display in front of each eye the user was brought into an immersive virtual environment and was able to interact with it through the motion of their head. Because of the lack for computational power and small high resolution displays the technology has then been unnoticed for a long time. Immersive virtual environments were meanwhile made possible with high resolution beamers that projected the images on up to six faces of a cube around the users. This technology is called CAVE which stands for: *Cave Automatic Virtual Environment*. In 2016, the first consumer suitable HMDs reached the marked and major game engines like Unity and Unreal started to support them.

VR became more and more popular in recent years. Today it is used in all kinds of fields from engineering via games and movies through to medicine. *Virtual Reality* provides functionality to look at huge models such as airplanes, cars or other kinds of machines and other products in a 1:1 scale before they are built, to communicate with

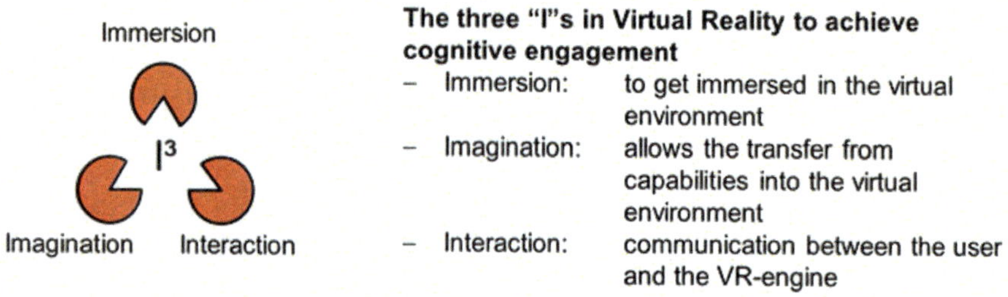

Fig. 13.2 The three "I"s: immersion, imagination and interaction

people who are on the other side of the globe with the possibility of spatial referencing or even to interact with the same object. Overall, Virtual Reality uses the three "I"s, immersion, imagination and interaction, to provide a realistic understanding of the virtual 3D models for the users (see Fig. 13.2).

13.1 Engineering Understanding of Virtual Reality

VR solutions provide powerful technology elements in the context of extended Virtual Product Creation scenarios. Engineers and IT experts integrate VR solutions into numerous industrial processes to achieve state-of-the-art digital connectivity and intelligent virtual support.

Primarily, VR technology for engineering is used in design reviews, production and assembly planning as well as in training. In general, VR can support tasks such as:

- Interactive visualization interfaces to support engineers, workers and costumers during prototype reviews or planning activities,
- Support of context-aware activities and of product user assistance as well as training,
- Visualization of a product in a fully computer simulated environment,
- Visualization of product and process-relevant virtual geometries and information.

Engineers are starting to appreciate *Virtual Reality* as a new convincing review and working environment where 1:1 scale representation of technical systems, products and manufacturing facilities can easily be comprehended and spatially be assessed. Designers and Engineers, however, are oftentimes still afraid how to become proficient in interacting and manipulating *Virtual Reality* scenes and, therefore, they often still depend on specifically trained *Virtual Reality* technology experts. Just recently, it becomes apparent that the future of engineering will need specific advanced VR support in order to better understand, study, analyze, assess and alter functional connectivity reasoning of technical systems and components as part of sharply rising model-based systems engineering and integration and data linkage engineering.

In addition to industrial VR applications, VR technology can add value in several further areas like sales (e.g. customer product presentation and configuration), healthcare (e.g. surgery training and support) and consumer market and culture (e.g. games and exhibitions).

13.1.1 Why Does an Engineer Use Virtual Reality?

An engineer uses VR applications e.g. to visualize digital CAD data no longer just on a 2D screen but in an immersive way allowing them to percept and interact with it in 3D and at original scale. This means that the engineer and other stakeholders can gain a more lifelike impression of the product before any physical prototypes have been created to help to comprehend the overall design and assembly situation of the product and to detect design issues earlier on in the product development process.

VR is also used as a decision-making tool to provide answers to questions in fields such as visibility and viewability evaluation, ergonomics and reachability analysis and review of the aesthetic qualities of the product. VR is also effective as a communication tool, which can be used for storytelling to explain use cases as well as a means to convey information between different disciplines such as engineering, marketing and design. One example at Bombardier is to conduct Virtual Reality based design reviews for train development as shown in Fig. 13.3.

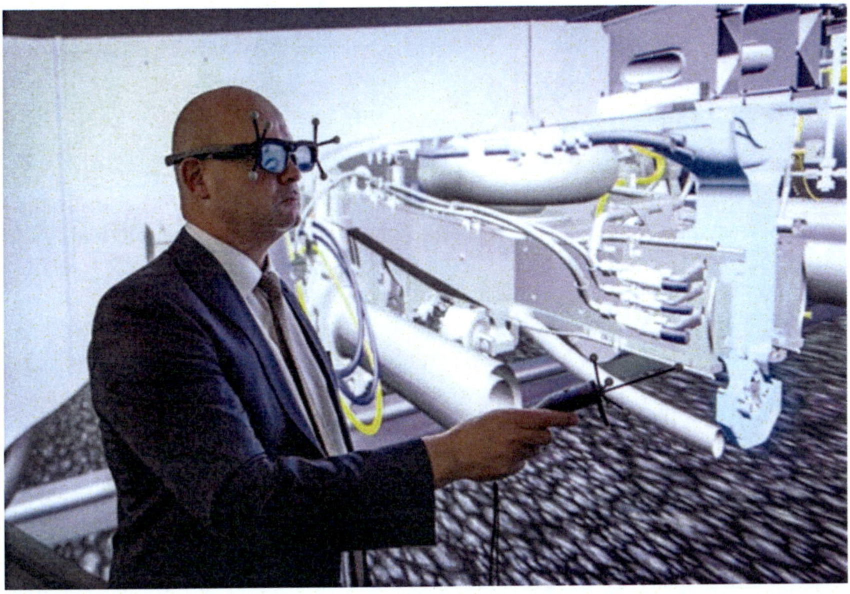

Fig. 13.3 Virtual reality based power wall design review at Bombardier with active 3D shutter glasses tracked by an outside in tracing

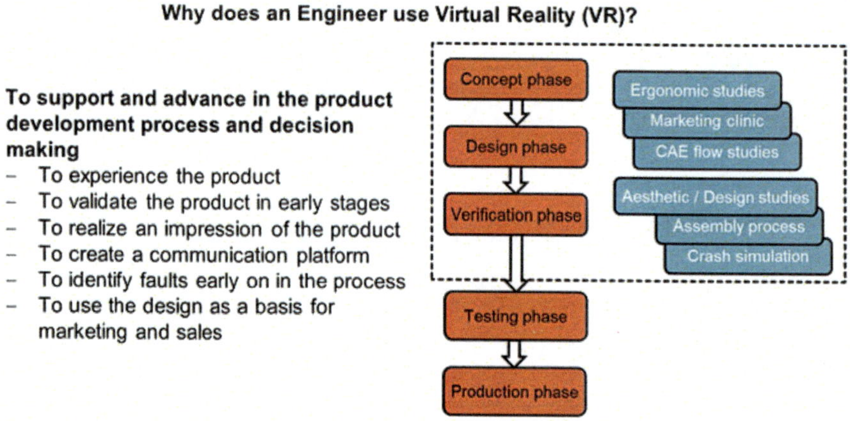

Fig. 13.4 Circumstances for engineers to use virtual reality

Virtual Reality is especially used in earlier phases in the product development. It enables engineers in the concept, design and verification phase to e.g. conduct ergonomic studies, assessments if the product can be assembled, CAE flow studies and simulations and it can also be used as a communication platform to inform others about the current state of the development status (compare Fig. 13.4).

In summary, *Virtual Reality* is a technology that enables engineers and managers to experience the product, to validate the product in early stages and to use it as a communication platform to identify faults in the product.

13.2 How Does Virtual Reality Work?

The virtual world is an imaginary space or a description of a collection of digital objects in space and immersion refers to the immersion into an alternate reality or point of view. The sensory feedback is the feedback from the overall VR tracking system e.g. based on the user's physical position and one form of the user's interactivity may be the ability to control the computer-based virtual world. A VR-scene consists of different objects/surroundings that are three dimensional and the experience is different from normal computer screens as the depth information can also be perceived. The digital objects in a VR-scene are basically computer-generated graphics that are designed in advance (compare Fig. 13.5).

One very crucial requirement from VR technology is that it should provide real-time conditions. Only then, an experience comparable with reality can be obtained. From a visual perspective, a motion can only be recognized by the eyes of a human being if the frame rate of the screen can deliver at least 24–25 frames per second. However, if the user interacts dynamically as well, i.e. is not in stand-still situation, the frame rate for pictures delivered by the computer should be processed by the

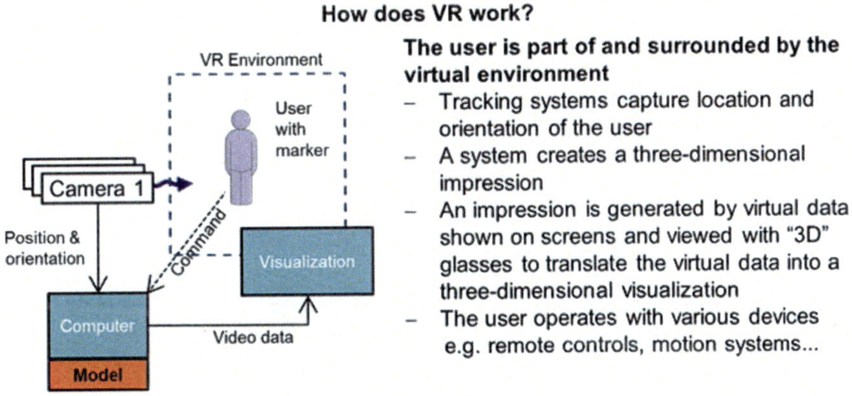

Fig. 13.5 The virtual reality user is supported by the VR technology stack

visual device with a frame rate of meanwhile 75–90 Hz. Why is that? The reason for a much higher frame rate (respectively a lower latency) is that a couple of different physical conditions of the images need to be distinguished with respect to the human eye and brain recognition: the refresh rate (of the same images) and the frame rate (of individually different images). According to [5] the critical refresh rate starts as slightly under 50 Hz and an absolute flicker free recognition for the human being needs a frame rate of almost 100 Hz. For HMDs the refresh rate plays a more important role which drives a fast reaction time of the HMD screen. The visualization latency, however, is just one contributor of the overall VR latency: in addition to the contribution of the human tracking latency, the VR scene simulation (computer) latency and the connectivity latency (of all sensors and the compute system) need to be considered and controlled meaningfully in order to avoid human recognition distortion.

In a VR Environment the HMD or the tracking glass is usually tracked by one or more cameras through infrared pulses. These cameras are connected to a computer which analyses the information and spots the two displays in the headset precisely. The matching stereoscopic perspective of the user referring to the location and orientation of his head can then be rendered and the video data is transferred to the displays with a cable or the information is transmitted wirelessly. This creates a three-dimensional impression for the user which has an immersive character if a frame rate of minimum 60 frames per second (meanwhile a higher frame rate is anticipated as explained earlier) and a field of view of at least 100° are used.

Furthermore, the VR environment can be affected by additional trackers and controllers. They can track the movement of the users' body to display it in VR or the user can manipulate the environment in an interactive manner. Some HMDs also support eye tracking. With an infrared controller, the direction of the user's eyes can be monitored in order to adapt the virtual world, save render power or direct additional mechanical parts of the HMD.

13.3 Virtual Reality Technologies

Virtual Reality requires a rich set of technologies and system architecture in order to provide a rich immersive and interactive engagement for the users with the 3D model scene. The following sections will step by step provide an insight to these technologies.

13.3.1 Setup of the Overall Virtual Reality System Architecture

A virtual reality system consists of different components: input devices translate and deliver the user inputs such as location of the user or interactions to the VR engine. Such input devices may be, for example, 3D glasses (passive or active), head-mounted displays and/or controllers with associated tracking system (see Fig. 13.6).

The VR engine then creates the visualization of the virtual environment based on the current location and orientation of the user. Engineering software customizes the VR-application, for example for haptic and feedback calculations. These calculations can be used to feedback the forces to the input devices to allow real-time force and ergonomic validations.

In order to bring an image of a virtual environment to the displays of the headset, a real-time rendering engine is needed. Major Virtual Reality application providers have developed their own real-time rendering engines. However, from 2015 onwards the influence of gaming environments has also been increasingly notable. The two major applications for such engines are currently Unity and the Unreal Engine. They are both game engines that had already been widely used for game development before the introduction of VR. At the beginning of the VR trend both engines decided to broaden their user base by making their software usable free of charge if

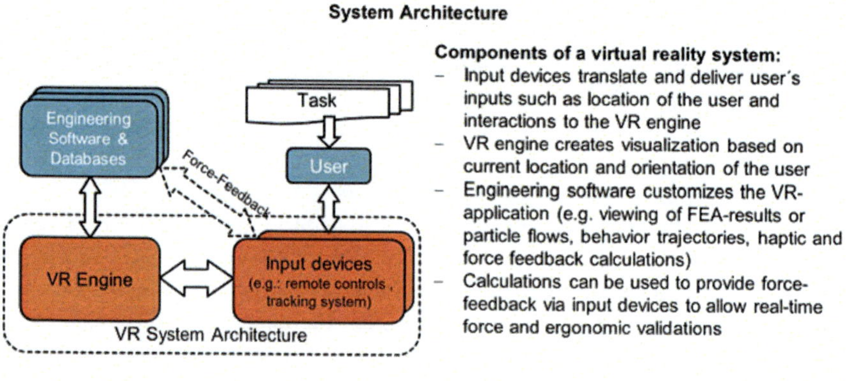

Fig. 13.6 System architecture of a VR system

the user is without revenue through game sales. This way, engineers from various fields started working and researching with virtual reality scenarios. The support for individual headsets, controllers and trackers is implemented by open source SDKs or handled through official partners like SteamVR. For developing VR experiences and interactions Unity is using C# where Unreal is based on visual scripting and C++.

13.3.2 Head Mounted Displays, 3D Glasses, Projection Displays

For a long time, the stereoscopic viewing for the user was exclusively enabled by passive filter or active shutter 3D glasses that produced two separate images for the right and left eye. This was achieved while looking at a display that offered two images (one right and one left, compare Fig. 13.3). Head Mounted Displays (HMD) were long time too heavy to allow for convenient carrying and were used in specific applications in industry with limited success (too heavy, field of view limited to 100–110° maximum).

A head-mounted display (HMD) is similar to a pair of glasses or a helmet worn on the head display device. This may or may not allow visibility to the outside world, i.e. be transparent or cover the entire field of view. The first HMD was developed by Ivan Sutherland in 1968 [6] (see Fig. 13.7). In his accompanying publication "A head-mounted three-dimensional display" Sutherland described the mode of

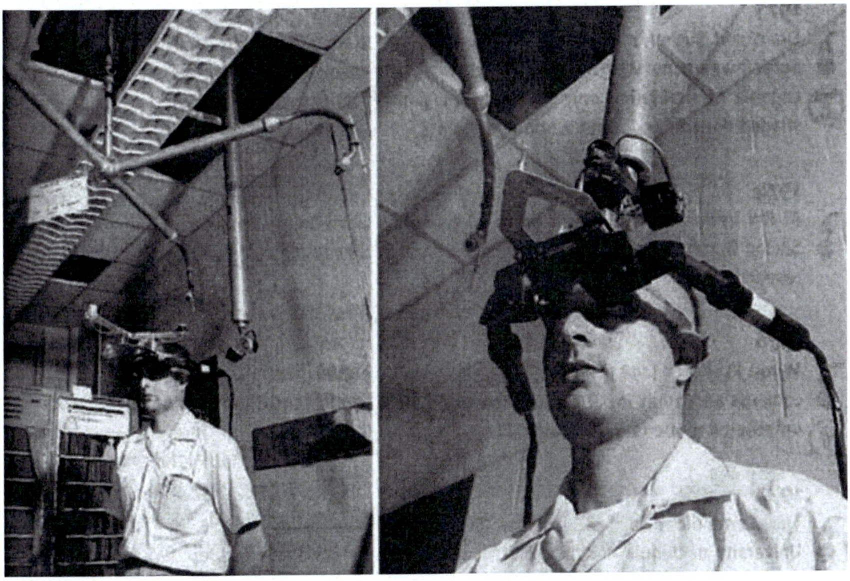

Fig. 13.7 First HMD developed by Ivan Sutherland in 1968 [6]

operation, whose basic concepts can still be found in today's HMDs. Thus, head-mounted displays have two (one eye each) built-in screens that represent images of a virtual environment or virtual objects so that the user gets the impression of a three-dimensional space. A so-called tracking system allows the tracking of the position and orientation of the user's HMD or head in space. By means of this information, the images displayed on the screens and thus the perspective of the user in the virtual space or on the virtual objects are adapted.

Since 2012, first modern HMDs were unveiled. Nowadays, many different models and approaches to VR headsets have hit the market. While the display of the first prototype (Oculus Rift DK 1) had a resolution of 1280 × 800 pixels, state-of-the-art headsets can now reach up to 7680 × 2160 pixels (Pimax 8 K). The increase from 640 × 800 pixels per eye to 3840 × 2160 pixels per eye makes the pixels almost imperceptible for the user. For the immersion the field of view is equally important. The first-generation HMDs in the 90s and 2000th used to have a FOV of 100–110° only, which corresponds to the stereoscopic binocular FOV of a human. Yet the borders of the screen are still visible in the two monocular fields of view. For a full coverage of the monocular visions a FOV of at least 200° is needed, which the newest models of Pimax already achieve. To bend the light according to the FOV typically different Fresnel lenses are used. The lenses are essential because the eye of an adult human cannot focus a display that is located three to seven centimeters away from their eyes. Virtual reality glasses are a form of HMD covering the entire field of view of the user. Such glasses allow the user through this technique to look around and move in the virtual environment analogous to the real world [3]. An example for such virtual reality glasses is the HTC Vive that was developed by the technology group HTC and the software company Valve in cooperation with each other. It has two built-in AMOLED 3.6 "screens (one for each eye), each with 1080 × 1200 pixels (together 2160 × 1200 pixels) and a refresh rate of 90 Hz. Together these screens offer the user a field of view of 110°.

In addition to HMDs, other technologies allow users to immerse themselves in virtual worlds. Such a technology is for example a CAVE (computer aided virtual environment). As shown in Fig. 13.8, a CAVE is a room in which three or four of the walls are illuminated by projectors in such a way that the illusion of a 3-dimensional virtual world is created for one person in this room who is looking through passive or active filter glasses [7].

Figure 13.9 provides an overview of the different visualization techniques for virtual reality.

13.3.3 Tracking

In the early days the tracking of the PC-based HMDs and shutter glasses changed from inside-out tracking with build-in motion sensors to outside-in tracking with cameras. It was possible to make such progress because a higher tracking precision could be achieved with tracking cameras from outside. Today's headsets usually use

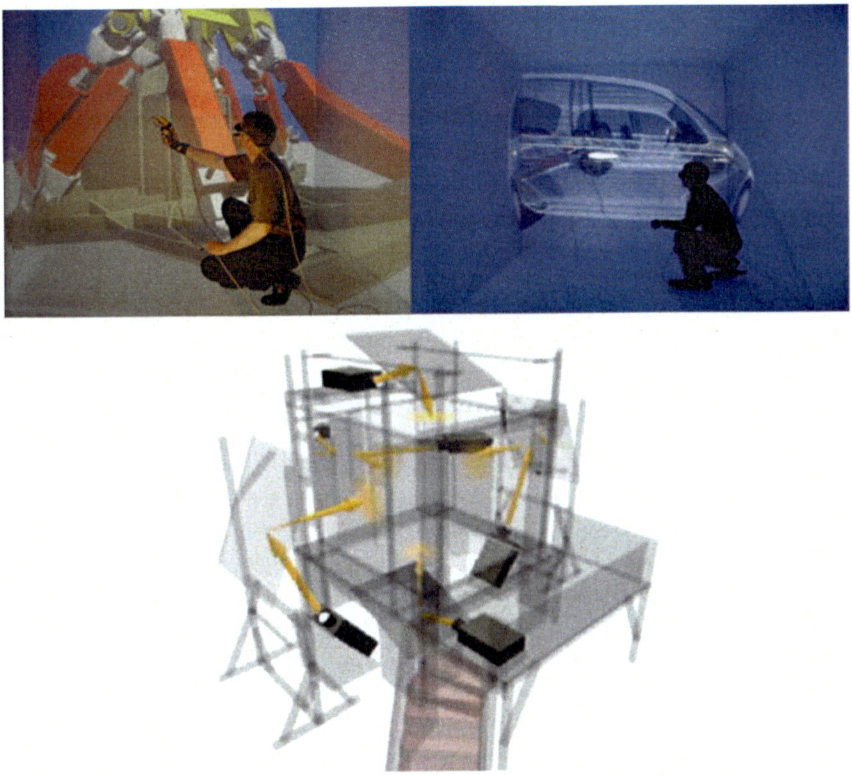

Fig. 13.8 CAVE examples: left: design review of a tooling machine; right: structure analysis of vehicle side door; bottom: 5-sided CAVE system (4 surrounding ones plus bottom)

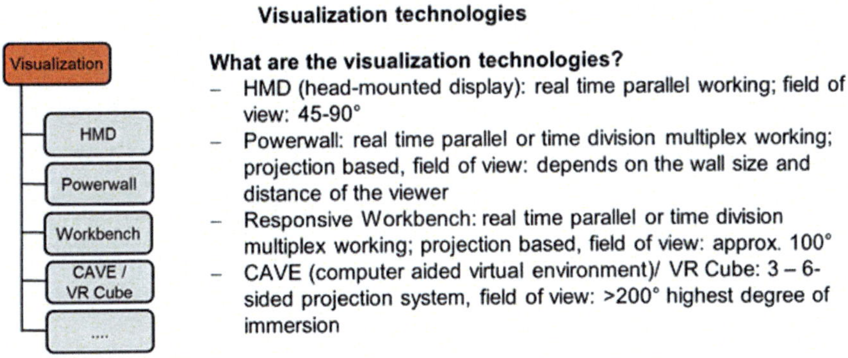

Fig. 13.9 Different types of visualization technologies to enable virtual reality

two tracking cameras that send and receive the infrared pulses with infrared LEDs and lasers. Nevertheless, the most recent developments from the HMD manufacturers are coming back to inside-out tracking because the VR setups can then be offered at a cheaper price. The tracking works with visible light and the base stations are not needed anymore. However, this means that the tracking will not completely work in the dark.

There are two different ways in which the detection of movements in the three-dimensional space of HMDs works (compare Fig. 13.10). With outside-in tracking, the position of the user in space is recorded via external cameras or signal transmitters [8]. As examples, the Oculus Rift CV1 or the HTC Vive can be mentioned here, since both systems require external hardware for position detection. The advantage of these tracking systems is the high tracking accuracy of six degrees of freedom and thus a stable measurement of the position in space. In contrast to this, there is the Inside-Out-Tracking. With Inside-Out-Tracking the HMD autonomously detects the position in space by internally installed sensors. This can be done based on stereoscopic camera images or by using sensors. Examples of HMDs with Inside-Out-Tracking are the Oculus Quest, or the Windows Mixed Reality glasses. The disadvantage is the lower accuracy, as well as a higher probability that the controllers required to use the HMD are not correctly captured. The advantage of this technology is that no additional tracking hardware, besides the HMD, is needed. It is also possible with this technology to use self-sufficient HMDs without connection to a computer.

Latest tests of tracking systems for VR systems have revealed that there exists a wide range of tracking accuracy between 0.1 and 5 mm [compare 8]: outside-in tracking systems provide the highest accuracy with 0.1 mm whereas the range of inside-out tracking systems vary between 0.69 and 5 mm incl. observations of a position drift cause by the motion controller.

What are the different tracking technologies?

All technologies use markers and sensors to detect the position and orientation

- The contactless optical technology detects the presence or intensity of light
- Electromagnetical: non-contact devices based on "time-of-flight" measurements of the outgoing and reflected signal
- Acoustical sensors send out an ultrasonic chirp and detect the return echo from the marker
- Mechanical tracking detects the position and orientation by using data gloves with a fixed position

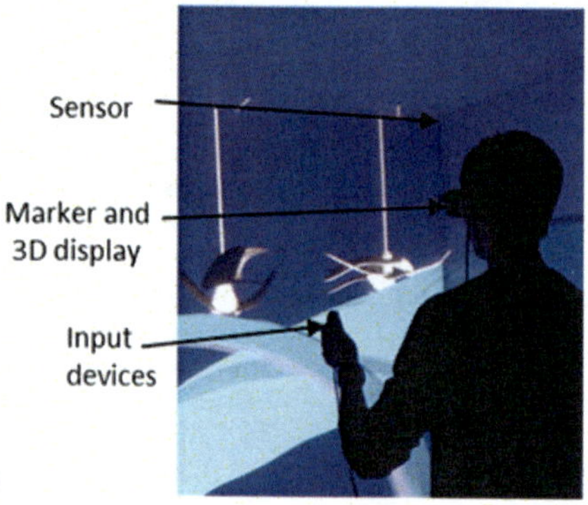

Fig. 13.10 Tracking technologies for virtual reality systems

13.4 Human Interaction with VR

Four main functionalities that are necessary to interact in VR are navigation, selection, manipulation and system control to change the system preferences (see Fig. 13.11).

Through various technical approaches, humans are able to interact with virtual environments. Early types of VR interaction used game controller devices that were already used for 2D games. For simple applications such as zooming, choosing & picking, selecting or discarding they fit well by offering simple motion and selection operation with e.g. a joystick or specific buttons (compare the over the air connected fly stick in Fig. 13.3 and a cable-based navigation and manipulation in Fig. 13.12).

As described by Stark et al. (compare [9]) new interaction research and demonstration solutions have been developed already in terms of modeling, sketching and designing technical products in Virtual Reality. The idea is to enable designers to

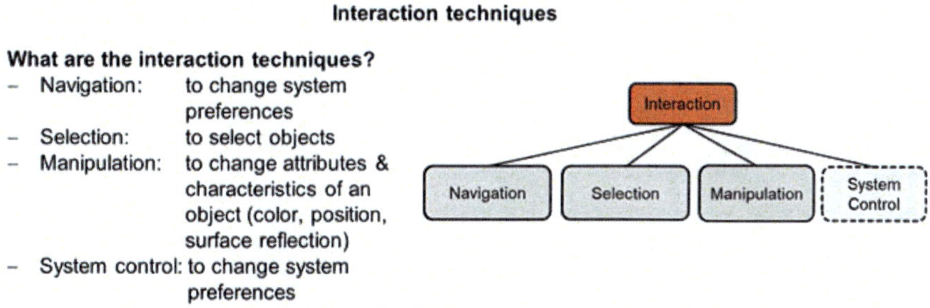

Fig. 13.11 Interaction techniques in virtual reality

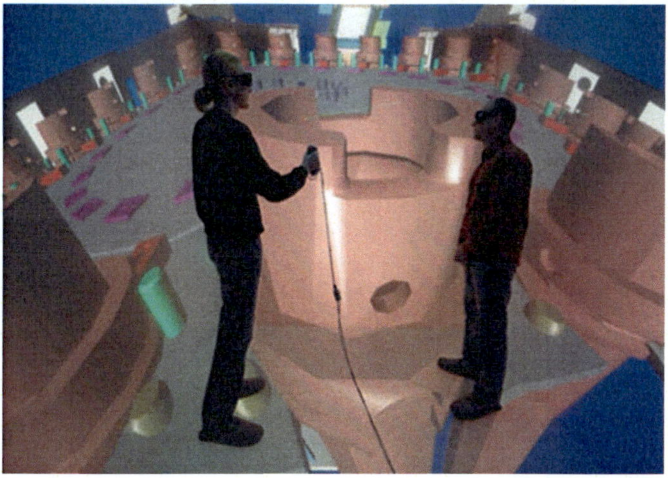

Fig. 13.12 Cable-based navigator/manipulator device in a VR design review

directly model in true 3D space by offering different types of interactive tools as part of the VR system environment. Figure 13.13 shows a range of VR based interaction technologies which have been developed at TU Berlin and Fraunhofer IPK in Berlin, Germany. These CAVE or projection screen-based interactions offer an interesting mix of direct visible physical interaction devices while being immersed in virtual reality.

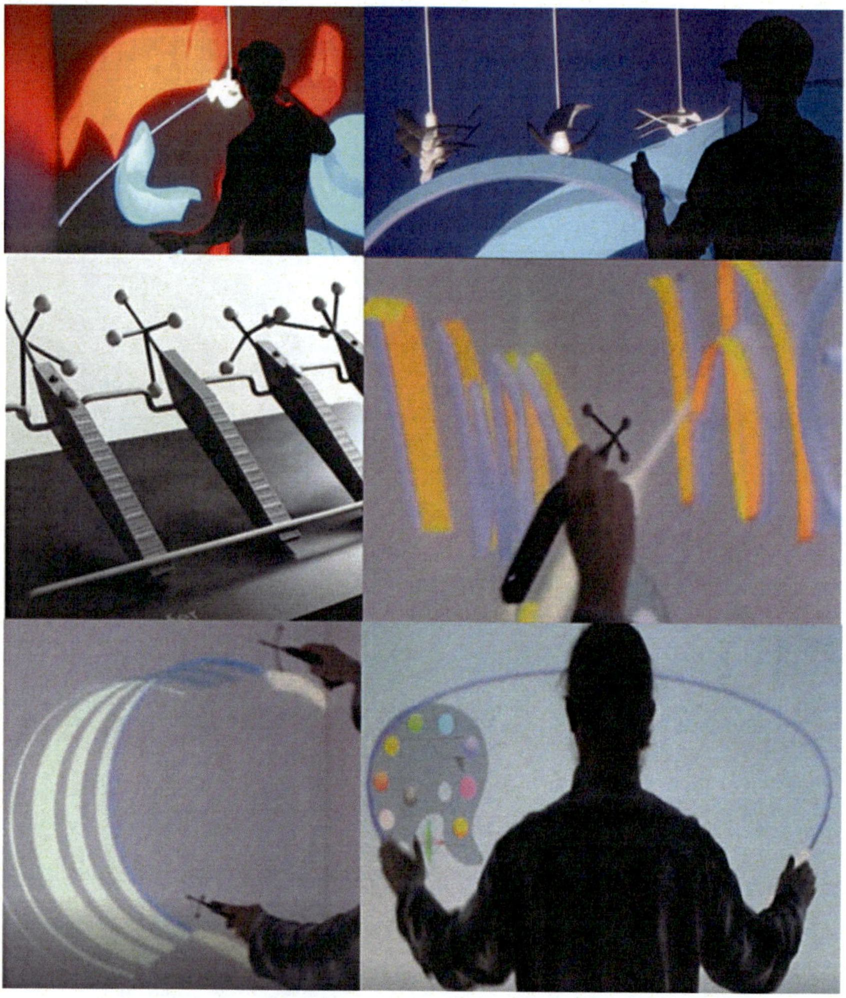

Fig. 13.13 Interaction devices for sketching and modeling in VR CAVE space *top:* sketching lamp design with line and surface modeling devices; *middle:* application specific input devices with function selectors; *bottom:* interactive surface modeling devices and bare hand modeling based on camera based technologies

However, the conventional and the tailored manipulators have the big disadvantage that the controllers are not visible while wearing a Head Mounted Display (HMD) and their position in 3D reality cannot be used. Special VR controllers for the consumer market solve this. They are tracked like an HMD and can be visualized as a virtual representation so that the user always knows their position and the software or system in which the controllers are used can also work with this position data. They can be used to point to things, control units only by moving them through the air and they still support the use of buttons and touchpads. With a simple (scalable) tactile feedback of the whole device, the user can receive a simulation of tactile feedback of their actions.

Even though this type of controller already enables simple intuitive gestures and gives the user the feeling of actively interacting with the world with their hands, the user is still not able to use their fingers naturally and individually. The entire hand is always grasped around a controller. The HTC Vive controllers have "grip buttons" on their sides, but they do not make a difference between grabbing and releasing the hand, because they have to be pressed much too actively and the controller still has to be held after 'grabbing'.

The next generation of consumer VR controllers further develops this grabbing behavior. For instance, the Valve Index Controllers as well as the upcoming Pimax Sword Sense Controllers support the recognition of single fingers and can be strapped to the hand, so the hand can be released and the controller recognizes this by its sensors.

In addition to the controllers that are widely used, there are systems that enable users to bring their hands into Virtual Reality. These systems can be divided into optical Systems like LeapMotion or Intel Realsense and haptic or non-haptic data gloves (see Fig. 13.14). Whereas optical systems have the advantage that the user does not have to put on additional technical equipment, they do not provide stable tracking data. This can result in wrong hand and finger poses. Data glove on the other side can provide stable data but the user has to wear them in addition to the HMD.

Furthermore, with the improvement of AI-technology, voice interaction is also a powerful interaction technique. With this, users can e.g. in design review situations document their decisions by just speaking them and adding them to the part. Thus, one has a direct relation between the part and the decisions of the design review, which makes it easier to implement the changes afterwards.

A solution for better haptic feedback are the Tactical Haptics Reactive Grip Controllers. They do not support finger detection, though. Two plates on the grip complete the haptic feedback. One of them lies under the fingers and the other one under the palm. These get shifted a little bit up and down, mostly in opposite directions to simulate shear and friction forces. In addition, the controllers are able to vibrate and they can also stick together with different magnetic attachment points on their upper area. This allows the user to combine them for several use cases, it combines the two one-handed grips to one two handed grip or it generally emulates devices with two grips that are arranged in a fixed ratio to each other.

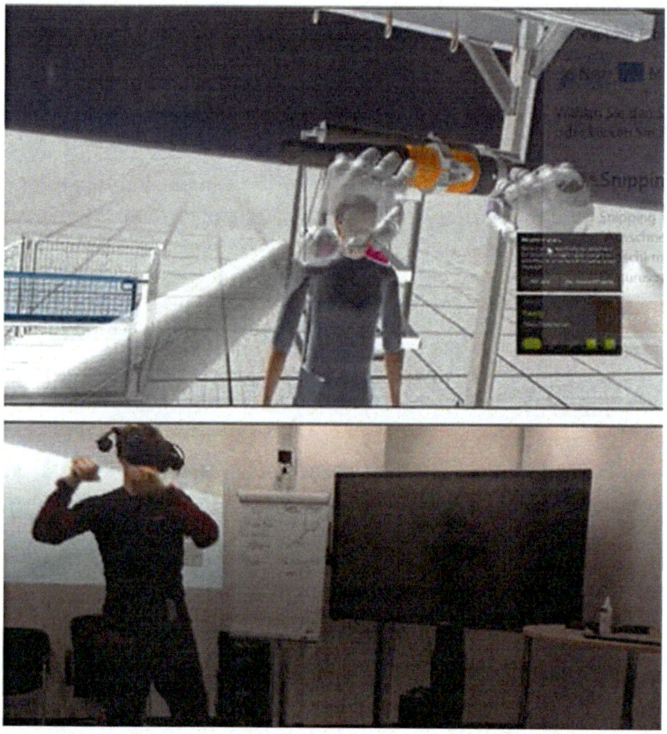

Fig. 13.14 Leap motion camera equipped HMD to track hands for VR based assembly planning; bottom: real person; top: virtual assembly scene

13.4.1 Development and Use of VR Applications

The development of VR applications is similar to those of traditional real-time 3D applications and mostly takes advantage of the same tools. The 3D interaction and perception constitute the biggest differences. This becomes apparent during testing VR applications. In traditional applications, a developer can typically use the same display for both development and testing and in many cases even the same input devices such as keyboard and mouse. In order to test a VR application properly, however, the tester or developer has to use a HMD or projection-based VR test environment and tracked controllers to closely resemble the experience of the end user. This also means that, depending on the hardware used, a tracking set-up may have to be present in the tester's workspace.

Applying a VR application to a specific use case is, in many cases, a process that still involves staff support to operate VR hardware and software and prepare the VR scene as well as the model that should be investigated. Depending on the complexity of the use case the time required for model and scene preparation can vary. Especially if complex materials and textures have to be applied or if kinematic constraints have to be set up, the process can take multiple days. When physical devices are to be employed in the use case, the tracking of these also has to be set up and they have to be aligned with the virtual content.

13.5 Use of VR for Engineering Working Tasks

In engineering working tasks Virtual Reality is primarily used in product-decisive phases for virtual and hybrid prototyping (see Fig. 13.15). It offers the possibility to experience the geometric, technological, interactive and physical characteristics of a product as well as its special impression.

Engineers can walk around a virtual 1:1 scaled version of their future product, look at special parts of it, evaluate it and change certain (predefined) features such as scale, detail, annotations on/off, position and orientation, cut-off plane, culling etc. Therefore, new ideas and engineering issues can be visualized rapidly and functionality experienced early.

Furthermore, it is possible to review products remotely together with other engineers around the globe or to show future product ideas to customers.

Virtual Reality can also be used to train new employees. Thus, an experienced engineer can record a workflow and new employees can watch the recording later and train the process using the virtual version of the actual product, e.g. the machine.

13.5.1 Technological Limitations

One of the main limitations in VR applications is the fidelity and level of detail of the simulation presented to the user. These limitations are mainly caused by the display hardware and computing power available to the user. The resolution of most

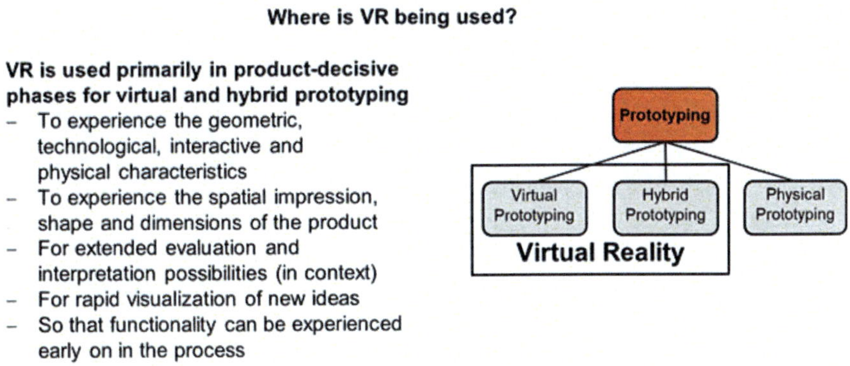

Fig. 13.15 Engineering working tasks in virtual reality

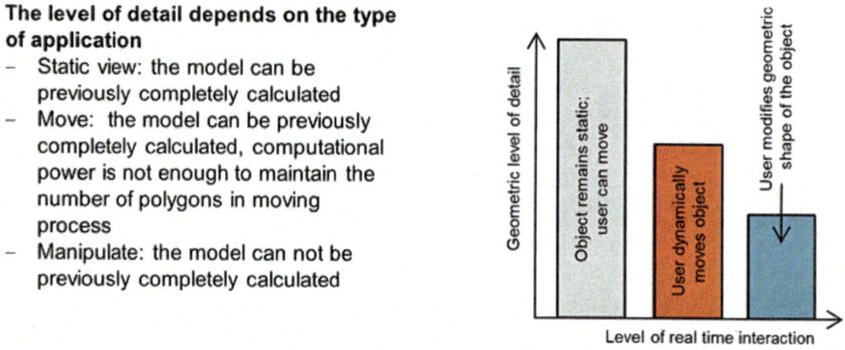

Fig. 13.16 Qualitative relation between type of VR application in respect to geometry level of detail over level of real time interaction

consumer level HMDs, e.g., is not high enough so that the user cannot notice the grid of pixels anymore, which decreases the sense of immersion. On the contrary, a higher resolution of the displays will increase the demand on graphics hardware, which can negatively affect the frame rate resulting in a decrease of immersion sense, too. To mitigate this, various methods are conceived. One of them is to take the optical properties of the headsets into account when rendering the scene to selectively decrease the resolution on the edge of the field of view. Another method is to project previously rendered frames again to artificially increase the framerate.

Other type of limitations to VR applications are present in non-immersive real-time 3D applications, which indirectly amplify and increase demands in resolution and framerate present in the final VR applications. This includes the geometric complexity of the scene, the visual quality of the rendering, physics simulations and collision detection. Especially when CAD models are to be rendered in VR, these limitations come into effect: compared to typical video game scenes CAD geometry has an extremely high level of detail and little optimization regarding its complexity which result in both, heavy demand on VR hardware fidelity and power. In addition, in a CAD context higher precision it is typically required in physics and collision calculations compared to games. Figure 13.16 provides an understanding of the trade-off necessary between geometry level of detail and level of real time interaction for VR applications.

Moreover, the size and weight of the HMDs or of other VR glasses constitutes another limitation to the user experience because these factors can cause discomfort in the user after extended periods of use.

13.5.2 VR Applications

Coming a long way from using an "expensive and difficult to use" Virtual Reality exclusively in research only or deploying it mainly for assessing 3D design shape and early virtual prototypes in industry, Virtual Reality technology and applications nowadays become step by step a *day-to-day* solution in engineering which will boost the understanding and interaction with 3D immersive models and other collaborative working meetings. The following sub-sections provide an overview of typical Virtual Reality application patterns.

Engineering Review Activities with Virtual Reality

The motivation to use VR for design and engineering review activities is to reduce time and cost since virtual models enable the avoidance of building the product or parts of the product physically with expensive tooling. Thus, the idea and goals for digital and virtual reality-based reviews is to study, analyze and evaluate if the reviewed product or parts of the product fulfill the rich set of requirements that were defined during the development project. To conduct an engineering review it is necessary that the relevant stakeholders participate in such meeting. Therefore, it is necessary that VR can be used in multiuser settings. The advantages of using VR in engineering review situations are manifold, e.g. it is possible to:

- create and review photorealistic designs
- review the product in a 1:1 scale
- review different design alternatives in direct comparison to each other
- include the product or factory line in the future environment which is embedded also as virtual model in the VR scene
- conduct acceptance tests of product uses, factory workers and service personnel due to the high sense of immersion.

Peer-to-Peer and Team Interaction with Virtual Reality

As already mentioned, in Virtual Reality it is necessary that more than one user can participate in a design review. When using CAVE systems, it is possible that more than one person is in the CAVE and use appropriate glasses so that they can see three-dimensionally. However, only the user with the tracking target on the glasses perceives the right perspective. Thus, the other participates in the design review have to try to get the same perspective as the main user with the tracking target. This can be achieved by standing very closely together.

In addition to that, by using Head Mounted Displays, it is not possible for others users to see in 3D what is displayed on the HMD. Due to the small amount of cost and existing software, it is possible to use them in multiuser-settings. This means, that several users wear an HMD and everyone can see the same models, products or environments. To see who else is in the multiuser VR-session the users are usually presented by avatars. The advantage of a setting in everyone using a HMD is also that everyone has controllers and can freely navigate through the VR-environment (see Fig. 13.17).

Consequently, contrary to CAVE and powerwall systems, the perspective is not limited to only one person. There exist several software applications, that support CAVE-systems as well as multiuser use cases, e.g.: IC:IDO, WeAre, STAGE, Virtalis and MiddleVR. Whereas Virtalis and IC:IDO are built on proprietary engines, WeAre,

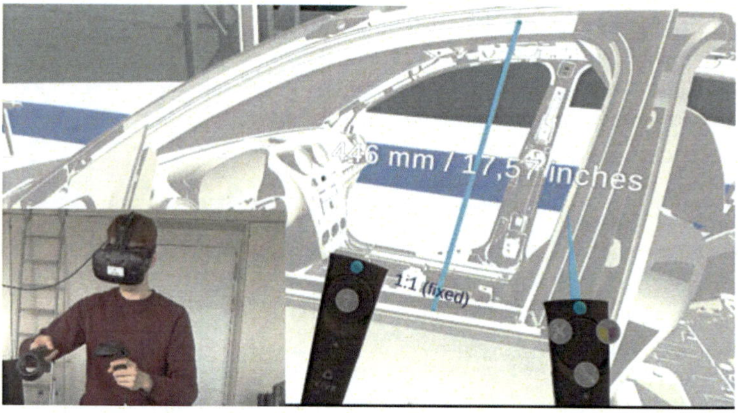

Fig. 13.17 Multi-user application of Fraunhofer IPK

STAGE and MiddleVR are built on GameEngines, which enable companies to rely on existing multiuser technologies from the gaming sector. As an advantage, those companies can integrate new technologies quickly as soon as the GameEngines provide new features.

In summary, these applications enable companies to:

- conduct meetings and workshops with long distances between the participants,
- reduce cost and time to travel, as it is possible to share more information than just on 2D screens (PowerPoint) like in ordinary 2D desktop viewing applications,
- increase the development time, as reviews can be conducted more quickly and with the relevant information which is needed to make decisions,
- improve the efficiency for interdisciplinary teams in complex projects, as it is easier and more intuitive to present and describe the current state of the project.

Due to the distance between the users, however, it becomes necessary to address the topic of data security. Before using collaborative VR/AR software in business contexts it is mandatory to investigate which data are shared between the different users and how secure these data are. This is especially important for 3D-models. In the future more powerful *computer hardware and streaming solutions* will reduce the risk for exchanging 3D models prior to Virtual Reality reviews and might also help avoid the process of preprocessing 3D geometry in order to reduce the geometry level of detail.

Besides using VR only for design reviews, it is also possible to use Virtual Reality technologies in creative product design processes. For example, it can be used to design models of any kind dependent on the functionality and aesthetics (see Figs. 13.13 and 13.18).

VR enables designers and all other users to design the products in a 1:1 scale in the environment immediately as the product will be used in the future. This is especially a big advantage in comparison to existing CAD or other design modeling tools (CAID or CAE modeler). The VR based systems are mainly used in early design processes to

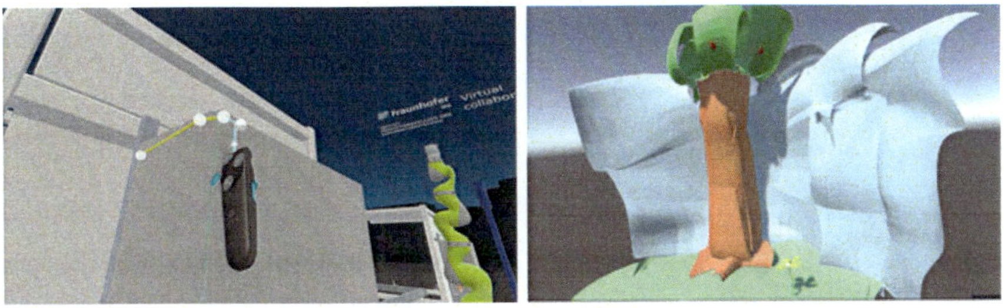

Fig. 13.18 Sketching at Fraunhofer IPK

generate different design alternatives and first sketches. These applications address different use-cases. Whereas some tools are mainly used to generate sketches with primitive objects, other tools address freeform surfaces.

Next to creating and designing new models, it is also possible to use Virtual Reality for factory layout planning: Fig. 13.19 shows a use case in which a user can design a factory layout in Virtual Reality based on a construction kit. The construction kit has also integrated the degrees of freedom of each part of the factory production line. Using VR for the factory layout planning, enables companies to assess immediately the path workers would walk or investigate ergonomics. This use case has been leveraged within the publically funded research project VIB-SHP ("Virtual Commissioning with Smart Hybrid Prototyping"), compare [10].

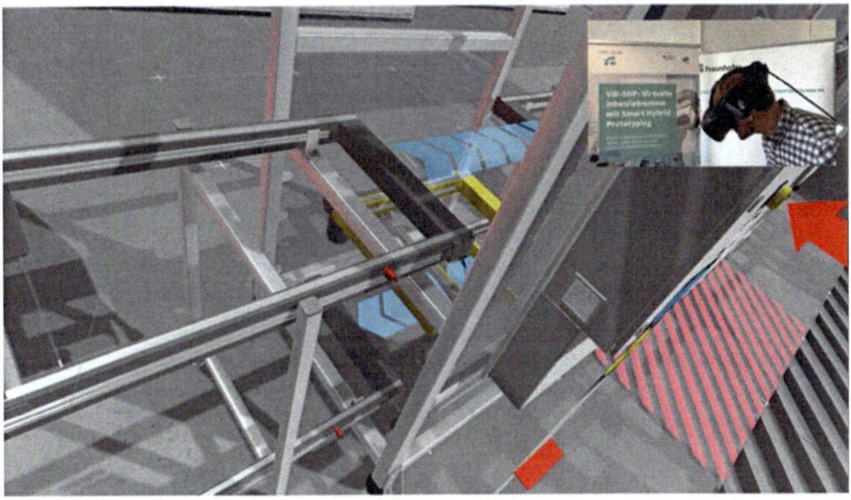

Fig. 13.19 Interactive VR-environment for factory layout planning [10]

13.5.3 Summary of the Technology's Benefits and Main Trends

In summary, Virtual Reality is a technology, which can be used in all phases of the product development process. The wide spread of use-cases as well as the low cost to start using the technology due the Head Mounted displays (in comparison to costlier powerwall and CAVE installations) leads to a more widely use in industrial applications. Nevertheless, one has to take into consideration that for different use-cases it is necessary to use different software or hardware tailored to the specific purposes.

Currently, the main trend is to increase the usability of VR-software so that it is easier for everyone to use VR applications. In addition, it is mandatory that the VR applications can be easily integrated into the product development process and into the IT infrastructure of the different companies. The latter is particularly nowadays an impediment to use applications that are based on game engines, as the data interfaces for industrial applications are still to be developed.

References

1. Milgram P, Takemura H, Utsumi A, Kishino F (1995) Augmented reality: a class of displays on the reality-virtuality-continuum. Telemanipulator and telepresence technologies. Int Socr Opt und Photonics 2351:282–293
2. LaValle SM (2017) Virtual reality. http://lavalle.pl/vr/. Accessed 19 Jan 2019
3. Sherman W, Craig A (2003) Understanding virtual reality: interface, application, and design. Elsenvier Science, San Francisco
4. Slater M (2003) A note on presence terminology. Presence Connect 1–5
5. Buhr M, Pfeiffer T, Reiners D, Cruz-Neira C, Jung B (2013) Echtzeitaspekte von VR-Systemen. In: Dörner R, Broll W, Grimm P, Jung B (ed) Virtual und augmented reality (VRA. Virtual und Augmented Reality): grundlagen und Methoden der Virtuellen und Augmentierten Realität. Springer, Berlin/Heidelberg, pp 195–236
6. Sutherland IE (1968) A head-mounted three dimensional display. In: Association for computing machinery (ed.): AFIPS 68 (Fall, part I): Proceedings of the December 9–11, 1968, fall joint computer conference, Part I. AFIPS fall joint computing conference. San Francisco, California, 9/12/1968–11/12/1968. ACM Press, New York, pp 757–764
7. Cruz-Neira C, Sandin D, Hart J, Kenyon R, Defant T (1992) The cave-audio visual experience virtual environment. Commun ACM 35(6):64–73
8. Head-Mounted Displays: Messung räumlicher Präzision bei VR-Trackingsystemen (Head-Mounted Displays): Measurement of spatial precision of VR-tracking systems). https://www.vdcfellbach.de/fileadmin/user_upload/Applikationszentrum_VAR_-_Werkstattbericht_04_-_VR-Tracking_-_Positionsgenauigkeit_v04.pdf. Accessed 13 Sep 2020
9. Stark R, Israel JH, Wöhler, T (2010) Towards hybrid modelling environments-Merging desktop-CAD and virtual reality-technologies.CIRP Annals—manufacturing technology 59(1):179–182. CIRP General Assembly, Pisa, Italien, 22–28 August 2010.
10. Stark R, Müller P, Grosser H (2018) ViB-SHP—Virtuelle Inbetriebnahme für Industrie 4.0 zukunftssicher beherrschen (Robustly controlling Virtual Commissioning for Industrie 4.0). Fraunhofer Verlag. https://doi.org/10.24406/IPK-N-509999

Chapter 14
Major Technology 8: Augmented Reality—AR

Executive Summary

This chapter deals with the following topics:

- Basics and advanced techniques of Augmented Reality
- Providing insight into how engineers benefit from using Augmented Reality (AR) technologies
- Describing functioning, benefits, and limitations of AR technologies in practice.

Quick Reader Orientation and Motivation

The intention of this chapter is:

- to give an overview of AR technology in Virtual Product Creation as driver and enablers for Digital Transformation in engineering
- to present AR technology as part of Virtual Product Creation from a practitioner's point of view to analyze the need and usefulness for day-to-day industrial work practice
- to give instructions on how to use AR technology
- to explain models, frameworks, and mathematical representations that help to grasp the internal working modes of AR technology.

In contrast to Virtual Reality (VR) applications, the Augmented Reality (AR) approach enriches *real world objects* with computer generated perceptual information *by means of an overlay*. As being part of the wearable computing research segment and industry, AR stands for a multimodal augmentation; it is often related to the visual sense to see both worlds simultaneously—an intelligently projected virtual image blended with the real three-dimensional environment. The benefit lies in the augmentation of the user's visual perception of his physical surroundings with additional and meaningful context-sensitive information. With advanced technologies such as computer vision, virtual objects become executable for visual consumption and are the base for extended user interaction within working scenarios.

R. Stark, *Virtual Product Creation in Industry*,
https://doi.org/10.1007/978-3-662-64301-3_14

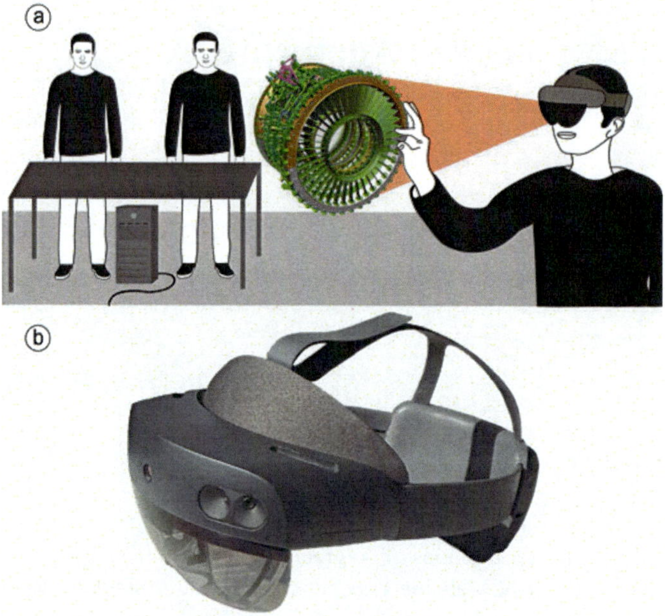

Fig. 14.1 **a** The AR principle of viewing digital content (example digital aero engine model) in the context of a real environment (e.g. office, test facility etc.); **b** Microsoft HoloLens 2 as an example for a Head Mounted Display [2]

The beginnings of industrial AR range back to the 1960s with first research prototypes, while the development ramped up steadily between the 1990s until today. Since then, the technology has evolved from being the subject of experimental research projects to being able to deliver effective and scalable applications that assist engineers during their daily work. Today, big tech companies like Microsoft, Google and Apple have identified AR as a key technology, investing heavily into it and already providing robust hardware and software solutions (see Fig. 14.1) [1]. Modern AR head mounted displays like the Microsoft HoloLens 2 shown in Fig. 14.1 (compare [2]) allow for up to 1–2 h of continuous interactive working meanwhile.

In 2018, AR was stated as one of the ten most strategic technology trends [3] with the potential to influence the technical development long-lasting. A recent study conducted by Deutsche Bank predicts that until 2020, the global market for AR will increase from 500 million Euro to 7.5 billion Euro [4]. Improved hardware and tracking algorithms make AR increasingly attractive [5]. Similar assessments of the technological and application related positioning of AR solutions, e.g. as part of advanced virtual assistants, are yearly assessed by the Gartner Research Group.[1]

[1] https://www.gartner.com/en.

14.1 Engineering Understanding of AR

AR solutions provide powerful technology elements in the context of extended Virtual Product Creation scenarios. Engineers and IT experts integrate AR solutions into numerous industrial processes to achieve state-of-the-art digital connectivity and intelligent virtual support.

Primarily, AR technology is used in planning or execution of assembly, operating and maintenance activities as depicted in Fig. 14.2a. In general, AR can support tasks such as overlaying digital information sets onto real physical objects, assisting humans in understanding and executing specific tasks according to in-situ needs or prescriptive workflow seqeunces and to provide explanations as part of interactive sessions:

- Interactive visualization of interfaces to support engineers, workers and costumers during prototype reviews or planning activities
- Support of context-aware activities and product user assistance as well as training,
- Object localization in the factory or field,
- Visualization of product and process-relevant virtual geometries and information (e.g. geometric differences between a real component and another version of it via virtual overlay).

In addition to industrial AR applications, the technology can add value in several further areas like sales (e.g. customer product presentation and configuration), health-care (e.g. surgery training and support) and the consumer market (e.g. games, services such as navigation or tourism, head-up displays). Other fields of AR usage exist within education and training as well as in guiding people within tourism and as part of cultural exhibitions.

14.2 Why Does an Engineer Use AR?

Within the overall range of AR use pattern, engineers are increasingly leveraging the following capabilities of Augmented Reality within their task portfolio. Please note the most common ones:

- Early detection of design errors by overlaying CAE results on top of real proto-types in order to compare high stress or strain related areas of a component or assembly.
- Descriptive and meaningful documentation of products and technical systems (and machines) in order to provide "on the fly" direct information sets at a physical product area or feature in order to enhance in-situ checks and understanding of operation.
- Simple communication between all planning stakeholders during reviews of physical objects and prototypes.

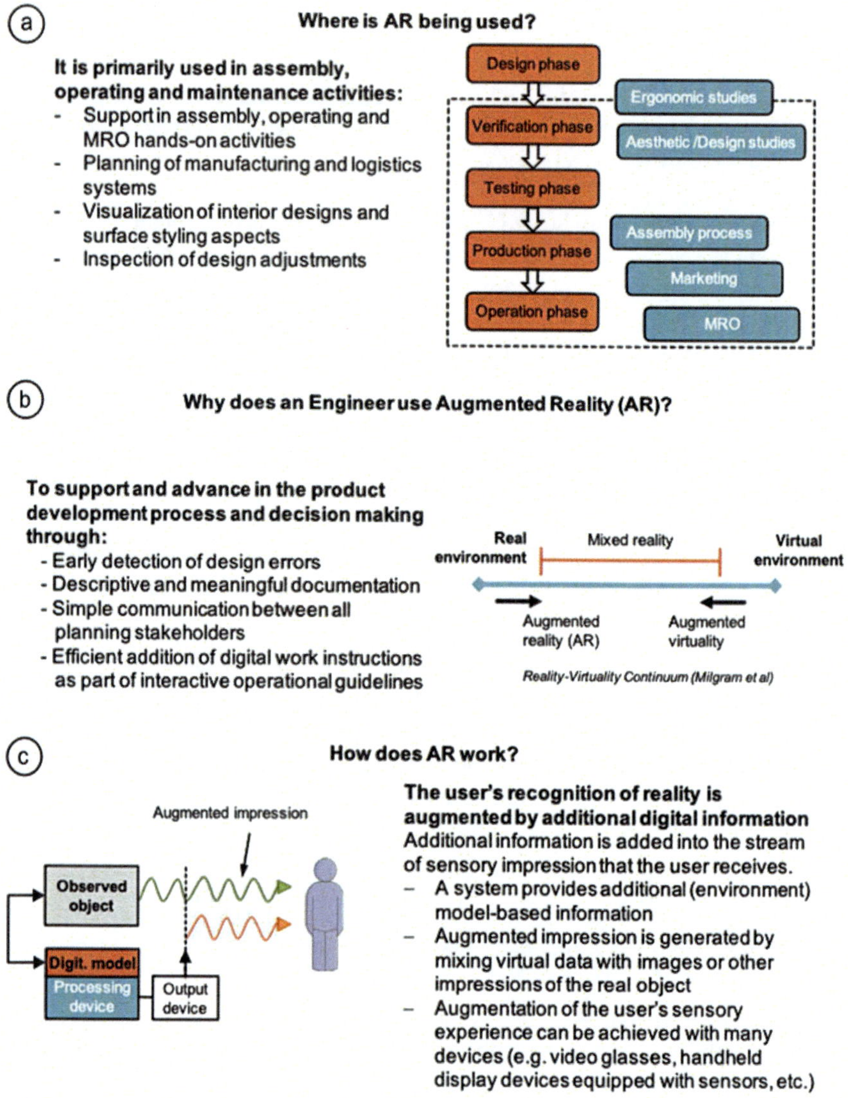

Fig. 14.2 Engineering application of AR (subfigure **b** based on [6])

- Efficient addition to operational guidelines with the help of static and dynamic digital information sets on top or in the nearness of the physical areas of a real asset, gadget, product or component within the MoL (mid of life) lifecycle phase.

Engineers are using Virtual Reality (compare Chap. 13) during the pure digital engineering phase of product creation when no physical objects and components exist. Augmented Reality, on the other hand, is utilized to overlay digital information with physical objects. For a better understanding and differentiation of the related technologies, the continuum between reality and virtuality is depicted in Figure 14.2b. The limit on the left hand represents real world physical elements such as persons or physical objects. The limit on the right hand designates sole virtuality. Here, the virtual environment only models fictional objects, i.e. there exist a completely computer-generated environment that can be arranged in a full immersive set-up. Apart from the controllers' inputs, this "pure virtual" environment is not connected to the physical reality of the context. In between the two opposite sides, the range of mixed-reality expands. Coming from the left, one could think of superimposing lightweight virtual information in Head-Mounted-Displays (HMDs) while seeing the real world that is considered in a context-sensual software application. Continuing to the right, almost complete virtual scenarios that only partially include real components (e.g. real seats, steering wheels or humans). Due to AR's key feature, there is still the chance to see the real surrounding like persons and objects and the scenes are less immersive compared to VR. However, the ability to superimpose digital information is more attractive for many companies and engineers because the real and the digital world can be perceived simultaneously in a specific arrangement. Even though today's virtual and augmented reality applications differ in many ways, experts see a merge of the two technologies in the near future. Indicators of this are development efforts in the sector of video-see-through devices that can switch between the two concepts.

14.2.1 What is AR Doing for an Engineer?

Augmented Reality provides technology support for Engineers in many occasions if the overall AR solution set has been set up for it. The following capabilities of AR are described and illustrated in order to create incentives for engineers to request such AR technologies within their personal engineering solution set.

14.2.1.1 Capabilities of AR in Product Design and Manufacturing

Product design requires many design iterations that consist of both, synthesis and analysis. The latter requires validation tasks of the specific prototypical design stages that AR can support within an intuitive way. A technical example of this is the comparison between virtually investigated simulation results and its real-world counterpart's crash deformation during real tests for validation. This correlation test use case can also be helpful between digital CAD models and complex prototypical components that are built for product integration evaluation. Furthermore, AR enables the chance to reduce prototypical designs to a minimum while replacing certain equipment with virtual holograms. By doing so, time as well as budget efforts are reduced and

Fig. 14.3 **Leftside** superposition of a crashed door with the crash simulation result [7]. **Rightside** augmentation of the virtual engine compartment with the real body-in-white for product validation. (*Source* BMW Group)

concepts are becoming mature in an earlier design stage. The two example use cases are depicted in Fig. 14.3.

In order to ensure manufacturing and assembly feasibility for prototype builds and series production AR offers capabilities known as *Augmented Visual Inspection*: it is, e.g., possible to check and inspect the released product data with the real physical product situation (see Fig. 14.4).

Meanwhile, first applications are under development to improve the interactive positioning and orientation of machines in factory and power plants shop floors. As shown in Fig. 14.5, a CAD-model of a machine is projected into the real world with a HoloLens 2. The user can then place the machine via hand gestures in the room. The translation and rotation will be returned back to the native CAD-program (in this case *Mechatronics Concept Designer, MCD* from Siemens). The AR application provides the engineer a better understanding on how the machines will be placed in

Fig. 14.4 AR based visual inspection of the physical member floor side with its digital product data based on the TWYN system of Visometry GmbH. (*Source* Porsche Leipzig GmbH)

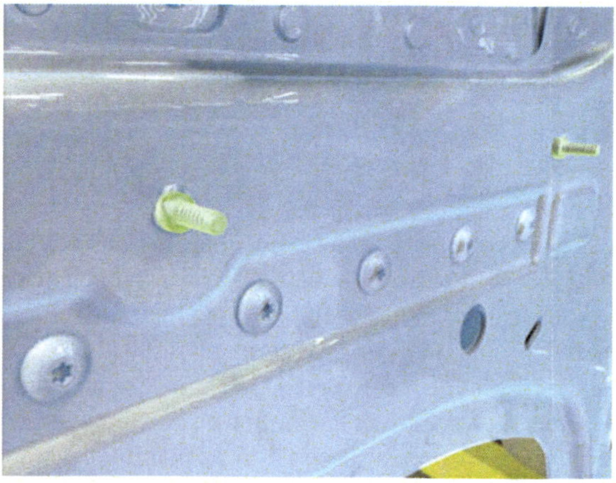

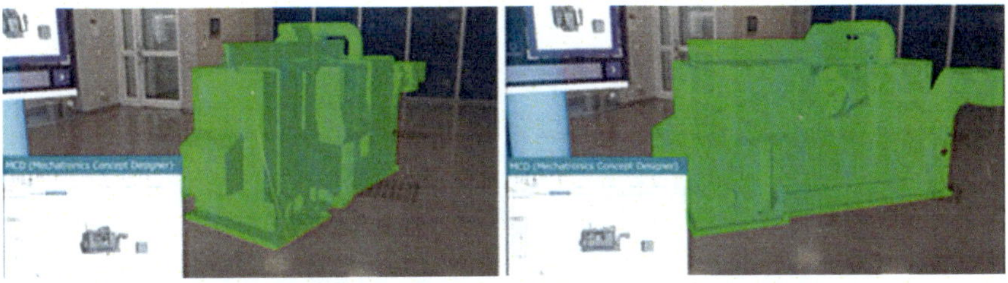

Fig. 14.5 AR application to position an electric drive on a factory shop floor. (*Source* TU Berlin, chair of Industrial Information Technology)

the real context. This will be advantageous for factory and power plant floor design of the future.

Figure 14.6 shows a similar AR application realized on the shop floor of the production site in Leipzig (Germany) of the vehicle manufacturer Porsche.

Such AR solution enables robust digital factory integration planning and layout sign-off between digital planners and station designer with the operational factory floor experts.

Figure 14.7 shows a new type of interactive AR prototype by Fraunhofer IPK to be used in conference room based digital reviews.

Fig. 14.6 AR based factory shop floor layout integration of a fixture resource based on the SuPAR system of CDM-Tech GmbH. (*Source* Porsche Leipzig GmbH)

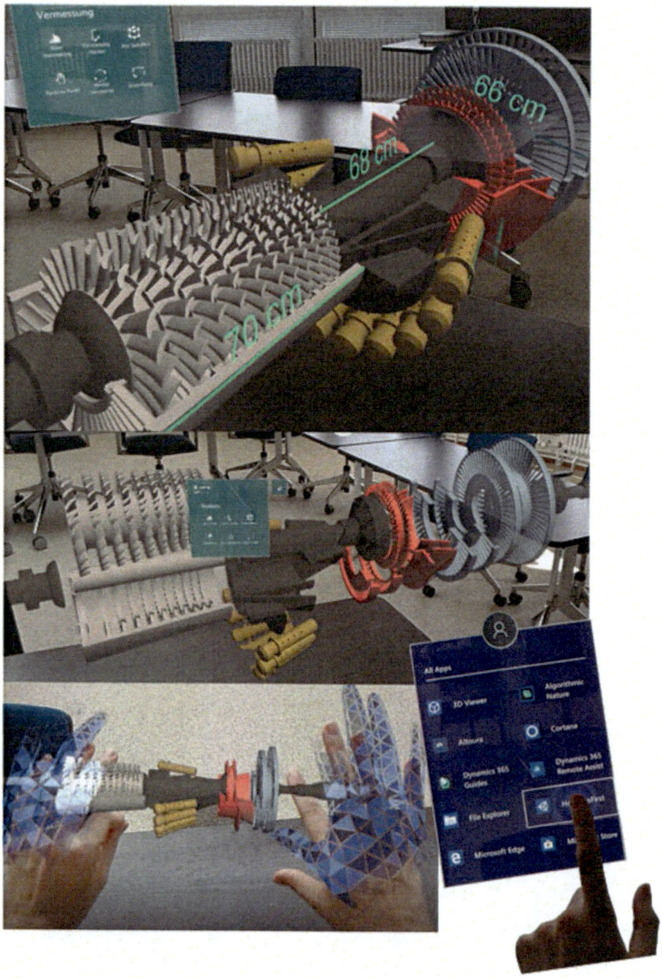

Fig. 14.7 Interactive AR prototype as part of a digital conference room review (example of an aero engine, courtesy by Rolls-Royce)

14.2.1.2 Capabilities of AR for Interactive CAE of Physical Objects

With increasing capabilities of handheld computers—such as tablets and smart phones or of see-through glasses with edge computing devices (or coupled to handheld computers via Bluetooth or Near Field Communication (NFC)—it will be possible to allow interactive AR based CAE analysis directly on the objects in the field, i.e. within the operation environment. As shown in an applied research solution of the University of Singapore by Wenkai [8], it is possible to provide an *AR front end* to the physical environment of a civil engineering object such as a bridge in order to interactively apply different load and boundary conditions to a predefined *back end CAE model.*

With the help of control sequences via the internet it is possible to modify key parameters of the CAE model in the back end and to invoke an instant CAE run for such problem (see Fig. 14.8). The advantage is that non-CAE expert users are enabled to use an appropriate CAE visualization environment directly at the location of the real object in the field. This helps to apply an on-the-fly CAE calculation and visualization of possible alternative solutions of the bridge pillar or trust construction as well as on modified load assumptions directly at an existing bridge that needs to be overhauled or re-engineered.

14.2.1.3 Capabilities of AR in Maintenance and Service

Typically, maintenance tasks are associated with high cost and a greater risk of errors because service technicians do not work in common workflows. To create efficient AR systems and applications, the cooperation between the Bosch Common Augmented Reality Platform and REFLEKT ONE provides several solutions. As an example, AR is used to support the complex repair of passenger vehicles. In this example, a tablet-based AR app supports the technician by highlighting the necessary work steps. Additionally, the application includes an instruction video and a list of required tools and components [9].

Another research-project using and testing AR is the multi-disciplinary joint project ALUBAR [10]. The process of turbine maintenance is supported with a head-mounted display. The project aims to support older workers in their daily activities and to ease the new or re-entry into the employment. The user is provided with relevant information to perform the maintenance, while the findings (e.g. damages) are documented automatically upon a simple command. Due to the interaction with the AR system by voice commands, the user has free hands to work safely and it is not necessary anymore to carry protocols and paper-based information material inside the narrow turbine. As a special feature, the developed AR system is adaptive and responds depending on the user's physiological state. If a certain stress threshold has been exceeded, the system automatically adapts the augmented information. By doing so, expensive errors can be avoided and work accidents are prevented.

14.2.1.4 Capabilities of AR in Commissioning

An example of how AR can support workflows is the use of data glasses applications in industrial logistics during picking processes on shop floors. To reduce error rates and to increase more effective workflows, the user gets relevant information about component parts, e.g. its location in the warehouse and the required part ID. Supported by a hands-free AR system depicted in Fig. 14.9, the user can make use of relevant information in situ via AR-based visual aids. Additionally, data glasses provide visual feedback that informs about the task correctness. A hand-mounted

Fig. 14.8 Parameter update in a mobile AR-FEA system (compare [8]). **a** the set-up consists of one natural feature image tracker and target outdoor structure; **b** initial state of the FEA result; **c** stress distribution after loading is added; **d** switch to deformation results display

Fig. 14.9 Picking and commissioning process on shop floor in automotive industry. **Left** Photograph of the commissioning task from a third-person-perspective, **Right** Augmented view through data glasses. (*Source* BMW Group)

camera scans automatically the bar code of the component the user reaches for. If the worker takes the correct one, a green coloring appears; if it is wrong, a red visual feedback shows up on the display. User research studies with comparable motivations succeeded and led to high user acceptance and the desired result of a more efficient workflow [11].

Figure 14.10 A shows the virtual commissioning and inspection of an electric drive with the help of an AR based application (developed by TU Berlin, chair of Industrial Information Technology). This AR solution adds 3D-models with attached simulation-data (drive shaft in combination of temperature distribution and bearings in combination with rest-useful-lifetime-estimation) as well as the position, name and index value of the sensors that are used for the inspection.

Through better visualization, this method eases the workflow of the inspection and has an entertaining side effect since it can also be used and followed out from home or the engineering office.

In the real-word, such an electric drive is of very large scale (up to several meters per dimension), so it needs to be inspected from different views. For this demonstrator two mobile phones perform AR-functions from different angles, which need to be synchronized across a network, showing e.g. rotation speed, temperature, sensor positions (compare Fig. 14.10b).

Over another network link, the views of the AR-devices are digitally streamed to a web browser, so the inspection can be conducted from another location. If the electrical drive is not ready for inspection yet, the inspection-workflow can be shown on an AR-model of the machine, so the customer and contractor can be prepared for the real inspection and check if negotiated inspection steps will be performed.

According to [12], other fields of research include Gamification approaches. An example for such capability would be a gamified application for picking processes as part of pre-commissioning tasks in industry.

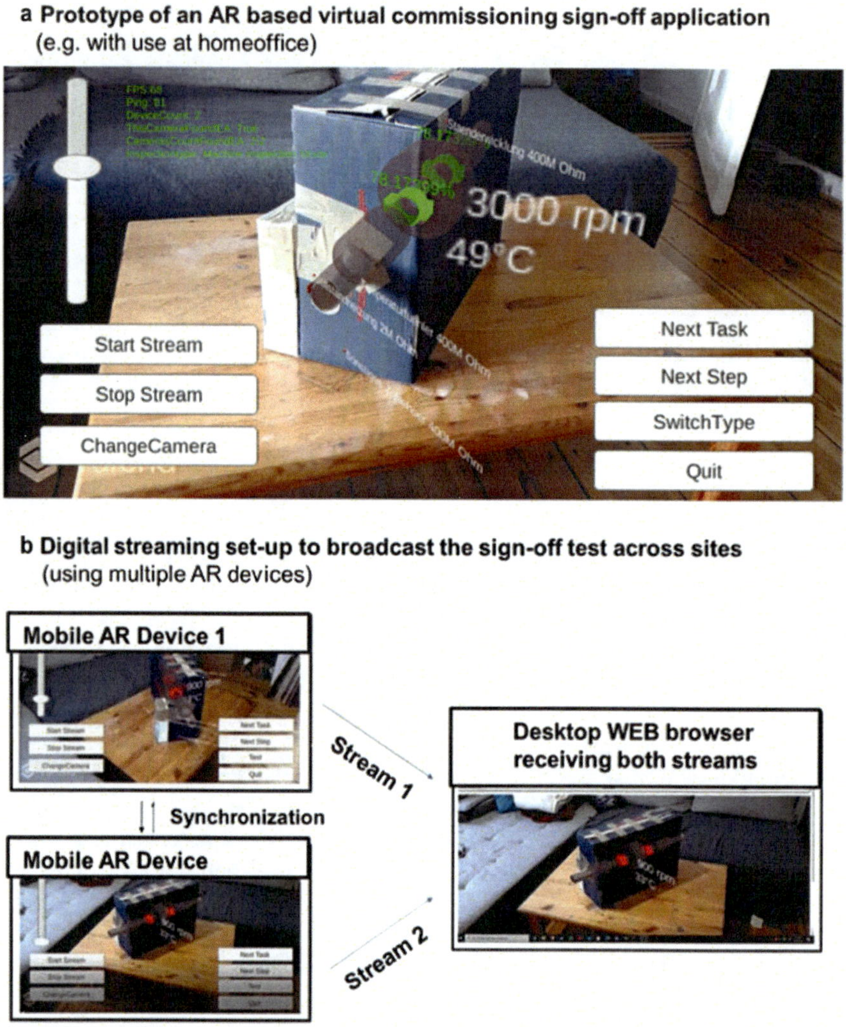

Fig. 14.10 AR application to support far-distance virtual commissioning and inspection

14.2.1.5 Capabilities of AR in Training

The possibility to add instructions to the real environment makes AR attractive for maintenance and training [13]. Based upon the composed sensor data acquisition, AR can be an assistance providing the user appropriate instructions, geometrical paths, work steps or further information. For instance, in a training scenario, an amateur can get information on every single working step (e.g. about tool use, navigational data) to practice a task and prevent mistakes due to the lack of expertise. The handling of particularly complex systems (e.g. aircrafts, industrial plants, etc.)

requires knowledge and expertise that can be supported by AR devices, thus reduces possible damage to expensive systems, and prevents work accidents.

14.2.1.6 Capabilities of AR in Generic Quality Assurance

Quality assurance either prevents or indicates missing, misplaced or defective components. To provide an example, "Werklicht Pro" by Extend3D is a project that develops spatial AR. Digital information about a construction unit (e.g. construction plans) are projected directly on a work piece. Furthermore, projected CAD data and work instructions can support employees in manufacturing and assembly. For enhanced error detection, spatial AR can reveal slight deviations by super positioning of the target state. Thus, quality can be improved and a communication basis about quality development is created. During internal or external product or process audits, AR can support the persons in charge with highlighted positions so that they can be aware of where to lay specific focus on.

14.2.1.7 Capabilities of AR in Remote Collaboration

Due to highly distributed facilities and international plants in automotive manufacturing, remote collaboration plays an important role. For example, when it comes to error detection and remotely supported maintenance of vehicles or manufacturing systems, experts do not need to waste time on expensive business trips, but rather share the captured video stream of an AR device with the colleagues on-site. Annotations can be made in the augmented field of view or voice instructions can be transferred by the sound system.

14.3 How Does AR Work?

As depicted in Fig. 14.2c, AR technology enriches the naturally sensed visual impressions by the human with virtual information based on either a wearable or statically mounted output device. In order to finally render the digital information on a display at a realistic position or at the right time, a computer needs to sense the reality for localization or interaction purposes. This enables the system to react to the users' environment, activities or work steps. The quality and quantity of sensor inputs depends on the use case and the application goal. Today, AR devices are shaped in many different ways depending on the specific use case. More detailed information on the technical side is provided in the following sections.

Figure 14.11 depicts a simplified systems data handling process to show the main functionality and the concept architecture.

In reality, the simplified loop of the system shown above is realized in different sub systems with separate processes run in parallel. Consequently, there is a chance

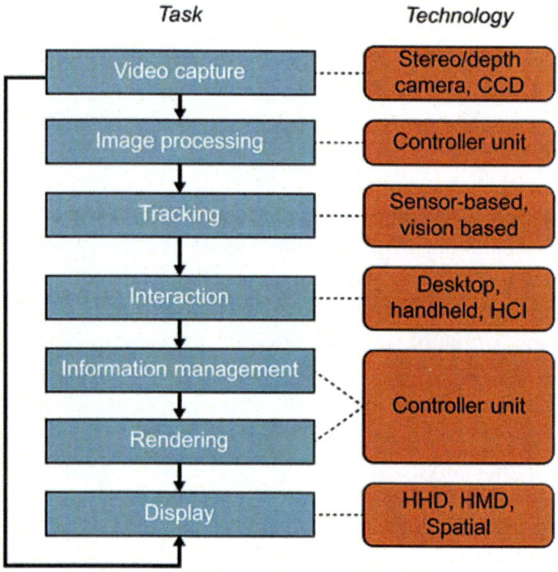

Fig. 14.11 Simplified AR pipeline according to [1]

that tracking rate and frame rate diverge, which can lead to noticeable instability of a 'static' virtual object like a virtual chair on the real floor while moving the camera (humans head).

If the tracking rate drops below the frame rate (typically minimum requirement of 60 Hz) or high tracking latencies occur, the object seems to wander during the cameras movement, which decreases immersion [13].

14.4 AR Technologies

AR hardware technologies can be distinguished in mobile and static applications (partially based on [1]). Figure 14.12a explains the three categories of visualization techniques that are used for the mobile and static applications. Mobile applications are supported by the first two categories:

Head mounted/head-up displays:

- *Smart Glasses*, also considered as data glasses or personal imaging system, are used like common glasses and add digital information, e.g. from the internet. Compared to HMDs, Smart Glasses are smaller and less powerful, typically without virtual depth perception.
- Examples: Google Glass Enterprise Edition, Vuzix Blade Smart Glasses, Bose Frames.

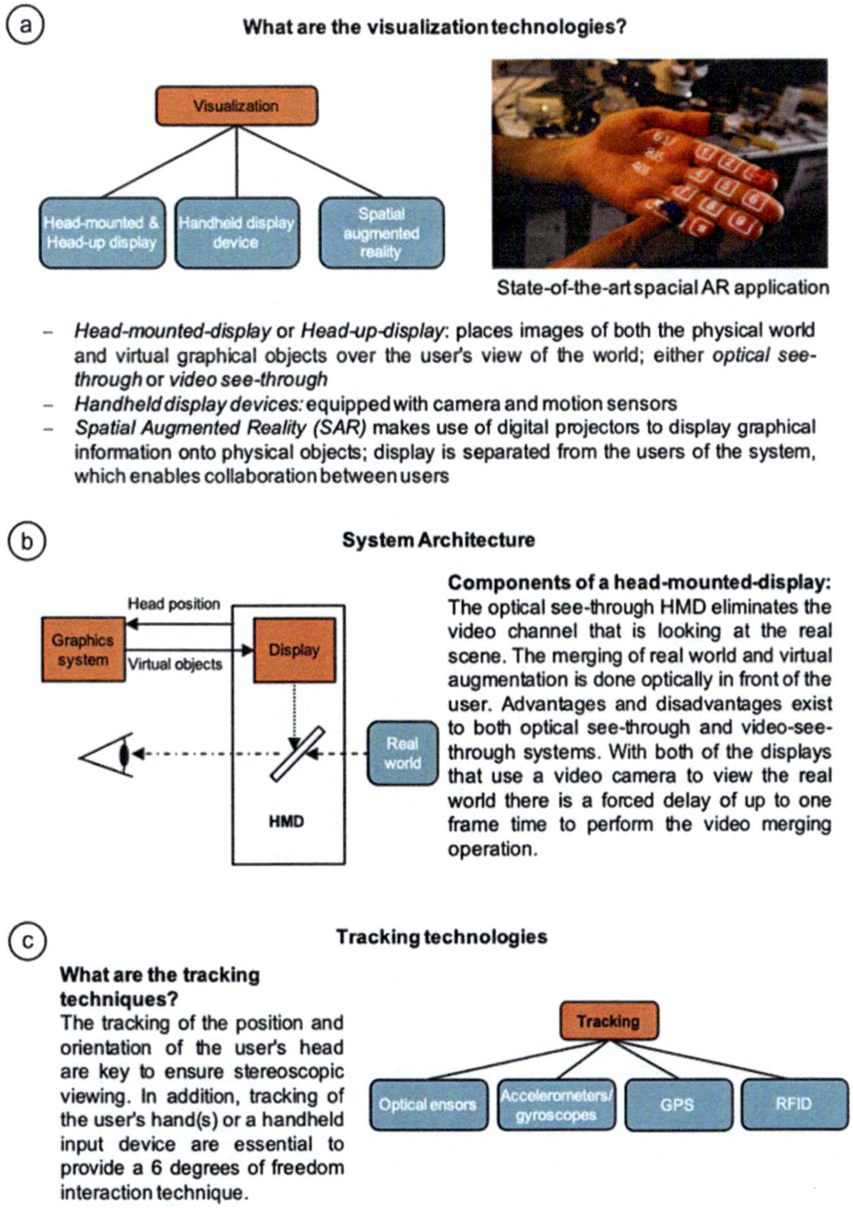

Fig. 14.12 Basis IT technology of AR

- *Advanced HMDs* designed as video see-through or optical see through, see Fig. 14.1.
- Examples: Microsoft HoloLens, Magic Leap (or VR/AR hybrids) and.

Handheld display devices such as:

- *Smartphones, Tablets* (hand held devices) used with AR apps.

Static applications use **spatial augmented reality (spatial AR)** solutions with the help of:

- *Projection based displays*, which consist of an optical projection on real objects. Example: Extend3D Werklicht.

Today's tasks in manufacturing and service require mobile hands-free devices, which is the reason for the recent development effort in the segment of smart glasses and HMDs. The requirement to have the devices as small and lightweight as possible limits the amount of the embedded systems computational power to some extent.

To overcome the problem of computationally expensive renderings combined with small hardware, applications can benefit of streaming the data from another device to the HMD. A drawback of this infrastructure technology are high requirements to the available bandwidth of the wireless connections. Ongoing research is conducted on the use of contact lenses. The desired retinal projection is achieved by directly projecting the virtual object into the eye with the help of a small wearable device. In addition to the wearable concepts, development efforts are taken also in the field of spatial projection. This is realized with the help of static projectors or displays in a dedicated room, as depicted in Fig. 14.13.

The software side can be distinguished in the following low level (embedded programming) and high-level technologies:

- Dedicated operating systems (memory management, hardware drivers)
- Tracking algorithms
- Rendering approaches

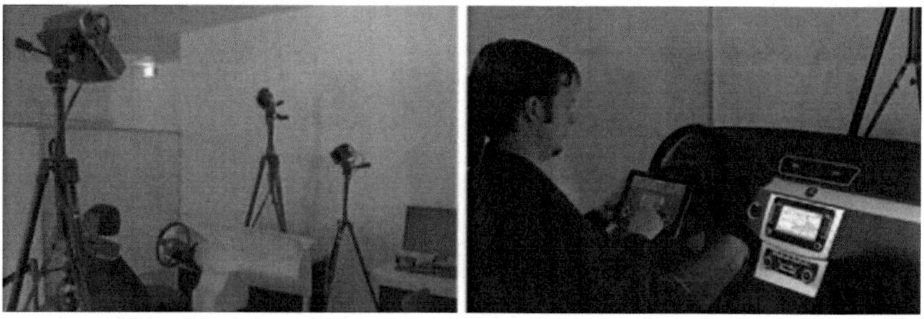

Fig. 14.13 Projection of the cockpit texture onto a rapid prototyped design mock-up apart of the automotive interior design process [13]

- Applications e.g. based upon development kits and game engines as frameworks for logic and visual implementations

14.4.1 Setup of AR HMDs/System Architecture

Today, head-mounted AR devices consist of optical components for the real-time virtual projection and of a central processor unit that runs programs and thus receives, analyzes, reduces and sends data. Network capabilities for inter-device communication often are integrated as well. The display typically is either designed as a 'video see through' or 'optical see through' concept (compare the explanations of AR technologies above).

The transparent display visualizes the augmented virtual information in case of an optical see-through-device. Figure 14.12b shows the components of a head-mounted display (optical see-through). In case of a video-see-through device, the collected camera image is rendered on an opaque display. Sensors as cameras (charge-coupled device, stereo or depth sensing), accelerometers, GPS, solid-state-compass or microphones provide the processor with real-world information. The human–machine-interface can be implemented in different ways: for example, via gesture and voice commands to input and to return visual information via display. The experience of the virtual content is realized with the help of intelligent transformation and projection based on the head movements. Digital content either is added based on internal storage or it is pulled from connected devices, such as a database server, an Enterprise Resource Planning System or the internet. Finally, the physical device then is used by an ergonomic element to hold or mount the device on the forehead.

14.4.2 Tracking

Tracking and environment registration approaches extract the spatial arrangement of objects and thus enable superposition of a virtual representation at the desired real-world location. Besides tracking the users head, applications also often require hand tracking, other object tracking and environment tracking. As shown in Fig. 14.12c, tracking can be based on mechanical, electromechanical, optical, acoustical, and inertial sensors or a combination of the mentioned techniques (sensor fusion towards hybrid systems). Today, typically, optical and inertial sensors are used due to their cost efficiency and scalability [7].

Optical tracking approaches can be distinguished into two different concepts (see Fig. 14.14):

- *Outside-in*: Based on the concept of photogrammetry, statically mounted sensors acquire the positions and orientations of pre-defined sticked markers (active) or marker less features (passive).

Fig. 14.14 Different tracking concepts with examples of corresponding markers. **Leftside** inside-out-marker [7]. **Rightside** outside-in-marker [14]

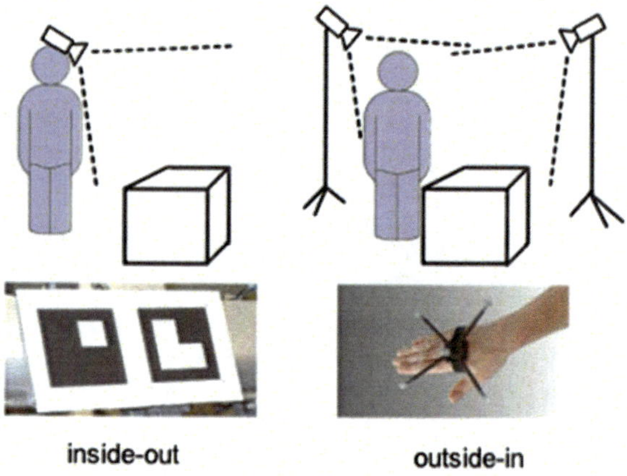

- *Inside-out*: A single camera is not fixed statically, but can be moved with the user or object and calculates its position and orientation based on environment features, such as 2D-markers (active) or object features such as specific geometric properties (passive).

The passive feature-based tracking approaches mentioned above are achieved with the help of geometric feature extraction. Beside edge or corner point detection, more abstract features of 2D or 3D objects can also be used for a camera-based identification. Since only a defined percentage of the features are required to successfully track an object, it is faster, more robust (e.g. during the exposure to disturbance such as partial covering) compared to marker-based tracking. The 'Simultaneous Localization and Mapping' (SLAM) approach adapts these feature-based concepts. Without any knowledge of the surrounding, the system incrementally maps the environment and localizes its position and orientation [13]. There exist a substantial global scientific and coding community for markerless AR using algorithms based on line and feature segmentation principles and edge detection methods.

Registration describes the calculation of spatially arranged coordinate systems for each object of interest, so a realistic and congruent perspective can be rendered even when the camera (the user) moves.

It is important to note, that an AR system also requires special "viewing" features and algorithms in order to achieve realistic occlusion in accordance to the rules of line of sight, i.e. it needs the ability to hide virtual objects behind real objects. The challenge stems from the fact that the display used to project the virtual content usually is closer to the eye than the physical environment. A possible solution is to introduce phantom models that overlay the occluded virtual objects (see Fig. 14.15) by using tracking sensors and computer vision algorithms. When these phantom objects are rendered black on AR displays with additive color composition, they will appear transparent to the user [5].

Fig. 14.15 Solving the problem of virtual objects (bin) behind real objects (folders) by introducing phantom objects (center figure) [13]

14.5 Human Interaction

For HMD devices, AR enables tremendous possibilities to realize human–machine interfaces for human input and output.

In order to make use of AR's potential, it is beneficial to rethink old window-based, two-dimensional UI paradigms by introducing 'spatial interfaces', which are much more natural to the human brain. The goals are to reach natural interactions based on previous experiences made in the real world. This makes a challenging learning curve of artificial workflows almost obsolete.

To reach this purpose, user interaction in the context of AR is considered an open field of research [15]. This is, however, similarly applicable to VR-interaction (compare Chap. 13).

Due to the diverging device characteristics of AR HMDs, a high variety of UX implementation concepts, which integrate the sensors and software logic, exist to interpret users' input:

- *Tangible interfaces*: physical devices (e.g. buttons, scroll wheels, virtual pens)
- *Haptic user interfaces*: tactile UI (e.g. touch or vibration feedback)
- *Camera-based interaction*: gesture control, ray-cast pointer of the camera, eye-tracking, object identification, depth camera
- *Audio-based interaction*: voice commands via microphone, environment sound
- *Interaction based on other sensors*: location tracking (e.g. GPS), Infrared sensors, etc.

As an example, virtual buttons can be placed wherever the user wants, e.g. on a virtual object to form an interactive control panel. Hand gestures and voice commands could round off the natural experience by looking at a previously identified real object to directly operate it, open its interaction menu or make it display the operating status. Combining the concepts mentioned above results in hybrid approaches that enable state-of-the-art intelligent multimodal interaction.

14.6 Development for AR Applications

After the 'Make or buy' question was clarified critically with regards to the commitment to individually work on AR software, developers coming from the traditional desktop development may need to leave old paths to engage in the new spatial thinking and interaction.

14.6.1 System Selection for Industrial AR

To successfully select and implement an AR app in an industrial environment, the following application-oriented considerations provide support in analyzing and determining important requirements (partially based on [16]).

Problem/application:

- Performance requirements: e.g. amount of objects/polygons
- Depth information requirements (2D: data glasses, tablet; 3D: HMD)
- Level of real-world consideration (tracking, object detection, etc.).

Environment:

- Infrastructure for data provision and exchange
- Need for mobility (local computation/rendering on external client, network connection)
- Consideration of environment variables (physically robust device, battery runtime, exposure to wear, etc.).

User:

- Desired degree of immersion (display resolution, field of view)
- Time and frequency of personal AR exposure
- Physical characteristics (weight, ergonomics, etc.)
- Acceptability of situational distraction by the virtual content.

Implementation (framework selection):

- Complexity: Static content (low), dynamic 3D-content (intermediate), interactive experience (high) helps to decide on possible software frameworks
- System integration (e.g. data bases)
- Deployment (app, program, internal service via locally installed software or remotely hosted-service via mobile web)
- Scalability
- Reusability.

Software development kits, such as ARKit (IOS), ARcore (Android) or Game Engines are helpful to avoid reinventing the wheel for high level AR application development. Efforts for standardized data formats are made for example with the

'Augmented Reality Markup Language'. This format describes the AR scene, such as its contents locations and appearance for geographic annotations in AR browser applications.

14.6.2 Implementation Design

Once the hardware and development platform is selected, the implementation can be initiated. Implementation of interactive AR applications is a multi-domain composition of three engineering professions as shown in Fig. 14.16a:

- *Systems engineering* (which type and degree of technical system or consumer/game environment need to be constructed?)
- *Software engineering* (how can such system environment be implemented by algorithmes and data?)
- *Usability engineering* (how can the user interaction be most proficient?).

To guarantee an effective and user-friendly user experience, the main points to be considered during software implementation can be clustered in three main categories (based on [17]):

- *Environmental design*:
  To be aware of the users' surrounding they are engaged with, such as space and situational context like public or private environment. This needs to be considered for example for the amount and size of virtual information, or device options as display brightness.
- *Interaction design*:
  To choose the right way of interaction regarding input options, feedback, which is at least partially also related to the context mentioned in the first bullet point, and other factors such as device capabilities or ergonomics.
- *Visual and audio design*:
  This cluster covers content, size and type of visual information, such as 3D objects or 2D information that was projected in the spatial real world. Shadow and Lightning play an important role in terms of immersion. In addition, it may be enhancing usability to implement visual or acoustic cues helping the user to find and understand possible ways of interact with specific objects.

Typically, industrial AR applications are to be integrated into the engineering design process and require connections and interfaces to the existing main system pillars such as CAD, PDM, PLM or ERP systems.

Fig. 14.16b depicts data sources together with data processing efforts before auxiliary relevant information is displayed effectively in the AR device. Beside the simplification of parametric CAD data towards tessellated visualization geometry, it is important to keep in mind, that the use of AR e.g. in the context of virtually augmented training also typically requires a newly referenced product data structure.

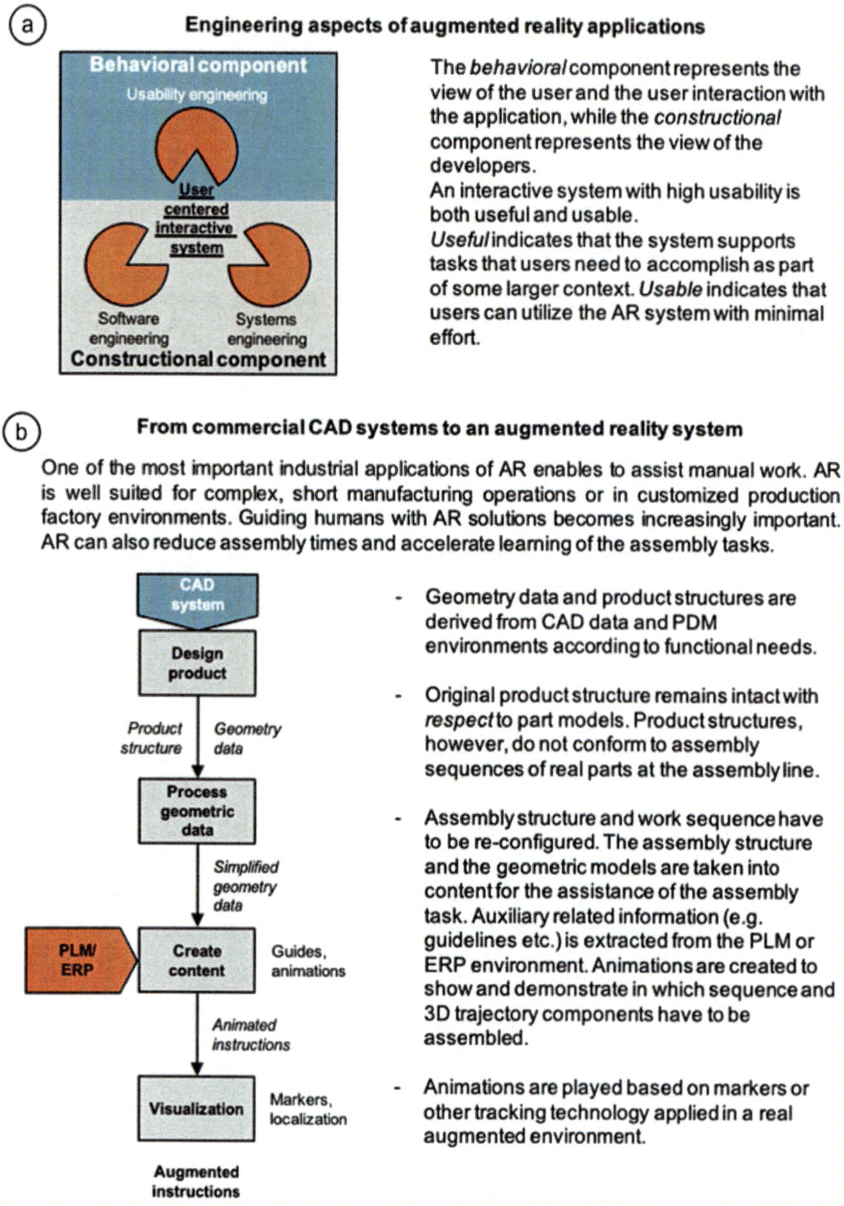

Fig. 14.16 Advanced technologies in AR

With the input of ERP systems, the AR content needs to be enriched and reconfigured according to processual information to create training animations

14.7 Technological Limitations to Overcome

Major development efforts are ongoing to overcome some usability limits, such as the form factor, limited field of view as well as display resolution, battery lifetime, processing performance, hardware weight and ergonomic shape [18]. For the everyday use with CAD objects out of a PDM system, the preparation-effort still leaves space for improvement. This can be at least partially traced back to the growing diversity of platforms that go along with incompatibility due to the lack of non-standardized software formats [19].

As with Virtual Reality (VR), a common problem of current AR devices is the *Vergence Accommodation Conflict (a human perception issue)*: the way that the lenses of human eyes focus on an object is very different from the way that human eyes physically aim themselves at the object the user is trying to focus on. Such phenomena lead to experience disturbance due to nausea and fatigue caused by divergences from physical laws.

A totally different challenge is given by the practical use of AR technologies in industrial and business environments: especially in product development and manufacturing, corporate security issues have to be dealt with in an appropriate manner (to avoid espionage and leaks of intellectual property). While engineers or workers are wearing the devices during long tasks at work, during important meetings or for communication, the use of camera, microphone and other accessible sensor data does not only go along with chances, but also with risks. Security risks exist e.g. in the case of using "sniffing" (network) malware or AR data streams in order to obtain intellectual properties concerning design features of future products, functional parameters of prototypes and/or engineering process knowledge.

Today, many industrial devices have radio connection to the outside world. Due to limited bandwidth and imminent interferences with high-priority systems on the shop floor, security and malfunction issues are to be conside red organizationally.

14.8 Summary of the Technology's Benefits and Main Trends

Overall, AR has high future potential for industrial use because the real world can be perceived simultaneously with helpful digital features. The new technology offers a wide range of potential benefits: context-specific visual support of tasks (e.g. maintenance and servicing), timesavings (e.g. by eliminating intense information searching by technicians), cost savings (e.g. by reducing training expenses and saving paper-based documentation), quality enhancement (e.g. by visually checking work steps), or worldwide availability of experts e.g. via AR-based remote systems.

Especially AR HMDs seem to be a technology with growing relevance for industry due to its scalability, mobility and the benefit of operating it hands-free. Continuous advances in hardware or marker less tracking algorithms further increase usability and flexibility.

Especially with a view to affecting peripheral technologies such as the advances in the sector of 5G, remote rendering as part of cloud-based computation will enhance the AR technology in a substantial manner and path its way towards further disrupting the engineering sector.

References

1. Fraga-Lamas P, Fernandez-Carames TM, Blanco-Novoa O, Vilar-Montesinos MA (2018) A review on industrial augmented reality systems for the industry 4.0 shipyard. IEEE Access 6:13358–13375. https://doi.org/10.1109/ACCESS.2018.2808326
2. Stein S (2019) Microsoft's HoloLens 2 feels like practical magic. https://www.cnet.com/news/microsoft-hololens-2-feels-like-practical-magic/. 24 Feb 2019
3. Gartner Inc. (2019) Top 10 strategic technology trends for 2019. https://www.gartner.com/doc/3891569?srcId=1-3931087981. 8 Feb 2019
4. Heng S, Hörster A-K, Karollus A (2016) Augmented reality: Bei Spezialanwendungen sollte Deutschland von dynamischem Zukunftsmarkt profitieren können; (Augmented Reality: Germany should benefit from the dynamic future market in the context of special applications); Deutsche Bank Research
5. Runde C (2017) Virtuelle Techniken 2017—Wo stehen wir heute?", Stuttgart. VDI-Fachtagung "Virtuelle Techniken in der Fahrzeugentwicklung. Virtual reality, augmented reality und mixed reality im praxiseinsatz, 31. March 2017, Stuttgart/Germany, VDI Wissensforum GmbH: Düsseldorf. https://www.researchgate.net/publication/327057800_Virtuelle_Techniken_2017_-_Wo_stehen_wir_heute?channel=doi&linkId=5b7594614585I546c90a0c03&showFulltext=true
6. Milgram P, Takemura H, Utsumi A, Kishino F (1995) Augmented reality: a class of displays on the reality-virtuality-continuum. Telemanipulator Telepresence Technol Int Soc Opt Photonics 2351:282–293
7. Nölle S (2006) Augmented Reality als Vergleichswerkzeug am Beispiel der Automobilindustrie (Augmented Reality as a comparison tool in the automotive industry). Dissertation (Ph.D. thesis), Technische Universität München, München
8. Wenkai L (2017) Integration of finite element analysis with mobile Aug-mented reality. Ph.D. Thesis, National University of Singapore
9. RE'FLEKT GMBH (2017) Highly-efficient augmented reality applications for automotive, industry and real estate. http://www.re-flekt.com/archive/en/products/158-augmented-reality-software-platform-one. 19 May 2017
10. Brandenburg E, Woll R, Stark R (2016) "Portable Hands-Free Process Assistant for Gas Turbine Maintenance", world academy of science, engineering and technology, international science index 118. Int J Comput Electr Autom Control Inform Eng 10(10):1702–1706
11. Pluta W (2017) VW führt Datenbrillen ein. https://www.golem.de/news/augmented-reality-vw-fuehrt-datenbrillen-ein-1511-117627.html. 19 May 2017

12. Korn O, Schmidt A (2015) Gamification of business processes: Re-designing work in production and service industry. Procedia Manuf 3:3424–3431. https://doi.org/10.1016/j.promfg.2015.07.616
13. Dörner R, Broll W, Grimm P, Jung B (2013) Virtual und augmented reality (VR/AR). eXamen.press. Springer, Berlin. ISBN 978-3-642-28902-6. online available under: https://doi.org/10.1007/978-3-642-28903-3.pdf (springer.com)
14. ART GmbH (2021) Internet material for AR-tracking technologies of the company Advanced Real-time Tracking GmbH. https://ar-tracking.com/en, 28 Feb 2021
15. Meta Company in 2019, the company declared itself insolvent), "Spatial-Interface Design. Augmented Reality and Neuro Science". https://pages.metavision.com/meta-augmented-reality-design-guidelines, 17 July 2017. Still available under: https://titanwolf.org/Network/Articles/Article?AID=f36d1179-06ac-4507-8d1f-9686e22440ee#gsc.tab=0
16. Tümler J (2009) Untersuchungen zu nutzerbezogenen und technischen Aspekten beim Langzeiteinsatz mobiler Augmented Reality Systeme in industriellen Anwendungen", (Analysis of user centric and technical aspects during long term usage of mobile augmented reality systems in industrial applications); Dissertation (Ph.D. thesis), Otto-von-Guericke-Universität Magdeburg, Magdeburg
17. Wilson T (2019) The principles of good UX for augmented reality. https://uxdesign.cc/the-principles-of-good-user-experience-design-for-augmented-reality-d8e22777aabd, 5 Feb 2019
18. Wiesend S (2017) AR- & VR-Lösungen im Unternehmen (AR & VR solutions in enterprises). https://www.computerwoche.de/a/ar-und-vr-loesungen-im-unternehmen,3217983,2, 19 May 2017
19. Ro YK, Brem A, Rauschnabel PA (2018) Augmented reality smart glasses: definition, concepts and impact on firm value creation. In: Jung T, tom Dieck MC (eds) Augmented reality and virtual reality. Progress in IS, Springer International Publishing, Cham, pp 169–181. ISBN 978-3-319-64026-6

Chapter 15
Major Technology 9: Digital Factory—DF

Executive Summary

This chapter deals with the following topics:

- Basics and advanced techniques of Digital Factory
- Providing insight into how engineers benefit from using Digital Factory (DF) technologies
- Describing functioning, benefits, and limitations of DF technologies in practice.

Quick Reader Orientation and Motivation

The intention of this chapter is:

- to give an overview of DF technology in Virtual Product Creation as driver and enablers for Digital Transformation in engineering
- to present DF technology as part of Virtual Product Creation from a practitioner's point of view to analyze the need and usefulness for day-to-day industrial work practice
- to give instructions on how to use DF technology
- to explain models, frameworks, and mathematical representations that help to grasp the internal working modes of DF technology.

Virtual Product Creation has the task to prepare for successful fabrication and production of products, which have been developed beforehand with the help of virtual models and tools. Similar to the digital modelling approach for products all involved elements of the production environment—as part of the overall factory—are treated in the same way. This means, that the factory layout, manufacturing resources, machines and tools, manufacturing process plans and tasks as well as the interaction with production workers and with the different states of the products during the manufacturing progression are subject for digital modelling, simulation, modification and maturation up to their final releases. This chapter, therefore, explains all major elements of umbrella solution framework *Digital Factory*, in some countries also referred to as (overall) *Digital Manufacturing* approach.

R. Stark, *Virtual Product Creation in Industry*,
https://doi.org/10.1007/978-3-662-64301-3_15

15.1 Engineering Understanding of the Digital Factory

The term *Digital Factory (DF)* has been driven from the German understanding of establishing an equivalent term to the ordinary, physical factory in order to describe all critical digital modeling and simulation capabilities and opportunities as part of digital manufacturing engineering.

15.1.1 Why Does an Engineer Use Digital Factory?

The *Digital Factory* results from many years of dynamic change in economic conditions. Due to globalization, different areas such as economy, environment and culture intertwine, which has substantial effects on products and their life cycles. As a consequence of the transformation from a seller's to a buyer's market, products are becoming increasingly diversified and individual and manufacturing companies face major challenges, as they have to demonstrate increasing quality, and simultaneously, falling costs in shortened periods of time [1].

In order to master this dynamic progression, computer-aided tools and methods have been used since the 1980s to increase planning efficiency and shorten the implementation time of products and production facilities [1]. Yet how does the *Digital Factory* relate to these tools and methods? The next section explains this term and introduces the underlying principle of the *Digital Factory*.

The Definition:

Research and industry have been concerned with the definition of the term "*Digital Factory*" over many years. In practice, however, the different attempts to explain the concept have often led to divergent interpretations. In order to counteract the growing misunderstandings, the VDI Guideline 4499 has developed a generally applicable definition that brings together the views of experts and supports a cross-industry understanding of the term [1, 2]. It reads as follows:

> "The Digital Factory is the generic term for a comprehensive network of digital models, methods and tools—including simulation and three-dimensional visualization—that are integrated through integrated data management.
> It aims at the holistic planning, evaluation and continuous improvement of all essential structures, processes and resources of the real factory in connection with the product."

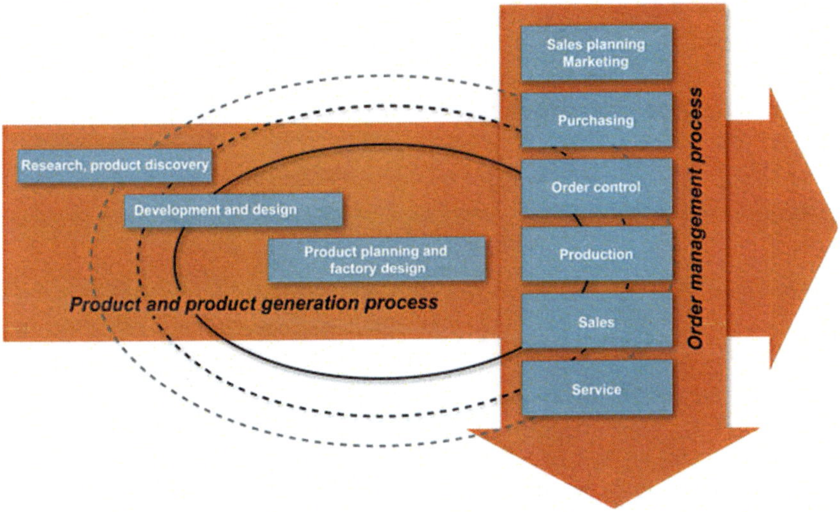

Fig. 15.1 Focus of the *Digital Factory* at the heart of corporate processes [2]

The *Digital Factory* is not solely a software-related topic, as the term may initially suggest. It rather focuses on production planning and factory design, which must be considered at an early stage in all business processes.

Figure 15.1 displays this focus in more detail. According to [2], production planning entails the planning of processes and production systems. Both, the requirements for the development and construction as well as those for the operative production process must be already considered in the early phases of the process in order to provide appropriate methods and tools that adapt to it. In addition, real production is to be continuously checked and improved by means of virtual instruments.

The *Digital Factory* is widespread in various industries. There is, however, a great difference within the implementation of its tools and methods. According to [1], automotive engineering and the aerospace industry play a pioneering role since the potentials were recognized early on and were incorporated accordingly into the individual processes. Taking an automobile manufacturer as an example, the progressive networking of planning processes can be illustrated. While in the 1990s isolated solutions with few interfaces were frequently used, the digital landscape has changed considerably over time. Processes that were not digitally supported could be transferred to the digital environment with new tools.

Tools used so far have also been further optimized and new interfaces have been added so that the overall picture of the *Digital Factory*, as shown in Fig. 15.2, is built up piece by piece.

Yet there are still hurdles to overcome, for instance, the implementation of standardization and data management. According to [1], the implementation of standardization is very advanced with regard to products, processes and sample solutions, since they represent a basic prerequisite for the introduction of IT tools and methods.

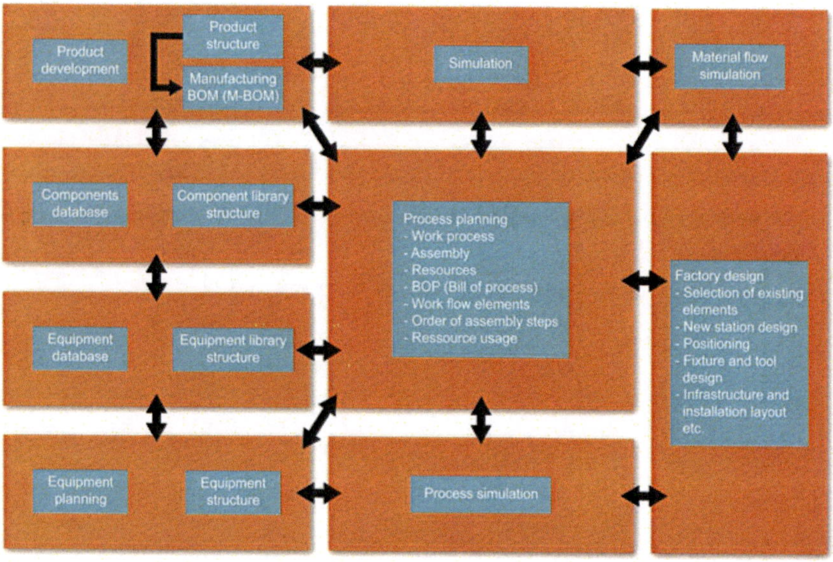

Fig. 15.2 Interaction of planning processes in the *Digital Factory* [1]

However, new requirements play an increasingly important role in the planning phase so that they can be determined and coordinated in every planning step. This creates a clear structure with which tools and methods can be introduced successfully.

On the topic of data management, significant efforts are required in the areas of change management and inventory data maintenance. Although interfaces between individual processes already exist, planning changes must still be recorded and maintained in the relevant systems so that all those affected receive information about any changes that occur. The completeness and consistency of the data and the relevant data management workflows and methods are critical to ensure high-quality manufacturing planning. The maintenance of inventory data appears as another major challenge, since it indirectly affects the planning data. In the first place, it involves the operational data that is available in the real production facilities and must be transferred to the *Digital Factory* repository. Due to the fact that a plant can always be modified during its operating life, it is important to record these changes. Primarily, those roles and individuals responsible for the data update in the information system have no direct advantage from this digital data maintenance. An analysis of these modifications can be only carried out effectively when these changes are transferred back to the planning systems, where process planning and manufacturing system design is carried out as part of modifications of the existing and the set-up of future plants.

In the entire value chain, not only the automotive industry strives to use the potential of the *Digital Factory*, but also the machinery and plant-engineering industry does so. In order to ensure that external plant manufacturers (suppliers) also support the internal planning processes of an automobile OEM manufacturer, the relevant data structures and formats need to be compiled. They are already specified in the invitation to tender for the specifications [1] and are finally detailed out as part of the statement of work. As a result, suppliers are faced with the challenge of meeting the requirements of different customers who, in turn, use different digital modeling, simulation and planning IT systems and resulting native modeling and data formats. However, this also holds the potential that machine and plant constructors will be short-listed for order placement if they have already been able to demonstrate support for the specified digital tools and methods in previous projects reliably.

In addition to the various possible uses of the *Digital Factory*, the question arises as to who works with digital tools and methods. In [2] target groups are identified, which can be enumerated as follows:

- External planning participants
- Internal planning participants
- Management of the company.

The external planning participants are primarily parts and module suppliers, plant suppliers and planning service providers. Different planning results are generated according to the specifications of the company-internal planning groups, which can be integrated in the subsequent processes. The internal planning participants consist of the areas involved in factory planning. They create, store and structure different types of planning information. They also use the results of the external planning participants in further planning steps in order to initiate the plant construction. A special target group is production staff, who are asked to incorporate their practical expertise into the planning process. Capturing such heuristic knowledge, however, is still limited within today's manufacturing and factory models and are subject for further research. Company and factory management refers to the people to whom the planning results are presented and to whom complex issues are explained so that they can drive the most relevant decisions. Company and factory management is the key stakeholder group who needs to be overall responsible and accountable to the overall planning results and associated budgets. It is still a dilemma, today, that most of the times digital data management tasks are not yet (fully) taken serious enough by company and factory management. This dilemma oftentimes is caused by the fact that company and factory management has a shortage in basic understanding and a lack of personal experience in such new digital engineering tasks.

Having positioned the role of the *Digital Factory* within the overall manufacturing planning situation and within the business scenarios of companies it is now time to explain the mission of the *Digital Factory* as key enabler within the *Virtual Product Creation* solutions portfolio. Figure 15.3 shows the major three mission elements:

- Part a explains the change of digital efforts upfront as part of the virtual front loading with the help of the *Digital Factory*
- Part b outlines the different layers of the *Digital Factory* (from the factory level down to the product level)

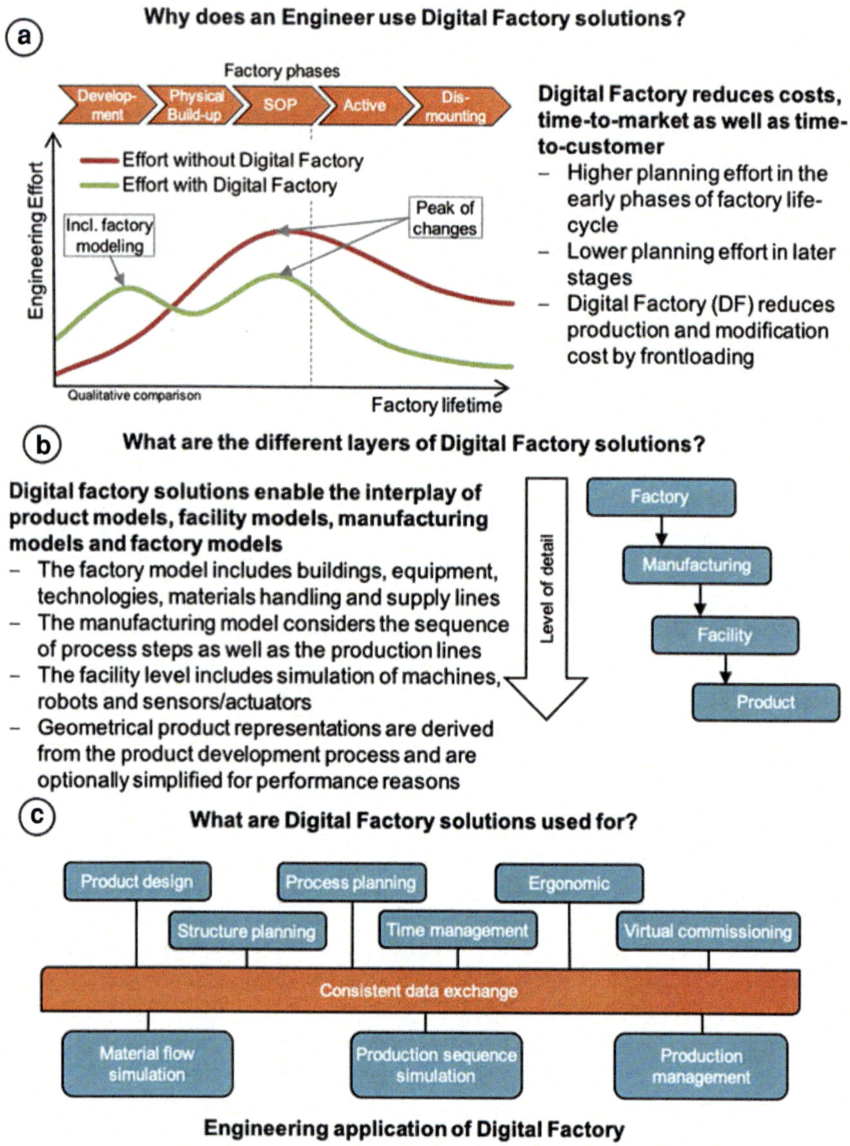

Fig. 15.3 The mission of the *Digital Factory* with motivation, layers and applications

- Part c shows the individual engineering applications of the *Digital Factory* including the critical integration element consistent data exchange (which requires a solid data management strategy and capability).

It should not be underestimated that the establishment of this *Digital Factory* mission does require substantial buy-in and support from both, corporate and factory management as well as the manufacturing planer and workers. It takes between five and ten years with a dedicated *Digital Factory program* to introduce and consistently set-up the engineering capabilities of a successful *Digital Factory* within enterprises.

15.1.2 What is Digital Factory Doing for an Engineer?

The *Digital Factory* influences the basic organization of projects with its tools and methods. As shown in Fig. 15.2, it causes planning processes to be networked, which in turn have various effects on the profitability of the company, the quality of the product and communication between the parties involved. Whereas in former times it was common to view product development and production planning as two separate, sequential project steps, the *Digital Factory* made it possible to increasingly parallelize these steps (simultaneous engineering). This not only positively affected reduced timeline of development projects, but also cost targets (reduction of change cost due to earlier feasibility and compatibility reflection). Nevertheless, it was and it is still necessary to invest continuously into digital engineering competence and operation, which in many cases remains a tedious effort in companies due to long living business habits of physical prove out approaches. The parallelization and shortening of the individual processes in early planning phases shown in Fig. 15.4 makes it possible to create and evaluate different variants with minimum effort.

Such "digital frontloading" leads to an intensive communication between product development and production planning, allows errors detection early on and reduces

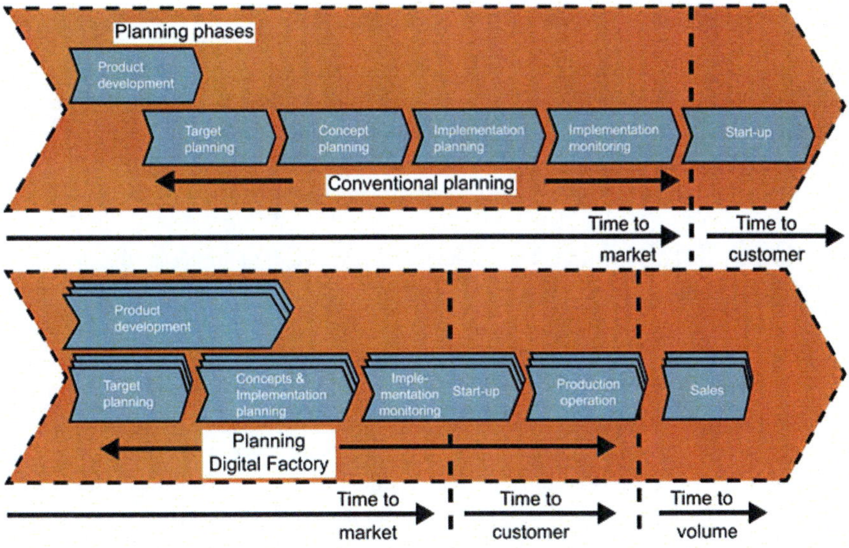

Fig. 15.4 Parallelization of planning processes through simultaneous engineering [2]

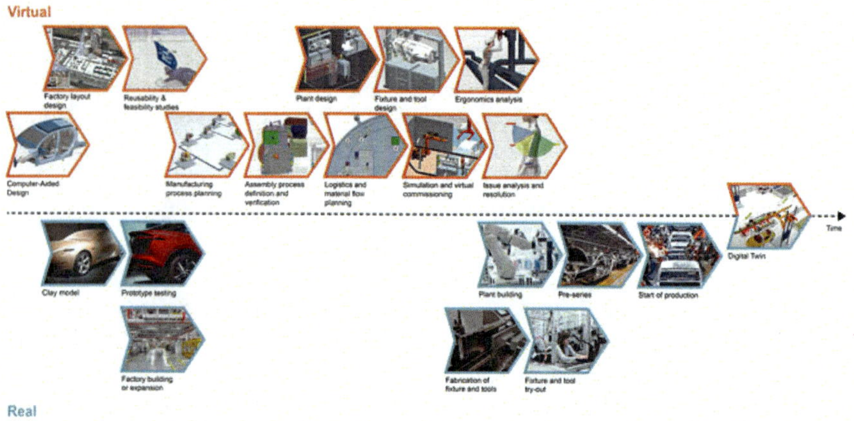

Fig. 15.5 Connecting *Digital Factory* virtual tools with real-world implementation

change efforts [1]. The *Digital Factory* offers possibilities of quality assurance of products already in the digital planning process as shown by the example of automotive virtual prove-out steps in Fig. 15.5.

Before manufacturing and assembly of the product takes place in the real physical factory, all process steps are virtually verified and product design as well as manufacturing systems might get improved along the process chain progression.

15.2 How Does the Digital Factory Work?

The *Digital Factory* is regarded as the "generic term for a comprehensive network of digital models, methods and tools" in the definition already presented. In order to describe this network, Fig. 15.6 presents an overview of different techniques of DF. Product creation is divided into three phases: product development, production planning and production. In each one of these phases, different methods and tools are used.

According to [1], a method is a systematic target-oriented approach that leads to a meaningful solution for a large number of problems. The methods shown in this context mainly describe organizational processes and procedures, which describe how an employee should use a certain tool at a specific time in principal terms. In addition, the detail digital working methods of using IT tools to create and use *Digital Factory* models are also considered. Simultaneous Engineering as an overall development approach represents an exemplary organizational procedure to enable collaboration between different functions such as Product Development and Manufacturing Engineering. The specific method how to model a virtual manufacturing assembly station belongs, however, to the portfolio of IT based digital working methods and is part of the *Digital Factory* solution.

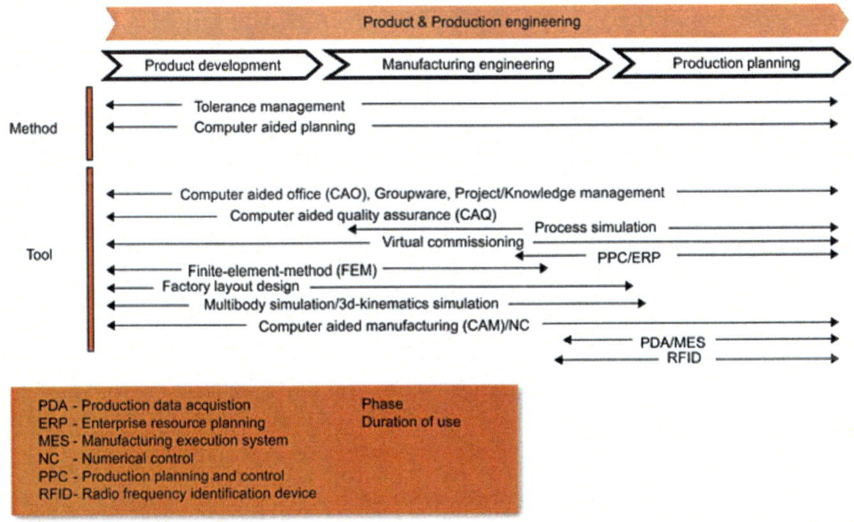

Fig. 15.6 Techniques of Digital Factory during product development (compare [1])

According to [1], an IT-supported tool is the software-technical implementation of a method or a combination of several methods, so that these can be used computer-supported, i.e. that a working method can be digitally executed. One example of such a digital working method is the Finite-Element-Method, which is used to carry out virtual tests on the load-bearing capacity of various machining components prior to the actual implementation of such a resource in a production line.

Another example is the RFID (Radio Frequency Identification) technology, which is suited to support manufacturing process execution. The simulation of such a RFID sensor as part of the overall manufacturing process execution can be used, for example, to locate vehicles manufactured in the automotive factory. In order to control cycle time and individual vehicle tracking in the *Digital Factory*, it helps to simulate the throughput from body construction up to final quality inspection.

In the *Digital Factory*, there exist many different methods and tools that differ in usage as part of manufacturing and assembly build feasibility, manufacturing process plausibility and overall factory management optimization. Increasingly, the degree of software integration into manufacturing resources are subject for latest innovations in *Digital Factory* capabilities. The author cannot present all of those methods and tools in the following sections. Rather, the focus will remain on those methods and tools that have already proven themselves through widespread practical implementation and are therefore safe in being deployed successfully.

15.3 Process and System Implementation of the Digital Factory

The methods and tools of the *Digital Factory* can be divided and categorized into three major areas. In order to implement factory planning digitally and to apply it throughout the *plant development process*, the areas of automation technology (1), plant logistics (2) and manual assembly (3) must be considered. These can be digitally mapped mainly in phases 3 to 5 of the plant development process, i.e. from the phase of *concept planning* to *preparation for implementation and final installation* (compare Fig. 15.7).

In addition, they can also be leveraged as reference for virtual commissioning and during the product launch phase. Therefore, it can be achieved digital plant-engineering activities can rely on digital layouts and simulation models created beforehand as part of the *Digital Factory*.

In addition to the integration into the plant development process, it is necessary to establish the technical set-up of the Digital Factory with respect to the IT system architecture and deployment framework, see Fig. 15.8.

In order to comprehend the importance and power of *Digital Factory* solutions to enable digital manufacturing and to conclude the specific application areas, each company has to analyze the relevant application and target areas, with corresponding use cases. Figure 15.9 depicts three potential areas and explains their major digital model arrangements.

Please find below the explanation of these three application/target areas in more detail.

15.3.1 *Logistics- and Production Flow Simulation*

In the area of logistics planning, the entire material provision for the production process of the product to be manufactured is defined and secured based on either a general or a detailed layout. In addition to the commissioning of production and purchased parts, station concepts and conveyor technology as well as their individual performance are calculated, simulated and validated against the target definitions

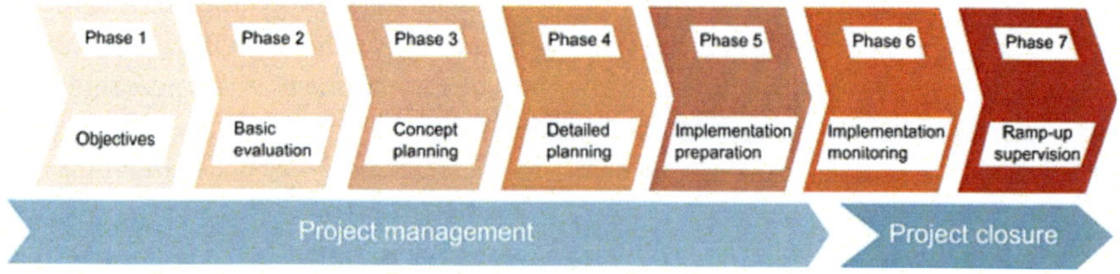

Fig. 15.7 Phases of the plant development process, according to [3]

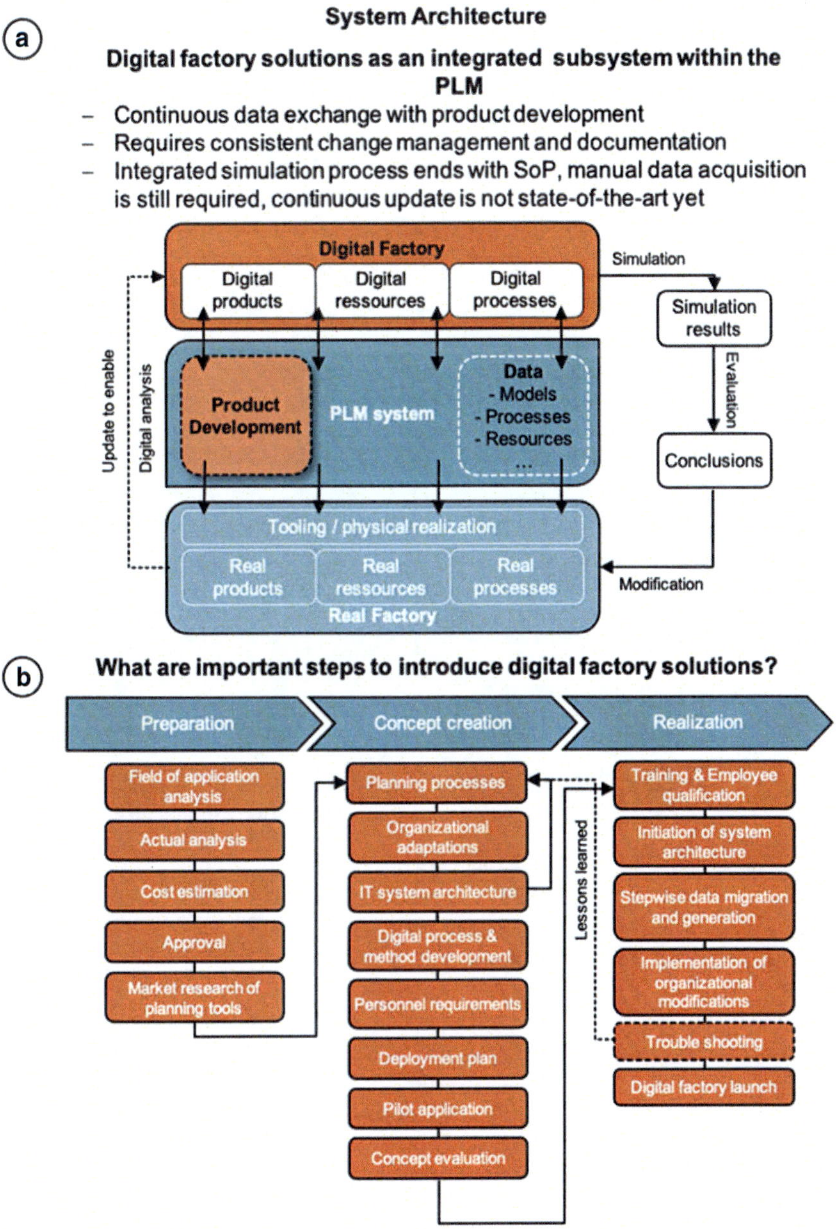

Fig. 15.8 The technical and deployment set-up of the *Digital Factory*

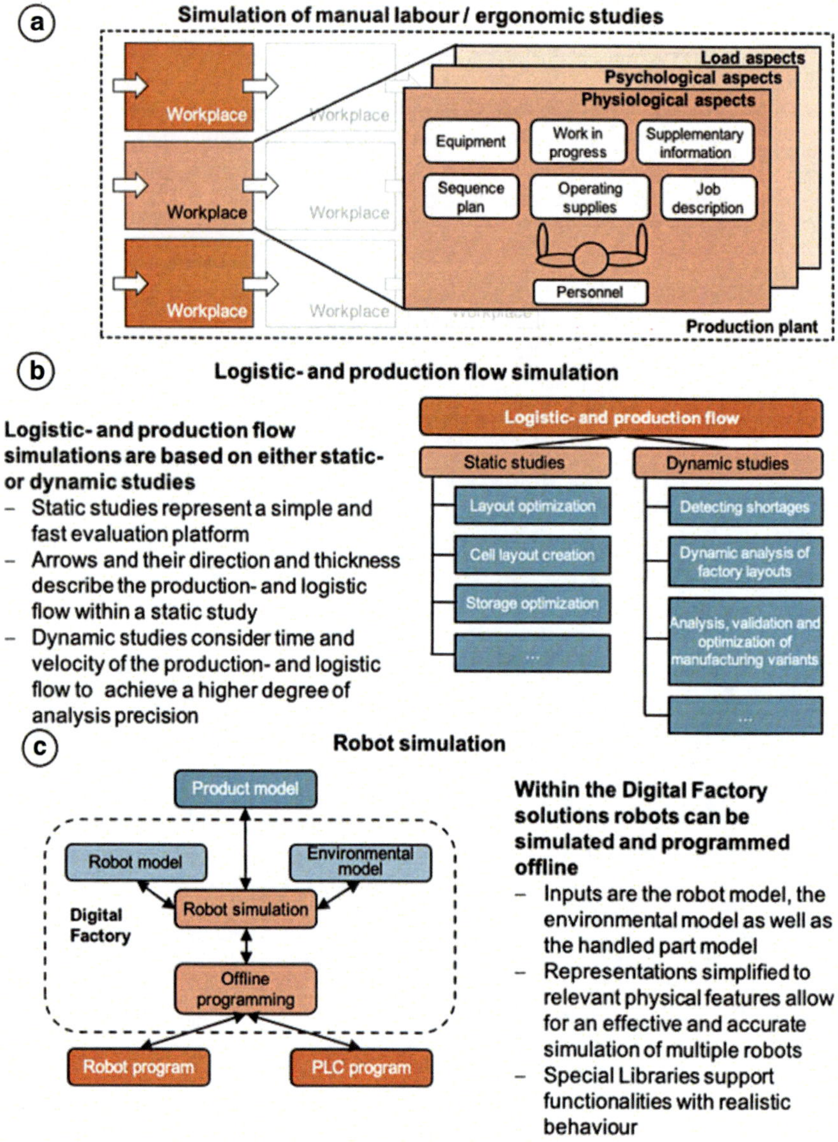

Fig. 15.9 Examples for application and target areas of the *Digital Factory*

(Phase 1). In particular, the material flow simulation also allows assisting the start-up support according to the simulation results.

15.3.2 Automation Technologies/Robotics

In the context of automation and robotics, any handling and manufacturing processes that have to be secured based on highly automated manufacturing processes are considered, simulated and validated. The goal is to optimize the production flow as much as possible with the help of tools from the *Digital Factory*. In terms of logistics and material flow simulation, these tools serve as input and output variables for automation and robot planning. The goal of robot planning is, for example, to handle components within a workstation in an optimized way (e.g. according to the limited space of the work cell) or to optimize their integration into the production processes (e.g. matching to the overall cycle time).

15.3.3 Simulation of Manual Labor/Ergonomics

In addition to automation technology protection, manual workstations must also be considered in production planning. *Digital Factory* tools support the production planner in the design and validation of workstations where manual assembly activities are carried out. In particular, simulation software is used in the following areas: route analysis, installation and removal studies, and ergonomics studies at virtual assembly workstations.

15.4 Digital Factory Technologies

In order to execute *Digital Factory* investigations, one should be aware of the different modeling approaches of the individual production environment and systems as well as understanding the associated simulation opportunities. This section will introduce the major modeling and simulation technologies in order to provide a solid understanding for typical *Digital Factory* investigations and deliveries.

15.4.1 Digital Factory Basic Modeling Technologies

One of the main tasks of the *Digital Factory* is the planning of the entire factory. According to [1], there is a distinction between a completely new planning (Greenfield) and the re-planning of already existing structures (Brownfield). In order to achieve an optimal result, in both planning cases an ideal planning without restrictions should be assumed, so that an optimized production process can be examined.

An important basis in factory planning is the plant model. In this model, support structures, walls, windows, roofs and other building elements are drawn according to

[1], so that it serves as a basis for the next planning steps. Based on the plant model, various participants can conceive and plan their ideas. Different structures can be created in a factory because different focal points exist. In logistics planning, for example, a low-crossing material flow can be tracked, whereas fire protection will focus on rescue and escape routes. Each of these participants can also use special software for modeling or commission corresponding planning service providers, so that the factory data can be available in different 3D formats for the planning results. This makes it necessary to transfer the data into a common Digital Factory model based on specified 3D CAD and/or 3D visualization data formats.

15.4.2 Layout Planning

In order to meet the challenge of an appropriate *Digital Factory* model, the required 3D CAD model is designed using the common factory layout. The main purpose of such a factory layout is to enable planners and decision-makers to visualize the respective results quickly so that they can understand and verify the functioning of the factory. Consequently, not all design details are relevant, as only the major geometric outer shape of the objects is required. Figure 15.10 shows a section of a 3D CAD model of a robot cell as part of the Body-In-White (BIW) production line of an automotive manufacturer that is used as a plant layout in planning.

According to [1], data reduction is necessary when creating such a model. This enables a fast display of the entire model on PC systems and handheld devices without much waiting time.

Fig. 15.10 Extract from a 3D model of a robot cell in BIW production [5]

15.4.3 Factory-Digital Mock-Up (DMU)

As mentioned earlier in this sub-chapter, different groups (internal and external planners) are involved in factory planning. Therefore, according to [1], it is necessary that a regular coordination between these individual areas takes place so that the respective results can be incorporated into the 3D overall model. This procedure is shown in Fig. 15.11 and it is called Factory Digital Mock Up (Factory DMU)—please also compare details of a DMU in Chap. 12.

In the first step of this procedure, the 3D data is checked in advance using checklists. Here the focus lies on the compliance with drawing regulations, such as for element design or specified colors. This step is followed by a collision check, so that it can be statically and dynamically examined whether there is contact between the production plants and building structure. An example of a static collision check is shown in Figs. 15.12 and 15.13 shows how dynamic movement space can be represented by a space placeholder for the clash analysis as part of a Factory DMU.

Through the visual control, a collision can be detected early in the planning process. The third and last step of the examination is the technical evaluation. With the help of such checklists, for example, it is examined to which extent there exists free

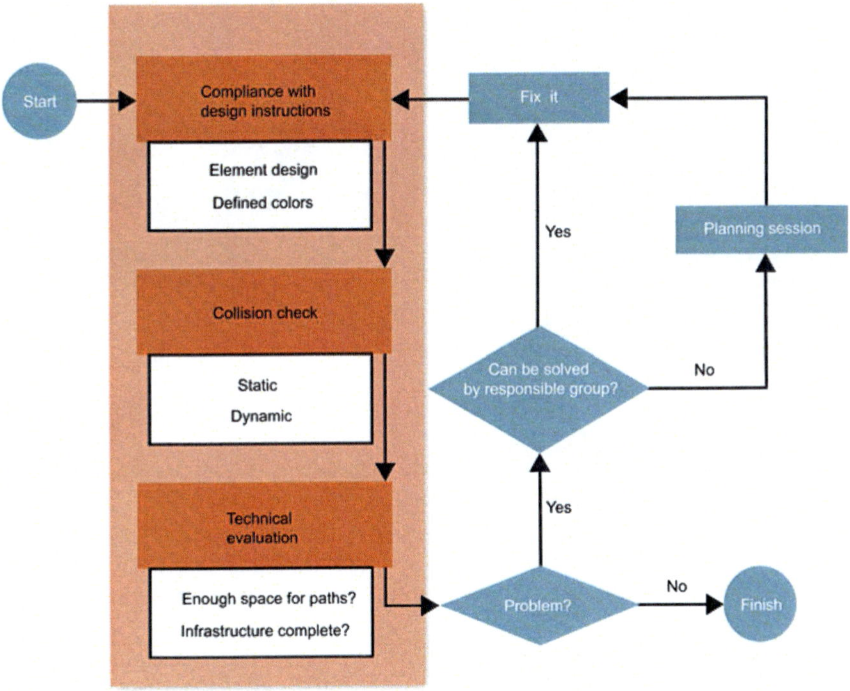

Fig. 15.11 Procedure of using a Factory-DMU for creating new 3D-CAD factory models, see [1, 4]

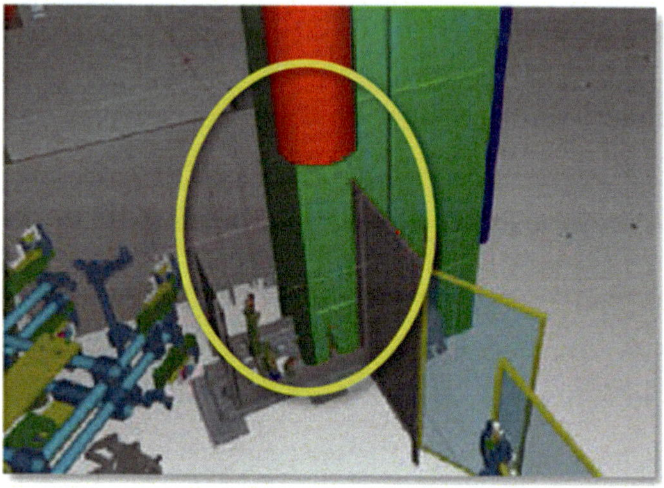

Fig. 15.12 Example of a collision between a pillar and a safety fence [4]

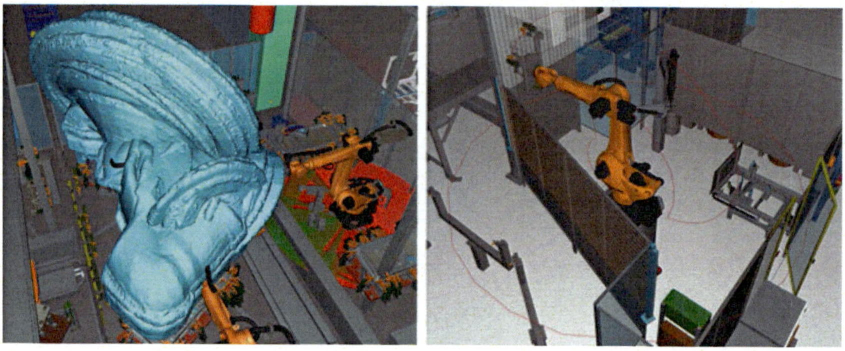

Fig. 15.13 The shrink-wrap volume of the roboter dynamic movement in the working cell as a static representation in the Factory DMU clash analysis [4]

space for tool and logistic paths and whether all necessary infrastructure installations are complete and accessible for maintenance. If problems arise in a particular area during this procedure, the responsible group is asked to rectify them. However, problems can also occur in different areas and therefore several groups could be asked to solve them together. In such cases, appropriate planning team meetings are scheduled in order to solve all issues and to agree on the appropriate industrial engineering solutions.

The presented process is iterative and with every iteration step, the planners and engineers improve the 3D factory model continuously. If all specifications have been met and there are no ambiguities any longer, the procedure is considered complete. In the construction of the entire factory, this final 3D overall model offers many advantages. It serves as a master source of information to derive various layout dimensions from it. In addition, it is leveraged to export certain points from the factory model into different manufacturing application tools. For example, it is used to display quickly the positions of drill holes for fixing plant components with a laser marking at the construction site as part of the physical set-up.

15.4.4 Behavior Models

In addition to purely spatial design modeling of a factory and the 3D layout mapping of the corresponding production stations, behavioral models of factory equipment are required as a basis for an accurate validation of manufacturing process and operations. Based on the *kinematization of geometric model data*, these behavioral models map dynamic and time behavior of plant components. With the help of the associated modeling steps, the entire behavior of the production plant is mapped according to previously planned cycle times. Manufacturing Engineers and Factory Planners use behavioral models for various aspects of production planning. In the context of the *Digital Factory*, the focus is usually placed on time behavior and its associated cycle patterns in order to map the simulation of material flows between the station concepts. In robot planning, the use of behavioral models is also very useful, especially during end effector path planning. The traverse paths of the robot body, arms and end factors as part of their work fulfillment (positioning, welding, screwing, painting etc.) and the sum of all cell robots' trajectories in their interplay are checked for mutual clash avoidance, clearance safety, are optimized in addition towards overall cell cycle time as well as wear minimization.

15.4.5 Electronics and Controls

For a holistic illustration of the planned production processes, an interdisciplinary modeling approach is advantageous. In addition to the 3D models created in the factory DMU, kinematization and behavioral modeling of the entire production plant, it is also necessary to describe it from the perspective of electrical project planning and control programming. The aim is to achieve consistent data continuity, starting with the factory DMU, through the complete wiring diagrams, to the initial control system design. Compared to the factory DMU described above, in electrical planning and control development the focus lies less on the three-dimensional representation of production plants. The wiring and pin assignment of the production plant as well as its function and performance are primarily secured. In the field of electrical planning, this can be ensured by means of circuit diagrams and I/O tests based on such diagrams. In the area of control development, first the function of the control code is developed. Then the real-time-capable response behavior of the planned production plant is considered and further optimized.

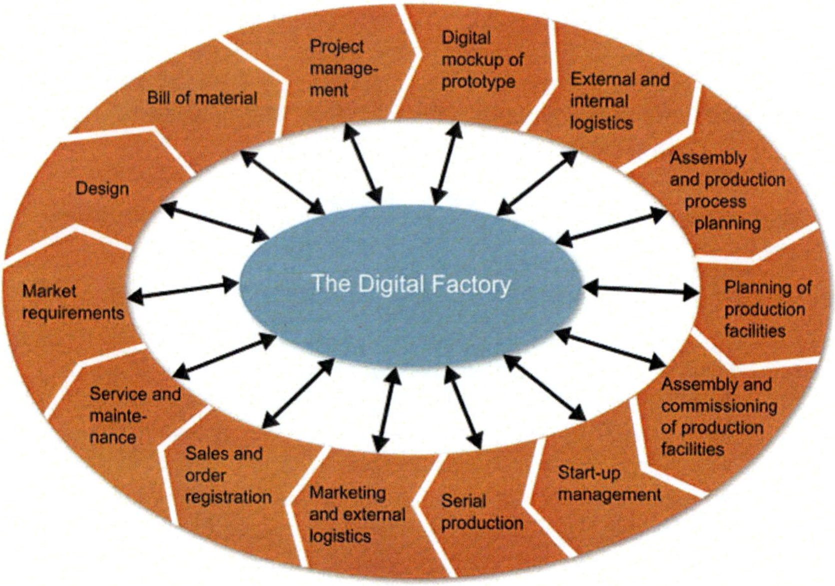

Fig. 15.14 *Digital Factory* operations as a part of the *Digital Factory* process (see [2])

15.4.6 Basic Simulation Technologies

During the last ten years, *Digital Factory operations* offer a wide range of digital solutions to cope with the increasing product diversity of companies as part of the overall *Digital Factory* process solution. As shown in Fig. 15.14, the operations mainly deal with the subordinate processes of production process planning, assembly and commissioning of plants up to series production, so that, according to [2], different methods and tools of the *Digital Factory* can be leveraged. Obviously, the *Digital Factory operations* have numerous contact points with the other Virtual Product Creation technologies as described and explained in this book. This is necessary since the *Digital Factory* simulations are dependent on the existence of and linkages between digital models and information databases of products, (manufacturing & material) process and logistics as well as all involved resources (PPR concept).

15.4.7 Virtual Commissioning and Robotic Simulation

Before a product can get started in series production, it is necessary to digitally (instead of only physically) anticipate, check, prove-out and analyze all manufacturing process in the individual plants in order to avoid all possible problems that usually can occur during production operation. Since such thorough check for manufacturability (or manufacturing feasibility) is applied to test the start-up of plants and their equipment as well as the execution of real manufacturing processes, it is also referred to as *commissioning*.

With the large number of product variants, as it is usual in many industry branches, many companies face severe economic and time line challenges: exceeding cost targets, delayed time line to launch the product and quality issues during ramp-up of production. In order to overcome these challenges, *virtual commissioning* has established itself as an important sub-process in *Digital Factory* operation over the past few years. With the help of *virtual commissioning*, a significant higher number of digital tests can be mastered at the same time, which results in shortening of factory start-up time. As shown in Fig. 15.15, *virtual commissioning* is located before the start of physical plant realization.

To carry out *virtual commissioning* analysis, it is necessary to provide the corresponding data available in a mechatronic library. Figure 15.16 shows an example of how such a library can be set up. In order to generate the digital plant model, mechanical, electrical and fluidic designs as well as robot, PLC and NC programs must be created in advance. Since plant designers as well as internal and external plant engineers deal with these tasks, the data of the individual participants are usually combined. To build up this library, referred guidelines have to be taken into consideration.

In the following step, the created plant must be linked to other plants as well as processes and resources. According to [2], the interdisciplinary cooperation of different trades as well as consistent data management can enable a holistic view of the entire life cycle of a plant. In addition to data management, change management also poses a major challenge since changes made during the operation of a plant should also be fed back into the digital model.

As illustrated in Fig. 15.17, planned changes to the digital model can be tested in advance prior to commissioning; nevertheless, they can also occur during plant operation. There exist several ways in which a *virtual commissioning* can be carried out. Figure 15.18 displays these possibilities.

In a system simulation, the *virtual commissioning* takes place with the help of a computer model. In the simulation environment, a complete virtual test can be

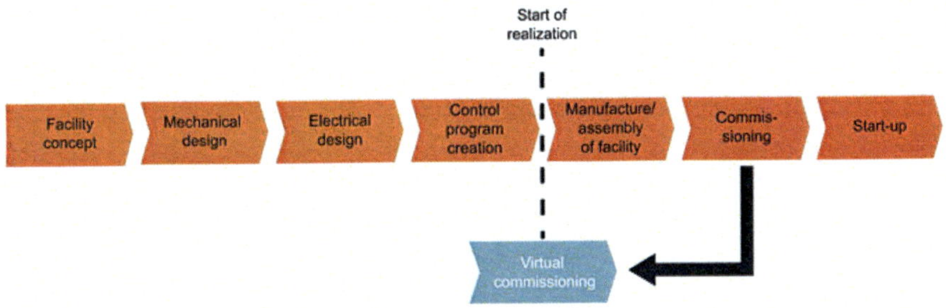

Fig. 15.15 Frontloading of commissioning into the virtual engineering process (see [2])

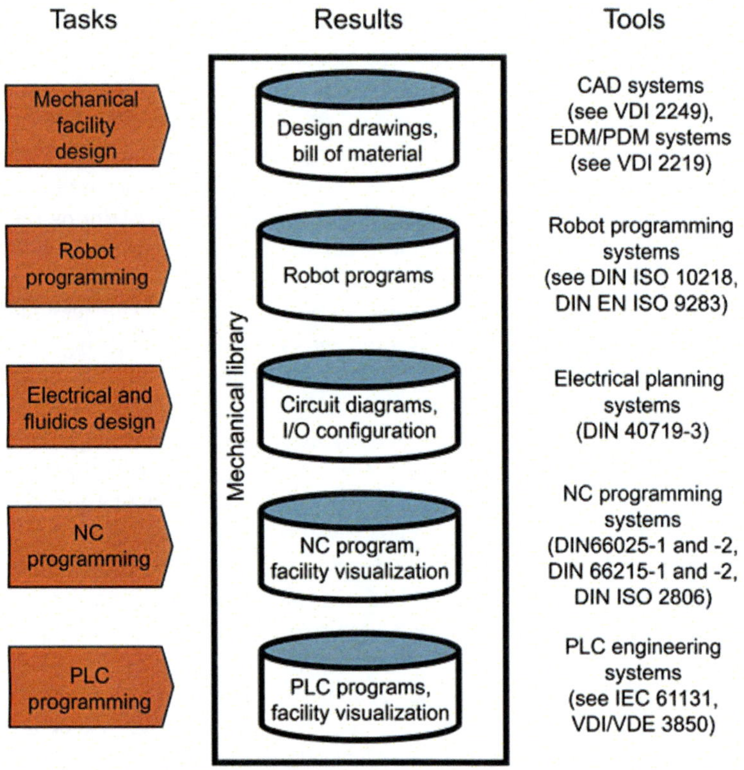

Fig. 15.16 Interdisciplinary factory planning tasks and tools [2]

carried out based on the computer model, so that analysis results can be obtained quickly with respect e.g. to function, timing and clash [2].

Another possibility would be a *Software-in-the-Loop simulation* (SiL), in which control and regulation algorithms under development are linked to the real plant and tested in real time. The advantage of such a simulation is that coding can be checked independently of control hardware [2].

As opposed to a SiL simulation, the created algorithms are imported into the control hardware and tested on the virtual plant model during the execution of a *Hardware-in-the-Loop* (HiL) simulation. This way, statements about the functionality and safety of the planned plant can thereupon be made [2]. A HiL simulation from the automotive industry for the plant area "body-in-white construction" is deployed meanwhile as a standard case. A very high degree of mechanization prevails in the body-in-white construction, since many production steps can be processed by robots and other tools. Typically, the control hardware is available for *virtual commissioning*, so that only the digital plant model has to be procured.

For a simplified representation, an exemplary plant model consists only of a single work cell. This work cell contains a robot that performs a welding task. In step one,

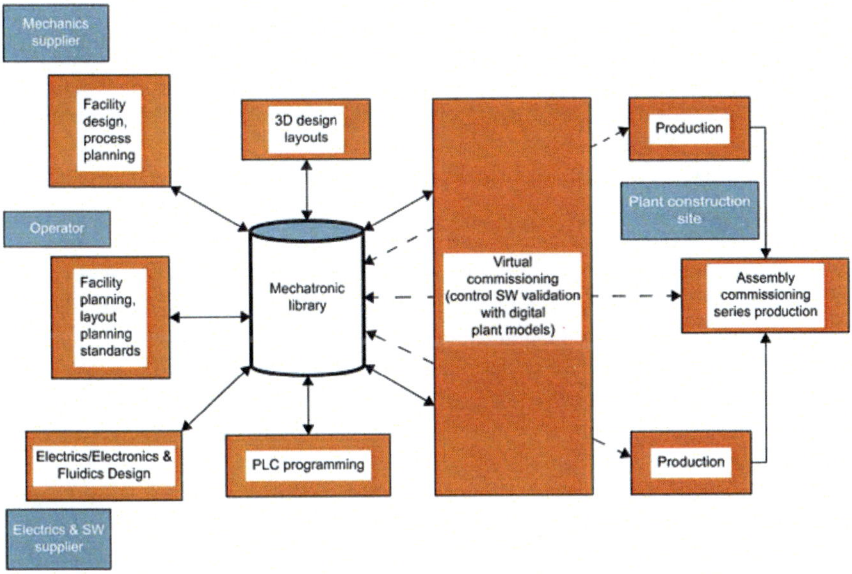

Fig. 15.17 Shared planning and development model (see [2])

		Production facility	
		Virtual	Real
Controller	Virtual	System simulation	Software in the loop (SIL)
	Real	Hardware in the loop (HiL)	Real operation

Fig. 15.18 Types of virtual commissioning (see [2])

the robot picks a sheet from the first feed and places it into the fixture. In the next step, it picks up a second sheet and sets it down next to the first sheet. The fixture serves as the robot's workbench so in the meantime the robot changes his gripper from gripping tool to welding tool. The fixture also has clamping elements to hold the sheets in their defined places during the welding process. After welding, the gripping tool is selected again and the welded sheets are transferred to the next work cell.

In order to reproduce this manufacturing process in the *virtual commissioning*, the specification of the robot and the kinematic 3D model are first required. With this information, the robot's processes can be calculated and visualized. After that, another essential element, the layout of the plant (work cell) and the building, is required in order to detect possible collisions during the robot movement. In a further step, the 3D data of the objects to be welded (sheets) and the fixture are requested, so that the gripping, depositing and welding processes can be analyzed in detail. As a final step, the control model is integrated into the simulation, whereupon the plant model for a HiL simulation contains the required data.

This simplified example already shows the complexity of a larger plant model. In actual practice, the body-in-white construction is made up of many plants, whereby these consist of several work cells. Each plant and its work cells must be secured before series production. In the case of real commissioning, short-term changes, e.g. geometric optimization of sheet metal, are associated with a high expenditure of time, as the real work cells must be changed and secured on site. In a virtual environment, however, the modified 3D data of the plates are already available, so that only the corresponding 3D models of the tools as well as the layout and the control model of the plant have to be optimized. This shows how promising *virtual commissioning* capabilities are.

This will also be evident in the following sections concerning material flow simulation and ergonomics investigation. In order to explain those digital technologies consistently a work cell example is taken from the collaborative research project VIB-SHP (compare [5, 6]) that concentrated between 2015 and 2017 with the future virtual engineering commissioning capabilities; the use case was take from automotive industry and is illustrated in Fig. 15.19.

15.4.8 Material Flow Simulation

By means of the material flow simulation, the delivery of the components to be manufactured can be mapped, simulated and secured along the production process. The material flow simulation is often based on flow charts in which the manufacturing and joining sequence is stored. In the initial rough planning, this sequence is transferred from a textual or graphical representation to a 2D or 3D representation, which is known as the layout plan. The layout plan can be used to display both the production and auxiliary goods to be considered and the associated assembly and manufacturing stations. Different station concepts, production processes as well as their material flow can now be simulated and optimized based on an allocation of time-based behavior models to manufacturing stations and on the material flow of the production goods. Material flows between different work stations as well as within a station can be modeled and they can also conveniently be combined with the method of *virtual commissioning* by means of the simulation of virtual signals and the positions of the production goods resulting from the behavioral models, see Fig. 15.20. In such a process, a real controller is connected to the material flow simulation and controls it so that errors in the programming of the plant control system can be detected at an early stage and can, at the same time, be optimized when compared to a plant that does not yet exist in reality (see also [6]).

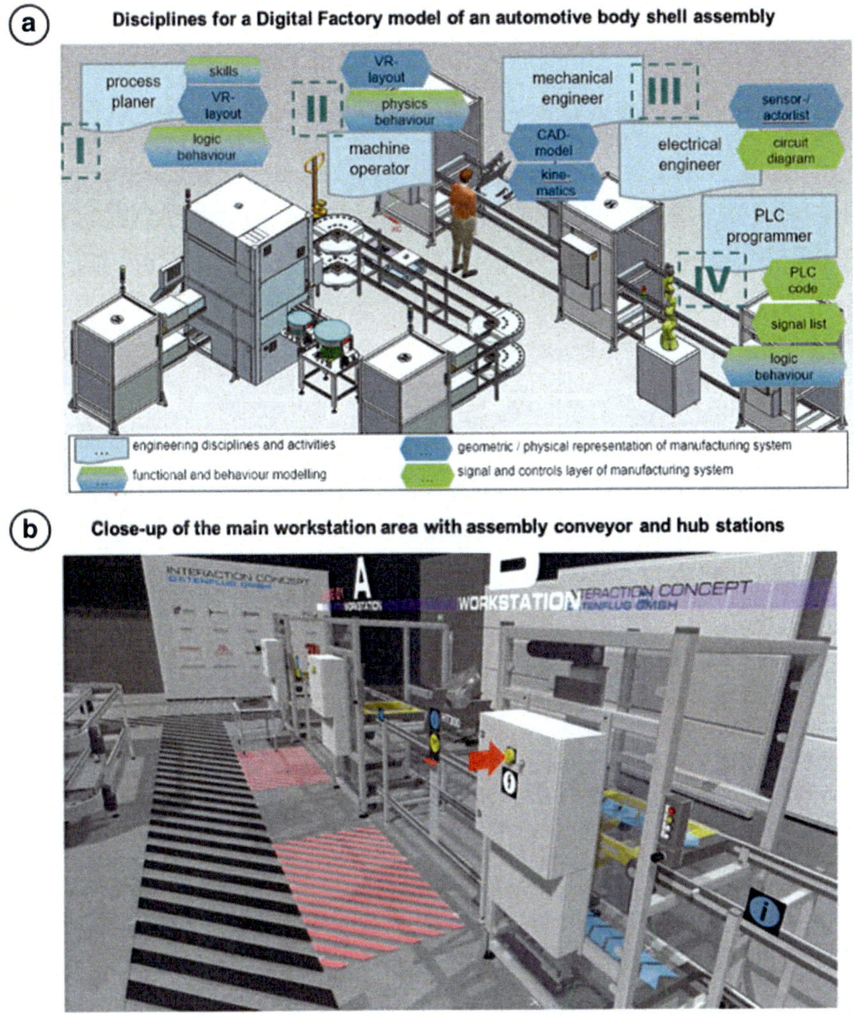

Fig. 15.19 *Digital Factory* work cell model [5, 6]

15.4.9 Ergonomics Validation

In addition to the material flow, as described above, each individual assembly and manufacturing station of the production process must be secured before it is started up. Especially for manual assembly stations, this represents a high challenge for the installer concerning the requirements for operational safety and for the long-term execution of the assembly work with no health risks.

During assembly validation, the arrangement of each assembly workstation is examined in terms of a number of influencing factors. Such factors can range from

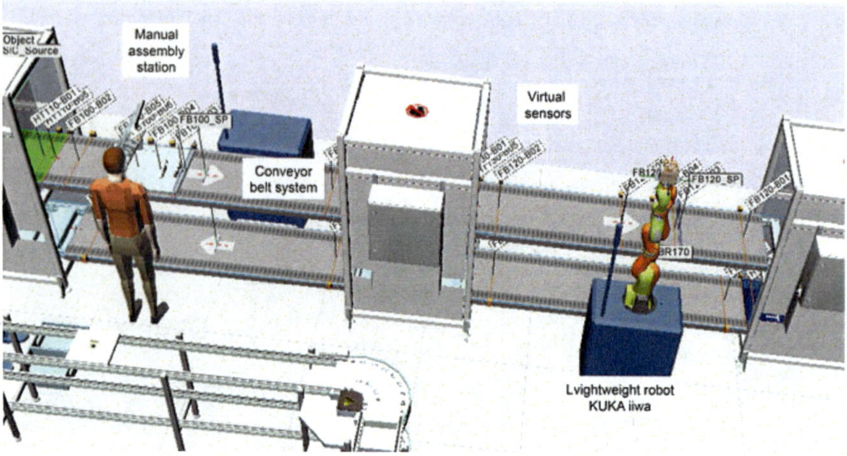

Fig. 15.20 Material flow simulation of manual assembly task

the spatial arrangement of the assembly workstation to the positioning of auxiliary equipment and tools for the manual assembly process. For a precise assessment of the real assembly activity, visual and accessibility analyses can be carried out during the validation process. This allows the temporal and spatial performance of the workers to be optimized and the assembly process to be more intuitively conducted. In this area, it is common practice to use so-called digital human models (DHM), which can be employed to simulate and analyze the installation of the production goods to be assembled in combination with the posture of the assembly worker. These ergonomic load analyses take into account influencing factors such as the mass of a component to be lifted, the posture of the technician and the frequency with which the technician has to lift this component during their shift.

15.5 Advanced Technologies

The increasing individual needs of customers for the product as well as the expectation of a high degree of adaptability of production planning to short-term changes before and during the manufacturing process poses new challenges for the *Digital Factory* methods and tools. In order to cope with such challenges, new perspectives on the upcoming problems are emerging. The mechatronic plant model already presented, which is used for virtual commissioning, is an example of them. The model represents a detailed digital copy of the plant until the phase of the real commissioning arrives. Therefore, there exists the possibility to link the digital model with the real plant in order to represent the real conditions during the running production in real time.

15.5.1 Consistent Data Modeling and Exchange

The prerequisite for a consistent, *Digital Factory* model is the cross-disciplinary model description and simulation of the planned production plant. For this purpose, it is necessary to understand the production plant as a mechatronic system and to build the corresponding factory data management upon it. Decisive interactions between the disciplines, such as in drive and sensor technology, must be consistently mapped across the boundaries of model data of the individual disciplines. In the above-mentioned example, the spatial positioning of a sensor within the layout of the production plant can have an influence on its function as well as electronic influencing variables and vice versa. Especially in the field of model data exchange and change management, these cross-disciplinary interfaces are of great importance.

15.5.2 Virtual Reality Used in the Context of Digital Factory

The use of virtual reality or augmented reality tools in the *Digital Factory* is still relatively low. Advantages of these technologies are particularly evident in the visualization of an entire production plant. In Design Reviews, the planned plant can be inspected spatially and on a scale of 1:1. For the observer such 3D Virtual Reality (VR) technology enables a realistic perception of distance, in which, for instance, the accessibility of plants and work stations can be checked, see Fig. 15.21.

Figure 15.22 shows the difference between a traditional expert observation of a 3D *Digital Factory* model which has been set-up for replay on a 2D screen and a

Fig. 15.21 Layout of an assembly station in virtual reality (VR) [5]

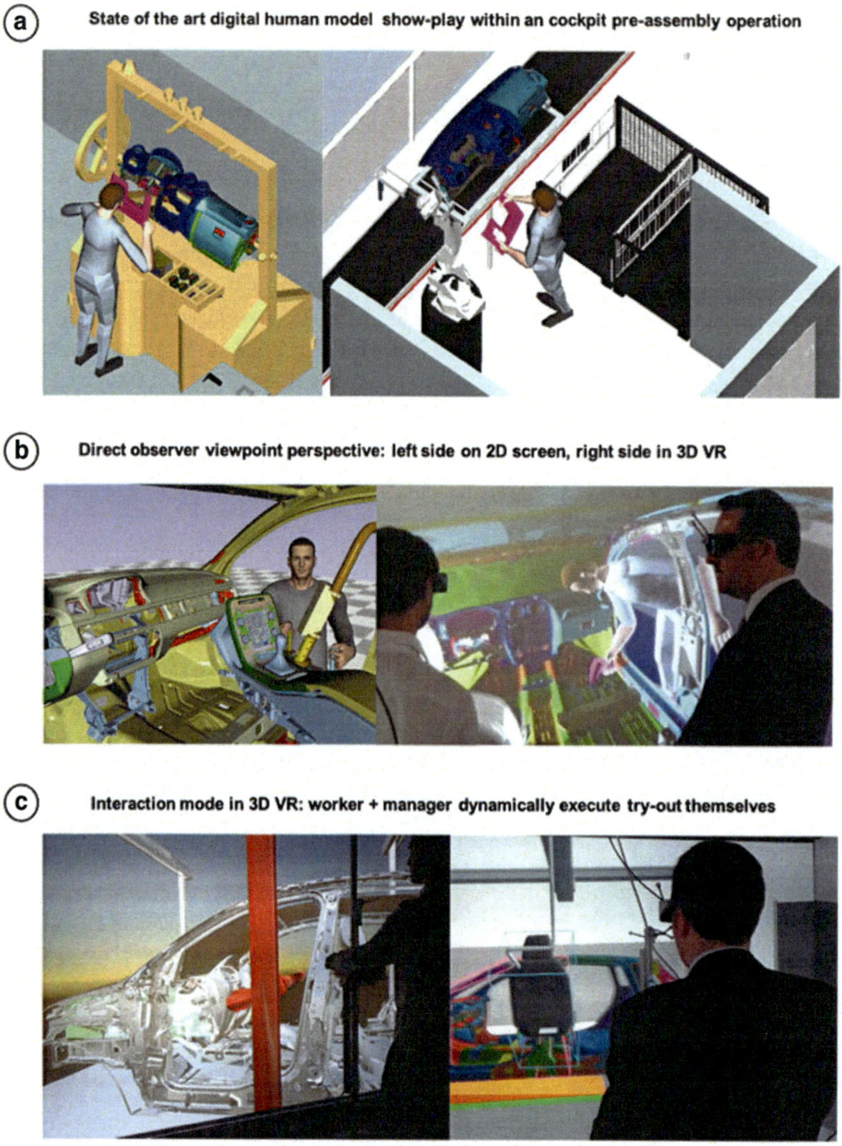

Fig. 15.22 Different observer and interaction possibilities for assembly tasks; comparison between 2D screen mode versus 3D Virtual Reality mode

3D *Digital Factory* model which can be either passively reviewed in 3D VR or even interactively used in 3D VR.

Figure 15.22a shows a perspective from a 3D *Digital Factory* model which has been set-up with the Tecnomatix process simulate software from Siemens. It shows the interaction of a digital human model (DHM) from an appropriate upper camera position for two different cell and process proposals of an automotive cockpit pre-assembly.

Figure 15.22b outlines the difference between two alternative analysis approaches of an automotive assembly line example. On the left side the dynamic digital human operating activities to assemble a console at the center are of a body shell compartment is shown in state-of-the-art 2D screen view. The 3D model has been created with the software IMMA, the Swedish Digital Human Modelling software of Fraunhofer Chalmers in Gothenburg.

On the right side, a 3D *Digital Factory* model created with the Tecnomatix tool suite of a similar task but enabled to be viewed and observed within a 3D Virtual Reality Cave. The 3D Factory model is enhanced by transporting the graphic stream in real time onto the VR screen with the help of the software TECHVIZ. The advantage is that the assembly experts together with Management easily can follow all analysis details in a 1:1 immersive view, which boosts the engineering understanding.

Figure 15.22c shows the innovative way of *Digital Factory* engineering of the chair of Industrial Information Technology of TU Berlin. With the help of the *Smart Hybrid Prototyping* technology, it is possible that a worker or a manager can try-out the interaction of an assembly resource like a semi-automated assembly fixture with the product to be assembled into another product (the example shows the installation of a cockpit on the left side and of a seat at the right side). The term *Smart Hybrid Prototyping* designates in this case the direct interaction of a 3D (Virtual) scene with a real physical device supported by real time contact collision and logic process control.

In contrast to virtual reality tools, augmented reality applications are nowadays mainly used for the visualization of operating data or the remote maintenance of a plant. This allows an untrained service employee to be instructed remotely by an expert, for example. Based on a consistent actual model (CAD model data) of a production plant, corresponding sections of the *Digital Factory* must be visually prepared and presented to the management, e.g. within the scope of a design review.

15.5.3 Human–Robot-Collaboration

The purpose of human–machine collaboration is to combine the advantages of an automated assembly or manufacturing station with the advantages of a manual workstation. On the one hand, the flexibility of an automated workstation should be increased and on the other hand, the repeatability of a manual station should be optimized.

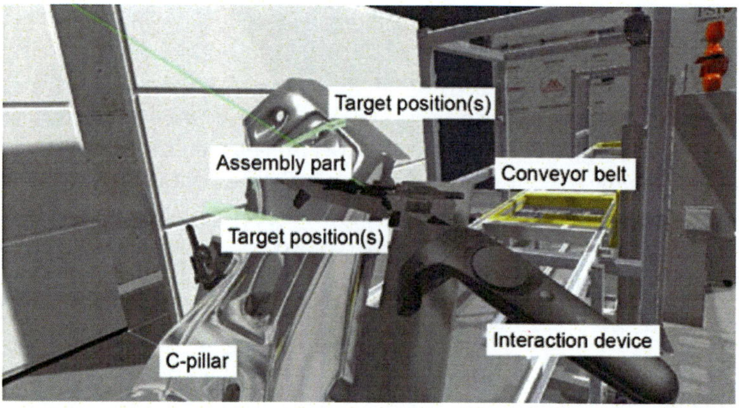

Fig. 15.23 Assembly validation in virtual reality (VR) [6]

The symbiosis of these two approaches can, for example, reduce the error rate in production and increase the number of variants of the production goods that are manufactured at the same station. With the combination of worker and automation technology and robotics, however, increased risk potentials also affect the operator, see Fig. 15.23.

The hybrid workstation must therefore be equipped with particularly high-performance safety mechanisms to allow workers and robots to collaborate effectively and safely. Digital security tools in the field of human–robot collaboration are often used with digital human models, but are not yet widely available in the *Digital Factory* or only as specialized simulation software.

References

1. Bracht U, Geckler D, Wenzel S (2018) Digitale Fabrik—Methoden und Praxisbeispiele. Springer, Berlin
2. VDI (2008) VDI Richtlinie 4499—Digitale Fabrik; VDI guideline 4499—Digital factory
3. VDI (2011) VDI richtlinie 5200—Fabrikplanung—Planungsvorgehen; VDI Guideline 5200—Factory Planning—planning procedure
4. Kerber S (2016) Prozessgestaltung zum Einsatz digitaler Fabrikgesamtmodelle; Process set-up for the application of overall factory models. Springer Fachmedien Wiesbaden, Wiesbaden
5. Stark R, Kind S, Neumeyer S (2017) Innovations in digital modelling for the next generation manufacturing system design. CIRP Annal Manuf Technol 66
6. Stark R, Müller P, Grosser H (2018) ViB-SHP—Virtuelle Inbetriebnahme für Industrie 4.0 zukunftssicher beherrschen. (Mastering virtual commissioning successfully). Fraunhofer Verlag. ISBN 978-3-8396-1416-7

Chapter 16
Major Technology 10: Artificial Intelligence (AI) in Virtual Product Creation

Executive Summary

This chapter deals with the following topics:

- Basics and advanced techniques of Artificial Intelligence in Virtual Product Creation (VPC)
- Providing insight into how engineers benefit from using Artificial Intelligence (AI) technologies in VPC
- Describing functioning, benefits, and limitations of AI technologies in VPC practice.

Quick Reader Orientation and Motivation

The intention of this chapter is:

- to give an overview of AI technology in Virtual Product Creation as driver and enablers for Digital Transformation in engineering
- to present AI technology as part of Virtual Product Creation from a practitioner's point of view to analyze the need and usefulness for day-to-day industrial work practice
- to give instructions on how to use AI technology
- to explain models, frameworks, and digital representations that help to grasp the internal working modes of AI technology in Virtual Product Creation.

Artificial Intelligence (AI) is not a new concept or technology. It first appeared in the 1950s, when several scientists came together with the dream to build machines as intelligent as humans. Afterwards, this field has experienced several hype cycles, including the so-called AI Winters, in which many research organization and companies failed to deliver their extravagant promises [1]. In the 80ties and 90ties of last century a wide variety of rule- and knowledge-based AI systems have already been introduced in industrial engineering work and in technical system operations (e.g. as part of damage analysis tools or engineering assistant systems in design synthesis). In the 2010s, the term of AI rose again, especially the sub-field of *machine learning*, due

R. Stark, *Virtual Product Creation in Industry*,
https://doi.org/10.1007/978-3-662-64301-3_16

to the development of next generation of computing power (e.g. Graphics Processing Unit (GPU), Clouding Computing, etc.), the increased amount and variety of data and the advances in algorithms, especially in *Deep Learning* (e.g. Artificial Neural Networks). Before delving into the definition of Artificial Intelligence, the definition of intelligence will be hereinafter introduced.

16.1 What is Intelligence? What is Artificial Intelligence?

Intelligence can be defined in many ways and that is why it may be controversial to try to find a unique comprehensive definition of the term [2]. From the perspective of psychologists, intelligence can be defined as the ability to solve problems, to create products that bring values within cultural settings [3]. From the perspective of AI researchers, it can be defined as the ability to process information properly in a certain environment. In order to define the criteria for an appropriate definition of intelligence, it is required that information is processed by corresponding experts [4].

Accordingly, AI owns a significant variety of subfields, ranging from general (learning) to specific tasks. Such as playing GO, writing lyrics, face detection, self-driving cars, diagnosing diseases, etc. Figure 16.1 [1] illustrates eight definitions

Thinking Humanly "The exciting new effort to make computers think ... *machines with minds*, in the full and literal sense." (Haugeland, 1985) "[The automation of] activities that we associate with human thinking, activities such as decision-making, problem solving, learning ..." (Bellman, 1978)	**Thinking Rationally** "The study of mental faculties through the use of computational models." (Charniak and McDermott, 1985) "The study of the computations that make it possible to perceive, reason, and act." (Winston, 1992)
Acting Humanly "The art of creating machines that perform functions that require intelligence when performed by people." (Kurzweil, 1990) "The study of how to make computers do things at which, at the moment, people are better." (Rich and Knight, 1991)	**Acting Rationally** "Computational Intelligence is the study of the design of intelligent agents." (Poole et al., 1998) "AI ... is concerned with intelligent behavior in artifacts." (Nilsson, 1998)

Fig. 16.1 Artificial intelligence explanation, organized into four categories [1]

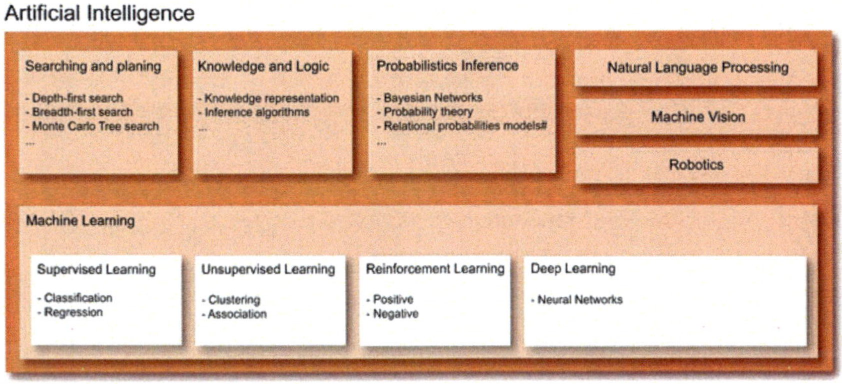

Fig. 16.2 Categories of AI based on [1]

of AI, divided into two dimensions. The above definitions are about *thinking* and *reasoning*, whereas the ones below are about *behavior*. Definitions in the left column measure success of AI according to how they are similar to human behavior, whereas the ones on the right column concern about the ideal performance of AI systems, called *rationality* [1].

Research about AI has the initial objective to build machines that could help to improve our understanding of intelligence. The technologies of AI can be broadly divided into the following types [5, 6]:

- Knowledge-based systems: explicit modeling with words and symbols
- Computational intelligence: implicit modeling with numerical techniques.

With the renewed rise of AI in the 2010s, terms such as are Machine Learning, Deep Learning and Neural Networks also increased in popularity. However, AI is a much broader concept and consists of many more subfields than these ones. Figure 16.2 illustrates different categories of AI.

16.2 Knowledge-Based Systems and Their Application in Industry

Knowledge-based systems are designed to answer complex questions within specific domains. They include techniques such as rule-based, model-based and case-based reasoning. They were among the first forms of AI and remain in a major position until now. In the simplest case, knowledge-based systems contain three modules: knowledge base, inference engine and user interface (as in Fig. 16.3). In knowledge base, the declarative description of problems is stored, e.g. some rules, facts and relationships, without the details about *how* or *when* to apply them. These details exist in inference engine. Since knowledge is explicitly described in knowledge base,

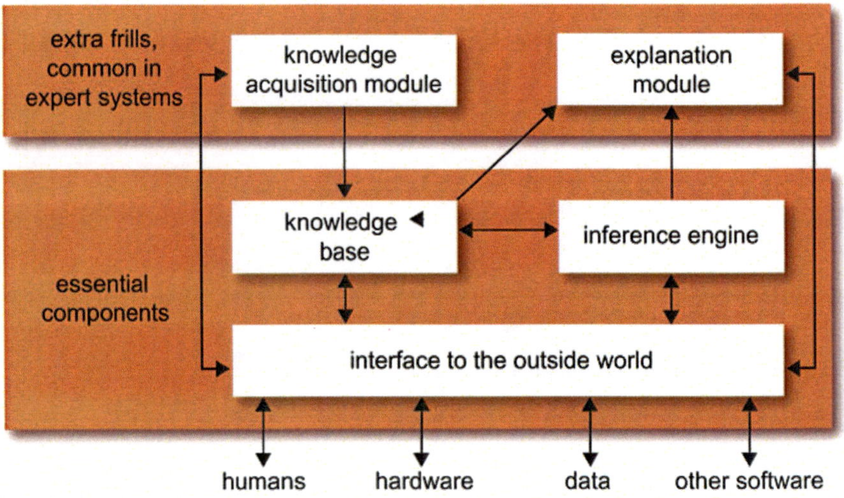

Fig. 16.3 The main components of a knowledge-based system [5]

rather than implicitly embedded in the structure of a program, domain experts can relatively easy update knowledge without any programming skills [5]. An example that displays the problem explicitly is "*If the pressure is high and the release valve is closed, then the release valve is stuck*" [6] On the other hand, the way an inference engine uses knowledge base is similar to the way a conventional program calls a data file [5].

Expert system is one type of knowledge-based system, which is designed to integrate the expertise into a specific domain, such as medical diagnoses and technical diagnoses. It is intended to act as a human consultant that could offer answers according to their domain expertise. Normally, the user interacts with the expert system by describing the problem through dialogues. Then the expert system offers answers, suggestions, or recommendations. Typically, the expert system shall be able to justify the current line of inquiry and explain the reasoning of conclusion, and this is the function of explanation module in Fig. 16.3 [5].

Knowledge based systems are one of the first AI applications created. In 1969, a program called *DENDRAL* [7] was initiated by Ed Feigenbaum, Bruce Buchanan and Joshua Lederberg, with the purpose to deduce molecular structure using information provided by a mass spectrometer. It was also the first expert system, written in programing language LISP,[1] which automated decision making and problem solving processes for chemists. It reached significant success at that time, since it clearly separated rule-based knowledge from reasoning component, which mapped knowledge from a general form to special forms, like cookbook recipes [5]. With lessons learned from DENDRAl and the objective to prove, the methodology of expert systems could

[1] LISP: short for List Processing, a favored programming language for artificial intelligence, which is based on lambda calculus. Works good for computation associated problems.

also be applied to other sort of human expertise like MYCIN, which was developed by Ed Feigenbaum, Bruch Buchanan and Dr. Edward Shortliffe to diagnose blood infection. MYCIN was able to perform as well as some experts, and even better than junior doctors. With the growth of applications of expert systems with the aim of facing real world problems, different representation and reasoning languages were developed, e.g. Prolog[2] [5].

With the success of commercial expert systems, there was an AI boom during 1980s and 1990s in which companies invested millions to billions of dollars in building expert systems, vision systems, software and hardware to implement AI systems. This success raised great optimism for AI, but only for applications for specific narrow domains. Applications for more broad-based representations of human intelligence were still difficult to achieve [8]. Typical rule and knowledge-based AI applications ("first generation AI industrial applications") were introduced to support the following tasks and application fields back then and are still effective today:

- Failure and damage analysis and explanation (reasoning).
- Model design synthesis and concept classification.
- Design knowledge templates to support design automation.
- Checking routines in engineering design and release as well as in technical system maintenance and overhaul.
- Business case calculation, cost estimation and financial assessment.

16.3 Machine Learning—The Most Widely Used AI Subfield in Industry

In this sub-chapter, the author focuses on Machine Learning, since it is currently the most widely used category of AI in industry. However, before approaching this topic further, some basic Machine Learning (ML) terminologies are in the following explained:

- **Attribute** [8]: also known as an independent variable or feature, which describes an observation (e.g. height, color, etc.). Generally, attributes are divided into the following two types:
  - **Categorical**: discrete values, which can be divided into two subtypes: *nominal*, in which there is no ordering between the values, such as last names and colors; *ordinal*, in which there exists an ordering, such as low, medium or high.
  - **Continuous** (quantitative): subset of real numbers, which means there is measureable difference between values.

[2] Prolog: a logic programming language, which is widely used in artificial intelligence and computational linguistics. It works well for rule-based logical queries.

- **Hyperparameter**: a high-level property of an AI model, which decides the learning rate and complexity of the model.
- **Model** [8]: also known as *classifier*, it is a structure or interpretation, which summarizes or partially summarizes a set of data, for description or prediction purpose. The result of most AI algorithms is such kind of models.
- **Knowledge discovery** [8]: the process to identify valid, novel, potential, useful and understandable patterns in data. This concept was first used in "Advances in Knowledge Discovery and Data Mining", 1996, by Fayyad et al. [9].
- **Training data**: the subset of data, which is used to observe, to learn and to train a model.
- **Test data**: the subset of data that is used to test the performance of a model after the model has been trained with training data and validated with validation data.
- **Validation data**: the subset of data apart from the training data, which is used to adjust the hyperparameters of a model.

Learning is the process in which the AI System is able to improve its performance on future tasks after making observations about the world [1]. *Machine Learning* is a subfield of AI that became an extremely popular term in the last decade. It is the science to get computers/programs to learn from experiences rather than programming with specific rules. A detailed definition from Tom Mitchell reads as follows [10]:

A computer program is said to learn from ***experience E*** *with respect to some class of* ***tasks T*** *and* ***performance measure P*** *if its performance at tasks in T, as measured by P, improves with* ***experience E***.

This definition of *Machine Learning* also defines the general guidelines to start any new projects in this field: before starting any *Machine Learning* project, the task (objective) T, the performance measure P and the experience E should be defined.

According to different learning styles, *Machine Learning* (ML) could be grouped into the following four types:

1. **Supervised learning**: the training data fed to the algorithm is labeled, i.e. the samples are marked or augmented with a meaningful tag which represents information. The algorithm learns the relationship (a function), which maps the given input data to the given output data [1]. According to the output types, supervised learning can further be grouped into *regression* and *classification* problems. In *regression* problems, the output is a continuous numerical value, such as 'weight' of a constructive part. For instance, in the absence of an analytical equation, if the radius of a part is 4.9 cm and the weight is *predicted* to be 200 g, then 200 g is the output of a *Machine Learning* system. Figure 16.4a illustrates a regression problem. In *classification* problems, the output is one of the labels/categories of the input dataset. Figure 16.4b illustrates a classification problem.

2. **Unsupervised learning**: the training data fed into the algorithms is *unlabeled*, i.e. no additional information exists for the samples. Algorithms learn the pattern

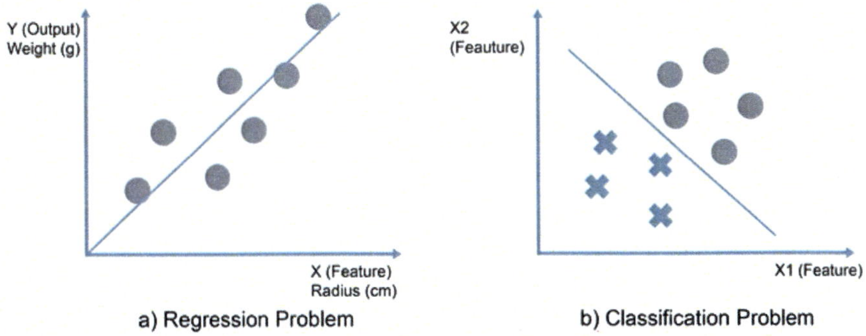

Fig. 16.4 Supervised learning

in the training data without being given an explicit output and model the underlying structure or distribution in the data. According to the *output types*, unsupervised learning can be divided into *clustering* and *association* tasks. In *clustering* tasks, the main focus is to detect potential useful clusters of the input sample [1]. Figure 16.5 is an illustration of a clustering problem. In *association* tasks, the main focus is to discover rules which describe the large amount of the training data.

3. **Semi-supervised learning**: In practice, the differences between supervised and unsupervised learning are not so obvious. In semi-supervised learning, the input is a mixture of labeled and (a lot of) unlabeled data. And even the labeled data may not be 100% correct [1]. Unlabeled data is easier to acquire, compared with labeled data, and the labels may require support from experts or special devices/software.

   The most common application about semi-supervised learning is the photo hosting services, such as Google Photos and Apple iOS Photo Stream: when photos are uploaded to the service, it automatically recognizes that the same

Fig. 16.5 Unsupervised learning—clustering problem

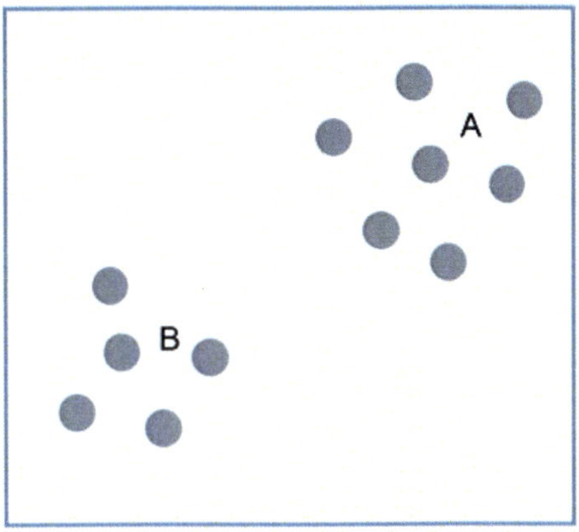

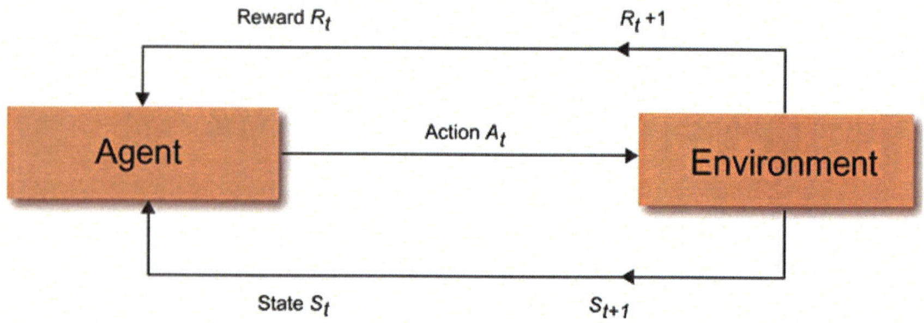

Fig. 16.6 The agent-environment interaction in reinforcement learning based on [12]

person A appears in photo #1, #3, #5 and person B in photo #2, #4, #6. This is also an unsupervised learning problem—clustering. If the system is informed about who these persons are, just by means of being labeled in one of the photos, then the system is able to name everyone in every photo. The same applies if pictures are taken from cracks of mechanical structures and certain types of them are categorized towards certain types of failures such as coating hairline crack, full surface crack or substantial (volumetric) fracture.

4. **Reinforcement learning**: the learning system makes *observations* in an *environment*, takes *actions* and in return, receives *rewards*. The learning system must learn the best strategy by itself by maximizing the rewards, this is called a *policy*. A policy describes what action the agent should choose in a given situation [11]. When there is no sufficient training data or the only way to learn about the environment is to interact with it (i.e. the ideal state is not clear), then reinforcement learning could play the biggest role. Figure 16.6 is an illustration of the agent-environment interaction of reinforcement learning.

   A simple example is a robot (agent), which applies reinforcement learning to learn to walk in a case when there exist two routes in front, a route A with fire and another route B with water. It firstly observes the environment and constructs its own representation of the environment (state), then it takes an action. If it chooses route A, it will get burned (next state) and will get negative reward. Then, it knows it should take fewer actions that lead to such a result (updating policy). On the other hand, if it chooses route B, it will get positive reward and it knows it should take more actions that lead to the result in the future. The robot will repeat the process until it finds a policy (what to react to under different circumstances), which maximizes the rewards.

Similarly, in manufacturing, the Japanese company *Fanuc* [13] has applied reinforcement learning to improve the efficiency and precision of industrial robots. A robot learns to train itself by picking up objects (*actions*) while capturing video footage of the process. After every success or failure, it records how the object looked like and all the relevant features, which are the *state* of the process. The robot gets a positive reward when it puts the parts into the correct container; otherwise, it gets a negative *reward*. The goal is to come up with a *policy* which tells the robot, which kind of part should be put into which container.

16.3.1 Deep Learning

Deep Learning is a subset of machine learning, which can also be divided into the learning types of unsupervised, supervised, semi-supervised and reinforcement learning. The major difference between standard *machine learning* and *deep learning* is that in standard *machine learning* the training data is described by a set of fixed-length features or attributes, whereas the features or attributes are to be extracted from the *raw input data* in *deep learning*. In other words, *deep learning* can process a large amount of data and at the same time requires less *data preprocessing* time. This is accomplished by utilizing one to many interconnected layers (hidden layers) of calculators, an input layer and an output layer, which form a basic structure of *neural network* (Fig. 16.7). This architecture is inspired by the brain, which is why the calculators are also known as 'neurons'.

The input layer of a *neural network* processes a large amount of raw input data. Then the hidden layer(s) in between learn(s) to increase the details of input features. The output layer is responsible for making a determination about the input data and afterwards, when the neural network is applied to new input data, it will make a prediction based on what it has learned. For instance, in order to recognize if the same person has appeared in the new picture.

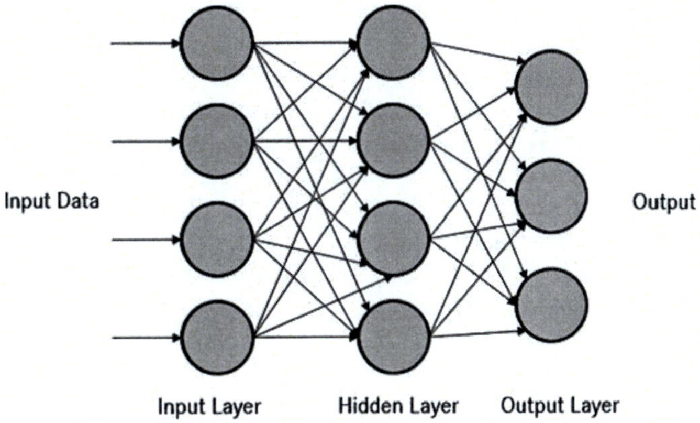

Fig. 16.7 A simple architecture of neural network

16.3.2 Standard Process for Machine Learning Projects

Machine Learning itself is just a core step of the complete methodology to deal with AI projects. Without a deep understanding of the existing problem and available data, it is difficult to achieve the objectives.

With the promotion of Industrie 4.0, comprehensive data is collected in companies within connected machines and systems. According to the *Wissenschaftliche Gesellschaft für Produktionstechnik* (WGP) in 2019 [14], the innovation and competitiveness of manufacturing companies is based to a large extent on the technological knowledge of engineering processes, machines and systems. The key question is how to link the knowledge with the new development of AI systematically and methodically in order to increase the efficiency and added value of processes, machines and plants in addition to the value creation in engineering [14]. There are in general two kinds of approaches to apply AI in the area of engineering: *data-driven* and *process-driven.*

For the *data-driven* approach, companies first collect a large amount of data by applying data analytics to find useful information from it. This relies more on an information technology perspective, which does not require much knowledge with respect to engineering processes. The disadvantage is that normally the collected data is not gathered consequently to existing engineering processes. Therefore, only limited possibilities to get valuable information from the data with respect to the specific engineering process steps are available [14].

The *process-driven* approach is generally aimed at monitoring, controlling or optimizing the process. It highly depends on the type of steps, the machines, the environment, the material and the people involved in the process. Therefore, in order to answer the questions of which data are needed and how to collect such data, an extensive knowledge on the domain is required and such knowledge has a high influence on the results of the AI project. Compared to the *data-driven* approach, the *process-driven* approach systematically extracts more valuable knowledge from engineering processes [14]. In Fig. 16.8, a new value creation potential is shown in the form of 'sensorisation'. The learned model can be applied to optimize the complex process, to evaluate the reliability of results and to improve the visualization the results for better decision making [14].

Cross-Industry Standard Process for Data Mining (CRISP-DM) [16] is the most popular model designed to orient researchers for a standard AI Project. It fits for both, *data-driven* and *process-driven* approaches, which are already mentioned above. The

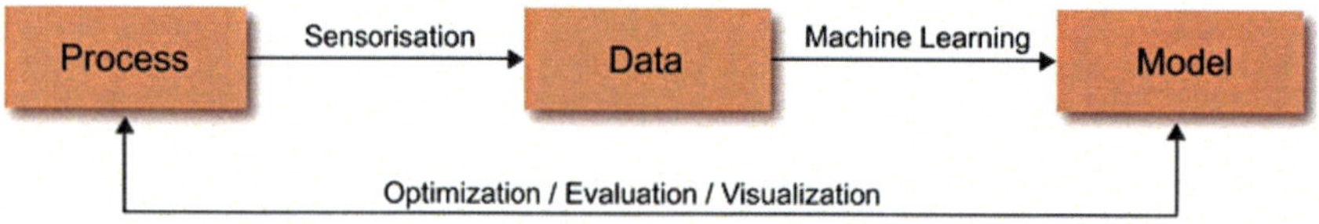

Fig. 16.8 Production technical process-driven approach to apply machine learning [14, 15]

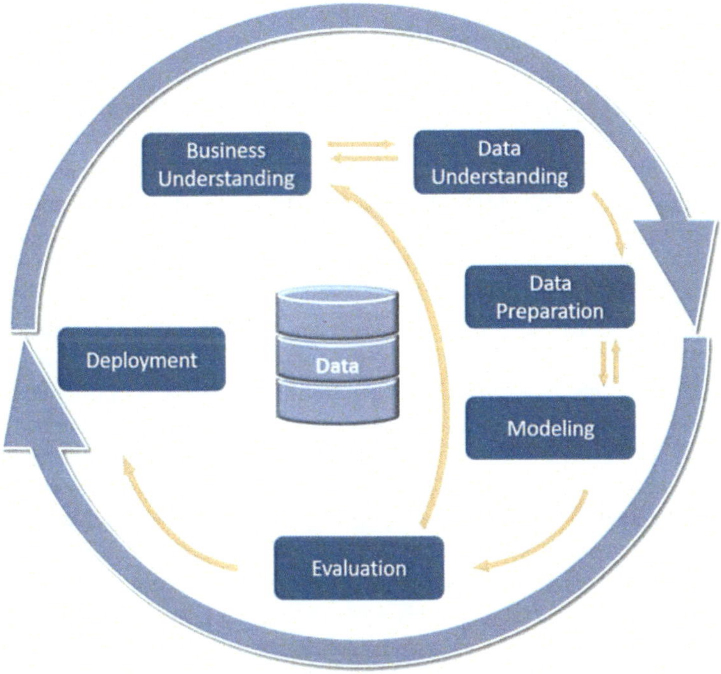

Fig. 16.9 The six phases of the CRISP-DM process [16]

value of CRISP-DM lies in that it involves data science steps that range from business needs to deployment and indicate how interactive the process is. CRISP-DM (see Fig. 16.9) in general describes six phases:

Business understanding: this step begins with an enterprise/industrial or academic need for learning new knowledge or improving current processes. Then the objective is defined, followed by assessing the current situation that needs technical/process knowledge. Afterwards, a plan for finding such knowledge is defined, such as how to collect data, analyze and report data. And then it will be transformed into the objective of the AI project [17]. For example, an automotive company wants to improve the quality and efficiency of constructive design work in CAD system by developing a CAD assistant system. Goals such as "According to the engineers' current design work, what are the next best *features* to be used?" "What kind of *parameters* should be chosen?" are needed.

Data understanding: after a clear description of the problem, the relevant data should be identified and collected. This is followed by exploration and quality assessment of the data [17]. For instance, in CAD systems, there are *log files* which record the used command (feature) combined with parameters and default geometries. According to the logged history, the design behavior can be learned by means of AI algorithms.

Data preparation: the purpose of this step is to select and clean the required data for better quality. This includes: integration, filtering of outliers, filling the missing values, etc. Besides, the selected data may also have different formats [17]. For instance, the CAD system log files need to be converted to a readable format for AI algorithms, e.g. csv. The feature name sequences need to be converted to integers in order to fit AI algorithms.

Modeling: depending on the problem, the appropriate modeling techniques will be selected. Different techniques could be applied in this step, results will be compared and the most appropriate technique will be decided. In the example of CAD assistant system, algorithms like *Random Forest*[3] and *Multilayer Perceptron (MLP)*[4] are applied and compared in the beginning, and *MLP* is chosen due to a better performance.

Evaluation: the results of modeling need to fit the business purpose and should be evaluated in context of business success criteria. In this step, an interaction of data analyst, business analyst or (virtual) engineering experts and decision makers is mandatory [17]. For instance, for evaluation of the CAD assistant system, design engineers, data analysts and managers get involved to assess the degree to which the model meets the business objectives.

Deployment: the results of modeling will be distributed as a usable representation and integrated in an organization process/system. In the example of a CAD assistant system, it is integrated into a CAD system such as *Siemens NX* or Dassault Système CATIA as a plugin and running in parallel with it. The performance will not be mutually affected.

16.4 (Big) Data in Product Lifecycle Management

To achieve better performance in *Product Lifecycle Management (PLM)* [18] with the support of *Big Data* and *Artificial Intelligence*, it is necessary to clarify which type of data sets are involved in which phase of the lifecycle of a given product, machine (both represent technical systems) or service. Generally, the product lifecycle could be divided into: Begin of Life (BOL), Middle of Life (MOL) and the End of Life (EOL) [9]. In BOL, the product concept is generated, designed and physically tested and its production is being prepared. In MOL, products are produced, distributed, used and maintained by customers or engineers. In EOL, products are prepared for re-use and/or recycled by manufactures or disposed by customers [19, 20].

Begin of life, BOL: According to Jun et al. [20], the most essential steps involved in BOL are: market analysis, product design and production preparation. In phase

[3] The algorithm *Random Forest* is based on a combination of decision trees. To classify a data sample, each decision tree provides a classification result for the input data. *Random Forest* then collects the results from each decision tree and choose the most voted one as the prediction result [18].

[4] *Multilayer Perceptron* (MLP) is the simplest neural network, sometimes also referred to a feedforward Artificial Neural Network.

market analysis, the target is to meet customers demand. There exist a variety of data formats, e.g. comments on blogs, videos that customers upload on the Internet, websites in which customers mark their purchasing behaviors. Besides, the information from MOL and EOL, for instance, customers' complaints and sales performance of similar products can also contribute to provide the goals for product design [19].

In the phase of product design/development and manufacturing engineering, the data involved can be the descriptions of needs, requirements, description of specific product functions, detailed design specifications—e.g. drawings or product configurations, the accurate programming codes for the automation of manufacturing equipment, and all kinds of technical parameters. Furthermore, the maintenance and failure information from MOL, like the records of breakdowns and root causes can also contribute to efficient and reliable product design [19] as part of "feedback-to-design".

Mid of life, MoL: In the middle of the product lifecycle, the product exists in its final form. The main issues and influence factors can come from production and from service [19].

In the production phase, while some data might be stable, other data are dynamic and change along the phase of product manufacturing. The data from product design will be regarded as standards for production processes and operation, and data from monitoring and testing of products are used to check whether all standards are reached and met [19].

In the logistics phase, warehouse management and transportation need efficient decision strategies to solve complex issues. Based on the order information, here considered as input data, the manufactures are able to find optimal arrangements. One of the main tasks along this line is to transfer order information into intelligent arrangement within a global view and supply chain network [19].

In the utility phase, customers operate products based on the information from user manuals or from heuristic knowledge. In this process, product status information are generated and potentially transferred back to manufactures: traditionally, for most of the products in field usage only failure modes are recorded, nowadays, due to internet technologies, the actual (positive) use data become decisive for new business models of manufactures during the utility phase. In addition, the field usage information is monitored and recorded to provide guidance for the product maintenance [19]. In the maintenance phase, by combining maintenance supporting information with product status information generated from utility phase, faults can be predicted and prevented. The adjusted maintenance plan with root causes and solutions is taken into account as output data during this phase [19].

End of Life, EoL: In the end phase of product lifecycle, lots of decisions have to be made regarding EOL product re-use (or partial re-use), recycle or disposal. With the help of data from MOL the following decisions can be supported: maintenance history information, product status information and usage environment information, the degradation status and calculation of remaining value of individual components. The purpose in EOL is to maximize values of products. Depending on the status of the product, suitable options such as recycle, re-use, remanufacturing, and disposal should be decided [19].

16.5 Internet of Things

The majority of current industrial products are mechatronic. With the evolution of micro embedded devices and software within mechatronic products, their intelligent capabilities, such as autonomy, real-time interaction, self-organization, etc. and the capabilities to communicate and network with other products have been improved. This type of product is now defined as 'cyber-physical systems (CPS)' [21].

The term "*Internet of Things (IoT)*" was first suggested by Kevin Ashton [22] in 1999. At that time, he viewed Radio-frequency identification (RFID) as the essential to the internet of things. Literally, IoT means "*...all about physical items talking to each other ...*" [23]. Nowadays, IoT carries a much broader designation since the term IoT is oftentimes also referred as a term to describe daily used gadgets and objects with internet connection such as TVs, smart watches, cellphones, ovens, refrigerators, cars, etc. All of them, however, handle data sets created by sensors in those objects and gadgets of daily live as well as in machines of smart factories.

Making products 'smart' means connecting and sharing data between them. On the other hand, it means capturing the huge amount of data, ingest, process it and then mine it as the business requires. Enabled by IoT, CPS could not only communicate and network with each other, but are also capable to perform a required functionality by integrating the available internet services. These products are called '*Smart Products*' [21].

There are many design challenges faced by the developer and engineers of smart products. Among many issues, such as availability of internet, the IoT is entirely dependent on the development of Wireless Sensor Networks (WSN) and Radio Frequency Identification Devices (RFID). Mukhopadhyay [23] has summarized the many challenges of IoT as follows:

- Availability of Internet everywhere and at no cost
- Security issues
- Low-cost smart sensing system development
- Energy
- Computational ability
- Scalability
- Fault Tolerance
- Power Consumption.

In 2014, a framework of CPS was proposed by Lee et al. [24], which provides a guideline for applying CPS to industrial use cases. This architecture consists of 5 "C"-levels:

- **Connection**: this level consists of properly selecting sensors and data sources, transferring protocols, and seamlessly transferring data to the central server [25].
- **Conversion**: in this level, intelligent algorithms and data mining techniques can be applied to various raw data to extract valuable information, which is also known as features in most Machine Learning projects. Then the calculated information along

with other machine state data is being sent through Ethernet or Wi-Fi Network to a cloud server, in which the information is managed and stored [25].

- **Cyber**: information from every connected machine will be gathered and analyzed in this level. The results of the analytics provides machines with self-comparison ability—performance of the individual machine can be compared and rated among the fleet [24] and with historical information of similar machines to predict the future behavior of this machine.
- **Cognition**: in this level, the acquired knowledge is presented as comparative information as well as individual machine status to experts, in order to support better decision making. Therefore, information visualization techniques such as graphics, tables are necessary to transfer the acquired knowledge completely [24].
- **Configuration**: this level acts as the feedback from cyber space to physical space. It applies the corrective and preventive decisions that have previously been made to the cognition level to the monitored system [24].

Case study: Cyber-physical system-based smart machine

So far, the application of AI based algorithms have become popular in real physical application cases, such as in manufacturing. The following case study of the "sawing material" example explains the approach and the appropriate measures which are necessary to apply the five "C"-level approach of Lee. Manufacturing processes start with sawing raw materials into designed sizes, therefore, speed and quality of sawing affect the whole production. Errors in sawing will propagate to the following steps and further affect the quality of product. Accurate sawing requires slowly cut but since it will affect the productivity of the production, an optimal balance between quality and speed need to be achieved [25].

In the **connection** level, data is collected from sensors and controller signals. Data, such as vibration, acoustic emission, temperature, blade speed, cutting time and blade height, etc., provide working status of each machine and will be processed in the industrial computer connected to each machine [25].

In the **conversion** level, the industrial computer performs feature extraction and data preparation. For instance, frequency domain features such as RMS (Root Mean Square), kurtosis, frequency band energy percentage, etc. are extracted from vibration and acoustic signals. At this stage, however, it is crucial to use manufacturing processing know-how and process knowledge (compare Fig. 7.10.8). Calculated features together with machine state data are sent through Wi-Fi network or Ethernet to the cloud server for storage and management [25].

In the **cyber** level, an adaptive clustering method [26] is performed on the cloud server to segment the historical performance of blades into discrete working regimes based on the difference of features comparing to normal baseline and local noise distribution. The clustering method (see explanations to unsupervised learning and Fig. 16.5) then compares the current features with the baseline and historical working regimes and identifies the appropriate cluster to match with the current working condition. If no appropriate cluster is found, a new cluster is generated [25].

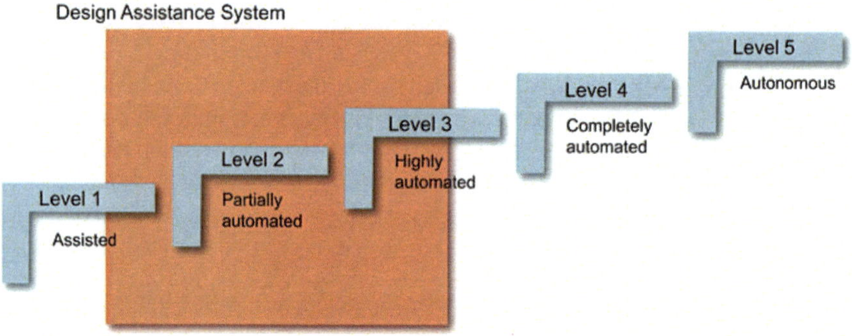

Fig. 16.10 The vision and levels of design assistance systems at Daimler (based on Daimler internal project material)

In the **cognition** and **configuration** level, decisions will be made based on the health information of each connected machine, which is visualized by Web or iOS-based user interface. For instance, for a new blade, a higher cutting speed will be chosen for high productivity without affecting the quality of production. After a certain amount of degradation, a more moderate cutting should be applied to ensure production quality [25]. Please refer to Chap. 20 for more details on "*Internet of Things (IoT)*", especially with regards to "*Industrie 4.0*".

16.6 Example of a Virtual Product Creation AI Application

Industry leaders have recognized the widespread of digitalization. Instead of been changed by the digital wave, many companies have decided to react to changes and be a game changer by implementing new technologies supported by agile working methods. AI is definitely one of the promising technologies which helps companies on the way to digital transformation. And it has a place in the future of Computer-Aided Design (CAD), as one AI example in Virtual Product Creation.

Recently, Daimler AG has developed a design assistant system—NeuroCAD[5] with the objective to support CAD Data construction in Siemens NX CAD System by means of *Artificial Neural Networks*, which are a sub-discipline of Artificial Intelligence. Similar to the five levels of autonomous driving, the vision of NeuroCAD is to enable highly automated design (see Fig. 16.10). There exist three assistance systems components of NeuroCAD: the (design) feature assistant, the structure assistant and the parameter assistant. Meanwhile, NeuroCAD has reached the capability to partially automate the CAD design work, which goes beyond the "high end template based" design automation approach from the first and second decade of this millennium.

[5] NeuroCAD is a separate program which runs in parallel with Siemens NX. The performance will not be influenced mutually.

16.6.1 The Main Function Description

This sub-chapter explains the main functional elements of the design assistance system NeuroCAD, which employs a range of AI elements.

Feature assistant: this first functional element learns the typical command sequences (NX functionalities) and then supports design engineers by selecting the next best commands in NX by suggesting the three most likely commands the user could use next (as shown in Fig. 16.11). Each click will recall the corresponding command in NX. If all suggested commands are not appropriate, the user can still choose commands directly in NX. Feature Assistant in this case, provides only suggestions as part of design assistance instead of automating the design work.

Structure assistant: the second functional element is the traditional feature-based modelling (compare the sub-chapter "Feature based Modeling", part of the Chap. 7 "Computer Aided Design—CAD"). The CAD System (in this case *Siemens NX*) keeps the history of each command (feature) with the used parameters and the default geometries in a structure tree as part of the traditional CSG based modeling paradigm (compare Chap. 7 "Computer Aided Design – CAD").

Each feature can be modified later and all subsequent features of the design will be recalculated. These geometry construction features build high interconnectivity of data. It is difficult to master the complexity if the data is not further structured. Designers thus use '*Feature Group*' function in NX to group the features applied to a specific geometry. Some typical constructive commands will be repeated with variants. Thus, the creation and extension of similar components can be suggested by using the historical logs and AI technology. The *Structure Assistant* can support in such a case: it makes suggestions during the creation of feature groups and recommends the appropriate feature groups based on the position of part in the Part-Navigator.

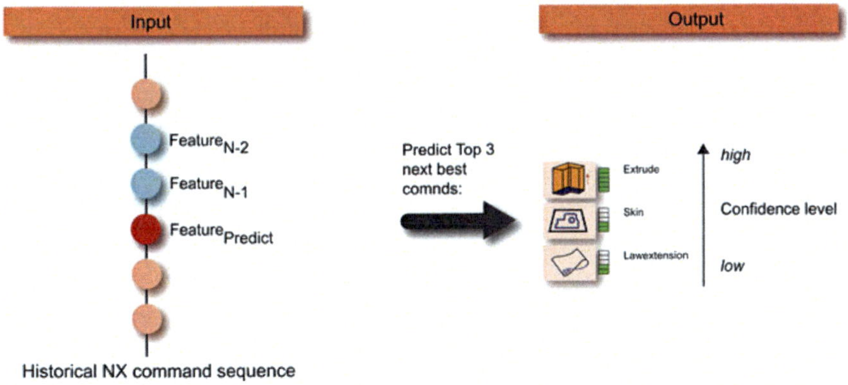

Fig. 16.11 Feature Assistant: by learning from historical NX command sequence, the next best 3 commands are predicted and displayed in confidence level from high to low (on the right side)

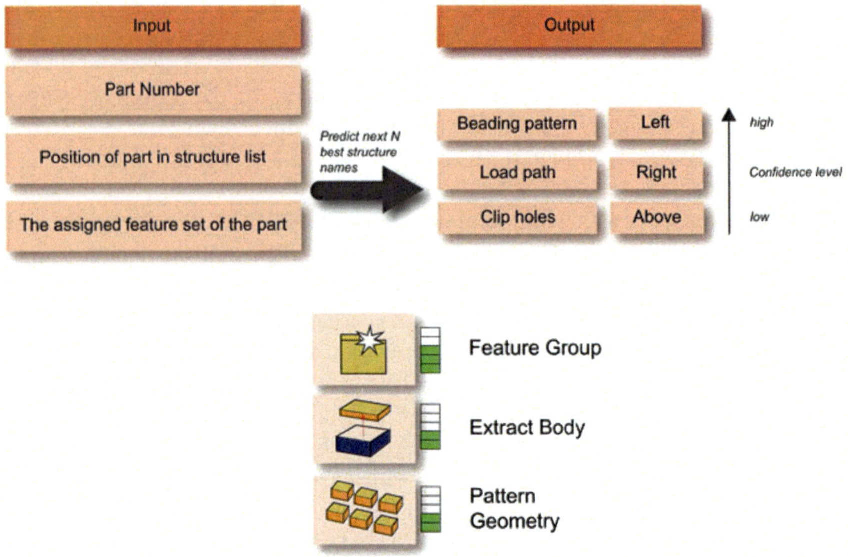

Fig. 16.12 Structure Assistant: by learning from information about part (e.g. part number, position of part in structure list and its corresponding feature set), the next best structure names are predicted and displayed on the right side

The structure assistant uses a dictionary with around 3000 terms from CAD data construction process in Daimler AG. As shown in Fig. 16.12, based on the structure level of part, part number and the assigned feature set of the part, the first prediction list—*Sickenbild (bead layout)*, *Lastpfad (load path)*, *Cliploecher (clip holes)*, etc. is provided. Regardless of the prediction list, the names can also be filtered by entering the first letters of the word. For instance, when 'ves' is entered, then only the names start with 'ves' will be listed.

Parameter assistant: for different features, the third functional element, the parameter assistant, suggests the meaningful initialization values, based on the data from *the start-part information* and the *name of Feature Group*. For instance, for a feature *CYCS* (Absolute Coordinate System), the given input will designate:

(Part Number), (Name of feature group).

The suggested coordinate value will be given in the following form:

(X, Y, Z).

Together with the feature assistant and the structure assistant, the parameter assistant makes the contribution to simplify daily design work and to offer the possibility of semi-automating design work.

16.6.2 Best Practice

NeuroCAD has learned the features from more than 21,000 CAD parts (with around 2.8 million features) and was widely rolled-out in thousands of workstations within Daimler AG. The current version (beginning of 2020) has reached 92% accuracy.[6] There will also be a mobile version in the future.

The development and deployment of *NeuroCAD* has been supported by Agile Software Development Methods [27] which will become the norm in continuous DEVS/OPS (Development & Operation) type of Engineering of the future following three essential paradigms:

- Individuality and interactions over processes and tools
- Working software over comprehensive documentation
- Customer collaboration over contract negotiation responding to change over following a plan.

The *NeuroCAD* team has provided some best practices when implementing AI in industry:

Think big, start small. *NeuroCAD* has the vision to enable high-automated CAD design. Instead of starting with several functions in the beginning, they divide this big vision into small realistic problems and start with the one with high data availability. Implementation time of the first prototype takes only 2.5 months, with one person with 100% capacity and 3 persons with 20% capacity. Building quick prototypes will help earning the confidence from stakeholders in the early phase and thus it will very important to the success of the project.

Involve stakeholders from the beginning onwards. As already introduced in sub-chapter "*Standard process for Machine Learning projects*", a deep understanding of the existing problem and available data is an essential step in an AI project. In the kick-off phase, the *NeuroCAD* team organizes several workshops to communicate with key users and to deeply understand their potential challenges during the implementation phase. This ensures that the final digital product is delivered according to the actual business needs.

In-house development. Many companies tend to hire external consultants or developers and SW-coders to deploy new technologies. This will be difficult or at

[6] It is assessed based on the correctness of the top 3 recommendations from feature assistant.

least challenging for AI projects, since the first challenge they encounter will be the data access problem. Yet, today's companies have oftentimes not found an internal policy and security way to open their data repositories for outside SW-development companies. Besides, the lack of enterprise's own knowledge will also be a barrier during the implementation process. Therefore, the development team of *NeuroCAD* at Daimler all stems from inside the company, with a high degree of programming skills and knowledge of AI. The team was supported by *SCRUM* method and Speed Coach, one local team without bureaucratic organization, and constantly exchanged experiences with local AI experts within the company. This ensures not only the high development speed, but also the fully utilized existing enterprise knowledge. In other circumstances, however, especially in smaller and medium sized companies, where such critical in-house skills are not available or cannot be mobilized easily, outside help from research institutes and SW companies are necessary and useful.

References

1. Norvig P, Russel S (2009) Artificial intelligence: a modern approach, 3rd edn. Prentice Hall, Upper Sadele River, NJ, USA
2. Legg S, Hutter M (2007) A collection of definitions of intelligence. Front Artif Intell Appl 157:17–24
3. Gardner H (1993) Frame of mind: theory of multiple intelligence. Basic Books, New York
4. Nakashima H (1999) AI as complex information processing. Mind Mach 9:57–80. https://doi.org/10.1023/A:1008322730047
5. Hopgood AA (2005) The state of artificial intelligence. In: Zelkowitz MV (ed) Advances in computers, vol 65. Academic Press, London, pp 1–75
6. Hopgood AA (2003) Artificial intelligence: hype or reality. Computer 36:24–28
7. Feigenbaum E, Buchanan BG, Lederberg J (1969) Heuristic DENDRAL: a program for generating explanatory hypotheses in organic chemistry. Mach Intell 4:209–254
8. Kohavi R, Provost F (1998) Glossary of terms. Machine learning—special issue on applications of machine learning and the knowledge discovery process. Mach Learn 30:271–274. https://doi.org/10.1023/A:1017181826899
9. Fayyad U, Piatetsky-Shapiro G, Smyth P Uthurusamy R (eds) (1996) Advances in knowledge discovery and data mining. American Association for Artificial Intelligence, Menlo Park
10. Mitchel TM (1997) Machine learning. MIT Press, McGraw-Hill Science, Cambridge
11. Géron A. (2017) Hands-on machine learning with Scikit-Learn and TensorFlow. O'Relly Media, Sebastopol
12. Sutton RS, Barto AG (2018) Reinforcement learning: an introduction. MIT Press, Cambridge
13. Ciaburro C (2018) Keras reinforcement learning projects: 9 projects exploring popular reinforcement learning techniques to build self-learning agents. Packt Publishing, Birmingham
14. Krüger J, Fleischer J, Franke J, Groche P (2019) WGP Standpunkt: KI in der Produktion. Künstliche Intelligenz erschliessen für Unter-nehmen. WGP Wissenschaftliche Gesellschaft für Produktionstechnik e.V., Hannover
15. Agrawal A, Gans J, Goldfarb A (2018) Prediction machines. Harvard Business Review Press, Boston
16. Chapman P, Clinton J, Kerber R, Khabaza T, Reinartz T, Shearer C, Wirth R (2000) CRISP-DM 1.0, CRISP-DM Consortium
17. Olson DL, Delen D (2008) Advanced data mining techniques. Springer, Berlin, Heidelberg
18. Stark J (2011) Product lifecycle management. In: Product lifecycle management. Decision engineering. Springer, London, pp 1–16. https://doi.org/10.1007/978-0-85729-546-0_1

19. Li J, Tao F, Cheng Y et al (2015) Big Data in product lifecycle management. Int J Adv Manuf Technol 81:667–684. https://doi.org/10.1007/s00170-015-7151-x
20. Jun H, Shin JH, Kim Y, Kiritsis D, Xirouchakis P (2009) A framework for RFID applications in product lifecycle management. Int J Comput Integr Manuf 22(7):595–615. https://doi.org/10.1080/09511920701501753
21. Abramovici M (2015) Smart products. In: The international academy for produ, Laperrière L, Reinhart G (eds) CIRP encyclopedia of production engineering. Springer, Berlin, Heidelberg. https://doi.org/10.1007/978-3-642-35950-7_16785-1
22. Ashton K (2009) That 'Internet of Things' thing, in the real world, things matter more than ideas. RFID J 22:97–114
23. Mukhopadhyay SC, Suryadevara NK (2014) Internet of things: challenges and opportunities. In: Mukhopadhyay S (ed) Internet of things. Smart sensors, measurement and instrumentation, vol 9. Springer, Cham. https://doi.org/10.1007/978-3-319-04223-7_1
24. Lee J, Bagheri B, Kao HA (2014) A cyber-physical systems architecture for industry 4.0-based manufacturing systems. Manuf Lett 3:18–23
25. Lee J, Ardakani HD, Yang S, Bagheri B (2015) Industrial big data analytics and cyber-physical systems for future maintenance & service innovation. Procedia CIRP 38:3–7. https://doi.org/10.1016/j.procir.2015.08.026
26. Yang S, Bagheri B, Kao HA, Lee J (2015) A unified framework and platform for designing of cloud-based machine health monitoring and manufacturing systems. J Manuf Sci Eng 4:137. https://doi.org/10.1115/1.4030669
27. Beck K, Grenning J, Martin RC, Beedle M, Highsmith J, Mellor S, Bennekum AV, Hunt A, Schwaber K, Cockburn A, Jeffries R, Sutherland J, Cunningham W, Kern J, Thomas D, Fowler M, Marick B (2001) Manifesto for agile software development. Available via Agile Alliance. https://agilemanifesto.org. Accessed 6 Nov 2020

Chapter 17
The Hidden Demands of the Engineering Community

Executive Summary

This chapter deals with the implicit expectations and requirements with respect to Virtual Product Creation solutions by the engineering community, their users and the underlying core foundational principles of engineering. In many realization projects of new PLM functionalities, modified or newly introduced virtual product creation working solutions and associated responsibilities of individual roles and job functions those hidden demands are oftentimes not known, understood and/or not taken seriously into account. This oftentimes leads to unwanted delays, limited progress and a high number of substantial problems in operational digital engineering business.

Quick Reader Orientation and Motivation

The intention of this chapter is:

- to describe and illustrate principles and expectations of the engineering community with respect to virtual product creation solutions
- to explain when and why such explicit and implicit expectations can create conflict potentials with typical PLM and virtual product creation (VPC) development, deployment and implementation approaches
- to provide advice and assistance how to avoid such conflicts
- to explain strategic and practical advice for modifying engineering principles and behaviors which are in better alignment to Virtual Product Creation innovations.

A rich mix of creative, routine, responsible, collaborative and self-reflecting elements, both with analog and digital engagement styles, characterizes engineering work patterns. As a reward for engineers, a thing, a sketch, a design, a calculation, a model, a mock-up, a test, a prototype etc., get accomplished with the goal to contribute to the next or modified product, production line, technical system for costumers or factories. Engineering is thrilled and committed and wants to get measured by this ultimate rewarding scale.

R. Stark, *Virtual Product Creation in Industry*,
https://doi.org/10.1007/978-3-662-64301-3_17

With this engineering DNA in mind, many product and manufacturing engineers raise the expectations as creators of the technical universe to be best suited to also define, develop and deploy the corresponding engineering tool-sets, methods and test capabilities. This ambition, however, is nowadays difficult to achieve due to specific needs in digital competence that are necessary to build digital engineering tool-sets and applications. Consequently, irritations, implicit expectations and frustrations arise if other experts are assigned and contracted to define, develop and deliver new digital working solutions for engineering communities. In many cases, it is a mix of pride, personal choices and indeed the best engineering knowledge that create mixed feelings of the engineering community about the appropriateness and best fit of digital solutions sets as part of Virtual Product Creation.

In principle, it is beneficial to know which expectations of the Engineering Community exist and what they are driven by. Only then, it will be possible to counteract it appropriately in order to provide IT based virtual product creation methods and tools and to suggest the right data and model mix for it.

The typical innovation situation

Product and Manufacturing Engineering has changed significantly from the mid of the 90ties of last century onwards: rather than typical line engineering progression with routine work contributions to product, machine or product system developments, more and more a stage gate-oriented project engineering approach is meanwhile the norm in day-to-day engineering progression. Consequently, a typical engineer has barely the possibility to use "special time" in order to specify, develop or even experiment with new and better engineering tool sets. In addition, in most of the companies a rather awkward and lengthy process of setting up innovation projects has been established. Generic product, tools or process owners undergo a rigorous review process, typically starting from June up to November of the previous year. The goal is to receive an official approval in January or February of the following year to conduct an 8–10 months prove of concept (POC) type of engineering innovation project with a tangible task to deliver a solid return of investment evidence within one year a maximum of 2 years. Special teams are assigned inside the companies, which are usually not in operational engineering work responsibility. They are supposed to interact closely with outside research institutes, vendors or other IT service providers and with internal key users from the engineering operational teams in order to investigate, demonstrate, explain and conclude with respect to potential future digital engineering solutions.

The operational engineering teams remain in their daily hassle and cannot devote any substance time into future engineering workplace solutions. Occasionally, they are invited to awareness sessions of innovation project outcomes or to hype demonstrations like "agile accelerators", "co-working boot camps", "Scrum for all, let's do it" and similar type of generic solutions frameworks. In only very few cases those operational engineers are rotated out of their daily routine in order to work with substance and foundational reasoning on the new way of digitally assisted and enabled engineering of the future. In terms of their own challenges like.

- sharply increased product and process complexity,
- overwhelming *day-in-and-out* risk mitigation actions as part of increasing intensities of reviews as part of gateway reviews, boot camp crisis engagements or simply direct reporting meetings
- growing technical uncertainties about fast growing system-of-system interactions of technical systems and growing legislative regulations as well as
- missing data and information infrastructures and
- manageable digital model and data working methods to cope with all of the above challenges, the gap even widens with accelerating speed.

Engineers who have served longer than 6–7 years in such an environment remain sceptic about improvements and are even cautious to become members of new engineering improvement initiatives. They rather prefer to stay passive and observe the "innovation arena" from the sideline or from the spectator stand with mixed feeling in terms of behaving silent or in "protesting behind the scenes". The needed pro-active engagement in order to represent and drive the voice for improved digital engineering solutions has oftentimes no longer a healthy culture in modern enterprises!

This is a severe problem, since creative ideas and technical competence on the one side, but also the hidden demands about implicit expectations of the engineering community no longer or with limited visibility and clarify come to surface for such critical digital transformations in industry.

The author of this book, therefore, wants to present the hidden demands, as he knows them from his own fourteen years of industrial experience as well as from recent and still active research and development projects and consulting engagements with more than 50 companies around the globe.

17.1 Hidden Engineering Demand #1: Intra Company Competence to Drive the Digital Future

Innovation and steering committees are established in industry to reflect on top down type of changes to the existing engineering operating system. By far, it is easier to invite expert groups from the outside to provide innovation impulses and re-engineering ideas rather than to deal with a bottom-up type of approach within the internal engineering community. This, however, is in clear conflict with engineering leaders and experts inside the companies as it is described in "hidden demand #1" in Table 17.1.

The hidden demand rationale

The specific knowledge of both, the technical expertise within a specific industry branch and its products, services and manufacturing circumstances, and the internal engineering operation principles is definitely a core asset that needs comprehensive consideration for defining future Virtual Product solutions. Engineers, therefore,

Table 17.1 Intra-company competence should be driving the digital future

Hidden demand #1	Background and explanation
Engineers know better, what is needed for the future, why should consultants from outside the company be contracted and engaged?	Engineers need substantial amount of expertise and experience on the job in order to deal with the development of technical systems in a fully res-ponsible way. Knowing this fact, Engineers do not understand and appreciate at all why outsiders from IT, digital vendor and consulting companies drive the future solutions. It should rather be the other way around!

expect that this basic rule can be seriously considered in the reflection of the as-is-status, the identified areas of improvement and the basic logic for future digital solutions.

The hidden demand conflict potentials

Following this expectation bears a couple of conflicts, both internal the company and with regard to potential partners and experts from outside the company. To find the right approach to balance internal expertise on best professional practices for engineering work with "fresh-eye" or "new innovation know-how" from the outside, the right technical leadership needs to be applied. Otherwise, the risk will be too high that digital opportunities from research and technology offering companies do not find its way into the intra company "new opportunity hitlist of business improvement candidates". Other risk often kicks-in, if different camps of technical conviction within a company block themselves mutually: neutral advice and new digital solutions opportunities from the outside are not recognized or they are even denied due to this internal fight of the different schools of thoughts.

The approach to mitigate the hidden demand conflicts

The competence of independent technical leadership within the company can help to mitigate such a risk. The set-up of a serious technical career paths in companies, however, is oftentimes difficult and do not resonate with traditional management leadership set-ups. The alternative or additional element of "innovation funnel processes" can help to integrate members of different leadership levels and departments within the future capabilities track within enterprises. Such process can help to invite outside knowledge into the early phases of new digital program definitions, POC (prove of concept) work and within resulting request for proposal rounds. A third approach is to free-up internal SMEs (Subject Matter Experts) part time or as part of short sabbaticals in order to give them the chance to extend and broaden their technical knowledge and technology immersion with partners at Universities, research institutes and technology think tanks.

The resulting need to change engineering principles

The engineering principle of "determine, model, build, test, measure and modify" should be principally remain intact but needs to be expanded and enriched by the following elements of equal relevance:

- Observe and reflect unknown solutions, regardless whether they are proposed, advised or even advocated by outsiders or insiders
- Conduct active reasoning about what could be changed and how, rather than concentrating on how to continue best with the existing solution principle by just adding another new or cool technology
- Pause and interrupt before making a half-baked decision just to follow a stringent time plan driven by an internal—and potentially artificial—innovation project circumstance situation.

17.2 Hidden Engineering Demand #2: Robust and Professional IT Application Integration

Engineers are meanwhile highly frustrated and no longer willing to accept "half-baked" digital applications streams, which make it necessary to enter multiple times data and information inputs manually. Due to oftentimes-overwhelming time commitments to satisfy application and database driven data inputs, engineers are increasingly disappointed about the lack of consistent and single source of truth information authoring within the company IT enterprise architecture.

The hidden demand rationale

Obviously, the increasing levels of Virtual Product Creation execution have been resulting in a high number of different IT applications with individual specific usage purposes as well as diverse sets of data and model generation scenarios. This is the reason for the hidden engineering demand #2 (see Table 17.2).

Hidden engineering demand #2

Consequently, many Engineers and other digital workforce actors feel themselves urged and forced to equalize the isolated data model regime of individual digital applications via manual re-entries in order to guarantee consistent digital engineering workflows. Since traditionally IT departments have limited know how only about underlying engineering processes and task execution, they depend on key users and their determination of use cases to provide at least a minimum level of data flow consistency in between IT applications series.

The following examples show the diverse set of hidden expectations concerning this second type of hidden engineering demand:

Table 17.2 Robust and professional IT application integration

Hidden demand #2	Background and explanation
Engineers expect seamless data and information logistics in between IT applications rather than multiple data entry authoring tasks at multiple points of the IT application line-up	Engineers see themselves as responsible technical system archi-tects executing design, analysis, validation, verification and test tasks. Anything, which becomes part of the corresponding digital engineering solution set, should offer a minimum of consistency of data input / output across and within engineering IT solutions. It is, therefore, not the tasks of engineers to execute as "dumb" data entering subjects in order to compensate for not planned, limitedly implemented or even wrongly de-signed data and information logistics within virtual product creation solutions

- Users expect a single log-on access to all engineering applications and engineers expect the offering of digital tools with respect to engineering tasks rather than scattered around divers' sets of IT systems.
- Engineers expect high service operations times of all digital applications, no matter whether these are locally hosted, network operated or even mainframe based and which types of back-up regimes are entertained.
- Nobody in a producing industrial company likes jobs of back-office workers such as physically re-entering the same data from one IT application into another one. Such type of work simply represents waste, creates error potentials and provides no valuable work ethics or even professional success. Therefore, the deeper question why such manual (re-)entries remain a "hidden" task within the digital engineering workspace has to be transparently clarified: e.g. IT integrations might be deemed too expensive, might not be possible to be realized in time or simply might not be wanted due to non-strategic IT application integration situations! For such sub-optimal situations, however, RPA (Robotic Process Automation) or DPA (Digital Process Automation) applications are available, which favor Bots (SW-robotics) to carry out such routine type of mapping re-entries across IT applications based on reliable rules according to human knowledge. If this is also not possible then the productivity of highly paid engineers might only be temporarily acceptable since midterm perspectives will cause substantial issues and conflicts.

The hidden demand conflict potentials

Still today, engineers oftentimes remain passive and negative if their digital work environment does not reflect core essential work labor requirements such as smooth data transition, workflow and import/export to other applications. Acting as "human disk jockeys of multiple IT applications" in order to perform official digital engineering workflows constitutes not only troublesome and error prone waste of engineering efficiency but also invokes the feeling of not being valued by the company. Unfortunately, short-term accepted workarounds –as such deficiencies are called in official tech language– usually remain for many months in industrial operations if not years as part of half-finished and never completed virtual product creation architecture solutions. In most of the cases, this deficiency is caused by limited IT project funding and by no end-to-end process thinking in IT vendor solution selection and contracting. Consequently, such situation will lead eventually to work refusal and high frustration of the engineering workforce, limited and risky work results as well as to quality problems of the final products in the market and in operational use.

The approach to mitigate the hidden demand conflicts

It all depends on the integrity and close cooperation between the engineering workforce (supposed to be represented by department leaders and chief engineers), the internal IT organizations and potential digital project leaders. The company culture has to provide an appropriate spirit and work forum for such kind of sensible future work simulation prove outs. If this is not established proactively within the development, deployment and implementation phases of digital transformation projects, there is little hope for disappointments and deep clashes afterwards.

In addition, IT integration does not end with API (Application Programming Interface) offering and opportunities. Rather the correct anticipation of the owner and the usage pattern of data and models are finally decisive to conclude explicit needs and number of formal and informal transactions and information exchange within digital engineering workflows. Therefore, it must be clarified early on, who in the engineering community really represents this decisive part in the solution development. The engineering community in most of the cases remains shy in expressing clear demands due to unclarities about ther own digital knowledge as well as due to fears to request the wrong or non-sufficient data gives and gets. Nevertheless, the engineering community needs to be pushed constantly by digital experts to open up, to engage on digital elements early on and to find agreements and compromises within their own community to represent their clear needs and service requirements consistently with one voice!

The resulting need to change engineering principles

For too long, the engineering community within enterprises remained passive and not interested in taking care about their own future digital work environment. Engineers tend to love their product related work success too much in order to stay open and determined about the changing circumstances of their own digital workspace and of the overall digital enterprise. Unfortunately, in many companies, engineers are meanwhile already used to being represented by Engineering IT groups who are established within Research, Development and Manufacturing divisions to serve as the digital face-off for Engineering due to their deeper IT skill sets. However, this remains critical over time since risks will increase substantially in terms of being too far away from new product technology and collaboration perspectives of the engineering community itself.

By accepting the important role of a driving stakeholder for new digital solutions, the engineering community needs to accept ownership of responsibility for future digital engineering solutions as well. By being part of the definition and development group of such new Virtual Product Creation architectures and solutions, engineers have to associate themselves as driving force behind the new digital solution approach rather than criticizing such commonly developed solutions in hind side as critical customers only! This mindset is unfortunately not yet common in industry and limits digital transformation progress substantially. Management has to become much more determined about it and needs to prepare the engineering community for such new role (compare Chap. 18).

17.3 Hidden Engineering Demand #3: Digital Simplicity and Joy

Dealing with digital tasks usually demands a split in mental working mode. On the one hand, it is necessary to follow a methodological game plan within the individual's brain in order to understand structure and determine the engineering task execution sequence and to decide profoundly at each decision point of the engineering assignment. On the other hand, much brain capacity is simply needed to follow the right digital methods steps within the broad spectrum of functionality offered within the individual digital engineering IT applications. Therefore, engineers have started to expect better-designed and implemented digital applications, which also suits the desire for application simplicity and joy of interaction (see Table 17.3).

Table 17.3 Digital simplicity and joy

Hidden demand #3	Background and explanation
Engineers, like many other digital users, meanwhile expect application interfaces that are simple and intuitive and provide a joy of use!	Unlike the early days of computation, where working with computers were pure expert's jobs, working with digital applications nowadays consumes more than 70% of the effective engineering work time. Such working pattern demands high mental and cognitive concentration in terms of ensuring correctness, completeness and accuracy of digital work deliveries. However, the hedonistic well feeling in using digital engineering applications and the joy of interacting with them becomes a critical enabler for job satisfaction. This is exacerbating in younger generations who have been grown up with cool gaming and collaboration apps on their smart phone and tablets. Tedious and old-fashioned digital interfaces, therefore, create a feeling of not being valued as important workforce. For difficult digital assignments, support functions are expected to keep the individuals task execution as simple as possible

The hidden demand rationale

IT tools and supported digital applications are meanwhile accepted mainstream working elements of modern engineering work. Consequently, such digital working environments should best possible support all engineers and other involved working personnel in their personal working ethics and in their highest possible working effectiveness and efficiency. Digital tool interactions, therefore, are meanwhile treated to an increasing degree as potential positive contributions to the daily work satisfaction.

To drive it such way, however, IT tool providers and internal IT organizations need to work seriously on improved task execution based on simple instructions in order to guarantee smooth execution steps as well as to provide a professional level of interaction joy with the application interfaces themselves. Engineers and other users will appreciate such ease and joy in working with the tools, which will result in less cumbersome help desk calls and time delays in digital engineering workflows. For controllers, however, this rationale might not be good enough, since they do expect rather hart facts in (digital) operational cost. Consequently, it will be necessary in many instances to conduct a digital value sequence assessment. This, however, remains a rather unknown digital transformation activity in today's company digital realities.

Hidden engineering demand #3

The hidden demand conflict potentials

Traditional heavy loaded and function rich interfaces of engineering IT applications without any application of scientific results on information fairness and effectiveness for individual users contribute substantially to error prone work execution and to rather frustrating moments of application usage. Engineers were heavily trained to cope with such information and function offering for a long time since they could derive a certain level of expert knowledge from being able to master such user-unfriendly digital interfaces. Starting from generation Y and more heavily represented in generations Z and Alpha, engineers no longer accept such a heavily disciplined approach of information offering and usage and do demand new type of working joy and associated support assistance levels by their employers and hence also by the company IT applications.

The approach to mitigate the hidden demand conflicts

Companies and IT vendors need to put much more focus on the following aspects:

- Simple navigation of software functionality—offering of engineering task oriented functionality offering and down-selects (engineers do know and expect similar capabilities as they are also offered by mainstream digital solutions such as smart TV or smartphones)
- Different degrees of professional usability of IT application functionality—times are over to simply offer "the one serves all" functionality to all
- Work with analogies to mainstream office and social media applications, e.g. excel type of interfaces for baby boomers and Generation X, Facebook type of

interfaces for Generation X and Y, Tik Tok type video instruction interfaces for Generation Z etc.

- Provide gaming type of rewards by counting of already successfully mastered digital task interactions and by providing extra gadget support opportunities ....

This might sound like a lot of extra IT effort, but it is not really. Why is that? Meanwhile IT applications can be configured much more easily for specific roles and tasks and do not necessarily need re-programming but structures and a professional approach within the engineering departments, themselves. In addition, it is necessary to employ skilled digital engineers and human–machine interacting experts to work closely with the engineering workforce in the engineering departments. The new generation of Engineering Managers seriously need to establish such a competence as core element of digital engineering and of an excellent Virtual Product Creation.

The resulting need to change engineering principles

A lot of engineering knowledge needs to be re-written to make it digitally consumable and executable. For instance, it is still not common or not even considered as part of engineering studies at universities today that machine element type of engineering expertise is provided as part of digital learning infrastructures and basic modeling and simulation environments. Such bottom-up digital learning and task resp. project execution level foundation, however, is needed in order to take such critical basic digital-technical know-how task away from technically limited IT solution vendors. It is time to bring it back to where it belongs: to digital enabled engineers, technical experts, professors and managers who want and need to install such new digital enabled technology competence in engineering and business operations. This would also help tremendously to counteract sympthoms that many engineers still have today, fears or little interest in learning competence in Virtual Product Creation tools and their functionalities on top of all the other engineering knowledge that is needed to do the job.

17.4 Hidden Engineering Demand #4: Personal Assistance to Avoid Failure Intrinsic Work

Engineers received traditionally direct working instructions and support by the superiors. The experienced boss or colleague was able to provide the right level of advice for the next level of engineering task execution. With the excessive use of digital working solutions, this personal advice has increasingly diminished and individual engineers need to use multiple, oftentimes non-aligned hints for engineering work assignments, the right level of digital execution expertise and the organizationally correct process deliveries.

Hence, there exist a gap of personal assistance from the point of view of best digital data flow, appropriate digital engineering workflow and best suited digital modeling and simulation preparation for individual engineering tasks (see Table 17.4).

Table 17.4 Personal assistance to avoid failure intrinsic work

Hidden demand #4	Background and explanation
Engineers expect personal digital assistance to cope with the broad and intertwined task executions such as modeling, analyzing, assessing, communicating, deciding and documenting	Engineers are nowadays heavily involved in digital tasks and digital engineering workflows across functions, disciplines and locations/regions of the world. This does require the simultaneous execution of more than 10 to 15 IT applications within daily working routines. In order to concentrate on rather difficult and complex technical engineering tasks engineers have meanwhile reached the point in which they raise concerns about being better personally assisted to use all those tools within specific situations and circumstances. This personal assistance becomes in-dispensable within their engineering task execution in order to avoid unnecessary error modes and failures within digital engineering workflow executions

The hidden demand rationale

Empowerment and down-delegation from Project and Department Leaders to Engineers and project members do transfer substantial amount of responsibilities and even accountabilities to the engineering work force. Timely accelerated task execution and oftentimes cross-enterprise and cross-regional if not global engineering operations lay another level of stress onto the daily engineering roles, responsibility and duties. At the same time, fast changing new and all-encompassing digital capabilities as part of PLM and Virtual Product Creation environment set-ups push the engineering workforce even further into challenges of failure intrinsic work patterns.

Personal, case-based assistance as know from the analog times of engineering can no longer be provided, hence digital assistance types are required and need to be compiled and offered. In the first twenty years of this century, this type of missing personal assistance created invisible inhibitors for many companies to find acceptance in their own workforce to get to the next possible levels of Virtual Product Creation.

Hidden engineering demand #4

The hidden demand conflict potentials

Engineers often doubt which type of digital progression fits to which level best to the team-based advancing in engineering clarification, work delivery as well product validation, verification and release. Sometimes there exist high demands from the management into the delivery capabilities of the engineering workforce with respect to time, cost and quality, independently of the degree of realized digital engineering excellence, which is oftentimes closely linked to newly set-up Virtual Product Creation processes and solutions.

However, what can no longer be provided in many companies is the appropriate level of personal assistance in order to apply the digital solution element to the type of engineering collaboration, model execution and validation or virtual product verification correctly or at least most meaningfully. Therefore, there exist either very different ideas of how to cope best with such digital situations as part of a team approach or of how to advice each member of such a virtual team to best leverage digital solutions and methods at hand for a given task or collaboration circumstance. Consequences resulting from this are team confusion, unnecessary delays or even development errors and/or super cautious workforce members who no longer carry out proactively their digital engineering tasks but wait reactively for a minimum set of coordination clarification by some "authorities". Such lack of appropriate personal digital assistance for complex task execution can even lead to company partial stand stills in its engineering progression in development phases where otherwise smooth digital partner interaction would guarantee successful virtual series deliveries and gateways.

Please note the following specific examples for personalized digital assistance to mitigate conflicts of typical failure modes:

- Simple, robust and high success driven data and information search across data bases and model libraries
- Specific views on digital tool functionalities to enable and improve engineering task execution in context and according to in-site circumstances
- Minimization of tedious data management and structuring (incl. data and model configurations) work through personal advisory and butler services (e.g. via bot support); as an example, PLM system should not always ask engineers for technical attributes that are always the same. Why does the PLM system still not recognize that many engineers in the same department conducts authoring of components with the same type/material/supplier origin?
- Explanation and "semantically" conversion assistance for model and data interpretations from other organizational functions and disciplines ....

The approach to mitigate the hidden demand conflicts

First, such desire for personalized assistance needs a climate of open and transparent digital work execution within the organizational environment, otherwise there will be no room for appreciation of personal actor needs to improve personal and team work advancement! Engineers need to be enabled to seed information, data and model elements for their personal work environment and pattern environment easily. Only if, such capabilities are entertained in companies there will be a chance for collecting a rich set of "supporting engineering intelligence" in digital engineering execution. To which extend, high level engineering rules as introduced in the late 90ties and the early 2000th are still sufficient to provide enough adequate support for today's highly digitally dispersed work patterns needs to be discussed and investigated in specific digital focus groups within the companies. In any case, the competence of personal digital support needs to be based on a companywide new engineering initiative that gets the right support by IT departments, IT vendors, and digital solution providers. However, such approach is still new for most of the companies and bears a clear sense of ownership within the engineering departments themselves!

The resulting need to change engineering principles

Engineers, themselves need to open up and need to pro-actively invest time and efforts in training its personal digital assistance support tools, bots and avatars as they might have learned from their private live digital assistance gadgets. Management and engineering teams should start building up personalized digital assistance with implicit top-down knowledge on lessons learned as well as different engineering development cases and systems. Digital experts and IT departments should leverage bottom-up data & information syntax rules as well as data analytics patterns.

The right level and type of personal assistance should be started for specific job roles, most cumbersome digital task executions, and most critical digital engineering workflows in order to gain trust and highest return on invest for such new digital capabilities. The digital assistance realization itself should also encompass the findings of the hidden engineering demand #3 “digital simplicity and joy”.

17.5 Hidden Engineering Demand #5: Self-modifiable Personal Digital Working Environments

In history, engineers were used to having a high influence on their own work environment and could, therefore decide directly, which techniques, process steps, tools and methods they needed in order to develop, test and produce new products, build prototypes and complete technical systems for service operations. This situation has been deteriorating systematically for engineers during the age of Virtual Product Creation and as part of the increasing digital engineering. In most of the cases, they no longer have the knowledge and the expertise to decide which information technologies are sufficient for their own engineering tasks and how such digital work elements can be modified to make them a best fit to the anticipated digital modeling, analysis and collaboration working modes (see Table 17.5).

In the beginning of digital computation, only a few advanced thinking and skilled engineers were able to develop new digital solutions by themselves since they needed a high degree of computational and programming skill set based on mathematical and physical laws and solutions. Today, engineers are highly depending on IT & digital solution providers, on IT organizations and software coders to provide to them digital computer solutions that allow for easy configuration of digital applications and data engineering solutions.

Table 17.5 Self-modifiable personal digital working environments

Hidden demand #5	Background and explanation
Engineers expect multiple ways to customize and self-organize their digital working environments and applications	Today, there exist limited capabilities for engineers to modify their own digital working environment. They might usually have the choice of arranging their virtual desktop with a sub-selection and zone oriented positioning of icons and information boxes. IT organizations and digital solutions providers in most cases do not support anything beyond such simple visual arrangements due to the reason that it would be non-supportable by 1st level helpdesk per-sonnel. Engineers are unhappy about such situation and feel limited in their own capabilities how to best arrange engineering content, calculation routines, data analytics and information channels as well as their own best practice routines and algorithms

The hidden demand rationale

If somebody is responsible to deliver product and technical systems solutions, which have to fulfill many demanding requirements and system capabilities, such individual must have the legitimation and the professional duty to influence the way such deliverables are to be developed and to ensure that those can be validated and verified in specific ways.

It includes the conceptual proof, the overall design layout, detailed design and component solutions, overall system behaviors, system integrations and final test and sign-off. Engineers, therefore, expect to have a direct influence if not to be a decisive authority, on how such activities are carried out, regardless of whether this requires physical or virtual tools and/or testing facilities.

In order to align best to the overall Virtual Proudct Creation environment—an environment that needs to follow requirements from a broad range of stakeholders– engineers should have personal opportunities to accommodate their own digital working environment to carry out their own digital development duties (see above). This includes the choice of a best fitting personal digital tool and data/model environment.

Hidden engineering demand #5

The hidden demand conflict potentials

The more engineering users a company has to support with Virtual Product Creation capabilities, the more stringent the overall IT architecture needs to be. This oftentimes is the argument for "OOTB" (*Out of the Box*) or "COTS" (*Commercially over the shelf*) configurations for companies as they are made available by digital solution providers (or PLM vendors) in their offering of engineering applications to their customers. The decisive question, however, is, how flexible such IT architectures can be digitally configured: rather than *customization* that requires a certain degree of special software code just to make it usable for a costumer, the digital solution configuration allows for flexible bundling of various solution elements within a certain spectrum of independent tool capabilities and virtual desktop arrangements. Due to the sharply evolving digital transformation, it becomes increasingly indispensable in the future, however, to provide a wider range of personalization of digital engineering tools and data/model fidelity building, analytics and fidelities for the individual engineering user in and across specific industry branches.

Today, the following options are principally possible if they are agreed on at an early stage during vendor selection and Virtual Product Creation architecture definition:

- Virtual desktop arrangements to provide tool icon down-selection, position and look & feel, window size and tiling, background color and desktop theme definitions
- Configurations of information feeds, alert messages and routing options to receive notifications and inform other users
- Priority listings of tasks and daily routines, etc.

In the future, a much stronger content rich personalization will become essential and needs to be designed and implemented with deeper modes of engineering data and model knowledge, modeling and simulation schematics as well as situation aware and process mining related intelligences. Please note the following three examples for this upcoming digital personalization service standard of the future:

- Engineering semantics driven information analysis to screen and present available data sets of competitors, existing simulation data, physical test data and engineering change management related tasks
- Model based engineering assistance configurations to establish trace links between model artefacts and other data sets based on historic knowledge and or preferred personal engineer's rationale
- Personalized bot services to help finalizing virtual series gateway deliveries according to team or process related configurations for BOM (Bill of Material), product structure and CAx model deliverables completeness, communication clarities and "difference picture" documentations.

The approach to mitigate the hidden demand conflicts

Obviously, companies need to start appreciating such personalization requests and demands by their existing and future engineering workforces. Engineers need to start proposing most useful and stringently desired customizable services with the opportunities to aggregate them to similar type of IT operations and data service algorithms. If an alignment becomes possible on such a level, it will be much simpler to arrange and prove out examples in order to assess the opportunities and limitations of existing and future virtual product creation architectures. Each personalized virtual desktop capability and service should pay for the overall effectiveness and efficiency of the individual engineers with respect to his/her own digital work profile. Measuring such effects will provide evidence of trust and will contribute to a self-inducing new digital work policy.

The resulting need to change engineering principles

Engineers and their management should investigate to which degree personalization of digital tools and data/model services delivers personal and/or team efficiencies and working robustness versus risks concerning divergence of no longer compatible engineering interactions and collaboration amongst team members. If the process re-engineering thinking of the 90ties of last century and the early 2000th is applied to the new world of fully digitalized working environments of the 2020th and beyond all engineering methods need an overhaul and a rather natural interpretation capability for each active engineer. Times are over, that engineering approaches can persist for many years or even decades. Constant reviews of team and personal duties and development skills will have to be reflected within personalized digital working environments. The better such new digital asset will be understood and appreciated amongst professionals, the earlier companies, agencies, management and engineers will be able to contribute successfully to the digitalization challenges of the future.

17.6 Hidden Engineering Demand #6: Quick and Continuous Improvement

All digital applications are dependent on a process, which transforms engineering principles, procedures or model/data assumptions into executable algorithms and software code in order to provide the digital processing of engineering value creation.

This process, however, is oftentimes flawed and not straightforward due to the necessary separation of duties between agreements on engineering approach, determinations of digital working modes, functional specification for digital task execution, reductions to specific uses cases as minimum set of software intelligence, design specifications for specific data models and software functions and various degrees of test cases. Such an approach widely differs from the traditional engineering approaches that are based on solid understandings of technical physical effects, physical behavior principles and a wide range of existing technical system solutions based on mechanical, pneumatic, hydraulic or electrical/electronic principles.

Engineers envy the quick implementation routines of software development and hence expect quite naturally that quick and continuous improvement for digital engineering tools are simply possible and should be realized wherever possible.

The hidden demand rationale

Engineers have learned the hard way during the last 30 years that the only development capability left in the new digital world is based on software based digital and analytical toolsets. Since engineers are in most of the cases not capable of building such digital capabilities, tools and toys themselves, they are usually fully dependent on others to provide, correct, improve and further develop such capabilities. This means in consequence, that Engineers meanwhile have become dependent on others in order to carry out pure basics and all advanced type of (digital) engineering.

After one or two decades of a rather devoted attitude towards such situation, the situation has changed towards a more demanding, professional attitude in comparison to the "ordinary" product and technical system world where clear and stringent business relations exist with high demands of product quality and delivery consequences. In natural denial to the very specific circumstances of software and digital application development and to the necessary Enterprise Architecture Integration of IT applications such as Virtual Product Creation solutions into company environments, engineers nowadays do insist on an increased service level with respect to quick and continuous improvements of digital solutions by the responsible development groups (see Table 17.6).

Hidden engineering demand #6

The hidden demand conflict potentials

The first conflict potential exists during the customer acceptance testing in terms of new IT application rollouts and deployments. If key users of engineering teams find fundamental and annoying failures of the software or if the logic of the digital

Table 17.6 Quick and continuous improvement

Hidden demand #6	Background and explanation
Engineers who are users of digital applications of the Virtual Product Creation tool suite expect from responsible IT application suppliers—inside the company as well as from outside partners and digital solution providers (such as PLM vendors)—a professional and proactive way to provide bug fixes, to constantly conduct application improvements and to listen to other costumer suggestions	Since the end of last century, engineers have been heavily educated and trained to be sensible to customer feedback of their own products. They have accepted and appreciate that professional set-ups have been established in companies in order to ensure product quality through new methods and development procedures as well as to measure customer satisfaction constantly. The goal of such endeavors is to provide quality and product improvements in a responsive and professional way. Software Engineering, however, still experiences today high failure rates in implementing costumer demands into executable software applications. It is still common in 2020 that software is delivered to customers with sig-nificant amounts of errors and non-working or non-intended functiona-lity. Engineers cannot understand such business approach in comparison to their own business. Hence, they raise concerns on such questionable business practices and demand quicker improvement turn-around cycles

approach creates flaws in daily digital engineering or at least in single groups, the implementation might be stopped, delayed or implicitly no longer pursuit. The traditional battle within PLM projects, IT implantation activities and overall Virtual Product Creation business improvement undertakings foresees lengthy and tedious negotiations between companies and their tool providers with respect to bug fixing prioritization. In such cases, their own company or internal digital project teams force engineers to allow a certain degree of annoyance and workaround willingness in order to keep the high priority bug fixing opportunities within a limited size according to the money value that was reserved for such cases during the initial contract agreement. Engineers have no understanding for such opportunistic approach.

The second conflict arises during the actual use of the digital applications in the course of engineering operations. If many help desk calls, user feedback based on questionnaires and other complaints are not really taken seriously into considerations by company and project authorities, engineers start acting negatively and call a crisis. This, however, might cause bigger churn and stress within the entire engineering team and within the development projects where such flawed digital capabilities provide clear inhibitors to deliver quality development results.

The approach to mitigate the hidden demand conflicts

The dilemma usually starts already at the beginning, when the basics of a Virtual Product Creation architecture and solution are assembled. By that time, oftentimes-wrong high level or non-future oriented carry-over assumptions form the basis for the relevant use cases which are then treated as contractual baseline for software application customizations and new developments. Consequently, it becomes essential to involve a higher number of lead engineers and digital competent method engineers into that 2–3 months initial phase for such a digital transformation initiative. It is necessary to get them at least half-time, if not full-time involved, during such phases in order to provide consistency into the bottom line assumption of the digital architectures. Those assumptions serve as base for any next level development work.

After full or partial implementation of new digital capabilities, a group of digital competent and well-respected key engineers should act as conduit between user groups of engineers, IT organizations, project managers of the digital delivery project and/or to the real business owners of such digital capabilities (if they have been identified clearly enough with all duties beforehand!). It is their important responsibility to discuss short-term work-around opportunities and new improved solutions with IT application engineers. They also need to keep the pressure on IT organizations, company stakeholders and digital solutions providers to deliver quick short-term solutions and profound long-term digital solutions. Similarly, such key engineers need to organize, with the help of IT application engineers, the professional level of digital operations with work-around if during such unpleasant project times.

The resulting need to change engineering principles

Engineers need to understand that digitalization is not just a “temporary thing or journey” that will be soon over. On the contrary, it becomes critical for engineers to understand that digitalization remains a constant journey with different episodes and timely phases. Therefore, universities and companies need to invest into such digital skills and transformational developments much more heavily, both with respect to strategic long-term evolutions and to short-term initiatives. Engineers need to engage directly and with all engineering wisdom at various levels on a personal and/or team level such as:

- Ideation of new digital principles for the future of the company and for the future of products and technical systems,
- Engagement with digital solutions providers to understand the challenge of developing and delivering appropriate and execution robust IT applications,
- Reflection of data and model consistencies and richness, which form the base for algorithmic support levels and needs of digital engineering applications,
- Encouragement to request persistently the right level of professional support for bug fixing, IT application improvements and new ways of digital engineering rather than just following old levels of (traditional) engineering practices.

Efforts will pay back if mechanisms are established to constantly and quickly review, judge, improve, implement and review all IT applications, especially in their interplay to each other!

17.7 Hidden Engineering Demand #7: Flexible Digital Test Beds in Production IT Environments

Virtual Product Creation (VPC) no longer constitutes just a collection of individual tools to create, change and save files and documents in order to describe models for representation, assembly, simulation, control and storage of engineering content and machine & product behaviors. It has been further developed to a digital engineering competence in order to exchange, release, collaborate with, improve and optimize ideas, new designs, technical solutions and full products/systems amongst co-located or dispersed team members and companies (compare Chap. 4, "Virtual Product Creation—what is it?", and Chap. 6 "The set-up of Virtual Product Creation in Industry"). Thus, Virtual Product Creation is like an active eco-system consisting of many digital technologies, extensive data and model management solutions as well as a high range of different engineering processes, methods and collaboration procedures. Any change in such a VPC eco-system has to be carefully planned, carefully reviewed, simulated, actually tested, finally assessed and judged, and potentially iteratively optimized. In order to do so, it becomes mandatory to provide various test beds and environments to enable professional validation, optimization and verification of Virtual Product Creation architectures, solutions and business operations.

The hidden demand rationale

After having experienced painful implementations and deployments of VPC solutions with many workarounds, flawed software components and process-wise unclear

Table 17.7 Flexible digital test beds in production IT environments

Hidden demand #7	Background and explanation
Engineers expect that new features, tools, methods and process flows of Virtual Product Creation solutions can be used and tested like a "digital prototype" in flexible test beds as part of the production IT environment. They offer all intended future capabilities including all needed data, models and workflow services in order to judge the readiness and appropriateness of the new digital solutions	In many digital transformation or innovation project tedious discussions crop up between engineering functions and IT organizations to clarify whether test labs or test beds are supported, or not. IT organizations are often frightened by engineering and manufacturing user requests to get dedicated test areas provided within operational IT production systems or at least test labs populated with realistic production data. The reason is that control regulations of IT administration might have to be managed, which differs from the traditional role of databases that serve for many applications across the enterprise IT architecture. Offering testing environments in a *live IT production environment* demand specific regulations and responsibilities, which are oftentimes avoided due to substantial efforts associated to them. Engineering departments underestimate the risk of such approaches and do not provide enough budget for the build and maintenance of such testing environments

collaboration and working patterns as well as overwhelmed users many companies and PLM project management members became cautious just to rely on unit test and high level key user tests to sign-off new digital solutions for business operations (compare new demand in Table 17.7). In addition, many engineers meanwhile remain passive and none willing any longer to use new digital tool integrations as part of existing or new Virtual Product Creation environments without thorough tests involving key players and *key engineering case scenarios* (rather than only the potentially associated individual *use cases* that are expected in today's agile IT DevOps approaches[1]).

Unfortunately, PLM and digital solution providers in most of the cases are just treated as IT vendors rather than as VPC partners and cannot directly influence the integration of their own tools and digital method solutions into the overall company IT architectures and infrastructure. Consequently, the pressure has been growing significantly on internal company IT organizations to seriously separate out test beds for intensive end-to-end testing by engineers within live production IT environments.

[1] DevOps is a set of practices in informatics and software engineering intended to reduce the time between committing a change to a system and the change placed into a normal IT production environment. DevOps architectures, therefore, improve the software development process by introducing agility and cross-functional team works. Integration and automation in the build and test process with supportive tools reduce manual IT work and, therefore, increased the speed of software implantations.

The set-up of professional test beds will benefit implicitly from the growing number of *microservices*[2] of software deliveries based on the Service Oriented Architecture (SOA) approaches. Establishing *microservices*, however, needs mid to long-term commitments and a new type of skill set within IT organizations independent of special IT innovation or digital transformation projects.

The alternative to entertain separate test labs usually lacks a consistency of getting live production data (or at least a representative frozen snapshot of it) into such separated IT server environments. Furthermore, old traditional approaches exist for those cases, which forces method development experts to manually download certain data and model examples rather than relying on automated snapshot freeze downloads in a professional digital set-up.

Hidden engineering demand #7

The hidden demand conflict potentials

If no agreements can be reached on such strategic important approach for smooth and forthcoming validation of digital solution environments, the following attitudes, mindsets and cause of actions will typically prevail and will cause unnecessary trouble situations:

- In case of lengthy and late discussions, time is running out to provide help for specific release dates, which puts significant pressure on individuals to sign-off production readiness of digital solutions without any thorough tests.
- Engineering teams might deny any buy-in to the digital solution suggested for production and can blame easily digital project or improvement authorities for all upcoming issues. This, however, will cause major irritations, churn and stress which will lead to expensive workarounds and higher resource demands to keep timing of the actual product or technical system development program where the new digital solution is supposed to be used!
- Short and mid-term misalignment between IT organizations and functional engineering and manufacturing areas will be counter-productive and eventually detrimental for mutually trusted digital future initiatives across organizations. Therefore, mutual understandings of needs are to be explained openly and should be appreciated in order to find compromises to reach digital robustness.
- Engineers are likely to cease from any future digital engagement once they recognized that quality testing is not to be supported seriously by their own company organizations –this, however, might stall any healthy digital transformation projects in the future.

---

[2] The term "*microservices*" is defined as small, self-contained applications, that can be implemented, deployed, scaled and tested independently from other applications. The *architecture of microservices* involve packing up software code and all its dependencies in a container so that the application in the container can run on any infrastructures. For each microservice, a separate database is designed with specific control rules.

The approach to mitigate the hidden demand conflicts

Such a delicate hidden demand needs to put on the official agenda at an early stage for a comprehensive view on all digital project circumstances. The validation and testing of modified and new digital product and manufacturing engineering solutions as part of Virtual Product Creation needs to be treated as seriously and as important as possible, similarly to any technical product and technical system development in comparison. Hence, the project leaders need to sensitize such hidden demands officially and need to get clear understanding and commitment of all Engineering Leadership behind it.

After having clarified such fundamental requirements within the official project planning and execution line-up, all details of representative test data and test conditions derived from officially reviewed requirements have to be analyzed and elaborated in a close three way engagement between key engineers, key method experts or key users and experts for IT operations and innovations. Only then, it will possible to size the efforts and approaches more easily on how to realize such overall test bed environment within the company, across locations or even with the interplay of suppliers and partners. Costing and budget provision are critical to get a professional set-up ensured. To be on the safe side, approximately ten percent of the overall PLM/innovation budget should be reserved for such set-up and test operations support.

The resulting need to change engineering principles

The biggest change deals with the recognition in Engineering and IT Management that such test bed undertaking is necessary and not just a nice to have item! Accepting thorough engineering principles in ordinary product, production and technical system development projects should be taken as a role model to find and establish the right validation and verification environments as well as for digital engineering innovations and optimizations. Engineers need to open up their personal believes and expert attitudes in order to provide the right support for the determination of testing activities and procedures and must support it as a “mission critical” activity. New digital working approaches for difficult and sensitive engineering tasks require a fundamental and well-thought-through approach of test scenarios: they consist of a new mix out of traditional and world class engineering capabilities combined with forthcoming new digital mechanisms and collaboration patterns based on data, information and model fidelities. Only if such new commitment can be established across functional organizations and within the individual engineering departments, it will be possible to guarantee smooth test, validation and verification as well as operational excellence later on.

17.8 Hidden Engineering Demand #8: True Appreciation for Digital Responsibilities

Changes in values and forms of work appreciation usually take a long time before they are transferred into daily routines and even longer before they are integrated in official job roles & responsibilities and company cultures. Digital transformations and their associated projects and initatives are still today mostly "controller driven" with the clear expectation to serve for a typical business case within a period of amortization (usually within one year, sometimes two years). Mid and long-term transformations of employee skills, core competencies of organizations and teams and associated responsibilities and accountabilities are missing, or at least do not play a vital role in typical two to three year's assignments of modern management.

The hidden demand rationale

In professional life, engineers might follow their personal calling and enthusiasm for quite a while unless both collide with stringent company rules, regulations and procedures as well as with career relevant personal achievements incentives and needs. For many years, they have received training to cope with specific tool functionalities and process-related development procedures. One important thing, however, has been never addressed (yet): which dedicated responsibilities are transferred to engineers in their job roles with respect to digital data in general, to their timely creation and maintenance and the appropriate storage, management and transferring to others! All digital activities that are usually hidden behind high-level process boards created by business and process consultants are by no means spelled out, valued correctly with respect to their influence to company success or estimated concerning the necessary time involvement (see Table 17.8). Engineers, nowadays, have started to sense this

Table 17.8 True appreciation for digital responsibilities

Hidden demand #8	Background and explanation
Engineers expect and demand an honest appreciation of the digital working activities and all associated digital responsibilities for data, information and models across the full range of officially established digital tools, databases, workflows and processes	Unlike to the past, where engineers where mainly responsible for their own solution know-how and for their rationale and contribution to specify the right design attributes for a given product or technical system, they have received step-by-step during the last 20 years major responsibilities for many digital authoring, storage, management and collaboration ac-tivities. The time necessary for such new and extra digital tasks has been never granted to them officially. On the contrary, it is expected by ma-nagement that such digital working steps are "automatically" absorbed through personal or team efficiencies and process optimizations by engineers

and wonder why companies and management are not capable or are at least shy on this dilemma.

In addition, engineers have meanwhile doubts whether digital engineering work capabilities and achievements are well recognized for career opportunities or whether it is simply assumed and treated as a hidden mandatory "must have experience" with no further critical skill potential. It is high time to establish *Digitalization Capabilities* as technical career boost!

Hidden engineering demand #8

The hidden demand conflict potentials

Engineers ask themselves to which extent it "pays back" to fulfill all implicit expectations of Engineering Management to care personally about all digital assets such as sets of data, information or models, entries in databases or workflow management systems and as part of interactive design and systems reviews. In many cases, there seems to exist a situation in which your immediate boss somehow simply does not know very much about those digital tasks. Why would a great fulfillment of those duties help and motivate you in making good impression on your superiors? In other cases, there does not exist true appreciation for it and it is downplayed by phrases such as "just get this "bloody" thing done and do not make a "big fuss" out of it!

Following such industrial situations, please note one of the following fundamental and brutal wisdoms of today's digital transformation business:

> Nobody (yet) values if you maintain your digital data, model and information sets in good shape in 99% of your daily routines. However, within the 1% range of your work when you might not have completed a specific data entry in a given situation or when you might have done a mistake, then this is noticed at once and you are put on the spot immediately!

Engineers, designers and analysts are increasingly disappointed about these ambiguities and non-acceptable situations and consequently start requiring a clear commitment to digital work, values and achievements by their professional members and management.

The approach to mitigate the hidden demand conflicts

A first necessary step deals with a clear and precise listing of engineering tasks that are officially recognized in the companies' working regulations, engineering processes and job roles & responsibilities and of the type of digital skills and activities they require. With such a base listing, it is then possible to rank the digital activities and tasks concerning the following value items:

- Level of skilled digital competence to carry out such a digital activity
- Urgency of applying such digital skill
- Degree of mutual intellectual combination of such digital activity with traditional engineering, design and analytical competences and know-how
- Scarcity of such digital skill competence in the organization
- Criticality of digital skill competence even as role-model-related for management and future executive positions.

In any case, consistency has to be applied for constant and regular reviews of digital achievements of each employee and engineer. Simple appreciation feedbacks in given situations and pro-active reward types after a series of demonstrated digital competence and duties should be considered, too.

The resulting need to change engineering principles

Principles of valuing engineering delivery, achievements and competence need to be extended to or even consistently shifted towards digital tasks, skills and achievements in order to motivate, support and excel on such new critical competencies! The new generations of Engineers and Engineering Management have to follow-up such important change of values within their personal agendas and career behaviors and life cycles. It also requires, however, a much higher degree of digital commitments by all daily engineering operations and task assignments. It remains doubtful why especially very traditional companies in machinery, vehicle technology and aerospace still consider two old-fashioned classes of digital skill: one, as part of IT plumbing and operations, i.e. far away from traditional engineering, and the second one as auxiliary and/or clerical job of supporting jobs for engineers. The newly recognized sill set of software coders for embedded software as integral functional part of products, machines and production systems has been recognized, too, but this is not associated to engineers. This is at least an oversight or even a wrong perception since technical systems will only benefit from digitalization if engineers are valued and educated in digital competencies of Virtual Product Creations technologies and solutions!

17.9 Hidden Engineering Demand #9: Upfront Simulation of Digital Engineering Collaboration

In today's business operations, line and project management are based on the principle assumption that either well founded business heuristics or process related planning experience form the basis for successful working environments and working procedures. Since implicit iterations and intense collaboration patterns are part of Virtual Product Creation behaviors, they are not really envisioned and, therefore, not seriously considered (although many process descriptions do flag them out as one of the theoretical situations), see Table 17.9.

The hidden demand rationale

Based on the experience that Virtual Product Creation solutions use quite naturally control loop operation and iterations (compare Chap. 6 and Fig. 6.5) engineers and engineering management have meanwhile understood that a realistic prediction of digital engineering progression does need additional forecast capabilities and not only experience and process assumptions. Engineers have lost their personal estimation capabilities that were traditionally based on localized and mostly analog discussions, reviews and sign-off procedures. Hence, it is difficult for them to cope with the new

Table 17.9 Upfront simulation of digital engineering collaboration

Hidden demand #9	Background and explanation
Engineers and Engineering Management would like to know upfront how certain digital solution sets can be set-up, altered and combined and which effects such elements will deliver for overall digital engineering collaboration based on specific workflow interactions as an entire new or modified Virtual Product Creation module	One of the most unsatisfying aspects of Virtual Product Creation is the rather vague upfront understanding and missing evidence of simulating VPC. Thus, leadership is often lost in determining critical success factors during the planning and conceptual development phase of new or modified Virtual Product Creation solution. Therefore, better, more reliable and reproducible simulations of personal and team interactions with respect to modeling, analyzing, reviewing, exchanging and collaborating as part of Virtual Product Creation are anticipated and requested by Engineers and their Management

challenges in the age of high information technology enabled services for digitally documented, modeled and mastered engineering collaboration across departments, sites, countries and regions. This is the reason why they desire and need realistic upfront simulation of digital engineering collaboration. Furthermore, they do not really understand why such Virtual Product Creation behaviors and progression is not professionally offered and established (yet) although much progress was made for the simulation of product and component behaviors already!

Hidden engineering demand #9

The hidden demand conflict potentials

Process re-engineering consultancy and agile working coaches have introduced a series of business process analytic review styles in order to find agreement, conviction and leadership delegation to introduce new business logic and behaviors. Please note just three examples:

- US driven "mission and control war room set-ups" to provide an effective "military type command steering board environment" to review, assess and determine mission critical steps for changed or new business approaches
- Japanese driven "stand-up style obeya room set-ups" to link things together, like strategic objectives to metrics, to planned work and to potential problems.
- California style "business case-oriented design thinking" approaches and associated stand-up style story telling demonstrations to clarify and motivate future team and business directions and capabilities ....

Unfortunately, all of the above approaches end up on a fairly high level where digital progression as part of the digital transformation foundations do not get involved, at all. Senior Management, however, often believes that such approaches will quite naturally lead to find a mission and order for the internal digitalization in companies and might even influence Virtua Product Creation behaviors directly. This however, is a major misapprehension!

Consequently, (Senior) Management has to rely on digital experts to provide their assessments that they cannot factor into the overall business process equation. This is one of the most obvious, but hidden conflicts as part of digital transformation projects and initiatives of today!

The approach to mitigate the hidden demand conflicts

The scientific community has already offered a couple of promising approaches, which did not find their full ways into mainstream process management, Virtual Product Creation workflows and behaviors and engineering factory operations. Please note, as examples, the following three approaches:

- Business process modeling notations such as BPMN (Business Process Modeling Notation) or EPC (Event driven Process Chains) with associated workflow engines to simulate overall process behaviors
- System related linked network models such as Petri Net, SADT (Structured Analysis and Design Technique) or SysML (System Modeling Language) to capture technical system or process related dependencies
- System Dynamics as an overall methodology and mathematical modeling technique to frame, understand, and discuss complex issues and problems as they also might occur in digital collaboration of technical systems development within Virtual Product Creation.

In the past Senior Management and their implicit process owners in companies called in business process re-engineering consultants to analyze business metric oriented organizational behaviors and to drive change via top-down approaches. Unlike this past success model, nowadays companies have to establish core competences in observing and modeling digital process behaviors, digital engineering progression and Virtual Product Creation collaborations in order to transform such measurable data flows and digital collaboration behavior patterns into meaningful models as already indicated above. Similarly, PLM and digital solution providers and digital consultancy agencies should establish deep expertise in such modeling and simulation behaviors in order to justify and explain better their own VPC solution elements for a given company situation and mission.

The resulting need to change engineering principles

As one of the first principles engineers, designers, analysts and all other digital workers need to become ready to measure themselves with the help of tools and scientific guidance how they perform digital tasks. Such measurements are mission critical in order to understand current individual and team working patterns and behaviors. With such voluntary group of individuals—approximately 25–30% statistically representative members of an observation group would be enough to conclude seriously the overall population behavior—digital simulation experts would receive reliable assumptions for their predictive process and VPC control simulation models.

The resulting digital value stream mapping can then deliver core advantages for both, Engineering Management and engineering individuals:

- Predict overall digital team performance, bottlenecks and improvement opportunities
- Receive advice on changes to personal digital behaviors or to train personal bot or avatar support (see also hidden demand #4).

17.10 Hidden Engineering Demand #10: New Advanced Human Interfaces

If humans need to interact with IT systems, direct visual, tactile and/or auditive user interfaces become essential. Research and science have delivered key evidence that the user performance, which includes elements such as:

- Task solving time
- Number of solutions found
- Degree of natural or intuitive interactions
- Degree of (information) immersion
- Learnability
- Memorability
- Effectiveness and
- efficiency.

Depends on the following influence factors:

- The user itself with the user experience, the cognitive, perceptive and motorized capabilities, the personality as well as personal likes
- The task itself with its selection, positioning and orientation and
- The IT system with its capabilities of interaction technology, degree of immersion, visualization and degree of freedom.

Unfortunately, the existing and prevailing doctrine of Virtual Product Creation stakeholders put a lot of burden to the re-definition and representative use case oriented sub-selection of digital tasks (often not oriented towards the engineering thinking) and on awareness resp. training sessions of the final users. Sometimes, users are at least grouped according to role and task categorization. Efforts towards target oriented interacting technologies as well as immersion and special visual analytics are barely considered or seriously implemented as a key capability (see Table 17.10).

The hidden demand rationale

Modern people experience new type of interface capabilities in private life by interacting with modern technologies and app-oriented software solutions on tablets, smart phones or even voice control gadgets and services like amazon echo/Alexa or google voice/assistant, car OEM interfaces, etc. Meanwhile engineers ask themselves

Table 17.10 New advanced human interfaces

Hidden demand #10	Background and explanation
Engineers expect cool and exciting human machine interfaces as they can experience them already with smart phone apps, gaming tools and other digital interactive devices of daily life	For a long time, PLM and digital solution providers deliver most comprehensive functionalities, as they are official demanded by company VPC and PLM project requirements. To serve many costumers and user types, internal IT interface environments of major office tool vendors mainly drive the underlying technologies embraced for viewing functionalities such as: forms, listed views, indented views, final delivery-oriented WYSIWYG (*W*hat *Y*ou *S*ee *I*s *W*hat *Y*ou *G*et) viewing panes etc. VPC/PLM vendors refrained from providing advanced visual interfaces for joy, fun, visual efficiencies and effectiveness and better cognitive arrangements. Also, audio, VR and AR interactions are not common yet

why such sophisticated interfaces are not offered yet and getting qualified for digital engineering type of interactions. Experiencing convenience support and hedonistic fun and joy in interacting with digital tools will become a "must have" characteristic of next generation Virtual Product Creation solutions, too.

Hidden engineering demand #10

The hidden demand conflict potentials

Younger generations such as Y, Z and Alpha generations will no longer be as patient as the former generations in demanding modern, more immersive and intuitive interacting interfaces in a creative technology mix including:

- Dynamic and interacting viewing analytics,
- Game oriented self-exploring levels of expertise for a giving subject area,
- Ubiquous viewing immersion including Virtual & Augmented Reality interfaces and viewing devices and
- Voice activated and controlled routine services for daily engineering tasks and advisory.

Engineers, in general, will expect and insist on more efforts to provide the latest technology related interaction support for their duties and tasks, not only to receive better guidance and visual comfort for apprehension but also to experience more joy and fun in delivering digital results constantly and with high personal motivation and quality ambitions. Companies are therefore asked to change their attitude towards a much more forthcoming perspective on those engineering satisfaction levels rather than downplaying it constantly via arguments such as IT integration cost burden and difficult support models.

The approach to mitigate the hidden demand conflicts

Meanwhile IT technologies like micro services and widely used WEB services and interface technologies exist to serve better such demands on the technological side. What remains difficult is that Engineering and IT Management in industry as well as PLM and digital solution providers still have difficulties to employ Human Factors Experts and Human Machine specialists. They are indispensable in order to gain more insights to the most effective ways of offering such new interface solutions in combination with the given engineering cases and tasks and the individual user profile characteristics (see explanation of user performance above).

The resulting need to change engineering principles

Digital Engineering tasks can no longer be just treated as mandatory advice activities that simply have to be followed up according to a prescriptive method. Digital Engineering activities (compare Chap. 6) are to be treated as socio-technical efforts that need to factor in joy, interaction success, tool usage satisfaction and openness to provide feedback to digital assistants in order to help others as well. Therefore, it is adviced to start such a journey with Senior and Middle Management since they need to get convinced about such new and important influence factors and technologies. To get them closer acquainted to more advanced digital technologies in order to trigger closer engagements to the ordinary (digital) engineering work will convince them to establish the right skillset for wider developments and implementations.

In summary, the described ten hidden engineering demands in this chapter are critical for the success of next levels of establishing digital competences and work patterns as part of next generation Virtual Product Creation in industry. All stakeholders, Engineers, Engineering Management, IT experts and IT Management of developing and producing companies, Management and Application Engineers and Consultants of digital solution companies need to deal seriously with them. It is not a question of desire, it is a question how quickly and thoroughly those demands can be met in a fruitful and professional manner, not just occasionally but consistently across all digital activities!

Chapter 18
The Challenge of Modifying Management Leadership Behavior Towards Virtual Product Creation in Industry

Executive Summary

This chapter aims at explaining the challenges and typical behavior types of Management in enterprises within the new competence field "Virtual Product Creation" (incl. all aspects of Digital Engineering, Product Lifecycle Management, Advanced and/or Model-based Systems Engineering, Digital Manufacturing and related IT-technologies etc.). Management as an organizational task and opportunity to change and develop new digital engineering principles, processes, methods, tools, data models and engineering model types, is a rather new skill set which is often missing to drive digital innovations and transformations in industrial companies. This chapter is motivated by the author's broad industrial experience: with many enterprises. This new type of digital leadership in Management is often missing, or at least, not yet equally established or even well anchored as a valid career opportunity like in traditional engineering leadership positions or specific IT management ranks. Without solving this dilemma, no new fundamental approaches will be achievable in enterprises and today's operational flaws in digital product delivery will continue to exist!

Quick Reader Orientation and Motivation

The intention of this chapter is:

- to derive the essential leadership needs for Virtual Product Creation
- to explain approaches to develop Management skills for successful digital transformation in engineering and to train Management in digital leadership skills
- to describe the Do's and Don'ts of Management behaviors in digital business and associated decision making
- to provide advice for Senior Management in new digital leadership.

Many industries and enterprises suffer from a lack of leadership in determining, defining, developing and implementing fundamental changes associated with digital innovations and transformations of their business operations. Figure 18.1 provides an understanding of this dilemma in comparing the *traditional business objectives*

R. Stark, *Virtual Product Creation in Industry*,
https://doi.org/10.1007/978-3-662-64301-3_18

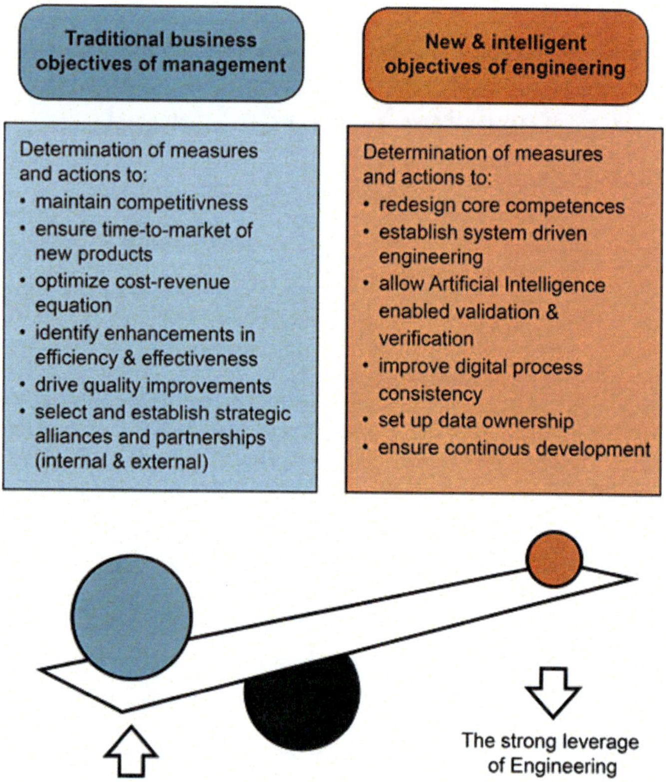

Fig. 18.1 The drift in management objectives driven by digital transformation

of Management with the *new intelligent objectives of Engineering*, which are not yet anchored and integrated into the roles, responsibilities and yearly objectives of (Business & Engineering) Management.

In the past, individual digital tools were introduced to transfer a manual task into a digital worksheet or model instance. Nowadays, major business approaches need to be transformed into a new digital operations set-up. These new demands require establishing end-to-end digital process continuations, data and information threads, new digital product intelligence, fully digitally enabled business practices as well as new digital business models and value creation elements. Even in areas where digital penetration has evolved substantially already during the last decades like in Engineering, it is now necessary to reach the next level of full digital, IT-based working environments amongst all roles and stakeholders. In addition, new (digital) engineering approaches have to be introduced to get complexity and dynamics under control with respect to digitally connected products, which require a constant feed of digital IoT (Internet of Things[1]) type data retrievals, analytical interpretations and subsequent operational execution.

Traditional Engineering, Business and Project Managers continue to rely on getting such digital change management transformations primarily addressed by IT Management and/or by high-level process re-engineering. This, however, is no longer sufficient and successful, since fundamental new business approaches in digital operations require new and additional leadership beyond typical IT (hardware and software) and process skills:

- Socio-technical systems require new type of digital interactions,
- Product features and functions rely on software-enabled control algorithms,
- Continuous product and technical system validation and verification are dependent on new transparent digital information traceability,
- new digital engineering quality assurance approaches and mechanisms are missing in order to mitigate and control risks caused by Artificial Intelligence (AI) driven drifts in product and system intelligence and operations,
- Traditional business models lack capabilities to reflect and enable digital platform-based data value deliveries.

Some companies tried to solve this by introducing additional CDO (Chief Digital Officer) positions. This new set-up might have helped to encourage companies to pay more attention to digital business models and associated processes but failed in delivering solutions to the fundamental evolutions and revolutions as described above.

Which sort of knowledge and decision making capability does the new type of "digital literate" Managers need to apply in order to drive today's digital presence and tomorrow's digital future? This chapter will provide insight to this fundamental challenge. However, before providing an insight into it, let us understand which leadership decisions are necessary in today's digital engineering business and Virtual Product Creation (VPC) solution sets.

Please note some examples where today's Management teams (in Engineering, Business, Project Management and IT) often fail due to missing background and technology understanding, working experiences and skills as well as limited digital business acumen or missing data and cost/benefit estimations:

1. How to invest into future VPC competencies with respect to new digitally executed incremental verification of technical systems (products, production lines, infrastructures)? Which investment distribution in time, effort and budget amongst candidates suits the purpose the best, with respect to:
    a. IT technologies
    b. analytical and digital engineering methods and working principles responsibility processes
    c. new virtual test facilities and prototyping environments?

[1] Please compare all IoT (Internet of Things) related subjects and explanations in Chap. 20.

2. Whom to assign inside and outside the company to define and design the new digital engineering core competences for the future? Which approach to take to recognize shortfalls in the existing engineering set-up with respect to the necessary next level of digital sign-off rules and methods of highly connected, automated and/or autonomous technical systems? Will the company still be in business in 5 years from now, if current engineering practices are still in daily use, especially with respect to cover new products and systems with high degrees of SW-enabled and data driven system behaviors?
3. Which time is left to make the final decisions to establish the new hybrid symbiosis between classical HW-prototypes, dynamically updated digital prototypes and appropriate virtual prove-out environments? Which degrees of new digital prototyping is best suited to allow for integrated approaches that follow *model based (systems) engineering* and to leverage *smart service/IoT based* development patterns?
4. How to retrain the current engineering workforce and to merge in new digital talents and specialists from the outside (e.g. data scientists or analysts, computational engineers, system architects) in order to cope with the growing challenges of delivering "error-free or resilient" intelligent products in the global market places? How to establish new ways of Dev/DesignOps for technical systems?
5. Which criteria and arguments should be used to finally decide whether new information standards need to be developed, integrated into IT systems, and trained to the workforce as part of the next crucial digital transformation? Which of the existing information standards are no longer sufficient or fit for the future, how can such "short term non-productive" but "long term strategically critical" new digital capability be justified? How to take the lead in it? Which investments are necessary?

The next section will clarify which responsibilities of Management are essential for the digital engineering capabilities and intelligences of today and for the future.

18.1 Needs for Improved Digital Leadership of Management in Virtual Product Creation

The internal and outreach recruitment system to fill Manager Positions in product and manufacturing engineering relies on the classical two-fold T-model shape skill assessment approach:

(a) **The "vertical depth":** deep dive knowledge in a specific classical technical field (such as automotive engineering, mechanical engineering, electrical engineering etc.) with an evidence of ~5 years of operational experience "on the product" in industrial practice
(b) **The "horizontal breadth":** integrated or additional knowledge and engagements with respect to project management, simultaneous engineering and

collaboration, international business understanding and assignment and cultural leadership.

Once being part of the Management system individuals grow internally with their experience, performance, internal reputation and network connections as well as with their capability to comply with the internal leadership culture. External hiring benefit from unique, complementary skill sets but it has to adopt to the companies' behavior style very quickly to stay on the fast track for promotion.

Unlike new digital start-ups, pure software companies or leading GAFA[2] type tech and digital data corporations, the traditional industrial corporations still rely heavily on HW (hardware) product centric leadership capabilities. Those leadership skills do include the specifics of many years of product hardware development and prototype practice within a specific business (factory and OEM-supplier network) and innovation (material and production technology) environment. Adopting management skills from one industry branch to the other remains difficult still today!

Which Challenges Exist to Build and Drive the Digital Transformation?

The new world of sharply advancing digital driven business approaches and working pattern makes it necessary to rethink this traditional approach and also to introduce new career elements and skill-sets into traditional industrial enterprises. This cultural and career relevant change as part of the overall digital transformation in most of the cases happens slowly. In the best case, this transition happens evolutionary. In some cases it is caused rapidly by crisis modes that are resulting from declining market equations and major technological system changes (e.g. like the migration from combustion powertrains to electric powertrains). It is also increasingly driven by a sharp growth of SW enabled product intelligence which leads digital conversion approaches such as replacements of single engineering control units (ECUs) to overall digital operating systems for the management of embedded software intelligence in products. Those drivers are often recognized too late in order to change pro-actively the development approach in companies. Many traditional industries cannot change as fast as they should and, therefore, the development approach is oftentimes not handled as part of a disruptive full redefinition of the company. In order to provide more opportunities for business segments to develop in such a dynamic phase of transition, companies allow a separation of new skills as part of a spin-off. This helps to provide leeway for quick transformation apart from the rather static set-up of the parent company. The reasons behind such an approach are the human individuals themselves since they do not like to leave their own comfort zone that allows them to run business based on well-established experience. Consequently, they do not drive self-motivated new and unknown business behaviors and success patterns. The question remains, how do drive digital transformations more

[2] The acronym GAFA stands for the leading western world digital tech corporations such as Google (G), as most used internet search engine), Apple (A), as a leading digital technology company, Facebook (F), as most used social media platform, and Amazon (A) as the world leading online dealer. Their new platform based business capabilities builds up on unique digital leadership mechanisms beyond the traditional strongly hierarchical organizations.

dynamically even in segments of well-established industries like mobility and vehicle technology, machinery, aerospace and aviation?

The fundamental base challenges for digital transformations in enterprises are the following ones that are difficult to achieve by ordinary management principles and behavior with a focal point on operational control:

1. The first base challenge is to get the whole organization behind a *new digital approach* and not just specific teams. As a pre-requisite, a core management team has to drive consistently the associated culture, architecture and operational spirit forward. It cannot just be down delegated to others! Experts from inside or outside might help but cannot substitute the key management leadership drive.

   The hidden dilemma: unfortunately, progress with respect to the whole organization is only as quick as the slowest hitter is and the landmark where to hit the ball needs to be clear for all in advance! This, however, needs dedicated commitment and encouragement by the teams and their leaders themselves. Business as usual needs to be openly dismissed and new ground rules based on new principles need to be established.
2. The second base challenge is all about knowledge, understanding and motivation. In order to achieve an organization shift towards new digital working principles the organizational members need to reach a similar comprehension level of the new approach.

   The hidden dilemmas: in case of long standing and well as established industries and enterprises, it is difficult to get all members motivated and activated to learn new digital styles and technologies of working and operating, and also to trust them and to adopt them for the personal working system and environment!

   From human perspective, it becomes essential for Management to provide practice zones (even for themselves): *if you had not hit a ball for a long time, you need to create motivation to (re)start again and you need simply practice to resume professional levels in order to show off in your "new digital neighborhood*".

Digital transformation needs a new open mindset with trust and transparence to equally comprehend and appreciate the different dimensions and influence factors. As it was already introduced and explained in Chap. 6 ("*The set-up of Virtual Product Creation in Industry—best practices, error modes and innovation speed*") the dimensions of the Engineering Operating System (EOS) provide an excellent understanding of challenge to harmonize the success factors for digital business and working solutions:

- *Process and organization* (the classic strength of Management)
- *Tools and IT system integration* (typically foreign to the majority of traditional Business & Engineering Management, and therefore, delegated to the IT organization)
- *Virtual models and digital data sets and information* about physical objects (usually not known by Business and Engineering Management, not even well known to IT Management!)

- *Operational activities* of designers, engineers, analysts, process planners etc. (some knowledge exit due to personal background and practices; this, however, is in most of the cases outdated in Management)

In order to rebuild, to extend and to newly set-up Virtual Product Creation as a key engineering discipline in industry, Management has to provide step by step a vision, a mission and leadership (in person and as a team) with a passion to design and determine such new and extended capabilities.

Virtual Product Creation in industry needs the following five critical proactive contributions by Management Leadership:

1. Leadership in vision & mission of the new Engineering objectives and intelligences. Product Development and Manufacturing Engineering need appropriate digital enablers and solution elements. Management has the task to enable and support modified business objectives by appropriate new engineering approaches and solutions. Thus, Management has to start paying more attention to robust systems engineering and integration of continuous and agile SW development as part of embedded sub-systems with modular interface driven hardware architectures and component deliveries of products and technical systems. Do not talk about data, argue with data! Never accept fuzzy meta data reports about development status, ask for the evidence of functional fulfillment by the individual elements, and do not hesitate to view it live!
2. Achieve personal commitment of individual Managers as well as aligned Management team engagements in defining needs, alternative analytical capabilities and specific target settings of the Virtual Product Creation target architecture (which comprises new engineering principles, digital processes and workflows, new synthesis and analytic capabilities and overhauled or even enhanced digital tool sets and methods). Work pro-actively together with research institutes and consultants, but get entrenched yourself—become part of conceptual run-throughs already in POC (prove of concept) work, rather than getting show cased late deliveries with flaws.
3. Drive change management to modify the organizational and cultural set-up and building the core fundamentals of the new full digital engineering approach by:

    a. deriving new engineering system thinking and integration principles,
    b. establishing new types of design reviews and decision-making,
    c. allowing and requesting constant reflection of working and collaboration practices incl. lessons learned) and
    d. ensuring appropriate scaling up of the new-piloted digital behaviors.

4. Incentivize Managers to work directly with data and information sets offered by business intelligence tools in engineering business rather than just requesting passive Excel- and PowerPoint File documents and presentations. The management & control of Virtual Product Creation development, deployment and business integration as well as daily digital engineering operations (rather than just steering a project set-up of VPC) needs to become an active management type of tasks of modern and future Managers!

5. Establish dynamic and continuous leadership control of digital engineering delivery (architectures, software, digital models and data, digital prototypes etc.). Digital Engineering activities need to be driven actively and in a continuous (agile) way by Management beyond the classical stage gate concepts, rigid milestones and gateways of existing project management methods and toolboxes.

From the end of the 90s of last century, Management Re-Engineering programs, Business Schools and MBA type of knowledge have heavily influenced Upper and Senior Management.

In addition, Project Management has been introduced intensively in most of the industrial enterprises. Both elements have been driving the way of working of Management towards regular operations reviews. Such operating meetings have been "optimized" with respect to the overview of actions versus tasks and initiative targets.

As a consequence, (Upper and Senior) Management step-by-step migrated into their own world of documents, which was increasingly decoupled from the rising stage of digitalization in the engineering world. Figure 18.2 illustrates such situation:

- Since Managers are no longer able to work directly with digital tools as part of Virtual Product Creation, they request a down cascade of relevant information from a full 3D analytical model environment into an easy to use PowerPoint presentation with a pre-filtered reduction of digital content and information (step 1 from a to b)

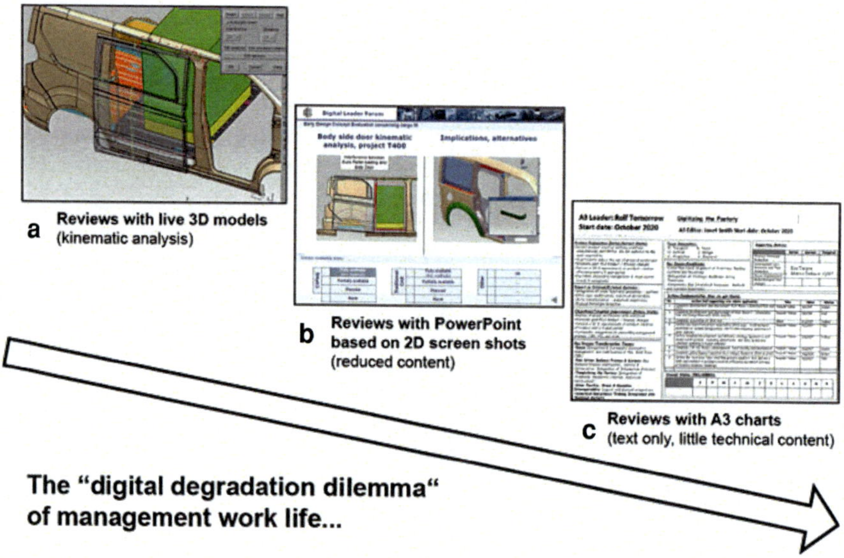

Fig. 18.2 Reduced "*digital content environment*" specifically prepared for upper and senior management

- In addition, especially for Upper and Senior Management, this type of information is further transformed into overall Management report sheets like an A3 documents (compare the background of A3 Management thinking in [1]) which are introduced to keep an overview about lessons learned, new business practices and management initiative and program description and status. From a digital point of view, those documents are nothing else than text verbatim and structured time & status elements without any trail or traceability back to the potential high value full digitalized materials (step 2 from b to c).

It shows that the digital transformation still has to change traditional Management practices in order to use digitalization consistently throughout all organizational working patterns. VPC and PLM solutions companies for a long time did not pay enough attention to this dilemma. Consequently, still today, there do not yet exist traceable "digital live/life interaction" dashboards for the diverse needs of management. It will be a major research and development task to build up *Engineering Intelligence Charts* that leverage diverse sets of digital and analytical data and models in engineering and production (similar to "Business Intelligence" charts). The complex socio-technical product-service systems of the future depend on a rich mix of partial system parameters and their dynamic control interaction via rules and/or data-based analytics. The resulting demand to observe, review and dynamically change their interdependence and traceability will grow substantially. This drives the need for live ("*exactly now*") and life ("*forecast, rewind of lifecycle periods snapshots*") interaction dashboards.

Traditionally, Management is powerful if it assumes responsibility and accountability of "something important" and hence feels ownership in command and control of it. Unfortunately, for a long-time engineering management in traditional manufacturing companies allowed itself a way out of direct digital responsibilities by simply leaving it with or transferring it towards IT organizations. The thinking behind such an attitude is simple: "digital means that software is engaged and software should run on computers, on databases and across digital networks. IT departments are responsible for such operations". The actual delivery elements created by digitalization were considered within this context as a storage element within the IT environment rather than as an engineering asset! The creation responsibility to change the underlying engineering working system towards new digital capabilities, elements and solutions is, therefore, oftentimes not proactively considered by engineering management. Consequently, many companies created *project type or "extra" organizations* to drive those digital elements within enterprises rather than integrating digital responsibilities directly into existing engineering departments. This even exacerbated decision making in assessing, deploying and execution new digital working solutions in industry. As outlined in Fig. 18.3, management is supposed to lead the digital transition overall and to have ownership for its robust set-up.

It starts oftentimes with a flawed ownership of scouting for, assessing of and proving out new forms of digital solution capabilities (see element 1 in Fig. 18.3). Only few companies have already established ownership for this including regular capability reviews to understand and drive such new digital engineering solution

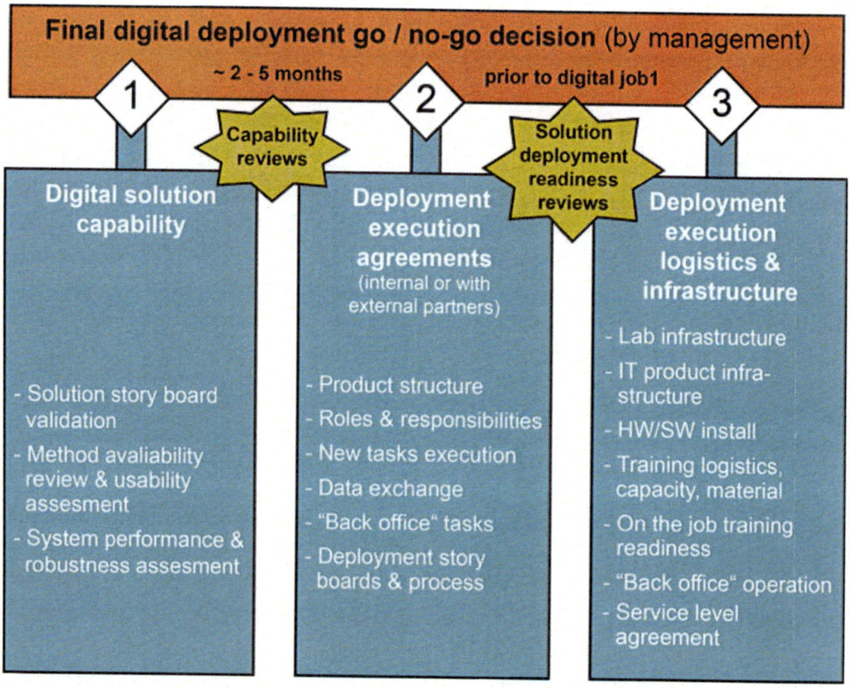

Fig. 18.3 The 3-pillar staggered digital management leadership responsibility

elements. Endless budget competing rounds as part of preparing for annual innovation funds are meanwhile exhaustively entertained in industrial companies to allow certain digital prove out work for the following business year. This painful waste of *creative enthusiasm and power* are significantly contradictory in order to establish meaningful ways of allowing continuity in digital innovation and progression.

The second challenge, however, is even bigger. As shown in Fig. 18.3 as second element, the critical task of developing, arbitrating and establishing agreements amongst all stakeholders for digital solution deployment, implementation and execution is key for any further digital success. It needs strong personality and leadership with direct contact to senior management to become powerful enough and successful. Most of the companies still experience constant failure modes with this element due to missing ownership and leadership for this task. Engineering managers sometimes are thrown into such leadership role as part of a sideway career step without having the necessary technical skill set or at least meaningful understanding of it. Hence, they act cautiously without strong mission and might look already for other or next career opportunities. In many cases, such ownership is not placed at the right level on the management hierarchy. Therefore, these managers do not have the right power to negotiate and determine necessary changes in digital engineering work practices across engineering and manufacturing disciplines.

For many companies the deployment readiness reviews and gateways for a new set-up of digital engineering roles and responsibilities as well as new working solutions are still foreign. Oftentimes, companies are unable to establish such transparent reviews due to their traditional set-up of concentrating on department internal rather than cross-departmental digital progression. Other inhibitors are created through a low-profile set-up of digital solution responsibilities within the overall management set-up.

The third major leadership task within industry digitalization and digital transformations is the responsibility of the digital solution execution and all associated logistics, infrastructures, reporting systems, steering and escalations mechanisms. Co-ownership between business engineering leadership and IT-leadership remains difficult due to different interests, success factors and associated business metrics. Business engineering management focus on getting delivered fully tested, robust digital technology for their work force with *no* or *exceptionally low* levels of bugs in digital application and digital workflow solutions.

Engineering management, thus, have major problems in accepting prioritized and force ranked software bug & error listings (and accepting implicitly that minor bug fixes might not get delivered at all!). In addition, training and competence set-up and progression is favored via the help of "on-the-job trainers" (OJT) who provide direct solution help at the engineer's work desk on the office or factory floor.

IT departments are more concerned about their responsibilities to provide a stable IT server and application factory according to certain service level agreements (SLAs) which need a buy-in and sign-off from business engineering leadership. This does include back-up operations, server uptimes and rapid data conversion and delivery services to the IT solution user base.

Overall, it is still a major challenge in companies to identify, shape and authorize the right management team, which assumes ownership and leadership for the ultimate decision and operation responsibility of digital solution deployment readiness and the final "*go or no-go decision*" with respect to full digital operation in business.

18.2 Management Behavior Do's and Don'ts in Digital Leadership

This section provides two examples of good and bad practices of management behaviors in leading digitalization, Virtual Product Creation and PLM in industrial companies. Those reflections should help companies, their management and future leaders to apply the right personal attitude, business acumen, digital technology assessment capability as well as appropriate motivations, judgement calls and leadership skills in the context of digital innovations and transformations.

The first example shows how political pressure is misused in order to rush for quick moves in a complex development context with immature digital solutions just to deliver success for personal career purposes.

Virtual Product Creation Experience in Industry
(*Management don'ts in digital leadership*):

Aggressive rollout plans for immature digital solutions in a complex development environment

The overall situation

The enterprise *Future Automotive* is under pressure: the scalability of their platforms and technologies needs significant improvements (new powertrain types, new connectivities and intelligent functions, location-based service integration etc.) in order to achieve higher margins in solution offering on the future mobility market. Therefore, a merger & acquisition strategy has been followed thoroughly. In consequence, the new partners and brands of this "fusion" must now work tightly together in delivering such new intelligent products, architectures and technologies. Unfortunately, as it is still usual in the twenty-first century, the individual company digital engineering solutions sets are diverse enough to prevent easy, efficient and effective engineering collaboration. Enterprises within company networks are, therefore, forced to entertain additional costly *digital bridging solutions* to "translate and deliver" engineering work results across the diverse sets of virtual product creation solutions. This accounts for approx. 20% extra cost which usually is covered by "hidden pockets" of the overall engineering budget spending. Nevertheless, the new enterprise transformation program "*Digital Innovation Edge*" has been funded and set up meanwhile to deliver within 4 years the new engineering capabilities of the future. This digital program has a similar size and funding schema as a typical full major vehicle platform delivery.

B. Mr. Tanterelli leads the global transformation program *"Digital Innovation Edge"* and has a group of seven managers reporting to him to ensure the appropriate development, delivery, deployment and daily execution of the new future digital engineering solution. Overall, approximately 400 heads are involved in this mission critical program.

Assessing digital solution readiness and deployment start

Mr. Tanterelli has invited all seven managers and various technical experts from five locations in the world to his digital headquarter facility in Shanghai, China. Overall, 4 days of common work in a team of approximately 40 persons are on schedule.

The first two days of technical deep dives are used via the help of *Digital Solution Capability reviews* (Fig. 18.3) to assess the true situation of the development status of the new digital platform (after 2 years of work). This is done together with three major digital solution providers and PLM vendors in order to prove, test and potentially sign-off the various digital capability degrees and to outline significant risks. The third and the fourth day are used to translate these findings into a status for generic deployment readiness in order to potentially start negotiations with vehicle and technology development program management.

Now it is Thursday afternoon and Mr. Tanterelli has summoned a 2 h "*final conclusion and decision*" meeting with his seven managers and support staff (overall more than 20 people in the room). Mr. Tanterelli starts the meeting as follows:

"Team, it is now time that we deliver our solution to the company units, no matter what it takes. Last week, I already promised this to the board and you have enough time to assemble everything this week here in Shanghai. You know, that we have to to move fast and therefore, I expect this pro-active behavior from you all! Are you on board?"

Nobody said a word, although everybody knows that the digital solution still has major flaws and that even the digital solution providers and PLM vendors could not yet recommend the full business use of this platform. The risks that development projects would fail due to missing robustness, non-delivered functionality and limited scalability have been commonly assessed as far too high.

Mr. Tanterelli continues the meeting by addressing each one of the managers 1:1 in front of the entire crowd. He starts with his local manager who has been assigned to deliver this *new digital platform* to a local (China only) derivative of an existing technology platform: "Mr. Li, are you ready to use it and will you be successful?" Obviously, Mr. Li quickly nodded his head and confirms with a silent yes. Mr. Tanterelli continues this approach 3 times more with managers from other regions in the world who have easy circumstances with only limited content development projects which make it easy enough to potentially revert back to the legacy solutions. In the 5th term, B. Tanterelli directly turns to Dr. Ryan who has the unfortunate task to support a full-fledged global platform delivery program out of Europe amongst three brands within the first new fusion commitment of the global enterprise *Future Automotive*:

"So, Dr. Ryan, how is it with you? Tomorrow night, during your flight back to Europe, I will be participating after midnight local time in a leadership call with the US and Europe. In this meeting, I have to explain whether we will shift the important platform program "*EU fusion excellence*" to the new global digital platform, or not …? So, how will you advise your Vice President in Europe Monday morning on your return? I need to know this now!"

Dr. Ryan recognizes that now all 23 persons in the room were looking at him knowing that he will now have a hard time to say no to his international boss Tanterelli, who has earned a well-known repu-tation to get quickly hot tempered in situations where he does not get answers he likes. So, Dr. Ryan waits for 5 s before he answers firmly and clear so that everybody in the room can hear it.

"Mr. Tanterelli, of course I will tell you now exactly what I will tell my Vice President back in Europe Monday morning. I WILL NOT RECOMMEND USING THIS DIGITAL SOLUTION YET; it is not ready for productive use and would cause major problems to the *EU fusion excellence program.* Sorry, but I have to be honest with you and to the company and I am more than happy to explain it to you."

Mr. Tanterelli immediately changes his facial expression, stands up and starts to take full control again of the meeting.

The next moment, Mr. Tanterelli starts to shout to Dr. Ryan: "I do not accept your position, this is against the commitment I had given already and you have to follow my decisions! This will destroy your career …".

Immediately, all other members start to leave the meeting room and after one minute, Dr. Ryan is alone with Mr. Tanterelli. Even the overall pan-European superior of Dr. Ryan, Joe S., has left. It seems now that the meeting degenerates to a performance report meeting for Dr. Ryan. Nevertheless, Dr. Ryan stays calm and answers back to Mr. Tanterelli: "Please stay professional and calm with me, I would like to explain how I came to my negative conclusion and I would like to fill you in since this will be important for your meeting tomorrow and potentially also for your future career. Let us go to the board, I will explain …".

Mr. Tanterelli somehow understood that he should better listen now and he allows Dr. Ryan to provide this interactive briefing to him explaining the snapshot of findings from the various reviews on the days before, incl. all risk assessments. After Dr. Ryan had survived the next 5 min after this clash, half of the other members return back to the meeting room and become part again of the review.

Mr. Tanterelli at least commits himself that he will think about it until the next day meeting. On the next day, when Dr. Ryan is in the air on his flight back to Europe, Mr. Tanterelli reverts back to his prior position and promises the use of the new digital solutions to the "*EU fusion excellence*" program. For Dr. Ryan it becomes again a difficult meeting in the office Monday morning explaining to his Vice President what the real situation is. The following days were full of clarification meetings and excuses. Bottom-line, however, it is understood why the solution can not yet be used. Such promise on wrong digital capabilities happens again 6 months later and again it is too early. The "EU fusion excellence" program finally has migrated 4 years later to the new digital platform after it was proven out with another global platform initiative 2 years after the Shanghai meeting.

Progress in new digital solution offering takes time and integrity!

Lessons learned, bad practice and successful re-action:

- Coming physically together to interactively study progress and shortcomings of upcoming digital solutions is critical for success and needs to be executed professionally; being in one location helps to grow together in understanding and common position, however, it does not prevent misuse of power!

- Making up a conclusion session to ask for non-critical buy-in to a pre-commitment, which was given by the most senior leader already far before such a review week is worst practice and should not be supported at all. Such bad management eagerness and misuse behavior destroy all trust levels of digital commitment and encouragement and adds negative damages to partnerships for successful digital operation.
- Putting individuals on the spot in a meeting amongst equals with the help of super power from a senior person constitutes bad and non-constructive leadership behavior.
- Other teammates also in management should help each other and should stay united to protect the team against such behaviors, leaving the room is the wrong reaction!
- Staying professional and calm in reaction in such a situation is best re-action and might lead to at least neutral moments in decision-making; however, it cannot achieve mindset shift at the other end.
- Overall, integrity of individuals are noticed by others and help changing the digital culture and management approach; however it takes time and needs similar behaviors by others in order to make positive impact.

Four years later, in an occasion of general reflection, Mr. Tanterelli thanked Dr. Ryan for his constant integrity and honesty in technical assessment and for his encouragement to speak up even in critical situations. He admitted that he did not find enough managers around him who offered such an attitude. Dr. Ryan accepted this late praise!

It should be noted that positive digital leadership is rare in management. Finding good role models remains essential to positively influence the digital culture and attitude of an entire organization. However, it is encouraging enough that such individuals exist even amongst management members who did not have the opportunity to deep dive in Virtual Product Creation technologies or specific digital engineering solutions.

The second example describes a digital leadership behavior characterized by a range of positive and successful general management attitudes: trust, capability to listen and comprehend, good personal preparation prior to important decision-making meetings, support of individuals in difficult situations like protest and rejection in critical meeting situations.

Virtual Product Creation Experience in Industry (Best Practice)

How digital solution implementation can rely on excellent senior management behaviors even in critical situations

The development situation

Senior Management had endorsed a major digital innovation and transformation program called ***"Digital Intelligence Future (DIF)"*** 3 years ago in order to reduce time to market by another 15 percent whilst increasing efficiency in delivering new powertrains, highly automated driving solutions for next generation interconnected mobility services by at least 30 percent! Those numbers are already booked as contributions within the future cycle plan. Now, after 3 years of global prove out in pilots of the new digital solution architecture the first major technology architecture program for level 3 highly automated driving is about to start in 3 months. The new *DIF* solution architecture comes with a new integrated PDM/FIE (product data management/functional intelligence elements) solution for hardware and product intelligence offering, incl. a dynamically coupled ALM (software application life cycle management) solution with an integrated SW delivery development (DevOps) and delivery (OTA, over the air) pipeline. At the same time, all (SW & HW) engineers need to undergo a new *Advanced Systems Engineering* training curriculum to change to a 70% MBSE (*Model-based Systems Engineering*) digital development solution. Management is targeted to become 50% more efficient by directly working with all digital elements in specific *Digital Management Browsers* (DMB).

The new *DIF* solution leverages engineering intelligence brokers with AR/VR (Augmented & Virtual Reality) interaction gadgets as well as IoT data platforms for data analytics. With the help of *DIF* each product in the field can be operated with a range of up to 100 digital twins, which themselves can be configured dynamically for different business, functional, safety and environmental purposes. The dream of digital thread and digital continuity, which became popular around the early 2020s, finally becomes reality. To be able to sustain leadership, each manager will have to train his/her AI enabled Digital Bot Assistant in order to keep abreast about the information and model inflation of the new world (individual have to work daily with at least 50 digital models.

Approaching the digital solution deployment agreement

Dr. Ryan knows that such transformation will be stressful and will cause fears and potential mistrust, especially within middle management. Those Managers are under enormous pressure to deliver the future new product types for the company with respect to the technology architecture for level 5 autonomous driving. Therefore, only selected individuals have been chosen, all of them with a strong record of excellent development and collaboration skills. Dr. Ryan knows that there are even 2 managers amongst them who had the opportunity some years back to develop "their own project" *pragmatic digital development solution* for one of the internal "*Beat the competition fighter*" development projects. Such important projects to that time have all been designated with Greek goddess names such as Athena, Artemis, Aura, etc. to underline the "epochal dimensions" of such a project or the future product line-up. However, Dr. Ryan also knows that both projects have sunk 80 million cash for such short-term digital solutions that were neither architected to work robustly nor to be scalable for the entire company. Now, with the company initiative ***Digital Intelligence Future (DIF)*** it is the other way around: a team of 100 experts has been co-located in a central location with 50 satellite collaborators to pilot the new way of digital engineering of the future for 2 ½ years. A non-cycle plan listed experimentation vehicle was developed with this new *DIF* solution in 15 months in order to be sure that *DIF* would deliver all relevant digital capabilities. The new digital working styles, types and tasks could be tested, observed, measured and improved to achieve deployment readiness. Dr. Ryan now has the task to introduce all managers to this new digital future!

Preparing for the deployment readiness meeting with management

Dr. Ryan has agreed with Denise K, Vice President R&D that it will be essential to prepare the management team in a dedicated, mandatory meeting under the leadership of the Vice President in 4 major steps:

1. Understand the core changes and the urgency of action
2. Explain the plan for first 12 months of deployment and training
3. Align the motivation, provide faith by showing pilot results, explain the steering set-up and the explicit manager tasks
4. Wrap-up with and next steps for the next 3 months.

The day before the management cascade meeting Denise K. invited Dr. Ryan for a late meeting at 8 pm to her office to get last things prepared. After having run through all slides, statements and tasks, Denise K. concluded to Dr. Ryan: "Tomorrow's meeting will be a difficult one; I wonder how many slides you will be able to show without intervention from some of the "alpha" managers. Just be prepared for it and please stay calm. You simply cannot avoid it! I will stand-by and will react appropriately to cover the situation. Trust me, we are one team!".

Deployment D-day with Management

It is 10 a.m. and other than normal, everybody is already seated in the management oval room, 25 managers, 12 at the local site in Germany, 7 at the remote side in the UK and 6 in China. Denise K. starts this special meeting by explaining shortly but precisely what the goal of the meeting is: to get every manager pro-actively on board to fulfil his/her leadership role in the most significant digital transformation in R&D in history of the company ever. She expects that everybody comes prepared for taking on personal assignments since everybody had received a preparation booklet by Dr. Ryan already 3 days ago. She looks around the local and virtually plugged-in management team and then turns to Dr. Ryan, who is seated directly besides her, and she finally says: "Dr. Ryan, please take us through our tasks step-by-step and explain them in all clarity, please".

Dr. Ryan switches on the newly 3D immersive Digital Presentation Streaming which has been replacing the formerly known PowerPoint presentation slides. This way, he can insert all live data from the new *DIF* environment piloted with the experimental car whenever needed.

Dr. Ryan starts—according to the agreed 4-step approach—by pointing out major drivers for the new digital development approach in the company. In minute eight of his presentation, on the fourth slide with the title "The plan for the first 12 months of deployment and training", one of the most experienced "alpha" managers, Gilbert G., suddenly and firmly interrupts him. He raises the question: "Dr. Ryan, are you serious, that we as managers have to really undergo a 2 weeks training …for these new digital gadgets? We have more important things to do than wasting our time with IT tools. Our engineers, designers and analysts are supposed to use them, but not us. We are the managers and not the digital workers!" Dr. Ryan looks around to all the other faces and they signal similar unwillingness and fears to him. However, all of them remain in tense silence. The internal voice of Dr. Ryan reminds him of the mission he is on and provides solid encouragement to him. Dr. Ryan responds clearly to Gilbert G.: "It might be unusual to you and your peers to make such a bold step to a new digital working environment of the future, but yes, I am serious about this step. Together with our global expert group and all management stakeholders in the ***Digital Intelligence Future*** team we have exactly decided this and do expect it also from you!" After 2 s of stunning silence across all rooms, major mumblings crop up everywhere. Another 10 s later another key manager, Andy S., raises his voice and directly addresses to the Vice-President: "Denise, it does not make sense at all to continue this waste of time here, this is unreal. It does not make sense that we get educated by a Digital Leader on how we as managers should digitally work in the future. It is up to *us* to decide what we need and want. Then we advise the digital experts what they have to deliver to us. This is the way, as we have done it in our "*Beat the competition fighter*" development projects some years ago. Please stop this here, immediately!" This speech killed all mumblings immediately. Again, stunning silence across all rooms.

Dr. Ryan has noticed that Denise K. has begun to scratch her finger knuckles during the last minute and has been changing her face color slightly every second onwards. The moment has come now as Denise K. had forecasted it last night in the preparation meeting. Consequently, Denise K. starts to talk silently but *very* firmly to all managers.

"Team, you better think before you talk. You know what is expected from us. We have to lead and we are supposed to be the role model for our engineers, planners, controllers and teammates. The company made a commitment by heavily investing into the future and a whole team has worked hard on the *DIF* solution for a couple of years, the ***Digital Intelligence Future*** solution. It has been proven already that this new solution does deliver these long times awaited new digital capabilities. Now it is up to us to finally make a serious commitment. Let me ask you, as my direct reports: "Who of you does not believe in this solution or is not able to make a commitment to learn and adapt to this new digital working environment? Please speak up know before it will be too late!" Denise K. pauses for ~10 seconds, looks everybody in his/her eyes, also to those in the virtual rooms, finally turns to Dr. Ryan and says in a very calm tone: "It has been cleared up, everybody understands what it takes, please continue with your explanation and plan for all of us, Dr. Ryan!"

The meeting goes on in a very professional way; all managers get "pro-actively" interested in understanding the plan forward and to learn their personal role in the upcoming leadership events with their own teams.

After the meeting, Dr. Ryan was approached from almost every manager. The following weeks and months obviously were cumbersome and stressful; the productivity went down by 30% for the first two months. Before working efficiency went up again to normal it took 5–6 months. The new potentials started to pay back already after 10–12 months and the rollout extensions to other teams were signed-off 14 months after the first productive introduction! Digital transformation can work.

Lessons learned, best practice and encouragement elements

- Despite major investments into new digital development environments, many managers still have difficulties to accept their new leadership role and personal working activities in it!
- Timely, mandatory meetings to align management teams as part of major digital transformation initiatives are key!
- Management needs to engage with digital experts to become digital drivers, to mitigate criticism upfront and to provide trust!
- Team leadership and digital passion drive motivation!
- Management needs help in learning digital leadership!
- Digital future needs reactiveness and not concern behaviors!

Comparing the two lessons learned examples the question comes up to how industrial management approaches and measures the capabilities of new digital engineering solutions and their readiness for deployment (i.e. start of usage) and for scaling it up in ordinary or new digital engineering activities? This, indeed, is a major

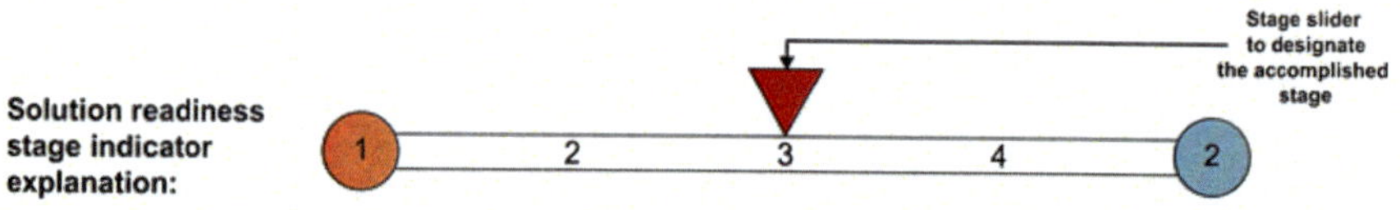

a) Concept, solution, method offering	• VPC concept developed, • Base capabilities in IT applications identified • Feasibility given • Capabilities exist to demonstrate the main way of working with the new concept	• VPC concept refined and calibrated with day to day engineering examples • Needs for method & tool functionality indentified and agreed to be part of specific digital solution release	• VPC concept and Conceptual Information Model (CIM) finalized • Agreed way of using the solution for real life work • Dedicated project team + plan exist to develop the solution for specific job1	• All VPC working solutions are demonstrable in test labs • Method usability assesment done to point out critical areas for end user readiness • End user testing started	• All VPC solutions tested by end user • All methods and workarounds developed, defined and documented • Critical enhance-ments & priority 1 & 2 elements identified • Performance leads completed
b) Business needs, storyboard, integration, R&R commitment	• Business need for new/revised VPC concept and digital solution identified by experts and endorsed by management	• High level business goals and process integration model established • Identification of which functions & departments and job oles are affected and in which development phase	• Deployment story board development finalized and agreed by key leaders from the functional activities (and respective management)	• Final agreement within organizations on usability of the solution delivery & related story boards • Business impact and value analysis completed • R&R agreed	• R&R and resource estimates finalized • Formal commit-ment by senior management to get individuals trained to start productive use of the new solution in a given develop-ment project

Fig. 18.4 Stages to describe and measure digital solution readiness

shortfall in industrial companies and is not (yet) treated serious enough by a robust management framework. The overall framework introduced by Fig. 18.3 in many cases is not executed thoroughly and are not underpinned by explicit management tools.

Dr. Ryan in the industrial examples of this book obviously uses a detailed "*solution readiness stage indicator*" in order to drive, validate and sign-off precisely the readiness of digital capabilities before they are treated as verified candidates for deployment. Figure 18.4 shows the two levels of readiness assessment and declaration:

(a) The level of concept, solution and method offering
(b) The level of business need, engineering storyboard integration and commitment for roles & responsibilities in the organization.

The stages 1 through 5 provide a precise readiness clarification with a gradual progression of individual characteristics such as principle solution understanding, conceptual fit to engineering examples, match to enterprise needs, working demonstrations and detailed method and digital application readiness. Those stages also address various aspects of business goal fitness.

Due to a missing understanding of such a management control approach, the majority of companies leave it up to individuals to find ways of how digital application functionality can be used without any stringent fit to the digital innovation and transformation goals of the individual company. It is somehow comparable to the situation in the beginning of the twentieth century in the physical world with the *non-factory like* individual machine shop working habits: at that time there did not exist any consistent factory approach for effective and efficient production line readiness levels.

Which Other Digital Challenges Exist for Management and How to Build the Digital Transformation?

The following general observations might help management to find their own way to set-up, drive, architect and steer digital innovations and digital working transformations within their organizations:

1. The challenge is to get the whole organization behind a new digital approach and not just specific teams; unfortunately, progress is only as quick as the slowest hitter is!
2. In order to achieve an organizational shift towards new digital working principles the organizational members need to reach a similar comprehension level of the new approach.
3. Without charismatic leadership of individual senior managers, it is almost impossible to drive the digital transformation. If a senior leader personally buys-in, shows trust, willingness and also a certain degree of dependency to/on technical expert teams, then this time of bond serves for the entire organization.
4. To drive appropriate communication and sensibilization of *digitization* in management, it is necessary to know the structure of the company and its management schools and behaviour types. Depending on the size of the company and the management structure, it is essential to adapt the communication strategy accordingly when communicating *digitization*. Different situations exist in different types of enterprises, not all can handle *digital innovation and transformation* in the same way:

    - ***Mega-companies: 200.000+***: these enterprises can set-up special digital departments in order to prove out new technologies before making them matured enough for large scale application; due to their size and global footprint with regional business and culture differences many efforts are necessary just for alignment and reconciliation of local digital working practices. In these types of companies, the internal structure of the company and the strategies of certain functional areas or managers associated with it add to the challenge. In order to stand out from the crowd of other areas, departments try to distinguish themselves from others with the help of digitization projects. This can lead to unexpected headwinds when attempting digitization, even despite the fact that it makes sense. Consequently, this might cause extra efforts again in terms of re-alignment and reconciliation to fit to a common enterprise approach.
    - ***Major companies with 50.000+ employees (often tier 1–2)***: big enough to tackle new digital capabilities once it fits to their business models and technology roadmaps. These companies are often highly dependent on technology shifts at and on alternating business equations of OEMs. These

companies increasingly compete with OEMs on new technical system technologies and the resulting product intelligence leadership (especially in the mobility sector). As a result, a higher versatility of innovative digital solutions and future Virtual Product Creation capabilities are necessary. Speed of realizing digital innovations is significantly higher compared to mega companies.
- ***Medium to Large Size companies with 500 to 5000 employees or even beyond)*: In many cases, still owner lead, where digital innovations need explicit senior commitment and convictions; digital transformation can be handled quite consistently if the business equation allows for it. It is highly dependent, though, from overall technology level and from willingness of the owner group to recognize digital innovations as key enabler for their future. In most cases, digitization is targeted at a specific area of the company. Therefore, it is important to address the manager responsible for this specific area of the company, in a double sense. The respective executive must be convinced that digitization is either necessary or delivers added value (measurable benefits). The path to corporate management (or owner) leads through this executive.
- ***Family-owned or owner-managed SMEs*:** such companies are usually strongly owner-oriented with regard to important business decisions and with respect to the right type of digitalization. Finally, the decision on where and how to invest into digital capabilities is made by the owner. In such companies, it is often the owners or their children who have built up the company and/or made it successful. Therefore, the introduction of digitization requires special sensitivity on work ethics and work attitudes and needs to respect the internal (analog) spirit & soul of the company.

It must be communicated to the company resp. its management briefly, precisely and transparently with detailed information at hand, where the advantage of digitalization for this company lies. Neither an international "fad" nor digitalization purely for the sake of digitization will lead to a positive decision in the management of the company.

The company's benefit from digitization can have several manifestations, either one of the manifestations alone or as a combination of these:

- Cost savings (product, internal processes)
- Reduction of development time for a new product
- (reduce time to market)
- Reduction of production time for a product
- Increase in product, process and service quality
- Shortening innovation cycles and ability to react to new consumer and technology trends
- Achieving competitive advantages, if applicable (incl. "white space" products)

For managers to be able to decide on a digitization project, it is important to show them which of the above points is to be achieved. Based on this, a decision can be made about a digitization project and its direction according to the company's priorities.

The following section is devoted to the question of how the next generation of digital leaders can develop as part of future management teams.

18.3 Development of Future Digital Leaders in Management

In order to meet the expectations and needs described in the previous two sections of this chapter, it becomes obvious that new digital leaders should be appointed, trained and supported for their development in engineering and manufacturing management. Independently of specific structural organization particularities of individual company specifics, it is essential to establish a new management skill-set, motivation and desire & passion to lead in comprehensive digital business and work situations. The following eight elements of management skills for Digital Leaders are key to achieve the goal of pro-active and competent digital innovation and transformations in industry. Those skills obviously need to be pooled with and cross-linked to the ordinary technical skill-set and experience in engineering and manufacturing (according to the industry branch necessities).

The following eight critical digital leader capabilities need to become part of any professional Management Development Program in industry:

- **(New) Digital business attitude**

Still today, there exists a puzzled picture on digitalization and digital business, especially within (traditional) industries. The first school of thoughts, mainly represented by the traditional management groups, considers "digital business" either

- as *new type* of "internet related" data business supporting services such as social media, internet browsing or other ordering and payment services or
- as a *traditional type* of internal company information technology-oriented service business to keep computers, workstations, voice-over-IP and data base servers up and running.

In any case, this first school of thoughts does not see itself in digital driving position; this is the task of others!

The second school of thoughts, mainly represented by the CDO (Chief Digital Officer) and related digital business consultants, favors a viewpoint that digital data should be treated as assets that must be used as new value creation elements on their own and are core elements for any digital business model in the future.

The technical IT related management and expert groups mainly represent the third school of thoughts. Here the belief is that digital business is closely connected to algorithms, software and software applications.

Digital Leader Management in the future needs to stay away from such single sided views on digital business and has to comprehend digital business on all levels in intra and intercompany business. Hence, it will be indispensable to teach and train the next generation of *digital leaders* in feeling and executing personal ownership to lead and guide new digital business understanding in traditional R&D, engineering and manufacturing operations. In addition, such new management needs to understand and live the difference in the digital approach: active usage of digital assets and provision of positive incentives for all participants to act that way will become the norm for successful leading companies of the future. The lagging ones will continue the obsolete set-up of three different schools of thoughts.

As it was envisioned already by the French existentialist Antoine de Saint-Exupery in the first half of the 20th century, which is reflected by his quote:

> If you want to build a ship, don't drum up people to collect wood and don't assign them tasks and work, but rather teach them to long for the endless immensity of the sea.

It is now the time to teach all management members the endless immensity of the digital elements, irrespective to whether they are compared to clouds, lakes, atmospheres etc. The desire to design the new world with such new digital business perspectives will also boost the positive and successful usage of Virtual Product Creation and its future capabilities.

- **Personal skill-set and experience in digitalization**

If you become responsible for something you have been working with for a long time or at least for a while, then you feel comfortable with it since you are a successful practitioner, even a true expert or at least an insider of this skill. If you receive responsibility for something that you have no real working experience with, you feel unsure und not capable of leading and driving it into the future. This describes a phenomenon that unfortunately still is in the way for more progressive and natural drive for digitalization in many companies. Today, many managers in execution power have limited and not up-to-date working experience with digital engineering tools of Virtual Product Creation. Consequently, they are heavily dependent on merely reflecting the associated digital innovations, actions and transformations on a digital meta level rather than on a digital execution level.

Investments are necessary into their own digital skills and into their competence to seek and find future ways of digital working. In addition, as digital leaders, they need to be trained to fight against their own traditional controller law that follows the "compulsive return of invest" syndrome: only invest if you are 100% sure that your "digital dividend" will pay off. In management, digital leaders should be supported in their development and given the necessary encouragement, competence and capabilities in order to provide "innovative digital start-up environments" in the company eco-system together with a reliable "safe harbor" commitment. This means, that the risk to fail with new digital working solutions needs to be pro-actively managed and not necessarily eliminated to zero.

On a personal level, all management members should accept to learn the digital groundwork situations by being actively trained and introduced to such situations at least one day per month. The digital leaders amongst them should be motivated enough to do so once per week! Overall, managers need to be trained in getting familiar with new digital technologies on a personal level rather than getting presentations only along the digital hype cycle enthusiasm by consulting companies.

- **Digital Leader influence to executive management levels**

Very few individuals exist high up in the management hierarchy to combine the following three capabilities: digital technical competence, digital business acumen and digital strategic thinking.

Therefore, it becomes crucial that digital leaders in middle management have good personal relations and contacts to the executive level. In many European and Japanese companies, there exist a connectivity gap of such desired close relations. Many proactive plans around digital innovations and transformations consequently do not find their ways at all to the top of the company via the ordinary vertical management meetings circles, or they get "brainwashed", "skewed" and "compromised" heavily on their way up. This slows down the overall digital transformation capability, quality and speed.

Therefore, it is highly recommended to train Digital Leaders in management to establish different kinds of "short-cut" channels, interactions and briefing types with and to Senior Managers, either on a personal mentor type basis, in their explicit role as a Digital Leader for a certain group of senior managers, or as part of specific senior executive digital review board. At the same time, it is crucial to learn a narrative explanation style for "rather difficult to understand" digital technologies and data/model realities in order to increase the chances to reach out to the comprehension, interest and motivation level of senior management. Digital Leaders need to have the skill set to explain the most relevant elements of Virtual Product Creation and digitalization in management language and show cast them with respect to executive buy-in comfort zone.

- **Dedicated digital training elements for management**

The times in which Management only works with office applications on a computer are over. Meanwhile, management rather gets actively involved in digital engineering work streams and digital sign-off of product functions, risk assessments, verification and production release.

The future will demand that Management gets intensively trained in data analytics, artificial intelligence, driven assistance and resulting digital intelligence assessments as part of the new world of decision making in Management.

nIn terms of comprehending and driving this new intensive style of digitalization, it becomes essential for Management to be technically well educated in core digital solution elements such as data, algorithms, engineering models, databases, information models, all as part of the company specific digital environment.

New digital training classes for Management should, therefore, cover the following core elements rather than software application functionality only:

- In our company, in my area of responsibility, what are the data, what are the underlying information models for them and what do they stand for? Where are they used (process and activities), in which way (e.g. digital methods) and by whom (job roles)?
- How to change and manage data and information successfully inside the company and across partners and the supply chain?
- How to establish a sufficient information model for digital business, whose views can and should be offered based on the different business perspectives?

- **Ownership for Virtual Product Creation capabilities**

One of the most critical challenges in industrial companies are expressed by the following question: who in the company receives which type of ownership to care about the digital future, the opportunities for different working practices, the associated engineering data and digital models as well as the digital business set-ups (incl. digital business models)? In most of the companies, it is difficult to funnel all relevant capabilities for such important range of ownerships into just one person. Thus, a well-aligned cross-functional team approach might rather be more stable to drive forward all of those necessary ownership set-ups and responsibility accountability duties.

Hence the Digital Leader training and education for management needs to provide the following critical elements which are supposed to be part of the final capstone team course with the help of practical assignments amongst the management training participants including their specific "home departments and digital responsibility areas":

- Technical competence to understand and assess how ownership should be established for new digital technologies and practices.
- Business experience to judge about the impact on existing and future digital business practices and analyzing the different forms of ownership around it (ownership for data, information, knowledge, engineering activity, digital model preparation, validation and verification, transfer and logistics to other users and customers, review and archiving, knowledge creation and fusion, etc.).
- Collective team ownership and split-up of responsibilities to ensure a consistent way of operational digital practice and the right strategic balance for the future.

- **Driving digital value creation and digital business benefits**

The strength of traditional successful engineering management is to oversee the situation, determine the performance and keep a good outlook. Based on these observations and use of management methods and tools, the future and the target of the future state can be determined, long before you realize it.

Unlike this traditional management approach, the existing way of analyzing, developing and driving Virtual Product Creation and its digital capabilities is not based at all on any solid *digital value* and *digital business benefit* theory foundation or underlying *digital value creation* model yet. This makes it difficult if not impossible to synthesize and develop the appropriate justifications for new improved digital engineering environments.

In order to make enough progress in this digital capability dimension Digital Leaders in industrial management need to be trained to collaborate closely with research institutes. They are currently in the process to create industry compatible models for *digital value definition* and the alignment to corresponding *digital business benefit* models (beyond the traditional value stream analysis of the physical world). The author of this book suggests the definition of digital value creation as described in Fig. 18.5 and the definition of digital business benefits as described in Fig. 18.6.

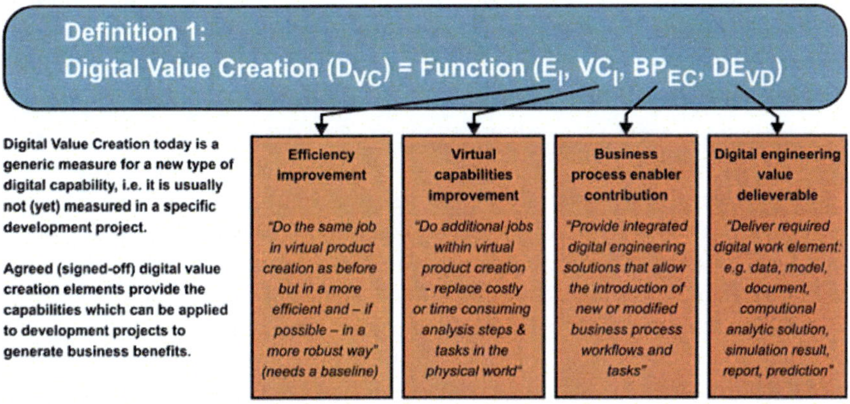

Fig. 18.5 Definition of *digital value creation*

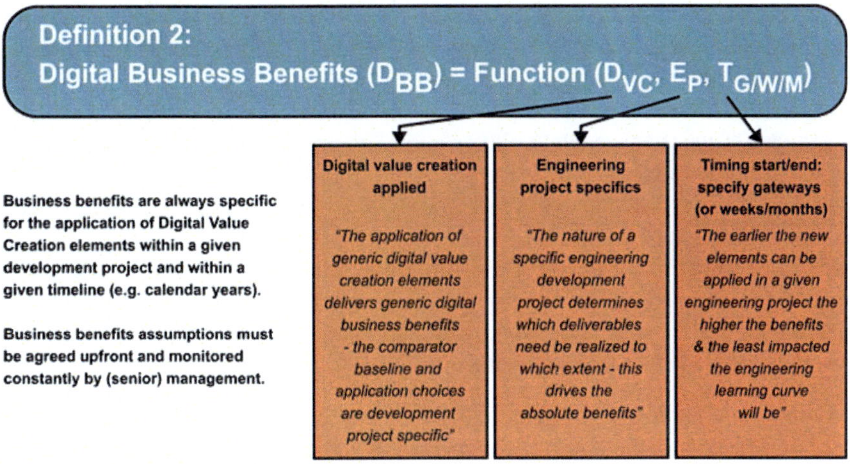

Fig. 18.6 Definition of *digital business benefits*

Digital value creation depends on the individual company specific absolute value factor "*digital engineering value deliverables*" that need to be assessed and determined first. Such determination does not exist in industry today yet. Based on this anchor element, additional relative factors such as "*business process enabler contribution*", "*virtual capabilities*" *improvements* and "*efficiency improvements*" come into play to increase or decrease the value. Whether there exists a certain time dependency drift (or not) in the generic core anchor element "*digital engineering value deliverable*" is still subject for fundamental research. No industrial standards exist yet for such digital value determination. Industrial Digital Leaders, therefore, need to apply their own skills to further drive such a digital value system within the various development streams of companies. The "*digital business benefits*", as second measurement scale, are built up on the "*digital value creation*" determined before, adjusted by the relative factors "*engineering project specifics*" and "*timing*" (usage related). Those foundational academic models Virtual Product Creation research can be now used to create first value assessments in industry in order to improve the understanding on how to drive digitalization.

- **Leading and managing digital transformations**

Digital Leaders in management will only be successful with a solid understanding of how to unlock the organizational rigidity and to drive changes in the business and technology culture. This new digital transformation capability becomes decisive as a personal and collective management skill. Training programs and respective personal education advisory are key to teach and train the new generation of digital leadership in building up and using the right leadership network to establish the targeted lobbying for systematic digital transformation steps. Leading for new digital set-ups and working activities require appropriate change management attitudes of managers. The personal motivation to drive and live the new digital value creation elements and to modify long lasting procedures of the traditional analog world into the new digital spectrum becomes essential. This becomes even more critical in order to convince the peers in middle management leading to a more transparent digital data-based management routine.

Cloning of the same character for a successful manager, as it was done in the past as part of the management cultures and associated management development programs in enterprises, is no longer the right approach. It leads to non-diverse leadership attitudes, which creates risks for broader digital mindset integration into business organizations.

Digital leaders in management need to embody a positive mentality and attitude, which clearly articulates opportunities and successful ways forward rather than pondering on pessimistic outlooks and preaching cautious ways of acting only. These new attitudes need to be learnt and practiced:

- Start with the positive message, the goal, the incentive, the aspired target, the motivation and explain the way forward on how to get there. Traditional professional management skills such as control, risk assessment and progress monitoring will kick-in as needed.

- Achieve a clear and precise explanation of digital elements, their need and their usage. Describe the digital transformation with specific examples from digital engineering workflows on the office and shop floor.
- Use the leadership instruments in meaningful ways: "give the right amount of cocaine, provide the right drug" (i.e. bring incentive and motivation into play), control the media with which you want to report about it in a responsible way and deploy "military and police" to help keeping order during the digital transformational steps.
- Deliver the right target setting for management and technical leadership in order to provide clear orientation like the ordinary Engineering Management is used to: e.g. in car development the product attributes such as vehicle acceleration from "0–100 km/h" in x seconds …, maximal weight of…, etc. Everybody needs to know the targets!
- Listen, understand, engage, support, judge and decide.
- Set-up ownership, trust and empowerment rather than check and control only.
  Never start with only the pessimistic and trouble related picture!

- **Setting vision and mission for the digital future**

Establishing a clear vision and mission is essential to get momentum behind the digital transformation and digital future. Such vision and mission should not be too complicated. Both are important to convince all members about the way forward and the business rationale behind it. Vision alone does not makes sense without a mission how to accomplish it. Mission needs more explanation and most importantly a solid budget for its realization! Individuals, teams and organizations will sense this difference quickly and will follow pro-actively if the future perspectives have been cleared, especially in regards to the personal future circumstances.

Providing an appropriate period for vision and mission operations is essential. Allow for a minimum of 2 years, if not a minimum of 3 years in bigger organizations, avoid going beyond 5 years into the future (in Europe, America and Australia, in Africa and Asia longer time periods might be possible from cultural point of view but less from digital technology point of view!) Be mindful to allow for internal digital lab and pilot experimentation and experiences with new digital solutions (PMTI, i.e. process, methods, tools and information standards) in close interactions with specific business and engineering initiatives. Digital Leaders should always reflect on known dilemmas when they design the digital future prove out environment:

- Existing future evaluation approaches often just provide lab activities on synthetic use cases as playground to keep a safety net against failures ("do not make your hands dirty" in case of occurring problems and negative results")
- Companies would like to create fast moving digital islands with a kind of start-up mentality within their organizations but do not provide to them the right base financing to establish full technology solutions. Consequently, many new digital solution principles do not get scaled up within enterprises and might even gain bad reputation unnecessarily! Successes get socialized, failure get privatized…

Digital tool driven transformations and those solely based on tool migrations or harmonization are not sufficient at all to declare a new digital future! Digital leadership also knows how to avoid part time hobby type of digital transformation approaches (accounting for another additional 5% leadership objective task for the year).

New digital transformation programs need a solid understanding of the following two challenges and limitations of today's digital business in industry:

- The change of IT departments in industry: the core of IT gets step-by-step reduced to the core responsibility of running the infrastructure and the associated server and network utilities in a robust manner (operational service). For the transformation aspects, it becomes more and more critical that IT departments accelerate their work on enterprise architecture frameworks and interconnected data bases and lakes to be offered as scalable services to all functional activities in the organization
- New digital capabilities like data analytics (based on data science principles) and data engineering need to be established in the functional activities of engineering and business rather than in IT organizations (compare also Chap. 21). Today, this is usually not the case yet since new digital approaches are oftentimes treated as a skill or task in the IT department per default.

Every organizational change is accompanied by a certain amount of unrest in the beginning and, depending on the extent of the digitization, a decline in productivity of varying degrees until the new processes have become established (compare the examples given in Sect. 18.2). Management must be clearly informed about the advantages of the proposed digitization (cost savings, competitive advantages, securing the future ability to work etc., see also digital business benefits above) and need to have the certainty that digital business processes will not be disrupted during the changeover phase.

A parallel strategy with a demonstration environment in which digital business processes are mapped without influencing day-to-day business is helpful but requires extra efforts. In such a demonstration environment, the new digital work processes can integrate down to the last detail and finally it can be assured whether the digitization meets the expectations of the desired process optimisations, or just to a certain degree, or not at all.

It is also important for Management to be able to stop digitization projects at any time if the expectations are not met ("emergency exit") without causing damage to the day-to-day business. This is only possible in a separate and encapsulated demonstration environment. Such demonstration environment provides management the certainty of encountering few surprises during a changeover and of ensuring protected business operations. A step-by-step approach is advisable and facilitates acceptance by management and employees. However, once being started with the full rollout in production environments all preparations steps need to be finalized according to the staggered approaches shown in Figs. 18.3 and Fig. 18.4.

Reference

1. Sobek DK. Understanding A3 thinking: a critical component of Toyota's Pdca management system. Smalley, Art. ISBN: 9781563273605

Chapter 19
The Role of Digital Technology Vendors

Executive Summary

This chapter deals exclusively with the hidden champions in digitalization, the Digital Technology Vendors (DTV).[1] They act usually in a triple role, first as digital innovators for new VPC capabilities, second as suppliers for PMTI (Process, Methods, Tools and Information Standards) market solutions and third as partners and technology consultants for industry. Their product offerings are oftentimes associated with the following terms:

- *Virtual Product Creation* (VPC) and *Virtual Engineering*,
- *PDM/PLM* (Product Data Management, Product Lifecycle Management) and *ALM* (Application Lifecycle Management),
- *Digital Engineering, Digital Thread, Digital Continuity, Digital Factory, Digital Twin,*
- *Computer Aided Design/Engineering/Manufacturing (CAD/CAE/CAM), Virtual Reality (VR), Augmented Reality (AR),*
- *Modelbased (Systems) Engineering* (MBE, MBSE), *System Design and Simulation*
- *Computational Analytics, Data Analytics, Data Contextualization and Semantics, AI based Engineering* and
- Others (e.g. *model/software in the loop, mathematic modeling, collaborative and streaming engineering* etc.).

Digital solution offerings for engineers and their integration into company IT enterprise environments are both essential to enable appropriate digital technology foundations, IT architectures, best suited digital engineering applications as well as process and methods expertise for enterprises and their value creation networks.

[1] The author prefers this term since it is neutral with respect to the offered type of digital technology. DTVs are also known as *PLM or System Vendors*, *CAD/CAM/CAE vendors* and *IT suppliers/companies*. The group of DTVs in this book does not include Internet Service Providers (ISP), Computer & IT infrastructure and equipment vendors or ICT network providers.

R. Stark, *Virtual Product Creation in Industry*,
https://doi.org/10.1007/978-3-662-64301-3_19

Quick Reader Orientation and Motivation

The intention of this chapter is:

- to explain the principle set-up and competence of Digital Technology Vendors (DTVs),
- to describe the role of DTVs in Virtual Product Creation and the projects to develop, customize, integrate and deploy associated digital technologies,
- to discuss the variance of partnerships between DTVs and industrial companies,
- to forecast likely changes of future DTV directions and associated business models.

19.1 The Set-Up of Digital Technology Vendors

Digital Technology Vendors (DTV) have been emerging from the 80ies of the last century from different technology directions and business backgrounds (please also compare Chap. 5: *The technology history of Virtual Product Creation*):

- Mathematical, geometric and systems modeling
- Computational analytics
- Data management and exchange
- Process modeling and simulation
- Research and innovation
- Technological spin-offs from OEMs in aerospace and automotive.

Therefore, the companies of the DTV group are typically not older than 40 years; the majority might only exist between 10 and 20 years,[2] many new companies emerge every year. Due to capitalization needs, however, many young spin-off or start-up DTVs, which might have been founded based on a new research & innovation idea, get (continuously) acquired by major DTV vendors such as Siemens Digital Industry Software (formerly known as Siemens PLM), Dassault Systèmes, PTC Inc, Autodesk Inc. and others. In some cases, multiple merger and acquisitions have evolved in a period of 10 years or more before such DTVs were finally formed into one legal entity.

The core business of each Digital Technology Vendor is software-based revenue for digital applications in product and manufacturing engineering and planning. In addition, a sub-group of DTVs do also offer digital tools for factory operations, field operations and product maintenance. Over the last years, DTVs have extended their typical PLM-type market penetration from the classical manufacturing industries also into other business sectors such as life science, health & medical, retail, insurance and finance.

[2] One of the best overviews of the spectrum of Digital Technology Vendors (DTV) can be found in the annual report of the prostep ivip associations which publishes all member names under (example 2020): https://prostep.epaper-pro.org/annual-report-2020_english/#6. Approx. 80 DTVs are listed there under the term IT companies (System Vendors).

Typically, DTV consists of the following core functions:

1. Leadership, strategic planning (incl. merger & acquisitions) and finance
2. Research and Development (R&D) with the following principal departments:

    a. Core (internal) research and new technologies (usually just a small core team, ~5%)
    b. Business application knowledge streams (~5%)
    c. Software (SW) development divided into several program lines (50–60% of the R&D headcount and expense)
    d. Test & verification and packaging (30–40%)

3. Technical sales, incl. pre-sales activities and POV (prove of concept work), application engineering, software distribution and customer support
4. Marketing and public affairs.

The following two drivers mainly influence the development of new digital SW applications at Digital Technology Vendors (DTVs):

- General IT technology trends and digital research innovations (mainly from Universities and application-oriented research institutes),
- Needs and demands from industry and other business sectors as well as new legislative requirements with high impact on digital applications.

The longer DTVs exist in the market, the more they usually enlarge their SW application portfolio. This expansion requires additional knowledge in the SW application related business process knowledge and application expertise in order to consult and advise the customer based adequately. The present also applies for the SW applications themselves with respect to the individual functions and features. For the lifecycle of the software application (also known as ALM, Application Lifecycle Management), DTVs need to establish and keep internal efficiencies for ongoing software maintenance. They need to bundle many functionalities originally requested by and co-created with different costumers within a common application architecture. This high number of different software functionalities, however, makes it increasingly difficult for the DTV application experts facing to the customer base to keep abreast about the diversity of the SW functions and features. It remains a constant challenge for DTVs to find a best knowledge fit and balance of the offered SW functions and features to the high number of engineering working scenarios and digital method compliance of their clients, customers and users.

Digital Technology Vendors typically rely on the following three innovation pipelines to drive their strategic future digital capabilities and offerings:

1. *Internally funded new business development projects*: Market observations and business model transformation in different business sectors might be the motivation point for internal product owners of existing SW application portfolios (e.g. data management or product modeling and simulation, business intelligence) to use internally available skills and resources to develop new application prototypes. Such prototyping could be achieved via *user journeys*, *interactive application storyboards*, *conceptual click demonstrators* and *minimal viable (software)*

products. They are intended to be used for internal and external customer focus groups and in clinics and workshops with potential future users. Such internally funded projects are also partially done together with external universities and research institutes and are limited to a period of 3 to 6 months.

2. *Intellectual property (IP) related innovation projects together with trusted partners*: Such partners are typically long-standing customers who seriously geared up to invest together with the DTV into new ideas and applications types including the assessment of business models and internal business case development. In many cases such business-critical topics are based on engagements that run for at least 1 year, in some cases even up to 2 years. It depends on the type of IP contribution and resource working split when the final step is made with respect to a go/no-go decision for further industrialization under different terms. The following options are common:

   a. further co-funded development towards a production ready solution for the IP partner with an exclusive use period of 2–3 years (compare PDM/PLM customization in Chap. 11) or
   b. finalization towards a market ready solution by the DTV including the IP content by the trusted partner (with special conditions for the first years of license use for the trusted partner).

3. *Externally co-funded pre-competitive research and development* with other industrial and university/research institute partners in publicly funded collaborative research projects: There exist different types of such public funded research programs such as *Horizon Europe* in the European Union or various research programs by federal organizations in all major industry countries. In contrast to the other two-preceding innovation schema, the DTVs are obliged in this case—as any other consortium member—to publish major project results (by keeping all rights of using new findings and prototypes for themselves). DTVs usually like such projects since they can be used also as a pre-cursor for standardization work afterwards (together with other industrial companies) and to trigger additional internal development work to accomplish the finalization of market ready architectures and applications.

Some of the Digitial Technology Vendors put a lot of emphasis in collecting and describing best practice process & methods dossiers in order to be better prepared to engage with business partners in Virtual Product Creation pre-sales and solution integration work. Whenever the pre-sales and application engineering workforce conduct workshops with (potential and existing) business customers, research institutes and tech start-ups summarize and compare the findings with their internal intelligence in order to merge new facts, trends, opportunities and engineering methods into their internal knowledge management solutions framework.

The revenue side of Digital Technology Vendors (DTV) differ significantly from the majority of their typical customer base in industry. This fact usually creates a certain degree of churn and unhappiness. It gets usually addressed in special negotiation rounds of IT responsible management and purchase professionals at the industrial company side. The evidence of the intended software usage, however, need to be analyzed, estimated and justified by business and engineering functions of the industrial customers. It helps to compare the situation of the different revenue model types in order to understand the controversial debate around that topic.

Any DTV, from start-up, tech company, or digital business supplier, may operate with multiple revenue sources and, consequently, with different revenue models. Depending on the industry and the product/service type, the revenue model will look differently in order to convince (industrial) companies to accept the inherit rules of the software business.

The group of traditional enterprises in industry is used to transaction-based revenue models. A *transaction-based model* is a classic way business can earn money. The revenue is generated by directly selling an item or a service to a customer. The customer can be another company (B2B) or a consumer (B2C). The price of the product or service constitutes production cost and business margin. Increasing the business margin, the business is able to generate more income from sales but run the risk to become non-competitive. Traditional industries in some sectors like automotive industry have modified this transaction model by adding options of finance leasing or renting models to it. Nevertheless, everything is related still to the product price that provides the framework also for the various financial models. During the 2010th decade industrial companies in the long-term investment sector also introduced industrial product-service systems (IPSS): They combine the provision of products with a certain operation or maintenance service. The business models associated with such type of IPSS also offer new revenue models, which might be linked to delivery-oriented business metrics. Nevertheless, such kind of business models are still rare in traditional industries.

Selling software products or services entails using different pricing tactics. It has to do with the core basic that software (once coded, compiled and implemented as running application) does not follow the principle of "piece price" as any material product has (due to the physical element *material*). However, in many cases certain services of maintenance or update might be included with software and their usage.

Some of the following software related revenue models might be considered as separate ones; however, they also come often used in combined packages. Today, the following revenue models and pricing types are common in software business:

1. **Develop and deliver software based on a fixed or dynamic budget for unlimited use**. Negotiation about SW maintenance is handled separately and might even handed over to other companies. Such as model has been popular by industrial companies in the beginning of Virtual Product Creation in the 70ies and 80ies of last century. This model, however, has been getting more popular again during the last years in large enterprises due to the fact the industrial companies have recognized the importance of specific software packages in the context of their own context and hence have started developing their own software

internally (or together with specific SW coding companies) *without* any DTV engagement again!

2. **Licensing/one-time purchase.** This entails selling a software product by license that can be used by a single user or a group of users. The general idea is to offer a product that requires making only one payment for it, e.g. Microsoft Windows, Apache Server, a majority of video games.
3. **Subscription/recurring payment.** Unlike licensing, a user receives access to the software by paying a subscription fee on a monthly/annual basis, e.g. Netflix, Spotify, Adobe products or Autodesk Fusion.
4. **Pay-per-use.** This pricing tactic is mostly used by different cloud-based products and services that charge you for the computing powers /memory/resources/time used. Examples are Amazon Web Services, and Google Cloud Platform.
5. **Freemium/upselling.** Freemium is a type of app monetization in which a user may access the main product for free, but will be charged for additional functions, services, bonuses, plugins, or extensions, e.g. Skype, Evernote, some video games.
6. **Hybrid pricing.** Sometimes pricing plans are a mixture of more than one. E.g. a *freemium plan* might morph into some form of pay-per-use tiered plan. After passing some limit in computation or resources, a user can be forced or offered to use another type of pricing, for example platforms such as Mailchimp, Amazon Web Services, and SalesForce.

Digital Technology Vendors in the PLM and VPC sector have primarily used the second software revenue model so far and have modified it towards specific customer wants as part of the negotiation rounds. DTVs. Offer the following most popular modifiers of the traditional license deals in the PLM and VPC business (also in custom combinations):

- *Peak license usage*: A certain cap of how many licenses can be used simultaneously in order to limit the number of licenses needed; in such a model, a trusted relation is necessary in order to rely on a robust license count server solution.
- *Named user licenses*: With the introduction of cloud-based Software-as-a-Service (SaaS) platform offerings DTVs have started to charge SW licenses by identifiable users rather than by generic licenses, which do not recognize specific users but only a used license per time slot.
- *Token flex licenies*: With the introduction of cloud-based Software-as-a-Service (SaaS) platform offerings DTVs have also started to offer companies the choice which type of application is selected by individual organizations and users and measure it by different kinds or different numbers of tokens. This provides more flexibility in using new digital applications and follows the thinking of office applications packages (e.g. Mircosoft Office).

For each of the license models the maintenance and update circumstance are added to revenue model. This again is different to the traditional industry business rules where maintenance models are not necessarily constitute a decision point at the point of purchase: DTVs follow the model that customers need to decide on the maintenance conditions upfront for a couple of years. Usually 2–5 years of maintenance contracts are common, but might differ between customers and application types. From 2015 onwards first DTVs have started to offer their VPC and PLM digital application offerings as part of cloud platforms introducing with modified license models (incl. careful extension to revenue models #3, #4 and even #5).

Due to the successful software license revenue model most of the Digital Technology Vendors (DTV) deliver significant higher business margins compared to traditional industrial companies. Successful DTVs easily reach a profit level of 15% in relation to the net revenue, whereas DTV leaders might deliver business margins of 25% and even higher. Questions and doubts, however, get louder whether such "software favorable" situation for DTV has the potential persist and to resist against open and free software in the mid and long-term range. DTVs have meanwhile recognized that they have to invest significant foresight thinking into the fundamental issue which future business conditions will still be supportive to support this traditional software revenue model. Hence, it becomes evident to ask the following question: Will DTV of the future no longer have a chance to survive in its current shape and needs to be adapted to a new sustainable future?

19.2 The Role of Digital Technology Vendors in Virtual Product Creation

Digital Technology Vendors (DTV) meanwhile play a core and crucial role within digitalization of industry. Without the solution offering of DTVs industry could neither introduce, modify, extend and professionalize overall Virtual Product Creation environments nor optimizing, constantly updating and regularly overhauling company specific Digital Engineering and PLM architectures. With the help of their internal functions (compare Sect. 19.1 they host enough internal digital capabilities to tackle a significant range of tasks within the overall spectrum of Virtual Product Creation business, but also have limitations and short falls. This section will explain the strengths and weaknesses of DTVs in Virtual Product Creation.

One of the hidden success factors in Virtual Product Creation business is the ability to establish close and trustful connections to Senior Management of industrial enterprises. This majorly enhances the chances that DTVs are not only treated as a vendor for digital/IT solutions but as a trusted partner to establish new and reliable business and engineering solutions to guarantee today's and future (digital) value creation. Oftentimes, such senior connections are pivotal to enable and support core senior management decisions for or against major digital investment initiatives and programs in industry. Digital Technology Vendors, therefore, put a lot of emphasis to establish solid matching lines of their core leadership personnel with the appropriate levels of enterprise in industry:

- DTV CEO to match with potential industry board members and/or vice president levels of industrial enterprises
- DTV Vice President or director level with influenceable business division leadership and CIO (Chief Information Officer) and/or CDO (Chief Digital Officers) in industrial companies
- DTV regional and key accounting leadership with OEM/supplier digital core initiative and program leaders in IT and in major business functions
- DTV technical architects and research evangelists with technical leadership and specialists as well as IT architects in industrial companies.

DTVs, which follow such a stringent line-up even without having already a solid customer base within specific industrial accounts, will be able to double the chances to be recognized and considered for *future requests for quotation (RfQ)*.[3] The interaction of the appropriate DTV management members with the equivalent management levels of industrial companies are critical to help industry management with their challenges and problems as explained in Chap. 18. However, DTV had to learn appreciating the "hidden role" of senior leaders in VPC research and education (such as directors of chairs and research divisions) who not only deliver the next generation of digital literate employees to industry but also have the advantage to be in a true neutral role regarding digital technologies and working methods. Unlike consulting companies who leverage current know how with best future projections, VPC research leaders have their fingertips constantly (!) on future technologies and overall new digital approaches and ways of working. Consequently, successful DTV run ongoing R&D programs with such VPC research leaders and university institutes driven by senior DTV management.

Moreover, DTV play a major role in transforming current business practices into new future VPC technology options. As explained in Chaps. 7 through 16 in great detail, such evidence has been demonstrated in the last 20 years like in:

- modern and automated CAD and CAE technologies,
- KBE (Knowledge Based Engineering) template approaches,
- new data management capabilities (from vaulting towards full digital configuration spaces for any kind of data, model or information) and
- easy-to use visualization environments (e.g. DMU) and virtual interaction solutions (Virtual/Augmented Reality).

The new challenges ahead in ASE (Advanced Systems Engineering) and MBSE (Model based Systems Engineering) as well as the need for AI (Artifical Intelligence) assisted engineering intelligence will require new ways of DTV solutions deliveries (compare such type of future digital solutions in Chaps. 20 and 21).

---

[3] A request for quotation (RfQ) in VPC and PLM business is a business process/method companies or public organizations use to request a quote from a supplier (or a preselected group of potential suppliers) for the development and/or purchase of specific digital architectures, applications, products or services. *RfQ* generally means the same thing as *Call for bids (CfB) and Invitation for bid (IfB)*.

POC (Prove of Concept) work engagements between DTV and industry represent a typical interaction element of today's VPC innovation streams. DTVs are expected to understand companies needs and to transform it into first workable "digital solution snippets". The following types are common for such POC demonstration environments (usually a mix out of agile working elements and problem/heuristic-based reasoning sessions):

- conference room pilots using brown bag paper and obeya room illustrations with *process cards* and *process run through swim lane type of notations*
- *Design Thinking* type of digital problem structuring and solving sessions allowing explicitly for open and creative rationales
- Highly interactive digital demonstrator environments with clicking type method run-throughs
- Fully equipped digital lab prototypes and minimal viable solution-based exercises even together with process owners and digital project and product owners.

DTVs have earned credentials within the VPC community to serve within an expert technical tole in digital solution customizing and solution integration. Such credentials evolve through constant engagement within company POC, PLM projects and Virtual Product Creation futuring initiatives and are usually not referenceable by official publications and peer-reviews or neutral expert checks like in product homologation in classical industries. This special digital circumstance shows that Digital Technology Vendor business runs in a special mode that is characterized by a reputation mixture out of personal experience and feedback, constant innovation delivery evidence and successful business longevity.

The Virtual Product Creation industrial business often relies on DTVs to take on implicit architecture roles, both on a strategic side and for data model development. This is clear for the DTV own software applications and information models and needs long standing partnerships between DTV and the industrial company if this should be extended to other non DTV-owned information models. In that business, it is even common that senior data professionals acting as freelancer experts execute well-paid data architecture roles in PLM projects. They are frequently contracted by DTVs on a project as needed base and they have to closely work with internal architects belonging to the DTV or industrial company permanent staff.

However, there exist also business areas in Virtual Product Creation and PLM operations where Digital Technology Vendors have only limited expertise and are not accepted as appreciated partners. Due to the unfavorable cost structures of DTV application, engineers compared with digital service agencies and due to limited knowledge of technologies from other DTV competitors, the DTV consulting support accepted and ordered by industry customers has been significantly reduced in the second decade of the twenty-first century. As a subsequent consequence DTV meanwhile have serious difficulties to aquire and keep profound knowledge in industrial digital solution usage and true operational work implementations in industry. Some DTV were really successful in establishing such tight working relations to their customer bases for many years due to solution deployment projects and even in on-the-job implementation work. Such model was especially successful for US based

DTVs. Meanwhile the belief has changed since such high numbers of pre-sales and application resources are often too expensive, according to the internal cost structure, and oftentimes are no longer regarded as critical enough to ensure software revenue as DTV core business. This evolution, however, is somewhat controversially discussed within DTV Senior Management and enlarges the risk of no longer *digital business grounded* enough to understand and reach out to realities of industrial companies.

In looking to new business segments of DTVs, another opportunity crops up and need significant commitment and investment: limited knowledge in engineering model content, collaboration and interpretation and analytics know-how. Traditionally, DTVs concentrated on generic digital model template and structures. Only few DTVs owned core engineering knowledge within their own company and were successful to maintain it. Examples are the LMS knowledge pool within Siemens Digital Industry SW or the recent acquisition of ISKO Engineering by Contact Software. Dassault Systèmes always had the advantage to be an integral part of the overall Dassault Corporation, which has a strong stake in aviation and aerospace industry.

Digital Technology Vendors have no role, no responsibility and no day-to-day working experience in direct product related digital engineering delivery responsibilities, tasks and tactics. This situation makes it difficult, if not impossible, to them to truly understand, support and/or modify the following characteristics of daily Virtual Product Creation work hazzle (compare more detailed explanations in Chaps. 6, 17 and 18):

- The *true level of pain and joy* experienced by ordinary users of their software
- The *business pressure* scenarios under which digital solutions and software do not function or require major work arounds
- The *operational burden* to guarantee error free digital deliveries within the VPC working framework and environment despite of high numbers of software glitches, digital architecture limitations and limited software integration levels into the overall digital solution architecture.

In summary, DTV represents an indispensable pillar of today's and future Virtual Product Creation business solutions. They will most certainly remain in such strong position; however, they have to heavily invest into new core and future service capabilities. Digital Technology Vendors (DTV) most certainly have to change their digital business role and their business directions more significantly compared to the last 20 years. The following section will analyze these scenarios of the future in detail.

19.3 Transformations in Digital Technology Vendor Business

The future of Virtual Product Creation with its future IT technologies and new digital engineering capabilities will be introduced and discussed in the following two Chaps. 20 and 21. However, in order to make already meaningful assumptions within this chapter regarding most likely transformational steps for the group of *Digital Technology Vendors* (DTVs), certain fundamental issues need discussions and elaborations. One fundamental element is indeed related to the base assumptions of software offerings for the various degrees of digital engineering tasks and activities as part of the overall Virtual Product Creation environment.

Unlike the initial days of Virtual Product Creation (see Chaps. 4 and 5), when industrial companies developed their own software applications, nowadays the majority of industrial enterprises do rely on the development and provision of such software by DTVs. Nevertheless, this situation is not necessarily carved in stone for the future. In certain industrial sectors (like automotive and high tech), OEMs and system suppliers as well as private–public partnership organizations (e.g. in aerospace) have already started to return to internally owned VPC software development.

To the same time, there exist significant competition in between the Digital Technology Vendors. Basic VPC and PLM applications such as CAD, CAE or PDM often no longer differ significantly if it comes to basic functional foundations. Therefore, the amount of digital application commodities are growing and will increase the pressure on DTVs to defend their market share with more attractive license and associated service offerings. The situation will be exacerbated quickly once the efforts to migrate and transform today's diverse set of vendor specific models and data types will be significantly reduced. This could happen already mid-term (3–5 years) but most certainly will happen in the end. It will be accelerated, however, if it will be directionally pushed by more stringent measures of openness and standardization by the customer base, i.e. by the industrial companies (compare e.g. the CPO, Code of PLM Openness in Chap. 11 PDM/BOM). Today, DTVs still have the luxury to bond high numbers of legacy customers who do not easily take the effort to migrate to other vendor solutions. They are discouraged by high efforts associated to it. If digital commodity cost pressure and likelihood to migrate existing line-ups of company specific data and models into any other DTV eco system grows, then DTVs will have to be ready for new business models. Those business models will favor different or additional revenue offerings than just selling software licenses for digital tools and applications. DTV, therefore, need to prepare themselves to other and higher degrees of Virtual Product Creation engineering excellence than today.

Figure 19.1 shows the one of the author's outlook perspectives on how today's Digital Technology Vendors (DTVs) might transform themselves towards *Digital Engineering Excellence Provider (DEEP)*. DEEP, will represent a new class of *engineering competent digital technology vendors* in the future. The transformation from a pure DTV, i.e. being solely dependent on software centric revenues, towards a full capable DEEP, requires various transformation steps to offer high reliable Virtual Product Creation service levels. Up to a full degree of ownership for full service delivery on digital engineering core elements with high value creation potential and associated service rates. The nature of a DEEP, however, should not be mixed up with typical Digital Engineering Service agencies as known today. DEEP run their own software for future digital engineering work and still will benefit from the software

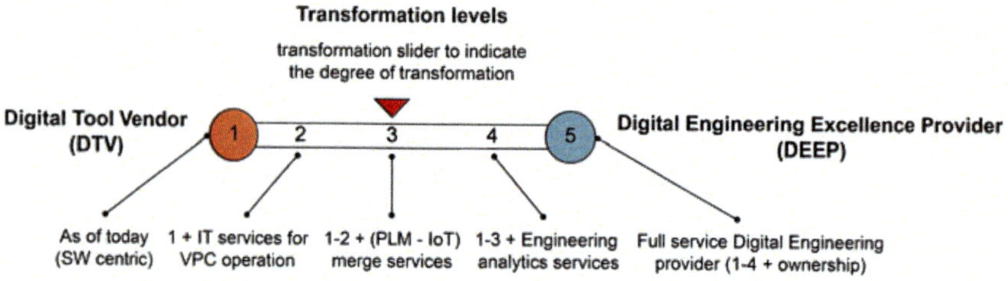

Fig. 19.1 Transformation levels from DTV towards DEEP

revenues of the traditional DTV business. Acquiring new valuable business without losing profitable earnings in the traditional role will be the leadership profile of the future!

At level one, still operating as pure DTV, a couple of business changes are expected and necessary, too. The leadership role of Autodesk Inc. in migrating away from a "*license type*" *software revenue model* to a full "*subscription based*" *software revenue model* has shown major advantages, both, for the customer base and for the DTV itself. Autodesk Inc., e.g., has achieved this type of transition between 2015 and 2019 by converting to almost 5 million subscribers whilst delivering a solid (non-gap) operating margin of 25% via constant and solid subscription earnings and giving the chance for divestment options to profile the future.

According to a specially selected content from the editors from *Catalyst* Autodesk has sponsored a report about their subscription model and the advantages for the customers (subscribers) on the Web[4]; please note the following core characteristics of the Autodesk subscription model (as of May 24th of 2021):

> Subscribing to Autodesk products streamlines the software management workflows, including licensing compliance, software updates, and upgrades. Every subscriber receives direct access to Autodesk support specialists via online chat, by appointment, or through e-mail, as well as direct access to a vast and growing knowledge base of documentation, tutorials, training videos, and community support forums. Remote desktop assistance offers secure hands-on troubleshooting. For collection subscribers, benefits do not end with access to software and support. A subscription license allows home use and use of software on the road. Most products in the collection also include access to prior versions. CAD managers and IT staff can use new administrative tools to simplify managing software licenses and account use. Reporting and analysis tools are available to monitor product use, spending, and productivity, and estimate future needs. Having a collection is much like having Microsoft Office: Not every employee needs Excel or PowerPoint every day, but the software is there when needed, especially as needs change over time. Companies can now reduce operational costs (IT and procurement spend) by standardizing on a collection of technology flexible enough to suit the needs of a majority of users rather than managing unique software deployments for each employee.
>
> In the past, fluctuating staff size could mean companies had to make tough decisions about whether to add or drop software licenses — and if so, how many? Subscribing to a collection, however, makes it much easier to manage such change: Acquiring software for

[4] https://damassets.autodesk.net/content/dam/autodesk/draftr/1759/spotlight-autodesk-collections-final.pdf.

new employees has a lower cost of entry and you have more flexibility around who uses each subscription.

… for cloud-based Autodesk services, including certain rendering and analysis tools in the AEC (Architecture, Engineering & Construction) Collection, users choose whether to store data locally or in the cloud.

Another major transition towards DEEP offerings according to Fig. 19.1 will achieved by the capability to act as *digital platform provider* across all levels. Autodesk Inc. and Dassault Systèmes have been pioneering to *digital platforms* for their product portfolio (partly together with digital platform technology provider such as AWS or Azure): the successful platforms are called *Fusion 360* by Autodesk and *3D Experience* by Dassault Systèmes. In order to provide best services to users on such platforms DTVs have changed the *serial number license* to a *named user license model*. By doing so, each user has a personal login from any capable client. For the DTV it becomes easy to analyze software impact caused by individual users and—to the same time—to consult users more professionally in best software usage questions.

Figure 19.2 explains the first two transition steps of the transformation of DTV on their way to a DEEP (Digital Engineering Excellence Provider). Many Small and Medium (SME) companies in industry have severe difficulties to establish core groups of Virtual Product Creation (VPC) competence, both in IT departments and within engineering and manufacturing functions. SMEs, therefore, will majorly benefit from the stepwise offerings since they can leverage this capability at least

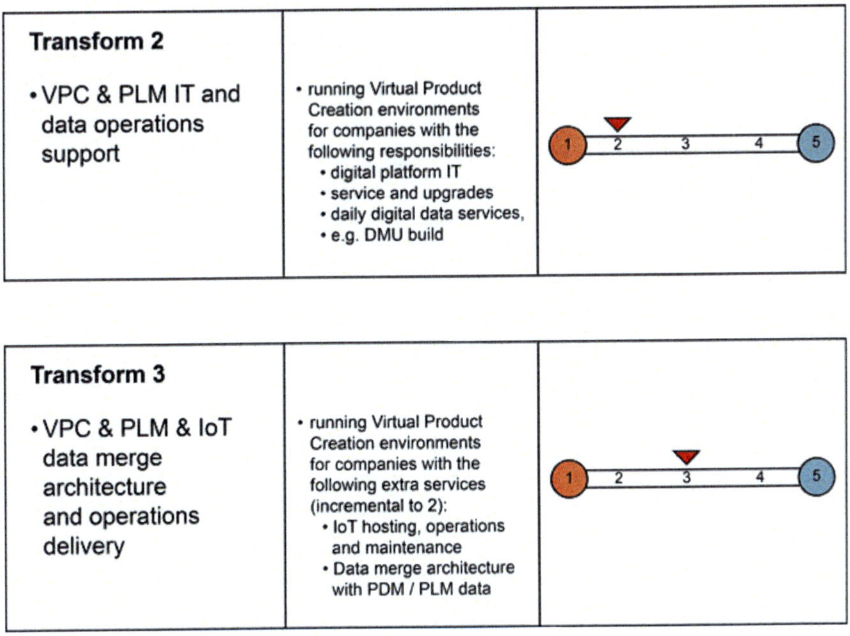

Fig. 19.2 Characteristics of DTV-DEEP transformation levels 2 and 3

in the first years by DEEP service levels. In step 2 they could already leverage the competence of DEEP for a professional set-up of the underlying IT and data operations support for the wide range of VPC solutions. This does include from today's perspective, e.g.,

- operations of special server and services to run DMU clash & clearance detection,
- population and maintenance of voxel-based geometry finder applications such as QPL (Qiuck Part Locator) functions or
- operating robust linking mechanisms for data, document and model traceability across databases including SOA (Service oriented architecture), REST (Representational State Transfer) and graph-based technologies.

Bigger companies already today implicitly rely on such competence by DTV but realize this options trough expert leasing rather than by DEEP service levels.

Step 3 would open the "merge connection" between the IoT (Internet of Things, compare Chap. 20) active data world reaching out to the factory resp. field and the PLM type of engineering backbone models and data. This connection is critical to start establishing pre-requisites for *personalized* or *technical system driven* engineering analytics including Digital Twins (compare Chaps. 20 and 21). Industry would benefit from such offering by robustly realizing new "intelligent data lake" types of information environment providing different degrees of data contextualization. This capability will enable and boost *no-code* and *low-code* development—i.e. without specific computer language knowledge—of specific data analytics applications by engineers. Additionally, such level 3 service would push new robust ways to connect and integrate AI-type of IT services (incl. running special lambda type IT architectures[5]) to support new types of engineering support and intelligence.

Stages 4 and 5 (see Fig. 19.3) represent offerings of digital engineering capabilities for future types of Virtual Product Creation. They are based on IT infrastructure knowledge, IT services capabilities, data structures and network fidelity levels, model content and hierarchies as well as digital application functions.

Digitally assisted Virtual Product Creation does require high robustness of data and model synthesis, analysis and linkage, especially in the context of growing system (or even system-of-system) interactions. Such interactions engage model and data flows between partial technical domain architectures and digital artefact types (such as requirements, functions, system partitions, behaviors, physical structures). The level 4, therefore, does address engineering service capability that delivers meaningful engineering exploration, analysis and synthesis results. Functional engineering activities of industrial companies are the direct recipients for such digital engineering deliveries. DEEP capability does require substantial systems and domain knowledge and competence. Those DTV who retain or have acquired direct engineering knowledge will have an advantage to reach such level. Level 5 constitutes the highest capability in delivering *and* owning digital engineering content with the obligation (and associated highest value proposition) to serve as integral partner

[5] Lambda architecture is a data-processing architecture to process massive quantities of data by leveraging both, batch and stream-processing methods.

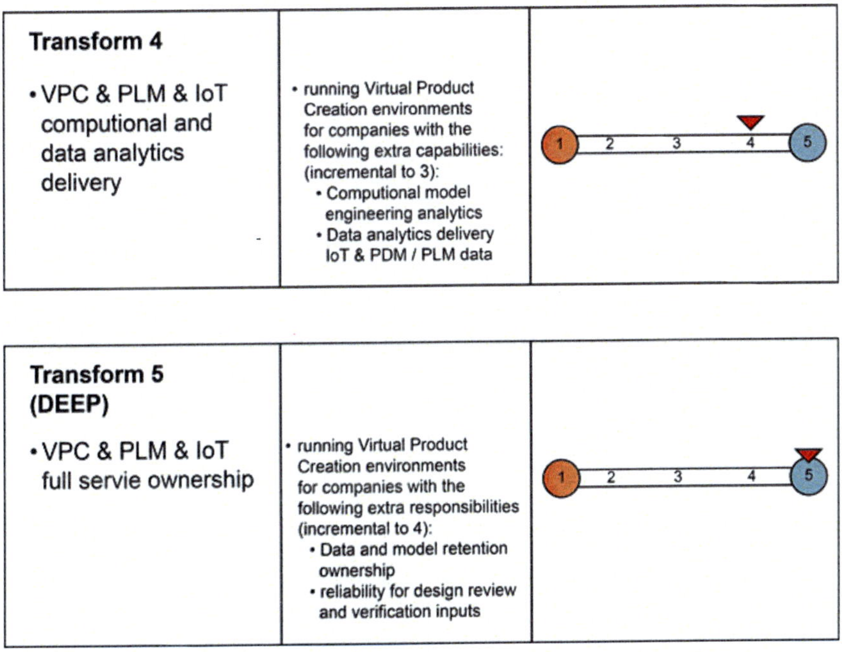

Fig. 19.3 Characteristics of DTV-DEEP transformation levels 4 and 5

in the engineering progression amongst "system/product" owners and stakeholders. Important tasks such as model fusion deliveries with associated predictions and technical assessments are covered by this degree of DEEP capability. DEEP, therefore, no longer sell generic models' types and data analytics as digital templates to industrial customers but actually use them in real project context to deliver engineering reasoning, analysis and model/data content for their industrial customers. This will be digital engineering service levels, which only a few DTV of today would be able to achieve due to missing active operational engineering experience. It will be interesting to see to whether Engineering Service Provider (ESP) might come from the other end to take on work tasks of DEEP level 4 and 5, or whether strategic alliances between DTV and ESP will even accelerate aligned DEEP offerings in the market!

From technology point of view, Digital Technology Vendors have to heavily invest into cross model and cross digital architecture solutions to be able to fulfil the needs of industry regarding networked technical systems and IoT based intelligences. Democratization trends of model and data usage will force DTVs into easy to use data environments based on IT micro-service architectures. To the same time, however, the desperate need for new extended information models across all disciplines and lifecycle areas will demand new ways of highly standardized but flexible data elements or information objects. Those new levels might even include self organization of information objects (which does not exist yet)!

Today's mostly transactional oriented database schema are no longer capable enough to support the automated engineering reasoning of data and models. IT technologies of today cannot deliver on data semantics. DTV need to become much more active on the data and model side, both in terms of semantic standards and with respect to open their own data and model architectures for flexible exchange mechanism. The future will put much higher emphasis on the value of data and model. Such massive change in value understanding in industry will boost the acceptance of today's DTV to develop themselves toward the revenue models according to new DEEP responsibilities.

The following new scenarios need serious attention by Digital Technology Vendors (DTV) in the future to be able to survive and play a successful business role in Virtual Product Creation of "tomorrow". Those scenarios become true in certain industries, but not in all. DTV no longer can afford the one and only revenue model of the past. The pressure to act rises!

Scenario 1 (Likelihood: High)

- Software feature & function differentiation for basic digital engineering applications no longer plays a decisive role due to high similarities amongst DTV offerings
- DTV face severe risks of declining digital commodity business
- Competition from open source digital engineering apps increase steadily and does erode classical software revenues
- Differentiation to cloud offerings from digital platform providers as part of SaaS, PaaS or IaaS set-ups becomes increasingly difficult; pure cloud provider become too powerful to accept high license, maintenance and subscriptions fees from DTVs
- DTV need to increase their footprint in high value DEEP business

Scenario 2 (Likelihood: Medium)

- Industry has fully picked up on MBE (Model based Engineering) and on MBSE (Model-based Systems Engineering) due to immense societal and political pressure caused by error prone and non-reliable verification and homologation of highly networked products and technical systems in the field
- Industry has, therefore, recognized that reliable digital models and data become the prime asset in Virtual Product Creation of "tomorrow", a major shift away from the SW tool and application-oriented asset viewpoint of the first 20 years of the millennium
- Such change triggers heavy investments and new revenue models (incl. leasing) for digital engineering data and model generation incl. contracts of co-ownership of engineering content deliveries
- The new business roles "engineering solution provider" emerges with core method competence in certain engineering disciplines and engineering methods; DTV see the value of transforming to DEEP
- Competition from open source digital engineering apps increase with the growth of the model market

Scenario 3 (Likelihood: Medium)

- Industrial companies steadily increase their investment into internal software development and are no longer or only partially dependent on DTV competences and offerings
- Such new digital competencies and model-based software development capabilities in industry allow agile development approaches for engineers to compose their own digital development and validation environments
- VPC data and models are highly standardized; VPC, therefore, follows the software engineering approach (high reuse of software code) to guarantee high degrees of model and data reuse
- Such new sets-up make it possible to offer easy *configurable no/low code* digital application which becomes a must for DTV of the future

These potential new scenarios show that DTVs might undergo through substantial if not disruptive or even radical transformations. Therefore, it is important to have a closer look how today's Digital Technology Vendors (DTV) asses their own position in the market and their roles to drive the future of Virtual Product Creation. The following section will provide a positional overview giving by six Digital Technology Vendors that are competitors to each other but also offer complementary software applications. In their presentations, they explain their self-understanding, their focal points moving into the future with their technical solutions and the challenges ahead in term of driving innovation forward.

19.4 Perspectives by Digital Technology Vendors

To be able to describe the strategic positioning of Digital Technology Vendors (DTV) today and for the future the author of this book has asked several well established and long-standing digital technology companies as well as one start-up to lay out their own views on the following five questions:

1. How do we describe ourselves?
2. What is our role as Digital Technology Vendor (DTV)?
3. What is our vision to drive the future digital technology portfolio?
4. What are our biggest challenges being a digital technology vendor?
5. Which inhibitors exist for robust integration of our digital capabilities in industry? Why?

Please follow their valuable statements and views in alphabetic order of their company names.

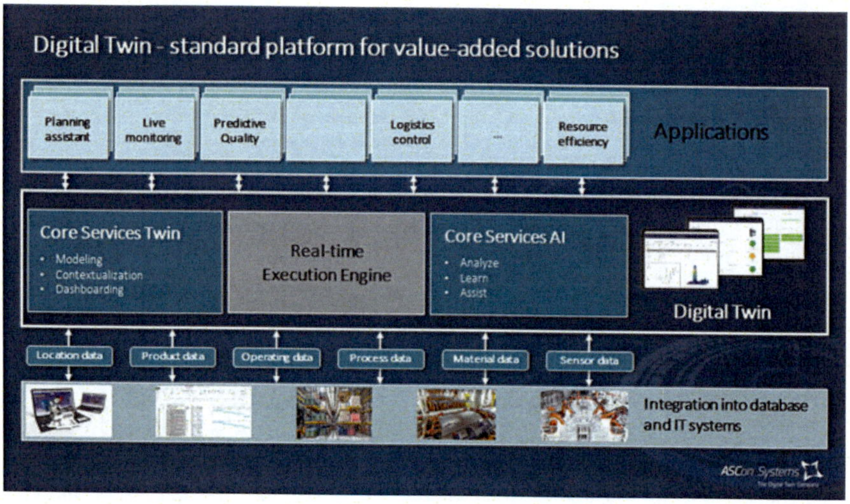

Fig. 19.4 Real-time execution engine as core block of the ASCon digital twin platform (*Source* ASCon system GmbH)

ASCon Systems GmbH—The Digital Twin Company[6]

How do We Describe Ourselves?

ASCon Systems is a provider of real-time, context-based digital twins for manufacturing. The high-tech start-up was founded in January 2017 and employs over 90 people at four locations. Its unique selling point is a real-time kernel for continuous data acquisition directly in a context-based behavioral model for the synchronization of product development, planning and production as the basis of a twin platform and building innovative solutions. The ASCon Digital Twin closes the gap between PLM (product development, planning and virtual commissioning), analytics (Big Data/AI) and the real production world.

What is Our Role as Digital Technology Vendor (DTV)?

Our role as Digital Technology Vendor is focused on the implementation of digital twin-based solutions for the industry.

The ASCon Digital Twin is based at its core on a, in Europe, the USA and Japan patented, real-time, discrete-event, non-time-clocked process execution engine, which is called *IoT Execution Engine*. This is the core of a general-purpose platform that also includes modeling and direct signal coupling. This integrated overall environment allows control processes to be defined and executed without having to program for them (no-code), from modeling to connectivity to execution in the execution engine. The platform approach is shown in Fig. 19.4.

[6] The *ASCon System* technical contribution has been created and delivered by Prof. Dr.-Ing. Uwe Winkelhake and Mathias Stach.

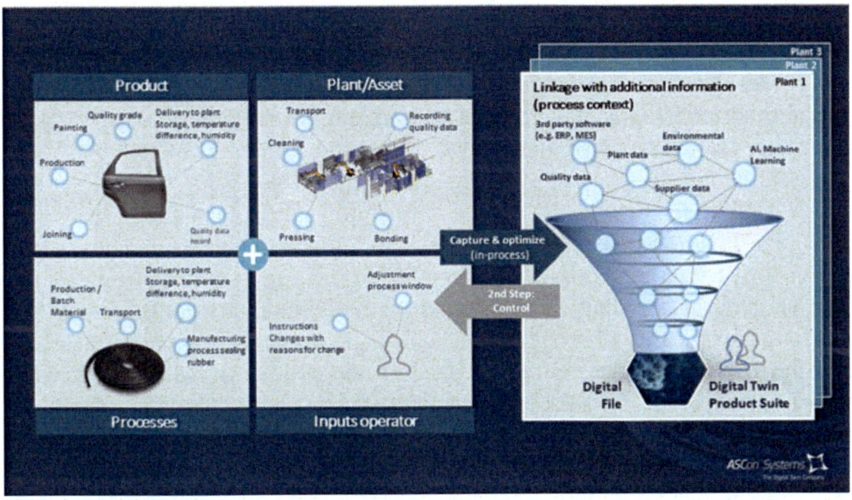

Fig. 19.5 Reference project: rolling head plant (*Source* ASCon systems GmbH)

The Digital Twin platform is based on solution modules that enable AI-based optimization in the virtual model. A typical application example is shown in Fig. 19.5, which deals with the operational optimization of a door rubber assembly line in the final assembly area of an automobile manufacturer. Six industrial robots work together to automatically install the foam rubber seal in the door areas of the vehicle bodies.

The task is to collect a large amount of data in real time and to continuously incorporate it into the virtual plant model. The plant model has been configured in a no-coding approach on modular elements and the control level has been connected via an ASCon device level. Based on the ongoing situation analysis, the digital twin-based solution continuously provides operating instructions and shows the key figures of the overall plant. By using the solution, it was possible to significantly reduce the operating effort of the plant and, by implementing preventive measures, to reduce downtimes and thus increase output. Overall, this project has a very short payback period of less than 10 months. Currently, the rollout of the standard solution in other plants and an adaptation to further plants is planned.

What is Our Vision to Drive the Future Digital Technology Portfolio?

ASCon Systems GmbH was founded with the ambitious goal of revolutionizing the way of planning and manufacturing in the future.

With our digital services, we monitor, analyze and control production and create an efficient ecosystem of modular, powerful and connected solution modules for assisted production. In this way, we are already laying the foundation for future autonomous production.

In the future, factories will be able to efficiently manufacture highly individualized products in very small batches. They will control themselves, anticipate malfunctions in automated equipment and in the operational process, and initiate AI-based measures to avoid failures. These smart factories are adaptable and can also respond immediately to changes in production and products. Digital Twin solutions offer great potential for the planning and operation of such factories [1].

In analogy to the established vehicle navigation, test drives in virtual images of plants or planning projects are made possible. Similar to the navigation systems used in cars, advice is given on how to avoid traffic jams and bottlenecks on the route and thus arrive safely and predictably at the destination. Similarly, there are hints for plant operation or even the automatic implementation of planning measures [2]. Good overviews of reference projects, research projects and also providers are available for this topic area, for example, in [3, 4].

The vision of ASCon is shown in Fig. 19.6. Today, the projects are about supporting the users. Concrete measures for improvement are proposed during plant operation or even line planning. In the future, these optimizations will be autonomously incorporated directly into the control systems, thus achieving continuous improvements automatically.

What are Our Biggest Challenges Being a DTV?

As an innovative and disruptive company, ASCon Systems GmbH faces the challenge of penetrating existing and established market structures. Furthermore, it is necessary to break up grown organizational structures. The responsibilities for planning, controlling and operational execution often lie in different areas of the companies. With the integrated twin platform of ASCon, these areas grow together and new processes have to be adopted. To achieve these changes, not only the technology

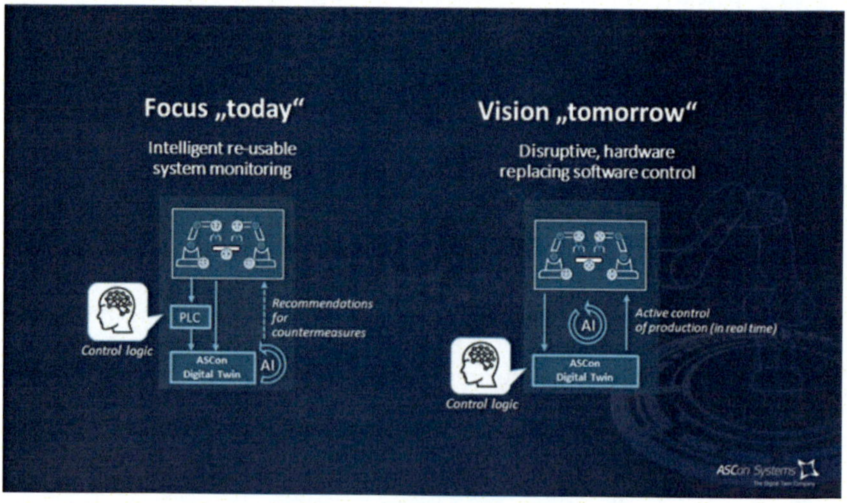

Fig. 19.6 ASCon vision of future production (*Source* ASCon systems GmbH)

must be convincing, but also the general approach to concrete projects. Integrated control and optimization on the basis of a digital twin is still in its infancy today with initial pilot projects. There is a lack of overarching, platform-oriented solutions for which basic requirements are named in the implementation recommendation on Industry 4.0 [5]. According to this, it is necessary to address the entire value chain across all company organizations and also across company boundaries in horizontal integration, to enable a dialog between the company control level and individual machines in vertical integration, and to continue to establish digital continuity and continuous integration of engineering across the entire product life cycle and production system. This is where the unique ASCon solutions are positioned. Based on the real-time kernel or the twin platform, further solution modules and fields of application will be quickly established and thus inspire existing as well as new customers and partners with innovative solutions.

Which Inhibitors Exist for Robust Integration of our Digital Capabilities in Industry? Why?

The prerequisite for powerful digital twins is the near-real-time recording of all relevant information. On the one hand, these are plant signals from a wide variety of sources, such as actuators, sensors and controllers, and on the other hand, product and process information as well as data from the factory environment, such as quality, weather and logistics information. This information is often not directly usable data, but only context-free measured values. All this information must be captured and assigned to a logical context. For this purpose, ASCon Systems has developed a unique technology, the real-time kernel, which enables the construction of powerful digital twins (compare Fig. 19.7).

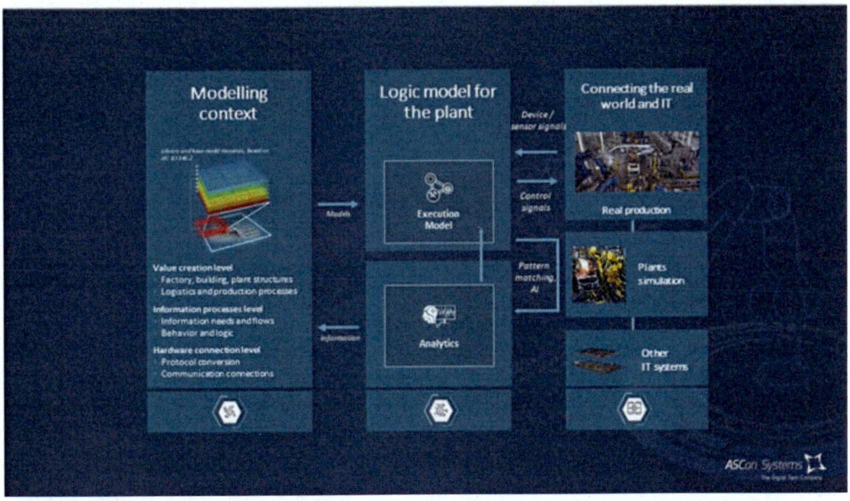

Fig. 19.7 Emergence process digital TWIN (*Source* ASCon systems GmbH)

For the implementation process, it must also be ensured that the necessary data for the digital twin in operation can be mapped by appropriate sensors and processed at near-real-time speed. For larger data volumes, it may be necessary to supplement existing database systems or to replace them with an edge computing approach, for example. ASCon Systems has the corresponding process knowledge and the resources to solve these challenges through its constantly growing partner network.

What is Our Way for Successful Partnership with Research and Industry?

ASCon Systems has a number of cooperation with universities. Thereby, we focus on the use of state-of-the-art technologies and their application possibilities in the industrial environment. In addition, we participate in research projects to bring innovations to industrial maturity together with the participating companies and universities. Development partnerships with strategic customers complement our approach of bringing state-of-the-art technologies to market in a very agile manner.

CONTACT Software[7]

How do We Describe Ourselves?

CONTACT Software is a software vendor offering a portfolio of information management and collaboration solutions for industrial and enterprise customers. Operating mostly out of Germany, with subsidiaries and offices in several European countries, customers all over the world are served through a network of partners in the Americas and Asia. CONTACT and its partners complement the software products with implementation and consulting services.

Our approach is based on our scalable low-code platform *CONTACT Elements* and a range of composable application building blocks ("apps") running on top of that platform. Application modules are either off-the-shelf standard software provided by CONTACT and third parties, or customers and service providers can additionally build custom apps on top of *Elements.*

The portfolio of 100 + ready-made apps covers a broad range of functional areas, from PLM core activities like CAx data management, bill of material, variant and requirements management, to traditional and agile project and process management, to IoT and manufacturing execution. The platform includes means for efficient global operations through data replication, and aims for market-leading user experience to improve user acceptance and facilitate process improvements (see Figs. 19.8 and 19.9).

The *Elements* platform is also marketed to other software vendors that re-brand the technology and build their own software products on top of it. OEM customers include global vendors of industrial software.

CONTACT itself is bundling its platform into differently branded offerings: *CIM Database*, a comprehensive PLM system, *Project Office*, a collaborative project management software, *CONTACT Elements for IoT*, targeting the industrial IoT, manufacturing operations and maintenance processes, and *Collaboration Hub*,

[7] The *CONTACT Software* technical contribution has been created and delivered by Frank Patz-Brockmann.

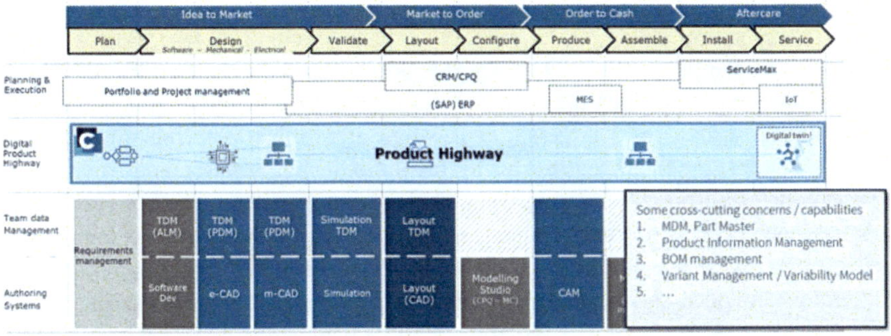

Fig. 19.8 A customer scenario of the system landscape for the digital thread' (*Source* CONTACT software)

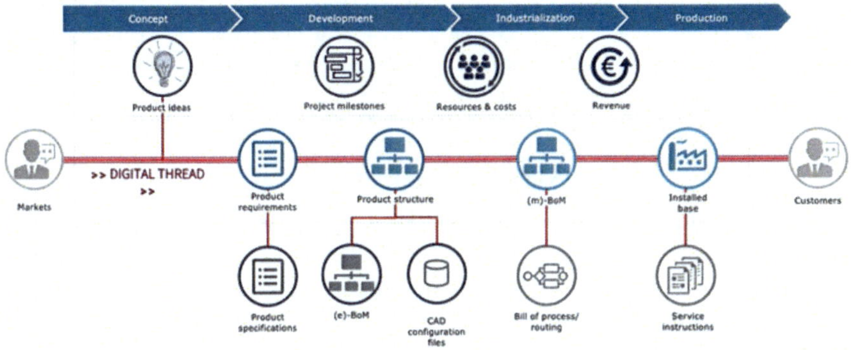

Fig. 19.9 Key information models involved in the digital value chain (*Source* CONTACT software)

supporting cross-company collaboration in engineering supply chains. The "apps" are available in either offering, and can be freely combined to build a targeted solution.

CONTACT *Elements* solutions can be run on-premises or in software-as-a-service (SaaS) and platform-as-a-service (PaasS) models.

What is Our Role as Digital Technology Vendor (DTV)?

Following our vision of "energizing great minds", we aim to be an enabler for digital value chains inside and beyond the boundaries of the enterprise. As a partner for our customers, we want to understand the customer's challenges and goals, and contribute our software products and industry knowledge to build solutions that put the customer in the best position possible. Although we believe that every company eventually requires "digital" skills, it is our role to be our customer's specialist partner for lifecycle software and digitalization. Through our standard software offerings and the tailor-made solutions augmenting it, we enable customers to build efficient digital processes, that seamlessly integrate with other software tools and the broader "digital environment" of connected devices and the internet.

What is Our Vision to Drive the Future Digital Technology Portfolio?

Processes for designing, manufacturing and operating products supporting certain business models vary widely in their underlying patterns—from engineer-to-order to configure-to-order, from large-series to single-item manufacturing. In addition, products are designed and realized using technologies and materials as diverse as those products' purposes—ranging from airplanes and smart phones to steel mills. To satisfy the need to differentiate itself in the market place, nearly every company's lifecycle processes have unique properties not found elsewhere—not accidentally but deliberately and necessarily. On the other hand, many abstractions and paradigms are similar or the same and can be handled by standardized software if it is sufficiently flexible.

It is our vision to build a software portfolio for supporting lifecycle processes with information technology that embraces the diversity of users, methods and tools in use. Business models, the resulting processes, and therefore the landscape of methods and tools are subject to an ongoing discourse in any organization and are ever changing, raising the need to constantly re-organize software systems. We want to ensure the digital sovereignty through a holistic combination of customizable standard software and effective consulting.

What are Our Biggest Challenges Being a Digital Technology Vendor?

The biggest challenges in being a digital technology vendor revolve around making customer success sustainable. Against the backdrop of trends like the opportunities of digital business and the risks of being outpaced by competitors that embrace digital business models, customers more than ever need consultative approaches from vendors: we increasingly have to help customers discovering their needs. Once a project is in place, it is in some situations challenging to maintain the motivation for continuous improvement and moderate the definition of meaningful priorities.

We see a growing gap in the market place between leaders and laggards in digitalization. Supporting organizations from both ends of the spectrum requires a vendor to support both, stability and a higher cadence of innovations.

Which Inhibitors Exist for Robust Integration of our Digital Capabilities in Industry? Why?

A major inhibitor for lifecycle digitalization initiatives, specifically where those aim to close loops between design, manufacturing and operations, is the inability of organizations to untangle and resolve the inherent complexities. This applies to complexities in business requirements, changing established processes, but also in re-configuring and changing existing IT systems and adopting new digital technologies like Artificial Intelligence. Software vendors and service providers need partners on the side of the customer at equal footing to align potentials and expectations, and build and evolve the system landscape.

Finally, as competition is intense, and although every vendor is claiming the opposite, some vendors try to force customers into lock-in situations by inhibiting interoperability like imposing contractual restriction or concealing information in proprietary data formats (e.g. CAD).

Occasionally, product lifecycle initiatives lack top-management support, resulting in missing strategic momentum, inappropriate priorities or insufficient funding. A significant number of industrial companies is not aware of the risks of being marginalized by digital leaders or becoming locked-in by less well-intended vendors—to gain digital autonomy it is essential to properly understand and manage risks and opportunities and not just "buy the necessary tools".

What is Our Way for Successful Partnership with Research and Industry?

Regarding research, we are maintaining a structured research roadmap, that is informed by a "trend radar" aligned with our strategic goals. Our in-house consulting staff is well connected in industry and academia, and a competent dialog partner for the research community. Actual projects are put to work with the support of a network of research partners. Strategic research partners in Germany and Asia are funded directly through long-term agreements. The *Elements* development platform is a powerful enabler for research prototypes and demonstrators, and also facilitates efficient exploitation.

CONTACT maintains a dedicated team for acquiring and governing alliances and industry partners. We are participating in partner programs of all major players in our market. Strategic alliances are in place with globally acting companies like Mitsubishi.

Dassault Systèmes[8]

How do We Describe Ourselves?

Dassault Systèmes, the 3D EXPERIENCE Company, is a catalyst for human progress. We provide business and people with collaborative 3D virtual environments to imagine sustainable innovations. By creating virtual twin experiences of the real world with our 3D EXPERIENCE platform portfolio, our customers push the boundaries of innovation, learning and production. As of 2021, Dassault Systèmes brings value to more than 290,000 customers of all sizes, in all industries, in more than 140 countries.

For more information, visit www.3ds.com.

What is Our Role as Digital Technology Vendor (DTV)?

Dassault Systèmes has already led the charge in transforming how products are designed, developed and supported, with 40 years of digital technology innovation. The accelerated pace of innovation required in the three sectors of the global economy we are serving, Manufacturing, Life Sciences & Healthcare, and Infrastructure & Cities, can only be achieved by the continued platformization of industries, where companies can leverage the social enterprise to support their innovation

[8] The *Dassault Systèmes* technical contribution has been created and delivered by Philippe Laufer.

processes throughout their value chain and across all disciplines, to drive successful end-customer experiences (see Fig. 19.10).

Today, consumers (whether a corporation, small company, individuals or government entity such as a city), make purchase and usage decisions, not based on the product or service itself, but on their experience with it. Our objective is to help our clients create, test and evaluate these experiences to make sure they are rewarding for their users, ensure that the product manufactured or the service provided meets expectations, and use this information to drive further improvements in the end-user experience.

Our 3D EXPERIENCE platform, which pioneered the category of „business experience platform", provides a collaborative environment that empowers businesses and people to innovate in entirely new ways and create these products and services using the virtual world. We are positioned to help customers become platform-driven through:

Fig. 19.10 Experience in the context of human, nature and technologies (*Source* Dassault systèmes)

- **A system of operations** coupling Modeling & Simulation (Mod-Sim) with extensive data capabilities
- **A business model** acting as a marketplace or trading platform that connects service providers and buyers

The 3D EXPERIENCE platform enables customers to reveal real-world data, from many disparate sources, elevated to a consistent, actionable semantic, and activate them into virtual twin experiences of the products, manufacturing facilities, or even enterprises themselves. The offer is built on a rich portfolio of data-driven industry processes and roles, spanning a wide spectrum of domains from high fidelity modeling and scientific simulation to production and logistics optimization. It is applicable in sectors such as natural resources, cities, transportation, buildings, smart products, consumer goods, as well as biological systems and chemistry.

This strategy focuses on *Human Industry Experiences*:

- **Human**: centers on online, mobile and ease-of-use, for collaborative innovation and for bringing 3D to consumers. For example, our HomeByMe solution helps millions of people all over the world imagine, easily create and place furniture in rooms, and experience them in virtual reality.
- **Industry**: centers on creating the knowledge and know-how needed to ensure that our solutions match closely the needs of the industries we address. Large clients have a strong focus on deep transformations to adapt to the respective challenges of their industries. In all these industries, new entrants have appeared with small teams focusing on sub-segments of their markets and proposing high-value experiences with products. Our solutions appeal to industry leaders and startups, both of which are shaking up industries.
- **Experiences**: Being able to model experiences is how companies can innovate and create new categories of products and solutions that will drive new, better experiences for their consumers. But this use isn't limited to companies. Our work with cities demonstrates that we can do this at the most demanding level thanks to the 3D EXPERIENCE platform's capabilities to model city experiences to improve the lives of citizens.

What is Our Vision to Drive the Future Digital Technology Portfolio?

Dassault Systèmes 'key driver is *sustainability*. Today, we are supporting the United Nations Sustainable Development Group (UNSDG) and its 17 Sustainable Development Goals such as Industry, Innovation and Infrastructure, Good Health and Well-Being, and Responsible Consumption and Production, through our purpose to provide virtual universes to businesses and people to imagine sustainable innovations that can harmonize product, nature and life.

To do this, we provide industry with the technological capabilities to create real-time virtual representations of a product, platform or ecosystem. These virtual twins can be used to model, visualize, predict and provide feedback on properties and performance, reduce operational costs, and drive end-to-end disruption in value chains, making them a key enabler of disruptive and sustainable innovation as well as more circular business models that would be prohibitively expensive, risky, and complex to develop and test in the physical world.

A virtual twin is not only defined by its physical representation, but also by its operational, functional and logical representation, hence the importance of the objects dictionary (Requirements, Functions, Logical items, Physical items, Processes, Resources, Models, Scenario, Scenes, Results, etc.) provided by the platform and all related services such as Lifecycle, configuration, change, ... Dassault Systèmes is the only company delivering these unified services.

Today, many of the items people use on a daily basis—from a shampoo bottle to a car—have been designed, engineered, improved and developed using virtual twin technology. The technology has enabled disruptive solutions that can positively impact society, from smart city initiatives and driverless vehicles, to record-breaking solar-powered aviation, hydropower plants and wind turbines.

The next frontier of virtual twin technology involves healthcare, which is why, as a company, we have expanded our focus „*From Things to Life*".

What is the difference between things and life? Life is not made of parts: the human body is one piece and hyper connected. Life doesn't do standardization: it's personalized design, production and usage. And life isn't "used" but lived. Life is an experience. Therefore, to improve life, we have to invent new ways of representing reality. We have to invent the virtual twin experience of life.

A virtual twin experience of the human body with the 3D EXPERIENCE platform integrates modeling, simulation, information intelligence and collaboration (see Fig. 19.11). It brings together biosciences, material sciences and information sciences to project the data from an object into a complete living virtual model that can be fully configured and simulated. By combining art, science and technology, it makes it possible to understand the invisible to represent the visible. Industry, researchers, physicians and even patients can visualize, test, understand and predict what cannot be seen—from the way drugs affect a disease to surgical outcomes—before a patient is treated.

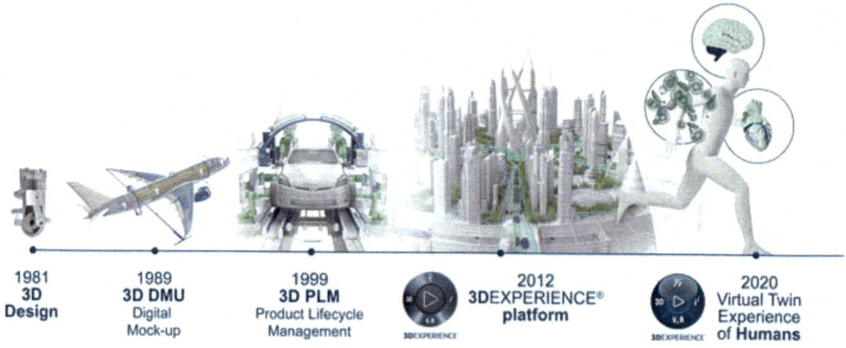

Fig. 19.11 The evolution from *3D design* to *virtual twin experience of humans* (*Source* dassault systèmes)

What are Our Biggest Challenges Being a Digital Technology Vendor?

Accelerating the understanding and implementation of virtual twin experience concepts, and making them inclusive for everyone in a company, whatever its size, are some of the key challenges Dassault Systèmes is addressing and meeting everyday.

Global GHG emissions are projected to reach around twice the *IPCC*- and *UN*-recommended CO2e target by 2030 in order to limit global warming to 1.5 °C by the end of the century. Achieving a carbon-free, circular economy in which waste is eliminated and resources are continuously used cannot be done incrementally and requires radically disruptive innovations by industry that are only possible through a molecular-level understanding of each product, material or process through its end-of-life.

Virtual twin technology is a non-negligible opportunity to stay within the recommended global carbon budget. Yet the technology has only been adopted by 10% of the companies that should be using it due to a number of barriers such a limited understanding of technology use cases and benefits. This requires advancing the thinking on the potential for virtual twin technology to accelerate this sustainable transformation towards a more circular economy.

Which Inhibitors Exist for Robust Integration of Our Digital Capabilities in Industry? Why?

Dassault Systèmes is already working with disruptive startups that are using the 3D EXPERIENCE platform to invent new industries. We're helping to facilitate the transformation of global OEMs into leaders of what we call the *„Industry Renaissance"*, a digital revolution that is transforming every aspect of industrial business: how product experiences are conceived, developed, tested, made, sold and serviced; how supply chains form, operate and braid themselves between industries; how consumers and manufacturers interact; how the real and virtual worlds inform and reinforce one another; how value is created; how the workforce is trained; and the very nature of work itself.

But the *Industry Renaissance* is powerful, and its transformations occur fast. Companies must either embrace it today or disappear tomorrow. This was further revealed during the pandemic, when companies needed to continue to work with their data and maintain business, operational and digital continuity. This was possible only by extending their communities, project management and business to a virtual environment on the cloud.

Cloud adoption, although now accelerating, is key for robust integration of "out of the box" solutions to implement virtual twin experience concepts. It eliminates costly and unuseful customizations that systematically hinder inclusive usage, through normalized processes and data.

Lastly, digital platforms, intelligent 3D models that embed knowledge and know-how in context, and digital marketplaces that can transform traditional supply chains into value-creation networks are critical components of success, but are not yet widely integrated in the standard thinking and decision process in industry.

What is Our Way for Successful Partnership with Research and Industry?

Dassault Systèmes has embarked on many initiatives to put our technology and knowledge at the service of research, education, culture, and artistic creation, as well as to drive our commitment to sustainability.

We advance our purpose through collaborations with larger, leading companies in key industries that have given us the means to solidify our pioneering position, as well as with the „movers and shakers" that are catalyzing the creation of new categories of products and reinventing industries.

Our research and development is conducted in close cooperation with users and customers to develop a deeper understanding of the unique 'To be' business processes of their industries and their future product/experiences directions and requirements.

We also have long-standing scientific and technical collaborations with key partners to maximize the benefits from available technology and increase the value for our shared customers. These alliances are established with three objectives: to cover end-to-end solutions with holistic offerings; to participate in shaping the future structure of industries; and to integrate the most advanced features of these technologies into our solutions. Dassault Systèmes participates in several hundred public–private projects including ones with the FDA, with prestigious universities such as Harvard, MIT, or Berlin University of Technology (TU Berlin), and with world-leading institutes such as INRIA, INSERM, and Fraunhofer Institutes. We also collaborate with renowned scientists including Nobel Prize winners.

We are also working very closely with innovators within the 3D EXPERIENCE Lab. This open innovation laboratory and startup accelerator program was created to foster entrepreneurship and strengthen society's future of creation. First established in France and now located in the U.S. and India, it has yielded successful projects aligned with the UN SDGs, including large-scale additive construction using robots, 3D printing of personalized organs for simulation of surgery, and unmanned long-range solar drones. The Lab has grown to include more than 25 incubators, accelerator, educational, entrepreneurial, technology and fab lab partners worldwide, 1200 mentors, and collaborations with multinational companies on co-accelerating promising projects in specific industries.

La Fondation Dassault Systèmes is dedicated to transforming the future of education and research through 3D technology and virtual universes. The foundation provides grants, digital content and skillsets in virtual technologies to education and research initiatives at academic institutions, research institutes, museums, associations, cultural centers and other general interest organizations throughout the European Union, the U.S. and India. In 2019, La Fondation supported 35 projects across these regions.

In education, Dassault Systèmes partners with institutions worldwide such as Berlin University of Technology (TU Berlin) closely with Prof. Dr.-Ing Rainer Stark, to jointly develop enhanced teaching methods that help transform science, technology, engineering and mathematics (STEM) education, and to define and implement policies and initiatives that contribute to preparing the workforce of the future. It is one of the founders of key academic associations such as the Global & European

Engineering Deans Councils, the International Federation of Engineering Education Societies, and the Cartagena Network of Engineering. It also organizes and supports competitions for science and technology students worldwide, to get young generations interested in the domains, anticipate and meet skills requirements, and boost their employability.

The strength of Dassault Systèmes lies in its inclusiveness, and its ability to reach all audiences and offer specific solutions to fulfill each need.

MathWorks[9]

How do We Describe Ourselves?

MathWorks is the leading developer of mathematical computing software for technical computing and system development. Founded in 1984, MathWorks is a privately held company with a staff of over 5000 people in 34 offices around the world. Millions of engineers and scientists worldwide rely on its products to accelerate the pace of their discovery, innovation, and development. MATLAB, the language of engineers and scientists, is a programming environment for algorithm development, data analysis, visualization, and numeric computation. Simulink is a block diagram environment for simulation and Model-Based Design of multidomain and embedded engineering systems.

Building on those two platforms, the company produces over 100 additional products for diverse applications such as signal processing, control system development, machine learning and deep learning, optimization, automatic code generation for production use, communication systems design, and computer vision.

What is Our Role as Digital Technology Vendor (DTV)?

Our role spans three key areas: education, research, and industrial applications.

In education, our tools enable students to learn and master engineering and scientific concepts, by applying them in real-world applications and problems. More than 6500 colleges and universities around the world use MATLAB and Simulink for teaching and research in a broad range of engineering and science disciplines. Our tools integrate with learning management systems, can be leveraged for massive open online courses (MOOCs), and work with more than 2000 textbooks that present theory, real-world examples, and exercises in engineering, science, finance, and mathematics.

In research, the open-system architecture of MATLAB and Simulink enables researchers to create and share leading-edge techniques as community toolboxes, built on the impeccable MATLAB numeric foundation and the ability to integrate with a broad range of professional toolboxes developed by MathWorks. Researchers can also integrate with techniques and libraries created in other languages such as C/C++, Java, and Python.

In industry, MATLAB and Simulink are used throughout the automotive, aerospace, communications, electronics, and industrial automation industries as

[9] The *MathWorks* technical contribution has been created and deliveredby Jim Tung.

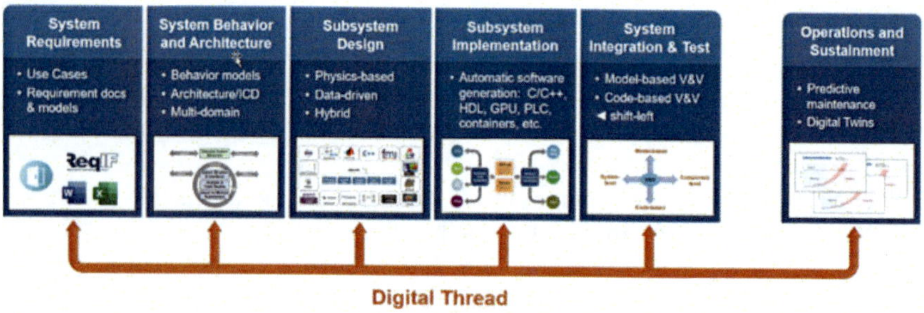

Fig. 19.12 MathWorks solutions along the digital thread (*Source* MathWorks)

fundamental tools for research and development. They are also used for modeling and simulation in increasingly technical fields, such as financial services and computational biology. Our capabilities span the system development lifecycle, from requirements management, to system architecture, multi-domain modeling and design, code generation, model-based/code-based verification and validation, and the use of models as a digital twin (compare Fig. 19.12). Our capabilities also enable powerful data analytics, with the ability to deploy to embedded devices, edge systems, on-prem/HPC, and cloud.

MATLAB and Simulink are designed to interoperate with other aspects of the digital ecosystem, including traditional PLM systems, modern systems such as Jira and Git, more than 100 other modeling tools and languages, CI/CD (continuous integration and continuous delivery), and enterprise systems based on Spark, Hadoop, and other Big Data frameworks.

Our tools support most popular development approaches, including the traditional V-model, Agile, and DevOps (compare Figs. 19.13 and 19.14).

What is Our Vision to Drive the Future Digital Technology Portfolio?

Our vision is to enable the systematic use of data and models so organizations can create and deliver superior value to their customers throughout the entire product/service lifecycle. The data include experimental data, production data, test data, and operational data. The models include dynamical and physics-based models, AI models, and other data-driven models. Value is obtained by enabling decisions in the asset, at the edge, on-premises, and in the cloud.

What are Our Biggest Challenges Being a Digital Technology Vendor?

In companies across all the industries that we serve, "digital transformation" initiatives are dramatically changing their strategies, workflows, people, and systems. In some cases that creates uncertainty, and sometimes conflict, between engineering groups who want to use model-based approaches and digital/software groups who want to use data-driven approaches. A key challenge—and major opportunity—is to work with those groups to define synergistic approaches that leverage the strengths of both perspectives.

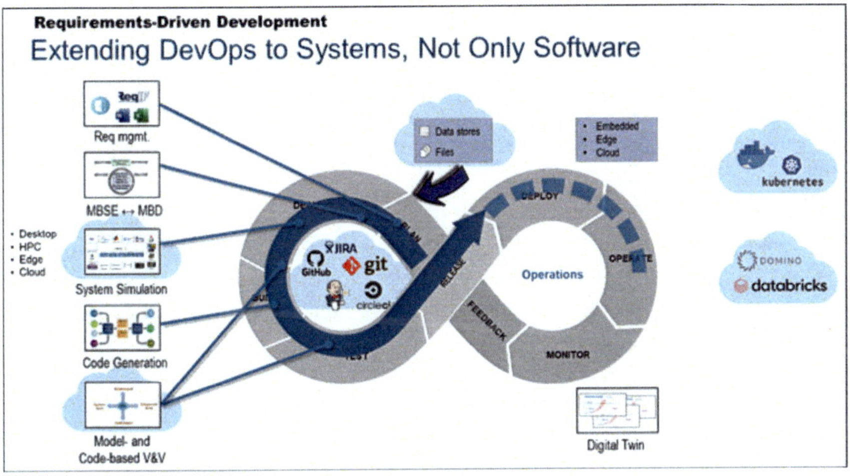

Fig. 19.13 *Requirements* driven DevOps to (technical) systems (*Source* MathWorks)

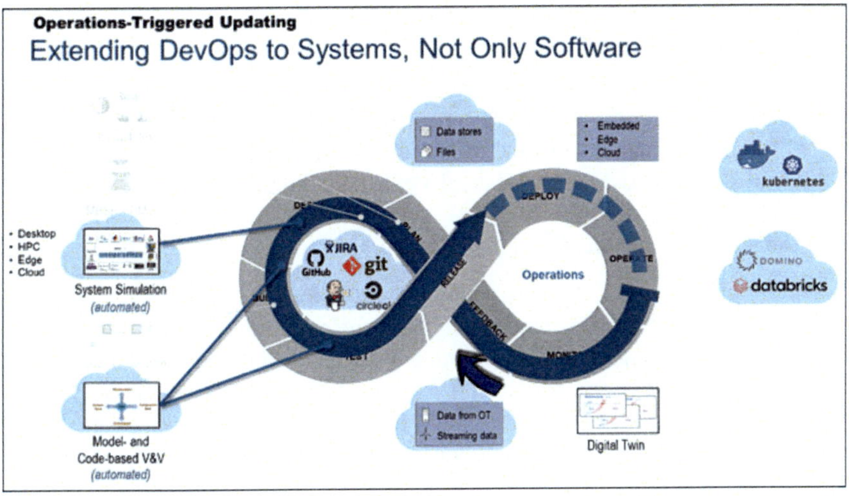

Fig. 19.14 *Operations triggered update* driven DevsOps to (technical) systems (*Source* MathWorks)

Which Inhibitors Exist for Robust Integration of our Digital Capabilities in Industry? Why?

In general, MathWorks has a proven track record enabling robust integration of our capabilities with other tools. We do that by carefully designing APIs (application programming interfaces) in our products, then documenting, supporting, and maintaining them as features of our products. We rely on well-proven protocols and standards for the given workflow, such as COM, .NET, and DDS.

An inhibitor can exist when a standard is not well proven or lacks rigor needed for a particular workflow. One example is the XMI interchange specification in SysML, which is not able to support rapid exchange and roundtrip workflows due to the specification and SysML tools' support for it.

What is our Way for Successful Partnership with Research and Industry?

The MathWorks approach for successful partnership with research and industry relies on our robust, performant, and flexible technology foundation. With that in place, we can enable research and industry customers to self-serve their way to success, and we can also define ways to partner, extending and customizing our capabilities when appropriate.

PTC Inc.[10]

How do we Describe Ourselves?

PTC Inc. (formerly Parametric Technology Corporation) is an American computer software and services company founded in 1985 and headquartered in Boston, Massachusetts. The global technology company has over 6000 employees across 80 offices in 30 countries, 1,150 technology partners and over $1bn in revenue.

The Company offers a portfolio including.

- IoT products, such as ThingWorx, KEPServerEX,
- PLM products, such as Windchill, Integrity, Navigate, Creo View and Arena,
- CAD products, such as Creo and Mathcad and Onshape,
- AR/VR products, such as Vuforia and Vuforia Studio.

What is our Role as Digital Technology Vendor (DTV)?

IT is now playing a more prominent role in organizations' decisions to adopt Industrial IoT platforms as part of broader digital transformation initiatives [6]. Importantly, IT's involvement is being met with enthusiasm from OT groups at the organization. The historic divide between IT and OT has been well documented, but as the Industrial IoT platform market has matured, so has the concept of IT-OT convergence. Platforms continue to be the dominant option for Industrial IoT functionality, but CIOs should also be aware of the emerging selection of Industrial IoT solutions that address specific use cases. Together, a platform and a solution built on that platform are a compelling option for digital transformation projects [7].

In this context, PTC has consistently been named a leader in Industrial IoT by all major industry analyst firms, e.g. [8], over the last several years, see Fig. 19.15. PTC can deploy its Industrial IoT offerings on-premises, in the cloud, or in a hybrid environment. PTC has developed a broad partner ecosystem in the Industrial IoT space, including strategic partners, global and regional systems integrators, and technology partners.

[10] The *PTC Inc.* technical contribution has been created and delivered by Dominik Rüchardt and Dr. Erik Rieger.

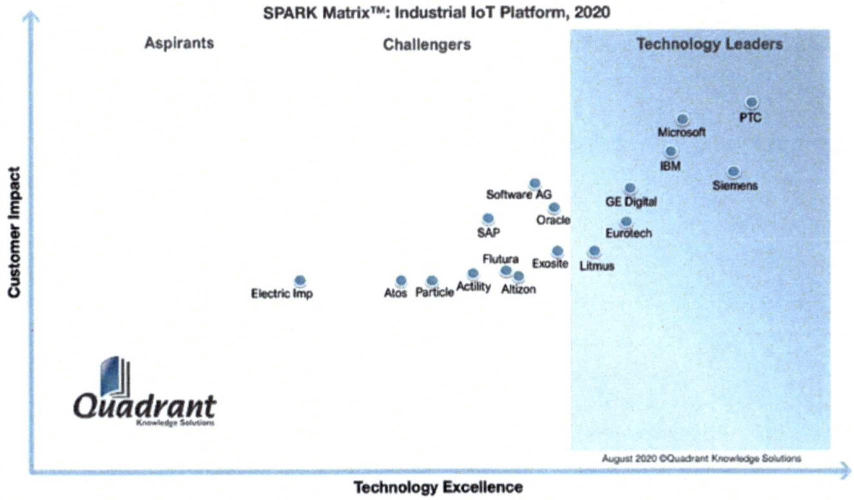

Fig. 19.15 Quadrant of IIoT leaders [7]

PTC brings its industrial IoT offerings to market through this partner ecosystem, a growing channel network, and a global direct salesforce. Complementing core IIoT functionality like application enablement, analytics, and device management, PTC offers a range of device connectivity offerings, including some of the broadest support for industrial protocols and drivers. Moving forward, PTC will offer more of its Industrial IoT functionality in a pure software as a service (SaaS) model.

What is our Vision to Drive the Future Digital Technology Portfolio?

The increasing capabilitites of smart connected products not only reshape competition within industries but expand industry boundaries. This occurs as the basis of competition shifts from discrete products, to product systems consisting of closely related products, to systems of systems that link an array of product systems together [9]. Also, in contrast to traditional product development, in the future product development is expected to continue along its entire life cycle [10].

Industrial value creation will benefit from a convergency of the physical and the digital world which creates powerful new capabilities as Digital Twins, Augmented Reality based interactions, any many more.

The fundamental step to these capabilities is the digital thread—the seamless connectivity along the lifecyle an cross all functions and views in connection to industrial products and services (Fig. 19.16).

The Digital Thread is the logical next step coming from PLM and Systems engineering. An open digital architecture connecting things and software systems along the value chain and supporting applications and solutions for the business-oriented interaction. One is the digital twin concept which becomes key for managing products in the future. The management of the digital version of the real product instance is basis for leveraging the benefits of the digital transformation such as

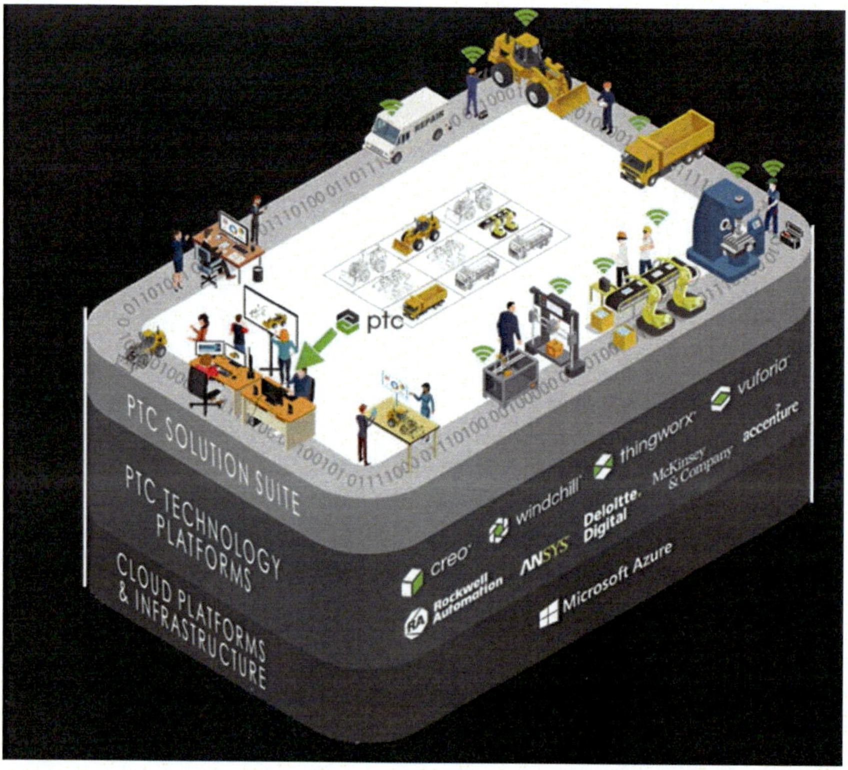

Fig. 19.16 The digital thread connects the entire lifecycle and is the prerequisite for the realization of connected business models and digital twins (*Source* PTC Inc.)

remote product management, performance control, failure prediction, continuous improvement, 'Product as a Service' offers and flexible recomposition.

This comes together with an evolving IT focus on moving IT infrastructure into the cloud, but also leveraging SAAS models. Product related data is now available across the entire cycle from all locations. Data is gathered from different systems, but also along operations (SCO) and directly picked up from the product, while it is in use or in maintenance (SCP).

The comprehensive data is used to establish the digital twin and thus forms as basis for AR and VR embedding. It opens up multiple differentiation paths. It can create companion experiences that expand the capabilities of products, give customers more information, and increase product loyalty. PTC supports all pathes based on the described technology offerings together with the experience and vision as leader in the magic quadrant.

What are Your Biggest Challenges as a Digital Technology Vendor?

With the increasing degree of connectivity value creation shifts more and more from a well fenced product- or user group to distributed user groups and processes across functions and enterprises. The traditional "Customer-Vendor" relation tranforms in that context into a partnership which inludes many stakeolders contributing to the final value add. Vendors have to deliver into complex value networks which continuously change.

At the same time new business models like SAAS (software as a service) and cloud operations change the way of payment, the way of evolution of vendor-customer relations, the way of maintaining software and the way of supporting user communities.

Which Inhibitors Exist for the Robust Integration of our Digital Capabilities in Industry? Why?

A digital transformation investment of a customer is much more than buying software. It includes changes of behavior and the need to integrate with all interfaces along value chains. A digital strategy which includes business and stakeholder evolution is absolutely necessary for short- and long-term success. In addition, standardization and orientation on industrial reference architectures is key to benefit from the fast ecolution of digital capabilities.

How do You Achieve Successful Partnerships Between Research and Industry?

Research need shifts from technology to the question how to apply technology in a given or changing business context. Research results in reference models and processes. As a large vendor of commercial software for the industry the highest responsibility is in ensuring the robustness and long-term reliability of the software. Research consortiums should be reshaped in that sense and follow long term architecture concepts.

Following these principles PTC partners successfully with many universities with a special academic program and runs a large industrial experience center in the RWTH Aachen Business Campus. PTC also supports public funded research in a broad manner with technology and advisory. PTC is active member in asssiciations as OMG, ProStep iVIP, Bitkom, Plattform Industry 4.0 and many more and supports initiatives and consortiums for sustainable business innovation.

Siemens Digital Industries Software[11]

How do we Describe Ourselves?

Siemens Digital Industries Software is world's leader in industrial software. We offer an integrated portfolio of software, services, and collaboration across a broad spectrum of engineering domains called *Xcelerator*. Xcelerator accelerates digital transformation for companies of all sizes and in all industries. The Xcelerator portfolio supports three key facets of the digital enterprise: the comprehensive *digital twin*, *personalized, adaptable solutions*, and an *open, modern ecosystem* (see Fig. 19.17).

[11] The *Siemens Digital Industries Software* technical contribution has been created and delivered by Brenda Discher.

Fig. 19.17 The Xcelerator portfolio to accelerate digital transformation (*Source* siemens digital industries software)

What is our Role as Digital Technology Vendor (DTV)?

Siemens uniquely enables customers to create a better future with state-of-the-art technology so that anyone can turn today's ideas into tomorrow's products and experiences. By providing all the tools to connect the virtual and real worlds of product development and production processes, it creates a closed-loop environment leading to continuous optimization possibilities.

What is Our Vision to Drive the Future Digital Technology Portfolio?

1. *Comprehensive Digital Twin* (compare Fig. 19.18)

Merging the virtual and physical worlds with a complete digital representation of the product and its creation process allows engineers, designers, production engineers and even end users to experience every facet of a product or system long before a prototype is available, a chip is manufactured, or a factory is built.

The Digital Twin enables integration of the entire product lifecycle with the factory and plant lifecycle, along with performance data. Siemens is the only DTV whose Digital Twin concept encompasses the integrated circuit, embedded software, electronics, electrical systems, mechanical design, physics, and the actual system in operation. Given that the worlds of integrated circuits and systems and products are converging more and more—with automakers, aerospace companies and other manufacturer starting to design their own specialized chips—this ability is becoming increasingly important.

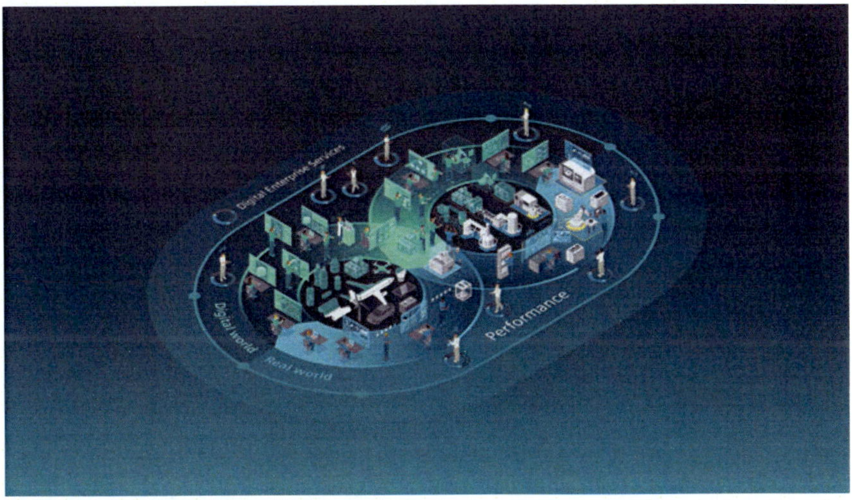

Fig. 19.18 The comprehensive digital twin (*Source* siemens digital industries software)

2. *Personalized and Flexible*

We believe all users of software will require personalization. This means the users will also define the development of the application to their individual needs. IT systems should assist the designer/engineer, customer workflows, digital threads, and industry specific customer solutions. Flexibility means the ability to work across engineering domains and across design, manufacturing, utilization allowing for ubiquitous engineering 24/7, from any device and ramping up new engineering capabilities on demand. Siemens acquired Mendix in 2018, allowing business users to create and tailor our software applications without having to be software developers.

3. *Open Ecosystem*

In the complex product lifecycle world, it is important to have an open architecture so our customers can allow data to easily flow into existing third-party applications. By securing collaboration and sharing data and IP with partners, suppliers, or developers, an open ecosystem enables innovation with low-code and cross-platform compatibility. Siemens enables over 4 million users to leverage its 3D modelling engine Parasolid, and over 130 members utilize our visualization tools. In addition, customers can choose from a strong low-code community of over 190,000 developers[12] to help them. This creates an industrial network effect to speed up innovation, enhance customization efforts and promote partnerships across the supply chain.

[12] All data as of April 2021.

What are our Biggest Challenges Being a Digital Technology Vendor?

The top challenges are:

1. to support our customer's transition into new business models; the disruption happening across most industries,
2. to enable digitalization in the face of a significant skills gap. Currently, design, engineering and manufacturing jobs worldwide go unfilled due to the growing gap between the skills employers need and those recent graduates have.

Meanwhile, digitalization changes constantly and Industry 4.0 drives the need for skilled workers that understand data-driven, AI-powered, connected design and manufacturing. As the industry landscape evolves, the need for highly skilled technologists has never been greater.

How do you democratize and knowledge and experience? We believe digital technologies are already starting to play a major role in accomplishing this goal. Siemens' low code platform Mendix helps to "encode" knowledge and experience within apps, leveraging and exposing the capabilities of the whole Xcelerator portfolio for commercial users as well as for academic institutions. Xcelerator provides the bridge across engineering disciplines to enable more cross/inter-disciplinary collaboration which supports new business models that rely on the combination of sophisticated technologies, services and monetization models.

Which Inhibitors Exist for Robust Integration of our Digital Capabilities in Industry? Why?

Those reluctant to change will point to many factors preventing them from embracing digital transformation. A company's existing investments in technology, regardless of its potential pitfalls and perceived or past successes, may be setting back innovation. Having already invested has translated into a refusal to evolve.

We see that investment in digitalization separates the winners from the laggards, especially as startups enter established industries. Other factors that drop barriers to innovation include the democratization of design and supply chains and the lower thresholds of infrastructure cost fueled by cloud technology.

Most progressive companies in terms of digital transformation initiatives see the highest levels of return. But they must be all-in. We have seen the biggest gains come from customers who have leveraged technology to have different disciplines working in a more united way. An example of a company embracing digitalization and being able to pivot is VinFast. VinFast wanted to be competitive in Vietnam and globally right from the beginning, so the company relied on Siemens' expertise to utilize the latest technology. This resulted in a closed-loop manufacturing system, which uses digital twins of the products, the production, and the performance of production and product. The fully digital factory was built in 21 months, 50% faster than usual, and is designed to be easily scalable for future expansions.

What is our Way for Successful Partnership with Research and Industry?

For Siemens, research and industry cooperation is essential. The company's R&D expenditures reached €4.6 billion in fiscal year 2020 and issued 2493 patent applications with more than 25% of the patents related to "Industrie 4.0" and digital technologies.

Proven and new forms of bilateral partnership allow for the greatest possible sharing of knowledge, which is why Siemens partners with academia on several levels. These partnerships include supplying of our products and solutions for education, free training, certifications, and modular curricula content for educators as well as joint research and development projects. Siemens works closely with universities and research institutions on many research and innovation projects.

The dialog with university students includes idea competitions and hackathons, industry-sponsored doctorates for graduates at Siemens, and university teaching by Siemens employees. Siemens also serves on various academic bodies, where we seek to better fuse industrial and academic requirements.

Research "co-locations" and "living labs" allow professors, Siemens experts, and students to collaborate intensively. Siemens shares its industry expertise and its knowledge of industry needs in the development of new curricula in the emerging technologies of artificial intelligence, digital twin, additive manufacturing as well as model-based systems engineering. Siemens is also actively driving new partnerships with academic incubators.

Siemens serves an ecosystem of 4000 + schools and 1,500,000 + students, as well as over 4000 technology partners.

References

1. Deuter A, Pethig F (2019) The digital twin theory—Eine neue Sicht auf ein Modewort (The digital twin theory—a new perspective on the buzz word); Industrie 4.0 Management 35. https://www.industrie-management.de/sites/industrie-management.de/files/pdf/deuter-The-Digital-Twin-Theory-IM2019-1.pdf
2. Winkelhake U (2021) Die digitale transformation der automobilindustrie (The digital transformation of the automotive industry) 2nd edn. Treber—Roadmap—Praxis, Springer
3. BMBF (Federal Ministry of Education and Research), BMWI (Federal ministry for economic affairs and energy), Plattform industrie 4.0: https://www.plattform-i40.de/SiteGlobals/PI40/Forms/Listen/Downloads/EN/Downloads_Formular.html?https=1&__site=PI40&cl2Categories_Typ_name=veroeffentlichung&nsc=true
4. Bitkom (2020) Industrie 4.0 platform industrie 4.0: vorschlag zur systematischen klassifikation von interaktionen in industrie 4.0 systemen (Proposal for the systematic classification of interactions in industrie 4.0 systems), Bitkom, Berlin, https://www.bitkom.org/Bitkom/Publikationen/Whitepaper-Vorschlag-zur-systematischen-Klassifikation-von-Interaktionen-in-Industrie-40-Systemen. Accessed 20 March 2020
5. BMBF (Federal Ministry of Education and Research), Forschungs-union, acatech: recommendations for implementing the strategic initiative INDUSTRIE 4.0. Final report of the industrie 4.0 working group; https://en.acatech.de/publication/recommendations-for-implementing-the-strategic-initiative-industrie-4-0-final-report-of-the-industrie-4-0-working-group/. 8 April 2013

6. Goodness E, Velosa A, Friedman T, Berthelsen E, Kim S, Havart-Simkin P, Thielemann K (2020) "Magic quadrant for industrial IoT platforms," Gartner. https://www.ptc.com/en/resources/iiot/white-paper/gartner-mq-for-iiot
7. Porter M, Heppelmann J (2017) A manager's guide to augmented reality. Harvard Business Review, Boston
8. Quadrant-Knowledge-Solutions (2020) "SPARK Matrix™: industrial IoT (IIoT) platforms". https://quadrant-solutions.com
9. Porter M, Heppelmann J (2014) How smart, connected products are transforming competition. Harv Bus Rev 92(11):64–88
10. Erler S, Rieger E (2016) Product evolvement through entirely scheduled lifecycles. Procedia CIRP 48:79–83. https://doi.org/10.1016/j.procir.2016.04.077,Berlin

Chapter 20
Industrie 4.0 and IoT Technologies

Executive Summary

This chapter deals with the following topics:

- Understanding which new *Cyber-Physical System (CPS)* concepts and *IoT (Internet of Things)* technologies will drive future Virtual Product Creation capabilities,
- Explanation how *Industrie 4.0 solutions* and related digital technologies change product offerings
- How data services become essential in future operations of smart products and product service systems.

Quick Reader Orientation and Motivation

The intention of this chapter is:

- To describe the major *Industrie 4.0* concepts and solutions
- To help learning what *CPS (Cyber Physical Systems)* and IoT (Internet of things) are all about

---

The specific term *Industrie 4.0* (instead of Industry 4.0) as specific German expression for an entire strategic program of the German government will be used throughout this book to indicate the concepts, solutions and realizations of the platform Industrie 4.0 and its research board. This strategic program has established cross-links to similar initiatives in other countries (such as *Smart Manufacturing* in China or *the Industrial Internet Consortium* in the USA). This chapter introduces and describes core concepts, technologies and solutions of *Industrie 4.0, Internet of Things, Cyber-Physical Systems, Smart Products, Digital Twins as well as Cloud, Edge and Digital Data Platform architectures*. The chapter also explains Industrie 4.0 applications in industry. The relevance of these new Industrie 4.0 technologies and solutions increase steadily and they influence not only future *smart factory solutions* and *digital manufacturing* aspects but also many new *"intelligent" (smart) product and product service systems* solutions. This chapter provides critical insight to which extent and in which areas *Virtual Product Creation* needs to evolve in order to cope with these new Industrie 4.0 technologies and solutions.

R. Stark, *Virtual Product Creation in Industry*,
https://doi.org/10.1007/978-3-662-64301-3_20

- To provide an understanding of *Digital Twins* and their potential roles in future technical systems development and operations
- To explain necessary technology stacks in order to build up IoT type solutions in industry
- To describe hardware and software infrastructure for edge, cloud and platform computing
- To derive future Virtual Product Creation (VPC) capability needs based on the above-mentioned new solution elements of technical systems.

Industry is undergoing a major transformation, which in Germany is referred to as Industrie 4.0. It is a high-tech strategy of the German federal government to secure Germany's future as an industrial nation [1]. Novel digital solutions and technologies, which this chapter is dedicated to, drive the change in industry and are challenging and significantly shaping future *Virtual Product Creation* capabilities. Therefore, the first section of this chapter gives a strategical overview of the changing industry, its historical roots, arising concepts and required research. The *Digital Twin* concept and *IoT* technologies are majorly characterizing the change and are addressed in the second and third section. In turn, cloud and edge technologies give the computational foundation for novel solutions and will be introduced in Sect. 20.4. To illustrate the solutions in application, exemplifying industrial scenarios are presented in the last section.

20.1 Industrie 4.0

The origin of the Industrie 4.0 strategy lies in the extensive outsourcing-activities of production from established industrial nations to low-wage countries in the last decades. Germany, as one of the biggest industrial nations in the world with increasing wages, suffers more than competitive nations, since the industry share of the German economy is still the highest of all industrial nations. While the share of production, particularly in the USA, UK and Japan, has fallen significantly over the last 20 years, it has remained at a high level in Germany [2]. In order to compete with low-wage countries on a cost-covering basis, German industry has seen the new digital opportunity for a new type of "intelligent production" in order to achieve a new level of efficiency and sustainability. Based on new technical principles like digitally networked devices, objects and components throughout the manufacturing processes and product lifecycle Industrie 4.0 defines a new substantial digital foundation for the next generation factories and all machine and product operations through the global world.

Industrie 4.0, therefore, encompasses more than the automation of production itself, as Fig. 20.1 displays. Industrie 4.0 is the continuation of the last three industrial revolutions [3]. The first industrial revolution was aroused by the water and steam-driven mechanization of the loom in textile industry. The second one was characterized by the emergence of mass production and the electrification of production lines following the theories of the Taylorism and Fordism. The third and last

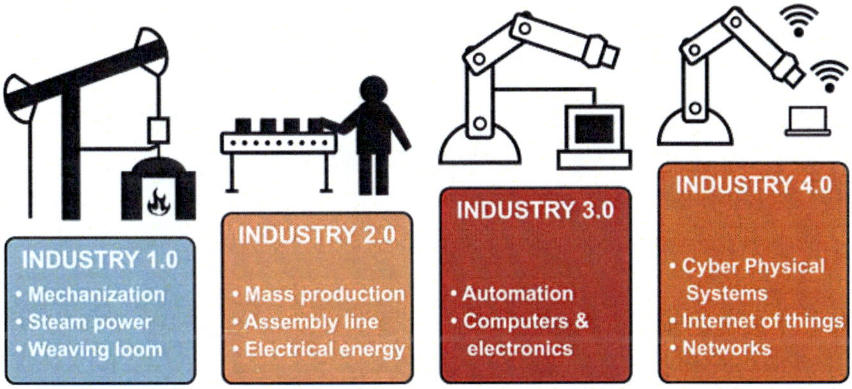

Fig. 20.1 The four industrial revolutions

revolution was dominated by the automation of production, based on new technologies in computing and electronics. Industrie 4.0 stands for a new stage with the fully digital organization and control of the entire value chain over the entire product life cycle [4, 5 and 6]. New technology is i.e. applied by embedding software with respective hardware in almost all products which turns them into *smart products*[1] [5].

As result of Industrie 4.0, the development and manufacturing of products reach new levels of complexity [2] while at the same time showing the potential of a higher degree of efficiency of up to 18% [7], as current studies examined.

20.1.1 The Concept, the Initiative and the Vision

The revolution of IT in industry drives radical transformation. Nowadays, the combination of smart products and cyber-physical systems[2] (CPS) herald the fourth industrial revolution [6]. Industrie 4.0 aims at digitally orchestrating whole enterprises, their networks and the associated value creation streams (no longer fixed chains!) over the entire product lifecycle (PLC). This ambition, however, requires full digitalization of the industrial processes. It is, therefore, no longer just the digital control of machines and production lines which is in scope for digitalization. The goal with Industrie 4.0 is to reach out to the embedding and the effective interplay of all major value creation relevant processes, activities and services to enable new capabilities such as:

- New in process quality data driven manufacturing builds,
- Dimension and shape deviation-based sequence and work plan rescheduling of components and assemblies to be manufactured,

[1] Smart products: Smart products are CPS with services through internet connection. Those products can, for instance in production, communicate to other products, systems and the environment with their embedded intelligence 5.

[2] Cyber-physical system: CPS are systems with physical parts (manufacturing resource) and cyber parts (cloud). See Sect. 11.3.1 for a more detailed explanation of the term.

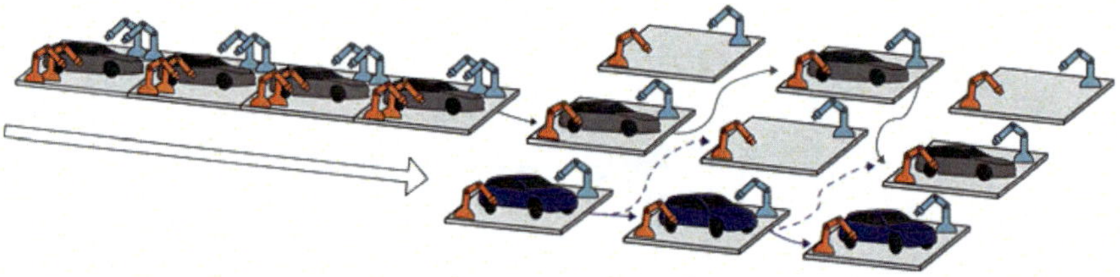

Fig. 20.2 Conventional assembly line (left) and autonomous modular assembly system (right)

- Dynamic rescheduling of product orders and build sequences due to material availabilities and material batch test records,
- Context aware optimization of human–machine and human–robot interactions, both to drive assembly process yield and to deliver higher quality results.

This new type of industrial digitalization creates large amounts of data caused by sharply increased sensor measured signal detection. Those signals and digitally "shadowed" measurements of physical states are to be exchanged and forwarded through IoT-technologies to "analytic engines". Those "analytic engines" serve as "intelligent cognition nodes" to enable thereby the "new intelligence" and subsequent interaction of smart products as part of Cyber Physical Systems (CPS) with their physical and digital twin interactions. Hence, complex digital networks arise and, on this basis, every single decision for the orchestration is made. This decision-making is based on advanced and versatile algorithms in conjunction with AI and simulation techniques.

The left side of Fig. 20.2 shows a robotic assembly line. Nowadays, production lines of this type are used in automotive industry, e.g. in manufacturing stages such as the body-in-white. The assembly line is highly-automated, the overall process is sequential, and the control of the assembly line is carried out in a highly deterministic way by IT-systems. The right side of the figure visualizes a highly flexible, changeable and above all autonomous assembly system. The CPS is based on the new principles of Industrie 4.0. It consists of individual assembly stations that adapt in a flexible way to the product to be produced. Smart products operate autonomously and move through production autonomously with the help of autonomous guided vehicles (AGV). The AGV find their path through the assembly stations by communicating with each other, the environment, the CPS and a cloud-based scheduling and orchestration system. In smart factories of this kind, each product can be localized at any time and the current status and requirements to fulfill the production are known at all times [6]. The assembly systems are vertically integrated within the automation hierarchies of the enterprise and horizontally connected to value chains and can be managed in real time. Hence, the moment a product order is placed, new routes through the production system, or rather the all-encompassing orchestration of the entire enterprise is calculated, and eventually specific manufacturing work plan changes might even be triggered on-the-fly.

20.1.2 The Platform Initiative

The Platform Industrie 4.0 is one of the worlds' largest networks for Industrie 4.0 and aims at coordinating the digital transformation of industry in tight interaction with research and science. The German network is composed of various members from mainly politics, research and industry. Its members are organized in working groups that focus on the major challenges of Industrie 4.0 [8]. Industrie 4.0 has become an international brand and the platform has established a large number of national and international partnerships. In addition, the platform leads the discussion of G7/G20 on the digital transformation of industry (c.f. [8]: Plattform Industrie 4.0, S. 1–3).

The platform's guiding principle is to shape the digital transformation of industry in a coordinated manner. Therefore, reference models and standardization are one of the platforms' key focus. The use of standards is a basic prerequisite for achieving the digitalization and automation throughout the entire value creation process. The Reference Architecture Model Industrie 4.0 (RAMI 4.0), defined by the platform members, depicts a three-dimensional model which aims at ensuring the vertical and horizontal integration of production objects from different perspectives [9]. Therefore, the model targets the IT-based description and integration of those objects, see Fig. 20.3.

The *first dimension* (vertical) shows the above-mentioned perspectives organized in layers and the objects to which they can be assigned, such as the functional description or the communication behaviour. Each layer is, in turn, described by the dimension two and three.

The *second dimension* (horizontal integration) describes the single phases and the respective data continuity of the product life-cycle phases such as the development phase and the production phase.

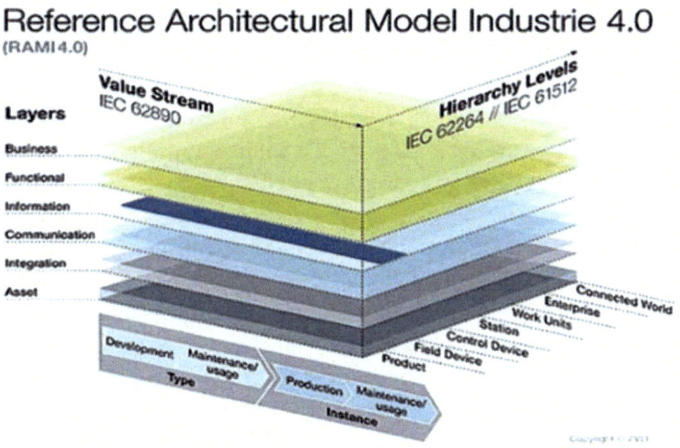

Fig. 20.3 RAMI 4.0—reference architectural model for industrie 4.0 [9]

And the *third dimension* (vertical integration) describes the factory hierarchy from product and field devices over workstations up the whole connected enterprise. In Industrie 3.0 the communication worked from hierarchy to hierarchy and the product has not been part of communication mechanisms. Industrie 4.0 creates a communication network which breaks the existing limitations of Industrie 3.0. Semantic knowledge is anticipated to be present and expressible in all segments of the RAMI 4.0 model.

Imagine the following scenario to clarify the model. A manufacturer of electric motors develops a specific motor (see base level "product" on the Hierarchy Level in Fig. 20.3) and sells it to an industrial engineering company designing and building assembly systems. The electric motor is then installed in an assembly system in a producing company and therefore categorized as field device (compare Fig. 20.3). The usage of the engine generates information, which brings up various Industrie 4.0 use cases, based on the information layer. The information layer sets standards for the communication with the field devices' information over the product life cycle, (compare to the dark blue highlighted area in the light blue information layer in Fig. 20.3). This creates several use-cases over the PLC for all involved companies. The producing company can use the information for preventive maintenance and anomaly detection. The industrial engineering company can use the information for advanced system modelling based on real shop floor data and after all, the electric motor company can use the information for further improving their products [9].

The RAMI-Model conjuncts with internationally recognized reference architectures such as the Japanese IVRA (Industrial Value Chain Initiative) with the *Smart Manufacturing Reference Architecture*, the Chinese IMSA (*Intelligent Manufacturing System Architecture*) and the American IIRA (*Industrial Internet of Things Reference Architecture*).

20.1.3 The Industrie 4.0 Roadmap Ahead

The RAMI model opens a solution space, in which research and industry identified necessary technologies and have developed application solutions. First products based on the RAMI model are successfully applied in industry towards fulfilling the Industrie 4.0 vision. This chapter outlines innovations and required research topics for Industrie 4.0 from a technological and methodological standpoint.

Industrie 4.0 underlies the digitalization of industry and makes the *Digital Twin* a central concept. The Digital Twin is a "*[...] digital representation of an active unique product or product-service system [...]*" [10] and depicts much more than the digitization of paper processes. For this reason, the following Sect. 20.2 is dedicated to the to the *Digital Twin technology*. One requisite for the implementation of Digital Twins is the Internet of Things (IoT), which will be introduced in the Sect. 20.3. Lastly those two technologies require novel flexible and cross-company IT-infrastructures which also make cloud and edge computing a central topic of Industrie 4.0 and is addressed in Sect. 20.4.

These solution concepts and technical underpinnings form the technological foundation of Industrie 4.0. Nevertheless, further research-topics will be necessary to drive the new industrial revolution. The German research council for Industrie 4.0 has identified the following four research and development fields for establishing a focused research roadmap for Industrie 4.0 [1]:

- Prospective technology trends
- Value creation scenarios for Industrie 4.0
- New methods and tools for Industrie 4.0
- Work and society.

Prospective Technological Trends

It is apparent that Industrie 4.0 will rely on new prospective technology trends, such as the latter mentioned technologies *Digital Twin*, IoT and *cloud computing*. Current progress in independent production technologies as well as information and communication technologies will be merged to interdisciplinary solutions and will lead to flexible, modular, context-aware and autonomous production systems. Those systems will be fully integrated in the overall product lifecycle (PLC). There is the need for new and disruptive machine concepts, applying capability-based and self-configuring modules. Those systems should not be based on the conventional automation pyramid,[3] rather a cross-system networking of the single components needs to be achieved.

Current engineering, planning and operation procedures cannot be applied to those new systems. The new complexity of such new type of devices and machines, generating and using a full range of additional new signal flows and data types, require novel and versatile algorithms supporting engineers over the entire system lifecycle. The algorithms processing the new data set volumes will be based on a combination of statistical, deterministic and AI-methods and will lead to mostly autonomous systems. The creation of such systems, however, need to be highly supported by engineering assistant systems and new model-based engineering approaches (compare Chap. 21). It has to be mentioned that nowadays the application of AI in engineering differs significantly from the application e.g. in a consumer environment. This gives research new needs towards the exploration of industrially useable AI. The terms *IIoT (Industrial Internet of Things)* has been created in the US and just recently the term AIoT (= AI + IoT) has been created by the technology company BOSCH for such new industrial sensor and data flow rich solutions across networks. Furthermore, it will be necessary to establish hybrid models linking *model-based* and *data-driven* approaches in product, manufacturing and industrial engineering [12].

Sensors and actuator systems are the link between the physical and the digital (cyber) world. Current and future developments aim at downsizing those components, decreasing their complexity and driving down prices. In addition, the generalizability and transferability of sensor technology will be key. Thus, a wide deployment of sensors in the field can be established and enabling the cross-system communication,

[3] See [11].

one of the bases of Industrie 4.0. The development of new communication technology itself will be another cornerstone of Industrie 4.0. The 5G-based communication, for instance, will fulfill the high requirements of autonomous, smart systems, regarding higher data-rates, level broadband and real-time communication. In this context, data security constitutes a significant caveat and, therefore, will remain an open-research topic (c.f. [12] p. 13).

Value Creation Scenarios for Industrie 4.0

Making Industrie 4.0 a reality, existing manufacturing processes will need to be improved. Moreover, innovative business opportunities will come up with data-driven and platform-based business models. The digitalization of industry enables businesses to create *Product-Service Systems (PSS).* PSS represent a combination of physical products combined with human-based or automatic services leveraging data services, called Smart Services. Those latest level PSS follow the new principles of 'Everything-as-a-service' (XaaS). Such modern PSS already found its way to industry via e.g. 'pay/power-by-the hour', 'pay-per-use', 'pay-per-load' models for machines. Those concepts extend the revenue generation over the entire product lifecycle and induce more advanced services such as software updates via 'over-the air' (OTA).

Still today, research must further investigate the scalability and the extension of such Smart Service based PSS solutions. It is likely, however, that Smart Services and Product Service Systems will demand and trigger additional big-data and AI-solutions supported through industrial data platforms and streams. Those might be leveraged for a customer-specific automation adaption or flexible price settings and revenue streams. Thereby, the digital twin becomes one of the key elements of PPS. In turn, however, such solutions have high technological demand and requires highly connected and distributed systems, significant compute power, high-speed and secure communication networks to enable high volume and real-time digital data transfer with low latency [12]. In 2019 the Industrie 4.0 platform has published their 2030 Vision for Industrie 4.0: *Shaping digital ecosystems globally* (see Fig. 20.4).

The 2030 vision declares three closely interlinked strategic fields of action are crucial for a successful implementation of Industrie 4.0: *autonomy, interoperability* and *sustainability (c.f.* Figure 20.4*).* Industrie 4.0 sustainability principles such as *circular economy* and *resource efficiency* are targeted to be achieved. Unfortunately, so far, the potential of sustainability is mainly untapped. Hence, methods are necessary, which guide industrial companies for the realization of sustainable-friendly tools and processes, i.e. methods that avoid physical prototypes by virtualization of products, or methods that consider options for the end-of-life use of products.

New Methods and Tools for Industrie 4.0

Pre-requisite for implementing Industrie 4.0 is the development of novel methods supporting all major engineering phases: Planning and design, development, validation, simulation and testing. It is obvious, that Industrie 4.0 causes a significant increase in the system complexity - while bringing the efficiency to another level. In particular, the autonomous orchestration of systems over the complete product

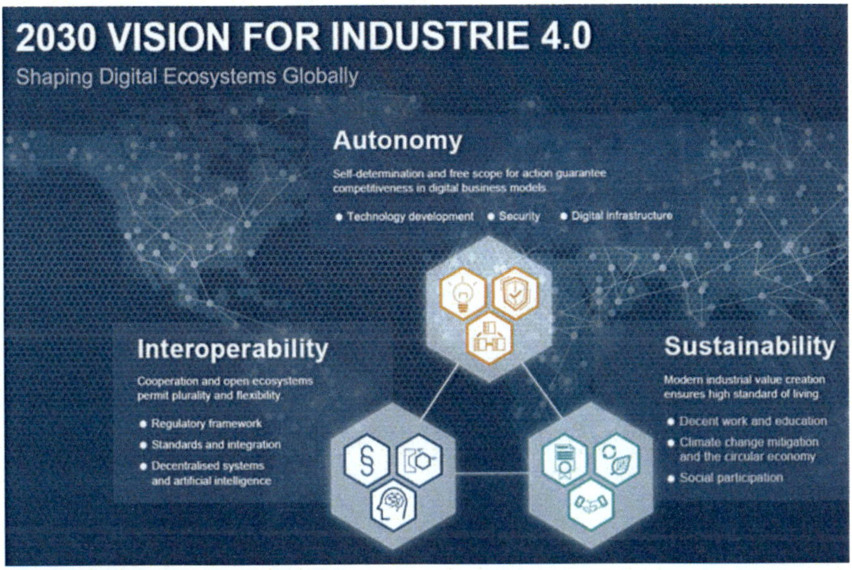

Fig. 20.4 2030 vision for industrie 4.0 by the industrie 4.0 platform[4]

lifecycle leads to an engineering challenge that has not been proven yet. Engineering practice, therefore, needs to change from focusing on the design of single products to the design of adaptable system networks instead. Today's system modelling, simulation and testing approaches insufficiently consider products' interplay and adaptability to the environment. In addition, interrelationships between value creation, products and services need to be considered already in the early phases of product design.

One approach to face this challenge offers the discipline of *Advanced Systems Engineering (ASE)*. In Germany, ASE is meanwhile an official national acatech[5] program and steers the research and development activities in the three Industrie 4.0 relevant fields *Advanced Systems* (new market solutions for future value creation and business models), *Systems Engineering* (manage the solution complexity consistently) and *Advanced Engineering* (innovations for intelligent engineering of tomorrow). While conventional system engineering approaches have been majorly driven by urgent industry needs, *Advanced Systems Engineering (ASE)* focuses on a mid- and long-term viable perspective and integrates the product, the service and the production system needs by explicitly including the holistic technical systems orchestration. In addition, the use of *MBSE (Model based Systems Engineering)*, *DSE*

---

[4] Publisher: Federal Ministry for Economic Affairs and Energy (BMWi), Public Relations, 11,019 Berlin, www.bmwi.de Editorial responsibility: Plattform Industrie 4.0, Bertolt-Brecht-Platz 3, 10117 Berlin, Germany. Accessible under: https://www.plattform-i40.de/PI40/Redaktion/EN/Downloads/Publikation/Vision-2030-for-Industrie-4.0.html.

[5] *Acatech* is the designation for the National Academy of Science and Engineering in Germany.

(Data Science and Engineering), *Artificial Intelligence (AI)*, different types of simulation algorithms and emerging digital and virtual technologies are key capabilities of this future engineering intelligence framework (compare chapter 21). New and combined algorithms will help engineers to predict the digital Industrie 4.0 reality and enable the offline self-learning of systems before operating them in the field and in factories. This, in fact, prerequisites the adoption from design and system models at any time and requires comprehensive standardization of underlying data models and standards (compare also MBSE in chapter 21).

4.0 Work and Society

Industrie 4.0 is bringing a major change to work and society. Research plays a key role in shaping this change and bringing success to the future industry ecosystem. In this regard, the following key-themes have been identified: *data protection and security*, *socio-technical definition of systems*, *work development and training*, *acceptance*, *participation* and *management culture*. Data protection and security topics have to fulfill legal challenges. The implementation of Industrie 4.0 will provide data sets, which are created by machines and humans. Data will enable the evaluation and behavior of humans to make single individuals traceable. Thus, new requirements towards data handling, protection and security arise. Nevertheless, exactly this data potentially improves operation procedures or can be used for feedback and training processes [12].

To design new ways of work as part of Industrie 4.0 in a human-oriented manner, socio-technical understanding and methods are required. This is the only way to actively shape and implement the new principles of increasing human–machine interaction. This new way of work requires life-long learning and skill-development. It requires new forms of learning. New learning approaches will be based on the practically-oriented and individual learning through the application of digital and smart solutions [12].

20.2 Digital Twin Concept

The *WiGeP*[6] position paper on Digital Twins states that "Digital twins will change the product creation process substantially and enable new business and value creation models." [13] So, what is a Digital Twin?

Digital Twin (DT) as a term is attributed to Michael Grieves and John Vickers, the former being the first to have described the underlying concept as early as in 2003 [14]. While the first decade after the inception of DTs did not see much research dedicated to them as a research area by itself, the underlying concept as well as the underlying principal IT and digital network capabilities have significantly evolved

[6] The WiGeP is the *Scientific Society for Product Development* in central Europe headquartered out of Germany with 75 Professors of leading university institutes with more than €100 million external research funds (public and from industry) per year.

since then. To the same time the need for new smart service business models have pointed the attention to DT use cases and DT engineering applications.

The original concept of Grieves and Vickers has been substantially expanded and Digital Twins as a research field of its own have been established substantially during the decade of 2010 through 2020. Use cases for DTs arise from a large and seemingly increasing number of disciplines/domains [15]. The potential roles of Digital Twins are versatile and first applications fields become prominent in long living investment good industries in the energy, manufacturing and aviation/aerospace sectors but also in the automotive and transportation industry as well as in the health sector. Jones et al. cite the following use case areas: Industrie 4.0, smart factories/smart manufacturing, learning, product design, model based engineering, communication technologies for factories, machine health monitoring, composite material optimization, smart cars, farming, human health and agriculture jones [16].

According to Grieves and Vickers the basic idea that has led to today's concept of DTs has existed since the dawn of CAD technology, but has only really begun to manifest its relevance and potential with the increased feasibility of simulating dynamic behavior of products/systems through digital models. For Grieves and Vickers the motivation behind DTs comes from a Systems Engineering and Product Lifecycle Management (PLM) perspective, specifically from the need to be able to anticipate behavior of a complex system[7] before a physical realization of the system is produced/deployed [17]. In this use case, a DT is a means to the end of foregoing physical simulations (with physical prototypes) and the associated costs, while still eliminating uncertainty about the real-world behavior of the future system. But this is just one of the many DT use cases discussed in literature nowadays. Other use cases range from mere representations of physical entities, i.e., for illustration or monitoring purposes,[8] all the way to DTs being artificially intelligent digital entities capable of autonomously controlling the behavior of their physical counterparts. Consequently, definitions of the DT also show a wide range of possible DT morphologies.

The following subchapters give an overview of the DT concept by recapitulating selected state of the art definitions, classifications as well as considerations for applying DTs in the context of Industrie 4.0. Technologies and solution elements for the implementation of DTs are presented in the context of Subsect. 21.4.

[7] In short, in this context a complex system is characterized by consisting of several objects (where at least some of them are physical ones, in contrast to digital ones), and by the behavior of these objects together as a system not always being easy to predict. See [17] for a more detailed definition.

[8] This construct would be called a Digital Shadow rather than a Digital Twin according to latest international definitions, see Subsect. 20.2.1.

20.2.1 Digital Twin Definition

A good starting point to grasp the Digital Twin concept is the definition by Grieves. In this definition, the concept has three main constituent parts [14]:

- A physical entity in real space, the "Physical Twin" (PT).
- A virtual entity[9] in virtual space—the actual "Digital Twin" (DT).
- A two-way information and data flow between the physical and the virtual space.

In [17] it is further explained that a "*... Digital Twin is a set of virtual information constructs that fully describes a potential or actual physical manufactured product from the micro atomic level to the macro geometrical level. At its optimum, any information that could be obtained from inspecting a physical manufactured product can be obtained from its Digital Twin*". Grieves and Vickers also go on to introduce the distinction between *Digital Twin Prototypes* and *Digital Twin Instances*, as well as the environment a DT "lives" in, the *Digital Twin Environment*. This distinction is further expanded in the classification of types and lifecycle phases of DTs proposed by [10], see Sect. 20.2.2.

The "optimum" described by Grieves and Vickers is, of course, an ideal vision of the DT. Real-life implementations of DTs will certainly fall short of this ideal, but still qualify as DTs. As analyzed in [16], there currently are no known DT implementations cited in literature that would come close to the optimum described by Grieves. "Lesser"—but feasible—DTs have many use cases and potential, nonetheless. In fact, literature does not discuss a lower threshold for the fidelity/accuracy with which the Digital Twin replicates the Physical Twin. Mostly, definitions are content simply saying that the Digital Twin needs to be as accurate and in sync with the Physical Twin as the use case for which it was designed requires it to be.

The Jones et al. paper provides a recent and exhaustive analysis of definitions and characteristics proposed and discussed in literature directly for the Digital Twin. Interestingly, Jones et al. also survey technologies/research fields that are not directly addressed under the term Digital Twin, but still have characteristics attributed to DTs nowadays. These technologies/research fields mostly predate the DT concept and have a narrower scope in terms of targeted industries and applications. In this sense, the following technologies/research fields can be considered to also fall within a broader scope of the DT concept: computer integrated manufacturing, virtual manufacturing systems, model-based predictive control, advanced control systems, machine health monitoring/prognosis (intelligent predictive maintenance also falls under this term) and building information modeling. Of course, this list may not be finalizing.

---

[9] Admittedly, the term *virtual entity* is very unspecific. Most scientific publications do not go into details about what a *virtual entity* actually is, but it can be assumed that digital entities in the context of Digital Twins are algorithms and programs (software)—with associated digital data—that run/are stored on any type of computer, such as embedded, desktop, edge or cloud server computer.

Table 20.1 contains the characterizing items common to DTs identified by Jones et al. in their systematic literature survey. The exact terms used for each characteristic may vary from publication to publication.

One characteristic in the list by Jones et al. is particularly interesting: The Virtual-to-Physical Connection/Twinning. According to Jones et al., many DT definitions do not contain the Virtual-to-Physical connection. Furthermore, literature does not seem to have a clear stance on exactly what characteristics this connection must possess or how the Virtual-to-Physical twinning process should go about for the digital entity in question to be considered a real DT. For instance, it seems to be up for discussion whether a human-in-the-loop that may take the role of an actuator triggered by the Virtual Processes of the DT, represents a valid Virtual-to-Physical connection. If one were to acknowledge such a process for a DT, one would also

Table 20.1 Common characteristics of DTs identified by Jones et al. in their literature survey [16]

Characteristic	Description
Physical Entity/twin	The physical entity/twin that exists in the physical environment
Virtual Entity/twin	The virtual entity/twin that exists in the virtual environment
Physical environment	The environment within which the physical entity/twin exists
Virtual environment	The environment within which the virtual entity/twin exists
State	The measured values for all parameters corresponding to the physical/virtual entity/twin and its environment
Metrology	The act of measuring the state of the physical/virtual entity/twin
Realization	The act of changing the state of the physical/virtual entity/twin
Twinning	The act of synchronizing the states of the physical and virtual entity/twin
Twinning rate	The rate at which twinning occurs
Physical-to-virtual connection/twinning	The data connections/process of measuring the state of the physical entity/twin/environment and realizing that state in the virtual entity/twin/environment
Virtual-to-physical connection/twinning	The data connections/process of measuring the state of the virtual entity/twin/environment and realizing that state in the physical entity/twin/environment
Physical processes	The processes within which the physical entity/twin is engaged, and/or the processes acting with or upon the physical entity/twin
Virtual processes	The processes within which the virtual entity/twin is engaged, and/or the processes acting with or upon the virtual entity/twin

have to consider traditional Statistical Process Control (SPC) paired with automated parameter measurement and manual maintenance activities (triggered by the SPC) a DT application. This could be problematic for the research field of DTs because it might lead practitioners to believe that the Digital Twin is just an arbitrary new term for applications that have existed since the dawn of industrial computers. Jones et al. dedicate a longer section of their analysis paper to the Virtual-to-Physical connection because it seems to be one of the key points that set the DT concept apart from "conventional" CAD and/or systems monitoring and simulation methods.

A term associated with the DT that should also be explained is *"Digital Shadow" (DS)*, which is a term that is often used synonymously with the term Digital Twin in practice. While the two are closely related, there is a significant difference between the two that could be described as follows: A DT always encompasses a digital information construct that describes the current/last captured/last measured state of its physical counterpart. This state information is what shall be called the *Digital Shadow* within the DT, and it can be used to create a digital representation of the PT [13]. Note that in this context representation refers to whatever characteristics/aspects of the PT are relevant to the particular DT use case, and that a DS does not necessarily refer to a human-visible representation (e. g. a 3D model displayed on a computer screen). Recalling the previously presented characteristics of the DT, it is thus clear that a DS alone—while being a fundamental part of every DT—does not make for a DT on its own, since it lacks e. g. the characteristic of virtual processes.

In the context of Industrie 4.0, it is apt to present the DT definition that has been adopted by the CIRP Encyclopedia of Production Engineering [10]. The short version of this definition is given by Stark and Damerau as follows:

> A Digital Twin is a digital representation of an active unique product (real device, object, machine, service or intangible asset) or unique product service system (a system consisting of a product and a related service) that comprises its selected characteristics, properties, conditions and behaviors by means of models, information and data within a single or even across multiple life cycle phases [10].

This is one example for a definition that does not name the Virtual-to-Physical Connection nor the Physical-to-Virtual one explicitly, albeit the latter can be implied since there can be no digital representation without it. However, in the rest of the publication by Stark an Damerau [10], the authors proceed to introduce an 8-dimensional model for classifying DTs. This 8D model goes well beyond the short definition in terms of richness of functionalities and capabilities that DTs may possess, and it does name the Virtual-to-Physical connection for more sophisticated DTs. Thus, the CIRP definition may be regarded as a definition for the lower functionality end of the DT spectrum, that is, the DT base line, while the 8D model also foresees DTs that may even surpass the ideal vision of the original DT definition by Grieves and Vickers in certain aspects and various dimensions.

While the Digital and Physical Twins and their connections lie at the heart of the overall DT concept, there also exists the notion of a Digital Master (DM), which can be interpreted as the source and blueprint for creating instances of pairs of twins (physical and digital). A Digital Master is not tied to a particular physical entity,

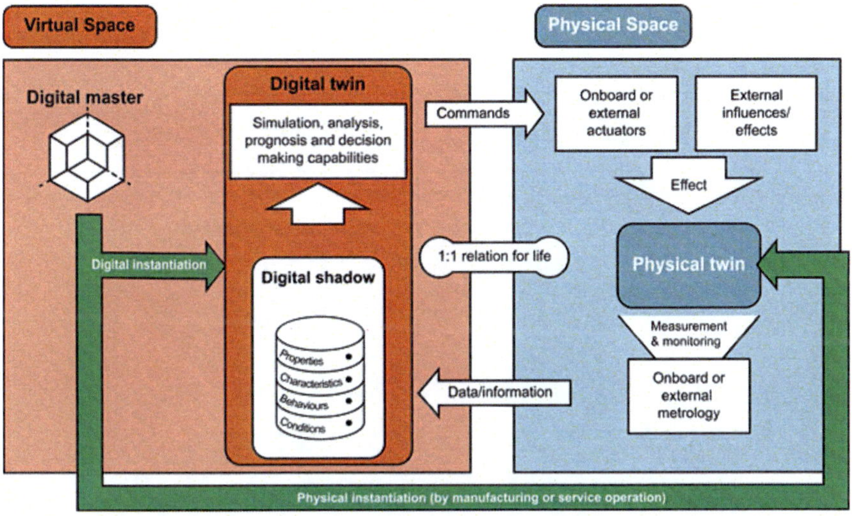

Fig. 20.5 Digital twin concept, interpretation of CIRP definition by stark and damerau [10]

whereas a DT is always tied to a specific singular physical entity. However, there may be different DTs involved with a physical entity throughout its lifecycle. A further overview of life cycle considerations for DTs follows in Sect. 20.2.2, but for now Fig. 20.5 summarizes the core of the DT concept as seems apt for the context of Industrie 4.0.

Figure 20.5 is roughly split into the Virtual Space (on the left) and Physical Space (on the right). It shows how both DTs and PTs have their origin in the Digital Master, which is the blueprint for their instantiation. DTs and PTs are instantiated in pairs that remain connected throughout their common life cycle phase(s). This connection is illustrated through the arrows that indicate data and information flow from the physical to the virtual space and possible commands that go back from the virtual to the physical space and may allow the DT to trigger effective changes in the PT. On the virtual space side, the figure illustrates that the DT is comprised by the state information that forms the Digital Shadow within the DT as well as the models and methods useful for describing and simulating the behavior and evolution of the PT. These models and methods are defined in the DM from which the DT was instantiated and which are "inherited" by the DT.

A DT can be regarded as a digital product by itself [Stark and Damerau CIRP definition], and as such it has a life cycle of its own, not to be confused with the life cycle of a PT instance. Figure 20.6 roughly depicts the different phases of a DT life cycle.

It starts with the development phase, during which the *Digital Master (DM)* is firstly created, in many cases via geometry models using CAD technology (compare Chap. 7) but also in form of mathematical and functional models. During product and manufacturing engineering possibly a rang of *Digital Prototypes (DP)* are

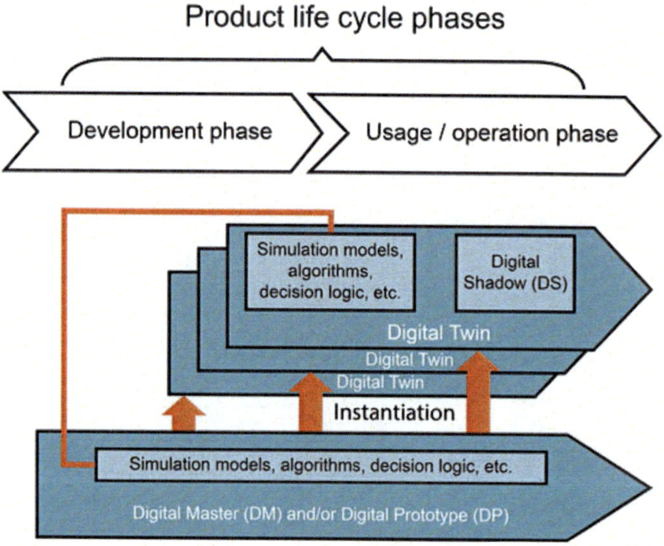

Fig. 20.6 The digital twin broken down across different phases of its life cycle, adopted and modified from [18]

created based on or derived from the *Digital Master (DM)*, possibly in an iterative manner, improving the DM with every iteration. Here, the major VPC technologies *CAP/CAM, CAE, DMU* and *Digital Factory* are mainly used (compare Chaps. 9,10,12 and 15). The DPs can already be used as upfront (anticipated by model assumptions) Digital Twins for the Physical Prototypes of the physical product in question and can be a helpful resource for gaining insights for possible design improvements for the overall CPS product. After virtual and physical product creation (product development and manufacturing engineering) the product life cycle phase production, usage, operation and service starts. In the first subsequent phase, during production, real products are produced and mated with Digital Twins during their respective instantiation. For the physical product, instantiation of course refers to the production/manu-facturing according to the build plan presented through the DM, and for the DT it refers to the installation/registration and persistence of a new DT instance. From thereon out the DT contains the *Digital Shadow* (DS) of the PT, which continuously twins its characteristics and conditions. Furthermore, the DT instance is "equipped" with the virtual process capabilities (simulation models, algorithms, etc.), which it inherits from the DM. The same applies for the twinning of products in their operation in the field.

20.2.2 Digital Twin Classification

Figure 20.7 shows the 8D model by Stark and Damerau, which differentiates levels of functionalities and sophistications of DTs. Of the 8 dimensions, the first three (*Integration Breadth, Connectivity Mode and Update Frequency*) concern the DT environment. Together with the 8th dimension (*Product Life Cycle*), these four dimensions characterize the environment and context of a particular DT, whereas dimensions 4 through 7 (*CPS Intelligence, Simulation Capabilities, Digital Model Richness* and *Human Interaction*) form the group of characteristics describing the behavior and capability richness of the DT. [10]

Although the model is not intended for measuring the maturity of different DTs [10], the scale of all eight dimensions in some dimension can be interpreted as ordinal, but no necessarily in all of them The lowest level (level 0) indicates the easiest implementation but also the lowest functionality of the DT, whereas the highest level in each dimension refers to the highest functionality but also to the highest implementation effort. However, this should not be read as "the higher the achieved level, the better", since functionality and implementation effort should have a reasonable ratio for the intended use case of the DT. The same notion is described as "application focus" in [13].

Looking at the 8D model, one can appreciate the large number of possible morphologies that DTs can adopt, which emphasizes that the development and implementation of a DT involves quite a few design choices that may heavily influence the implementation effort and the complexity of the task. In the end, a DT should be tailored to fulfil its intended purpose and must be able to function within its environment. This, in turn, may also put additional requirements on the integration environment of the DT as well as on the real environment of the PT. Therefore, it is recommended that practitioners consult the 8D model during the design of DT applications and carefully consider the implications of the implementation levels of the different dimensions.

The first dimension—*integration breadth*—could be summarized as describing different DT magnitudes in terms of size and complexity of the PT that the DT represents. The dimension ranges from DTs for single products or machines (i.e., clearly delimited devices/objects) all the way to DTs for representing the "whole world" (i.e., certain aspects of the world). Regarding *integration breadth*, DT designers should consider the effort and costs involved in modelling/measuring large parameter sets that may be necessary to twin larger, possibly complex physical systems with sufficient fidelity for the use case in question. At the same time, DTs could be particularly useful for gaining knowledge about such complex systems in an iterative manner, for example by continuously improving the simulation and prognosis capabilities of a DT by means of e.g., statistical, numerical and/or artificial intelligence methods. Such an application is well aligned with the use cases envisioned by Grieves and Vickers.

Connectivity mode is the dimension that describes the connections between the PT and the DT. The first possible level is *uni-directional*. The CIRP definition of the

Fig. 20.7 Eight ordinal dimensions to describe digital twin types and generic properties (c.f. [10])

8D model does not specify whether this uni-directional connection could be implemented in either of the two directions in question, but DT applications where the direction is uni-directional like Physical-to-Virtual seem the most probable, since this connection direction is necessary to get a DS. The two higher levels of *connectivity mode* are *bi-directional* and *automatic*. Seemingly, it is implied that *uni-directional* and *bi-directional* connectivity levels are not automated but need manual interventions to a certain degree (e.g. a sub-choice of priorities within the DS-analysis). The CIRP definition does not tell explicitly whether the third level shall only represent bi-directional automated or also uni-directional automated connectivity, i.e., sole automated twinning of the PT. Therefore, in this dimension it might be reasonable to toggle this 3rd attribute in conjunction with one of the first two.

The dimension of *update frequency* and its different levels are self-explanatory. The choice of the appropriate level may still pose a challenge, because the best trade-off between update frequency and cost might not be easy to find. Especially, one should consider that too low of an update frequency may mask out patterns in state changes that are only visible with higher update frequencies. The update frequency majorly depends on the application field of the DT and which type of "real time" need exist to deliver sufficient business value.

The dimension of *CPS intelligence* refers to the Virtual Processes that the DT performs/is engaged in. The lowest level is *Human triggered*, which should be regarded with certain skepticism for similar reasons as the unidirectional Physical-to-Virtual connectivity. The next level of *CPS intelligence*, where the DT is actually capable of performing virtual processes on its own, is *automated*, as in rule-based [10]. The remaining two higher levels *partial autonomous* and *autonomous* refer to virtual processes triggered by and/or performed using AI, which implies that the DT can learn and adapt its processes over time. Stark and Damerau acknowledge in the CIRP definition publication that the "full cognitive acting" and "human-like intelligence" envisioned for the level (fully) *autonomous* is yet to be enabled by "future artificial intelligence and cognitive solutions" [10].

The next dimension describes four levels of *simulation capabilities*. The first level, *static*, describes simulations with non-time-dependent input parameters, for example useful for "snapshot" type situation assessments at a single point in time. The next level is *dynamic*, which means simulation input parameters can be time-dependent, enabling the simulation of time-dependent processes. The third level, ad hoc, describes dynamic simulation models coupled with the possibility to use the current state of the DT as a continuously re-parameterized model for the simulation. As Stark and Damerau state, this makes the DT available for in-the-loop real-time simulation applications. Lastly, the highest level of *simulation capabilities* is *look-ahead prescriptive*, which refers to the ability of the DT to predict future states of the CPS based on the real current state and history of past states. This is an important capability for applications such as (automated) predictive maintenance services or bottleneck process predictions in manufacturing systems.

Dimension #6 describes three levels of *digital model richness*, referring to the extent with which characteristics of the PT are mapped to the DT. The three distinguished levels are *geometry and kinematics*, *control behavior* and *multi-physical behavior*. What these levels refer to might not be completely intuitive: Geometry and kinematics may refer to a model of the geometric shape and the degrees of freedom of movable parts of the model. A typical example would be a CAD or CAE MBS (Multi-Body-System) model. But, as Stark and Damerau state in the CRIP definition, *geometry* may also refer to abstract characteristics of the PT, so that "morphology" may have been a better term for this level. The second level of *control behavior* refers to models that describe the possible reactions of the PT to external influences and/or control inputs. An example would be a logic model describing the reactions of a manufacturing machine to changes in input parameters of its programmable logic controller. The third level is *multi-physical behavior*, which refers to models that comprise both multiple physical characteristics of the PT, e.g., mechanical and electrical morphology, and behavioral models.

The dimension of *human interaction* designates different types of user interfaces that the DT may offer. The first level is *smart device*, which according to Stark and Damerau refers to "digital twin interfaces tailored to commercial off-the-shelf hardware". As an example, they name smartphone apps, but it is not clear from the description in the CIRP definition what the term "smart" in the level's name refers to in this context. Computer programs that run on normal desktop or industrial computers can certainly provide interaction interfaces for human users of DTs, but they do not intuitively fit in the category of "smart devices" (nor any of the further levels). Therefore, it is safe assumption that smartness also includes the notion of different degrees of networked interaction capabilities to and with the Digital Twin (e.g. by using a mouse controller interface). The second level of *human interaction* for DTs is *virtual or augmented reality*, which provides immersive interaction possibilities for the user. Lastly, the level of *smart hybrid* describes immersive interaction methods involving advanced modalities like haptic technologies in addition to virtual or augmented reality.

The last dimension distinguishes DTs with regards to the portion of the PT's *life cycle* during which the DT is applicable. The life cycle is roughly split into three phases: Begin of life (BoL), mid of life (MoL) and end of life (EoL), please compare more detailed explanations in Chaps. 11 and 16. The three DT levels attainable in this dimension according to the 8D model are BoL, BoL + MoL and the entire life cycle, BoL + MoL + EoL. The authors of the 8D model explain that the levels are incremental, because it can be assumed that any DT capable of representing a later life cycle stage will automatically be able to represent all earlier stages, and the BoL stage should automatically be covered due to DTs being instantiated from DMs, which should in turn contain all models necessary for the creation of both DTs and PTs.

20.2.3 Digital Twin Use Case Examples

The following section presents examples of DT use cases, making references to the 8D model wherever helpful and briefly discussing certain DT design implications.

An example for a Digital Prototype that is used during the development phase of the product life cycle could be that of a passenger car. While the Digital Master of the product continuously evolves as development progresses, multiple different DPs can be instantiated throughout this process for testing different product design alternatives. For instance, a DP equipped with multi-physics behavior models (digital model richness level 2 in the 8D model), dynamic simulation models (simulation capability of level 1) and smart hybrid interaction modalities (human interaction on level 2) allows product testers to virtually test drive the digital car prototype in a realistic and immersive manner. The product testers can then proceed to give valuable insights into perceived consequences of different design alternatives to the development team. However, this example also serves well to emphasize that a DT/DP may be highly dependent on its virtual and physical environment. In this example, the environment needed would be that of an immersive haptic drive simulator, and the DP might need to be adapted to this environment and/or vice versa.

As products complete the development phase and enter the production phase, new use cases for the DT arise. Staying with the passenger car example, each car that is produced would be mated with a unique DT that represents exactly that car, starting from the very first step of production. The DT contains, for example, all information about the exact product features that the customer ordered, the bill of material needed for assembly, production job meta data such as planned and scheduled production and delivery dates, and possibly much more. By interfacing with the production control system, the DT can provide information to assembly workers, like for example visual representations of individual parts that need to be assembled and visual descriptions of the incremental morphological target state after each production step. This allows easy and efficient identification of deviations/errors through comparison of the as is PT with the prognosis of the target assembly state provided by the DT. If the PT is equipped with suitable self-metrology capabilities, quality assurance processes can even be carried out by the DT autonomously, without need for a human quality assurance operator.

As the car enters the utilization phase, the DT can provide convenient functionality for the customer, in our example the new owner of the car. For instance, the DT would allow the owner to inspect all relevant characteristics and parameters of his car remotely via a smartphone app. Examples of such parameters are the current fuel/battery level, cabin temperature, parking location and many more. If the DT is equipped with look-ahead prescriptive simulation capabilities (level 3), it may be able to calculate remaining mileage based on current tank/battery level, past driving style and planned journey route and it may even suggest when to take driving breaks (for example based on sensed fatigue of the driver) and when and where to make refueling breaks (partial autonomous CPS intelligence, level 2). Additionally, the DT would also provide the owner with remote interaction possibilities, like for instance remote controlling the air conditioning system to pre-heat the cabin before a journey or to control battery charging times according to daily energy price fluctuations (in case of an electric vehicle).

Another usage phase DT example could be that of a smart coffee machine with an app interface. The DT would be able to provide convenient information about, for example, how much coffee is left in the tank and how long this amount of coffee will probably last based on past consumption habits of its owner. Through the app, the owner could also be allowed to program the coffee machine to start making coffee in the morning autonomously.

An important aspect for DT developers regarding the usage phase of DTs to be considered—especially for consumer products—is that of data availability and asses. As Jones et al. aptly states, "If the aim of the Digital Twin is the exhaustive capture of all physical environment parameters, then there is a high possibility that those parameters can in some way directly or indirectly relate to aspects of people's lives, intellectual property, and everything in between. Determining how this information and associated data sets are shared between organizations and individuals poses a major challenge" [16]. This shows that DT design not only poses technical, but possibly also legal and ethical questions, many of which have not yet been answered by researchers and practitioners. Unlike physical assets where there exist a distinct legislative difference between *ownership*—the state, relation, or fact of being an owner ownership with the explicit right to share, lease, rent etc.)—and *possession*—control or occupancy of property without regard to ownership ant its rights—the digital world formally only differs between data availability, data access, data sharing, data exchange and data protection. Data ownership does not exist from a legal point of view. Consequently, there will exist a rather community and/or partnership-based agreement policy between the different stakeholder groups to allow and support data access and usage from physical objects supporting the Digital Shadow of the Digital Twin.

As a last example, consider the DT of a turbine in a power plant. A typical use case for a DT of such a piece of equipment would be that of predictive maintenance. In this case, maintenance is costly and must be carefully planned and scheduled to keep downtime of the turbine minimal. At the same time, maintenance must be carried out with great care because a failing of the turbine mid operation could pose significant safety threats to the power plant and its operating personnel. To enable the use case of predictive maintenance through the DT, the physical turbine is equipped with arrays of sensors that monitor each of its important operational aspects like vibrations, pressure differences, rotation speeds of all moving parts, etc. For such a use case, the update frequency of the DT should be close to real-time/immediate (level 3) and the digital model should have a richness that allows the modelling of multi-physical behavior (level 2). Furthermore, the DT shall be equipped with numerical, statistical, AI and other models for look-ahead prescriptive simulation capabilities (level 3) that allow foreseeing future wear out states based on past wear development in relation to operational parameters. This may also be achieved by means of using training data from other DTs of turbines of the same type. Furthermore, it makes sense to equip the DT with algorithms and functions for full autonomous acting (level 3 CPS intelligence) coupled with automatic connectivity (level 2) to the PT. In this way, the DT is able to shut down the turbine autonomously should it detect that there is a safety issue. Similarly, the DT can autonomously propose maintenance schedules to the maintenance personnel. An important point to realize with the use case in this example is that the metrology needed to provide the DT with the necessary data (reliably) must be considered during the design phase of the turbine. With such a complex and high-performance piece of equipment, it will probably not be possible to incorporate (machine-internal) sensors that have not been already foreseen during the development phase of the machine. This undermines the notion that Physical and Digital Twins need to be tailored to one another from the very beginning of their respective product life cycles.

This section did not explain how to develop the different type of Digital Twins with which kind of design elements. Please refer to Chap. 21.4 for more details on how to develop Digital Twins.

20.3 The Internet of Things (IoT)

The following sub-chapters give an overview of the Internet of Things (IoT) concept. To start off, the common Internet and its evolution is briefly explained, as it is the foundation that IoT technology builds upon. After explaining the IoT concept, relevant[10] technologies/concepts for IoT applications are roughly explained.

20.3.1 The Global Internet and Its Evolution

The technological roots of today's Internet (and consequently "the Internet of Things") lie in several research programs that took place in the US (DARPA), the UK (NPL network) and France (CYCLADES) in the 1960s. Most importantly, the DARPA (Defense Advanced Research Projects Agency) program would give birth to ARPAnet, a wide area network[11] (WAN) that was arguably the closest ancestor of today's Internet. ARPAnet began operation in 1969. The University of California, Los Angeles and the Stanford Research Institute were the first "nodes" to ever exchange a message transferred over a WAN. The first important use cases of the early WANs were file transfer and time-sharing[12] between computers, as well as electronic mail (Email) and sharing of information over Usenet newsgroups. Over time, ARPAnet grew to be a nation-wide network in the US and yet it only connected research and military facilities. During that time, due to the military usage, ARPAnet was mostly a classified network. One of the first prove of concepts for the Internet was performed by Vinton G.Cerf (today Chief Internet Evangelist at Google) and his colleague Bob Kahn in 1977: out of a van in California they could prove that it was possible to

[10] From today's point of view, that is, in the year 2021.

[11] Wide Area Network refers to a network of computers in which the computers are physically far apart from each other, as in different cities, states, countries or even continents.

[12] Time-sharing refers to the sharing of computation capacities of a computer between several computation tasks, e. g. when two researchers commonly use the same remote computer to perform calculation tasks. In the early days of computer science, when computers were expensive and scarcely available, time-sharing was commonplace.

connect various networks—wireless, satellite and APRAnet—via the base protocol *TCP/IP* (*Transmission Control Protocol and Internet Protocol),* both are standardized in the Open Systems Interconnection Reference Model (OSI) according to ISO (International Organization for Standardization since 1984 (compare more details in Sect. 20.3.3).

This fact changed in 1984, when the military portion of the network was split off into the MILNET. The now research oriented ARPAnet was eventually transformed into NSFNET (National Science Foundation Network), which would then proceed to be fully connected to other WANs that had been created independently, like CSNET (Computer Science Network) and EUnet (European Network). At that point, the joint networks began to clearly resemble today's Internet, and over time, further WANs were connected/expanded, until the computer network that is now known as the Internet span across the whole globe.

For more information on the history of the Internet, a comprehensive resource can be found at [19].

As it has already been explained, the Internet is a global network of WANs, a "network of networks". This begs the question of who operates and controls the Internet. The physical infrastructure of the Internet, i.e. the cables, facilities and supporting infrastructures are of course located within the legal jurisdiction zones of their respective geographical location. This infrastructure is, therefore, arguably in the main controlled by governments and operated by for-profit companies (which may in turn be regulated or directly controlled by governments). However, the Internet is intentionally designed and architected to not be dependent on the influence and control of any single country. For instance, most countries have several redundant connection gateways (with different neighbor countries) to the rest of the Internet. It is worth noting that access to the global Internet is also possible through wireless satellite connections. While this possibility adds another layer of redundancy and availability assurance, it is currently not apt to replace wired connections on a large scale in terms of latency and bandwidth.

Besides the physical infrastructure, the technological aspects such as ensuring global standardization and interoperability of technology stacks are just as important to keep the Internet operational and accessible all around the world. Arguably, this impressive feature is accomplished through continuous efforts of global cooperation between a multitude of non-profit and for-profit organizations. At the same time, some core functions of the Internet, such as ensuring a redundancy-free, global Internet Protocol (IP) address space (more on this in Subsect. 20.3.3), must be managed in a centralized manner. For this purpose, the US have a special role in that the globally centralized management and assignment of domain name system (DNS) namespaces (including IP addresses) is (as of today) carried out by the Internet Corporation for Assigned Names and Numbers (ICANN), a non-profit organization which is headquartered in Los Angeles and is, therefore, subject to US jurisdiction. The ICANN has an international board of directors whose members are chosen/nominated through several different mechanisms. This is achieved e.g. through internal nominations from within ICANN but also through an independent nomination committee, populated with representatives from further committees that represent world-regional non-governmental, technical, private-sector, governmental,[13] and other minor interest groups (see more details in [20]).

In summary, the Internet is factually governed/controlled by no single entity, but through the contributions and cooperation of many organizations and institutions, both non-profit and for-profit, which are assigned from all around the world, who work together for the public good that is the global Internet.

Closely related to the Internet is the World Wide Web, or simply "the Web". But the two are not to be confused with each other. While the Internet is a technical and physical infrastructure, the Web can be regarded as a virtual network of applications and information that runs/resides within this infrastructure, in parallel and, in turn, interconnected with other application networks of different types. Web technology was introduced to the rest of the Internet by Tim Berners-Lee from CERN.[14] It was the interconnection of ARPAnet and EUnet (which CERN was/is part of) that enabled this technology to spread and become as omnipresent as it is today. Web technology provides a method to distribute and consume information between users of the Internet using a standardized framework with rich possibilities regarding the presentation of such information. At the core of this framework lie the hypertext transfer protocol[15] (HTTP) and the hypertext markup language[16] (HTML). HTML introduced a formal language for the presentation of textual and other audiovisual information, as well as the possibility to interconnect different information objects via the novel "hyperlinks", which are navigation shortcuts that allow users of the Web to jump back and forth between different information objects. Information objects on the Web, such as texts, images, audio content, etc. are arranged/embedded in so called "websites". HTTP is the technical protocol that governs how websites can be provided to and accessed by users on the Web.

Compared to its early days, the Internet has vastly grown in both size and complexity of the network itself as well as of the technologies and concepts employed. The Web, on the other hand, while also having evolved in terms of complexity and functionality, has hardly changed its core mechanism: The Web poses a method to publish, share and consume information for human readers. Information and its presentation on the Web is designed for humans to read and interpret. While computers are employed to store, transport and present data across the Internet, the computers are oblivious to the information that lies within the data, that is, the meaning, the semantics of the data. But this might be changing in the foreseeable future, possibly giving us the next revolution in computer and information science:

---

[13] The governmental committee within ICANNs nomination committee does not get to vote on ICANN board of directors nominees but has advisory functions.

[14] In French: Conseil Européen pour la Recherche Nucleair. The European nuclear physics laboratory.

[15] See Fielding, R., Gettys, J., Mogul, J., Frystyk, H., Masinter, L., Leach, P., and T. Berners-Lee, "Hypertext Transfer Protocol–HTTP/1.1", RFC 2616, https://doi.org/10.17487/RFC2616, June 1999, https://www.rfc-editor.org/info/rfc2616.

[16] See https://www.w3.org/TR/html52/.

The "Semantic Web". The Semantic Web—a term coined by Berners-Lee—aims to transform the normal Web into something that not only links pieces of data in the form of documents and websites together, but also conveys their embedded information/meaning in a way that it becomes recognizable and interpretable by computers [21].

The Internet and the Web have been evolving and increasing their use cases and functionalities since their inception. It is hard to foresee the limits of their future potential. A novel phenomenon that has just emerged during the last decade or so is the so-called *"Internet of Things"*, which is one of the key enablers for Digital Twins and Industrie 4.0.

20.3.2 Internet of Things (IoT)

Researching for a definition for the *Internet of Things* is rather difficult because a simple Web search for the term will yield countless pages of links to websites with definitions (please also compare the first introduction of IoT in Chap. 16). However, what they all seem to have in common are two aspects: firstly, IoT refers to a network of computerized devices (the "things") that communicate which each other. And secondly, the communication can happen (but not exclusively) without any immediate human triggering/without a human being involved in the process. The choice of words regarding "computerized devices" is important here. It intends to convey that these devices are not primarily computers, but objects that have another main function/purpose besides being able to digitally communicate in a network. The main functions/purposes of the things shall not be purely digital, but (at least to some extent) physical, as in, taking place in the "tangible" world. Another term often used is "smart devices", but for a definition of IoT it seems too restrictive because it is usually associated with particular kinds of devices, primarily intended for communication and human interaction, like smartphones or smartwatches. The term "smart" is further misleading because it is often associated with AI (Artificial Intelligence) technology, which is not necessarily the meaning in the context of IoT.

A term closely related to IoT is *"Cyber-Physical System" (CPS)*. CPS is somewhat synonymous to IoT (-system), but each one of the terms is used more frequently in certain respective contexts: IoT is used more often in a general, i.e., not industry-specific, context, like e.g., in the public discussion about the digitization of society through connected consumer electronics. CPS seems to be used more often in the context of industrial applications and engineering domains. To take the terminology confusion even further, there is another related term, *"Industrial Internet of Things" (IIoT),* which simply refers to IoT applications in an industrial context, as is the case with Industrie 4.0.

The US National Institute of Standards and Technology (NIST) gives the following definition for CPS: *"Cyber-physical systems are smart systems that include engineered interacting networks of physical and computational components. [...] In addition to CPS, there are many words and phrases (Industrial Internet, Internet of Things (IoT), machine-to-machine (M2M), smart cities, and others) that describe similar or related systems and concepts. There is significant overlap between these concepts, in particular CPS and IoT, such that CPS and IoT are sometimes used interchangeably"* [22].

The CIRP Encyclopedia of Production states in its definition for CPS that *"[...] CPS can be generally characterized as physical and engineered systems whose operations are monitored, controlled, coordinated, and integrated by a computing and communicating core"* [23]. This definition is adopted from [24, 25].

In the following, the term IoT will be employed interchangeably with CPS. At the same time, when talking about the "Internet" of things, the term shall not be limited to describing a concept that is tied to the global Internet, but instead, to digital networks of arbitrary sizes. Thus, one may also read IoT as "IntRAnet of Things", if it better describes its application, for example when the IoT-system in question resides in a company-internal network without connection to the Internet, in which case the use of IIoT would also be apt.

The IoT is not defined by any single technology, nor is there any single specific problem that IoT targets. Therefore, it could be argued that IoT is a phenomenon, or perhaps the result of applying a certain kind of system architecture with certain characteristics, rather than a precisely defined method for predefined types of problems. There is no standard "cooking recipe" for applying of the IoT concept in arbitrary use cases. Nevertheless, being conscious about characteristics and implications of IoT may help practitioners steer clear of common pitfalls, manage expectations and leverage the possibilities of IoT technologies. The following list tries to summarize some typical goals that are associated with IoT in the context of Industrie 4.0:

- To improve efficiency in processes through increased availability of process-relevant information. → Speed-up of processes through faster information flow between the physical and digital domain,
- To improve effectiveness of processes through availability of process-relevant information. → Perform the right processes with respect to current circumstances,
- To gain knowledge about the manufacturing system through the analysis of relevant data collected through IoT techniques, and thereby, to identify improvement possibilities for the manufacturing system,
- To enable Digital Twin applications, for use cases both in the product creation process as well as for product service systems for the end consumer,
- To gain knowledge about the usage of a product or product service system, and thereby, to identify improvement possibilities for the product (service system),
- To automate data/information recording processes through IoT techniques,
- To rationalize data acquisition processes through the deployment of IoT hardware and infrastructure. Thereby, to enable more complex automated processes that would otherwise not be possible due to information lag/too high of an information gathering effort.

When building/setting up an IoT system, the developer/designer of the system will have to consider some architectural aspects of the system. After all, an IoT system is a cyber-physical network, and the functioning together of its components is typically not trivial, both from a systems engineering as well as from a computer science and engineering point of view. Apart from the immediate functionality of the system, strategic considerations for future scalability also need to be addressed. A good overview about IoT architecture and conceptual technology stacks can be found in [26]. It explains several especially important concepts that lie at the heart of any IoT system:

At the *"edge" of the IoT system*, there are so-called "constrained" devices/things. "Constrained" refers to the real-world requirements/limitations that many "things" used in IoT systems face: The need for the "thing" to be small (size constraint), to be cheap (cost constraint), often also to be mobile as in wireless and having a long battery life (energy-consumption constraint), or any combination of the above. Because of these requirements, constrained things often have limited functions/capabilities (often because of limited computing resources). Constrained things typically have the role of sensors and/or actuators in the IoT system. They represent the interface between the physical and the digital world (hence, they are referred to as being located at the "edge"). Another important kind of constrained things are devices/machines that were not designed having IoT capabilities in mind. In these cases, the limitations often refer to their connectivity capabilities.

The limited capabilities of constrained things bring the need to abstract further needed functions and capabilities away from the constrained "edge" into the rest of the IoT system. This is where IoT gateways come into play. IoT gateways are devices whose main purpose is to act as communication brokers between the constrained devices and the rest of the IoT system. "An IoT gateway will often offer processing of the data "at the edge" and storage capabilities to deal with network latency and reliability. For device-to-device connectivity, an IoT gateway deals with the interoperability issues between incompatible devices. A typical IoT architecture would have many IoT gateways supporting masses of devices." [26] Note that the boundaries between constrained things and gateways may not always be so clear in practice. In-between scenarios are possible.

Lastly, another important component of an IoT system is the *"IoT platform"*. The IoT platform represents the purely digital software infrastructure of the IoT system. In other words, it is the central nervous system of the IoT system, where all ties come together at some point. It provides services through which external applications may consume and use the information that was collected from the edge, as well as interfaces for triggering commands to actuators at the edge of the system. The platform may even directly host applications "natively" to form an autonomous system of sensors, actuators and active business logic. For this, data analytics functionality, possibly including AI methods, may also be provided directly by the platform. The platform itself is typically hosted on a cloud infrastructure (i. e., on a remote server infrastructure) or in an on-premises data center. IoT platforms will be explained in further detail in Sect. 20.4.3.

Figure 20.8 schematically illustrates the typical topology of an IoT system.

In the following, some important considerations for the planning of an industrial IoT system are compiled:

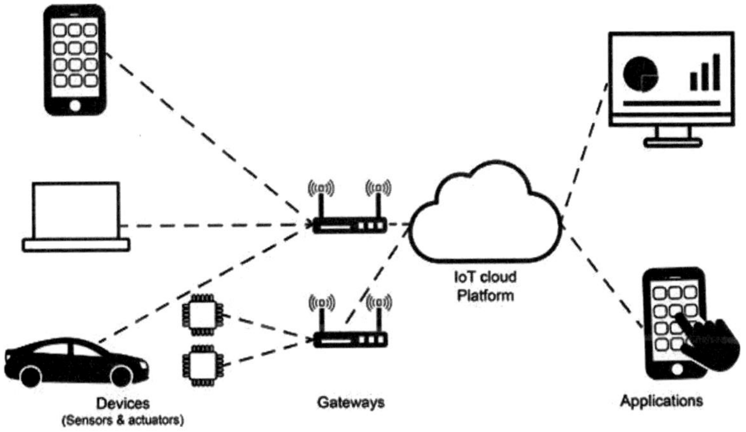

Fig. 20.8 Schematic overview of an IoT system differentiating constrained devices, gateways, platform and applications

- Is there legacy equipment (brown field resp. reuse) that should be incorporated and if so, what is its supported technology stack?
  → Gaps between interfaces/protocols that are supported by legacy equipment "as is" and state of the art IoT systems need to be overcome through gateway components, possibly requiring proprietary/special case solutions.
- Security aspects of technologies to be used.
  → Both security in terms of equipment/system failure and safety but also security against malicious intent. One major choice to be made, for instance, is whether platforms and data storage shall be hosted remotely or on-premises. See also Sect. 20.4.
- Which scalability is desired for the future in terms of supported protocols and interfaces but also for hardware scalability, e. g. scalability of wired and wireless connections at the shop floor, bandwidth, maximum number of clients, etc.?
  → The infrastructure and platform technology need to be chosen/implemented accordingly.
- Technical systems and solution requirements derived from engineering or use cases, e. g., latency requirements, equipment and service availability, system failure/down-time (in-) tolerance, etc., need to be considered. It is important to assess the system in its entirety and the characteristics that emerge from the "piecing together" (composing) of several IoT components. When assessing data transfer and computation latencies, the entire communication and processing network needs to be considered, not just the individual network object links. Special attention should be given to shared resources and services that might constitute bottlenecks.

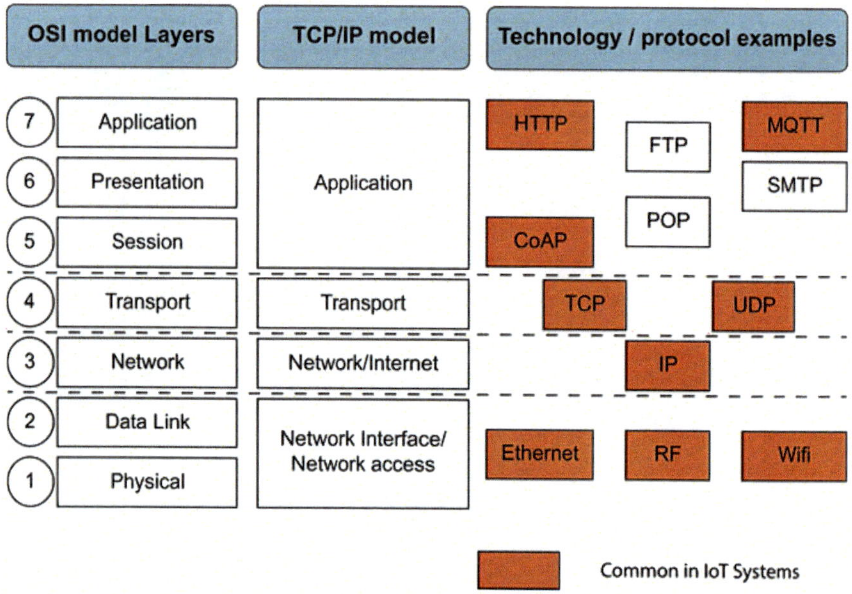

Fig. 20.9 OSI reference layer model, TCP/IP model and examples for wide-spread technologies/protocols

20.3.3 IoT Connectivity Stacks

Reference [27] state that IoT shall "[…] make use of an intelligent connectivity that relies on the consolidation of communication technologies." Further, it is noted that the range of requirements for IoT technology is growing. Consequently, "[…] the wireless connectivity for IoT deployments will continue to diversify with different communication protocols being used according to the needs and performance required by different IoT applications and services." The following subchapter aims to give a short impression of current IoT technology stacks. For further reading, [27] is recommended. Moreover, readers should know that most communication protocols and technologies are open standards, whose specifications and descriptions can be found on the Internet.

When it comes to comparing or referencing the employed communication technologies and protocols of different communication networks or components of such networks, the OSI reference model[17] [28], being a widely accepted standard, serves as a helpful common denominator. It segments the technologies and protocols that form the stack into seven layers, as shown in Fig. 20.9. The layers are organized from most "raw" (the *Physical* layer, layer 1) to most "abstract" (*Application* layer, layer 7). The technologies/protocols used on each layer fulfill specific functions for the next higher layer. However, the distinction of all 7 layers is not always possible/necessary

[17] Available at https://standards.iso.org/ittf/PubliclyAvailableStandards/index.html.

when looking at a specific implementation, in the sense that some technologies may take care of the functions of more than one layer at the same time. Typically, communication protocol stacks are modular, in the sense that some or even all of the layers have alternative protocol options that can be exchanged independently of the choices for the other layers.

The general protocol stack of the common Internet only distinguishes 4 layers (through abstraction of the 7 OSI reference layers), of which the second layer, *Network*, is populated with the Internets' most characteristic protocol: The *Internet Protocol (IP)*. On layer 3 (*Transport* layer), the *Transmission Control Protocol (TCP)* is the most widely used in the Internet protocol stack. Hence, the whole stack is also often referred to simply as the "TCP/IP" stack. While the original TCP and IP protocols have evolved into newer protocols (e. g., TLS and IPsec, which can be regarded as TCP and IP with additional security features), they are still referred to as the TCP/IP protocol family. Since TCP/IP is not only used for the communication in the global Internet but is also the most used protocol family within LANs (also in the industrial context), it is also an especially important protocol family for IoT applications. In Fig. 20.9, TCP/IP is mapped to the OSI reference model and exemplary IoT-relevant protocols are shown on the right.

The middle layers of the largest parts of IoT networks will typically be implemented using the TCP/IP protocol family. More thought shall be put into the *Application* and *Network Access* layers of the stack since choices of technologies and protocols on these levels may more heavily influence the characteristics of the overall system. In the following, first some common protocols towards the lower end of the stack are presented, followed by some application-level ones. This selection shall only serve as an exemplary, non-finalizing list, highlighting the differences that can exist between technologies and the different requirements of IoT systems that are addressed nowadays.

Notable *Network Access* layer (*Data Link* and *Physical* OSI layers) protocol/technology examples:

Ethernet: *Ethernet* is a standard for wired connections in LANs and, with limitations, also for WANs. The standard is published in [29] and derivatives. Currently, Ethernet supports data transfer rates of up to 400 Gbit/s (with optical fiber cables as physical medium) [29]. The speed, efficiency and security of Ethernet and its' establishment as a wide-spread standard make it one of the most used network technologies in LANs. The obvious drawback of Ethernet is the need for physical wires and network hardware like switches and routers, which makes network installation/setup and expansion take longer than with wireless connections.

WLAN: *Wireless LAN* is a family of technologies and protocols for wireless network connections specified in [30] and derivatives. The physical transport medium is the air and signals are transmitted via radio frequency modulation on specific bands. The most current standard, IEEE 802.11ax, also known as Wi-Fi 6, supports (theoretical) data transfer rates of up to 11 Gbit/s in optimal conditions and configurations by making use of the 6 GHz frequency band. Typical and effective transfer rates (in suboptimal conditions and configurations) from the point of view of the application layer are substantially slower. The advantage of WLAN is increased mobility of network devices because no physical wires are needed. The drawbacks are slower and less reliable connections and relatively high-power consumption compared to other wireless technologies.

IEEE 802.15.4: This is a standard for *Wireless Personal Area Networks (WPAN)*, which in contrast to WLAN is intended for shorter distances (hence personal area) and designed for energy efficiency. This standard is most relevant for networks that use certain application layer protocol stacks like ZigBee.

5G: *5G* is a new cellular network standard, and the planned successor of 4G. It is a radio access technology (RAT) with an increased bandwidth of up to 20 Gb/s downlink and 10 Gb/s uplink at latencies of about 1 ms. 5G is discussed as an important enabling technology for higher scalability and performance of IoT applications in industry, and it is therefore of high interest for the Industrie 4.0 agenda. "The 5G Alliance for Connected Industries and Automation (5G-ACIA) has been established to serve as the central and global forum for addressing, discussing, and evaluating relevant technical, regulatory, and business aspects with respect to 5G for the industrial domain." [31] According to the 5G-ACIA, there are still important challenges for the 5G standard to face before it can be deployed in industry on a large scale. Among other challenges, they name frequency spectrum licensing questions, operator models for provision of 5G to the industry, safety and security aspects for the use of 5G in industry and availability of 5G-enabled industrial components [31].

Notable *Application* layer (*Application, Presentation* and *Session* OSI layers[18]) protocol examples:

HTTP: The *Hyper Text Transfer Protocol* is the most important application protocol on the Web and can also be employed in LANs. Its specification can be found at [32]. It follows a request/response logic, where clients send parameterized requests (from a predefined catalogue of possible verbs) to servers, which in turn respond with a response, e.g. containing a requested piece of information/data or signaling the successful execution of a requested action. In the context of IoT, it must be said that HTTP is not an efficient protocol in terms of consumed bandwidth and latency. Furthermore, due to its synchronous one-to-one communication pattern and relatively high resource consumption on the application-hosting device, for many typical IoT applications it will not be a suitable protocol. However, since HTTP uses TCP/IP on the lower protocol stack layers, it represents a reliable and secure communication protocol, which is easy to implement and wide-spread, and thus well supported. Therefore, if the aforementioned limitations are not of concern, HTTP may still be employed in applications for IoT systems.

Industrial Ethernet: *Industrial Ethernet* refers to a group of standards/protocol families that use the Ethernet protocol at the physical layer of their protocol stacks and aim to make it suitable for industrial applications, where additional requirements regarding latency, connection security and durability of the hardware (including cables and plugs/sockets) exist. Among popular representatives there are Profinet (open standard) [33], EtherCAT (proprietary standard) [34] and, one of the oldest and the currently most wide-spread standard, ModBus (open standard) [35]. Most

[18] Not all three layers are compulsory for all of the presented standards.

Industrial Ethernet protocol stacks have a specific field bus protocol that they are compatible with, which is typically an important aspect when interfacing manufacturing equipment with PLCs (Programmable Logic Controllers) to an IoT system. The protocol families are listed in [36].

MQTT: The *Message Queuing Telemetry Protocol* is an application protocol based on a publish/subscribe logic, where client devices can publish and subscribe to so-called "topics". MQTT broker applications take on the role of receiving topic publications from clients (e. g., sensor devices) and broadcasting them to all subscribed clients (e. g., an IoT platform application). The protocol is lightweight in terms of implementation effort on the client side and has a low bandwidth footprint. MQTT is a wide-spread protocol for IoT applications. The basic version of the protocol uses TCP/IP on the *Network* and *Transport* layers, whereas MQTT-SN (where the suffix stands for "Sensor Network") is an adaptation for non-TCP/IP stacks, such as often the case with WPANs. See [37] for more information.

CoAP: "The *Constrained Application Protocol (CoAP)* is a specialized web transfer protocol for use with constrained nodes and constrained (e.g., low-power, lossy) networks" [38]. In other words, CoAP is an IoT application protocol for the Internet Protocol stack, specially designed for constrained devices, e. g. embedded sensor devices, enabling easy connection of these devices to the Web.

ZigBee: *ZigBee* is a communication protocol for WPANs that builds upon the IEEE 802.15.4 standard, meaning that it is not inherently Internet Protocol compatible. However, there is a ZigBee adaption for the IP stack, called ZigBee IP. ZigBee devices can form networks with mesh topologies, where participants can act as repeaters, effectively extending the reach of the network while maintaining low energy and fast network setup characteristics. Due to the mesh nature of the network, messages can take alternative routes through the mesh, making the network more robust against failure of individual network nodes. However, "orchestration" of a ZigBee network (device registration, ID/address allocation, etc.) is still performed by a coordinator device, which can be regarded as the vulnerable point of the network. Just like CoAP, ZigBee is designed for low power consumption, but also for low bandwidth consumption. See [39] for further details.

A special case of technology that should also be mentioned is *Bluetooth*. Bluetooth is a standard that comprises a stack of protocols which together represent a whole functional communication model. However, the protocol stack is not directly structured according to the OSI reference model, making it difficult to compare it to other protocol stacks. Bluetooth was originally developed for WPAN communication between mobile devices and peripherals such as wireless headphones. As such, it has only a short range of a few meters. However, most importantly, with the derived *Bluetooth Low Energy (BLE)* standard, it has relatively low power consumption and is a wide-spread and well-supported standard. Bluetooth technology is specified in [30].

For an extended list of current communication technologies and protocols, see [27].

20.4 Cloud, Edge and Platform Technologies

The Industrie 4.0 vision requires new technologies to enable the autonomy and flexibility of products and production in alignment with the holistic data continuity across the entire lifecycle. In this context the required computing effort is consistently increasing for engineering and production design, simulation, and operation. The complexity and administrative efforts to operate private or on-premise computer centers increases steadily in line with the growing compute demands. Cloud computing is one of the core IT technologies of Industrie 4.0 and aims at tackling those challenges. Cloud computing refers to the use of highly scalable computing power, usually not owned by its user. The computers are accessed remotely and for the use of the computers costs are charged [40].

However, in many cases, decisions have to be taken in real-time, e.g. for the autonomous driving of an AGV in production or a vehicle in traffic. For those, cloud approaches are only suitable to a limited extend, due to latency concerns, network bandwidth constraints, privacy rules, etc. [41]. This is exactly when edge computing technologies come into consideration. Edge computing means to compute locally—hence the opposite technology to cloud computing. Whereas fog (alternative terms: mist, dew) computing is a concept in which the computing happens in the edge of the network in small and decentralized computer centers. Fog computing leads to smaller latency issues due to less communication over the internet, due to the computation and communication happening at a shorter distance from the point of data/information usage. So-called fog-nodes decide whether to compute in the cloud or in the fog.

Both cloud and edge computing technologies will be introduced in this chapter. Finally, the interaction of these technologies is discussed, which is underpinned by example scenarios in the last Sect. 20.5.

20.4.1 Cloud Computing

The idea of cloud computing is not new: in 1960 John McCarthy envisioned that computing facilities could be provided to the general public as a utility [42]. Cloud Computing has been a result of the continuous availability of high-performance networks, low-cost computers and storage devices [42]. The National Institute of Standards and Technology (NIST) defines Cloud Computing as follows:

> a model for enabling ubiquitous, on-demand access to a shared pool of configurable computing resources, which can be rapidly provisioned and released with minimal management effort or service provider interaction [43].

Hence, cloud computing is a new operating model and allows consumers to pay only for the resources they actually require for their respective business case, known as 'pay-as-you-go/use'. In the past, heavy investments for infrastructure have taken solid shares and risks of new digitalization projects. Cloud computing transformed the act of buying computing resources in the act of buying computing services.

NIST [43] published the following five cloud computing characteristics:

- On-demand self-service: unilaterally and automatic (no human interaction required) provisioning of computing capabilities.
- Broad network access: availability over network and the access is ensured via standard mechanisms, enabling the usage with different kind of clients, such as computer, mobile phones, tablets.
- Resource pooling: computing resources, e.g. storage or memory, are pooled via multi-tenant technology. The pooling is based on dynamically assignment and reassignment of resources according to the current consumer demand.
- Rapid elasticity: capabilities can be elastically provisioned and released at any point. Capabilities appear to be unlimited for costumers and can be adapted at any time.
- Measured service: cloud systems automatically control resources. Resource usage can be monitored and controlled by providing transparency for both the provider and consumer.

There are multiple ways to implement cloud computing. Different standard architectures give guidance for the implementation. Figure 20.10 shows four different layers in which a cloud architecture can be sub-divided: the hardware, infrastructure, platform, and application layer. The hardware layer manages physical resources in data centers, such as physical servers, routers, switches etc. The infrastructure layer is responsible for creating virtual partitions of the resources. Key cloud-features such as the dynamic resource assignment are based on that virtualization. The platform

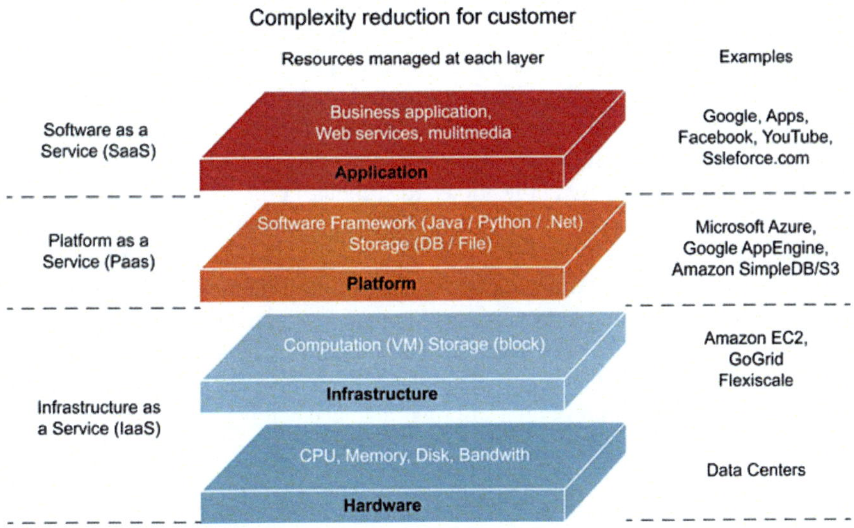

Fig. 20.10 Cloud computing architecture and business models

layer consists of operation systems and application frameworks or even tools for developing software. Thus, this layer can be used for software development, testing, deployment etc. Lastly, the application layer consists of the cloud-applications. Other than traditional applications installed on rich clients,[19] cloud applications can profit from the automatic scaling-features to improve the performance [42].

Based on these levels, services offered by major cloud providers have emerged, as shown in Fig. 20.10. It is the customer who decides which service suits the best to the respective business case.

Infrastructure as a Service (IaaS) is a service which provides the capabilities of the hardware and infrastructure layer, e.g. processing, storage, or other computing resources. Hence, the basis of the service are virtualized computing resources, in turn based on the respective hardware, both provided by the cloud provider. Users pay per use and depending on the chosen resources. The costumer has the possibility to deploy and run software of his choice. The underlying cloud infrastructure is managed by the cloud provider. Nevertheless, the user has control over the operating systems, storage and the deployed applications [42, 43].

Platform as a Service (PaaS) provides a platform to deploy onto the cloud infrastructure individual or commercial software applications. The platforms are flexible against various programming languages, libraries, and tools. The consumer does not manage the cloud infrastructure but can change configuration settings of the respective environment.

Software as a Service (SaaS) describes the capability to use the provider's applications, deployed on cloud infrastructure. The applications are accessible from clients through interfaces, such as a browser or a program interface. In the meantime, cloud providers offer SaaS for different types of industries, such as manufacturing, healthcare, finance etc. Besides some user-specific application settings, the user does not manage any of the beneath lying layers [44].

It should be noted that there exists a clear distinction to *Grid Computing*. Grid computing is a distributed computing organized by the "using organization (e.g. company) and not by a compute infrastructure provider. Grids of computers are set-up in the organization's own network whereby a "super virtual computer" is composed of many networked loosely coupled computers acting together to perform large tasks. For certain compute intensive applications, (distributed) grid computing constitutes a special type of parallel computing that relies on complete computers (with onboard CPUs, storage/hard disks, power supplies, network interfaces, etc.) connected to a computer network by typical network devices, like the Ethernet. Grid computing, therefore, is in contrast to traditional supercomputers, which have many processors connected by a local high-speed computer bus and differs from Cloud Computing by not being composed with the help of similar or even identical hardware as part of a scalable data (and compute) center offering rapid elasticity.

[19] A rich client is a computer that provides rich (i.e. comprehensive) functionalities independent from the central server.

20.4.2 *Edge Computing*

A significant part of today's data storing and processing capacity in the field of Industrie 4.0 and IoT is connected to cloud computing. This is how computing power is outsourced and centralized. However, in many cases, the transfer to the cloud requires time, which does not meet the requirement for real-time reactions in a multitude of use-cases in production, autonomous driving, etc. [45].

Edge computing aims at minimizing latency, prevent network congestion, and ensures the smooth functionality of cognitive systems and other latency-sensitive applications [45]. Thereby, the term *edge computing* points at the execution of computing in the edge of the network and encompasses the following implementation approaches [46]:

(i) multi access edge computing, MEC; (former term: mobile edge computing),
(ii) cloudlet and
(iii) fog computing.

In this context, it is not always possible to clearly differentiate between those terms. *Fog computing* is based on the so-called fog-layer, which in turn is based on gateway devices and wireless routers. Those devices are called *fog computing* nodes and are applied to store and process data from the edge devices, before sending them to the cloud [46]. The *cloudlet* approach functions similar but is based on the application of dedicated devices for the processing of data. It operates in the immediate proximity of the end device. The devices resemble data centers, nevertheless the size is significantly smaller and leads to the term **micro-cloud**. The *MEC* proposes the usage of devices with computing capabilities in the base station of networks and in this way enables *cloud computing* inside the radio area network [46].

20.4.3 *Interaction of Cloud and Edge Computing and Platform Technologies*

By 2025, connected IoT devices could lead to a yearly generated data volume of almost 79 zettabytes. The number of connected IoT devices could reach 75 billion globally [47]. Considering the scale of this annually produced data volume, efficient storage and processing is inevitable.

Figure 20.11 shows the three layers in which computing capacity can be provided: the edge, the fog, and the cloud-layer. As visualized in the figure the number of entities in the edge-layer is usually the highest, and the number decreases to a minimum in the cloud-layer. In turn, in reasonable architectures, the abstraction level of processed data increases from the cloud-layer to the edge-layer since it is not meaningful to transmit all data from the edge to the other two layers. The decision on which layer the processing and storage capacities should be provided, is highly dependent on the use-case. Nevertheless, parameters can give guidance in the decision process, such

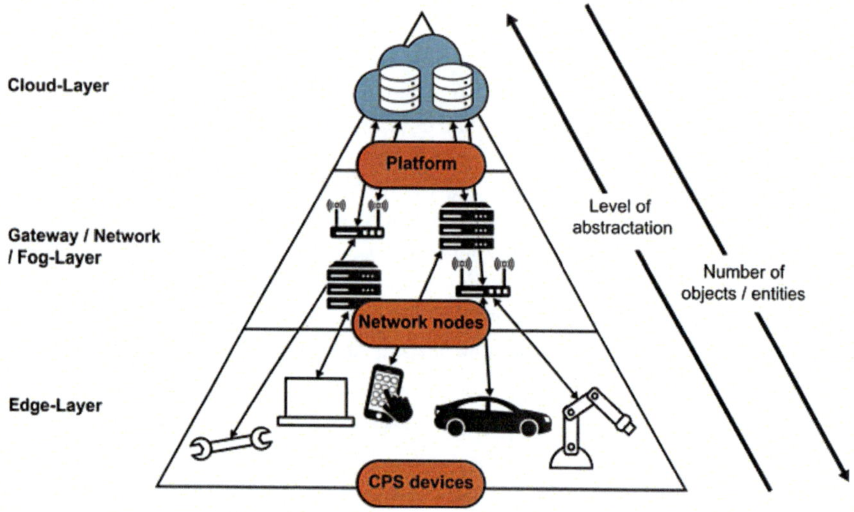

Fig. 20.11 Edge, fog, and cloud computing

as the logical proximity, access medium, context awareness, power consumption, computation time, level of abstraction, etc. [46].

The decision-making process is to be clarified exemplarily by means of the use-case described in paragraph 20.5.2. The orchestration of production systems depicts a particularly expensive computational process. Beyond that, it requires information from a high percentage of all involved devices. The information is to be gathered and processed at a central point. The orchestration is not time critical and must not be performed in real-time. Hence, the orchestrations' processing unit is usually located in cloud infrastructure. For the steering of the AGVs in production systems, in which human–machine interaction plays a major role, serious safety regulations impact the architectural decision. The AGV must react in real-time when it comes to avoiding collisions. The respective processing unit should therefore be located in the edge device itself.

As the previous paragraphs have shown, the importance of processing large data amounts from IoT-devices and the orchestration of systems will increase over time. Cloud-based platforms aim at tackling this challenge. Commercially available platforms such as Azure IoT, AWS IoT, SAP Leonardo IoT, and Siemens Mindsphere consist of different modules [48]. The so-called data collection module is supposed to collect data from different sources. Sources can depict IT-applications, Operation Technology Systems (OT-systems) or devices. Therefore, *Container Technology* enables the usage and the integration of different software types. As described before, the data can be preprocessed, filtered by edge or fog devices, before transmitted to the platform and therefore needs to handle different protocols (such as HTTPS, MQTT, AMQP, etc.) and devices from different vendors [49].

The second module of IoT-Platforms targets the data management. First to mention is the data ingestion, which includes the process of obtaining and importing data from the different sources by storing in a repository. Thereby, data can either be streamed or ingested in batches [49]. After storing the data, it can be processed, and semantic views are supposed to be created for the last module—the analytics module. This module is responsible for the data analyzing and modelling regarding future process planning, control, and optimization. The respective analytics software can be hosted on-premise, or in the cloud environment. The last module is dedicated to all security aspects within the platform. Please refer to Chap. 21 which will provide more details to the future capability needs in the new Virtual Product Creation discipline *Data Science and Engineering* (DSE).

20.5 Exemplary Industry Application of Industrie 4.0

As explained in the beginning of this chapter, Industrie 4.0 envisions production systems being Cyber Physical System, enabled through Internet of Things, Digital Twins and advanced digital methods like Big Data Analytics and AI, ultimately facilitating smart, efficient, flexible, and highly adaptable production and business processes in general. The following two sub-chapters try to demonstrate how the different technologies may work together for two exemplary use cases.

20.5.1 Efficiency in Manual Assembly Through Connected Processes

Manual assembly is an essential step in the production process of passenger cars in automotive manufacturing. Due to the dexterity and flexibility that is required, human assembly workers are still the most cost-effective resource in multi-model assembly lines. One challenge for manual assembly processes in continuous-flow assembly lines is attaining high efficiency.

Processes that are especially cumbersome and potentially inefficient are screwing processes where multiple parts need to be fastened by one worker, each with screws of different types/screwing parameters, possibly including a serial number registration of the assembled parts due to legal documentation requirements, as for instance in the case of safety–critical parts. A conventional method for a process with these characteristics could look as follows:

> The worker uses a barcode scanner to scan the serial number of a part. The scanned number is submitted to the manufacturing execution system (MES), which in turn instructs the control unit of the electrically controlled screwdriver to adjust the screwing parameters according to the requirements for the scanned part. With the screwdriver ready, the worker fastens the screws, and then proceeds to scan the next part, and the steps repeat.

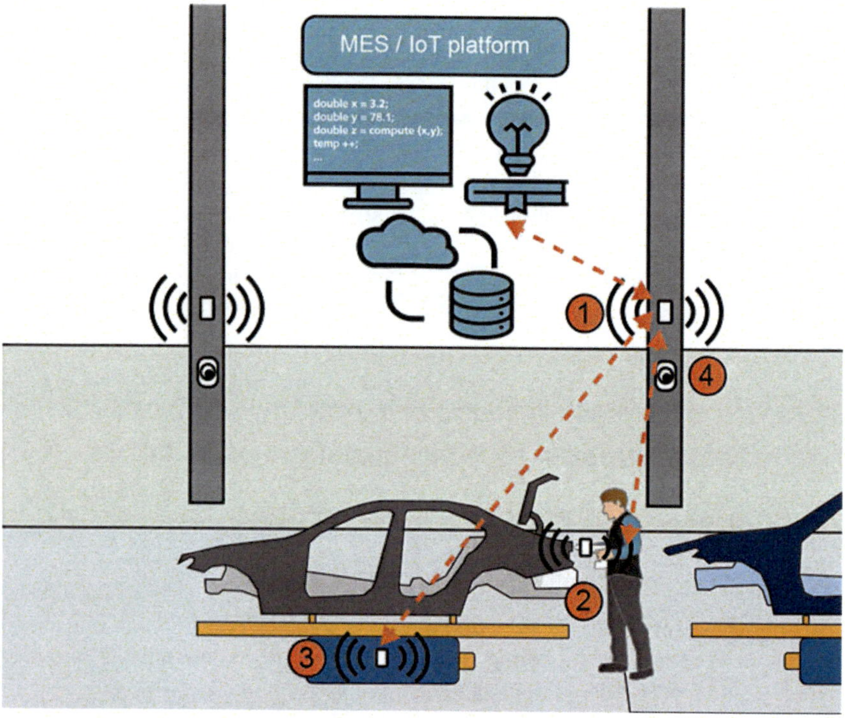

Fig. 20.12 I4.0-enabled manual assembly station for screwing processes in an automotive assembly line

The challenges or potential problems with such process are that

- operating the scanner,
- switching tools back and forth between the scanner and the screwdriver and
- conducting body movements of the worker between the point of scanning and the point of screwing.

all represent waste, as in non-value-adding process steps.

An IoT solution for improving this kind of process is described in the following. The assembly line shall be equipped with a so-called *Real Time Location System (RTLS)*. The system consists of a set of "beacons[20]" (see highlight #1 in Fig. 20.12), mounted in pre-defined positions onto the assembly shops' structural pillars (which stand evenly distributed on both sides of the assembly lines), and battery-powered "tags" that can be attached to any moving object.

The beacons and tags send out radio frequency and/or ultrasound signals, and through multilateration methods, the relative positions of tags and beacons can be calculated (with a certain error margin). To increase the accuracy of the position

[20] Beacons are small, wireless transmitters that use low-energy technology (e.g. Bluetooth) to send signals to other smart devices nearby as shown in Fig. 11.12.

tracking, the tags can be additionally equipped with Inertial Measurement Units (IMUs), which measure relative movement through integration of accelerations sensed over time by the unit. The beacons of the RTLS are connected to the IoT network of the assembly shop, and function as IoT gateways. They continuously report to the MES (which in this case also takes on the role of "reduced" IoT platform) the tracked positions of the tags, which in turn, represent a good example for constrained devices on the edge of the IoT system.

The conveyor units (possibly AGVs), which transport the cars through the assembly lines, and the screwdrivers of the workers are now equipped with RTLS tags (highlights #2 and #3 in Fig. 20.12). Thus, their relative positions within the assembly line are tracked at all times with an accuracy of, e. g., ±25 mm. Furthermore, the MES registers the vehicle production job IDs and tag IDs when the vehicles are put on the tagged conveyor units. In the same way, the screwdrivers (with model, parameters, etc.) are also mapped to their corresponding tracking tag IDs. Since the tags on the conveyor units are also placed in precisely pre-defined positions, and the relative position of the vehicle and the unit are also known, the MES can now effectively calculate the relative positions of screwdrivers and vehicles.

The next solution element is the *Digital Twin* of the product. Within the PLM system, a configured DT for every concrete vehicle (associated with a specific customer order) is instantiated. This specific configured DT contains a full geometric model of the vehicle-to-be (possibly in a reduced form, i.e. without complete and/or exact 3D element details), which not only serves as a basis for listing the bill of material (BOM) for production, but also tells the exact position and orientation of each part relative to the local coordinate system of the vehicle (including the screws relevant to our use case example). With this information, provided by the (configured and instantiated) DT to the MES, the MES can compare the positions of screwdrivers to the target positions of screws/parts. Furthermore, such a DT can provide the screwing parameters for each screw to the MES, so they do not need to be "hard-coded" in the MES or the control units of the screwdrivers. The MES then holds all information needed to automatically determine which screwing parameters need to be activated for which screwdriver based on the proximity of the screwdriver and a particular screw/part. The appropriate process logic and data connections, including proximity thresholds for triggering the parameter updates, is configured directly in the IoT platform interface of the MES.

The only "thing" missing for completing all Digital Twin readiness elements of the assembly process is the scanning of the specific parts (a specific type of Digital Shadow) for documentation and quality process purposes (after all, up to this point the MES only infers that a certain part *should* be in a certain position on the vehicle, but it has no certainty that this is actually the case). For this, in addition to the RTLS, an *Automated Optical Inspection System (AOIS)* is also deployed along the assembly line, consisting of several motorized cameras (highlight #4 in Fig. 20.12), placed at convenient spots (e. g. again the structural pillars of the shop, or any other suitable attachment point), so that all process-relevant parts for that particular segment of the assembly line are visible. The AOIS now automatically detects the barcodes of parts (through image processing algorithms), giving the MES certainty about the parts' presence in the respective station and vehicle. An improvement possibility for the described part presence detection process would be if the AOIS could detect parts through their appearance in images without the need for barcodes. While this is already possible with state-of-the-art machine learning image processing methods, setting up and automating the required data and processing tool chain is still a challenging task.

The human-observable, I4.0-enabled process finally looks as follows: The worker no longer needs to use a handheld scanner, and instead only uses the electronically controlled screwdriver. The process steps of switching back and forth between the two tools have been eliminated, as well as the scanning process including related body movements of worker. When the worker approaches a certain screw with the screwdriver, a signal light on the tool indicates to the worker that the screwdriver is ready for screwing, meaning that the MES has activated the corresponding screwing parameters. The worker performs the screw operation and can directly proceed to the next assembly process/vehicle. This example shows how a new cyber-physical system symbiosis can be reached between the technical manufacturing system and the human worker with the help of smart Industrie 4.0 solutions.

Figure 20.12 schematically illustrates the major solution elements of the described AGV and Digital Twin based Industrie 4.0 factory example.

20.5.2 Agility and Flexibility Through Autonomy—The Matrix Production

One of the core objectives of Industrie 4.0 is to enable most efficient and best possible intelligent production of a wide range of products. However, the focus of Industrie 4.0 not only lies on the flexibility to produce a variety of products, but also on the adaptability towards the production of new/additional products.[21] The concept of matrix assembly tries to accomplish these exact objectives. Matrix assembly is based on new production organization principles, in which the product finds its path through production in flexible, adaptable or even autonomous ways. Assembly stations are no longer interlinked with each other in a strictly linear way, as used to be the case in assembly lines since the second industrial revolution.

Figure 20.13 shows an assembly system used for demonstrating the matrix production. The assembly system consists of decoupled areas: the logistics areas (green) and the assembly area (blue). In the assembly system car bodies are assembled. The respective body parts to be assembled are stored in a parts store and can be transported via AGVs to the assembly stations. The stations in the production area are standardized and are not directly interlinked. For processing joining steps, matrix

[21] Flexibility and adaptability of production systems: production systems are *flexible*, when they can produce a fixed number of products based on predetermined configurations of the production system, whereas *adaptable* production systems are able to change the overall system set-ups (and related intelligences) in response to new products required to produce. The degree of intelligence makes up the difference between the notion of adaptability and autonomy.

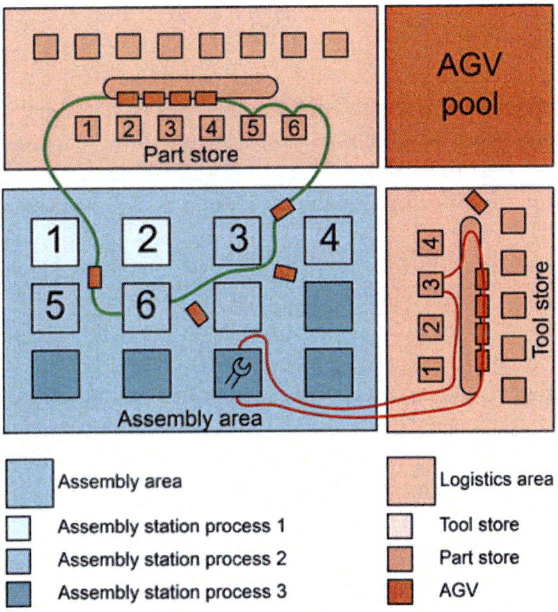

Fig. 20.13 Exemplary structure of a matrix production

stations can be equipped for one or more joining technologies. If needed, AGVs supply tools from the tool store to the stations for reequipping.

Figure 20.13 shows two different product-paths through the assembly stations. Those paths are flexible, and their determination is based on the assembly precedence graph of different product variants. Each path aims at performing one assembly step. The parts for each assembly step are collected in the parts store and are loaded onto the AGV. Afterwards, the path leads through stations in which the parts are joined. The selection and order of stations for each path is based on the overall status of the assembly system which is described by a high number of variables such as the availability of stations, the equipment and process capability of the stations, the length of the route of the AGV, etc. Hence, the orchestration represents an complex discrete optimization problem which aims to maximize the systems output in a huge solution space. Thereby, the solution space is composed by the sequence planning of the products, possible assembly sequences and possible paths through the station pattern. The orchestration of such novel assembly systems is still considered an open research question [50] in the scope of Industrie 4.0 and will require algorithms based on AI methods for predicting the most promising assembly scenarios and powerful real-time simulation methods for validating those.

Similarly, the planning and design of such novel assembly systems represents a complex problem and calls for new planning and production techniques. The planning of matrix production systems differs significantly from the design approaches of traditional assembly lines, since stations are not interlinked and cycle times are not necessarily fixed. The determination of the configuration of the matrix stations, AGV, etc. need to meet the required output. The configuration of such systems can no longer be planned manually (by fixed rules or heuristic knowledge by the planer), due to the high complexity of the holistic system. Novel planning algorithms and comprehensive simulation tools will be necessary to overcome this combinational complexity—incl. the high number of discrete event and prsoduct variety triggered simulation runs—and put the manufacturing engineers and production planer in a new role: from a pro-active planning architect and engineer role to a decision-making role, manufacturing analytics planer based on declarative engineering knowledge [51, 52].

Figure 20.14 focuses on a more technical perspective of the matrix production. It shows an AGV inside the already introduced matrix assembly system. The AGV is autonomously driving through the system, while taking the environment into consideration. Therefore, the steering and control of the AGV is processed on included edge controllers, while the current location and the status of the AGV is continuously streamed to the cloud. If e.g. a worker, as indicated in the figure, walks into the AGV's driving path and is at risk of provoking a collision, the edge controller gives the instruction to stop the AGV instantly. The corresponding algorithmic decisions are made in the edge since they have to be made in real time and therefore should not depend on transmission times of the network.

The target location that the AGV pursues is given by a cloud-based orchestration application. As illustrated in Fig. 20.14, the upcoming production execution decision

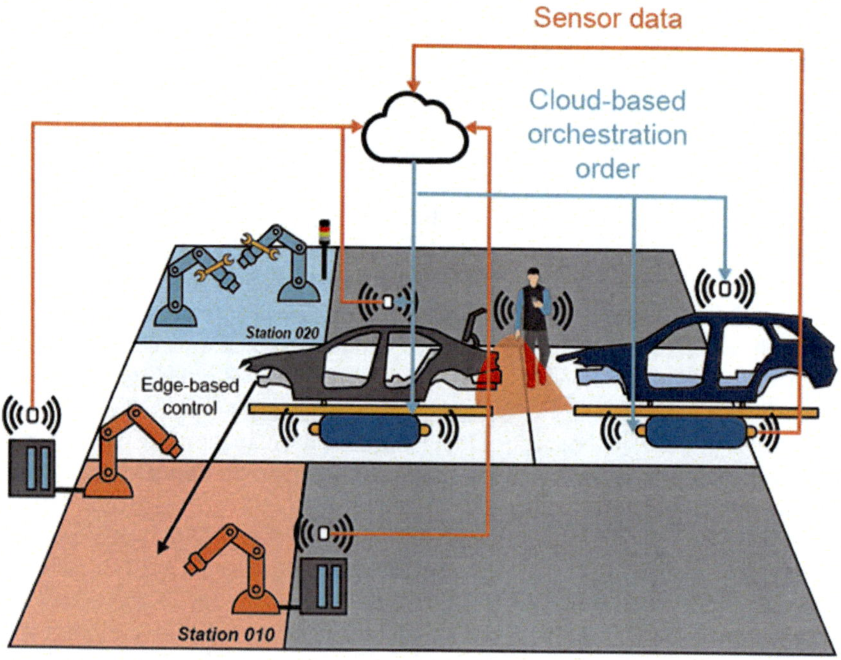

Fig. 20.14 Use of edge and cloud technology in matrix production

is visible, whether the targeted joining process should take place in station 010 or 020. It is likely that the decision takes place in the cloud back-end since the condition of all involved instances must be taken into consideration.

In the example shown in Fig. 20.14, the algorithmic decision has been made mainly based on the condition of station 020 as it undergoes maintenance routines. Once the vehicle enters station 010, the station recognizes the car variant to be assembled and the responsible control instances of the station initiate the respective robot programs. The corresponding robots then carry out the joining processes. At any time of the joining inside the station, current process and resource conditions are shared to the cloud applications for further decision making. In case of this vehicle variant, the next decision to make is to set the next target location. A number of open questions remain, e.g. with respect to which information elements might be of high interest to be allocated to the Digital Shadow data sets of the vehicle (product) or the assembly station (manufacturing resource).

This chapter has introduced a range of *Industrie 4.0* and *Internet of Things* concepts and technical solution options. It did not explain which new Virtual Product Creation disciplines and new engineering approaches are necessary to plan, design and validate/verify them in the future. The next chapter will provide more insights to such capabilities as part of the future Virtual Product Creation (VPC) solutions.

References

1. Acatech—Deutsche akademie der technikwissenschaften (2013) Deutschlands Zukunft als Produktionsstandort sichern—Umsetzungsempfehlungen für das Zukunftsprojekt Industrie 4.0. Abschlussbericht des Arbeitskreises Industrie 4.0
2. Sendler U, Baum G, Borcherding H, Broy M, Eigner M, Huber AS, et al (2013) Industrie 4.0. Beherrschung der industriellen Komplexität mit SysLM. Springer, Berlin
3. BITKOM, VDMA, ZVEI (2015) Umsetzungsstrategie Industrie 4.0
4. Bauernhansl T (2014) Die Vierte Industrielle Revolution—Der Weg in ein wertschaffendes Produktionsparadigma. In: Bauernhansl T, Hompel Mt, Vogel-Heuser B (eds) Industrie 4.0 in Produktion, Automatisierung und Logistik. Anwendung, Technologien, Migration. Wiesbaden: Springer, pp 5–24
5. Tomiyama T, Lutters E, Stark R, Abramovici M (2019) Development capabilities for smart products. CIRP Ann 68:727–750
6. Forschungsunion, acatech (2014) Recommendations for implementing the strategic initiative INDUSTRIE 4.0. Final report of the Industrie 4.0 Working Group
7. BMWi (2015) Industrie 4.0 und Digitale Wirtschaft
8. Plattform Industrie 4.0 (2019) Digital transformation "Made in Germany"
9. Wagner J (2020) RAMI 4.0—Ein Orientierungsrahmen für die Digitalisierung
10. Stark R, Damerau T (2019) Digital twin. In: Chatti S, Tolio T (eds) CIRP encyclopedia of production engineering. Berlin, Heidelberg: Springer Berlin Heidelberg, pp 1–8
11. ANSI/ISAANSI/ISA-95.00.01-2010 (IEC 62264-1 Mod) Enterprise-control system integration—Part 1: models and terminology
12. Acatech—Deutsche akademie der technikwissenschaften (2019) Key themes of industrie 4.0. Research and development needs for successful implementation of industrie 4.0
13. Stark R, Anderl R, Thoben K-D, Wartzack S (2020) WiGeP-Positionspapier (Position paper): „Digitaler Zwilling“ (Digital Twin). ZWF 115:47–50

14. Grieves M (2015) Digital twin: manufacturing excellence through virtual factory replication
15. Barricelli BR, Casiraghi E, Fogli D (2019) A Survey on digital twin: definitions, characteristics, applications, and design implications. IEEE Access, p 7
16. Jones D, Snider C, Nassehi A, Yon J, Hicks B (2020) Characterising the digital twin: a systematic literature review. CIRP J Manuf Sci Technol 29:36–52
17. Grieves M, Vickers J (2017) Digital twin: mitigating unpredictable, undesirable emergent behavior in complex systems. In: Kahlen F-J, Flumerfelt S, Alves A (eds) Interdisciplinary perspectives on complex systems. New findings and approaches. Springer, Switzerland, pp 85–113
18. Stark R (26 March 2019) Opportunities and obstacles in establishing and using digital twins. Initiative seminar: the mind of the machine—digital twins and processing intelligence. Gothenburg, Sweden
19. LivingInternet (2021) Internet history, design, advanced use, help, security, important features…. LivingInternet. https://www.livinginternet.com/. Accessed 16 May 2021
20. Internet Corporation for Assigned Names, NumbersICANN Homepage. One world, one internet. https://www.icann.org/. Accessed 03 Jun 2021
21. Berners-Lee T, Hendler J, Lassila O (2001) The semantic web: a new form of web content that is meaningful to computers will unleash a revolution of new possibilities. ScientificAmerican.com
22. Griffor ER, Greer C, Wollman DA, Burns MJ (2017) Framework for cyber-physical systems: volume 1, overview. National Institute of Standards and Technology, Gaithersburg, MD
23. Monostori L (2014) Cyber-physical systems. Arrazola Arriola PJ (eds) Modelling in cutting. Springer, Berlin, Heidelberg pp 1–8
24. Rajkumar R, Lee I, Sha L, Stankovic J (2010) Cyber-physical systems. In: Sapatnekar SS (ed) 2010 47th ACM/IEEE design automation conference (DAC 2010). Anaheim, California, USA, 13–18 June 2010. Piscataway, NJ: IEEE, p 731
25. Wani MA (ed) (2019) 18th IEEE international conference on machine learning and applications. ICMLA 2019: 16–19 December 2019, Boca Raton, Florida, USA: proceedings. IEEE, [Piscataway, NJ]
26. Eclipse IoT Working Group (2016) The three software stacks required for IoT architectures. https://iot.eclipse.org/community/resources/white-papers/iot-architectures/. Accessed 02 Jun 2021
27. Vermesan O, Coppola M, Nava MD, Capra A, Kornaros G, Bahr R et al (2020) New waves of IoT technologies research—transcending intelligence and senses at the edge to create multi experience environments. In: Internet of things—the call of the edge. Everything Intelligent Everywhere. River Publishers, p 168
28. International Standards Organisation (1994) ISO/IEC 7498-1:1994 open systems interconnection reference model. ISO-standard. https://www.iso.org/cms/render/live/en/sites/isoorg/contents/data/standard/02/02/20269.html. Accessed 03 Jun 2021
29. IEEE Ethernet Workin [g Group, IEEE P802.3bs 400 Gb/s ethernet task force
30. IEEE 802.15.1-2005—IEEE standard for information technology—local and metropolitan area networks—specific requirements—Part 15.1a: wireless medium access control (MAC) and physical layer (PHY) specifications for wireless personal area networks (WPAN)
31. 5G alliance for connected industries, automation, White Paper
32. Fielding R, Gettys J, Mogul J, Frystyk H, Masinter L, Leach P, et al (1999) RFC 2616—hypertext transfer protocol—HTTP/1.1
33. Pi (2021) profibus.com. https://www.profibus.com/. Accessed 11 Jun 2021
34. EtherCAT Technology Group (2021) EtherCAT technology group. HOME. https://ethercat.org/default.htm. Accessed 11 Jun 2021
35. (2021) The modbus organization. https://www.modbus.org/. Accessed 11 Jun 2021
36. IEC—international electrotechnical commission (2016) IEC 61784-3-18
37. (2021) MQTT specification. https://mqtt.org/mqtt-specification/. Accessed 11 Jun 2021
38. (2021) rfc7252. https://datatracker.ietf.org/doc/html/rfc7252/. Accessed 11 Jun 2021
39. (2021) Connectivity standards alliance. https://csa-iot.org/. Accessed 11 Jun 2021

40. Stradi-Granados BA (2020) Cloud computing for engineering applications. Springer International Publishing, Cham
41. Al-Turjman F (2019) Edge computing. Springer International Publishing, Cham
42. Wang L, Wang XV (2018) Cloud-based cyber-physical systems in manufacturing. Springer International Publishing, Cham
43. Mell PM, Grance T (2011) The NIST definition of cloud computing. National institute of standards and technology, Gaithersburg, MD
44. Alcácer V, Cruz-Machado V (2019) Scanning the industry 4.0: a literature review on technologies for manufacturing systems. Eng Sci Technol, Int J 22:899–919
45. Hassan N, Gillani S, Ahmed E, Yaqoob I, Imran M (2018) The role of edge computing in internet of things. IEEE Commun Mag 56:110–115
46. Dolui K, Datta SK (2017) Comparison of edge computing implementations: fog computing, cloudlet and mobile edge computing. Global internet of things summit: CICG, Geneva, June 6–9, 2017: 2017 proceedings papers. IEEE, Piscataway, NJ
47. IoT.Business.News (2020) Connected devices will generate 79 Zettabytes of data by 2025. IoT Business News
48. Hoffmann JB, Heimes P, Senel S (2019) IoT Platforms for the internet of production. IEEE Internet Things J 6:4098–4105
49. Moura R, Ceotto L, Gonzalez A, Toledo R (122018) Industrial internet of things (IIoT) platforms—an evaluation model. In: 2018 international conference on computational science and computational intelligence (CSCI). IEEE, pp 1002–1009
50. Burggräf P, Dannapfel M, Adlon T, Schukat E, Kahmann H, Holtwiesche L et al (2020) Modeling and evaluating agile assembly systems using mixed-integer linear programming. Procedia CIRP 93:1073–1078
51. Hagemann S, Stark R (2020) An optimal algorithm for the robotic assembly system design problem: an industrial case study. CIRP J Manuf Sci Technol 31:500–513
52. Hagemann S, Sünnetcioglu A, Stark R (2019) Hybrid artificial intelligence system for the design of highly-automated production systems. Procedia Manuf 28:160–166

Chapter 21
Future Virtual Product Creation Solutions with New Engineering Capabilities

Executive Summary

This chapter deals with the following topics:

- Understanding why and how *Virtual Product Creation* capabilities will be under significant change,
- Explanation which new *Virtual Product Creation* capabilities, technologies and solutions will arise and
- How those new Virtual Product Creation solutions will drive the *Engineering System* of the future.

Quick Reader Orientation and Motivation

The intention of this chapter is:

- To gain an understanding why *Model-based Systems Engineering (MBSE)* will be necessary as future engineering capability
- To learn what *MBSE* is, how it works and how *MBSE* will interact with traditional *Virtual Product Creation (VPC)*
- To explain and describe *Data Engineering & Analytics (DEA), Digital Platform Engineering (DPE)* in the context of VPC
- To provide knowledge how *VPC*, *MBSE*, *DEA*, *DPE* are key elements to enable *Digital Twin Engineering (DTE)*
- To motivate students, researchers as well as industrial practitioners and leadership to drive forward in *future VPC*.

The future challenges of our planet, our societies and our industrial value creation networks increases the pressure on accelerating the paths of both, *digitalization* and *virtualization*. In order to increase the efficiencies of our goods, gadgets, machines and products (incl. product-service-systems) in terms of materials, functional operation and maintenance operations it becomes critical to further drive the *virtualization* and the *virtual prove* out far before anything will get produced in factories and used by customers in the field. To the same time, the interplay of such objects in technical

R. Stark, *Virtual Product Creation in Industry*,
https://doi.org/10.1007/978-3-662-64301-3_21

systems and systems of systems becomes critical to enable circular, connected and fail safe systems. This new type of massive systems interaction and integration calls for new type of virtual system descriptions and simulations. If such *systems virtualization and operations* turned into reality, engineers and operators need to deal with a significant higher amount of digital models and digital data, both structured and unstructured, where the later one will prevail. In order to reveal the critical and important information elements out of such data streams, lakes and repositories the capabilities of *digitalization* need to grow, evolve and sustain. As a consequence, Virtual Product Creation (VPC) will have to evolve, too, and therefore, has to provide a range of new or altered digital disciplines in close conjunctions to already existing VPC major technologies (compare Chaps. 7–16). The following sections of this chapter will introduce and explain these new digital engineering capabilities as adjacent, coupled or integrated parts of future *Virtual Product Creation* solutions.

21.1 Model-Based Systems Engineering (MBSE)

In the past *Systems Engineering* has been leveraged as an overall management approach to coordinate all actors (not just the engineers) in the development of complex technical systems. As the Virtual Product Creation (VPC) solutions offer meanwhile a high number of different digital models to describe characteristics, attributes, logic connections and simulated behavior of technical system elements it becomes natural to transform System Engineering towards a Model-Based Systems Engineering (MBSE) approach. The MBSE approach is tasked to allow for higher degrees of dynamic digital model connections in terms of functional networks, critical path traceability, user and requirements validation and final technical system verification.

21.1.1 Motivation and Needs for MBSE as New Extension of VPC

Product creation is no longer focusing on simple combinations of mechanical parts that can be overviewed by single developers. Modern *Virtual Product Creation* focuses on systems that consist of many different sub-systems and components, which are interconnected which each other. Klaus and Liebscher [1] described this form of interconnection as complexity. This chapter concentrates on technical systems which might also inherently embody software systems as part of the overall system function and structure.

Tomiyama et al. [2] described the evolution of the smart product concept from mechatronic products, intelligent mechatronic products and Cyber-Physical systems (CPS) to smart products as highly interconnected form of products that extend their functionality through internet services. Even though, not all products are smart products, many existing products have meanwhile mechatronic characteristics with interconnections between their internal partial systems, e.g. the functional interplay between an *electronic stability control system (ESC)* with the *anti-lock braking system (ABS)* in the product *vehicle*. In extension, a dedicated group of smart products might even get dynamically interconnected, so that they are composed to a *systems of systems* (SoS). Examples for such SoS are connected systems such as *vehicle to traffic infrastructure* (via connectivity to road sensors in combination with vehicle onboard systems such as intelligent camera systems to detect traffic signs) or inter-vehicle systems (leveraging car-to-car communication). While a system can be defined as an "purposeful whole that consists of interacting parts" Walden et al. [3, p. 5], a SoS is a collection of constituted systems that fulfill functions not achievable by the systems alone [3, p. 8]. These collections of systems raise even more challenges than current interconnected systems.

To manage the complexity of such products *Systems Engineering (SE)* and its successor competence *Model-Based Systems Engineering (MBSE)* is currently heavily under development with many prove-out initiatives and are gaining traction in the industrial integration. MBSE describes the usage of modeling to support the development of complex systems [4]. As *Virtual Product Creation (VPC)* includes many modeling and information management technologies and solutions (see previous chapters) already and since VPC serves as digital engineering foundation for all kinds of products, the usage and integration of MBSE in the overall VPC solution set for *complex* products constitutes a natural and useful extension and a feasible mean. The toolbox of multiple methodologies—which includes tools, methods and processes [5]—that MBSE can offer, support VPC in a structured and focused way.

This chapter shall introduce MBSE in general as well as its combination potentials with VPC principles and major technologies.

21.1.2 MBSE Foundation on and Differences to Systems Engineering Principles

Traditional, "classical" (document-based) *Systems Engineering (SE)* and *Model based Systems Engineering (MBSE)* share a number of common features, but also exhibit some fundamental differences. SE is the transformation of a "problem" into a solution. It uses methods of systems thinking and techniques of project management. In addition, different procedural models might be used.

But what are the differences between classical SE and MBSE? While document-based SE focuses on static artifacts like texts, pictures, drafts etc. MBSE employs models, which are connected amongst each other via functional as well as non-functional relationships [6]. Classical SE focuses on the specification of the system. It is largely tool- and model-independent and allows the use of several virtual tools and methods *without* the need for an overall coordination or orchestration. Typically, different tools and descriptions are used, e.g., for a functional architecture and a physical architecture and the alignment has to be done in person, i.e. verbally or manually with the help of paper-based documentation. In document-based SE, the architecture specifications are available as a static artifact, which is treated not as a living and ever changing single-source of truth, but as a document with baselines and often multiple versions over time. This means that the traceability of problems is limited and there is a high risk of incorrect entries, which can and does lead to regular inconsistencies [7]. There is no system model that can be used across domains or disciplines, but documents contain the core of the system information and describe statically the progress of different aspects of system development.

MBSE leverages the use of comprehensive system models from which the required views of the system can be derived dynamically. Changes are automatically adopted and the documentation is a by-product of the modeling of the system aspects and components. In contrast to document-based approaches, requirements are mapped to functions in the course of development and are, therefore, *traceable* throughout the entire system, even if changes to design or requirements occur. It remains an expert task, however, to decide which type of traceability is necessary or desired for a given product or technical system within the context of engineering development (incl. validation, verification and change management) and homologation needs.

Table 21.1 lists the differences between the two systems engineering concepts. A direct comparison of the aspects is not always possible, because MBSE is not only an evolution of traditional systems engineering, but in some aspects, it is a fundamentally different approach [8] relying on an already well-established Virtual Product Creation foundation.

Applying MBSE for technical system development results in providing and establishing views for consistent requirements management and (virtual and physical) test specifications. In addition, overall system behavior simulation as well as (generic and specific) system architecture are made available in an integrated system model. This approach needs to be scalable on all levels of detail. Furthermore, MBSE processes should be standardized and/or modularized in contrast to ad-hoc processes, which are possible in document-based systems engineering and inevitably heavily rely on

Table 21.1 Comparison between document-based SE und MBSE [9]

Document based SE	MBSE
Derived from existing documents	Derived from a unified system model
Independent diagrams and blueprints	Consistent views
Static documents	Executable behavior
Static views	Dynamic views
Ad hoc Processes (High risk of inconsistencies)	Only replicatable Processes
Manual implementation/management of changes	Automated change management across the entire development lifecycle

implicit knowledge held by process owners and subject matter experts. This consistent MBSE approach requires, however, a solid preparation phase and the willingness of Engineering Management to invest steadily into such a new Virtual Product Creation capability.

21.1.3 Theory and Principles of MBSE

In order to gain a deeper understanding of MBSE, the term "system" must first be defined. The term "system" is used in a very general way outside the scientific/technical context without clearly defining what makes a phenomenon or groups of events, people or physical objects become a system. Prominent examples are terms like "solar system" or "ecosystem". There exist multiple definitions of systems, which are used across various domains. As mentioned earlier already, Walden et al. defines a system as "a purposeful whole that consists of interacting parts" Walden et al. [3, p. 5]. Haberfellner et al. mention that a collection of elements becomes a system because the parts it comprises are connected to each other [10]. Ropohl et al. [11] mention that systems can only be described by viewing them from three different perspectives, as shown in Fig. 21.1: functional, structural and hierarchical. While the structural concept describes that the system elements are interrelated to each other in a certain way, the functional concept describes the system and its reactions to various inputs with defined outputs as result of a change of states inside the system. The hierarchical concept describes that each system can consist of multiple other (partial) systems (type a) and to the same time can be part of an even larger system (type b).

In consequence of all these definitions, in the following a system includes related parts that form a whole and exhibit the following properties:

- Systems consist of *elements* (also often referred to as sub-systems and as components).

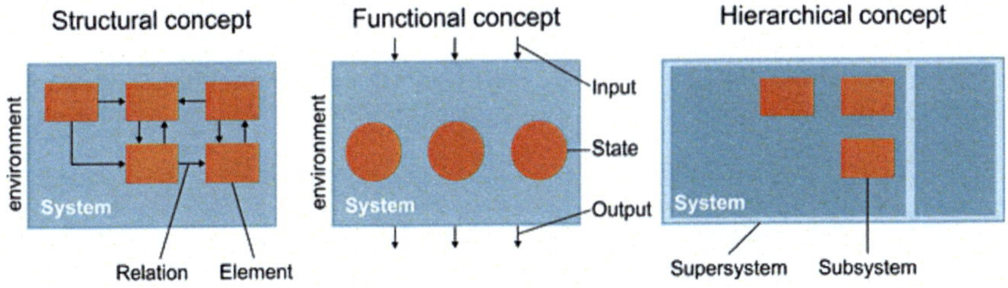

Fig. 21.1 System concepts of ropohl [11] (left to right): structural concept, (describing the relations of the elements inside of the system), functional concept (describing the behavior of the system as output to a specific input) and hierarchical concept (describing the system consisting of subsystems (type a) or as part of another supersystem (type b))

- Elements have *functions* and *properties*; whereas functions describe intended (sub) system behavior, properties contribute that intended (sub) system behaviors can be realized.
- Elements can be understood as *(sub) systems* again.
- Elements are connected to each other in form of *relationships*; those *relationships* carry important knowledge and are critical to deliver the "system intelligence" together with the individual elements of the entire system.

General system theory is based on the assumption that all open systems have commonalities that allow domain-independent analysis approaches. The pioneer of systems theory, Ludwig von Bertalanffy [12], postulated in 1968 that knowledge about a certain system thus allows conclusions about other systems. The elements of a system can be symbolic representations of organizational concepts or real objects. Systems that consist only of concepts are called abstract systems. Systems that contain at least two real objects are regarded as concrete systems. However, real objects do not necessarily have to be of physical nature, but can also be information (for example software code).

Explanation of Core Basics

A company group of different smaller enterprises fulfills the required properties of system mentioned above by consisting of different elements (enterprises) that can be systems on their own, are connected to each other and have various properties and functions. They can be seen as *symbolic representations of organizational concepts* and thus form an *abstract system.*

As second system, the turbine of an airplane can be considered, as follows: it consists of multiple *real objects* in either physical (e.g. blades, casing and shaft) or informational form (e.g. executable software code to control the fuel mix). Therefore, this system is a *concrete system.*

System theory assumes that a system always has a system boundary. This boundary is derived from how the relationships between the elements are evaluated and categorized, because it is from this categorization that the system membership is ultimately derived. The system boundary and thus the scope of the system is entirely dependent on the observer.

Context Example 1

Consider the development of a vehicle from an automotive OEM perspective. The top-level system composition can be defined by different aspects: the systems can be defined for example based on their functionality (e.g. *chassis system* or *environment perception system*) or their elements domain (e.g. electrical system, mechanical system and information system). The boundaries of these systems differ depending on the chosen aspect.

A technical system is usually the technical implementation of a defined solution concept, contains a defined number of elements and has certain properties. The system elements can be viewed from two primary perspectives. *Requirements* within a problem space and *architectures* within a solution space. The *system architecture* includes all system elements, their relations amongst each other and external interfaces. Therefore, the *system boundary* is an integral part of the *system architecture.* Through the relations of the system elements to each other, a system is given a *structure* and a certain *behavior*. The *system structure* describes thereby, how the elements are connected with one another and the *system behavior* describes the effects, which are produced, if a part of the system reacts with the system environment.

Context Example 2

The boundaries of the vehicle developed in the previous example (context example 1) shall be defined by functionality by having a closer look to the *environment perception system.* The architecture shall consist of a radar sensor, a lidar sensor, and a processing unit. The radar and lidar sensors are both connected to the processing unit but not to each other. This construct forms the *structure* of the system. Both sensors get information of the environment and forward them to the processing unit. This unit performs certain functionalities, e.g. calculating the distance of certain objects, and sends this information to external systems such as *human machine interfaces (HMIs).* This is the *behavior* of the system. The interconnection is shown as an *internal block diagram (ibd)* of the *Systems Modeling Language (SysML)*[1] in Fig. 21.2: *SysML internal block diagram of environment perception system to visualize the structure for a certain behavior*. This diagram is used to display the structural concept of the system of interest.

[1] The *Systems Modeling Language (SysML)* is standardized graphical modeling language to describe systems of interest. It aims to be an interdisciplinary, general purpose modeling language with focus on Systems Engineering. It uses nine different diagrams in the current standard v1.6, depicting the views on the system: *requirement diagram (req), activity diagram (act), sequence diagram (sd), state machine diagram (stm), use case diagram (uc), block definition diagram (bdd), package diagram (pkg), internal block diagram (ind)* and *parametric diagram (par).*

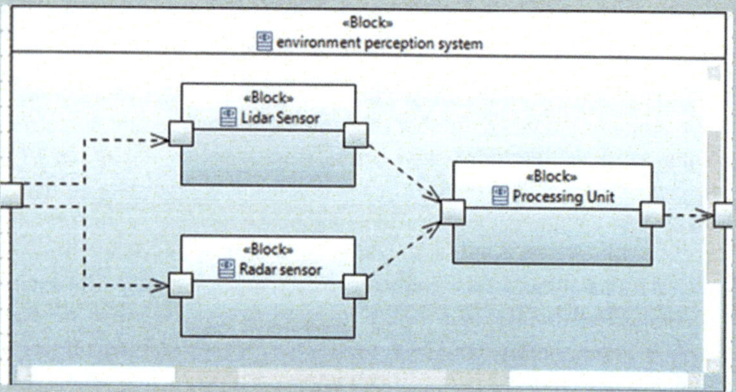

Fig. 21.2 SysML internal block diagram of environment perception system to visualize the structure for a certain behavior

Since most systems are dynamic systems—the relations between the system elements are changeable and depending upon configuration—these systems assume different states. The totality of the possible states is called state space of the system. This fact also plays a role in the development of technical systems, since the totality of all possible wanted and unwanted states can be used as a basis for planning system and component tests.

Further Explanation of Context Example 2

Possible examples of states within an environment perception system's *state space* are: processing input data, scanning input, inactive, defective, updating processing code.

Most systems are networks with a hierarchical structure, which in turn consist of elements that can be considered systems in themselves and are often described as *subsystems* or *partial systems* in technical systems development. These properties lead to the concept of *emergence* as a fundamental part of the definition of the term system. *Emergence* means, that the whole system has properties, which do not make sense or would not exist in relation to single elements. In relation to technical systems, this leads to the problem that a decomposition of the system can impair the understanding of the system, because information gets lost in the process. The dynamic nature of the interactions within a system and with elements outside the system can also lead to system behavior that cannot be captured deterministically.

Context Example 3

A radar sensor on itself is not capable of calculating distances to objects or of even giving feedback on the urgency of interaction. Neither is the processing unit capable of sensing the required input data for its distance calculations. These capabilities are the *emergence* of the environment perception system.

A model is a limited representation of reality. According to the US Department of Defense Modeling and Simulations Glossary, a model can be a physical, mathematical or otherwise logical representation of a system. A physical model is a concrete representation of the system, which is different from abstract mathematical and logical models. A model is always an illustration or representation of a natural or an artificial original, which itself can be a model. In general, models do not capture all attributes of the original, but only those that seem relevant to the originator or user [13].

Explanation of Core Basics

Exemplary models are clay models (physical), math-spring-damper models (mathematical) and CAD assembly models (logical representation of physical objects).

Oftentimes digital models are not clearly enough assigned to their "realizations" in the physical (real) world later on. In any case, however, digital models fulfill the important role to act as *substitute and development aid* during the engineering progression.

(a) for certain subjects (for whom?),
(b) within certain time intervals (when?) and
(c) under restriction to certain mental or physical operations (what for?)

The modeling of complex technical systems is an iterative process in which experimental findings are continuously compared with simulation results. In addition, the underlying theories can be adapted during the modelling process if fundamentally faulty assumptions are detected during the iterations.

From a *system architecture* model of a technical system, different sub-models and simulations can be derived, which are necessary for certain aspects of the overall technical system development. So (partial) system models can be used, in order to lay out the components of the targeted technical system and to interact with other (partial) systems according to the overall system architecture. In order to integrate these (partial) system models, which might get developed in the context of different domains, it is important that the system models are semantically compatible, which requires a high degree of up-front agreements, standardization and formalization. The ultimate goal of robust *systems integration* requires *semantic interoperability*.

Semantic interoperability means that data can be exchanged between models without loss and that each model can interpret and integrate the data unambiguously. In engineering practice system models must be able to exchange data amongst themselves, either manually triggered by System Engineers or even automatically according to new intelligent coupling and interaction mechanisms. This can be realized e.g. via a common database and only the development of *modeling standards* enables engineers to leverage the broad application of MBSE for systems development.

In any case, it becomes crucial to work within an appropriate model framework in order to provide a meaningful engineering interpretation of the (digital) engineering models in their overall system context to each other. This, by the way, is one of today's shortcomings in most of the industrial companies: such model frameworks have not yet been emerged from the individual modeling practices of departments and enterprise functions and are still far away from being standardized according to the individual company product spectrum.

The author of this book, therefore, recommends to establish such frameworks and would like to point to the reference *model cube* which has been worked out by Prof. Hick and his team at the Technical University of Graz in Austria (compare [14] and see Fig. 21.3).

Model-based development approaches rely on the extensive use of models in all phases. This requires model management and methods to structure related models. The concept and reference framework of the *model cube* addresses this need to structure and classify models, which are used in product development (incl. MBSE) and other phases of the product lifecycle. In this context, only digital models are considered, but the basic concept can be transferred to structure physically manifested models (e.g., prototype on test bed), as well.

One essential consideration with this *model cube* is the differentiation between *system models* and *specific models*. The main intention of *system models* is to incorporate multiple views in breadth and width in order to provide system relevant statements and to support *interdisciplinary collaboration*. *Specific models* focus on

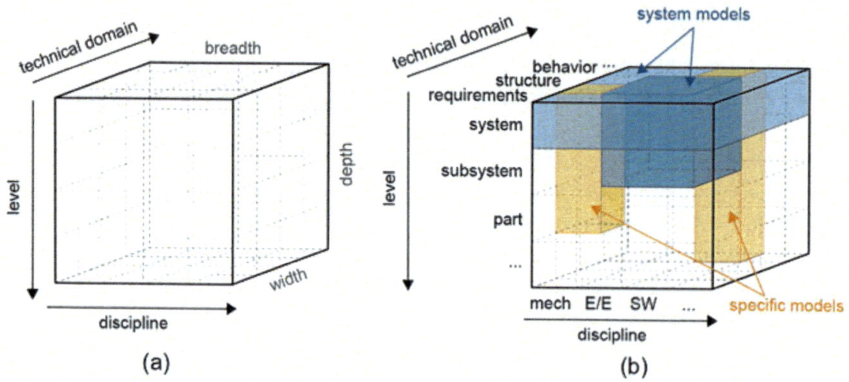

Fig. 21.3 The model cube according to [14]

specific technical aspects from the perspective of one discipline allowing in depth descriptions of (sub-system or part/component) model types in dynamics, kinematics, electronics, hydraulics, logic etc. Nevertheless, the differentiation of system models and specific models also depends on the point of view (e.g., OEM or supplier, position of the main technical system within the overall system hierarchy etc.). The basic structure, which is visualized in form of the model cube, is based on the following three dimensions:

- *Breadth of the model cube*: considering different disciplines involved in product development (in this case *mech* for mechanics, *E/E* for electrics/electronics and *SW* for software),
- *Width of the model cube*: considering different aspects of a system, that are described in models (requirements, structure and behavior),
- *Depth of the model cube*: considering the system structure (system, subsystem, component, etc.).

The *model cube* is a simplified illustration to structure and visualize models, which is helpful to address challenges originating from model-based development approaches. As the *model cube* represents an overview of the models used until a certain point in development, it supports interdisciplinary efforts by showing the broad range of different (discipline) specific models and the overlapping areas between these models. The overarching and integrative role of system models is highlighted by the illustration as well, while it should be emphasized that systems models are not limited to SysML or similar descriptive languages, but can be of quantitative nature as well (simulative character). In summary, such a model cube reference provides orientation and clarity in model usage and model connectivities.

21.1.4 Disciplines of MBSE

As MBSE is not tied to a specific method, process or tool, there exist multiple methodologies, which Estefan et al. [5] defined as a combination of the three aforementioned aspects. These methodologies give the user a sort of manual, how to develop the system of interest. While for some applications this might be helpful, it leaves the engineer with a sequential process and only a low level of adaptability. Since modern product development is most of the time of an iterative character, this leaves the engineer often with the need to jump back in the development process and redo the steps of the already beforehand executed process when changes occur. Current clustering approaches like the four pillars of MBSE (methodology, tools, language and data management) [15] or the five tenets of MBSE (organizational understanding, modeling language, model-based processes, structure/governance and presentation framework) [16] have a strong bias to the usage of graphical modeling languages

Fig. 21.4 System modeling approaches: direct linking of system elements, linking through a meta-model and linking of partial results [17]

like SysML or *Unified Modeling Language (UML)*[2] and thus for methodologies that support these languages. The survey of Estefan et al. [5] seconds this bias by mainly showing the application of graphical modeling languages. However, system modeling approaches can go beyond using a graphical modeling language as a meta-model. Stark and Schulze [17] differentiated system modeling into three approaches that are shown in Fig. 21.4 and consist of direct linking of system elements, usage of meta- models and linking of partial results of the system elements.

These approaches can be combined and may require the application of different methodologies depending on the user preferences and scope user. To allow this methodology-independent application of MBSE the research organizations of Fraunhofer IPK (division "*Virtual Product Creation*") and TU Berlin (chair "*Industrial Information Technology*") defined *five development capabilities* that are required to perform and master MBSE [18]. These development capabilities are designed as disciplines, as the current activities in methodologies have a sequential connotation that is not aligned with the iterative character of most product development projects, where MBSE is applied. Figure 21.5 depicts the MBSE capabilities in the 5D-model introduced by Stark and Auricht [18] and presented in Schmidt et al. [19].

Referring to the overall description in Schmidt et al. [19] the *5D MBSE* development capability include the following five development capabilities:

Systems Environment Analytics (SEA): This development capability focuses on the definition of the system boundaries and of the general interaction with surrounding systems. Even though, this development capability is important for an appropriate application of MBSE, it does not necessarily require the formal creation of models, however it represents an indispensable activity to clarify the (partial) system interaction landscape.

The early formal usage of e.g. different types of SysML diagrams such as *block definition diagrams, internal block diagrams* or *use case diagrams* might be useful and advantageous in already matured discussions, but could also lead to a by far too

[2] The *Unified Modeling language (UML)* is another standard of the *Object Management Group (OMG)*. It is agraphical modeling with focus on Software Engineering. In the current version of SysML (v1.6), UML forms the foundation for SysML. In the currently developed standard v2 of SysML, a new foundation will be used and UML might become a *domain specific language (DSL)* for the Software domain again.

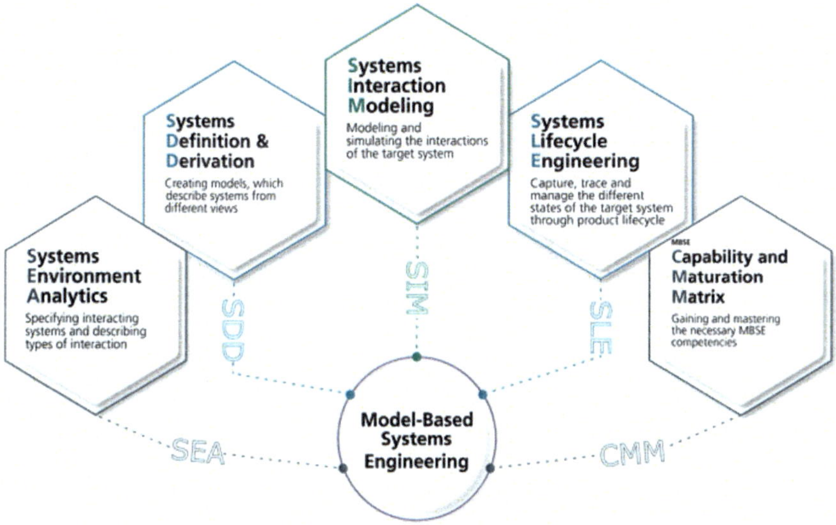

Fig. 21.5 MBSE development capabilities according to Schmidt et al. [19]

detailed discussion on specific system capabilities in the early development days of a systems development project. It will be possible in the future to use AI clustering methods (compare Chap. 16) evolving for this first MBSE discipline: using live (IoT type) data from already existing or similar systems in operation will provide statements of proposal or evidence for *system environment analysis (SEA)* in terms of level of interaction tightness of identified (partial) or envisioned (partial) systems.

In the following a turbine of an airplane shall be developed. A team of developers has to decide, whether they include e.g. the turbine control unit into their system or consider it as external system and outside their *system boundary*. In context of this example, it shall be considered as internal system element.

As part of the *Systems Environment Analytics (SEA)* the *interactions with external system elements* have to be investigated and determined sequentially. This includes the interactions with environmental influences like wind, rain and temperature, the interaction devices of the pilot or further power consuming systems. As the systems matures in the development process, the system boundaries have constantly to be checked and maybe adapted as well as the interactions with the external systems.

Right at the beginning, System Architects and System Engineers together with Data Engineers need to analyze which data sets are available under which circumstances and can be analyzed with the help of which criteria sets to drive an appropriate and useful system environment investigation.

Systems Definition and Derivation (SDD): SDD comprises high level but formal system specification and first mandatory modeling activities. As the main system elements are defined and modeled, further subsystems are derived over time. This development capability has the sequential character of most methodologies; it starts by the definition of the target system and moves on to the description of the overall system architecture and the derivation of the corresponding systems (incl. carry-over systems). The models connected to this development capability are usually rather static and focus on the structure of the system, such as system architectures defined in SysML with the help of *block definition diagrams (bdd)*, *internal block diagrams* (ibd) or *use case diagrams (uc)*.

Based on the *SEA development capability*, the *target system* can be defined. The turbine system's *core purpose* is to generate thrust and thus accelerate the plane. An additional purpose might be the generation of energy inside for electrical equipment inside the airplane. A possible *system architecture* for that purpose can be based on a functional decomposition for the core purpose. To generate thrust, energy has to be converted after it has been supplied to the system. This energy conversion can be used for the thrust generation or the sub purpose of energy generation for the equipment. Jet propulsion can be generated by different media such as water streams or gas. In the *system derivation,* different quantifiers (e.g. cost or expertise) are used to select the right solution elements for the system. In this example, we select an air-breathing jet engine to convert supplied energy to thrust.

All these elements are modeled to allow for a computer-supported evaluation and linking of the artefacts. A typical form of modeling in MBSE is the application of graphical modeling languages, e.g. SysML or the languages behind the *Object-Process Methodology (OPM)*[3] for this purpose, but system definition and derivation within MBSE is not limited to these modeling approaches. It can well be that even annotated concept CAD models might be used in this scenario or AI supported analysis methods in case of textual requirements or functional decomposition and network notations.

The system architecture can then be used as guidance for further system development: Königs [20], e.g., developed and used a *SysMT (Systems Modeling & Management Tool)* solution to define a system architecture that can be used as template for further system development, validation and verification.

Systems Interaction Modeling (SIM): While MBSE users could create most models upon mastering the previous development capability, these models are all

[3] The *Object-Process Methodology (OPM)* is combination of modeling languages and a methodology for modeling different systems, mainly automation system. It is standardized as ISO/PAS 19,450. It comprises a graphical modeling language, which uses *Object-Process Diagrams (OPD)* and a textual expression in form of the *Object-Process Language (OPL)*. It is mainly used to describe objects and their transformation or use by processes.

static and cannot be used for much more than descriptive actions. SIM focuses on models for the description and simulation of systems interactions. This may include timely interactions, causal reactions, signal flows, messages or full behavior-based interactions. In many instances those type of interactions trigger the need for target-oriented simulation models of interactions. The results of such interaction simulations can then for example be used to further derive system details under the SDD capability. Prime examples of such model types affected by the SIM development capability are.

- functional network models with causal reasoning analysis,
- logic flow models as part of state machine propagation simulation,
- behavior models of technical systems as part of 1D or 3D Computer Aided Engineering (CAE) in Virtual Product Creation or
- system dynamics models as part of Business Management and Operations Technology (OT) in case decision making (or cause-effect rationale) is treated in scope for technical systems interaction.

The system architecture defined in the previous step shows which elements are relevant and how they are connected to form the system, but cannot be used directly to perform trade-off analysis. Based on appropriate technical knowledge *Systems Interaction Modeling (SIM)* will provide a target model environment to drive the selection of the right fuel-mix ratio for a defined desired distance of the airplane. The input (system architecture model) generated in the SDD development capability can be used to create *simulation models*, which *depict the interaction of the system elements* energy storage and the jet propulsion engine and thus optimize the storage size for different distances.

Systems Lifecycle Engineering (SLE): As a system can obtain different states over its lifecycle, they have to be captured, traced and managed. SLE focuses on the activities connected to the states of the system from the development phase to the final system in use and in maintenance. Typical models in connection to this development capability are state machines and all kinds of models that consider multiple states of the depicted system. These states of systems are also often connected to PLM, e.g. in System Lifecycle Management (SysLM) as shown by Eigner et al. [21].

The turbine system can be active, inactive, out of fuel, in a rich or lean combustion or generating a power overshoot. These states have to be captured with other operational and lifecycle usage states in one or more *state models* for the system. The ways of capturing operations, usage and lifecycle states with appropriate data sets becomes a decisive competence as part of *Systems Lifecycle Engineering (SLE)*. The new emerging future *Virtual Product Creation* fields *Data Engineering (DE)* (compare Sect. 21.2) and *Digital Twin Engineering (DTE)* (compare Sect. 21.3) will serve as indispensable underpinnings for future SLE capabilities far beyond today's rather awkward and limited *System Lifecycle Management (SLM)* solutions or simple *lifecycle units* based on signal recordings.

In the future Digital Twin technology and solutions will become essential for Systems Lifecycle Engineering activities as well as IoT platform-based services (compare Chap. 20 to learn more about both technologies). SLE, therefore, will highly depend on the maturities and the evolution of those new data engineering and analytics-based solutions. The Digital Twin connection capability between operations data and engineering models will play the essential role to provide synthesis and analysis methods in *Systems Lifecycle Engineering (SLE)*.

MBSE Capability and Maturation Matrix (CMM): While the other four development capabilities allow a theoretical and practical application of MBSE, they have to be learned and applied by humans. The gaining and mastering of the necessary MBSE competencies is summed up in the CMM development capability. It measures how well the MBSE development capabilities are understood and how they can be taught and mastered. For this development capability, there are no clearly associated model classes, as it focuses more on the users of the models than on models themselves. However, it is possible to define a maturity model for model type definition and usage by engineers in an organization, network or development project.

Engineers of different technical disciplines rely on the information sets of the turbine system: e.g. electrical, software and mechanical engineering but also controlling, project management and customer communication. To fully understand where to look for the required information they have to understand how to read and filter the available models and have to acquire a basic knowledge on the development capabilities to understand certain model elements and their interaction intelligence. In such a way, engineers can meaningfully collaborate and generate additional information as part of the MBSE progression.

Even though these five MBSE development capabilities represent major current research topics, they did not yet make their way into industrial mainstream engineering. Zimmermann et al. [22] elaborated the use of those capabilities for the development of digital twins and Schmidt et al. [19] depicted their use for the development of an automated driving function. The basic idea and need for them is generally recognized, but only fragmentally established as industrial practice so far. Industry currently focuses mainly on the *SDD development capability* by creating system architectures with graphical modeling languages like SysML, although research work has already demonstrated that much richer solutions are needed (compare the SysMT work of Königs [20]). As organizations seem to notice the importance of learning and teaching MBSE [23], the industrialization of the other development capabilities is a major topic that will characterize the upcoming years.

Finally, it is important to note, that *system validation* ("are we developing the right system for the requirements in scope") and *system verification* ("have we developed the system right, i.e. does it work and get accomplished in time, budget and quality"), need to be primarily supported by the three disciplines *Systems Environment Analysis (SEA), Systems Definition and Derivation (SDD)* and *Systems Interaction Modeling (SIM). Systems Life Cycle Engineering (SLE)* is core for system surveillance and continuous improvement as well as for *Feedback to Systems Design (FTDS)* in order to improve the next generation system architectures.

21.1.5 Core Elements of the New MBSE Approach

After having introduced the theory and the major disciplines of the overall MBSE approach it is now necessary to identify and describe the core elements of MBSE which are subject of daily work of engineering within VPC.

21.1.5.1 Model Types and Digital Artefacts

The primary focus of system modeling is to use models supported by a well-defined modeling language and the appropriate model classes, instances and objects. While less formal representations can be useful, a model must meet certain expectations for it to be considered within the scope of MBSE. In particular, the initial classification distinguishes between informal and formal models as supported by a modeling language with a defined syntax and the semantics for the relevant domain of interest.

In order to apply systems modeling to the development of technical systems, well-defined modeling standards are required or at least useful. This often manifests as a formal modeling language that is used as an umbrella to encompass all domain-specific models and bridge the existing gaps between their specific modeling artifacts and approaches. Principal approaches of linking system elements and system results as shown in Fig. 21.4 are gaining traction with standards like the *Open Services for Lifecycle Collaboration (OSLC)*[4] and the *Specification Integration Facility (SpecIF)*.[5]

[4] The initiative *Open Services for Lifecycle Collaboration (OSLC)* aims at providing standardized interfaces between different applications to connect application data. It is based on the *Representational State Transfer (REST)* software paradigm used in web applications. Specifications are defined for the core of OSLC and different domains like PLM, ALM or Requirements Management. It misses, however, semantic and parametric interactions with domain specific engineering models such as CAD, CAE and mathematical models.

[5] The *Specification Integration Facility (SpecIF)* aims at a more artifact-centered exchange instead of a document-centered exchange. Its core is the extraction of semantic information of each model and thus the combination of different forms of models on semantic level.

While informal system representations can be useful, a model must meet certain expectations in order to be able to contribute to an MBSE approach that supports system models and connects relevant domain specific system representations employed during the development phase. This requires models to apply a defined syntax as well as consistent semantics [24].

During systems development, various types of models are used, from graphical modeling languages like SysML/UML and functional network reasoning, to logical control and behavioral models as well as geometrical models of system components. Models utilized during the development of technical systems can be grouped in two classes: descriptive models and analytical models. A descriptive model describes the structure of a system, for example a *system architecture* describing *system elements* and their relationships with each other and, therefore, represent the working artefacts of *Systems Definition and Derivation (SDD)*. These models can describe actual physical properties of a system or abstract functional properties.

Analytical models describe the logic control, the dynamic behavior or the cause-effect operation of a system, usually by describing mathematical relationships using equations or value tables and support the quantitative model-based analysis of system properties during system development. These kinds of models can describe time-varying states of a system and are used to explore performance characteristics of a system, for example speed and efficiency of a vehicle. Behavioral models can also be used for static systems analysis in order to determine a system's reliability, mass or cost properties. Applying MBSE for system development means that such models can also be utilized for system analysis, without the need for separate simulation models in order to obtain time-series data and be able to explore dynamic system properties. A model can comprise analytical and descriptive aspects and descriptive models might be used for analyses as well, but only allow logical analysis instead of quantitative analysis of a system, e.g. completeness checks. Analytical models play a decisive role in *Systems Interaction Modeling (SIM)*.

21.1.5.2 Model Content, Model Views and Model Linkages

Descriptive and analytical models created in various forms as part of the development with MBSE must be integrated across multiple domains to effectively leverage model-based approaches. System models become the focal point since they can be used to specify an overall system in *domain-independent* representations. Hence system models serve also as integrator to connect various domain-specific models with greatly varying contents and methods [25].

An example is given in Fig. 21.6 for the model network used in the TU Berlin development of the *Safe Door Opening System (SDOS)*, an automotive safety system to avoid accidents with bicycles during uncareful opening of vehicle side doors. Two different MBSE approaches have been used and studied within the engineering project: *Type A*, a structured and cross-linked RFLP (Requirement, Function, System/Logic, Physical) model approach integrated in the 3DExperience Platform of Dassault Systèmes (compare also Subsect. 21.1.7 *Examples of new MBSE methods*

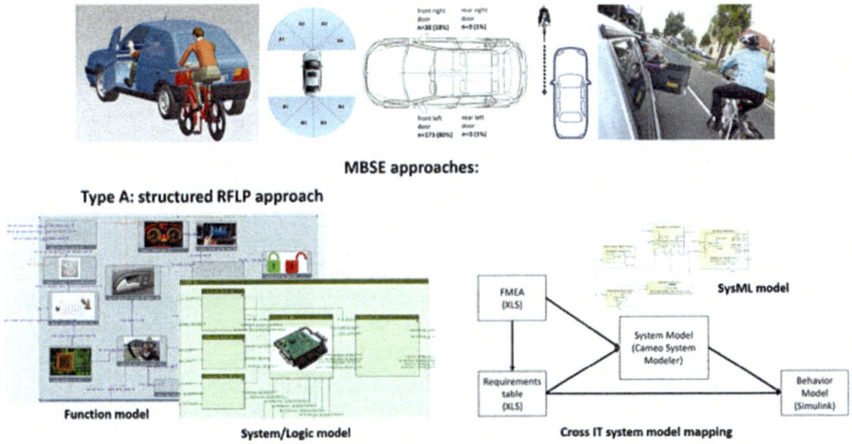

Fig. 21.6 Connected models of a safe door opening system (SDOS)

and tools). *Type B*, a loosely linked system model approach using a SysML model as core element as part of a cross-IT system model architecture. The SysML system model is connected dynamically with—in this case—static requirements and FMEA safety artifacts (in both cases represented in.xls files) as well as with configurable behavior models (Simulink models).

During the development of technical systems, consistency across all used models is a key aspect, since those models describe different aspects of the same system. Particularly challenging is the fact that models inherently reduce and simplify information, and depending on the specific type of a model, its usage and its domain, focus on certain aspects while reducing or omitting others [4].

Ideally, those models tie into each other and create additional value by representing a multitude of consistent aspects of a system's structure and behavior, but this also means that the interrelations and connections between different models are not trivial or always self-explanatory and in themselves might be driven by design decisions. Figure 21.7 depicts examples taken from the SDOS model environment shown in Fig. 21.6 (type B). It shows how parameters are connected across models to enable dynamical connections between multiple models. This still requires a high level of subject matter expertise and semantic structures and meta-models are only slowly beginning to be adopted to enable more automation in this area. In approach type B the Cameo System Modeler plug-in *Syndeia Magic Draw* is used to establish explicit linkages to Mathlab Simulink models. In an internal 3D Experience type A approach those connections are directly enabled through the internal data model. In the future, the connections will get increasingly realized through *REST* (Representational State Transfer) API (Application Programming Interface) technologies which have become the norm for WEB based digital platform environments.

MBSE capabilities are critical if it comes to enable multi-domain system functions, interaction and behaviors. In contrast to the specific geometry-based CAE

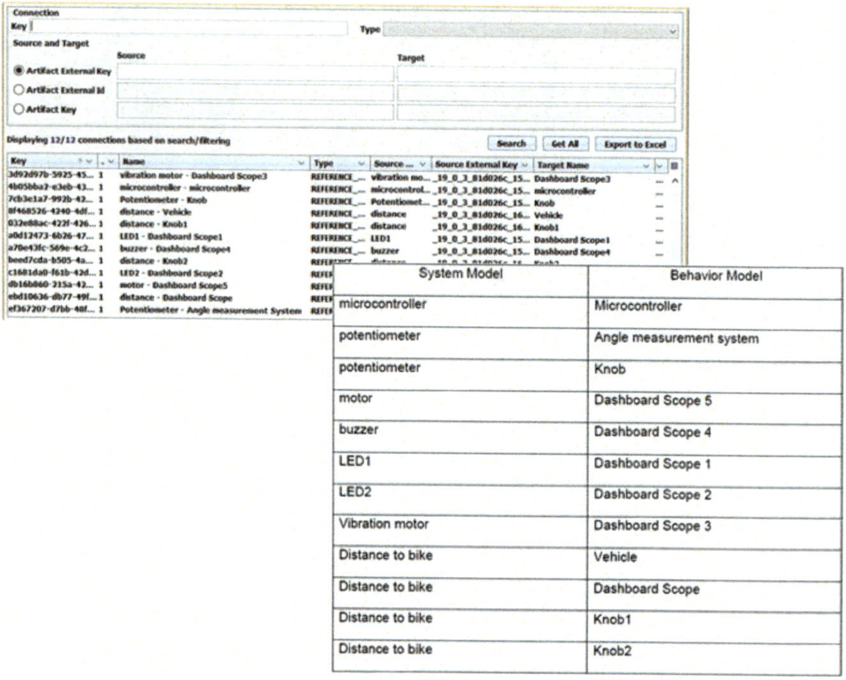

Fig. 21.7 Practical examples of the connection of parameters across models

models (compare Chap. 10 as part of the major technology chapters), MBSE primarily leverages 1D CAE models to represent the relevant technical cross-domain system physics and the mathematical equations of the boundary conditions and the control behavior types. Within this context *Modelica* has been developed as a language for modeling of complex physical systems. Based on the open *Modelica* language specification[6] a range of commercial tools (e.g. Simulation-X, Dymola, Maple Soft) have been developed to offer specific simulation capabilities based on the Modelic foundation. In contrast to Simulink models which are *causal representations* of the physics problem (i.e. each control object needs one direct predecessor as input) Modelica describes *a-causal models*, i.e. components and objects are described by their inner relations (as algebraic or differential equations) and their physical connections to other components.

If a system design focuses on solving a problem through means of mechanical parts, there will inevitably be a focus on those aspects of system modeling, while a system design solving the same problem employing controllers and software would obviously shift the modeling focus towards those development domains and simplify other aspects of the system. This means while all kinds of views can be generated from a MBSE system model, not all views would always provide the same amount and granularity of information about the technical system under development. Despite

[6] www.modelica.org.

this, all models must be sufficiently integrated to trace the actual system design elements to their respective requirements in order to validate the developed solutions. To integrate the used models a construct in one model must have the same meaning as a corresponding construct in another model, as well as across different modeling tools [24].

This can be achieved in practice by model transformations, formal modeling languages and data exchange standards. In that way, the three form of system modeling (direct linkage, linkage through a meta-model, linkage of partial results) as shown in Fig. 21.4 can be achieved.

21.1.5.3 Traceability Across Digital/Virtual Artefacts

Traceability in document-based Systems Engineering can be achieved by using links between documents or items within them to provide a trace from requirements to the system design as well as test cases, procedures and results in order to verify the design solution against system requirements and validate the system. This is typically accomplished by assigning unique identifiers to documents and items they comprise and by documenting their relationship in a matrix or a table, which today is usually done via relational databases. These connections maintain static and only provide information about the logic or the nature of the relationship between concerned document items. In MBSE, traceability is particularly important, due to the interconnectedness of different models utilized during system development. Figure 21.8 shows an example of interconnected models that have to be considered in system development. Despite the known advantages, traceability is still not established as a

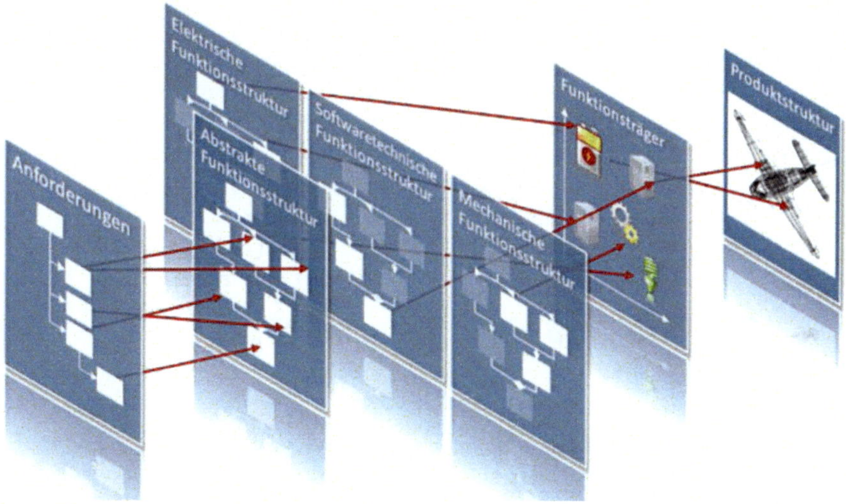

Fig. 21.8 Tracing concept along different artefacts shown in Beier [28, p. 35]

standard in digital engineering and Virtual Product Creation (VPC). Large effort for traceability creation and maintenance counts as one of the major obstacles against traceability [26].

Besides comparing the as-is status of a system against the to-be properties, traceability can be utilized in order to explore system characteristics and facilitate a deeper understanding of a system, by enabling engineers to explore logical links across the system and follow traces in order to analyze change impact or identify defects within the system design. However, traceability also plays into the system design, by driving linkages between system components and causal chains, therefore making the establishment of tracelinks part of the overall system design process. In some industries, traceability is mandated by a number of standards such as DO-254, ISO 9000ff, ISO/EC 15504 (SPICE) to show legal compliance for certification [27].

Significant research efforts are being invested in the development of improved automated tracing, to lower the amount of manual work required for creating and maintaining traceability. In automatic traceability creation, smart algorithms take over the decision-making process completely or partially to detect logical relationships in product development [26].

One of the major challenges of system development with MBSE is to transfer implicit knowledge about relationships and connections between different parts (both logically and physically) and aspects of a system into a formal network of descriptive and analytical models and therefore explicitly available to engineers.

Two major types of traceability can be distinguished based on their information content [27]. Qualitative tracing only allows the identification of whether objects are related whereas quantitative tracelinks enable the full potential of MBSE, by allowing for parameters and values within models to be explicitly linked and dynamically influence each other.

Quantitative tracing requires a higher modelling effort, but allows for a network of models, unified by an overarching inter-domain system model, which directly supports the exploration of time-varying states of the developed systems through co-simulations. However, it cannot be the aim to model all implicit relations in finest granularity since it would lead to an unmanageable effort both in terms of IT-infrastructure as well as human stakeholders, who would get overwhelmed with the sheer amount of information. Therefore, there is a challenge in identifying those dependencies that really benefit the development process [27].

21.1.6 Co-existence and Interaction with VPC Major Technologies

Even though (or even because) MBSE is an interdisciplinary approach that connects the various disciplines, it cannot stand on its own. It relies on existing VPC technologies and even has to leverage all associated technologies.

The most obvious and widely discussed interaction with VPC technologies is the interaction between MBSE and PLM. Even though PLM is a concept rather than a specific form of tools, multiple digital solution providers and PLM vendors call their product data management (PDM) environments PLM-tools in order to underline its potential for the entire life cycle (compare Chap. 11). Here the first commonality with MBSE becomes visible, as MBSE focuses on the full lifecycle of the system (compare the SLE development capability in Fig. 21.5), as well. Both concepts rely on *tracing* as an integral part of their underlying concepts. So far, MBSE mainly focusses on the systems states (SLE development capability) and PLM on the used artifacts. Some major PLM systems even integrate MBSE tools, like SysML modeling tools, into their backbone. Because model management is an important topic in MBSE, PLM tool vendors even try to integrate these models into their tools as well, in order to allow a sufficient management of these models (with all of their parameters and attributes!).

Even though these concepts can certainly benefit from each other, they also clash with each other when it comes to the question, which of them is the main driver in system development. While MBSE is focusing more on the methodology to support interdisciplinary development, PLM advocates are mainly trying to enable this interdisciplinary development through the advancement of interconnected and integrated tools. When both concepts are developed in synergy, the future product development can be supported even better. Current research projects and industrial working groups are trying to address these topics to reach the best synergies. As one example of this development, Eigner et al. [21] shall be mentioned, who developed *System Lifecycle Management (SysLM)* as a concept to integrate the two layers system models and PLM.

Examples of Transforming Stepwise Towards MBSE

To address current customer needs, a company focused on mechanical bicycles wants to develop electrical bikes. As this development introduces further domains and the complexity of the system rises, the company decides to use MBSE as their new development approach. As they already have developed the mechanical bike based on various CAD and CAE models and as they can reuse basic concepts of it, they can leverage those models in the PLM system for their model-based system design and validation. Through the step-by-step integration of these models into their MBSE approach, they might require less time in the future for the entire development execution and/or might reach a higher technical system quality and robustness at the end.

One of the most significant ambiguities of the interaction between VPC technologies and MBSE is the role of CAD models for or in MBSE. As CAD is wide spread in industrial use and since CAD (and closely integrated CAE) modeling has already been customized to serve in most mechatronic development approaches, it is one of the best-known digital models in product development. Even though, the pure usage of CAD models does not support the development of the overall technical system and thus is not yet to be equaled with Model-based Systems Engineering (MBSE).

Despite this misunderstanding, CAD and CAE are some of the main technologies for mechanical and electrical engineering, which are both part of the discipline specific design modeling. The digital models created in CAD and CAE show a structural description of one or more system elements and solve a specific physics problem by the appropriate simulation models and (mathematical & informatics) algorithms. If there exists a profound MBSE integration approach as part of the entire technical system architecture and design development, the CAD assembly generation, e.g., should be driven by the system architecture. The CAD assembly can then use the appropriate parameters defined for the overall system and can even be directly derived from the existing system models. In order to achieve this, there have to be specific model generation rules, that allow the implementation of given specific parameters of the parameter models into parametric CAD templates (compare CAD template technology in Chap. 7). When the foundation is automatically generated in such a template-linked way, the engineer can focus on the specific design aspects and optimizations. Modified physical, logic and technology parameters can then be used to update the system models and to perform automatic technical system design evaluations and consistency checks.

As part of his PHD thesis Köngis [20] researched intensively on MBSE template technology at the chair of Industrial Information Technology of TU Berlin: he extended CAD template to a full system development prototype platform by combining SysMT, a kind of SysML extension for active system architecture, design and simulation, within CAD, CAE and data management environments. This prototype was piloted successfully at Daimler in two engineering cases at the R&D center in Stuttgart immediately before Daimler decided to switch from the Dassault Systèmes CATIA solution to the Siemens Digital Industry Software solution NX. Figure 21.9 exemplary depicts this integration.

Examples of Transforming Step-wise Towards MBSE

In the development of an electrical bike, a mechanical engineer wants to design the layout of the drive train. The system engineer integrates a CAD model of a supplied motor that fits the system requirements for minimal and maximum torque into a CAD assembly for the technical sub-system drive train and starts the conceptual mechanical system architecture of the motor, the gearbox and the chain-drive. With the possible chain length options, the system engineer can update the maximum and minimum wheelbase and, therefore, updates this parameter of the entire drive train system model. Such drive train system parameters can then be used directly in simulations of specific chain drive dimension and critical sections.

When 3D geometries of system components are evaluated with the help of such driving system parameters with respect to certain product, sub-system attributes such

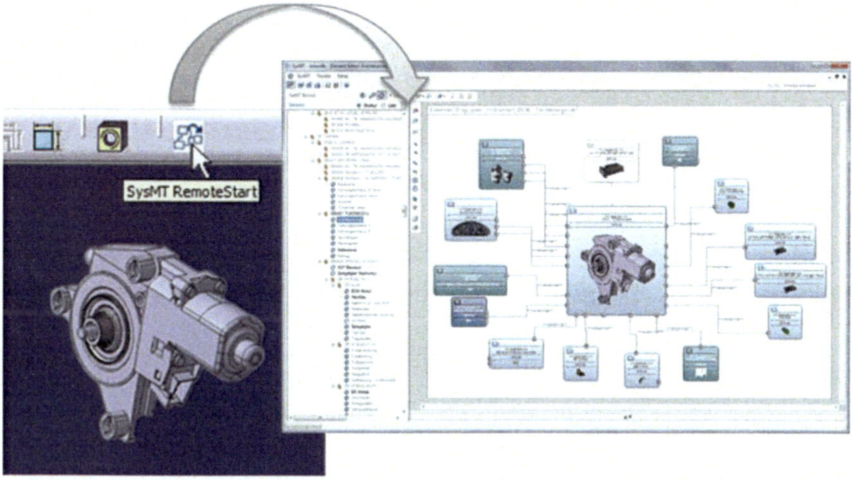

Fig. 21.9 Coupling of CAD/CAE and SysMT from Königs [20, p. 155]; the CAD component gets linked to the SysMT diagram which drives parameters in both tools

as durability, traditional 3D CAE, or specific 1D CAE solutions offer a promising engineering routine approach. The traditional CAE methodologies can use and optimize the parameters of (sub-) system elements concerning specific target durability behaviors such as critical component stress or strain. This kind of influences that have been captured through the SEA development capability can then be combined with other interaction types, to define the forces, torques and flows that affect the individual system element. Therefore, CAE models, e.g. FEA or CFD analysis, can be generated, updated and evaluated automatically for most parts of the system when MBSE is fully integrated into the system development (please compare specific CAE approaches in Chap. 10).

Examples of Transforming Step-wise Towards MBSE

The turbine of an aero engine has been designed based on fuel economy, heat performance, NVH and durability targets defined by airplane companies in the beginning of the project. As a prototype turbine is expensive and takes considerable time to get manufactures, the validation shall be performed model-based. Based on system models defined as part of MBSE aligned environment conditions (i.e. through SEA) have to be determined. The appropriately designed turbine model is converted and enriched towards a full CAE model in order to be simulated in the environment context through a specific CFD analysis. As the simulations might indicated, that the turbine needs a slightly higher fuel-mix ratio to fulfill the above-mentioned target spectrum, the parameters are adapted in the simulation, get directly updated in the system model and thus are usable by all involved technical system development domains.

While MBSE focuses on full system lifecycles, the lifecycle phase *system manufacturing* constitutes an important, but often neglected aspect of the overall product creation. When manufacturing systems are considered as an external system in the SEA development capability, the manufacturing system aspects have to be considered early on during the technical system development. The manufacturing system itself can also be developed with the help of MBSE. When the states and parameters of the manufacturing system are captured and maintained in system models, they can be used in CAM and CAPP activities for the manufacturing system developed with MBSE. When the system of interest is not a manufacturing system, the parameters can at least be used to optimize materials and geometry of the components in sense of *design for manufacturing*. Additionally, the system requirements can be used, to define manufacturing parameters, such as maximum temperature for injection molding or maximum roughness of the surface for erosive techniques.

Examples of Transforming Step-wise Towards MBSE

After having designed gas turbine mechanical system, its casing shall be manufactured. For a given gas turbine with its major systems rotor, blisk and turbine blades, various core parameters such as requirements for the roughness and thickness of the casing have to be captured in system models as well as GD&T (Geometric Dimensioning and Tolerancing) of the bearing support. The CAPP-CAM system takes such technology and part manufacturing information parameters and calculate the optimum process (incl. work holding fixture arrangement) and tool path generation for the machining of the casing. The tooling and its parameters can then be fed back into the system models of the manufacturing system to further optimize and maintain them, e.g. through the development of *digital twin technology* (and their underlying operational digital shadow data) based on the system models.

Manufacturing operations with specific machines should be transferred to the overall production system processes for the integration of multiple machines as part of the overall factory operation. Here, the interaction of CAPP and MBSE has to be considered in specific ways. CAPP defines processes for full factories and individual production lines for the manufacturing of products (compare Chap. 9 and 15). As with CAM and the individual machine system, the plant can be regarded as one system that can be developed with the help of MBSE capabilities. Some requirements can directly be linked from the system requirements of the manufactured product and then linked, to allow plant adaption for changed high-level requirements, such as the maximum production number per year or used materials, which influence the machine suitability for a given production order and product spectrum. In that way, CAPP can even profit from MBSE when the system of interest is not the entire factory. MBSE can benefit from CAPP by using the input of the available machines, to recommend the application of various machining and testing methods back to the production engineer, process planer and for the operation production planer.

Examples of Transforming Step-wise Towards MBSE

As machining of the casing from the previous example cannot be realized with one single machine only and since manufacturing of the additional components of the gas turbine should be aligned with the casing, the full factory has to be managed and treated as a system. For CAPP system models of respective manufacturing sub-systems such as in-plant logistics, material flow and human worker safety are to be linked to create system models for full factory operation. Overall system linkages and integration drive manufacturing of the full product.

One of the biggest inhibitors to use MBSE for collaborative engineering is the missing shared language for various stakeholders [29]. One feasible way to address this topic is the usage of 3D visualization. DMUs (Digital Mock-Up) have been used for years to allow discussions on the physical aspect of product, such as the dimensions or positioning of components and assemblies in the context of the entire product or factory environment (compare Chap. 12). DMU solutions can be used to visualize system elements and their relations in an "engineering understandable way", also for other stakeholders such as buyers or cost estimators. DMU solutions, therefore, can be used as basis for discussion and if properly integrated into an overall MBSE approach, e.g. as input medium to illustrate technical system models. The results of overall system type discussion, assessments, solution brainstorming and evaluation can be forwarded to system models and thus be integrated into the further system design. FMUs (Functional Mock-ups) leverage this approach by including interaction possibilities (see details and technologies in Chap. 10). All kinds of stakeholders, from customer to management, can interact with the current version of the system architecture core design and thus give their expert feedback in a comprehensible and understandable way. Such valuable input, e.g., can be integrated into the system requirements models and used for further refinements. The FMU can even be used for first system concept validation tests to drive system architecture completion as basis for the technical system design.

A supporting approach is the application of extended reality, such as VR or AR (compare Chaps. 13 and 14). This allows the user to interact with the system or product in early phases without the need to create physical prototypes (see Fig. 21.10). When properly combined with FMU, e.g. with the help of *Smart Hybrid Prototyping (SHP)*, the application can even include haptic interaction and thus gives the user an even more immersive experience. The interconnection of models can support these approaches, as they can be directly derived from the existing system models. Other

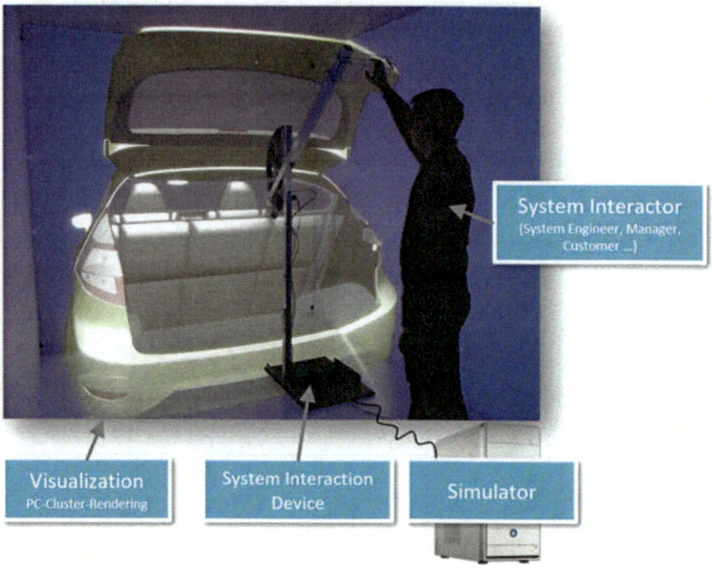

Fig. 21.10 Smart hybrid prototype of a vehicle tailgate, which uses connected, models to create an immersive user experience for the entire system (presented in Auricht et al. [31])

approaches to combine extended reality and MBSE are for example the usage of system models to define the VR scenes. Mahboob et al. [30] used behavior models created in SysML to update parameters of the VR visualization. The user can use an interaction device to update the SysML models, which are synchronized with a Physics engine and the VR-Software. In that way, the behavior does not have to be modeled in the VR software separately and can be updated more easily through the SysML models.

All of the approaches for the interconnection of MBSE and VPC solutions require the human to either create models or interconnect them beforehand in the overarching MBSE approach. As with general VPC technologies, the application of AI can leverage most aspects of MBSE. It can be used to support mechatronics design [32], functional decomposition [33] or the actual model generation from patterns [34]. In that way, stakeholder with little expertise in MBSE can be supported to participate in the development of complex systems. Experts in that area can benefit from AI as well, as it can reduce repetitive tasks to a limited degree and thus leverage the creative thinking aspect of the design. This supports the innovation process and thus the creation of better products. A core group of MBSE architects and system engineers, however, need to be in charge to oversee and approve such solutions sets in order to avoid the creation of MBSE induced risks to the overall technical system development, validation and verification. Consequently, major investments into such MBSE capabilities are unavoidable and need to be pro-actively driven by Senior and Middle Management in companies. This remains, today, a major challenge since the

MBA management schools have educated several manager generations rather into cost and efficiency driven business behaviors rather than into long-term capability investments for future business operations.

21.1.7 Examples of New MBSE Methods and Tools

With the introduction of MBSE, new forms of developing capabilities are at hand. While the application of these rather new overarching model-based development capabilities is not yet properly integrated into industrial practice, these methods and tools are continued to be developed without specific assignment by or input from applying companies. Even though, the examples of MBSE disciplines methods, tools and technologies presented here, shall be mapped to show the possible MBSE tool support for the development capabilities of the future.

When speaking of methods and tools, the definitions of Martin [35] shall be considered that have been the foundation for the survey of Estefan et al. [5]. These define a "method" as a collection of techniques to perform a specific task, so the basic elements of processes. Tools are then used to support these methods, but require certain expertise of the (engineering) user. However, oftentimes this is driven the other way around: based on a bundle of developed and offered MBSE tool functions *Digital Technology Provider (DTV)*—compare Chap. 19—start to advise implicit method competence as part of their service offering very closely linked to the capabilities of their own MBSE tool functions. If companies do not allow themselves to establish core knowledge about their own internal MBSE content framework and associated methods they will not succeed in reaching the next level of engineering capability and intelligence.

Typical tasks when applying MBSE can be seen in the V-Model presented by Buchholz et al. [36] and shown in Fig. 21.11. In addition to the specification and expectations tasks (incl. neutral functional type of analysis) of MBSE on the left

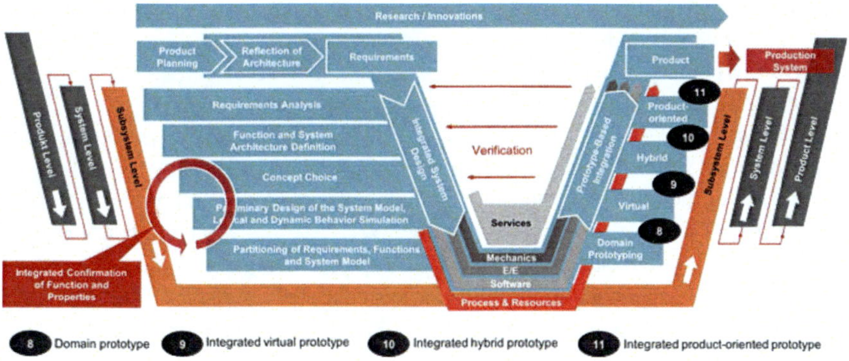

Fig. 21.11 MBSE process of Buchholz et al. [36]

	MBSE Type 1 Systems architecture	MBSE Type 2 Systems modelling	MBSE Type 3 Systems integration & traceability
Goal	Establish the in-scope composition of the overall system architecture (What?)	Gain the understanding about the system characteristics, functions and selected behaviours (What exactly, how?)	Validate and verify the overall system behavior, internal sub-systems integration and cause-effect intelligence (Ready for safe operation?)
Development approach	• Use of system meta modelling language such as SysML (various diagram types) • Indetification of and rationalefor sub-systems and system components • no further systems interaction modelling	• Use of detailed system modelling types with respect to functions, behaviors, information networks, logic control + casual effects • engagements of structural (models like SysML), function models, control models, behavior models with multi-physics, mathematical and time dependent stimulations	• Engineering case oriented systems interaction modelling to prove critical, system parameters according to pre-determined test and legislative boundary conditions • Use of interconnected RFLP moduls and of modul object and parameter tracelinks with extensive trace link rationale and intelligence
Required MBSE capability	Simple SEA + Medium SDD	Full SEA + full SDD + medium SIM	Full SEA + full SDD + full SIM + best possible SLE

Fig. 21.12 MBSE development framework with three major MBSE styles

side of the Engineering-V it shows in which steps the model-based systems prototyping as part of validation and verification should be applied on the right side of the Engineering-V. Some of them shall be used for the exemplary introduction of new MBSE methods and tools according to the MBSE development framework as described in Fig. 21.12.

The MBSE development framework of Fig. 21.12 differs three different types of Model-based Systems Engineering. MBSE Type 1, *Systems Architecture*, concentrates on the determination of the *systems architecture*. This type of MBSE focuses in addition to the normal engineering development approach on the determination of the overall composition of the (technical) system as guidance and steering of all engineering activities. It applies MBSE capabilities of (simple enough) *Systems Environment Analytics (SEA)* and concentrates on (medium intensive) *Systems Definition and Derivation (SDD)*. The development approach is characterized by SystML type of system meta modeling and does not include any detail systems modeling. This type of MBSE concentrates on clarifying the "WHAT?" question and hence determines the major system elements. The MBSE type 2, *Systems Modeling*, is targeted towards a gain of understanding about the systems characteristics, functions and selected resp. partial behaviors. Therefore, it takes into consideration a wide mix of different type of system describing models, from structural decompositions, functions, behaviors, multi-physics and mathematical equations incl. different degrees time dependent simulation set-ups.

It applies MBSE capabilities of (full) *Systems Environment Analytics (SEA)*, (full) *Systems Definition and Derivation (SDD)* and, in addition, (medium) *Systems Interaction Modeling (SIM)*. This type of MBSE concentrates on clarifying the "WHAT EXACTLY is part of the system and HOW will it work?" question.

The MBSE type 3, *Systems Integration and Traceability*, is aims at validating and verifying the *overall system behavior and interaction* as well as scrutinizing the oftentimes implicit *system cause and effect intelligence*. In order to achieve this MBSE goal, the approach has to explicitly prove evidence that the overall systems behavior and interaction comply with engineering cases, derived test cases and official legislative boundary conditions. Thus, MBSE type 3 has to carry out engineering tasks of interconnecting different type of system models (such as the RFLP types of models, which will be explained in the following, see also Fig. 21.14 later in this chapter) in order to establish tracelinks which can be used for systems reasoning for all activities within MBSE. MBSE type 3 applies MBSE capabilities of (full) *Systems Environment Analytics (SEA)*, (full) *Systems Definition and Derivation (SDD)*, (full) *Systems Interaction Modeling (SIM)* and best possible *Systems Lifecycle Engineering (SLE)*. The type 3 of MBSE concentrates on ensuring the "ENGINEERING READINESS for safe systems operation".

The increasing degrees and intensities of the three MBSE development types show that companies and individual engineers need to become aware of the differences in model engagement intensity in order to become proficient enough in model use. As a general guidance Fig. 21.13 shows one important separation of two different models' spaces of MBSE (there might exist additional ones). *Functional models* are at the

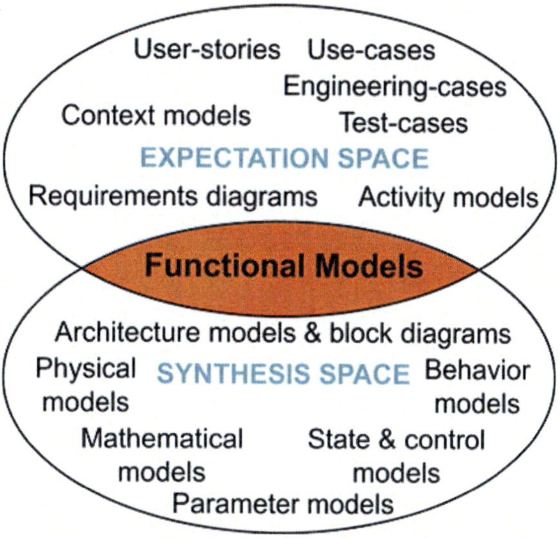

Fig. 21.13 MBSE model spaces (expectation versus synthesis)

direct interface between the *expectation space* and the *synthesis space* of MBSE models. Whereas the *expectation space* is mainly the focus for SEA (Systems Environment Analysis), the *synthesis space* becomes essential for the *SDD (Systems Definition and Derivation)*, the *SIM (Systems Interaction Modeling)* and *SLE (Systems Lifecycle Engineering)*.

As exemplary methods and tools used in MBSE, the *ARChitecture Analysis and Design Integrated Approach (ARCADIA)* in combination with the Capella tool and the functional modeling as part of the *Requirements Function Logic Physical (RFLP)* data structure with the 3D Experience platform as tool shall be described.

At first, the ARCADIA method shall be explained. It has been developed by the Thales industry group between 2005 and 2010 through an iterative process involving operational architects from all the Thales business domains. Since 2018, Arcadia is registered as Z67-140 standard by AFNOR, the French national organization for standardization. It is an approach to analyze and develop the architecture of a target system.

It consists of 5 primary steps that are depicted in Fig. 21.14: Operational analysis (1), System Need Analysis (2), definition of the Logical Architecture (3), definition

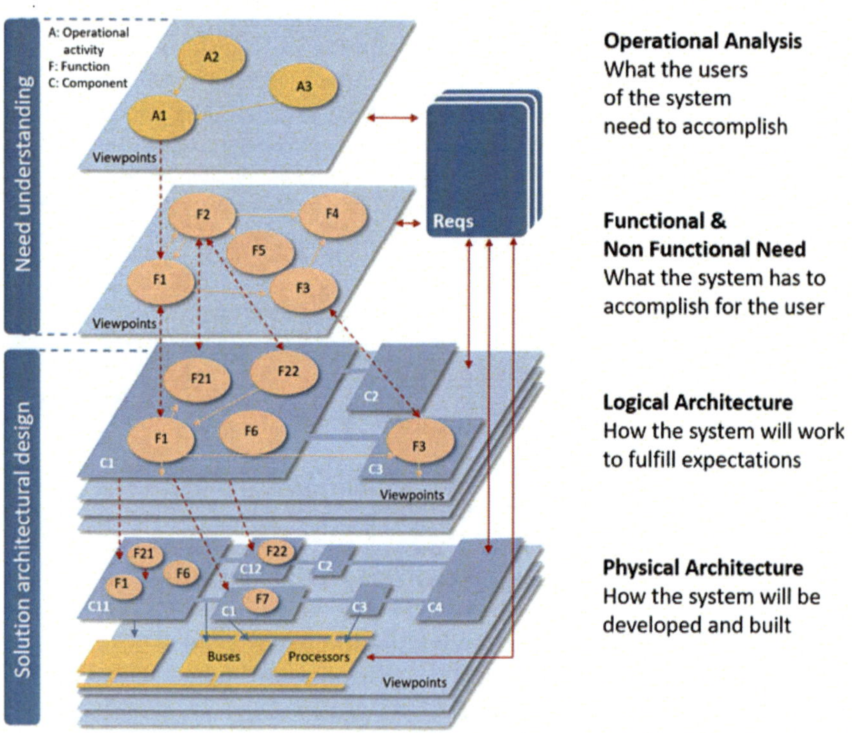

Fig. 21.14 Five steps of the ARCADIA method: operational analysis (1), System need analysis (2), definition of the logical architecture (3), definition of the physical architecture (4) and definition of the product breakdown structure (5). Adapted based on [38]

of the Physical Architecture (4) and definition of the product breakdown structure (5) [37]. Compared to the 5 MBSE development capabilities shown in Fig. 21.5 the ARCADIA method is primarily positioned in the MBSE capabilities *SEA (Systems Environment Analysis)* and *SDD (Systems Definition and Derivation)* and only to a limited extend in the *MBSE capability SIM (Systems Interaction Modeling).*

The first step of this method focuses on the *Systems Environment Analytic (SEA)* development capability. Operational entities and operational activities (A in Fig. 21.14) are collected and modeled. An example is the operator of a system or another system that the target system interacts with. Functions (F in Fig. 21.14) are defined for each of these entities and linked to each other. These functions are then used to define a logical architecture of the target system, which mainly describes the allocation of the defined functions and possible components for their implementation. This step is mainly used to describe architecture drivers. The physical architecture then defines physical components that include the components of the logical architecture. The last step of the product breakdown structure is mainly used for contracting and collaboration, as the physical architecture is clustered here in integrated components or assemblies that rely on each other or can be produced together and thus might be required from one subcontractor.

As a model-driven approach, ARCADIA has been developed with a good tool-support in mind in its conception. The open-source tool Capella has been developed to support this approach. As can be seen in Fig. 21.15, it guides the user through the steps of the ARCADIA method. Each step is broken down into activities for which specific diagrams are created to describe the current scope of the system. The

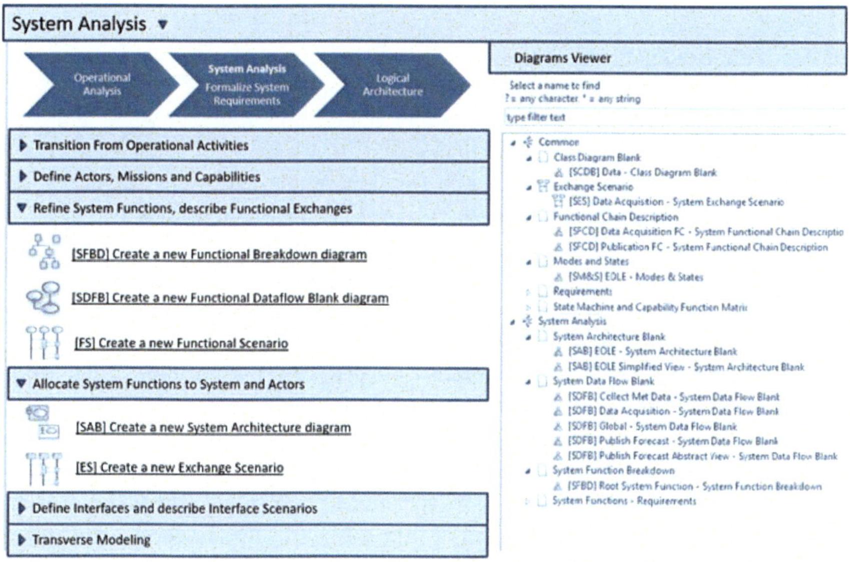

Fig. 21.15 Activity-browser type like the Capella tool, based on [37]

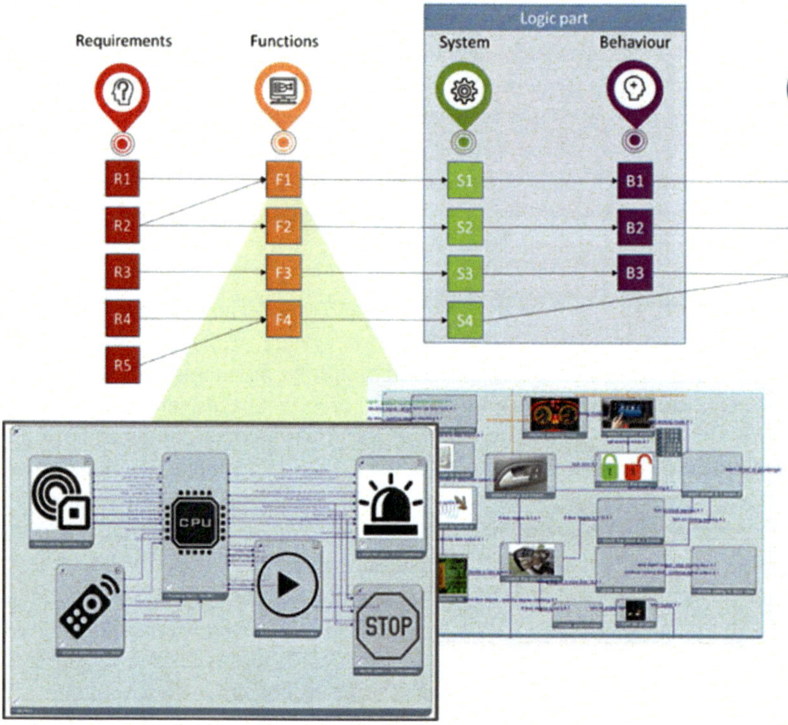

Fig. 21.16 Orientation of function modelling in a RFLP framework

element in these diagrams are integrated into an overall model of the system that can be used for further synthesis and analysis of the system.

A method used in nearly any methodology and is often used with other methods is functional modeling. Functional modeling (Fig. 21.16) is recommended to be used in projects to develop new products, achieve better modeling results, as well as to change or improve old product concepts, to develop collaboration systems [39].

The functional model is of great importance in the product development process. Regardless of the complexity of the product, the functional model allows clarifying the activities of the future product at the initial stage of development. The main idea of functional models is to describe future actions of the system based on its neutral (non-yet solution oriented) functions and to answer the question "What is the system doing?" In addition, the functional model provides guidance for the further development process and the search for solution implementation.

There is currently no holistic functional modeling method and each discipline has its own approach. Functional modeling in mechanical engineering is mainly driven by approaches defined from Pahl and Beitz [40]. In Electrical Engineering *Multilevel Flow Modelling (MFM)* is an exemplary approach to modelling the functions of complex processes, which corresponds to the development of electrical circuits. In

Software Engineering the *Rational Unified Process (RUP)* approach developed by Kroll and Kruchten [41] uses models to describe the functionality of the system.

The 3D Experience Platform from Dassault Systèmes has included the functional modelling aspect as one of the RFLP data model foundations. As in ARCADIA, it uses the functional model as foundation for the logical architecture of the system. It is depicted in Fig. 21.17. Other Digital Tool Vendors (DTVs) offer similar kind of solutions, such as Siemens Digital Industries Software with the integration of *System Modelling Workbench for Teamcenter* or IBM with *Rational Harmony* and *Engineering Systems Design Rhapsody—Developer.*

With these example methods and tools, it is reasonably possible to develop modern products in a flexible and yet guided way. However, such methods still have many limitations which are not yet solved and are still subject for many ongoing research and development projects:

1. As of today, there does not exist yet any scientific foundation on how to best integrate the different types of system related models such as functions, structural building blocks, logical and control interactions models, signal and information flow models, behavior models and process models. Consequently, the overall systems integration and traceability type 3 MBSE needs to entertain various sub-integration layers with limited reconciliation options with the tradition CAE model types.
2. Functional modeling is still understood highly controversially across the technical domains. In addition, the term function is quite loosely used within industry

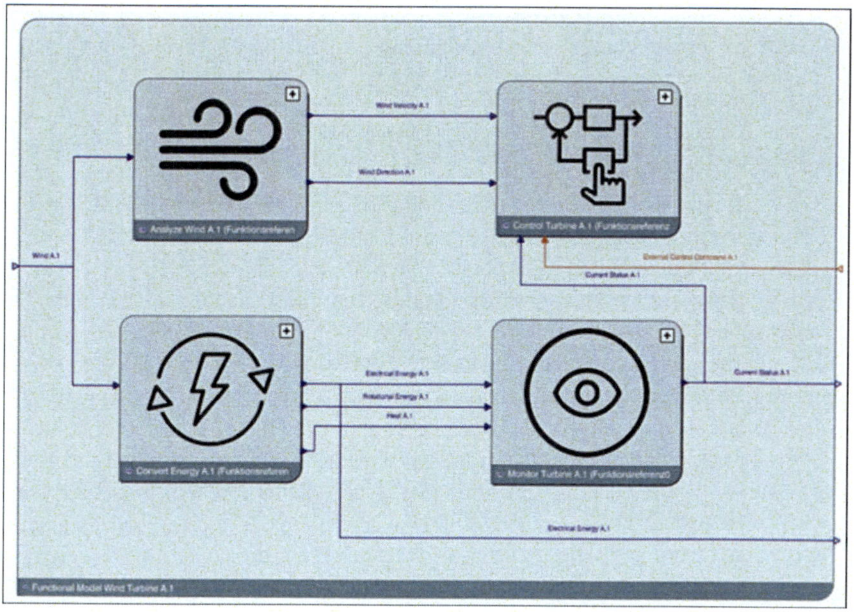

Fig. 21.17 Exemplary section of a functional model in the 3D experience platform

and in many cases are miced up with product features and sellable customer options. This makes it difficult to establish functional modeling (inkl. functional networks, intended but still neutral product property or behavior, function interactions etc.) as a full and precise MBSE method across departments, companies and industry branches or system boundaries.

3. So far offered MBSE methods concentrate on artifacts and objects to be modelled, rather than on the competence and knowledge to be applied to effectively carry out the methods. This, oftentimes, is DTV (Digital Technology Vendor) driven in order to be able to offer and sell an IT application for it. Industrial companies have not yet picked-up intensively enough this competence side of MBSE. In many cases they are overwhelmed to define their own MBSE technical framework before they become active with the appropriate digital tool and application landscape.
4. New MBSE practitioners have difficulties to work abstract enough in order to truly follow a top down system working style. As young discipline, MBSE currently employs a rather disjunct series of non-directly related engineering deliverables (requirements, functions, logic & control behavior, system simulations by CAE attributes). For MBSE synthesis proficient method are still missing to establish a stringent system driven way to call-up and integrate partial CAD system design and CAE system analysis model capabilities. System meta modeling (like SystML) is not yet grown up enough to find a seamless way down to neutral system behavior modeling and simulation (1D CAE, e.g. based on Modelical physics models or Simulink plat models) and to the subsequent FMU (Functional Mock-up) type of cross discipline system analysis with coupled 3D CAE co-simulation solutions.
5. MBSE as socio-technical working challenge still misses aggregated interaction models for Managers and decision makers. Proposals for such environments have been made already by research institutes but industrial companies did not yet pick them up since they still expect MBSE technology deliveries along that line by Digital Technology Providers (DTVs). This is highly disputable since DTVs usually have limited understanding of and interacting with operational Engineering Leaders and Management in industry. Unless this MBSE white space is not filled, MBSE will not be able deliver to its promises!
6. MBSE still suffers from awkward and limited data and model openness, linkage and exchange across the different digital technologies and applications within the internal portfolios of the various Digital Tool Vendors (DTV), and even more with respect to the digital application portfolios across DTVs. Due to the increasing need to deliver improvements for it the Digital Tool Vendors have started to embrace the term *Digital Thread* for this capability and have started to overhaul their portfolios accordingly. True openness which is desired for successful and cost efficient MBSE, however, remains difficult and needs much more push from applying industry, as they represent the technology receiving and using side for system driven Virtual Product Creation (VPC).

21.1.8 The Challenge of Integrating MBSE into Industry

MBSE is not yet broadly integrated into industry. Huldt and Stenius [22] concluded that the main hurdles for an effective MBSE introduction are cultural issues, clear value indicators, trained personnel, management support and a steep learning curve. This shows that especially the CMM development capability has to be improved for a proper integration into industry.

The research project AdWiSE [42] elaborated the current state of SE and MBSE and addressed similar issues. Additionally, the Integration in the existing IT-Infrastructure, the selection and adaption of existing methods and the usability of modeling tools and languages are mentioned as hurdles.

Looking at the publication landscape, most MBSE publications focus on requirements engineering and system modeling with modeling languages like SysML or OPM.[7] Digital Technology Vendors also mainly focus on these aspects and present their tools as full MBSE suites. This leads to non-converged definition of MBSE and some confusion, what MBSE actually is [23].

Cameron and Adsit [43] elaborated in a survey, that 88% of system data is still captured through documents and only 11% through models. This slows the integration of MBSE even more down, as the benefits raise, as more models are integrated in that approach.

Schmidt and Stark [29] scanned different literature regarding challenges for the collaboration in MBSE and listed a shared language/vocabulary, management of interactions, exposition of data, consistency of data, large collaboration networks and system compatibility. While all of them are mainly collaboration driven, this influences the integration of MBSE into industry as well, as MBSE is an interdisciplinary approach and thus depends on collaboration. A feasible means to address these issues is the usage of visualization as well as domain specific tools. Approaches like OSLC and SpecIF try to address this tool combination and thus allow a better collaboration.

With a better integration of the MBSE development capabilities, these hurdles could be addressed. Especially the CMM development capability has to be addressed for a better integration into industry. Here it is especially important to use people-oriented approaches for the MBSE knowledge transfer. In that way, MBSE can be introduces more intuitively and thus is easier understood and accepted. A feasible approach that has been tested in a set of workshops is the Design Thinking method in form of an interactive format, where the knowledge transfer is oriented on the stakeholder's problems and supported by interactive prototypes. With this approach, the most relevant skills for VPC can be transferred in an easily understandable way.

MBSE as a new discipline of Virtual Product Creation needs thorough preparation as a *new technical competence* and as *highly connected information network of digital applications and databases*. Starting the MBSE journey as top down *Systems Engineering* program only will fail since it cannot provide enough anchorage in today's

---

[7] *Object Process Methodology (OPM)* is a conception modeling language and methodology for capturing knowledge and designing systems, specified as ISO/PAS 19,450.

wide and complex *Virtual Product Creation* method, data model and tool application world. Treating it simply as an additional set of highly aggregated but isolated digital tool and process environment does not work either. The only way forward of successful implementation and integration of MBSE is to establish step by step the MBSE capabilities as explained and shown in Fig. 21.5 within a chosen and justified *MBSE development framework* (compare MBSE working styles in Fig. 21.12). Without help from outside most of the industrial companies will have difficulties to establish such new ways of working and might also be limited to establish the true value of *"technical system models"* and *"systems integration and traceability"* within their organizations.

Company internally, significant investments are necessary and justified to ensure system model technical correctness, engineering applicability and engineering usability in a balanced way. The author experience in teaching MBSE for more than 10 years indicates that most of the engineers are overwhelmed in the beginning with the following aspects of MBSE:

- Unclear combination of traditional digital engineering methods with new MBSE type methods. E.g., the traditional 3D system space and layout methods such as 3D space allocation, check & clearance analysis and proximity/neighbor topology analytics oftentimes do not get coupled with the MBSE SEA (Systems Environment Analytics) methods such as intended interaction modes, connectivity essentials and operational safety considerations.
- Within the interactive MBSE Design Reviews engineers use the individual MBSE diagrams and model's insolation instead of overlaying them with each other and establishing and explaining tracelinks due to cause dependencies between them.
- Limited knowledge how to recognize and efficiently establish tracelinks between different type of MBSE artifacts and objects such as requirements, overall product/system functions, CAE-type behaviors, CAD/PDM structural models, logical interactions and software function calls in order to reach an optimized synthesis of the overall system performance.

Since the need to collaborate on system development with partners and suppliers increases steadily companies are forced to agree on system modeling standards not only within but also across the company. The prostep ivip association is an international association with more than 180 members headquartered in Darmstadt, Germany, has committed itself to developing innovative approaches to solving problems and modern standards for product data management and virtual product creation. In its white paper "*Collaborative Systems Engineering on the basis of Engineering IT Standards*" from 2019 (compare [44]) it provides a guidance of MBSE related data and model standards across the Engineering-V (see Fig. 21.18).

Industry is dependent on MBSE data and model exchange standards since different IT implementations of MBSE IT environments as part of the overall Virtual Product Creation (VPC) architecture cannot guarantee open API and linking mechanisms like REST to be functionable across company IT environments. In the future it will become possible to use MBSE engineering space within specific platforms of Digital Technology Vendors (DTV) and possibly also across digital platforms with the help

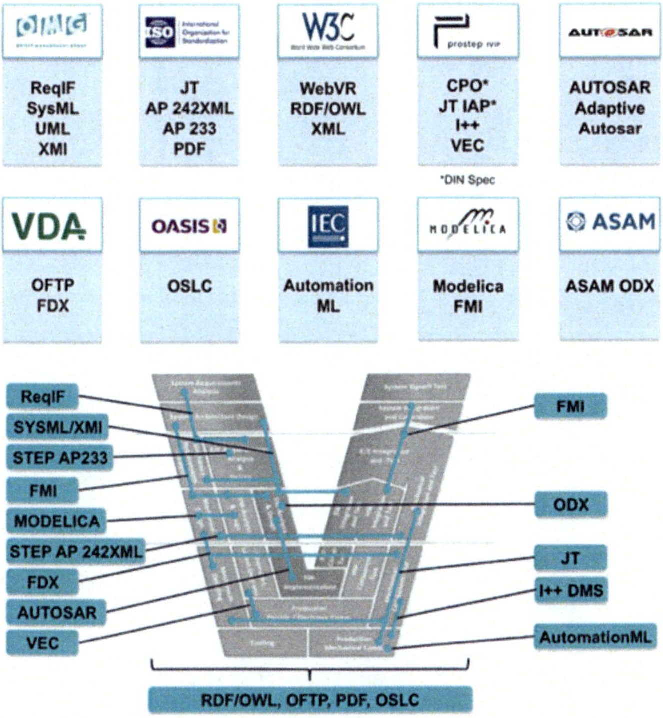

Fig. 21.18 Fact sheets by standards organization and MBSE relevant IT data and model standards under consideration in the V model [44]

of federated data infrastructures and cloud services such as GAIA-X in Europe. It remains, however, still unclear which type of industries and digital technology companies will take the lead to establish such type of extended data and model community for Model based Systems Engineering as part of the overall Virtual Product Creation data stream.

21.2 Data Engineering and Analytics (DEA)

Another key discipline of future Virtual Product Creation (VPC) is called *Data Engineering and Analytics (DEA)* and has to be integrated into the VPC framework and aligned to MBSE capabilities. DEA is the discipline of designing data environments for use in technical systems. VPC has the important new role to create new data enabled intelligent technical systems. Future digital engineering and VPC, therefore, requires new comprehensive skillset—especially data literacy (cf. [45])—for DEA to bring Engineering and Operations closer together. Major VPC technologies (see Chaps. 7–16) such as PDM/PLM/BOM and AI will get leveraged by the following

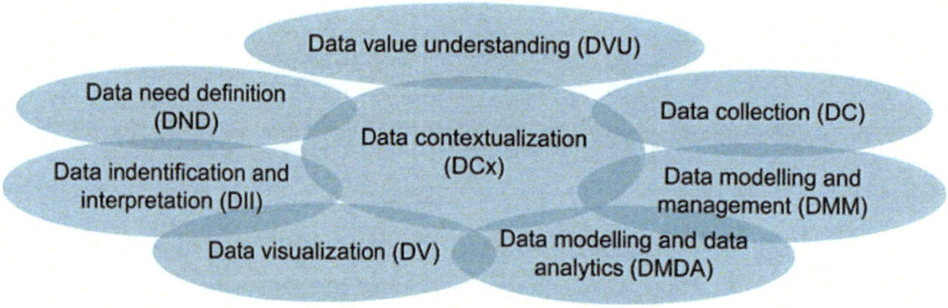

Fig. 21.19 Overview of DEA disciplines

eight DEA disciplines but are no longer sufficient to accomplish the needs and power of DEA. Figure 21.19 gives an overview of the eight disciplines of DEA as well as their relation and intersection. The sub-disciplines forming DEA are explained in the following sections.

21.2.1 Data Value Understanding (DVU)

Industrie 4.0 and the Internet of Things (resp. Everything) are leading to a sharp increase in the volume of data. These data volumes are getting increasingly analyzed using advanced data analytics techniques, creating a value creation pipeline with data as a source. (cf. [46, p. 11]). "As a result, heads of maintenance departments, and consequently business managers, can start to benefit from the value of data. Effective analysis of data enables equipment manufacturers and machines users to deepen their understanding of equipment, processes, services, employees, suppliers, and regulator requirements" [47, p. 77]. The discipline of *data value understanding* (DVU) aims at providing a framework to understand the value of data. It is important to note, that value is not equal to price. DVU is, therefore, not suitable to explain the pricing of data on the free market. This also implies that DVU will not compute numbers for "value of data". This is comparable to the value of PDM systems: There is no direct value stream in EUR (yet!) coming from the usage of a PDM system. Nevertheless, it brings value to the table by enhancing product quality, enabling handling of complex products, and providing traceability along with the PLM. As there are already data-driven business models in place—especially in the internet "search-present-broker" digital economy field with players like Google—it is clear, that data has value and can be used to enable data-driven business models. Still, a framework for the qualitative evaluation of industrial data within companies driven by mechanical products is missing. Therefore, a lack of understanding regarding the value of industrial data is present. Within this book, the value of data is understood as the potential of data to drive economic value. The higher the potential, the higher the value.

DVU is needed to be understood as a new discipline because other disciplines have gaps regarding the evaluation of data value for data-driven business models. For example, the field of data quality management aims at evaluation and improvement of data quality (e.g. [48]), but leaves the economic perspective aside. The field of business administration cannot deal sufficiently with the fact that the value of data often cannot be computed with quantitative methods. The field of data science sees the value of data in fulfilling a given business case and consequently provides different methods to increase the value of data (e.g. data fusion methods described in [49]). Nevertheless, there are approaches in all of the disciplines mentioned that are useful for assessing the value of data and thus for the new discipline of DVU. However, these need to be combined to consider the value of data holistically so that the full potential of new data treasures can be leveraged.

To estimate the value of data, the following core activities are essential:

1. **Data overview**: A systematic overview of available data is the start of any evaluation of data. This also requires the first categorization of data (e.g. master data, metadata, sensor data). The discipline of *Data collection (DC)* is needed at this stage.
2. **Quality estimation**: The quality of data needs to be estimated (not deeply quantified) to get a first approximation for potential sources of high-value data. A detailed quality assessment is neither necessary, nor possible at this stage of the data value estimation process, because the quality of data depends significantly on the use case, which is unknown at this point.
3. **Internal business and market opportunities**: Data with potential of high value needs to be mapped to potential usage within the current business processes and/or for new business models. By mapping data to the business perspective, the value of data becomes clearer and more quantifiable.
4. **Detailed value check**: While having identified potential business cases as well as potential high-value data, a detailed check on the data value is necessary. This is done by applying the methods and tools of the *data need definition (DND)* discipline (see Subsect. 21.2.2). As a result, there is a clear understanding of the value of data for a given business case.
5. **Realizing data value**: If the detailed value check finally results in an attractive business case, the business case itself needs to be realized using the methods and tools of data science, product creation as well as the other disciplines of DEA described in the following sub-chapters.

21.2.2 Data Need Definition (DND)

Data need definition (DND) describes the discipline of identification and formalization of data required to derive insights from data. Without a clear understanding of which data is needed, data collection and identification are nothing else than a best guess, try & error cycle. Data need definition (DND) needs a given use case to fulfill.

There are a variety of methods for describing data (e.g., [50–52]). However, to date, there is a lack of systematic approaches to systematically define data needs in the early phases of product development. Approaches currently used in practice are mostly data-driven. This requires, however, the existence of data, which can be problematic especially within the context of developing new, non-existing products and technical systems. This is, e.g., the case during the development of new product and smart service components since logically no field data exist of not yet built, tested and/or operated products. Therefore, insufficient data layers can be identified late in the product development process. The discipline of DND deals with the use of domain knowledge for an early, knowledge-driven definition of data requirements.

To enable an early definition of data requirements, the following activities are necessary: First, those domain experts must be identified who can contribute to the definition of data requirements through their expertise from the domain and for the use case under consideration. Then, systematic use of domain knowledge is required to describe which information the use case related data must contain. Once this is captured, *data needs* can be derived from the *information needs. Data need definition* should be carried out in the right system context: physical realized systems, as well as company-wide information systems, are a potential source for data. The information within the data is used to generate insights useful to the system. Here it is important to define a shared meaning of conceptual elements [53, p. 73f.] and define the system functions consuming the data [52, p. 160]. With respect to business roles, [54, p. 242] it is recommended to establish the appropriate „data definition owner“ role. “Theories of data modeling are useful for data definition” [55] and, therefore established data modeling techniques (examples can be found in [56–58]) should be used in the DND discipline.

The example of predictive maintenance for technical products and machines clearly shows that the early definition of data requirements is essential for the success of predictive maintenance projects. Already during product development, serious considerations are critical to determine which elements of a system are relevant for maintenance and which health indicators can be used to determine the condition of the system. This usually requires the integration of sensors into physical product components. If these sensors are missing due to an incomplete or missing definition of the data requirements, the state of the product cannot be determined in a data-driven way. The sensors must be retrofitted or maintenance cannot be data-driven.

The core result of the DND discipline is a description of the data requirements. This is the basis for the subsequent data collection.

21.2.3 Data Collection (DC)

Data collection (DC) is the discipline of acquiring those data sets which have been defined by the DND discipline (as explained in Subsect. 21.2.2). The scope of DC starts with defined data needs and ends with the final disposal of a product. This means, data collection deals with extracting data from IT systems and products as well as openly available data (c.f. [59]). The data collection phase is often time-consuming. “Empirical studies have shown that it constitutes around one-third of the total project time” in manufacturing operations projects [60]. The fog, edge, cloud IT architecture discussion (e.g. [61] or [62]) is also part of the DC discipline: DC needs to create a data collection pipeline to deliver data where it is needed while saving as much bandwidth, storage, and energy caused by computations as possible. This is always a trade-off discussion because technologies like data lakes and 5G make it tempting to always centralize all data in one place—making the data collection pipeline easy to architect. While those approaches are convenient for data collection engineers, they have drawbacks of creating more load on networks, IoT devices, and back-end servers than necessary—leading to higher costs and energy consumption.

Taking the example of predictive maintenance, data collection tasks are key to firstly decide whether the analysis of the remaining life of a component or systems takes place right at the IoT device (edge/fog) or if sensor data relevant for predictive maintenance analytics is transferred over the internet (cloud on far away geographical location) or maybe over a local network at the same site (local servers nearby) to back-end data servers. This decision goes hand in hand with identifying the relevant data (the edge, cloud, local server, etc.) and developing the data pipeline to ensure continuous monitoring and remaining lifetime forecast enabling predictive maintenance of an IoT device.

The discipline of *data modeling and management (DMM)* is closely related to DC, as DC and DMM are interdependent: The collected data needs to fit the data models and needs to be manageable in volume while having good quality.

21.2.4 Data Modeling and Management (DMM)

Data modeling and management (DMM) is the discipline of structuring data (modeling), making sure that the collected data is available and that the data keeps the quality (cf. [63]) after data collection. Due to the increased data volume caused by IoT devices, NoSQL approaches like graph-based or key-value data models rise in popularity.[8] While relational data models (cf. [64]) will not go away anytime soon and will continue to have a big market share, AI is centered around big data—and big data needs NoSQL databases (cf. [65]). Nevertheless, SQL databases are still relevant for big data (c.f. [66]). Therefore, DMM differs from traditional data management (cf. Chap. 11 PDM/PLM & BOM) especially regarding the three Vs of big data: Variety, Velocity, and Volume. Additionally, the storage of raw data and its context (cf. Sect. 12.1.5 Data contextualization) becomes more important. This is caused by the fact that raw data could be used for different use cases, whereas storing

[8] Rankings available on https://db-engines.com/de/ranking. NoSQL meanwhile stands for “not only SQL (Standard Query Langugae)”, originally referring to “non-SQL” or “non-relational”, and designates databases to provide a mechanism for storage and retrieval of data that is modeled in means other than the tabular relations used in relational databases.

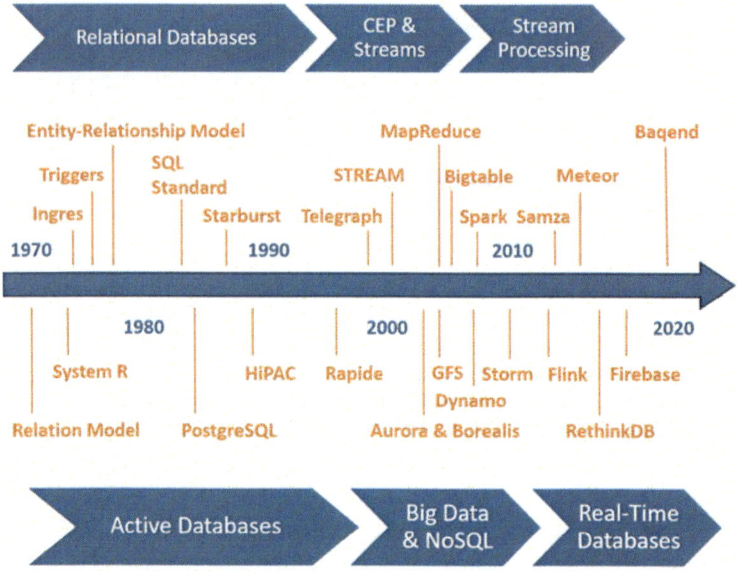

Fig. 21.20 Different classes of data management systems [67]

just analyzed data in the context of a particular use case limits the value of stored data for future applications that are maybe unknown during initial data collection. Since the database concepts available today are diverse (cf. Figure 21.20), there are also a large number of approaches to modeling data (cf., [64]). There are always new developments under way, which requires thorough expertise to decide whether they are suitable for the intended use case. For example, it makes no sense to try to manage sensor data of millions of IoT devices with a relational database approach. In this case, real-time stream data management (cf. [67]) is needed. It also makes no sense to manage a product structure with a database such as Apache Cassandra, since this database is optimized for the properties of big data and a product structure only represents small amount of data but in a well sorted relational or even hierarchical way.

Taking the example of predictive maintenance, DMM deals with the provision of data models and storage capabilities to manage data collected from IoT devices for their monitoring. In close collaboration with engineers responsible for data collection and the given data need definition, DMM engineers decide on what type of databases are used and how the data needed for predicting next necessary maintenance actions will be structured within the chosen databases. Additionally, DMM links DC and data contextualization (DCx). This is done by integrating the context of data *modeled* by DCx engineers into databases where the linking of this context to the *collected* data occurs.

21.2.5 Data Contextualization (DCx)

The discipline *data contextualization (DCx)* deals with the contextualization (semantic referentiation) of (field) data by digital model metadata of technical systems. This is an essential discipline for all other DEA disciplines. "Data contextualization becomes a base foundation for the design and operation of manufacturing systems" [68, p. 160], data-driven engineering and cyber-physical systems. Having the data context makes life easier for all other disciplines of DEA (e.g. better reuse of data possible by using it in different applications, reduce the probability of misinterpretation, or providing extra information for AutoML [69]. DCx is different from just storing metadata. It is about systematically formalizing the context of data and making use of domain knowledge that is necessary to interpret the data in a meaningful way (cf. [70]). For example, it makes a difference if the data model of a database just stores "This data set consists of integer values." versus "This dataset contains the number of test cycles of the ball bearing per day. The data was acquired on the test site Berlin using sensor model X".

DCx has two sub-fields: front-end and back-end contextualization. Front-end contextualization is about providing context-aware data analytics tools (cf. the DEA discipline DV). Back-end contextualization is about describing the context of data in a machine-readable manner, storing the context along with the data (cf. the DEA disciplines DMM and DMDA) and applying data analytics techniques if necessary to fulfill front-end data needs.

For making front-end contextualization work, related fields like UX design need to be involved. It is important to understand the context of the user and the use of context-aware data analytics tool. Therefore, a good description of the use case and the users is required as first step for front-end contextualization. Next, the data should be defined in order to fulfill the user requirements (cf. the discipline DND). The determination of "data needs" forms the interface between front-end and back-end contextualization. Back-end contextualization is in charge of providing the relevant data needed for front-end contextualization. After having established the data flow between front-end and back-end contextualization disciplines, front-end contextualization designs will get accomplished by implementing user interfaces leveraging techniques from UX design.

For making back-end contextualization to work, related fields like traditional database management, DND, DII, and DMM need to be involved. Technologies like semantic web (cf. [71, 72]), ontologies or graph databases are used. Based on the data need definition provided by front-end contextualization engineers, back-end contextualization engineers are in charge of fulfilling the data need. Therefore, the location of feasible data needs to be known (cf. the discipline DII). The discipline DMM needs to be consulted on which data models should be used to store the data. During this consultation, it is the responsibility of DCx back-end engineers to challenge the data models build by the DMM discipline regarding their feasibility to store necessary data context feasibly. Next, DCx back-end engineers carefully check whether the necessary data context to fulfill the data need definition is present. If not, DCx back-end engineers are responsible to make sure (e.g. by involving domain experts) that the needed data context is known and stored in the back-end. Last, DCx back-end engineers conduct a last check, whether data models, databases and available data context fulfill the data need described by DCx front-end engineers.

Taking the example of predictive maintenance, maintenance engineers want a dashboard showing the condition of relevant assets to plan maintenance actions. First, DCx front-end engineers describe the use case for this dashboard application and derive a data need definition. This data need definition is handed over to DCx back-end engineers. The DCx back-end engineers now check in with DII engineers to locate the data needed. After analyzing the available data regarding its feasibility, DCx back-end engineers conclude, that data context is missing. For example, there is data on each relevant component, but the data of the components are not linked in a machine-readable way and therefore cannot be automatically linked to fulfilling the data needs of the front-end. DCx back-end engineers decide to store the missing context data in an ontology. To design the data model of the ontology, they closely collaborate with DMM engineers. After having the data model in place, DCx back-end engineers systematically integrate the links of components into the ontology with the help of domain experts as well as available product structure data. After finishing the ontology creation, the data flow to the front-end is designed and implemented in close collaboration with DCx front-end, DII, and DMM engineers.

21.2.6 Data Identification and Interpretation (DII)

The discipline *data identification and interpretation (DII)* deals with identifying data which fulfil a data need definition (cf. discipline DND). To identify exactly that type of data, DII needs to deal with the correct interpretation of available and to be acquired data. Some tasks of data interpretation can be automated (e.g. [73]). However, the author of this book is convinced that domain knowledge of engineers will play a major role for data interpretation still for a long time to come. Therefore, DII has a close relation to the disciplines of DND (Which data is needed?), DC (How is the needed data acquired?), DMM (Where and how is the needed data stored?), and DCx (What is the meaning of the available data? cf. [74]). Finding the relevant data can be done using two complementary approaches: data-driven and process-driven.

Data-driven DII follows the data flow from the source (e.g. an IoT device) to target (e.g. a dashboard to interpret the data) or from target to source, depending on which direction is easier or more successful to use for data identification in the concrete scenario (use case, IT infrastructure, organizational structure, etc.). Additionally, the data-driven approach uses statistical analysis to identify relevant data (e.g. applying a clustering algorithm to all data available yields clusters of similar data helping to start from a single data point and traverse through similar data on a cluster). Here it is important to always be alert regarding spurious correlations: Correlation does not always imply causality. Therefore, it is good practice to check the identified correlations with domain experts and making use of the data context provided by DCx.

In contrast, the *process-driven DII* approach starts with analyzing the relevant business process regarding the data created and consumed (e.g. analyze which steps are required to solve the problem at hand and which data is typically used to do so). An important benefit of the process-driven approach is the fact that it is based on proven ways (business processes) to solve a problem. Additionally, the business processes provide context and, therefore, the process-driven DII approach is by far less dependent on well-established DCx. However, process-driven DII depends on the availability of business processes related to data needs. If no business processes with relation to data needs are available, these processes need to be created when opting for the process-driven approach. The data-driven approach has its benefits when dealing with new products or services without established business processes related to data needs. Additionally, the data-driven approach can be used by applying established algorithms from the field of data science (e.g., feature importance analysis, clustering, classification, correlation analysis, etc.). If the availability of data is expected to be good regarding the data need definition, data-driven approaches yield fast and inexpensive results. If the availability of data is expected to be bad (e.g. because the product in focus is not build yet), the process-driven approach making extensive use of domain knowledge is the only choice—in the end, data-driven approaches need data! Creating syntecial data might be an option, however can create severe problems in case no knowledge exists with respect to operational field or test experience of the system of interest.

Taking the example of predictive maintenance, DII engineers analyze results from DND, DC, DMM, and DCx. The result of such an analysis is an understanding of the needed data, their context and possible locations. Following this understanding, DII engineers conclude, that predictive maintenance is a common use case with well-established business processes. Therefore, they decide to use the process-driven approach for data identification and interpretation. First, the DII engineers identify relevant business processes. Once a company process is identified on carrying out predictive maintenance, this process yields the names of engineers involved in predictive maintenance analysis. These engineers are contacted to explain the company process for predictive maintenance and they can name the relevant data sources typically used. Additionally, the interviewed predictive maintenance engineers are asked for typical misinterpretations of the data and best practices for interpretation of data relevant for predictive maintenance decisions. After documenting the relevant data with their locations and guidance for interpretation of this data, DII engineer's handover to data modelling for data analytics (DMDA) engineers.

21.2.7 Data Modeling and Data Analytics (DMDA)

Data modeling and data analytics (DMDA) deals with preparing the data for data analytics purposes as well as applying data analytics techniques. DMDA—as well as data preparation in general—is essential, because data analytics approaches (cf. [75]) each require a fitting preparation of data (no one-fits-all data preparation solution available yet). Nevertheless, within this chapter, the author concentrates on data modeling for data analytics purposes, because the field of data analytics techniques is already quite mature: the field of Data Science, AI and statistical analysis already developed a wide range of data analytics techniques, e.g. [46, 67, 69, 75, 76] but approaches to fuel the required data pipelines in order to apply those techniques needs more research and development. Hence, DMDA needs to also encompass ETL (Extract, Translate, Load) processes and for AI applications, DMDA significantly intersects with data preparation (e.g., data cleaning, feature selection, dimensionally reduction, etc.) as described in the CRISP-DM ([76]; cf. AI Chap. 16). DMDA differs from established approaches like ETL or CRISP-DM data preparation regarding the links to other DEA disciplines: By linking DMDA to DCx and DII, domain knowledge is systematically used for data preparation. This reduces the risk of having undetected spurious correlations in the final data analytics models (DII tells DMDA how to interpret the data as well as the locations of feasible data), shortens the time needed for data preparation (DMM ensures comfortable accessibility), and data understanding necessary to conduct data preparation (data is described by DCx and DII).

Taking the example of predictive maintenance, the DMDA engineers start with analyzing the documents provided by DND and DII engineers. Those documents describe the relevant data, their location, and guidance on interpreting them. DMDA engineers conclude, that they need to build a neural network using a supervised training approach. One of the components to monitor for predictive maintenance is a ball bearing. To predict the remaining life of this bearing, labeled training data provided from laboratory tests are used. To train the neural net, the data identified as relevant by DND and DII engineers is selected as features. In this example, such feature could be filtered vibration data at the outer and inner ring of the bearing, rotations per minute of the outer ring as well as the temperature of the inner ring. Additionally, the remaining lifetime measured in laboratory experiments is selected as a label. These activities result in a table consisting of features and the label in columns and records in the rows. This table is a feasible data model to be fed into a neural network for the remaining life prediction. To understand the results of the trained neural network, it is necessary to visualize them using the discipline of data visualization (DV).

21.2.8 Data Visualization (DV)

Data visualization (DV) deals with graphical representations of data to support data analyses and communication of insights gained from data. The visualization of data has a long history when comparing to other disciplines of DEA (cf. Fig. 21.21).

DV uses technologies like visual analytics, dashboarding, and decision cockpits. (cf. [78]) Modern tools (e.g. [79]) automate many tasks creating visualizations, which

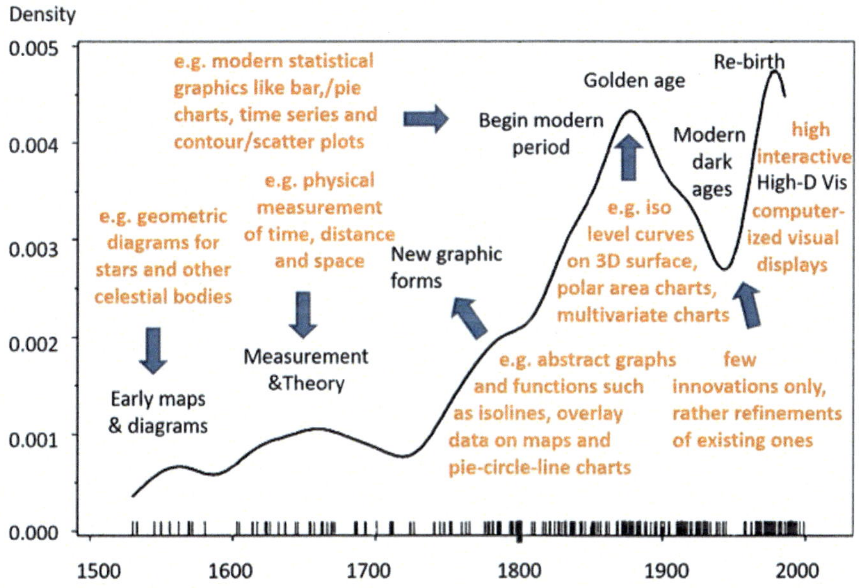

Fig. 21.21 The time distribution of events considered milestones in the history of data visualization (compare detail descriptions in [77])

creates room for DV engineers to concentrate on delivering the important information with visualizations. DV is essential for collaboration between the different disciplines of DEA. Visualizations make it easier to understand data, their meaning, and their context. DV is an established field, but needs to be extended in the context of DEA: DV is not just about making nice pictures out of data. It is about bringing together DEA disciplines, by linking the outcomes of each DEA discipline in a way that they can be easily understood by stakeholders coming from different professional backgrounds. As pointed out in the descriptions of the other DEA disciplines, there are intersections and handovers between the DEA disciplines. For example, DMDA engineers rely on the documented data needs (DND) and information on how to get and interpret the needed data (DII). Visualizations can help here by linking the data need definition graphically to the sources and interpretation of the needed data. This could be realized by simple boxes and arrows created in PowerPoint or by implementing a sophisticated web application showing an interactive graph with drill-down capabilities. This example illustrates, that DV in DEA is more than statistical diagrams—DV means delivering insights relevant for all DEA disciplines in a way, humans can easily understand. DV could be even placed at the core for the collaboration of DEA disciplines by providing tailored visualization templates for the documentation of insights relevant for DEA.

The following do's and don'ts should guide DV engineers in practice.

Do's:

1. Transporting the main message is more important than keeping all details. There will be always accompanying documentation that covers the details.
2. Set (internal) standards for creating visualization and stick with them. This will ensure that visualizations become comparable throughout different projects and use cases, which is important for the reuse of visualizations as well as for the familiarity and trust in using them for reviews, discussions and decisions.
3. Before creating a visualization, consult the engineers that are going to interpret the visualization. Truly understand their information need and design the visualization to fulfill this information need.

Don'ts:

1. Create visualization without legend, title, and axis labels.
2. Use different colors or markers that cannot be differentiated from each other.
3. Remove, modify or add data without consulting the receiving stakeholders of the visualization.

Taking the example of predictive maintenance, DV engineers closely collaborate with DCx front-end and DII engineers to create visualizations that are feasible to be used within a dashboard to plan maintenance actions. First, the information need by the maintenance engineers has to be understood. As it turns out, it is important to them having a complete overview of all assets under the responsibility of one maintenance engineer. This overview should highlight assets that need the attention of the maintenance engineer. Furthermore, assets to be monitored are located across different areas of the factory. The maintenance engineer, therefore, wants to get a visualization showing the locations of assets under his responsibility. To derive maintenance actions the maintenance engineer wants to get time series graphs on historical usage and important sensors of assets when clicking an asset within the dashboard. After analyzing the described information need, the DV engineers create a visualization positioning asset to be monitored on a map of the factory. The assets are visualized as small 3D representations. Assets that need attention due to results from automatic condition monitoring are highlighted by increasing the size of the corresponding 3D representation, increasing the color saturation of the relevant asset while decreasing the saturation of the 3D representations linked to the other assets. When clicking the highlighted asset, time-series graphs showing the condition of the asset become visible.

21.2.9 Conclusions

DEA is a discipline with eight sub-disciplines, namely DVU, DND, DC, DMM, DCx, DII, DMDA, and DV as explained above. While these disciplines all work together on fueling data pipelines for smart engineering, smart products, and smart services of the future, they all got their distinct area to contribute to DEA. DEA needs all these disciplines to be successfully implemented in practice. It is not necessary that engineers just own responsibility for a single discipline. In practice, DEA engineers can take on roles in different DEA disciplines at the same time. Engineers involved in DEA should have basic knowledge of each of the DEA disciplines to improve collaboration and avoid misunderstandings, especially in handover phases from one discipline to another. The capabilities of MBSE can nicely support DEA by providing.

- the context of technical systems and software (useful for DCx),
- the model data needs as well as
- the sources to fulfill those needs (useful for DND and DC).

The existence of MBSE might also be crucial to help establishing the new field of DEA by enhancing DEA chances that to become compatible with the approaches used in the engineering domains. DEA will become a necessary new Virtual Product Creation (VPC) discipline to robustly leverage AI (Artificial Intelligence) methods for technical systems development and operations. Because DEA provides the right mix of data synthesis and analytics methods for the correct engineering interpretations of right data types and the sufficient amount of data set occurrences. Similarly, to the introduction of quality engineering methods in the beginning of the 90ties of last century, *Data Science and Engineering (DEA)* as new Virtual Product Creation discipline needs new skillsets, education and training types as well as compatible integrations into engineering management and decision procedures.

21.3 Digital Twin Engineering (DTE)

Chap. 20 has already introduced the definition and the conceptual dimensions of the Digital Twin according to [80] in detail. The subsequent work of Stark et al. summarizes this Digital Twin foundation and provides the insight on which elements are essential for the new digital engineering capability to develop Digital Twins [81] (see Fig. 21.22):

> The area of DT environment and context is represented by the four dimensions integration breadth, connection mode, update frequency and product life cycle. The DT behavior resp. capability richness comprises the other four dimensions, i.e. the CPS intelligence, the simulation capabilities, the digital model richness and the human interaction. Each one of the dimensions provides three or four levels of realization: a higher level is not necessarily better than another but depicts a different and/or unique realization space. Four out of the eight dimensions, dimension 1 (integration breadth), dimension 2 (connectivity mode), dimension 7 (human interaction) and dimension 8 (product life cycle), however, do express with their increasing levels also an increasing degree of richness/fidelity (dimension 2 and 7) and of breadth/extent (1 and 8).
>
> The model allows describing major "behavior and context capabilities" to which a specific twin is designed for by allowing multiple target levels in each of the eight dimensions. Those eight dimensions are not exclusive or exhaustive but represent the most likely dimensions

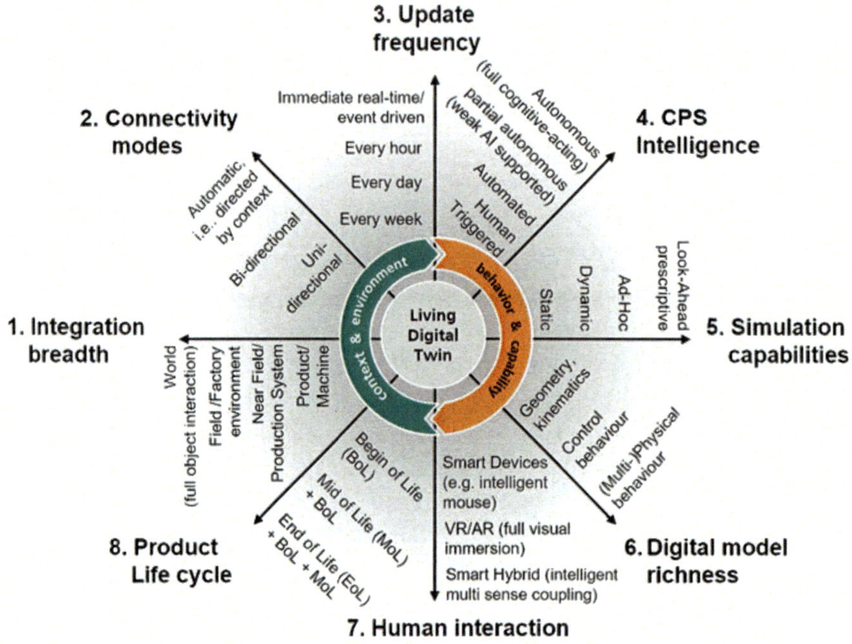

Fig. 21.22 The "Digital twin 8-dimension model" according to Stark et al. [81]

that are of importance to support the individual business context situations of the specific DT in scope. Engineers and business managers, therefore, are enabled to effectively use the "Digital Twin 8-dimension model" within the following situations:

- *As guidance in target setting in the development of completely new product by using an already existing DT,*
- *To extend existing products with the knowledge gained from their operational DT or*
- *To further develop DT as a product or service by its own (i.e. as a template), adding new functions along the eight dimensions as necessary".*

Since the role and the potential of Digital Twins will increase sharply in the context of providing lifecycle-oriented understanding and evidence of products and production equipment as well in controlling autonomous technical systems in the future, it becomes essential to provide a theoretical foundation for the Digital Twin design framework.[9]

Appropriate *Digital Twin design elements* constitute one of most critical capabilities within that Digital Twin design framework. Following the research results from Stark et al. [81], the following aspects are essential:

[9] In 2020 the Industrial Digtal Twin Association (IDTA) has been founded in order to standardize digital standards for Digital Twins: https://idtwin.org/en/.

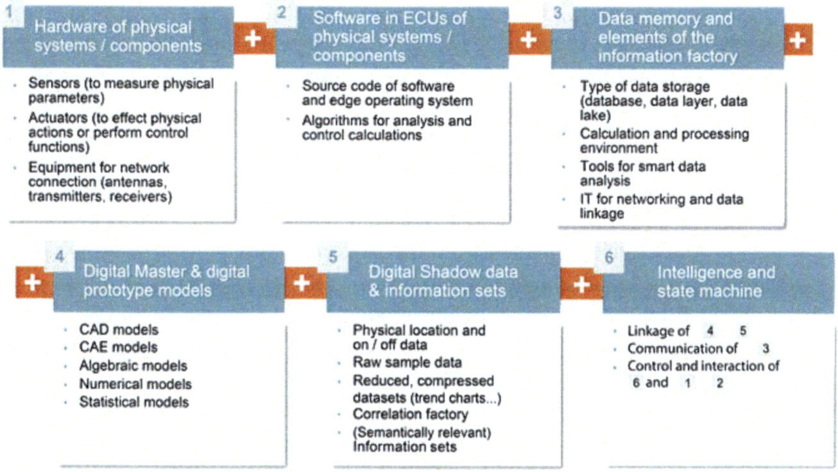

Fig. 21.23 Six major design elements of Digital twins acc. to Stark et.al. [81]

Considering design elements for DT the following two distinct DT use cases need to be protected in order to allow for most flexible applications:

- *The use of a DT on its own (i.e. without owning the physical product, object, gadget or machine) and*
- *The use of a DT in strict co-ownership with an in logical extension of the reference (physical) product.*

Treating those two basic use cases as equal and independent from each other the research team finally embarked on six major design elements, as shown in Fig. 21.23.

The design element 1 (certain hardware of the physical product) concentrates on those hardware components, which allow for analysis (sensor), control (actuator) and network interaction of the DT with the entire or certain subsystems of the physical product. The design element 2 (ECU SW of the physical system/component) ensures the description of the DT characteristics for the product or service on-board control algorithms and analytics. The design element 3 (data repository and core elements of the information factory) provides the capability to describe compute environments, associated data repositories, sets of analytic toolboxes as well as network connectivity information technologies. The design element 4 (Digital Master and Digital Prototype models) comprise all relevant digital models which form the base for DT (reference) capabilities. The design element 5 (Digital Shadow data and information sets) allows the integration of characteristics stemming from physical product or service operation, i.e. from measurements of data and related direct analytics. Last but not least, the design element 6 (intelligence and state machine) represents an interlink element between various other design elements and offers a wide variety of linkage richness and rigidity resp. flexibility."

The interlink of design element 6 enables the synchronization of the collected data with the to-be status represented by the digital master/prototype and may influence the overall digital twin target behavior in order to change the control of the physical system (physical twin) operation accordingly.

The major challenge of establishing successful digital twins lies in design element #6 (Intelligence and state machine, compare Fig. 21.23). Therefore, the following three new VPC capabilities need to be researched and developed to achieve robust *Digital Twin Engineering*:

1. How to engineer and establish robust linkages between *Digital Shadow* data (design element 5) and *Digital Master or Digital Prototype* models (design element 4)? This does inherently include data analytics on raw data (or signals) acquired and collected from the physical system (twin) to create most meaningful *Digital Shadow* data.
2. How to best reduce the *high amount of data sets and information elements* of the overall *information factory* in order to deliver most valuable Digital Shadow data for the target function of the Digital Twin?
3. How to engineer and to establish the *physical and logical connections* between the *hardware / physical components* (design element 1), *software capabilities of the ECU in the edge* (design element 2) and the *control and intelligence mechanism* of design element 6 (Intelligence and state machine)?

Let us start to reflect upon the first and the third new Virtual Product Creation (VPC) capability (see 1. above). In order to get closer to these new capabilities it is important to understand the different degrees of Digital Master/Prototype to Digital Shadow connections and the influence of this connection back to the Physical Twin (Physical System). It obviously depends on the specific business situation to determine which type of Digital Twin needs to be developed to work as part of the overall CPS (Cyber Physical System). Categorically, the following Digital Twin types are considered in industry:

- Product or technical system twin (which also includes e.g. machine twins),
- Process twin (engineering, manufacturing, operating, service, maintenance process etc.),
- Factory, production system or facility twin,
- Human worker or user twin.

Additional ones are thinkable but currently not yet highly marked. More important, however, is the rationale why an organization would like to operate a Digital Twin. The following business targets are meanwhile visible in industry and will most likely serve as major business drivers for Digital Twins:

- *Study and control performance* (incl. efficiency and effectiveness),
- *Leverage the Digital Twin information and extractable knowledge* to operate new business models and to generate new business offerings with respect to different types of attributes such as safety, convenience, new functionality and improved cost of ownership,
- *Gain feedback from and deeper insight* on actual operating patterns to streamline operations efforts,
- *Drive and control physical system (twin) behavior* in operational use with the help of Digital Twins,

- *Identify rest product life and across lifecycle potential* of living systems in real operation by monitoring and analyzing their Digital Twins,
- *Create feedback to design* evidence and use patterns in order to re-qualify and improve upfront digital master and digital prototype models.

Digital Twin Engineering has to provide new capabilities to plan, develop, validate, simulate, build and test Digital Twins based on the core element, the *twinning engine*. Figure 21.24 shows the twinning engine and the six DT design elements in their overall context of the Cyberphysical System (CPS) consisting of both, the *Digital Twin*, the *Physical Twin (Physical system)* as well as all other important interfaces and mechanisms.

The *twinning engine type* between *Digital Master* resp. *Digital Prototype* and the *Digital Shadow* determines which *Digital Shadow data* are necessary to interact with the *Digital Master/Prototype model*. The author of this book calls this type of Digital Twin Engineering approach the *Digital Twin inside-out development* approach. Once this type of *twinning engine intelligence* is determined based on the business rationale described above, all other elements and mechanisms shown in Fig. 21.24 need to be adjusted for it. This does include both, the CPS intelligence stream 1 (*Analyze and Understand, A&U*) and CPS intelligence stream 2 (*Drive and Control, D&C*).

The *CPS intelligence stream 1* requires the determination of the raw data elements and the associated *Data Modeling and Data Analytics (DMDA)* and *Data Collection (DC)* capabilities (compare details in Sect. 21.2). Based on this determination it will be possible to determine and engineer the right levels of sensor types, the overall sensor architecture, meaningful sensor locations, the sensor signal fusion regime of and the sensor *Data Acquisition (DAc) technology* for the *Physical Twin* (Physical System).

The CPS intelligence stream 2 first of all depends on the core intelligence gained from the *twinning engine*: how is the *Digital Shadow data* interpreted by the *Digital Master/Prototype model*? The conclusion taken from the interpretation of the *twinning engine* provides the decisive and appropriate signals and control pattern (algorithms) for both, the actuators and the embedded software of the ECU (Electronic Control Unit)—or in case of a production system—the PLC (Programmable Logic Unit).

In order to answer the first and the third question concerning the new VPC capabilities of Digital Twin Engineering (DTE) it helps to clarify the different *twinning engine types (TET)*.

The following Table 21.2 introduces the major *twinning engine types* with a short designation in the left column und explains the twinning mechanism as well as its potential role for the use as part of the overall CPS architecture (compare Fig. 21.24) in the right column. This table does not necessarily show all possible types but concentrates on the four foundational resp. essential ones according to the author's current expertise (summer 2021).

Figure 21.25 shows the Digital Twin set-up of a smart factory cell which produces configurable coaster. The overall factory cell consists of a conveying system with in-process quality control weighing to ensure the use of the right material uses as core a

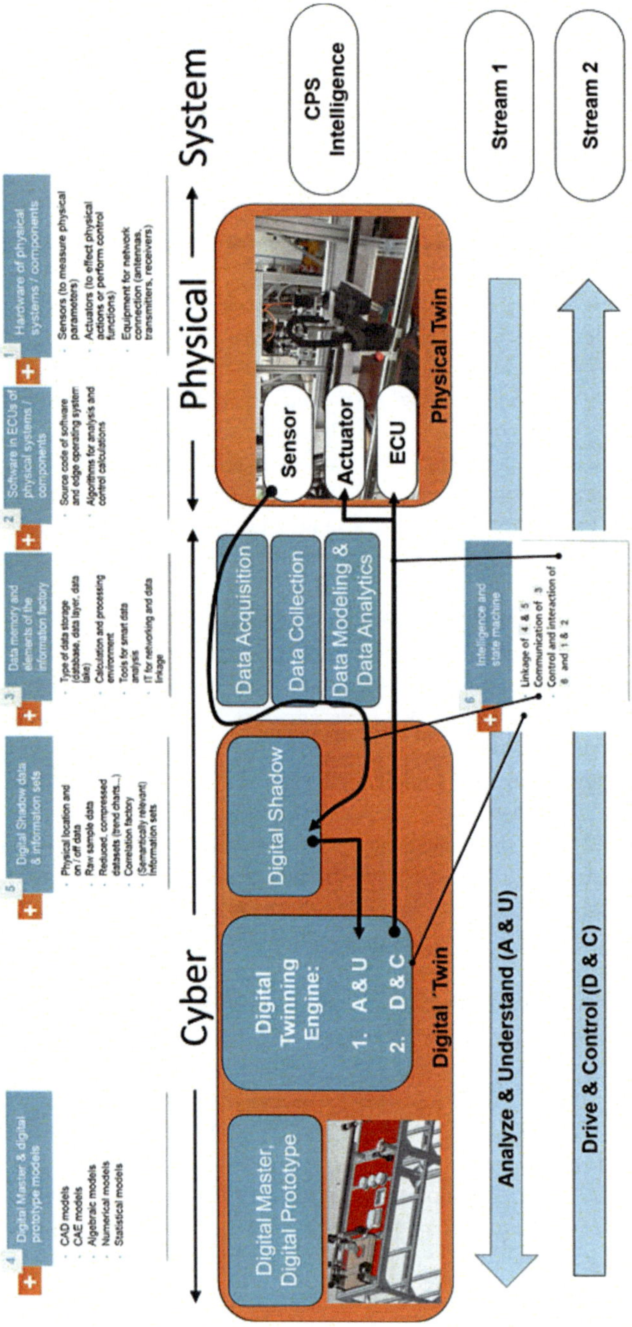

Fig. 21.24 Digital twin design elements within the CPS architecture

Table 21.2 Explanations of twinning engine types (TET) of digital twins

TET # and designation	Explanation and CPS use options
1. Nill—no twinning	There does not exist any twinning between the digital shadow data and the digital master/prototype model
	The physical twin (system) does not get driven at all through the digital twin but the digital twin might earn potential A&U intelligence through its digital shadow from the physical twin operation
2. Show/mark-up twinning	Digital shadow data (resp. smart data after analytics) is shown in digital master/prototype models (e.g. at 3D model or at mathematical equation variables) to allow better human interpretation and decision making
	The physical twin (system) does not get driven directly by the digital twin, however, human interpretation might trigger adjustments at the physical twin. The digital twin might earn potential A&U intelligence through its digital shadow from the physical twin operation
3. Model parameter twinning	Digital shadow data (resp. smart data after analytics) is used to influence digital master/prototype model parameters and variables (e.g. in behavior models or through other mathematical equations/models) to modify digital twin characteristics and behaviors
	The physical twin (system) might get driven directly by the digital twin as part of the D&C stream. In addition, the digital twin earns potential A&U intelligence through its digital shadow from the physical twin operation
4. Model rebuild twinning	Digital shadow data (smart data after analytics) is used to rebuild the digital master/prototype model. The analytics and quality of the digital shadow enables the reformulation of the digital master and/or digital prototype simulation models without any direct human interaction. This might be enabled trough trained machine learning analytics for rather non-linear and complex technical system behaviors. For safety critical technical systems this need to be carefully validated through deter-ministic models in addition
	The physical twin (system) might get driven directly by the digital twin as part of the D&C stream. In addition, the digital twin earns potential A&U intelligence through its digital shadow from the physical twin operation

milling machine to grave the individual name onto the coaster top plate before it will be processed for assembly in the second area of the factory cell. The Digital Twin of this smart factory cell consists of a 3D mechatronic system model in combination with a process simulation model. Both models can be triggered through signals from the physical twin elements (conveyor, gripper, milling machine).

For the CPS intelligence stream 1 a TET (Twinning Engine Type) type 2 has been chosen as is provides an energy consumption dashboard for the entire factory cell.

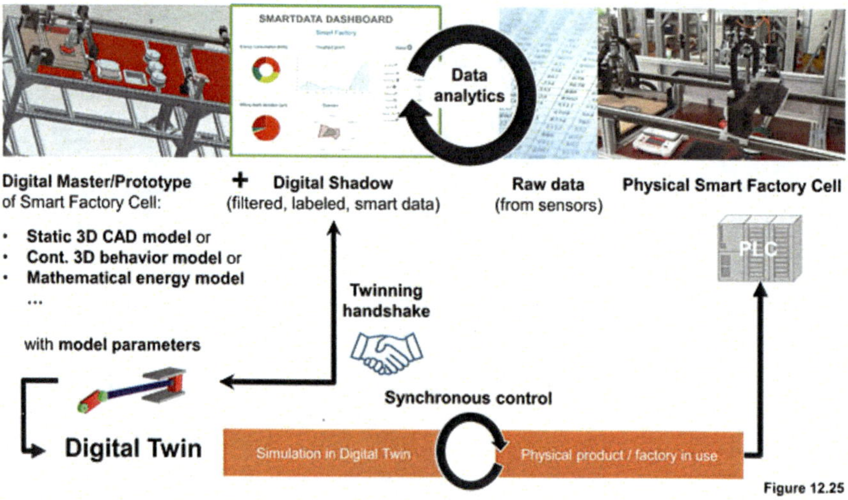

Fig. 21.25 Digital twin of a smart factory cell

In addition, it also carries the *PLC controller state signals as Digital Shadow data* in order to couple the physical production cell via the appropriate IoT stacks, i.e.

- via the communication protocol *CoAP* (Constrained Application Protocol using UDP as transport protocol) or
- via *OPC UA* (Open Platform Communications—Unified Architecture, using TCP as transport protocol)[10]

with the *continuous 3D behavior model of the Digital Twin.* This represents a TET (Twinning Engine Type) type 3. By doing so, it is even possible to use virtual sensors within the Digital Twin 3D behavior models to drive and control the physical smart factory cell within a given network latency. It becomes evident that the *twinning handshake* (to be realized by the twinning engine) between the *Digital Shadow data* and the *Digital Master/Prototype model* constitutes one of the core elements of *Digital Twin Engineering* as shown in Fig. 21.25.

Finally, the question "*how to establish new VPC capabilities for robust Digital Twin Engineering*" is discussed. It is key to analyze the ways of processing raw sensor type data to deliver the target oriented Digital Shadow data to influence the Digital Twin behavior. Figure 21.26 shows the core processing steps of *Data Modeling & Data Analytics (DMDA)*, as one of the core disciplines of *Data Engineering & Analytics*, introduced and explained in Sect. 21.2.

Based on the desired and/or needed Digital Shadow data types for the intended twinning handshake of the twinning engine with the Digital Master/Prototype models the sensor *Data Acquisition* and the following *Data Collection (DC)* and *Data Modeling & Data Analytics (DMDA)* processing will run through three major stages.

[10] Please compare Chap. 20 for the IoT protocol explanations.

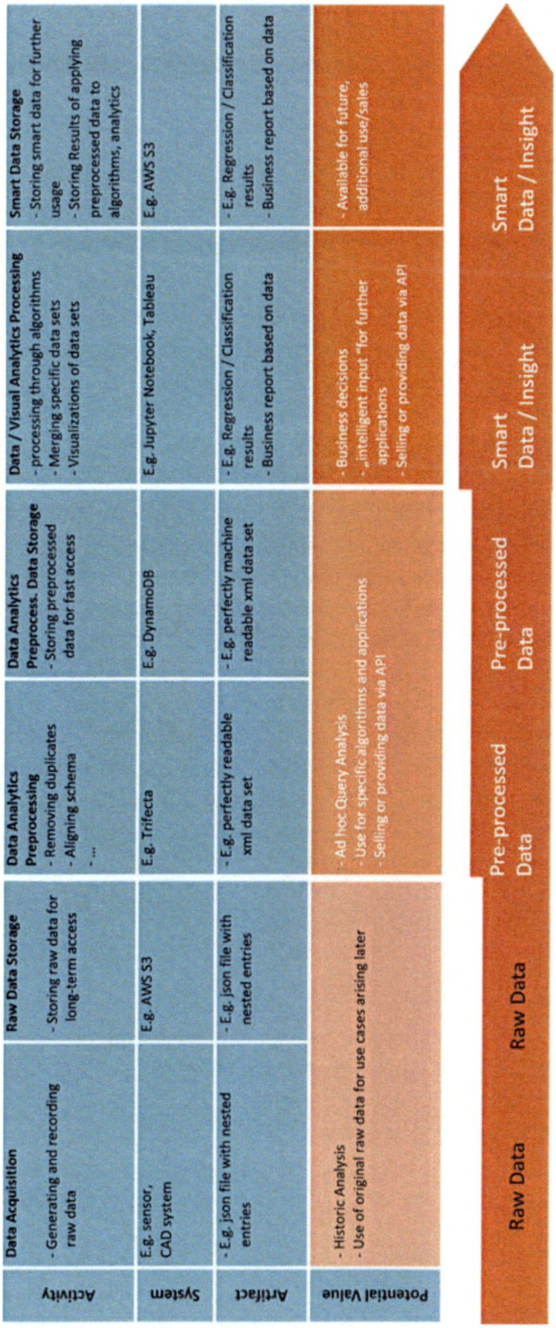

Fig. 21.26 Data analytics (DA) processing steps towards smart data

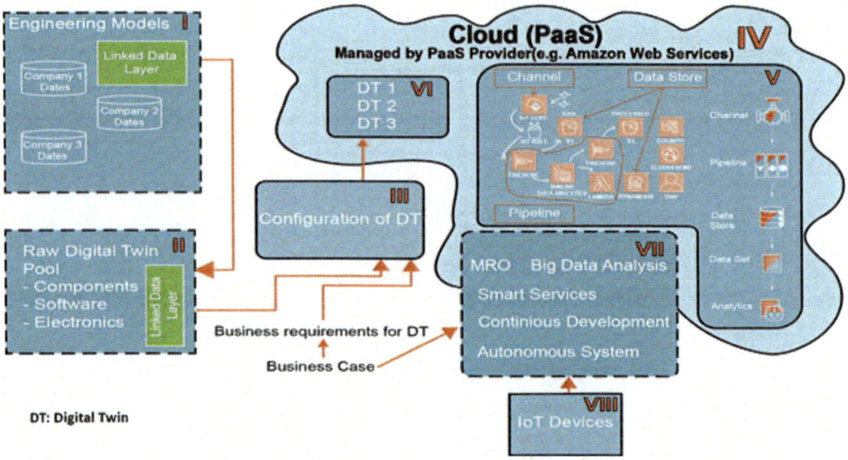

Fig. 21.27 Information architecture to configure and operate digital twins

The first stage *"Data Acquisition"* and *"Raw Data Storage"* (part of DC) ensures that all relevant data are stored in an appropriate database architecture[11] as part of an overall network data storage strategy (edge, cloud etc., c.f. IoT Chap. 20.4). On the second stage *Data Analytics Pre-processing* of data (belongs to the discipline DMDA) becomes essential in order to prepare them for their final stage, the *Data Analytics Process* and *Visual Analytics*. Both are necessary to reduce and interpreted data sets towards the intended smart data characteristics which are driven by the Digital Shadow needs. *Digital Twin Engineering (DTE)*, therefore, needs to leverage essential capabilities of *DEA (Data Engineering and Analytics)* in order to design the most appropriate data storage, processing and semantic interpretation for their use in the twinning engine (compare Table 21.2). Figure 21.27 illustrates the importance of the Digital Twin information architecture.

In order to configure and operate Digital Twins the different information technology elements of Digital Twins, such as the.

- engineering model environments (I),
- the raw Digital Twin pool with dynamic Digital Shadow data and twinning engine types in form of (data) components, software (analytic algorithms) and core electronics with the different types of controllers (II),
- the configuration engine of Digital Twins (III),
- the processing power of cloud services as part of PAAS, platform as a service (IV),

[11] Meanwhil NoSQL ("No" or "not only" Structured Query Language) data bases are designed for rapid data acquisition and storage capabilities which make them preferrable within edge networks. They are non-tabular, and store data differently than relational tables. NoSQL databases come in a variety of types based on their data model. The main types are document, key-value, wide-column, and graphs. They provide flexible schemas and scale easily with large amounts of data and high user loads and need special mechanisms to ensure database integrity.

- the data channels, pipelines, stores and sets as well as the libraries of data analytic algorithms and executable files (V),
- the individual Digital Twin instances (VI),
- the resulting Digital Twin services (smart service levels according to the CPS intelligence streams 1, *Analyze & Understand*, and 2, Drive & Control (VII) and
- the necessary IoT devices to transmit and communicate protocols, data and signal flows (VIII).

are to be executed in a linked and synchronized way to each other.

Digital Twin Engineering, therefore, heavily relies on a standardized and powerful orchestration of these IT elements, which are a pre-requisite for Engineers to develop, configure and operate Digitate Twins for technical systems. It is not clear at all, who will claim the leadership within that space, the applying industry, Digital Technology Vendors (DTV) or digital platform providers.

21.4 Digital Platform Engineering (DPE)

Ubiquitous data, information and digital model availability will become the core competitive advantage of all competing industries. This is not only due to the increasing entrenchment of digitization in future products and services in operation but also due to the need to stay in control about safe and robust development and validation of technical systems.

As a consequence, future Virtual Product Creation needs to get to the next level of virtualization: the virtualization of its own digital solutions! What is meant by this? After having established millions of companies specific VPC and PLM solutions with each a heterogeneous IT architecture and implantation set-up, the "logistics and streams" of providing digital data and models easily and even potentially in an ad-hoc fashion without a difficult regime of enabling IT services still remains a dream which finally will come true. How can this be achieved? New core fundamental algorithms as part of the next generation CPU cores will make it possible to "virtually" stream relevant digital data and model visualizations out of the database silos into live VPC of the future working spaces. The following sections will explain the fundamentals and will show first striking technology enablers in their functionality and application integration into today's and future Virtual Product Creation environments.

As it was already indicated in the previous Sect. 21.3 already with respect to the operation of Digital Twins, the targeted orchestration of appropriate information technology environments as part of digital platforms become an indispensable solution element of the engineering system of the future. There exists different motivation and rationales why digital platforms will become rather the norm than the exception for future Virtual Product.

Please note the following essential drivers and rationales for different types of digital platform engineering:

1. As first digital platform type, offered as a new extension from the traditional VPC perspective, proprietary *digital engineering platform environments* offered by Digital Technology Vendors such as *Forge* from *Autodesk* or *3Dexperience* from *Dassault Systèmes* provide an integrated VPC application environment with excellent data and data model interaction, integration and potentially even automated services across all platform applications. Those digital engineering platform solutions offer a high range of WEB service enabled application features which can be mixed in flexible ways in order to extend the traditional single application-oriented modeling, simulation and analysis activities of engineers. In addition, other type of capabilities are offered, e.g. with respect to project management, statistical analysis and cross-technology linkages like overlays of AR/VR visualization and interactions in combination with product modeling and Building Information Modeling up to complete factory and smart city environments.

   In addition, if offered through cloud infrastructure platforms such as Microsoft Azure, Amazon Web Services (AWS), SAP Business Objects, IBM cloud and others, rather than through on-premise enterprise installations, the administrative footprint is lowered and ITIL maintenance efforts can be dropped significantly.
2. The second type of digital platform focuses on *Industrial IoT* (IIoT) as a service. Here, data delivered by edge devices as part of IoT (compare Chap. 20) get stored and processed as part of data analytics and other data science processing patterns, incl. AI based predictions. Such data platforms are based on NoSQL databases to support the processing of raw data from IoT device raw data towards valuable Digital Shadow smart data as explained in the previous Sect. 21.3. It also helps different partners and companies to learn from the same data sources if used in a collaborative way. Typical examples of such proprietary digital platform are *Mindsphere from Siemens* or *ThingWorx from PTC* using advanced analytics from edge to cloud using and processing data from connected products, plants and systems. In order to make it possible for Engineers to create and configure their own applications *low code/no code* application frameworks (such as *Mendix* in case of *Mindsphere* or similar inbuilt applications in ThingWorx) are offered on those IoT platforms. Bosch as industrial equipment manufacturer has created the terms *AIoT (AI + IoT)* to designate the deep integration of AI application architectures (such as lambda architecture) within the IoT platform environment.
3. The third, *more generalized digital platform type*, is targeted to provide scalable compute power, storage and software environments according to general customer needs. The following cloud types, options and capabilities are distinguished:

   a. Cloud *Software as a Service (SaaS)* is a type of cloud that offers an application to customer or organizations through a web browser. The data for the app runs on a server on the network, not through an app on the user's computer. Software is usually sold via subscription.

b. Cloud *Platform as a Service (PaaS)* provides networked computers running in a hosted environment, and also adds support for the development environment. PaaS solutions generally support specific program languages and/or development environments. In general, it is possible to develop also digital engineering applications in such an environment by taking advantage of dynamic scalability, automated database backups without need to specifically code for it. However, general PaaS do not offer any specific digital data model options for it. PaaS is charged as an additional service on top of the IaaS charges (see below).

c. Cloud *Infrastructure as a Service (IaaS)*, provides the hardware and usually virtualized computer operating system to their customers. Software is charged only for the computing power that is utilized, usually CPU hours used a month.

As a general information factory approach this type of digital platform offering has the following advantages or incentives for companies if stable digital networks are available (!):

- *Storage*: Enterprise applications typically require on-premise storage, either from servers or hardware storage on a user's device. With the cloud, the storage is optimized for extreme performance with 99.9% durability. IT environment developers do not have to worry about creating and storing copies; the cloud does that for them.
- *Security*: The existence of cloud service providers depends on security standards often not understood well enough at the enterprise level. Competent vendors encrypt data with robust audit methods including knowledge who has accessed which information when.
- *Elasticity/scalability*: Digital cloud solutions enable applications to allocate resources as needed, easily scaling up or down when more compute, storage, or software instances are needed. A single user may require more or less compute power depending on current development project or program needs. Planning and acquiring hardware for the absolute maximum is not a cloud native thought process. Instead, the system automatically adjusts without oversight or downtime whether 10 users are interacting with the application or 1000 and more.
- *Performance*: The elasticity of cloud solutions ensure that users will not notice a change in performance. To the business and to users, a cloud solution, therefore, appears unlimited in terms of compute, storage and service power while maintaining consistent performance.

4. The fourth type of digital platforms provide a neutral layer of federated services, protocols and policies for many participants from industry, public or municipal authorities, non-profit organizations and private consumers to enable secured and open access to various data spaces. The project and initiative *Gaia-X* has been founded to establish such a trusted, sovereign digital infrastructure for

Europe. According to [82] one particular important aspect of Digital Sovereignty is Data Sovereignty:

Data Sovereignty is the execution of full control and governance by a Data Owner over data location and usage. By applying the core architectural principles outlined below, GAIA-X will enable Providers and Consumers to participate in a digital sovereign ecosystem. GAIA-X builds on a unique selection of technological approaches to bring digital sovereignty to life:

- *Federation: Supports standardized access to GAIA-X as well as multiple decentralized implementations. This way, a rich digital ecosystem is fostered.*
- *Self-Descriptions and Policies: The basic elements on a technical level for the selection, initiation and coordination of interactions between Providers and Consumers. Self-Descriptions represent GAIA-X offerings. Policies represent requirements. By matching both, Provider and Consumer can start to interact within the GAIA-X ecosystem.*
- *Identity and Trust: Helps GAIA-X Participants to verify that their engagement with others and the services they use are plausible, authentic and backed by Self-Descriptions and Policies.*

Figure 21.28 shows the intended Gaia-X ecosystem framework for services. The data ecosystem with advanced *smart services* (digital services based on smart, i.e. connected, products) and *data spaces* for various business streams are based on Gaia-x federation services. The infrastructure ecosystem as foundation provide various types of technical architecture elements and standards for commercial/public cloud providers to use. Please compare details of architecture guidelines, core architecture elements (e.g. services, nodes, data assets, catalogues etc.) in [82].

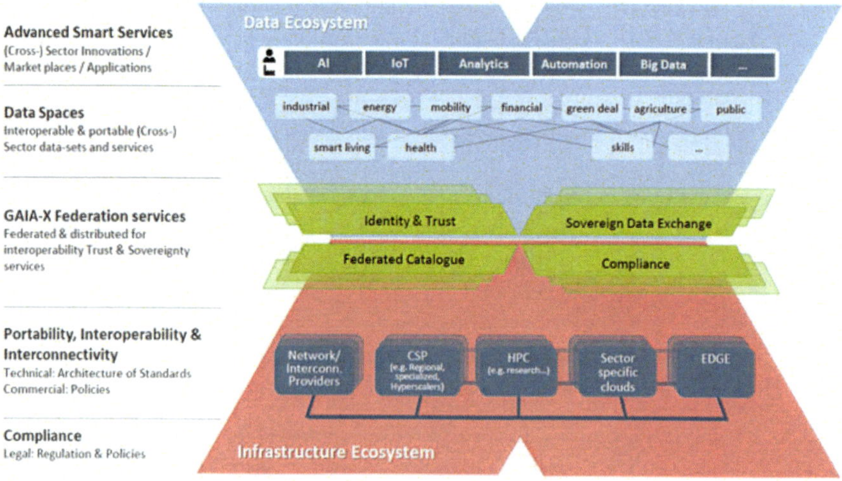

Fig. 21.28 The Gaia-X ecosystem of services and data [82]

The European Data Governance Act (DGA) offers the opportunity to define clear rules on how to use data responsibly as it is shared and exchanged between various parties. Together with the IDSA (Industrial Data Space Association)[12] a position paper [83] has been published which explains the practical steps towards realizing data sovereignty in a three-step approach through soft infrastructures:

1. *Develop functional, legal, technical and operational agreements that support the most pressing needs of people, businesses and governments in the various data spaces. These agreements should be co-created with the most eager participants, and form the initial version of the soft infrastructure. Much of the thought-work has been done already by researchers and business practitioners throughout the world. It is now a matter of agreeing on the optimal and coherent approach across all relevant disciplines.*
2. *The organizations which have created the agreements should roll out and implement the first version of the soft infrastructure.*
3. *The soft infrastructure should be extended across all sectors. And remember, soft infrastructures should be allowed to evolve over time. They are a 'living' form of standardization, and the common way of dealing with data must continuously respond to the needs of the market and its applications. This is secured through the set-up of a sound governance model which represents both private and public interests.*

Fiware [84] provides more insight to the concept of and the technology building blocks of Data Spaces: *A Data Space can be defined as a decentralized data ecosystem built around commonly agreed building blocks enabling an effective and trusted sharing of data among participants.* From a technical perspective, a number of technology building blocks are required ensuring *data interoperability, data sovereignty and trust* as well as *data value creation.*

The above described four different types of digital platforms, the policies for data sovereignties indicate that future collaborations between business partners and between IoT connected products and devices will provide different focal points of platform usages via related data spaces. Therefore, the number of data and model services to be enabled by network and cloud service providers, the power of high-performance cluster solutions to enable instant and ubiquous simulations and the penetration of digital analytics up to network edge devices will grow substantially and will demand different platform features. Hence, it will also be necessary to closely engage and follow new ways of establishing and developing new types of *Digital Platform Engineering (DPE)* offerings.

Let us have a look how these new DPE offerings and capabilities leverage creative and advanced Virtua Product Creation (VPC) solutions as part of the *Engineering Data Space*, which forms the base for the specific engineering business stream. In the following two examples will be used to explain the approach and the opportunities of *Digital Platform Engineering (DPE).* Obviously, the number of DPE offerings

[12] https://internationaldataspaces.org/.

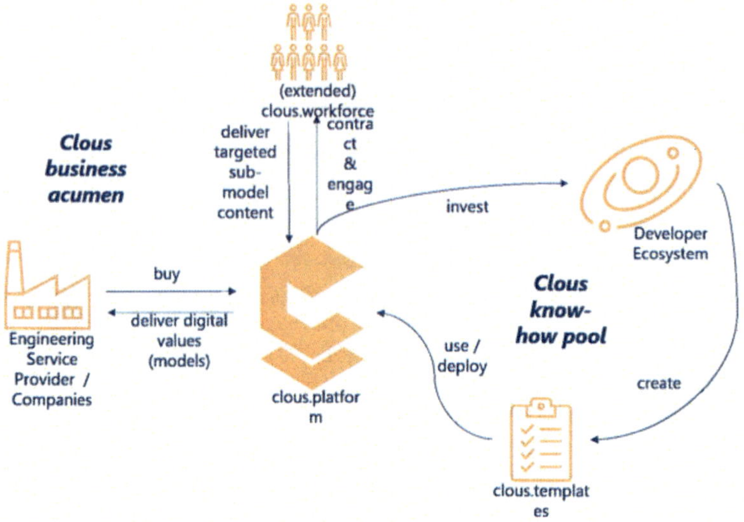

Fig. 21.29 Clous digital platform-based business and know-how model

will rapidly grow over time and will also influence significantly the way of Virtual Product Creation (VPC) of the future.

Example 1: Distributed Engineering of the Future by CLOUS
Digital Platform Engineering (DPE) will become a major technology and business approach for future Virtual Product Creation (VPC) if a successful mix between customer demands and solution deliveries can be organized and ensured by appropriate and easy to comprehend digital platform business models. Hence, it will be decisive to establish new ways of organized collaborations between industrial companies, their engineering departments and the new type of skilled digital platform operators who are able to offer new and additional engineering services with such a business model.

Clous GmbH,[13] a start-up in Germany, headquartered out of the German capital Berlin, offers such a unique and new digital service for engineering by addressing the needs to reduce lead times and costs through digital platform enabled simultaneous engineering and smart outsourcing. Figure 21.29 illustrates the new digital platform-based business and know-how model of Clous.

The core digital technology asset of Clous is design practice intelligence (*Clous know-how pool*) developed internally and with external developer teams. This design know-how for specific industry branches, type of products as well as digital information and model types is flexibly deployable *(Clous business acumen)* for new types of digital simultaneous engineering work patterns via a global network of designers (*the extended clous workforce*).

[13] https://www.clous.io/.

Clous has build up and orchestrates the extended workforce as part of a design practice democratization. This is achieved by splitting up complex design tasks into manageable and easy to deliver sub-tasks with limited design know-how needs.

The core solution of Clous enables such an approach via self developed intelligent algorithms and models. With such an approach, digital modeling and annotation design work can be offered via a global workforce. This can account for more than 70% higher cost efficiency compared to traditional high wage industry countries and can get even faster delivered to the industrial customers. Figure 21.30 explains the digital platform solution and delivery process by Clous.

Clous ensures via steps 2, 6 and 7 that IP (Intellectual Property) can be guaranteed and that final 3D CAD assembly model delivery fulfills stringent quality requirements of the industrial customers. Steps 3 through 5 (deployment with submission of tasks, distributed engineering workforce operation and the final model deliveries) leverage directly the digital platform capabilities.

The Clous offered digital engineering platform solution can be realized by all four digital platform types according to the explanation earlier in this sub-chapter. Thus, it depends how the different digital platform owners and operators respond to this new business model and which type of revenue sharing might be agreed with the new digital service offering company Clous.

This new digital business solution of Clous shows that engineering services will become one of the key innovation drivers for Digital Platform Engineering (DPE) offerings. It is safe assumption that a Gaia-X type of platform and service environment could boost this type of new digital value creation streams by guaranteeing sovereignty at all levels.

Example 2: Streaming Technology for Secure and Instant Collaboration
Digital collaboration has become essential in Virtual Product Creation due to an increased degree of involved development partners in engineering and manufacturing (local, regional and global). Exchanging and sharing data and models become quickly a bottleneck from perspective of digital technology resp. model availability and capability as well as of interactive collaboration methods. The Berlin VPC research team expressed this challenge with the following assessment [85]: *The rapid increase in information and the expectation of its global availability introduces a new field of information management that does not require a central distribution point but intelligent information containers that can manage the containing information and is able to route that information to systems and participants in a collaborative scenario that need them.*

For future Virtual Product Creation, it should not really matter any longer where the data physically resides to make them available for instant, secure and virtually composed collaboration. Streaming[14] technology comes into play to allow for intel-

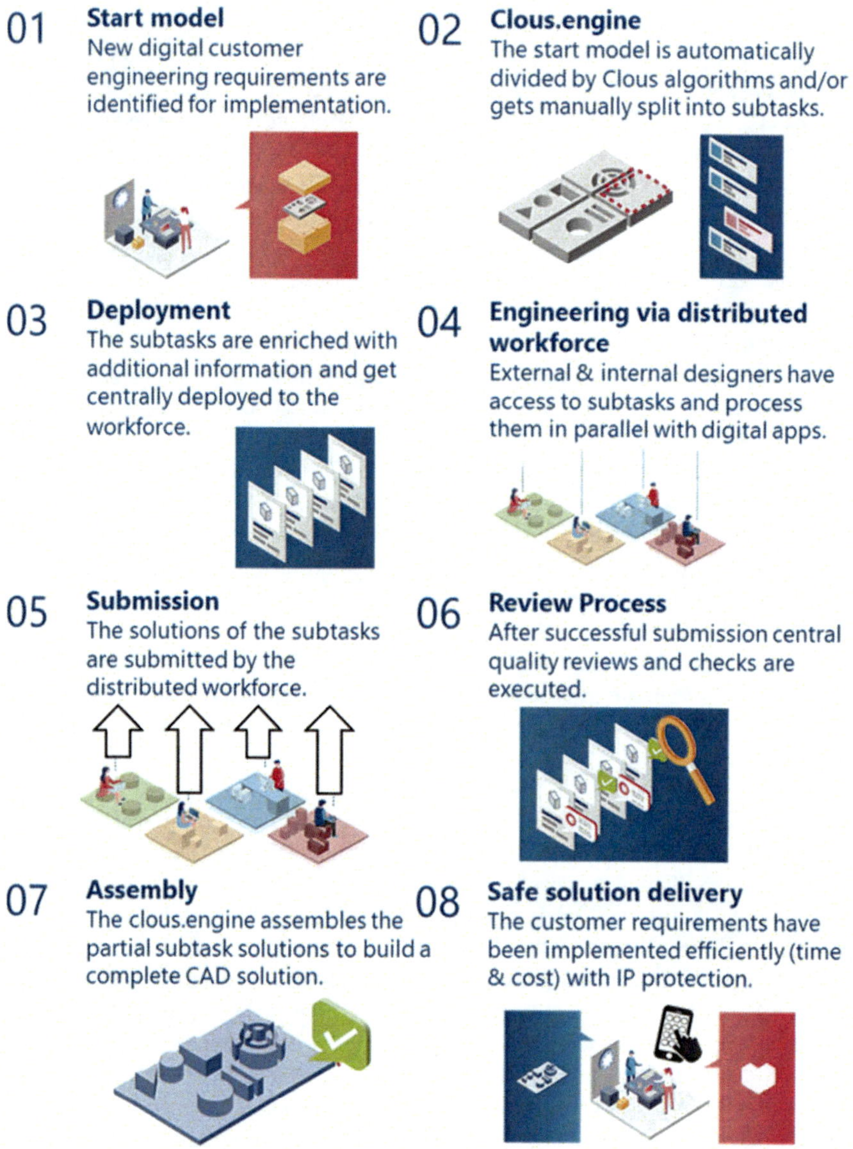

Fig. 21.30 Clous digital platform solution and delivery process

[14] The technology of transmitting audio and video files in a continuous flow over a wired or wireless internet connection. Streaming services, therefore, have a high need on network bandwith and latency requirements for dynamic interactions.

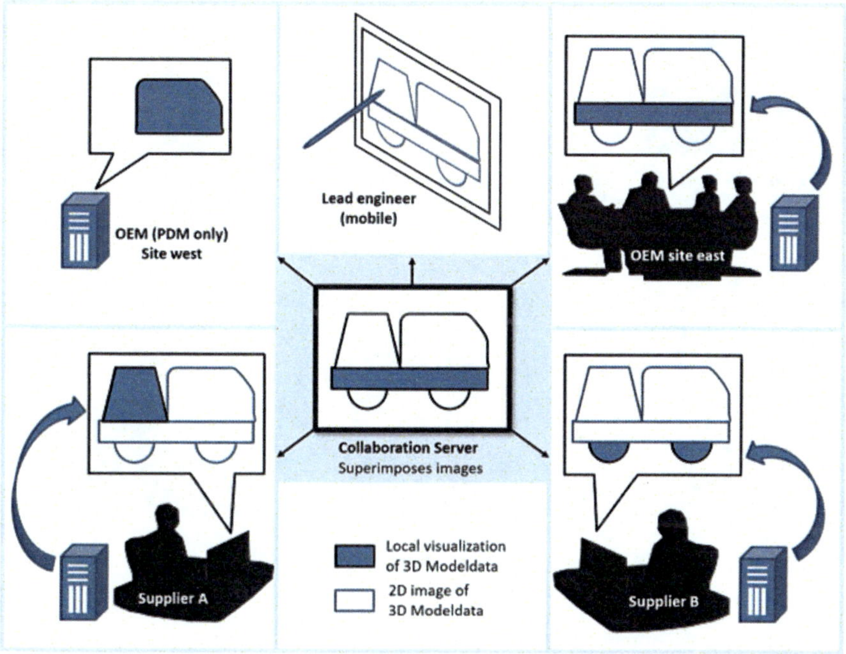

Fig. 21.31 Virtual collaboration server scenario across sites with heterogeneous local visualization file availabilities (see [85])

ligent overlays of different local visualization streams composed by a central collaboration server to stream (broadcast) the same dynamic model view to all participants (compare Fig. 21.31).

Reference [85] provides the detailed understanding of the technology interplay of the collaboration scenario shown in Fig. 21.31:

> The here presented solution constitutes a combination of screen sharing and the local visualization at each participant. In contrast to screen sharing not the whole program window or the whole desktop is being transmitted, but only the 2D image of the rendered 3D model, which is superimposed with the images of all participants. A correct superposition is necessary so that every participant can correctly perceive the visual impression of the complete product and properly interpret the correlations and distances between the components.
>
> This collaboration technique focuses on different scenarios shown in Figure 21.31. All participants shown in Figure 21.31 see a 3D representation of the object being reviewed, in this case a truck. The parts in blue are locally existent as 3D-Modells and are locally rendered on the computer and the rendered image is transferred to all participants. The gray parts of the model do not exist on the local computer. They are just 2D images streamed from one of the other participants. All views share the same point of view and orientation while looking at the truck. This information is also shared among the users and consists of a simple matrix. The scenario can also incorporate special participants like the mobile lead engineer which only needs a web browser to join the session. He is not supplying any 3D model, he just consumes the images. The opposing case is the PDM System at site B which just renders it locally stored data and sends it to the others. This participant does

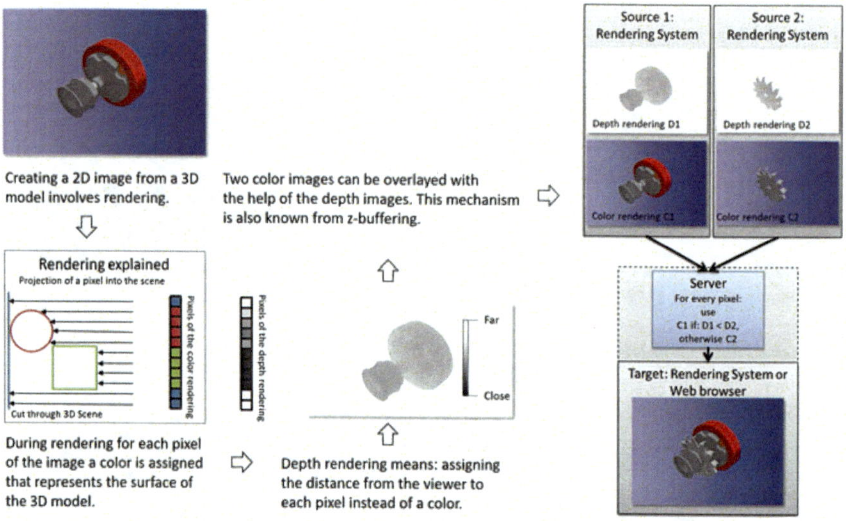

Fig. 21.32 In depth rendering technology to overlay different images

> not consume any information. All the other participants, the OEM at site east and the two suppliers deliver their own data and consume from the others. The OEM holds the 3D models of the chassis, while supplier A holds the cabin and supplier B the wheels. They only deliver their own property as images without being afraid that for example supplier B can steal the 3D model data from supplier A. To achieve the correct superposition a so-called depth image is transmitted additionally.

Unlike prerecorded infotainment like movies or audio broadcastings, the virtual collaboration scenarios shown in Fig. 21.32 do require an instant visualization stream creation triggered by dynamic interactions of the collaborating engineering actions within the session. The chair of Industrial Information Technology (IIT) of TU Berlin has developed a research demonstrator which uses in depth rendering technology to supercompose visualization streams from various 3D models (see Fig. 21.32) to merge them into 3D streaming collaboration solutions across different digital applications (see Fig. 21.33).

Meanwhile, Vertex Software, one of the premier visualization platform providers for 3D-powered digital model and twin on applications, headquartered in Des Moines, Iowa, US, announced on March 17 of 2021 that the U.S. Patent and Trademark Office issued U.S. Pat. No. 10,950,044, titled, "Methods and apparatus to facilitate 3D object visualization and manipulation across multiple devices". Vertex Software targets to offer a new innovative digital platform solution for instant and persistent visual collaboration and engineering without bothering their customers with any local 3D file footprint efforts.

Similar to the 3D streaming solution by TU Berlin, Vertex Software was founded by pioneers in manufacturing visualization and experts in order to leverage cloud computing for its core engine, the ultra-low-cost solution for remote 3D rendering.

Vertex Software explains their digital platform solution as follows:

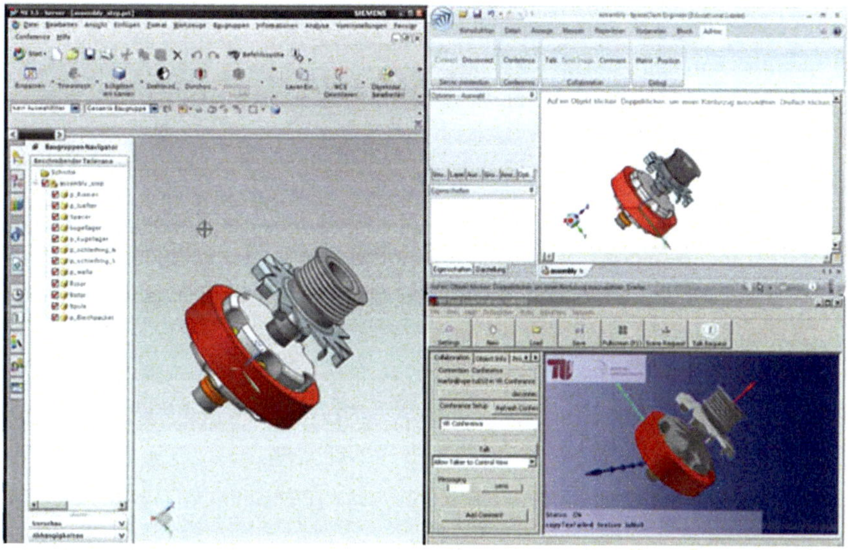

Fig. 21.33 3D streaming collaboration solution across the CAD systems NX (left) and Space claim (upper part) and a WEB browser (lower part) by TU Berlin

Vertex's patent describes a distributed computing method for interactive visualization of 3D models at scale. Specifically, Vertex has invented an ultra-low-cost approach for remote 3D rendering that is based upon the subdivision of 3D models, spatial indexing of geometric primitives, and a scalable fan-out/fan-in architectural computing pattern.

Vertex's architecture is built on four tenants:

1. Sharding subdivides 3D models into equally sized sets of geometric primitives based upon spatial proximity.
2. The shard datastore accommodates a large number of concurrent users rendering massive 3D models simultaneously. This makes the platform extremely responsive. Sharding also forms the cornerstone of the data security model.
3. Spatial indexing of shards provides for rapid lookup of shards needed to render a 3D scene. Rapid view frustum culling uses spatial indices to determine the jobs to be performed by the fan-out process. This supports fast and affordable rendering by reducing the scope of 3D shards to be rendered for a given scene.
4. High-speed laminate compositing assembles 3D images generated concurrently by a large network of CPU-based workers during the fan-in process. The resulting image is delivered to the end user device at a high frame rate to provide fluid interactivity and responsiveness for the end user.

The manifestation of this patent is the Vertex 3D Visualization Platform, a cloud-native digital twin platform that makes it easy to build and deploy low-code industry 4.0 applications. The Vertex platform runs on CPUs making it the most cost-effective approach for remote 3D rendering available. By rendering 100% of the 3D data in the cloud, Vertex securely delivers fully interactive digital twin experiences to any device, anywhere—instantly. This approach solves for the lack of specialized hardware and tools for distributed teams and customers outside of engineering.

Figure 21.34 shows the overall Vertex 3D cloud platform application environment as it is implemented and operated within companywide IT architectures and/or via (secure) public cloud infrastructures. The Vertex 3D cloud platform represents a core compute und distribution engine which offers continuous updates from source digital environments (such as 2D and 3D data from PLM, MES, ERP, CRM and IIoT) to the active *digital engineering activities* in the context of Digital Master/Prototype/Twin collaboration, assessment, analysis and decision making. The beauty of such approach is, that there does not exist a direct local footprint any longer at the client side. The pre-requisite, however, is a robust network access to the cloud computing source. Nevertheless, it is of striking evidence that running the Vertex 3D Cloud platform based in the mid-west in the US from an ordinary laptop client in mid-Europe offers full satisfying interactive response for all typical 3D viewing, annotating and interactive demonstrating digital activities.

Figure 21.35 provides an insight to the internal architecture of the 3D Vertex Cloud platform. It shows a combination of different types of Application Programming Interface (API) for internal and external applications and services, either offered by Vextex itself or developed by customers and companies based on specific needs and special know-how. From technology point either full Software Development Kits (SDK) are offered and REST (representational state transfer) type APIs are supported.

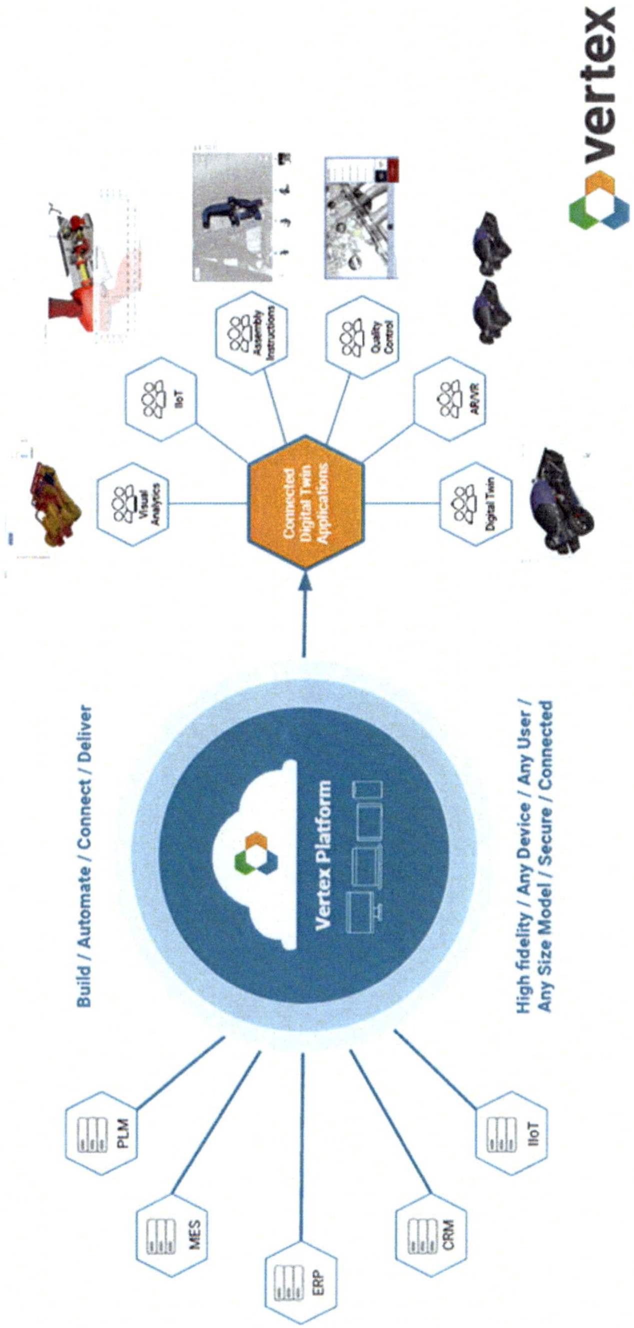

Fig. 21.34 The vertex 3D cloud platform application environment

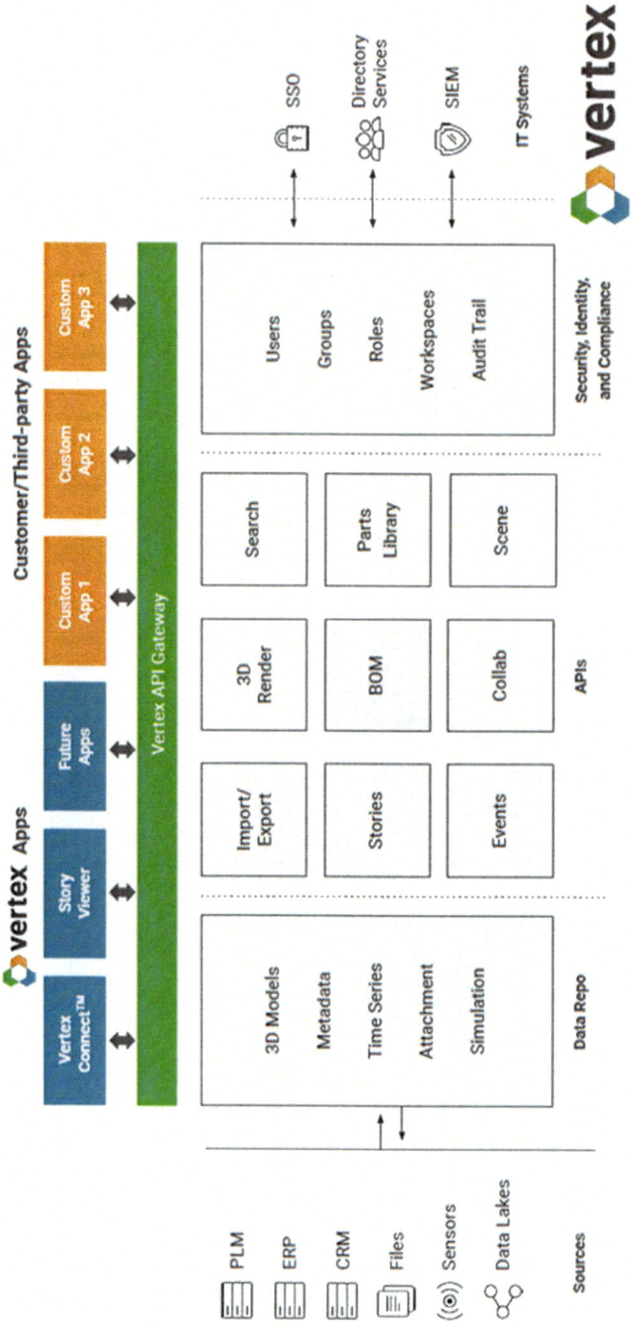

Fig. 21.35 The vertex 3D cloud platform architecture

21.5 Human Skill Sets for Future Virtual Product Creation

Information and communications technology (ICT) are penetrating more and more areas of product development, while discrete manufacturing companies increasingly provide more complex and interdisciplinary product and service offerings with new types of business models. This requires industry to understand and manage products and services throughout their whole lifecycle. Despite the increasingly significant role of ICT, people remain the core and most important asset of companies. Against the background of new and more complex products, services, systems, technologies, methods, tools and requirements, it becomes evident that the capabilities and skills of engineers also need to be updated accordingly. Key future skillsets of engineers include a holistic understanding of the developing systems and their lifecycle, as well as capabilities to apply value creating methods and tools. Furthermore, in today's engineering environment, a profound understanding of digital technology and ICT-engineering skills is required. Last but not least, the abilities to communicate effectively, manage projects and guide teams are becoming increasingly important. The future job of the engineer becomes increasingly interdisciplinary. While traditional products have emerged into mechatronic systems, CPS and Smart Products, the number of their components have increased and diversified (compare [86–89].

The work of the future VPE and PLM professional can include the tasks of engineering design, simulation and validation, certain elements of business analyst, data scientists and engineers, project managers, specialists for system engineering and operational improvements as well as capabilities in cultural change, and new business models. Therefore, future engineering professional must be able to master and apply expert knowledge from numerous disciplines and domains. In addition, they must be able to moderate conflicts within the company, reduce resistance, and communicate clear and target-oriented. The skillset of future Virtual Product Creation (VPC) is closely intertwined with the disciplines Product Lifecycle Management (PLM), see Chap. 4, and Cross-Lifecycle Engineering.

Four essential skill areas define the specific skillset and capabilities when it comes to future VPC (see Fig. 21.36). Firstly, the basics of PLM have to be understood throughout all areas of the business. Secondly, it is necessary to have a good grasp on how to use value-creating methods and processes throughout the product and system lifecycle. Here, the company's practitioners need to have a general skillset on how to engineer the respective products and systems (e.g. variant and configuration management, MBSE or methods of Digital Factory). The third skill area represents methods of information technology. To implement advanced engineering approaches, businesses need to be able to implement well-functioning IT- and engineering landscapes, which allow engineers to execute their work optimally. Last but not least, soft skills and project management skills are essential in order to design systems and organizations in a rapidly changing environment. In the following, the skill areas will be described in more detail.

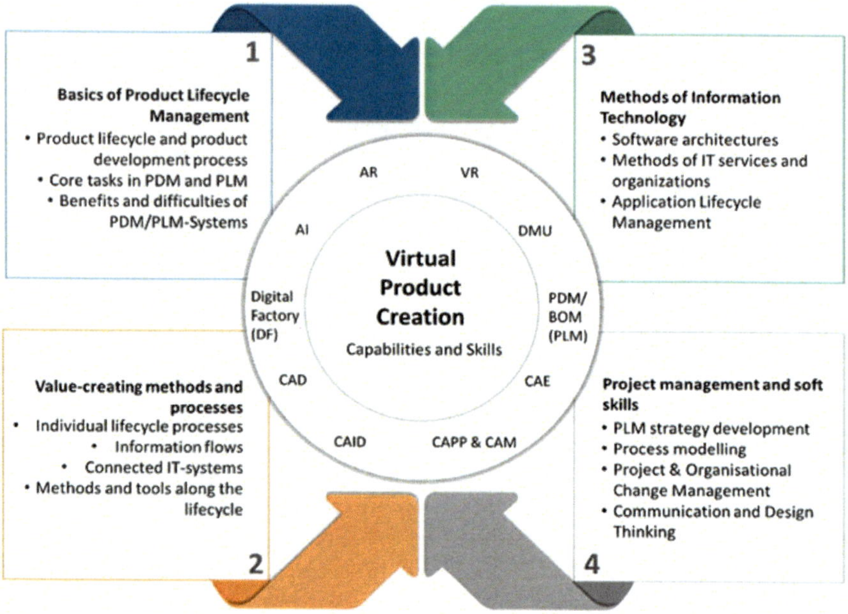

Fig. 21.36 Virtual product creation associated skills and capabilities

The Basics of Product Lifecycle Management

In the field of Virtual Product Creation, the principles of PLM have to be widely known and understood in order to tackle future complex product and systems development. In the first place, comprehension of the subject calls for precise knowledge of a company's Product Lifecycle (PLC) or Technical System Lifecycle. It then requires the design and execution of a concise Product Lifecycle Management (PLM), which includes identifying the potential benefits of the respective PLM- IT-system technologies and their impact on the product development process. Companies and their workforce need to have the ability to describe PLM applications with their core disciplines and functionalities, and distinguish between them. Moreover, future skillsets should include the description and assignation of tasks associated with project situations—the latter especially in the phases of the V-model of (Advanced) Systems Engineering.

In the future, successful businesses need to be constantly aware of megatrends, such as digitalization, new business models related to content platforms, internet-based service and storage concepts such as Software-as-a-Service (SaaS) and cloud-based services, Industry 4.0, Data Engineering and Analytics (DEA) technologies and IoT (compare explanations earlier in this chapter and in Chap. 20). Manager, Engineers and IT-experts need to know which impact these trends and tech stack options have on their products, development processes and IT-systems. To this end,

VPC and IT-departments need to monitor the respective advancements in PLM-Systems and to assess how they can be leveraged in order to improve the company's activities and address these megatrends.

Value-creating Methods and Processes

Successful future Virtual Product Creation depends on a solid understanding and the skill-set to establish and maintain basic phases of the product lifecycles, their specific processes and different types of cyclical information. The skill of being able to identify different types of information and business objects, models, IT-systems, and their specific relationships is crucial. Throughout the product development process, the definitions, flows and key processes of systems engineering, and the V-model in particular need to be fully understood and robustly executable by the engineering workforce.

Furthermore, it is indispensable that future Virtual Product Creation is able to match corresponding CAx system models with lifecycle data and Digital Twin feedback actively throughout the entire product creation process and ongoing DevsOps type continuous engineering. Professionals need to constantly review and modify information models and their PLM integrations. The future VPC skill set encompasses all major technologies (see Chaps. 7–16) and the new technologies as introduced in Chap. 20 and this chapter.

A basic understanding of articles and common parts management is also a component of the methodological basis for future Virtual Product Creation. It is fundamental to understand the nature and use of different terms and structure types (such as Bill of Materials) for product classification in PLM-systems. A well-trained handling in PLC requires the knowledge of concepts for versioning and release mechanisms throughout the product development process. In addition to the above understanding, an analysis of the goals, concepts and technical approaches of configuration and variant management is also crucial. Knowing the principles and goals of Engineering Change Management (ECM) is also essential in order to be able to react to changes quickly and ensure quality. Furthermore, it is important that information and changes in processes can be accessed and that these can be assigned to PLM-functions and -data structures. At present, engineers are required to have additional expertise regarding the importance and development disciplines of mechatronics, further expanding to cyber-physical systems and smart products with a high degree of connectivities.

Organizations need to develop a strategy for integrating customers, service providers and PLM-based cooperation. It is furthermore important to understand the principles of different views of different data structures, including their areas of application and objectives. In order to steer activities and processes better, companies should be able to implement workflows and manage them accordingly. It is necessary to be capable of describing the information flows and corresponding IT systems involved in the order processing. In addition, the core and cross-sectional tasks of Manufacturing Execution Management (MES) will need to be grasped. Furthermore, Enterprise Resource Planning (ERP)-Systems must be widely understood and managed properly. In addition to MES, the basics of the factory life cycle and the interactions between the different lifecycles (product, manufacturing, IT) must be properly understood. Here, it is again fundamental to have a clear image and understanding of the information flows and interactions between product development and production or manufacturing. Furthermore, companies need to be familiar with the methods, disciplines, tasks and benefits of the Digital Factory. For the individual Engineers a mandatory basic VPC professional training for 2–3 weeks will be necessary. Individual deep dive trainings for the specific T-shape in depth profile need to happen with assisted on-the-job training for a couple of months (incl. WEB-based self learning elements).

Methods and Processes of Information Technology

The knowledge about and the differentiation between Information Technology (IT), Information Logistic (IL) and Information Activity (IA) have been explained already in Chap. 6.5. Therefore, those skill sets become decisive for the training of the future VPC skills of Engineers.

Companies are meanwhile fully aware that the close interrelation between software, electric and electronical development in close collaboration with the data sets produced during the technical system operation can only be mastered professionally if the three information disciplines IT, IL and IA are completely digitally assisted. Therefore, software management of VPC solutions and within modern, complex products and technical systems need to correspond to each other! Consequently, it is fundamental to establish deep understanding of IT development processes and Application Lifecycle Management (ALM) in Virtual Product Creation. ALM aims to monitor, control and manage software development over the whole lifecycle of an application [90]. As shown in Fig. 21.37 tasks in ALM include application performance management, project management, customer/user experience and requirements, architecture and design, configuration management, coding, quality management and testing, as well as deployment and operation. All of these tasks have to be regarded in a software application context and, therefore, represent largely equivalent tasks compared to the physical and mechanical view of traditional PLM.

Practitioners should be capable of determining the importance and use cases of agile software development, and be able to define the terms and functionalities of ALM and PLM independently of each other and distinguish between them.

For the integration of software test management into the application development process and for the evaluation of error correction methods in the application lifecycle, the importance of test management should be included in the general software development process. Engineers should be able to distinguish between different types of test strategies and their fields of application, as well as understand the transition between error correction and change management.

Additionally, the capability of modelling IT-system architectures becomes increasingly important since products and technical systems are dependent on their own digital control units or even operating systems. Involved professionals must handle design patterns of software development—as well as views and examples of layers. Furthermore, they need to be able to explain heuristics, interfaces and drafts of software, as well as the specification of typical architecture documents. Information

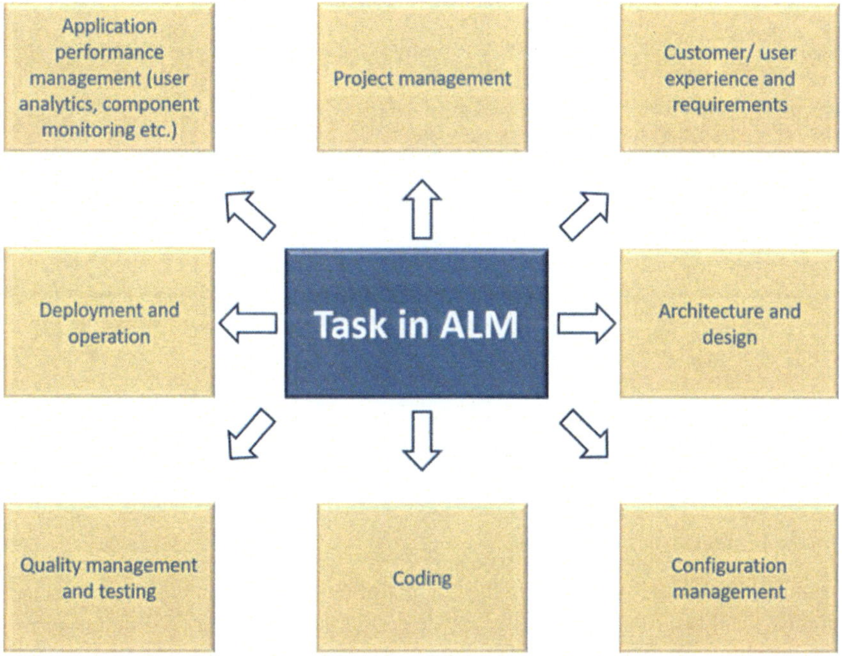

Fig. 21.37 Tasks in application lifecycle management (ALM), compare [91]

Technology Infrastructure Library (ITIL) as a reference process for IT operation and service is also worth mentioning at this point.

In this context, it is important to understand the basic idea of IT as a service. Creating optimal service level agreements and the basic performance parameters within the engineering environment is another key capability. Moreover, VPC Engineers need to be able to carry out tasks that occur after a business handover to ongoing operational services. Beforehand, they need to be able to judge and decide professionally to which extent an IT service organization/solution should be integrated into VPC digital engineering activities and become responsible for certain PLM documentation. Furthermore, compliance acquires greater significance. It is important to classify different documents, models and object types according to confidentiality levels and effects on the functioning of the IT-systems in use. Here, the protection of personal data according to the General Data Protection Regulation (GDPR) and effects on the service and operation of PLM applications is equally important. The GDPR is a regulation of the European Commission that protects natural persons with regard to the processing of personal data [92]. Since cloud and digital platform solutions are becoming a core element in future Virtual Product Creation, knowledge about their basics for on-premise solutions or hosting solutions are key future capability fields, too.

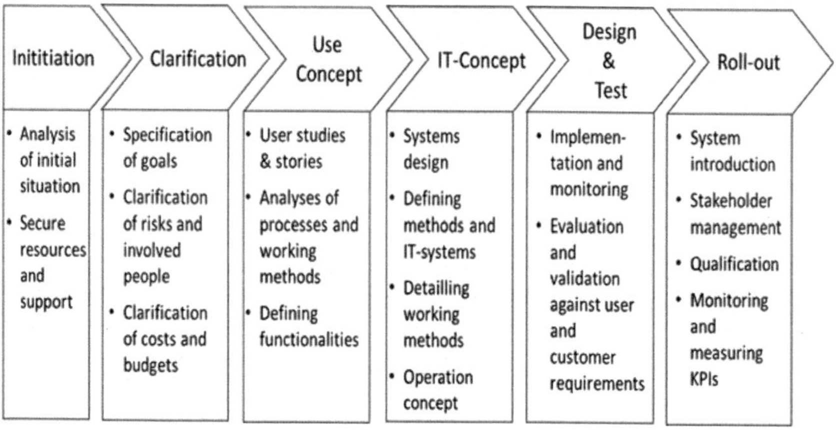

Fig. 21.38 General process for the implementation of VPE and PLM projects

Project Management and Soft Skills

Virtual Product Creation works in the context of a holistic strategy and is executed via individual projects, which will include a significant number of different disciplines (such as e.g. IT, mechanical engineering, software engineering, electrical engineering, marketing, production planning, management etc.). In this context, project management and soft skills are important capabilities for the practitioners involved.

Generally, there should be knowledge of how to develop a concise strategy and of how to pursue, implement and maintain it. This includes detailed analysis of the individual lifecycle phases and their corresponding tasks in the context of PLM and VPE projects. Additionally, executing IT-implementation projects systematically is an important capability. Here, projects need to be carefully planned, requirements to be analysed and individual roles and responsibilities within the company to get involved. Especially knowledge on how to roll-out IT-solutions has meanwhile become a key but rare enough skill! Fig. 21.38 depicts a general process for the implementation of VPE and PLM projects. First, the project should start with an analysis of the initial situation and the securing of resources and support.

In a next step, the project needs to define goals, risks, participants and budgets. A use concept specifies the target outcome of the project, while the IT-concept details the envisioned solution. In the phase development and testing, the system is designed and tested simultaneously. Finally, in the roll-out phase, the system is introduced to the user and monitored consistently. The capability to plan, execute and steer these implementation projects is key for a successful VPE. As explained in Chap. 18 ("*The challenge of modifying Management Leadership behavior towards Virtual Product Creation in industry*") Management has to take on a very strong role in proactively driving the set-up and execution of such projects!

In the context of outsourcing services and purchasing IT assets in a company's projects, professionals should be capable of handling the possible types of contracts and their requirements. When working in VPC and PLM environments, it is mandatory to contextualize the importance of *Intellectual Property* (IP) protection within the development architectures and processes, as well as part of the digital solutions and methods. Special attention is given to the awareness of risks and damages of loss of IP. An effective learning objective for the training of future engineers is the knowledge of success factors and typical reasons for failures in the context of IT- and engineering projects. In doing so, the professional critically examines their self-image and their areas of responsibility, can describe these, and place them in the project context.

It is also advisable to look at business case calculations and Total Cost Ownership (TCO) for projects. The technical experts should be familiar with the cost drivers and types so that they know how to calculate the project costs and the amortization. This also enables them to make well-founded recommendations for a project based on a business case.

For example, by means of process modelling, the engineers can identify the actual and expected situation in the company or project—within the framework of a process diagram. Third-party proposals for changes in the situation can also be evaluated in relation to these objectives. Considering the expected/target process, effective instructions are given with respect to the application of methods and to the resulting derivation of IT concepts and application environments. The responsible engineering teams must be capable of identifying precisely the advantages and risks of agile methods in such projects.

Finally, Organizational Change Management (OCM) is of increasingly high relevance for Virtual Product Creation. The introduction of new methods, IT-systems and tools and technologies changes the way a company and its employees work. PLM and VPE projects often require changes in the organization, which are sometimes unwanted by those affected. Concerns are raised and need to be mitigated if employees are afraid of poorer performance due to new ways of working, if individual employees lose their importance/power as a result, or if they are burdened with additional work or responsibilities as a result. Oftentimes, especially technical experts share and socialize fears when they have to pass on well-kept knowledge.

Next to these social challenges involving the employees concerned, management often expects that a clear Return of Investment (ROI) can be specified in advance for every project. However, this is often not possible. Next to these social challenges, also technical challenges of IT-solutions exist in parallel. They can include a lack of operability, compatibility or usability of the new systems.

All of these fears and challenges need to be properly addressed within a holistic OCM in order to make the most of the given opportunities of Virtual Product Creation and PLM. Engineers and professionals need to have knowledge of how to plan, implement and guide OCM. For example, general models on how to proceed in a given OCM project, such as the 8-step model by Kotter [93] should be familiar to professionals (Fig. 21.39).

Following this model, the urgency for change has to be increased first. After this first step, a guiding team has to be built and a vision should be developed. In order to create more attention and support within the organisation, the OCM project has to

Fig. 21.39 Model for the implementation of organisation change management (OCM) according to Kotter [93]

be communicated adequately. In further steps, action is empowered and short-term wins are created and made explicit. The project has to be continued until the goals are reached and finally the changes have to be made permanent by integrating them into an organisations culture [93].

In order to address some of the needs for new skills and capabilities, professional training and certification programs have been developed. One example includes the "PLM Professional" program offered by the Fraunhofer Gesellschaft, as outlined in Fig. 21.40.

The program is divided into two parts that can be taken separately. Level 1 of the course focuses on the basics of PLM in an eLearning format. The learning here is self-paced, using an eLearning platform. The essentials of PLM are taught here. Participants can take advantage of the learning opportunities consisting of videos and guided exercises in their own time from the comfort of their own home or at work.

In the first unit, "PLM Fundamentals", the basic concepts of PLM and product development are taught. In "Value Creating Methods in PLM", the focus is on methodical engineering work throughout the product development process. In the unit on "IT Methods in PLM", information technology aspects such as Application Lifecycle Management (ALM) and PLM architectures are covered, while "PLM Project Management" deals with management methods of PLM projects. In Level 2, participants learn more sophisticated skills and coaching in a face-to-face format and earn the personal ISO certification PLM Professional. PLM Professional Level 2 is conducted in a 3-day workshop with a certification exam. In further steps, Level

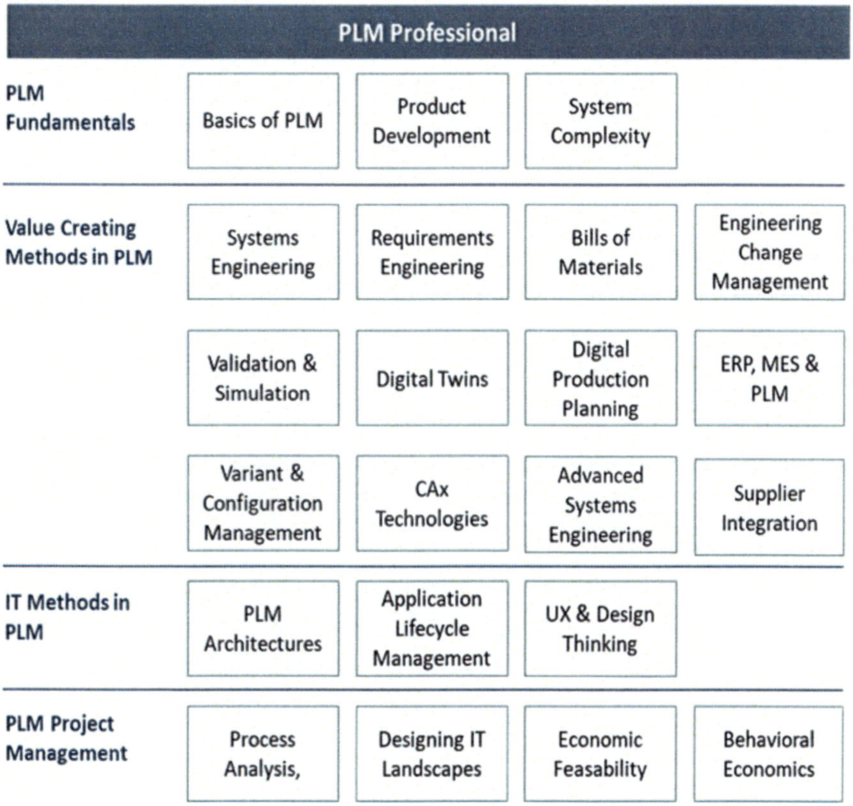

Fig. 21.40 Curriculum of the fraunhofer online certification program PLM professional

2 workshops will now be held in different European countries in cooperation with EIT Digital.

Such training programs should become the standard in educating engineers, planers and mangers for future Virtual Product Creation in industry. Master classes at universities need to stress these skillsets already as part of the curriculums in all technical faculties. Only with this dual path way forward, industry will be able to meet their challenges in future Virtual Product Creation!

21.6 The Engineering System of the Future

This book provides an insider view on Virtual Product Creation in Industry. In addition, it introduces a solid understanding of a number of new, next level skillsets and approaches which are critical in order to meet the requirements of.

- the next generation *Engineering System*, i.e. the new mix of digital engineering activities to deliver a blue print for successful development output beyond traditional high-level process descriptions,
- the underlying *Engineering Intelligence*[15] (core know-how to ensure reliable usage of models, data, algorithms and human heuristics) and
- *future Virtual Product Creation environments*, i.e. future working solutions based on processes, methods, tools and information.

Please note the following major changes in principles and fundamentals with respect to the future *Engineering System* of the future and its underlying core principles of *Engineering Intelligence* and *future Virtual Product Creation environments*:

1. **Principle of *Engineering Progression***
   Today:
   Engineering Progression is characterized as a loosely coupled, high level process sequence plan with engineering deliverables. It is up to teams and individuals to align and agree on appropriate meeting cadences and use of best practices to ensure digital compatibility.
   Future:
   Engineering Leadership owns the strategic task and operational duty to define, develop and operate an Engineering System which ensures a symbiosis between a cascaded model driven *Engineering Progression* and the engineering activities of humans and machines.

[15] *Engineering Intelligence* describes the ability of the Engineering System to reach its goals and target deliveries even under conditions of uncertainty.

2. **Foundation of the *Virtual Product Creation* solution**
   Today:
   Each engineering discipline develops its own *Virtual Product Creation sub-set solution* on a discipline specific theory and in isolation to the ones of other disciplines.
   Future:
   The Virtual Product Creation solution for the future Engineering System is built on a cross discipline theoretical foundation which makes its *Engineering Intelligence* (e.g. algorithms, core data sets and models) transparent and usable within the entire engineering community and their disciplines.
3. **Validity and aging of the *Engineering System***
   Today:
   Most of the companies 'internal development systems constitute a series of stage gate related milestones with specific engineering deliveries. Such development systems stay usually unchanged for 5 years or even longer. In addition, specific industry branch related process standards might get introduced on top of such development systems, e.g. the Automotive SPICE standard for embedded software and control unit development based on ISO/IEC 15504).
   Future:
   The Engineering System of the future needs *constant* change and evolution in order to meet the requirements for fast changing technology and business architectures. A core team in Engineering will be the active driver to add, delete and modify principles, rules and solutions with an integrated co-lead by the Information Technology department in order to provide appropriate modifications to the digital applications and related information models as well as method blue prints for engineering usages.
4. **The role of *Digital Data and Digital Models***
   Today:
   In today's engineering world digital data and model generation is nothing else than an implicit result of designers', engineers 'or analysts 'working time. There is no real value associated or even strategically incentivized to digital data and models in industry.
   Future:
   Digital data and models will receive official values and associated engineering & business relevance and uniqueness. Digital data and model price tags will substantially change the commitment to generate and use them as part of engineering tasks and activities. Conscious decisions on which data and models are most appropriate will be triggered and the demand for data and model sovereignty will grow substantially.
5. **Nature of *Decision Making***
   Today:
   Personal knowledge and heuristic experience of humans are still the prevailing sources and mechanisms for decision making and related engineering reasoning.

Future:
New types of *Engineering Intelligence* will be leveraged as main mechanism: a combination of equal rights of data analytics, model intelligence and human interpretation capabilities will form the new decision-making foundation.

6. **Approach of *Validation and Verification***
   Today:
   Validation and verification are determined based on past experience, are planned upfront but take place after the technical system development phases.
   Future:
   Validation and verification are conducted incrementally over the entire development time as a continuous and potentially even agile digital engineering activity ensuring quick and model centric system definition and prove-out.
7. **Differentiation between product *Hardware and Software***
   Today:
   Development of product hardware (HW) and product software (SW) typically are handled in separate development streams independent from each other using different, non-aligned digital development solutions.
   Future:
   Technical system development of the future expects closely coupled function and behavior-based development solutions that use integrated model capabilities for both, HW VPC models and SW development models.
8. **Engineer's *Digital Competence and Digital Support***
   Today:
   Development Engineers are expected to use their best understanding of methods and applications to build, design and analyze digital models and to finally interpret them meaningfully.
   Future:
   Development Engineers in future Virtual Product Creation will get general and personalized assistance system support to robustly carry out their engineering activities and to keep control over the high number of digital data and model types and interrelations.
9. ***Development Projects Versus Continuous Development***
   Today:
   A clear separation exists between new product development, *managed as a project*, and *ongoing engineering support* for product updates and improvements.
   Future:
   The Engineering System of the future will be able to allow *continuous development of different degrees*, even down to individual product instance modifications of product, devices and machines in the field or factory.
10. **Capabilities for *Lifecycle Engineering***
    Today:
    Feedback loops from *Mid of Life (MoL)* and *End of Life (EoL)* to the next product *Begin of Life (BoL)* only exist in a limited, non-consistent way.

Future:
The next generation Engineering System will deliver and use *Digital Twins* as synchronized conduit across different lifecycle phases (BoL, MoL, EoL).

11. **Capabilities of *Digital Working Experience***
Today:
Virtual Product Creation environments cannot detect, measure or preserve how engineers digitally work and hence do not take conclusions out of it.
Future:
Virtual Product Creation environments will be able to monitor, recognize, interpret and assist engineers personally and as a group (e.g. with the help of ongoing agent and/or AI-type training patterns).
12. **New ways and transformations of *Engineering Intelligence***
Today:
Engineering Intelligence is either encapsulated as upfront, pre-configured process or modeling template or afterwards as static lessons learned text document.
Future:
Engineering Intelligence will be continuously traced through dynamic analytics, data contextualization, model adaptions and knowledge dynamics (e.g. through active knowledge graphs).
13. **Extent and role of *Information Models***
Today:
No consistent information model approach exist for the entire Virtual Product Creation environment, instead highly fragmented silos of data and model repositories exist with non or limited semantic links.
Future:
A new type of connected and dynamically traceable *Intelligent Information Object (IIO) is* introduced to deliver engineering in-situ support, both, for engineering synthesis and analysis.

These are the most relevant changes in principles and fundamentals with respect to the future *Engineering System* of the future. It will be equally important to experiment with new ways of engineering assistance systems for future Virtual Product Creation (see [94]). Figure 21.41 gives a first impression of how *e-bots* could be used to support engineers in tomorrow's world of Virtual Product Creation and its engineering activities and tasks. It might still sound a bit spooky, but it will become real for good reasons. However, it needs serious and constant endeavors from all stakeholders to make it happen: experienced engineers, new digital nerds, IT specialists, data engineers and Management Leaders, who drive future Virtual Product.

Creation forward as a true digital engineering intelligence, capability and working discipline. Times are over to handle Virtual Product Creation and PLM just as IT enablers.

The innovation pace for *Virtual Product Creation* will accelerate further due to the following core drivers and potential negative consequences in case of just following the traditional steady state with no or limited progress:

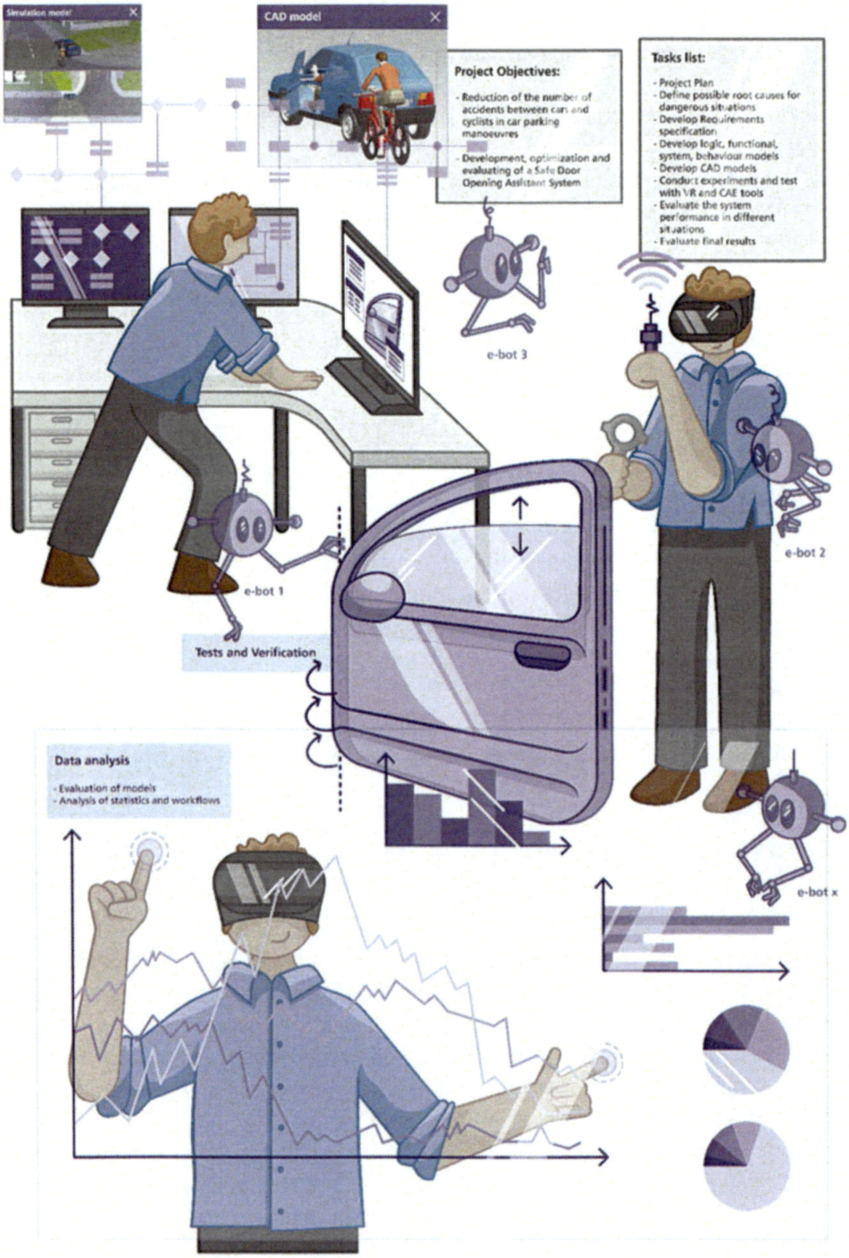

Quelle: Prof. Stark, TU Berlin (Vision, Mission und Plan, Dezember 2018)

Fig. 21.41 E-bot assistance for engineers in *future virtual product creation*

- Sharply *increasing pressure from legislative bodies and societal partners on transforming industrial sectors and consumer behavior* towards stringent *sustainable conditions* which requires substantial efforts to describe, model and simulate all (technical) systems and products (virtually) upfront and ongoingly in a true systems context,
- *New digital business values* getting created to justify the increasing dependency on digital value creation in the overall business; this will create a boost for Virtual Product Creation models, data flows and associated algorithms and tools.
- Necessity for humans (authorities, managers, engineers, planers, partners, clients, customers, consumers, citizens etc.) to receive *digital assistance and autonomous digital support in interacting with and controling digital/virtual models as well as data streams and informations sets*; otherwise resulting information overflows as well as lacks of digital consistency and traceability are unavoidable and the risk to erode the ability to design our future will grow sharply,
- Bright inventions, new scientific approaches and technology innovations will continue to boost upcoming *new ICT opportunities as foundational layer to allow new ways of "metaverse" type interactions and collaborations of humans, their avatars and digital twins*; Virtual Product Creation, therefore, will immerse gradually into new comprehensive cyber world solutions to allow for full *Virtual Living and Operations*.

Finally, let's stay optimistic and determined in transforming Virtual Product Creation to a new core engineering competence. It is worthwhile and will help us to build a hopefully "better" and sustainable world!

References

1. Klaus G, Liebscher H (1979) Wörterbuch der Kybernetik. Fischer-Taschenbuch-Verlag, Frankfurt am Main
2. Tomiyama T, Lutters E, Stark R, Abramovici M (2019) Development capabilities for smart products. CIRP Ann 68(2):727–750. https://doi.org/10.1016/j.cirp.2019.05.010
3. Walden DD, Roedler GJ, Forsberg K, Hamelin RD, Shortell TM (eds) (2015) Systems engineering handbook: a guide for system life cycle processes and activities; INCOSE-TP-2003-002-04, 2015, 4th edn. Wiley, Hoboken, NJ
4. I. T. Operations (2007) "Systems engineering vision 2020". INCOSE-TP-2004-004-02
5. Estefan JA (2008) "Survey of model-based systems engineering (MBSE) methodologies". International counsil on systems engineering (INCOSE), Seattle, WA, USA
6. Auricht M (2018) Entwicklung eines Validierungsframeworks zur erlebbaren Absicherung von Fahrerassistenzsystemen. Fraunhofer Verlag, Stuttgart
7. Saunders S (2011) "Does a model based systems engineering approach provide real program savings?—Lessons learnt
8. Pagnanelli CAG, Carson RS, Palmer JR, Crow ME, Sheeley BJ (2012) 4.5.3 model-based systems engineering in an integrated environment. INCOSE Int Symp 22(1):633–649. https://doi.org/10.1002/j.2334-5837.2012.tb01361.x
9. Scott Z (2011) "Model-based systems engineering". In: APCOSE—Asia Pacific counsil of systems engineering conference, Seoul, South Korea

10. Haberfellner R, (ed) (2015) Systems engineering: grundlagen und anwendung, 13th ed. Zürich: Orell Füssli
11. Ropohl G (2009) Allgemeine technologie: eine systemtheorie der technik. s.l.: KIT Scientific Publishing. [Online]. Available: http://www.doabooks.org/doab?func=fulltext&rid=15084
12. van Bertalanffy L (1968) General system theory: foundations, development, applications. Braziller, New York
13. Stachowiak H (1973) Allgemeine modelltheorie. Springer, Wien
14. Hick H, Bajzek M, Faustmann C (2019) Definition of a system model for model-based development. SN Appl Sci 1:1074. https://doi.org/10.1007/s42452-019-1069-0
15. Brusa E, Calà A, Ferretto D (2018) Systems engineering and its application to industrial product development. Springer International Publishing, Cham
16. Vaneman WK, Carlson R (Apr 2019) "Model-based systems engineering implementation considerations". In: 2019 IEEE international systems conference (SysCon), Orlando, FL, USA. pp 1–6
17. Stark R, Schulze E-E (2010) "Need and potentials to improve systems engineering by future PLM solutions"; prostep ivip symposium Berlin. Available through the prostep ivip association office: https://www.prostep.org/ueber-uns/geschaeftsstelle/
18. Stark R, Auricht M (2017) "Beschleunigung der Validierung innerhalb der Entwicklung von mechatronischen Systemen mit einer integrierten Plattform (Acceleration of the validation a spart of the development of mechatronic systems with the help of an integrated platform)". prostep ivip symposium essen; available through the prostep ivip association office: https://www.prostep.org/ueber-uns/geschaeftsstelle/
19. Schmidt MM, Schmidt S, Zimmermann TC, Stark R (2021) "Conceptual introduction of required disciplines in model based systems engineering". In: Drive train technology conference ATK2021
20. Königs SF (2014) Konzeption und Realisierung einer Methode zur templategestützten Systementwicklung. Stuttgart: Fraunhofer Verl. [Online]. Available: http://publica.fraunhofer.de/dokumente/N-292677.html
21. Eigner M, Dickopf T, Apostolov H, Schaefer P, Faißt K-G, Keßler A (2014) "System lifecycle management: initial approach for a sustainable product development process based on methods of model based systems engineering". In: Fukuda S, Bernard A, Gurumoorthy B, Bouras A (eds) IFIP advances in information and communication technology, vol 442, product lifecycle management for a global market: 11th IFIP WG 5.1 international conference, PLM 2014, Yokohama, Japan, July 7–9, 2014, Revised Selected Papers. Berlin, Heidelberg, s.l.: Springer Berlin Heidelberg, pp 287–300
22. Zimmermann TC, Masuhr C, Stark R (2020) "MBSE-Entwicklungsfähigkeit für Digitale Zwillinge". ZWF 115(special):51–54. https://doi.org/10.3139/104.112312
23. Huldt T, Stenius I (2019) State-of-practice survey of model-based systems engineering. Syst Eng 22(2):134–145. https://doi.org/10.1002/sys.21466
24. Wymore AW (2018) Model-based systems engineering. Boca Raton: Chapman and Hall/CRC. [Online]. Available: https://ebookcentral.proquest.com/lib/gbv/detail.action?docID=5389603
25. Schmidt MM, Zimmermann TC, Stark R (2021) "Systematic literature review of system models for technical system development". Appl Sci 11(7). https://doi.org/10.3390/app11073014
26. Sünnetcioglu A, Stark R (2018) Using transitive relations for automatic creation of consistent traceability in model-based systems engineering. Procedia Manuf 24:311–318. https://doi.org/10.1016/j.promfg.2018.06.022
27. Königs SF, Beier G, Figge A, Stark R (2012) Traceability in systems engineering—review of industrial practices, state-of-the-art technologies and new research solutions. Adv Eng Inform 26(4):924–940. https://doi.org/10.1016/j.aei.2012.08.002
28. Beier G (2013) Verwendung von traceability-modellen zur unterstützung der entwicklung technischer systeme. Zugl.: Berlin, Techn. Univ., Diss. Stuttgart: Fraunhofer-Verl., 2014. [Online]. Available: http://nbn-resolving.de/urn:nbn:de:kobv:83-opus4-51488

29. Schmidt MM, Stark R (2020) "Model-based systems engineering (MBSE) as computer-supported approach for cooperative systems development". In: Proceedings of the 18th European conference on computer-supported cooperative work: the international venue on practice-centered computing on the design of cooperation technologies—exploratory papers, reports of the European society for socially embedded technologies, Siegen
30. Mahboob A, Husung S, Weber C, Liebal A, Krömker H (2019) The reuse of SysML behavior models for creating product use cases in virtual reality. Proc Int Conf Eng Des 1(1):2021–2030. https://doi.org/10.1017/dsi.2019.208
31. Auricht M, Beckmann-Dobrev B, Stark R (2012) An interdisciplinary approach to validate mechatronic systems in early product development stages. In: NordDesign 2012. Aalborg, Denmark
32. Chami M, Bruel J-M (2015) Towards an integrated conceptual design evaluation of mechatronic systems: the SysDICE approach. Procedia Comp Sci 51:650–659. https://doi.org/10.1016/j.procs.2015.05.180
33. Folds DJ, McDermott TA (2019) "The digital (mission) twin: an integrating concept for future adaptive cyber-physical-human systems". In 2019 IEEE international conference on systems, man and cybernetics (SMC): Bari, Italy. October 6–9, 2019, Bari, Italy, pp 748–754
34. Holt J, Perry S, Brownsword M (2016) Foundations for model-based systems engineering: from patterns to models. Stevenage: IET
35. Martin JN (1997) Systems engineering guidebook: a process for developing systems and products. CRC Press, Boca Raton
36. Buchholz C, Tiemann M, Stark R (2018) "Durchgängiges prototyping mechatronischer systeme im MBSE entwicklungsprozess". In: Symposium design fox X, pp 128–140
37. Voirin J-L, Bonnet S, Normand V, Exertier D (2015) "From initial investigations up to large-scale rollout of an MBSE method and its supporting workbench: the Thales experience" In: 25th annual INCOSE international symposium (IS2015), Seattle, WA, July 13–July 16
38. Eclipse "Capella MBSE tool-ARCADIA". https://www.eclipse.org/capella/arcadia.html. last accessed 1 July 2021
39. Zafirov R, Kiefer J, Eigner M (2010) Functional modelling for efficient generation of mechatronic design and validation models of automated production installations. In: International design conference—design 2010, Dubrovnik–Croatia, May 17–20
40. Pahl G, Beitz W (2007) Engineering design: a systematic approach, 3rd edn. Springer, London
41. Kroll P, Kruchten P (2003) The rational unified process made easy. A practitioner's guide to the RUP, Addison Wesley Longman Inc., Amsterdam. ISBN: 0-321-16609-4
42. Dumitrescu R, Albers A, Riedel O, Stark R, Gausemeier J (2021) Engineering in Deutschland—status quo in wirtschaft und Wissenschaft, Ein Beitrag zum advanced systems engineering. Paderborn
43. Cameron B, Adsit DM (2020) Model-based systems engineering uptake in engineering practice. IEEE Trans Eng Manage 67(1):152–162. https://doi.org/10.1109/TEM.2018.2863041
44. White paper prostep ivip association: collaborative systems engineering on the basis of engineering IT standards. ISBN 978-3-9817958-6-8; Version 1.0, Februar 2019 https://www.prostep.org/fileadmin/downloads/WhitePaper_SSB_final_EN.pdf
45. Ludwig T, Thiemann H (2020) Datenkompetenz—data literacy. Inform Spektrum 43(6):436–439. https://doi.org/10.1007/s00287-020-01320-0
46. Lueth KL, Patsioura C, Williams ZD, Kermani ZZ (Dec 2016) "Industrial analytics 2016/2017: the current state of data analytics usage in industrial companies". IoT Anal 58
47. Jasiulewicz-kaczmarek M, Legutko S, Kluk P (2020) Maintenance 4.0 Technologies—new oppor-tunities for sustainability driven maintenance. Manag Prod Eng Rev 11(2):74–87. https://doi.org/10.24425/mper.2020.133730
48. Batini C, Cappiello C, Francalanci C, Maurino A (2009) Methodologies for data quality assessment and improvement. ACM Comput Surv 41(3):1–52. https://doi.org/10.1145/1541880.1541883
49. Liu J, Li T, Xie P, Du S, Teng F, Yang X (Feb 2020) Urban big data fusion based on deep learning: an overview. Inf Fusion 53:123–133. https://doi.org/10.1016/j.inffus.2019.06.016.

50. Diez-Olivan A, Del Ser J, Galar D, Sierra B (2019) Data fusion and machine learning for industrial prognosis: trends and perspectives towards Industry 4.0. Inform Fusion 50:92–111
51. Schroeder GN, Steinmetz C, Pereira CE, Espindola DB (2016) Digital twin data modeling with automationML and a communication methodology for data exchange. IFAC-Pap. 49(30):12–17. https://doi.org/10.1016/j.ifacol.2016.11.115
52. Peckham J, Maryanski F (1988) Semantic data models. ACM Comput Surv 20(3):153–189. https://doi.org/10.1145/62061.62062
53. Fürber C (2016) Data quality management with semantic technologies. Springer Fachmedien Wiesbaden, Wiesbaden
54. Hildebrand K, Gebauer M, Hinrichs H, Mielke M (eds) (2018) Daten- und Informationsqualität. Wiesbaden, Springer Fachmedien Wiesbaden. https://doi.org/10.1007/978-3-658-21994-9.
55. Tao F, Zhang H, Liu A, Nee AYC (2019) Digital twin in industry: state-of-the-art. IEEE Trans Ind Inform 15(4):2405–2415. https://doi.org/10.1109/TII.2018.2873186
56. Carlsson G (2020) Topological methods for data modelling. Nat Rev Phys 2(12):697–708. https://doi.org/10.1038/s42254-020-00249-3
57. Gadatsch A (2019) Datenmodellierung: Einführung in die Entity-Relationship-Modellierung und das Relationenmodell. Wiesbaden, Springer Fachmedien Wiesbaden. https://doi.org/10.1007/978-3-658-25730-9
58. Scheer A-W (1992) Embedding data modelling in a general architecture for integrated information systems. In: Pernul G, Tjoa A (eds) Entity-relationship approach—ER '92, vol 645. Springer, Berlin Heidelberg, pp 139–161. https://doi.org/10.1007/3-540-56023-8_10
59. Open data for development network and international development research centre (Canada) (2019) The state of open data: histories and horizons. Cape Town, South Africa; Ottawa, Ontario: African Minds: international development research centre: open data for development OD4D Network
60. Bokrantz J, Skoogh A, Lämkull D, Hanna A, Perera T (2018) Data quality problems in discrete event simulation of manufacturing operations. SIMULATION 94(11):1009–1025. https://doi.org/10.1177/0037549717742954
61. Das H, Barik RK, Dubey H, Roy DS (eds) (2019) Cloud computing for geospatial big data analytics: intelligent edge, fog and mist computing, vol 49. Cham, Springer International Publishing. https://doi.org/10.1007/978-3-030-03359-0
62. Lynn T, Mooney JG, Lee B, Endo PT (eds) (2020) The cloud-to-thing continuum: opportunities and challenges in cloud, fog and edge computing. Cham, Springer International Publishing. https://doi.org/10.1007/978-3-030-41110-7
63. International Standards Organization (2016) ISO standard 8000 part (61) version (1)—data quality management: process reference model. vol ISO 8000-6
64. Bühler P, Schlaich P, Sinner D (2019) Datenmanagement: Daten—Datenbanken—Datensicherheit. Berlin, Heidelberg, Springer Berlin Heidelberg. https://doi.org/10.1007/978-3-662-55507-1
65. García Márquez FP, Lev B (eds) (2017) Big data management. Cham, Springer International Publishing. https://doi.org/10.1007/978-3-319-45498-6
66. Meier A, Kaufmann M (2019) SQL & NoSQL databases: models, languages, consistency options and architectures for big data management. Wiesbaden, Springer Fachmedien Wiesbaden. https://doi.org/10.1007/978-3-658-24549-8
67. Wingerath W, Ritter N, Gessert F (2019) Real-time & stream data management: push-based data in research & practice. Cham, Springer International Publishing. https://doi.org/10.1007/978-3-030-10555-6
68. Chavez Z, Hauge JB, Bellgran M (2020) Advances in production management systems. Towards smart and digital manufacturing, vol 592. Cham, Springer International Publishing. https://doi.org/10.1007/978-3-030-57997-5
69. Elshawi R, Maher M, Sakr S (Jun 2019) Automated machine learning: state-of-the-art and open challenges". Accessed 15 Jun 2020. [Online]. Available: http://arxiv.org/abs/1906.02287
70. Gogineni S, Lindow K, Nickel J, Stark R (2020) Applying contextualization for data-driven transformation in manufacturing. Springer, Cham, pp 154–161. https://doi.org/10.1007/978-3-030-57997-5_19

71. Zaveri A, Färber M, Bartscherer F, Menne C, Rettinger A. Semantic web 0(0)1–53 1 IOS press linked data quality of DBpedia, freebase, OpenCyc, Wikidata, and YAGO." Accessed: 19 Jun 2020. [Online]. Available: https://www.cia.gov/library/
72. Woll R (2017) "Optimized data integration for tracelinking in product development through the application of semantic web technologies". Dissertation, Technische Universität Berlin
73. Prudhomme C, Homburg T, Ponciano J-J, Boochs F, Cruz C, Roxin A-M (2020) Interpretation and automatic integration of geospatial data into the Semantic web: towards a process of automatic geospatial data interpretation, classification and integration using semantic technologies. Computing 102(2):365–391. https://doi.org/10.1007/s00607-019-00701-y
74. Colace F, Lombardi M, Pascale F, Santaniello D (Nov 2018) A multilevel graph representation for big data interpretation in real scenarios. In: 2018 3rd international conference on system reliability and safety (ICSRS). pp 40–47. https://doi.org/10.1109/ICSRS.2018.8688834
75. Akerkar R, Sajja PS (2016) Intelligent techniques for data science. Springer International Publishing, Cham. https://doi.org/10.1007/978-3-319-29206-9
76. Shearer C (2000) The CRISP-DM model: the new blueprint for data mining. J Data Warehous 5(4):13–22
77. Friendly M (2008) A brief history of data visualization. In: Handbook of data visualization. Springer Berlin Heidelberg, Berlin, Heidelberg, pp 15–56. https://doi.org/10.1007/978-3-540-33037-0_2
78. Chen M, Hauser H, Rheingans P, Scheuermann G (eds) (2020) Foundations of data visualization. Springer International Publishing, Cham. https://doi.org/10.1007/978-3-030-34444-3
79. Waskom M (2021) Seaborn: statistical data visualization. J Open Source Softw 6(60):3021. https://doi.org/10.21105/joss.03021
80. Stark R, Damerau T (2019) Digital twin. In: Chatti S, Laperrière L, Reinhart G, Tolio T (Hg.) The international academy for production engineering, CIRP encyclopedia of production engineering. 2nd ed. 2018. Springer Berlin, Berlin, Heidelberg. ISBN: 978-3-662-53119-8. https://doi.org/10.1007/978-3-642-35950-7_16870-1
81. Stark R, Fresemann C, Lindow K (2019) Development and operation of digital twins for technical systems and services. 69th CIRP general assembly–Birmingham–UK–18–24 Aug 2019. CIRP Ann Manuf Technol 68(1):129–132. https://doi.org/10.1016/j.cirp.2019.04.024
82. GAIA-X: Technical architecture. Federal ministry for economic affairs and energy (BMWi) public relations division 11019 Berlin www.bmwi.d; release, June 2020. June 4th of 2020. https://www.data-infrastructure.eu/GAIAX/Redaktion/EN/Publications/gaia-x-technical-architecture.html
83. Position paper (Feb 2021) "Stimulating the digital economy by introducing the principle of "Data sovereignty". Position paper: stimulating the digital economy by introducing the principle of "Data sovereignty"—data sovereignty now
84. FIWARE Foundation e.V (June 2021) "Fiware for data spaces". FIWARE foundation e.V. Franklinstraße 13A, 10587 Berlin, Germany
85. Wolter L, Hayka H, Stark R (2013) Improving the usability of collaboration methods and technologies in engineering. In: Kovács GL, Kochan D (eds) Digital product and process development systems. NEW PROLAMAT 2013. IFIP advances in information and communication technology, vol 411. Springer, Berlin, Heidelberg. https://doi.org/10.1007/978-3-642-41329-2_9
86. VDI/VDE (2020) Entwicklung cyber-physischer mechatronischer systeme (CPMS). Development of cyber-physical mechatronic systems (CPMS). VDI Guideline 2206, Germany
87. Porter ME, Heppelmann JE (2014) How smart, connected products are transforming competition. Harvard Bus Rev: HBR 92(11):64–88
88. Abramovici M (2015) Smart products. In: CIRP encyclopedia of production engineering. p 1–5. https://doi.org/10.1007/978-3-642-35950-7_16785-1
89. Valencia A, Mugge R, Schoormans J, Schifferstein H (2015) The design of smart product-service systems (PSSs): an exploration of design characteristics. Int J Design 9(1)

90. Lacheiner H, Ramler R (2011) Application lifecycle management as infrastructure for software process improvement and evolution: experience and insights from industry. In: 2011 37th EUROMICRO conference on software engineering and advanced applications. pp 286–293. https://doi.org/10.1109/SEAA.2011.51
91. Rossberg J (2014) Beginning application lifecycle management: Apress. CA, US, Apress, Berkeley. 978-1-4302-5812-4. https://doi.org/10.1007/978-1-4302-5813-1
92. The European Comission (2016) The general data protection regulation. Online available under: https://eur-lex.europa.eu/legal-content/EN/TXT/?qid=1465452422595&uri=CELEX:32016R0679
93. Kotter JP (2012) Leading change: Harvard business press. ISBN: 9781422186442
94. Stark R, Brandenburg E, Lindow K (2021) Characterization and application of assistance systems in digital engineering. CIRP Ann. https://doi.org/10.1016/j.cirp.2021.04.061

Acknowledgement

I owe an enormous debt of gratitude to those who gave me constructive comments, assessments and future ideas on one or more *industrial practice* subject areas of this book *Virtual Product Creation in Industry*, including Mr. Ulrich Ahle, Dr. Thomas Bähr, Mr. Turgay Coruh, Dr. Michael Grieves, Dr. Siegmar Haasis, Mr. Stefan Harms, Mr. Armin Hoffacker, Mr. Frank Jelich, Dr. Markus Junninger, Prof. Alfred Katzenbach, Mr. Uwe Kloss, Mr. John Knight-Gregson, Mr. Thomas Kriegel, Mr. Pete Lamoureux, Dr. Jörg Lüddemann, Dr. Patrick Müller, Mr. Masakazu Ninomiya, Mr. Dieter Ottersbach, Mr. Steffen Müller-Urbaniak, Mr. Oleg Shilovitsky, Dr. Bernd Pätzold, Dr. Alain Pfouga, Prof. Oliver Riedel, Mr. Dave Russel, Dr. Dieter Schacher, Mr. John Scholfield, Dr. Dietmar Trippner, Dr. Armin Ulbrich, Mr. Christian Varnholt, Dr. Henrik Weimer, Mr. Philipp Wibbing, Mr. Christian Wilmers, Mr. Heiko Witte and Dr. Martin Witte.

In addition, I am very grateful for detailed, constructive and inspiring discussions on future *Virtual Product Creation* methods and technologies from my academic peers: Prof. Michael Abramovici, Prof. Reiner Anderl, Prof. Nabil Anwer, Prof. Alain Bernard, Prof. Martin Eigner, Prof. Detlef Gerhard, Prof. Hannes Hick, Prof. Eric Lutters, Prof. Frank Mantwill, Prof. Andrew Nee, Prof. Jivka Ovtcharova, Prof. Ralph Stelzer, Prof. Klaus-Dieter Thoben, Prof. Tetsuo Tomiyama, Prof. Sandor Vajna, Prof. Sandro Wartzack, Prof. Christian Weber and Prof. Klaus Zeman.

Anticipating the future of Virtual Product Creation and the associated new technologies of Industrie 4.0 and IoT (IIoT, AIoT) I appreciated very much the discussions and engagements with all members of the research council Industrie 4.0.

Furthermore, I fully appreciate that the following companies have supported this book with active contributions: ASCon System GmbH, AUTODESK Inc., BMW Group, Bombardier Transportation, Clous GmbH, Contact Software GmbH, Daimler AG, Dassault Systèmes, MathWorks Inc., KISTERS AG, PTC Inc., Rolls-Royce, Siemens Digital Industry Software, Vertex Software and Volkswagen AG.

I would like to express my special appreciation for the following scientific fellows and industry experts who have contributed personally to the following subjects covered by the book *Virtual Product Creation in Industry*:

R. Stark, *Virtual Product Creation in Industry*,
https://doi.org/10.1007/978-3-662-64301-3

	Special thanks of gratitude to Dr. Ulrike Völlinger for her participation, detailed support and commitment in delivering the subject "*History of Information Technology*". It was a valuable asset to count on her knowledge and experience in this field
	Special thanks of gratitude to Dr. Robert Woll for his participation, detailed support and commitment in contributing to the subject "*The technology history of Virtual Product Creation*". It was a true asset to count on his knowledge and experience in this field
	Special thanks of gratitude to Ludger Busswinkel for his participation, detailed support and commitment in contributing to the subject "*CAPP/CAM*". It was a valuable asset to count on his knowledge and experience in this field

(continued)

(continued)

	Special thanks of gratitude to Anton Dybov for his participation, detailed support and commitment in contributing to the subjects "*CAPP/CAM*" and "*MBSE*". It was a valuable asset to count on his knowledge and experience in these fields
	Special thanks of gratitude to Stephan Mönchinger for his participation, detailed support and commitment in contributing to the subject "*CAE*". It was a valuable asset to count on his knowledge and experience in this field
	Special thanks of gratitude to Dr. Carina Fresemann for her participation, detailed support and commitment in delivering the subject "*PDM/BOM*". It was a valuable asset to count on her knowledge and experience in this field

(continued)

(continued)

	Special thanks of gratitude to Dr. Kai Lindow for his participation, detailed support and commitment in contributing to the subject "*PDM/BOM*". It was a true asset to count on his knowledge and experience in this field
	Special thanks of gratitude to Dr. Patrick Mueller for his participation, detailed support and commitment in contributing to the subject "*PDM/BOM*". It was a true asset to count on his knowledge and experience in this field
	Special thanks of gratitude to Dr. Andreas Geiger for his participation, detailed support and commitment in contributing to the subject "*Virtual Reality*". It was a true asset to count on his knowledge and experience in this field

(continued)

(continued)

	Special thanks of gratitude to Bernhard Wolf for his participation, detailed support and commitment in contributing to the subject "*Augmented Reality*". It was a valuable asset to count on his knowledge and experience in this field
	Special thanks of gratitude to Nabil Hammouche for his participation, detailed support and commitment in contributing to the subject "*Digital Factory*". It was a valuable asset to count on his knowledge and experience in this field
	Special thanks of gratitude to Simon Kind for his participation, detailed support and commitment in contributing to the subject "*Digital Factory*". It was a valuable asset to count on his knowledge and experience in this field

(continued)

(continued)

	Special thanks of gratitude to Yuwei Pan for her participation, detailed support and commitment in contributing to the subject "*Artificial Intelligence*". It was a valuable asset to count on her knowledge and experience in this field
	Special thanks of gratitude to Simon Hagemann for his participation, detailed support and commitment in contributing to the subject "*Industrie 4.0 and IoT*". It was a valuable asset to count on his knowledge and experience in this field
	Special thanks of gratitude to Stefan Mengel for his participation, detailed support and commitment in contributing to the subject "*Industrie 4.0 and IoT*". It was a valuable asset to count on his knowledge and experience in this field

(continued)

(continued)

	Special thanks of gratitude to Marvin Manoury for his participation, detailed support and commitment in contributing to the subject "*MBSE*". It was a valuable asset to count on his knowledge and experience in this field
	Special thanks of gratitude to Thomas Zimmermann for his participation, detailed support and commitment in contributing to the subject *"MBSE"*. It was a valuable asset to count on his knowledge and experience in this field
	Special thanks of gratitude to Dr. Friedrich Halstenberg for his participation, detailed support and commitment in contributing to the subjects "*PLM*" and "*Human Skill Sets*". It was a valuable asset to count on his knowledge and experience in this field

(continued)

(continued)

	Special thanks of gratitude to Maurice Preidel for his participation, detailed support and commitment in contributing to the subject "*Data Engineering and Analytics*". It was a valuable asset to count on his knowledge and experience in this field

I owe an enormous debt of gratitude to the publishing House Springer for their patience and their unconditional and constant support. They encouraged me and allowed me to lead the progress and completion of the project in the manner I desired.

I do whole heartily appreciate the help of family members and friends who gave me different kinds of support in editing content for my book.

Finally, I want to thank my wife, Brigitte, for tolerating my incessant disappearances into my home office. A lifelong partner makes both the journey and destination worthwhile. Without her encouragement I would not have been able to sustain the efforts to write and finalize this book.